FOUNDATIONS OF PERTURBATIVE QCD

The most non-trivial of the established microscopic theories of physics is QCD: the theory of the strong interaction. A critical link between theory and experiment is provided by the methods of perturbative QCD, notably the well-known factorization theorems. Giving an accurate account of the concepts, theorems, and their justification, this book is a systematic treatment of perturbative QCD.

As well as giving a mathematical treatment, the book relates the concepts to experimental data, giving strong motivations for the methods. It also examines in detail transverse-momentum-dependent parton densities, an increasingly important subject not normally treated in other books. Ideal for graduate students starting their work in high-energy physics, it will also interest experienced researchers wanting a clear account of the subject. This title, first published in 2011, has been reissued as an Open Access publication on Cambridge Core.

JOHN COLLINS is Distinguished Professor of Physics at Penn State University. He has long experience in perturbative QCD. He has proved a number of the fundamental theorems that form the main content of this book, and has a record of formulating and deriving novel results in QCD. During his career he has received several awards, including a Guggenheim fellowship, a Humboldt Research Award, a Mercator professorship, and the JJ Sakurai prize.

FOUNDATIONS OF PERTURBATIVE QCD

JOHN COLLINS

Penn State University

CAMBRIDGE
UNIVERSITY PRESS

CAMBRIDGE
UNIVERSITY PRESS

Shaftesbury Road, Cambridge CB2 8EA, United Kingdom

One Liberty Plaza, 20th Floor, New York, NY 10006, USA

477 Williamstown Road, Port Melbourne, VIC 3207, Australia

314–321, 3rd Floor, Plot 3, Splendor Forum, Jasola District Centre, New Delhi – 110025, India

103 Penang Road, #05-06/07, Visioncrest Commercial, Singapore 238467

Cambridge University Press is part of Cambridge University Press & Assessment, a department of the University of Cambridge.

We share the University's mission to contribute to society through the pursuit of education, learning and research at the highest international levels of excellence.

www.cambridge.org
Information on this title: www.cambridge.org/9781009401838

DOI: 10.1017/9781009401845

First published 2011
Reissued as OA 2023

A catalogue record for this publication is available from the British Library.

ISBN 978-1-009-40183-8 Hardback
ISBN 978-1-009-40182-1 Paperback

Cambridge University Press & Assessment has no responsibility for the persistence or accuracy of URLs for external or third-party internet websites referred to in this publication and does not guarantee that any content on such websites is, or will remain, accurate or appropriate.

Contents

To Mary, Dave, and George

Acknowledgments

I would like to thank many colleagues for many comments on drafts of this book, including Emil Avsar, Markus Diehl, Gerardo Giordano, Aaron Miller, Ted Rogers, Anna Staśto, and the participants in an advanced QFT class in 2009.

I owe much gratitude to Dave Soper and George Sterman for our collaboration on many of the fundamental results in perturbative QCD. I also thank my department head, Jayanth Banavar, for his continued encouragement and support. Finally and most importantly, I thank my wife Mary for her continuous support and love during the arduous task of writing this book.

The drawings of Feynman graphs in this book were mostly made using the JaxoDraw program (Binosi *et al.*, 2009).

Work on this book was partially supported by the US Department of Energy under a research grant, and also, during a sabbatical at the Ruhr University Bochum, by a Mercator Professorship of the Deutsche Forschungsgemeinschaft.

1

Introduction

The theory of the strong interaction of hadrons – quantum chromodynamics, or QCD – is in many ways the most perfect and non-trivial of the established microscopic theories of physics. It is, as far as is known, a self-consistent relativistic quantum field theory. But, unlike the case of the electromagnetic and weak interactions, many primary phenomena governed by QCD are not amenable to direct calculation by weak-coupling perturbation theory. Moreover, QCD has few parameters.

To understand these assertions, first recall the classification of known microscopic interactions into strong, electromagnetic, weak, and gravitational. Precisely because the strong interaction is strong, it is useful to study QCD by itself, the other interactions being perturbations.

QCD is a quantum field theory of the kind called a non-abelian gauge theory (or a Yang-Mills theory). It has two types of field: quark fields and the gluon field. Particles corresponding to the quark fields form the basic constituents of hadrons, like the proton, with the gluon field providing the binding between quarks. There appear to be no states for isolated quarks and gluons; these particles are always confined into hadrons. This contrasts with quantum electrodynamics (QED), where instead of quarks and gluons, we have electrons and photons, which do exist in isolated single-particle states.

One key feature of QCD is "asymptotic freedom": the effective coupling of QCD goes to zero at zero distance. Thus short-distance processes yield to the highly developed methods of Feynman perturbation theory. Among other things, this allows a perturbative analysis to give a correct renormalization of the ultra-violet divergences of QCD. The theory therefore exists in a way that the electroweak theory may not. Thus QCD contains no hints of its own breakdown.

On the other hand, unlike the case of QED, where perturbative methods give (first principles) predictions of spectacular accuracy, many apparently simple phenomena in QCD are difficult and non-perturbative, for example, its simplest bound states, like the pion and proton. Although Monte-Carlo lattice calculations have made enormous progress, they are limited both in achievable accuracy and in the observables that can be predicted, and these include no hadronic high-energy scattering processes, for example. So QCD is highly non-trivial.

Moreover, its consequences are enormous. Of course, QCD underlies the whole of nuclear physics and it creates most of the mass of ordinary matter, as contained in the proton and neutron.

Despite the non-perturbative nature of the particle states in QCD, there is a vast domain where perturbative methods can actually be applied to realistic scattering processes in QCD. The purpose of this book is to give a systematic account of these methods and their justification.

At present these are the methods that show the power of QCD most strongly. They have an almost universal impact on experiments in high-energy physics, particular with hadron beams. This ranges from the long-range planning of experiments to the analysis of data, even when the primary subject of study is a non-QCD phenomenon, such as the weak interaction, the Higgs boson, supersymmetry, etc.

It is easy to state the characteristic method, hard-scattering factorization, that enables perturbative QCD to be systematically applied to these reactions. As an example, consider production of the predicted Higgs boson in proton-proton collisions at the Large Hadron Collider (LHC), which at the time of completion of this book (2010) was starting operation at the European Organization for Nuclear Research (CERN). In the factorization approach, the proton beams are treated as collections of so-called partons: quasi-free quarks and gluons, whose (non-perturbative) distributions, the parton densities, have been measured in other experiments, and are used at the LHC with the aid of the perturbative evolution equations of Dokshitzer, Gribov, Lipatov, Altarelli, and Parisi (DGLAP). The cross sections for collisions of partons of the various types to make the Higgs boson are calculated perturbatively given its expected couplings. A provable consequence of QCD is that many useful physical cross sections are predicted by convoluting measured parton densities with short-distance partonic cross sections. It is the proved universality (or, more generally, modified universality) of the non-perturbative parton densities, etc., between different experiments that gives perturbative QCD its predictive power.

Among other things, factorization allows an extrapolation of the physics by an order of magnitude or more in center-of-mass energy from previous experiments. Many cross sections of interest are so minute (down to femtobarns at the LHC) compared with the total cross section (close to 100 mb) that considerable quantitative understanding of QCD physics is necessary for a good analysis of experimental data.

There are many places where one can learn *how* to perform perturbative QCD calculations. But a newcomer or an outsider can be forgiven for questioning the logical foundations of the subject. For example, why should calculations be made with on-shell massless partons, as is commonly done? There seems to be an essential use of actual collinear singularities associated with on-shell massless partons; how can these be so routinely and cavalierly manipulated when we know that quarks and gluons are confined inside hadrons and are clearly not free particles and are therefore definitely not on-shell?

Therefore the purpose of this book is to give a connected logical account of the methods of perturbative QCD. The intended audience includes not only graduate students in high-energy physics, but also established researchers, both in high-energy physics and elsewhere, who want a clear account of the subject.

Readers are assumed to have a knowledge[1] of relativistic quantum field theory, up to non-abelian gauge theories and the elements of renormalization theory, together with a basic knowledge of elementary particle physics. Beyond this I try to keep the treatment self-contained.

There is a clear danger that the treatment gets bogged down in mathematical minutiae without getting to the practically applicable meat of the subject. But without a sufficiently clear and precise treatment, the concepts get muddied, further development is stymied, and the construction of new innovative and correct concepts is hindered.

Indeed, not everything in perturbative QCD is properly clear and established. One reason for such problems is the way in which much knowledge in perturbative QCD has been constructed.[2] It is common in science to induce theoretical ideas from a pattern perceived in a body of experimental data. But in QCD, we also often induce new higher-level theoretical ideas from calculations within QCD. For example, on the basis of a set of Feynman-graph calculations, one might see a pattern that can be formalized in the statement of a factorization property together with a definition of parton densities. The induced property can be tested both by further theoretical calculations and by comparison of its predictions with experiment.

Such a factorization property has the form of a statement of a mathematical theorem, and the soundest method of establishing the property is by proving it mathematically. Naturally researchers try to do so. But because of the difficulty of perturbative QCD, there are often interesting gaps in the proofs.

The theorems of perturbative QCD are supported not only by proofs, but also by a combination of agreement with the results of particular Feynman-graph calculations and agreement with experimental data. So a gap in a proof does not imply that a theorem is actually wrong. But the gaps can be frustrating to a newcomer learning the subject. They are suggestive of things that are difficult and not fully explicitly understood; the understanding in the collective consciousness of the workers in the field is quite non-verbal.[3] Such gaps could become particularly important in generalizations of the theorems.

In this book I try to make the gaps explicit. I will point out some of the danger areas, and suggest targets for research. I was also able to fill in or reduce many of the gaps.

1.1 Factorization and high-energy collisions

Since the idea of hard-scattering factorization is so central to the applications of QCD, it is useful at this point to formulate it quantitatively in a particular example. The results in this section are stated without any attempt at justification, the aim being to give a hint of the landscape we will explore in detail in the rest of the book. The section may be somewhat mysterious to a reader without any exposure to the general subject matter, and it can be skipped if necessary.

---

[1] Standard references include: Sterman (1993); Peskin and Schroeder (1995); Weinberg (1995, 1996); Srednicki (2007).
[2] These issues actually apply more generally in quantum field theory and in high-energy theory.
[3] A classic case of a similar situation is with Dirac's delta function.

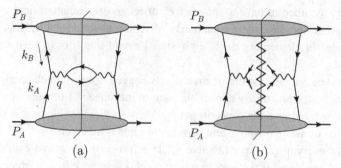

Fig. 1.1. (a) Drell-Yan cross section at lowest order in hard scattering. This represents an amplitude times its complex conjugate, with a sum and integral over the final state, at the vertical line. (b) Same as diagram (a) with interactions between the hadron remnants that fill in the gap between the remnants.

Let us choose the Drell-Yan process, the inclusive production of a high-mass muon pair, $H_A H_B \to \mu^+ \mu^- X$. Here H_A and H_B are incoming hadrons of momenta P_A and P_B in a collision with a large center-of-mass energy $\sqrt{s}$. The muon pair is produced through a virtual photon (or other electroweak boson).[4] The symbol X denotes that the rest of the final state is summed/integrated over, and is treated as unobserved.

In factorization, the $\mu^+ \mu^-$ pair is formed in the interaction of one constituent out of each hadron, with the lowest-order case being simple quark-antiquark annihilation, as in Fig. 1.1(a). In the center of the figure, a quark out of one hadron and an antiquark out of the other annihilate at a vertex with a line for a highly virtual photon. At the other end of the photon line are the detected muon and antimuon. The shaded blobs represent the remnants of the incoming hadrons.

Factorization applies when the $\mu^+ \mu^-$ pair has high mass. In this context the constituents are called partons, and the possible types correspond to the fields in QCD: the quarks of various types (called "flavors"), and gluons.

In a sense to be made precise in Ch. 6, the partons are approximately aligned with their parent hadrons, and we will define a momentum fraction ξ of a parton with respect to its parent. We will define "parton densities", $f_{i/H}(\xi)$. Here i labels the type of parton, and H the parent hadron. Then $f_{i/H}(\xi)$ is treated as the number density of partons of type i in hadron H. The factorized cross section is

$$\frac{d\sigma}{dQ^2\,dy} = \sum_{ij} \int_0^1 d\xi_a \int_0^1 d\xi_b \, f_{i/H_A}(\xi_a) f_{j/H_B}(\xi_b) \frac{d\hat\sigma(\xi_a, \xi_b, i, j)}{dQ^2\,dy}, \qquad (1.1)$$

where y is the rapidity[5] of the $\mu^+ \mu^-$ pair, and Q is its invariant mass. We have chosen to integrate over q_T, the transverse momentum of the pair relative to the collision axis. The errors in the factorization formula are suppressed by a power of a hadronic mass divided by Q or $\sqrt{s}$. We have a sum over the types of the parton that are involved, one out of

---

[4] Any other type of lepton pair may be similarly treated, e.g., $e^+ e^-$ or $\mu^+ \nu_\mu$.
[5] See Sec. B.3 for a definition.

each beam hadron, and we have an integral over the possible momentum fractions. The partons themselves undergo a collision, and this gives the $\mu^+\mu^-$ pair, with an effective cross section denoted by $d\hat{\sigma}(\xi_a, \xi_b, i, j)/dQ^2 \, dy$. This we call a "short-distance cross section" or a "hard-scattering coefficient". It differs conceptually from an ordinary cross section in that it is arranged to be not a complete cross section for partonic scattering, but to contain only short-distance contributions.

The short-distance partonic cross section can be usefully calculated in powers of the coupling $\alpha_s(Q)$, which is small when Q is large, because of the asymptotic freedom of QCD. The lowest-order hard-scattering for the Drell-Yan process is the tree diagram for quark-antiquark annihilation to $\mu^+\mu^-$, at the center of Fig. 1.1(a). At this order, the parton momentum fractions are determined from the muon-pair kinematics: $\xi_a = e^y Q/\sqrt{s}$, $\xi_b = e^{-y} Q/\sqrt{s}$, in the center-of-mass frame. Thus the Drell-Yan cross section gives a direct probe of the underlying quark and antiquark.

The correct definition of a parton density depends on a resolution scale, which can be usefully set equal to Q. There is an equation, the DGLAP (Dokshitzer-Gribov-Lipatov-Altarelli-Parisi) equation, which governs the scale dependence. It is a linear integro-differential equation – (8.30) below – whose kernel is also perturbatively computable.

There are many similar factorization formulae for a wide variety of processes, typified by the production of a high-mass electroweak system ($\mu^+\mu^-$ pair, and W, Z and Higgs bosons), or by the production of particles with large transverse momentum. There are, in addition, many further developments, characterized by more complicated kinematic situations.

We can now appreciate some of the issues that make the derivation and understanding of factorization very non-trivial. One is that in Fig. 1.1(a) there are beam remnants going into the final state. We can treat each of these as each being boosted from the rest frame of its parent hadron to a high energy, so that the final state appears to have two oppositely highly boosted systems, with a distinctive large rapidity gap between them. Such a state is in fact sometimes seen experimentally, and is called a "diffractive" configuration. But diffractive events are only a small fraction of the total. Therefore experiment suggests that the beam remnants interact with high probability, as notated in Fig. 1.1(b). We will need to understand whether or not factorization gets violated. For the fully inclusive Drell-Yan process, we will show that a quite non-trivial cancellation applies, so that factorization actually works: Ch. 14.

Moreover, the beam remnants in Fig. 1.1(a) appear to have quantum numbers, e.g., fractional electric charge, that correspond to beam hadrons with a quark or antiquark removed. This is not observed in actual diffractive events.

Another issue that we will solve is that the colliding partons are often treated as being free and on-shell, even though they are bound inside their hadrons. In addition, the partons are treated as having certain fractions of their parent hadrons' momenta, even though the parton and hadron momenta cannot always be exactly parallel; there is certainly a distribution over the components of parton momentum transverse to the collision axis. We will see how this is allowed for in defining parton densities, and how in some situations we need to treat it more explicitly. One standard example is the Drell-Yan process when we take the cross section differential in the transverse momentum q_T of the $\mu^+\mu^-$ pair instead of integrating over it as we did in (1.1).

1.2 Why we trust QCD is correct

QCD is generally regarded as the correct theory of the strong interaction. The reasons are not just that it makes successful quantitative predictions. Also important are structural arguments, as summarized in Ch. 2. These arguments start from high-level abstractions from data, and make a quite rigid deduction of QCD as the theory of the strong interaction. There does not appear to be an underdetermination of theory by data. The arguments are like those that led Einstein to the theories of special and general relativity, and to those that led Heisenberg, Born, Jordan, and Dirac to founding the current formulation of quantum theory. Once QCD and asymptotic freedom were discovered, there was a positive feedback loop where successful experimental predictions confirmed QCD. But the structural arguments are also critical in giving the realization that QCD applies generally to strong-interaction phenomena, including those that not yet derived from QCD, e.g., the bulk of conventional nuclear physics.

As with many modern theories of physics, QCD and its applications have many elements that do not in any immediate sense correspond to tangible real-world entities. Parton densities are a good example. These elements, and the associated mathematics, provide links between many different kinds of experimental data.

Indeed it is in the links between experiments that many of the predictions of QCD arise. Perturbative calculations alone do not predict any cross sections in hadron-hadron scattering and lepton-hadron scattering. Factorization gives us cross sections in terms of non-perturbative parton densities, which we currently cannot compute from first principles. But from QCD we deduce that the parton densities are universal between different reactions, with energy-dependent modifications of universality caused by the evolution of parton densities with a scale parameter. So we can fit parton densities from some limited set of data, and use them to predict other cross sections, with the aid of perturbatively calculated coefficients like $d\hat{\sigma}$ in (1.1).

No single experiment really provides a critical test of QCD. But a sufficiently large collection of experiments does simultaneously provide measurements of non-perturbative factors, tests of the factorization structure, and tests of QCD itself. There are some interesting issues in statistics and in the philosophy of science here, which do not appear to arise in such a strong form in other areas of science.

1.3 Notation

As regards normalization conventions and the like, I generally follow the conventions of the Particle Data Group (PDG) (Amsler *et al.*, 2008), since this forms a standard for our field. I point out exceptions explicitly. For a collection of many notations and standard results, see App. A. For acronyms and abbreviations, see Sec. A.3.

For the most part, I use the "natural units" conventional to the field, where $\hbar = c = \epsilon_0 = 1$, and the GeV is used as the unit of energy. See Sec. A.2 for common conversion factors.

I have found some variations on the standard symbol for equality to be useful. First, to flag the definition of something, I put "def" over the "=", as in

$$Q \overset{\text{def}}{=} \sqrt{-q^2}. \tag{1.2}$$

Quite often, it is convenient to explain the definition of some conceptually difficult object by first proposing a simplified candidate definition based on a naive picture of the physics, and then building up to the correct definition, e.g., with parton densities, starting in Ch. 6. So I use the notation

$$\text{Quantity} \overset{\text{prelim}}{=} \text{preliminary candidate definition,} \tag{1.3}$$

for the early incorrect definitions. This avoids having several apparently incompatible definitions of the same quantity, with the wrong ones prone to being taken out of context.

Similarly there are situations where I motivate proofs and statements of difficult results by first formulating them in a simplified situation. For example, before formulating and deriving true factorization theorems, I examine the parton model, an intuitive approximation to real physics that is relatively easy to motivate. So I flag these suggestive but ultimately incorrect results with a question mark over an equality sign, e.g.,

$$F_{\text{L}} \overset{?}{=} 0. \tag{1.4}$$

1.4 Problems and exercises

I have devised a number of chapter-end exercises. Some of these are relatively elementary and should be tackled by anyone learning the subject. Among these are exercises to complete derivations in the text; some others explore the conceptual framework by the derivation of further results. There are also harder exercises, rated with one to five stars. Those with three or more stars are really research problems. I do not necessarily have any answers or even suggestions for approaches to the research problems; a good solution could easily be suitable for journal publication.

2

Why QCD?

A possible approach to a theory like QCD is just to state its definition, and then immediately proceed deductively. However, this begs the question of why we should use this theory and not some other. Moreover, the approach is quite abstract, and the initial connection to the real physical world is missing.

Instead, I will take a quasi-historical approach, after first stating the theory. Such an approach is suitable for newcomers since their background in QCD is like that of its inventors/discoverers, i.e., little or none. There were several lines of development, all of which powerfully converged on a unique theory from key aspects of experimental data. Of course, we see this much more readily in retrospect than was apparent at the time of the original work, and my account is selective in focusing on the issues now seen to be the most significant. A historical approach also enables the introduction of ideas and methods that do not specifically depend on QCD: e.g., deeply inelastic scattering and the parton model.

I have tried to make the presentation self-contained, in summarizing the relevant experimental phenomena and their consequences for theory. The reader is only assumed to have a working knowledge of relativistic quantum field theory. Inevitably there are issues, ideas, experiments, and historical developments which will be unfamiliar to many readers, and for which a complete treatment needs much more space. I give references for many of these. In addition, there are several references that are global to the whole chapter and that the reader should refer to for more detail. A detailed historical account from the point of view of a physicist is given in the excellent book by Pais (1986). A good account of the phenomena is given by Perkins (2000). Standard books on quantum field theory also refer to them; see, for example, Sterman (1993); Peskin and Schroeder (1995); Weinberg (1995, 1996); Srednicki (2007). A comprehensive account of experimental results is given by the Particle Data Group in Amsler et al. (2008); this includes up-to-date authoritative summaries of measurements and their theoretical interpretation.

Naturally, QCD is not the whole story; there are known electromagnetic, weak and gravitational interactions, and presumably if we examine phenomena at short enough distances, beyond the reach of current experimental probes, we are likely to need new theories. But within the domain of the strong interaction at accessible scales, there is an amazing uniqueness to the structure of QCD.

2.1 QCD: statement of the theory

An expert in quantum field theory could simply define QCD as a standard Yang-Mills theory with a gauge group SU(3) and several multiplets of Dirac fields in the fundamental (triplet) representation of SU(3).

In more detail, QCD is specified by its set of field variables and its Lagrangian density $\mathcal{L}$. The Dirac fields $\psi_{\rho af}$ are called quark fields, and the gauge fields A_μ^α are called gluon fields. On the quark fields the indices ρ, a, and f are respectively a Dirac index, a "color" index taking on three values, and a "flavor" index. The gauge group acts on the color index. Currently the flavor index has six known values u, d, s, c, b, t (or "up", "down", "strange", "charm", "bottom", and "top"). On the gluon field, the color index α has eight values, for the generators of SU(3), and μ is a Lorentz vector index. The important role played by the color charge leads to the theory's name, "quantum chromodynamics" or QCD. Of course, the names "color" and "flavor", and the names of the quark flavors, are whimsical inventions unrelated to their everyday meanings.

To deal with the renormalization of the UV divergences of QCD (Sec. 3.2) we distinguish between bare and renormalized quantities (fields, coupling and masses). We define QCD by a Lagrangian written in terms of bare quantities, which are distinguished by a subscript 0 or (0). The gauge-invariant Lagrangian is the standard Yang-Mills one:

$$\mathcal{L}_{\text{GI}} = \bar{\psi}_0(i\slashed{D} - m_0)\psi_0 - \frac{1}{4}(G_{(0)\,\mu\nu}^\alpha)^2. \tag{2.1}$$

The full Lagrangian used for perturbation theory will add to this some terms to implement gauge fixing by the Faddeev-Popov method; see Sec. 3.1. The covariant derivative is given by

$$D_\mu\psi_0 \stackrel{\text{def}}{=} \left(\partial_\mu + ig_0 t^\alpha A_{(0)\,\mu}^\alpha\right)\psi_0, \tag{2.2}$$

where t^α are the standard generating matrices[1] of the SU(3) group, acting on the color indices of ψ. The gluon field strength tensor is

$$G_{(0)\,\mu\nu}^\alpha \stackrel{\text{def}}{=} \partial_\mu A_{(0)\,\nu}^\alpha - \partial_\nu A_{(0)\,\mu}^\alpha - g_0 f_{\alpha\beta\gamma} A_{(0)\,\mu}^\beta A_{(0)\,\nu}^\gamma, \tag{2.3}$$

where $f_{\alpha\beta\gamma}$ are the (fully antisymmetric) structure constants of the gauge group, defined so that $[t_\alpha, t_\beta] = if_{\alpha\beta\gamma}t_\gamma$. The Lagrangian is invariant under local (i.e., space-time-dependent) SU(3) transformations:

$$\psi_{(0)\,\rho af}(x) \mapsto \left[e^{-ig_0\omega_a(x)t^\alpha}\right]_{ab}\psi_{(0)\,\rho bf}(x), \tag{2.4a}$$

$$A_{(0)\,\mu}^\alpha(x)t^\alpha \mapsto \frac{-i}{g_0}e^{-ig_0\omega_a(x)t^\alpha}D_\mu e^{ig_0\omega_a(x)t^\alpha}. \tag{2.4b}$$

The quark fields have been redefined, as is always possible (Weinberg, 1973a), so that the mass matrix is diagonal:

$$\bar{\psi}_0 m_0 \psi_0 = m_{0u}\bar{u}_0 u_0 + m_{0d}\bar{d}_0 d_0 + m_{0s}\bar{s}_0 s_0 + \dots \tag{2.5}$$

---

[1] $t^\alpha = \frac{1}{2}\lambda^\alpha$, where the standard λ^α are given in, e.g., Amsler *et al.* (2008, p. 338).

Here separate symbols are used for the fields for different flavors of quark: $u_{0\,\rho a} = \psi_{0\,\rho au}$, etc.

The renormalized masses of the quarks are given in Table 2.2 below, along with the masses of the other elementary particles of the Standard Model. Large fractional uncertainties for the light quark masses arise because quarks are in fact confined inside color-singlet hadrons, which gives considerable complications in relating the mass parameters to data.

For their electromagnetic interactions, we need the quark charges:

$$e_d = e_s = e_b = -1/3, \qquad e_u = e_c = e_t = 2/3, \tag{2.6}$$

in units of the positron charge.

The only significant freedom in specifying QCD is in the set of matter fields, the quarks. At the time of discovery of QCD, only the u, d and s quarks were known; the c quark came slightly later. The discovery of the b and t quarks needed high enough collision energies to produce them. There have been many conjectures about possible new heavy quarks, both scalar and fermion, possibly in non-triplet color representations, but searches so far have been unsuccessful (Amsler *et al.*, 2008). The decoupling theorem (Appelquist and Carazzone, 1975) for heavy fields ensures that we can ignore the heavy fields if experimental energies are too low to make the corresponding particles.

A complete theory of strong, electromagnetic, and weak interactions is made by combining QCD with the Weinberg-Salam theory to form the Standard Model of elementary particle physics, summarized in Sec. 2.7.

2.2 Development of QCD

Why we should postulate the QCD Lagrangian and study QCD as the unique field theory for the strong interaction? An answer to this question should be at a high level and broad, since QCD is a high-level theory, intended to cover a broad range of phenomena, i.e., all of the strong hadronic interaction.

Starting in the 1950s, as accelerator energies increased, elementary particle physics gradually became a separate subject, distinct from nuclear physics. Several, not entirely distinct, strands of research led to the discovery of QCD in 1972–1973:

1. The quark model of hadron states.
2. The (successful) search for a theory of the weak interactions of leptons, including the weak interactions of hadrons.
3. Current algebra, i.e., the analysis of the currents for the (approximate) flavor symmetries of the strong interaction, including their relationships to the electroweak interactions of hadrons.
4. The theoretical development of non-abelian gauge theories.
5. Deeply inelastic lepton scattering and the measurement that the strong interaction is quite weak at short distances.

It is almost paradoxical that many of the key issues involved the weak and electromagnetic interactions; much of the research on pure strong-interaction phenomena was not critical to the discovery of QCD.

2.2.1 Quantum fields

Always present was the notion of quantum field theory. Soon after the discovery of quantum mechanics, it was apparent that quantum fields formed an appropriate candidate framework in a search for an all-encompassing underlying theory of known interactions.

First, the basic dynamical variables are local in a field theory, so that there are separate variables to discuss, for example, an experiment in Illinois yesterday and an experiment in Switzerland tomorrow. This happens even in non-relativistic quantum theory. A theory of interacting quantum Schrödinger fields is readily constructed; this theory can be shown (Fetter and Walecka, 1980; Brown, 1992) to be equivalent to a collection of ordinary quantum mechanical theories in terms of N-body wave functions, but now for any N and with specified inter-particle interactions. In contrast, an ordinary Schrödinger equation for a wave function concerns, for example, only one particular electron and proton. But a quantized field theory can be formulated to describe all possible electrons and nuclei. Thus it encompasses all of atomic and molecular physics, not to mention chemistry, etc. Of course to take account of radiative phenomena, one also needs the electromagnetic field.

Since quantum field theories are intrinsically many-body theories, they are suitable for the construction of quantum theories that obey Einstein's special relativity. Once sufficient energy is available in a collision, particles can be created, so that a framework where particles are conserved is wrong. Fig. 2.3 below serves as an icon of this: it shows the multiparticle outcome of one particular positron-proton collision.

Furthermore, it is natural in relativity that fields obey local field equations, written in terms of fields and their derivatives. A non-local interaction would involve action at a distance, and would require enormous conspiracies to avoid faster-than-light propagation, etc.

To obtain a quantum field theory, it is sensible to start by postulating fields that correspond to observed particles, and then asking what interactions, governed by non-linear terms in the field equations, give observed phenomena. This approach was successful for the electromagnetic interaction and gave us the theory called QED. With a long delay to allow the full formulation of the needed non-abelian gauge theories, this approach was also successful for the weak interaction. Considerable restrictions were applied to the candidate theories, concerning self-consistency and renormalizability.

But for the strong interaction, there was a failure of this obvious approach, where one searches for a theory written in terms of fields for observed hadrons, initially the nucleons and pions. In retrospect, the reason is obvious: hadrons are composite, with the size of the bound states (Hofstadter, Bumiller, and Yearian, 1958), around $1 \, \text{fm} = 1 \times 10^{-15} \, \text{m}$, being much less than the range of the strong nucleon-nucleon potential and the inter-nucleon separation in atomic nuclei (Hofstadter, 1956).

During the 1960s it became conventional, instead, to suppose that something other than a quantum field theory was needed for the strong interaction, an ultimately fruitless quest.[2] See the later chapters of Pais (1986) for a historical account.

2.2.2 Quark model

In strong-interaction physics very many unstable particle-like states, or resonances, have been discovered (Amsler *et al.*, 2008). They are generically termed hadrons. No fundamental distinction between the unstable and stable hadrons appeared to exist, stable hadrons simply being those that have no available decay channels. One natural hypothesis is that these states are bound states of more elementary particles, which turned out to be the actual case. The establishment of this view, starting in the 1950s, was quite non-trivial, however. Tightly coupled with these developments was the discovery that the strong interaction is approximately invariant under an internal symmetry, called SU(3) flavor symmetry; see, e.g., Gell-Mann (1962).

Within QCD, SU(3) flavor transformations are applied to the *u*, *d* and *s* quark fields, and would give an exact symmetry if the masses of the *u*, *d* and *s* quarks were equal. We get a useful approximate symmetry because the masses (and hence the mass differences) of these lightest three quarks are substantially less than a "normal hadronic mass scale", characterized by the proton mass. The *c*, *b* and *t* quarks (not known until after the construction of QCD) are singlets under SU(3) flavor transformations.

Flavor SU(3) symmetry is to be carefully distinguished from the later-discovered color symmetry group, which is also mathematically SU(3).

Gell-Mann (1964) and Zweig (1964a, b) constructed the quark model, in which baryons (like the proton and neutron) are bound states of three quarks, and mesons are bound states of a quark and antiquark. For the hadrons known at the time, they used three spin-$\frac{1}{2}$ quarks (*u*, *d* and *s*), with the fractional charge assignments of (2.6).

Now the *u*, *d* and *s* quarks are in the triplet representation of flavor SU(3). It follows (Gell-Mann, 1964) that baryons can be classified into multiplets that are singlet, octet and decuplet under SU(3), while the mesons are singlets and octets. Prior to the discovery of a satisfactory theory of the strong interaction, i.e., QCD, it was useful to investigate the consequences of the flavor symmetry abstractly, independently of any assumptions about a quark substructure or a Hamiltonian; see Sec. 2.2.4. Patterns of mass splitting within hadron multiplets can be understood quantitatively by using perturbation theory applied to symmetry-breaking terms in the strong-interaction Hamiltonian, with the hypothesis that symmetry-breaking terms are in an SU(3) octet. These terms are now identified with quark mass terms in QCD. See Amsler *et al.* (2008, Ch. 14) for a recent review and further references.

Each flavor of quark appeared to need three varieties (called "colors") in order for the spin-statistics theorem to hold. This is seen most easily for the $\Delta^{++}(1232)$

---

[2] Although the quest for a non-QFT theory of the strong interaction failed, it did lead to the invention of string theory, which now leads a prominent life as a candidate fundamental theory of everything including gravity.

baryon.[3] It is a ground-state baryon of spin $\frac{3}{2}$ consisting of three u quarks, so both the space and spin wave functions are totally symmetric.[4] But a side effect of the color hypothesis is that each meson (e.g., π^+) has an extra eight color states, which are not observed. An extra assumption is needed to prohibit the extra states.

Furthermore, there is a complete failure to detect isolated quarks in high-energy collisions, which requires the hypothesis that quarks are permanently confined in hadrons. Quark confinement obviously makes it harder to deduce from data the correct bound-state structure.

Thus there was a continued introduction of new hypotheses, which led to great scepticism (Zweig, 1980). Nevertheless, the situation was the unusual one of a correct general idea being forced by data into a unique implementation. In favor of the quark model, calculations with phenomenological interquark potentials allowed calculations for the energies of excited hadrons (non-ground-state hadrons), in essential agreement with data.

Around 1972, Fritzsch and Gell-Mann (1972) and Fritzsch, Gell-Mann, and Leutwyler (1973) had the inspiration that a non-abelian gauge theory, with an SU(3) gauge symmetry applied to the color degree of freedom, could not only give all these properties of the quark model, but could also solve other puzzles involving current algebra and the weak interactions of hadrons: Secs. 2.2.3 and 2.2.4.

Somewhat tentatively they proposed exactly the theory now known as QCD, missing only the heavy quarks, which in any event decouple from lower-energy physics and therefore do not affect the arguments. Understanding of the dynamics of the theory was still missing, in particular for the observations in deeply inelastic scattering: Sec. 2.3.

As regards the quark model, the unifying hypothesis suggested by the structure of QCD is that of "color confinement", that all observed states are color singlet. It simultaneously solves the quark confinement problem and the lack of extra meson states, and it is a natural conjecture, since gluons couple to color charge. Already in lowest-order perturbation theory it can be seen that the gluon exchange energy for a quark-antiquark pair is attractive for the color singlet state and repulsive for the color octet state: problem 2.1. Of course, perturbation theory for a generic strong-interaction quantity is at best a rough approximation. Even so, although a real demonstration of color confinement from QCD has still not been found, the hypothesis is consistent with all the evidence, theoretical and experimental.

The terms in the QCD Lagrangian that correspond to differences of quark masses give an operator in the Hamiltonian that transforms as an octet under *flavor* SU(3). This is exactly what had previously been assumed to explain mass splittings in the hadronic flavor multiplets.

---

[3] In this notation, the number denotes the mass in MeV, i.e., 1232 MeV, while the Δ^{++} denotes the quark and isospin content of the state (Amsler *et al.*, 2008, Ch. 8), which in this case corresponds to a baryon of isospin 3/2 with charge +2.

[4] The possibility that there are other types of particle statistics than Bose or Fermi was considered under the names of "para-statistics" or "quark statistics". But it was shown by Doplicher, Haag, and Roberts (1974) that all these possibilities are equivalent to ordinary Bose or Fermi statistics supplemented by selection rules on the allowed states. See also Drühl, Haag, and Roberts (1970). So the color solution is generic.

2.2.3 Weak interactions

By the early 1970s there was a leading candidate for electroweak interactions of leptons, the Weinberg-Salam theory (Weinberg, 1967; Salam, 1968). This theory used spontaneous symmetry breaking to give mass to the weak gauge-bosons. It became a genuine candidate theory after it was shown how to successfully quantize and renormalize non-abelian gauge theories, and has since become fully established. This work solved severe consistency problems of theories with massive charged vector fields.

How is the theory to be extended to include hadrons? We treat the situation perturbatively in the electroweak interactions, using a decomposition of the complete Hamiltonian as

$$H = H_{SI} + H_{0, \text{lept}} + H_{I, EW} + H_{SI-EW}. \tag{2.7}$$

Here H_{SI} is the full strong-interaction Hamiltonian, not yet known around 1970, $H_{0, \text{lept}}$ is the free Hamiltonian for non-hadronic fields, $H_{I, EW}$ is the interaction Hamiltonian for electroweak interactions, and H_{SI-EW} gives the coupling between hadronic fields and the electroweak fields.

We now do time-dependent perturbation theory with the unperturbed Hamiltonian including the *full* strong-interaction part, i.e., $H_0 = H_{SI} + H_{0, \text{lept}}$. Useful information can be extracted without either knowing or solving the full strong-interaction theory. The reason is that the couplings between the strong-interaction fields and the electroweak gauge fields were found from phenomenological evidence to involve currents for hadronic flavor symmetries. We write these in the form

$$H_{SI-EW} = \int d^3x \sum_A j_A^\mu W_{A,\mu} + \text{Higgs terms}, \tag{2.8}$$

where $W_{A,\mu}$ are the electroweak gauge fields $W^\pm$, Z and γ, while j_A^μ are the hadronic currents to which they couple. For consistency of the electroweak theory, the hadronic currents must be conserved, apart from the effects of their couplings to the electroweak fields. In fact, the currents, within the strong-interaction sector, are not quite conserved, which appears to be somewhat inconsistent. The inconsistency is solved retrospectively by the full Standard Model, where the non-conservation is caused by quark mass terms in QCD. Since quark masses arise from the vacuum expectation value of the Higgs field in the Yukawa couplings for the quarks, the lack of conservation of the flavor currents within QCD is essentially associated with weak interactions.

This form of perturbation theory, where the unperturbed Hamiltonian contains the full strong-interaction Hamiltonian leads to normal Feynman perturbation theory only for the electroweak fields (leptons, photon, etc.). In the strong-interaction part, the electroweak gauge fields are coupled to matrix elements of currents. For example, the decay of the neutron to $p + e + \bar{\nu}_e$ (Fig. 2.1) has an amplitude

$$\langle p, \text{out} | j_-^\mu(0) | n, \text{in} \rangle \frac{-ig_{\mu\lambda}}{q^2 - m_W^2} \bar{u}_e \gamma^\lambda (1 - \gamma_5) v_\nu \times \text{couplings}, \tag{2.9}$$

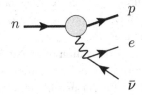

Fig. 2.1. Lowest-order weak interaction for neutron decay.

where j_-^μ is the hadronic current to which the W^+ couples, u_e and v_v are standard Dirac spinors for the states of the leptons, $q = p_n - p_p$ is the momentum transfer, and p_n and p_p are the 4-momenta of the neutron and proton.

2.2.4 Current algebra

For further references and for a more detailed historical account of the issues treated in this section, see Pais (1986, Ch. 21).

Initially, with no known theory of the strong interaction, and with no complete theory of the weak interaction, it was measured that the weak interactions of hadrons involved current matrix elements as in (2.9). This led to the subject of current algebra, i.e., the study of hadronic current operators. The current coupled to the W boson appears as one of the currents for an approximate symmetry group of the strong interactions. This group, a chiral $SU(3) \otimes SU(3)$ group, will be discussed further in the context of QCD in Sec. 3.8, together with its more exact $SU(2) \otimes SU(2)$ subgroup. Explicit breaking of the symmetry is caused by the relatively small mass terms for the u, d and s quarks in QCD.

It was found that the symmetries are spontaneously broken to a "vector" $SU(3)$ or $SU(2)$, with the pions being the Goldstone bosons for the $SU(2) \otimes SU(2)$ case. The explicit symmetry breaking by quark masses implies that the pion is not massless but simply much lighter than other hadrons. The residual vector $SU(3)$ symmetry is the one that is prominent in the quark model: Sec. 2.2.2.

Many consequences of the Ward identities for these symmetries were derived, in particular soft pion theorems. See, e.g., Treiman, Jackiw, and Gross (1972). One dramatic example is the Goldberger-Treiman relation that gives a relation between the matrix element in (2.9) and the long-distance part of the pion exchange contribution to the nucleon-nucleon potential; it thus relates a measurement of a weak-interaction quantity to an apparently very different quantity in pure strong-interaction physics.

Studies of symmetries require understanding of commutators of currents. This led to the study of matrix elements of two currents, like $\langle P \,|\, j^\mu(x) j^\nu(0) \,|\, P \rangle$, which are investigated experimentally in deeply inelastic scattering: Sec. 2.3.

A natural problem was now to find a theory that supports current algebra, i.e., a theory in which the currents are ordinary Noether currents and have the commutation relations postulated in current algebra. What excited Fritzsch and Gell-Mann (1972) and Fritzsch, Gell-Mann, and Leutwyler (1973) was that their proposed QCD Lagrangian not only could explain the quark model but naturally gave current algebra. The symmetry properties of the quark mass terms are exactly those used for the symmetry-breaking part of the strong-interaction Hamiltonian in current algebra.

Around 1970 it was found that the derivation of certain Ward identities for products of three currents fails in real field theories. It was found, moreover, that the resulting anomalies are correctly calculated within lowest-order perturbation theory; higher-order corrections are exactly zero according a theorem due to Adler and Bardeen. The methods of current algebra then enabled the decay rate for $\pi^0 \to \gamma\gamma$ to be calculated to the extent that the masses of the u and d quarks are small. Agreement with the observed decay rate is obtained if each flavor of quark has three color states. See Peskin and Schroeder (1995, Ch. 19).

Another line of argument related to current algebra was by Weinberg (1973a, b), who considered weak-interaction corrections to strong-interaction phenomena. In a generic candidate theory for the strong interaction, loop graphs have unsuppressed contributions from momenta around the W mass. The resulting violations of strong-interaction symmetries (e.g., parity) would be electromagnetic in strength, contrary to observation. Weinberg showed that this problem is avoided if the strong interaction is mediated by exchange of bosons whose symmetries commute with those for the electroweak bosons. This is the case for QCD, where color SU(3) commutes with the electroweak gauge group. The revolutionary consequence is that flavor symmetries were demoted from fundamental properties of the strong interaction to apparently accidental and approximate symmetries that occur because of the small size of the Yukawa couplings of the Higgs field to the light quarks.

2.2.5 *Non-abelian gauge theories*

The discovery of QCD needed a parallel track of purely theoretical work to formulate non-abelian gauge theories and establish their consistency. The initial formulation was by Yang and Mills (1954), who beautifully generalized the concept of local gauge invariance from the abelian symmetry of QED to a non-abelian group. Their attempt to apply their theory to the actual strong interaction foundered on the prejudice that the fields in the Lagrangian should correspond to observed particles, contrary to the now-known reality.

But the theoretical idea remained. With the discovery of the concept of spontaneous symmetry breaking, Weinberg (1967) and Salam (1968) found what is in fact the correct theory of electroweak interactions. At about the same time, Faddeev and Popov (1967) showed how to quantize such theories consistently. After this, it was quickly found how to derive Ward identities and thence to show that Yang-Mills theories, possibly including spontaneous symmetry breaking, are renormalizable.

With this, non-abelian gauge theories became fully fledged consistent field theories, setting the stage for the developments outlined in the preceding sections.

2.3 Deeply inelastic scattering

In parallel with work just described, the remaining developments that led to the establishment of QCD as the theory of strong interactions concerned deeply inelastic scattering of leptons (DIS). Since this process remains an important subject of study in QCD, we now examine those aspects that do not depend on knowing the strong-interaction Lagrangian.

We consider scattering of a lepton of momentum l^μ on a hadron N of momentum P^μ to an outgoing lepton of momentum l'^μ plus anything:

$$l + N(P) \longrightarrow l' + X. \tag{2.10}$$

The symbol X has a standard connotation, that we work with an *inclusive* cross section, i.e., a cross section differential in lepton momentum l', with a sum and integral over all possible states for the X part of the final state. Effectively only the lepton is treated as being detected.

There are a number of cases with different types of lepton for which there is experimental data: $e + N \longrightarrow e + X$, $e + N \longrightarrow \nu + X$, $\mu + N \longrightarrow \mu + X$, $\nu + N \longrightarrow \nu + X$, $\nu + N \longrightarrow (e$ or $\mu) + X$. When the momentum transfer at the lepton side is large, as we will see, we effectively have a powerful microscope into the initial-state hadron N. In actual data, N is either a proton or a heavier nucleus. Scattering on a nucleus is often approximated as scattering on an incoherent mixture of protons and neutrons. For more accurate work, "nuclear corrections" are applied to obtain cross sections relative to independent protons and neutrons.

In this section we will only treat the electron-to-electron case, for which the current state of the art for high energy is at the recently shut-down HERA accelerator at the DESY laboratory. There an electron (or positron) beam of energy 27.5 GeV was collided against a proton beam of energy 920 GeV, with a center-of-mass energy of $\sqrt{s} = 318$ GeV.

2.3.1 General considerations

Consider a wide-angle scattering of the electron in the center-of-mass frame, Fig. 2.2. The large space-like momentum transfer, $q^\mu = l^\mu - l'^\mu$, for the (essentially point-like) electron suggests that a short-distance scattering is necessary, which would naturally occur off a small constituent of the hadron. If we let the invariant momentum transfer be $Q = \sqrt{-q^2}$, then a natural distance scale is $1/Q$ (in units with $\hbar = c = 1$). At HERA there is data to above $Q = 100$ GeV, with a corresponding distance of less than 10^{-2} fm.

An enormous simplification occurs because, at high energy, the hadron is Lorentz-contracted and time-dilated[5] by a large factor, which is about 150 in the center-of-mass at HERA. A hadron like a proton has a size (Hofstadter, Bumiller, and Yearian, 1958) of around 1 fm, so it is reasonable to say that *in the hadron's rest frame* the constituents interact with each other on a time scale of order 1 fm/c. In the boosted hadron, as seen in the center-of-mass frame of the scattering, time dilation implies that the last interaction of the constituents typically occurred a long distance upstream. In the HERA center-of-mass frame, this is of order 100 fm, which is much larger than the scale of the electron scattering.

---

[5] These concepts are non-trivial (Gribov, 1973, p. 12) for microscopic particles in a quantum field theory, but that does not affect the motivational issues for this section.

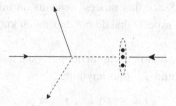

Fig. 2.2. Deeply inelastic scattering of an electron on a proton. The electron comes from
the left and the proton from the right. In the diagram, the electron (solid line) is depicted as
point-like and the hadron as a Lorentz-contracted extended object. The three dots inside the
proton symbolize the three quarks that are its constituents in the quark model. The struck
parton is indicated by a dashed line. Drawing a realistic Lorentz contraction would result
in a much thinner proton than shown here.

This suggests (Feynman, 1972) that in the short-distance electron-constituent collisions
it is a useful approximation to neglect the interactions that bind the constituents into a
hadron. Quantitative development of this idea leads first to the "parton model", to be
explained in Sec. 2.4, and then to the factorization theorems of QCD, which give a precise
and correct mathematical formulation of the intuitive ideas.

   In the original DIS experiments at SLAC, in the early 1970s, only the outgoing electron
was detected; there was no sensitivity to the rest of the final state. Moreover the electron
beam energy was at most 21 GeV on a fixed target. Modern experiments, like the ZEUS
and H1 detectors at HERA, can see the hadronic final state. An example event with
$Q = 158$ GeV is shown in Fig. 2.3. It supports the intuitive picture: an isolated wide-angle
electron recoils against a narrow group of particles, called a jet, which is reminiscent of the
scattered constituent. The scattered constituent (the "parton" in Feynman's terminology)
does not retain its identity as a single particle except at sufficiently microscopic distances;
this is of course compatible with the idea that quarks are permanently confined in hadrons
and never appear as isolated single particles. The standard view (Andersson, 1998) is that
many quark-antiquark pairs are created by the intense gluon field between an outgoing
struck quark and the proton remnant. These form into color-singlet hadrons, mostly pions,
that go in roughly the direction of the outgoing quark. The remnants of the proton continue
in motion, with excitation and only a small deflection: these cause hits in the detector
segments around the beam pipe, at the left of Fig. 2.3(a). Much of the remnant energy is
too close to the beam direction to be detected.

2.3.2 Kinematics; structure functions

We work to lowest order in electromagnetism and in this section we will ignore weak
interactions.[6] Then the amplitude for a contributing process is represented diagrammatically

---

[6] Unless Q is at least of order the masses of the W and Z bosons, weak-interaction effects are suppressed, by a factor
of Q^2/m_W^2. Higher-order electromagnetic corrections are smaller by a factor of roughly α/π, except for infra-red
dominated terms associated with the masslessness of the photon. It is conventional to present data "with the effects
of radiative corrections removed", so that higher-order electromagnetic corrections are effectively absent in published
data. The formalism is readily extended, with only notational complications, to deal with exchange of weak-interaction
bosons. See Sec. 7.1.

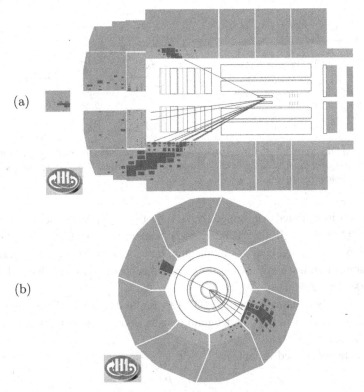

Fig. 2.3. Scattering event in an positron-proton collision in the H1 detector (H1 website, 2010) at a center-of-mass energy of about 320 GeV. The detector is approximately cylindrically symmetric about the center line which contains the beam pipes. Both a side view (a) and an end view (b) are shown. In (a), electrons come from the left, and protons from the right. One isolated track was identified as an electron, and there is a recoiling jet, approximately back-to-back in azimuth. The kinematic variables are $Q^2 = 25\,030\,\text{GeV}^2$ and $y = 0.56$ (see Sec. 2.3.2).

in Fig. 2.4(a), and is a product of a lowest-order leptonic vertex, a photon propagator, and a hadronic matrix element of the electromagnetic current, $\langle X, \text{out} | j^\mu | P \rangle$. The two independent Lorentz invariants for the hadron system are $Q^2 \overset{\text{def}}{=} -q^2 \geq 0$ and $P \cdot q$, both of which can be computed from the measured momenta l, l' and P, with the momentum of the exchanged photon being $q^\mu = l^\mu - l'^\mu$. The mass of the hadronic final state is then

$$m_X^2 = (P + q)^2 = M^2 + 2P \cdot q - Q^2, \tag{2.11}$$

where M is the mass of the initial-state hadron. A convenient combination of variables is Q and the Bjorken variable

$$x \overset{\text{def}}{=} \frac{Q^2}{2P \cdot q}. \tag{2.12}$$

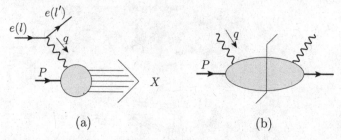

Fig. 2.4. (a) DIS amplitude to lowest order in electromagnetism. (b) Hadronic part squared and summed over final states. For the meaning of the vertical "final-state cut", see the discussion below (2.19).

Kinematically x is restricted to the range $Q^2/(s + Q^2) \leq x \leq 1$ (with fractional corrections of order M^2/Q^2 being neglected). In the parton model we will find that x gives an estimate of the fraction of the initial hadron's momentum that is carried by the struck parton. That the term "momentum fraction" has a useful meaning depends on the relativistic kinematics of the process: Sec. 2.4.

The term "deeply inelastic scattering" (DIS) applies to the region where both Q and m_X are large, so that there is a large momentum transfer and the hadron target is very much excited, inelastically.

Another commonly used variable is

$$y \overset{\text{def}}{=} \frac{q \cdot P}{l \cdot P}.$$
(2.13)

It lies between 0 and 1. In the rest frame of the hadron, this is the fractional energy loss of the lepton: $(E - E')/E$, so that it is simple to measure in a fixed target experiment. But it is not an independent variable, since

$$Q^2 = xy(s - M^2 - m_e^2).$$
(2.14)

The Lorentz-invariant inclusive cross section is then

$$E' \frac{\mathrm{d}\sigma}{\mathrm{d}^3 l'} \simeq \frac{\pi e^4}{2s} \sum_X \delta^{(4)}(p_X - P - q) \left| \langle l' | j_\lambda^{\text{lept}} | l \rangle \frac{1}{q^2} \langle X, \text{out} | j^\lambda | P \rangle \right|^2$$

$$= \frac{2\alpha^2}{s Q^4} L_{\mu\nu} W^{\mu\nu}.$$
(2.15)

In the prefactor, we have neglected the electron mass m_e and the hadron mass M compared with $\sqrt{s}$, while the fine-structure constant is $\alpha = e^2/(4\pi)$. The sum over X denotes the usual Lorentz-invariant sum and integral over all hadronic final states. The currents j_λ^{lept} and j^λ are respectively the electromagnetic currents for the leptons and for the hadronic fields. In QCD the electromagnetic current involves a sum over quark flavors:

$$j^\lambda = \sum_f e_f \bar{\psi}_f \gamma^\lambda \psi_f = \tfrac{2}{3}(\bar{u}\gamma^\lambda u + \ldots) - \tfrac{1}{3}(\bar{d}\gamma^\lambda d + \ldots).$$
(2.16)

In the second line of (2.15), we have separated out factors for the leptonic and hadronic parts. The leptonic tensor is obtained from lowest-order Feynman graphs, and in the unpolarized case is

$$L_{\mu\nu} = \tfrac{1}{2} \operatorname{Tr} \gamma_\nu l \gamma_\mu l' = 2(l_\mu l'_\nu + l'_\mu l_\nu - g_{\mu\nu} l \cdot l'). \tag{2.17}$$

The hadronic tensor is defined as a complete matrix element,

$$W^{\mu\nu}(q, P) \overset{\text{def}}{=} 4\pi^3 \sum_X \delta^{(4)}(p_X - P - q) \langle P, S | j^\mu(0) | X \rangle \langle X | j^\nu(0) | P, S \rangle$$

$$= \frac{1}{4\pi} \int d^4z \; e^{iq \cdot z} \langle P, S | \; j^\mu(z) \, j^\nu(0) \; | P, S \rangle \tag{2.18}$$

in the full strong-interaction theory. The normalization is a standard convention, and the variable S labels the spin state of the target. In general, this may be a mixed state, and the notation $\langle P, S | \ldots | P, S \rangle$ is a shorthand for a trace with a spin density matrix: see App. A.7, and especially (A.8) and (A.13), for details. For the usual case of a spin-$\tfrac{1}{2}$ target, the spin state is determined by its (space-like) spin vector S^μ, which obeys $S \cdot P = 0$. We normalize S^μ as in Amsler *et al.* (2008), so that $S^2 = -M^2$ for a pure state.

To obtain the last line of (2.18), we used a standard result for the transformation of fields under space-time translations:

$$\langle P, S | j^\mu(z) | X, \text{out} \rangle = \langle P, S | j^\mu(0) | X, \text{out} \rangle \, e^{i(P - p_X) \cdot z}. \tag{2.19}$$

This allows the conversion of the momentum-conservation delta function to an integral over position. Then we used the completeness relation: $\sum_X | X, \text{out} \rangle \langle X, \text{out} | = I$.

Diagrammatically, we use the cut-diagram notation of Fig. 2.4(b) to represent $W^{\mu\nu}$. There the vertical line is called a "final-state cut". It represents the final state $| X, \text{out} \rangle$, and implies a sum and integral over all possible out-states $| X, \text{out} \rangle$. The part of the diagram to the left of the final-state cut is an ordinary amplitude $\langle X, \text{out} | j^\nu(0) | P, S \rangle$; in perturbation theory it is a sum over ordinary Feynman graphs with the appropriate on-shell conditions. The part to the right of the cut is a *complex-conjugated* amplitude, in this case $\langle P, S | j^\mu(0) | X, \text{out} \rangle = \langle X, \text{out} | j^\mu(0) | P, S \rangle^*$.

The Particle Data Group's definition (Amsler *et al.*, 2008) of $W^{\mu\nu}$ differs in replacing $j^\mu(z) \, j^\nu(0)$ by the commutator $[j^\mu(z), j^\nu(0)]$, but I find it better to use the more obvious definition with the simple product. The other definition is a relic from the period when the commutator was a dominant topic of research. The second term in their commutator $- j^\nu(0) \, j^\mu(z)$ gives a contribution equal to $-W^{\nu\mu}(-q, P)$, so the commutator version can be reconstructed from knowledge of $W^{\mu\nu}$. In fact, for a given value of q, only one of the two terms in the commutator contributes, since, when P is the momentum of a stable single-particle state, only one of $P + q$ and $P - q$ is the momentum of a physical state that can be used for $| X, \text{out} \rangle$.

We now decompose $W^{\mu\nu}$ into fixed tensors times scalar functions. For this we observe that:

- The electromagnetic current is conserved, $\partial \cdot j = 0$, so that $q_\mu W^{\mu\nu} = W^{\nu\mu} q_\mu = 0$.
- $W^{\mu\nu}$ is linear in the spin vector, which is an axial vector.

- The strong interactions are parity invariant.
- $W^{\mu\nu}$ is a hermitian matrix, i.e., $(W^{\mu\nu})^* = W^{\nu\mu}$.

Then the most general form of the tensor is

$$W^{\mu\nu} = \left(-g^{\mu\nu} + \frac{q^\mu q^\nu}{q^2}\right) F_1(x, Q^2) + \frac{(P^\mu - q^\mu P \cdot q/q^2)(P^\nu - q^\nu P \cdot q/q^2)}{P \cdot q} F_2(x, Q^2)$$

$$+ i\epsilon^{\mu\nu\alpha\beta} \frac{q_\alpha S_\beta}{P \cdot q} g_1(x, Q^2) + i\epsilon^{\mu\nu\alpha\beta} \frac{q_\alpha \left(S_\beta - P_\beta \frac{S \cdot q}{P \cdot q}\right)}{P \cdot q} g_2(x, Q^2). \tag{2.20}$$

The scalar coefficients F_1, F_2, g_1, and g_2 are called structure functions. The invariant antisymmetric tensor $\epsilon_{\kappa\lambda\mu\nu}$ obeys $\epsilon_{0123} = +1$, i.e., $\epsilon^{0123} = -1$, a convention that is not universal.

2.3.3 Breit/brick-wall frame; helicity analysis

For much of our work, it will be convenient to use the so-called Breit frame, where the incoming proton is in the $+z$ direction, and the photon's momentum is all in the $-z$ direction: $q = (0, 0, 0, -Q)$. In the parton-model approximation, we will see that the struck quark gets its 3-momentum exactly reversed in this frame, which is therefore also called the brick-wall frame.

In the Breit frame we define structure functions with simple transformation properties under rotations about the z axis. These are the longitudinal and transverse structure functions:

$$F_L \overset{\text{def}}{=} F_2 - 2x F_1; \qquad F_T \overset{\text{def}}{=} F_1. \tag{2.21}$$

Then F_L corresponds to the components of $W^{\mu\nu}$ in the energy direction, while F_T corresponds to the components transverse to q and P.

2.3.4 Cross sections and measurements of structure functions

In the case of unpolarized scattering, which is the most usual situation, we set $S^\mu = 0$. Then (2.15) and (2.20) give

$$\frac{d^2\sigma^{\text{unpol}}}{dx\,dy} \simeq \frac{4\pi\alpha^2}{xyQ^2}\left[\left(1 - y - \frac{x^2 y^2 M^2}{Q^2}\right) F_2(x, Q^2) + y^2 x F_1(x, Q^2)\right]$$

$$= \frac{4\pi\alpha^2 s}{Q^4}\left[\left(1 - \frac{Q^2}{xs} - \frac{Q^2 M^2}{s^2}\right) F_2(x, Q^2) + \frac{Q^4}{xs^2} F_1(x, Q^2)\right]. \tag{2.22}$$

The errors in this formula are due only to the neglect of the electron and hadron masses with respect to $\sqrt{s}$, of the electron mass with respect to Q, and to the use of lowest-order perturbation theory for the electromagnetic interaction. The form of the kinematic dependence multiplying the structure functions is due to the established form of the electromagnetic interaction. Thus measurements of the structure functions are equivalent to measurements of the cross section. Without further knowledge of the strong interaction, a measurement at a single energy $\sqrt{s}$ only determines the x and Q dependence of a combination of structure functions, as is made clear on the second line. Measurements at a minimum of two different energies are needed to separate the structure functions. After that the cross section for all other energies is predicted for values of x and Q that are within the kinematic limits of the first measurements.

The remaining structure functions g_1 and g_2 can be measured with polarized electron beams on a polarized target; see Leader and Predazzi (1982, p. 256).

This finishes the summary of those results and definitions that apply independently of the theory of the strong interactions.

2.4 Parton model

The parton model was formulated by Feynman (1972) and formalized by Bjorken and Paschos (1969) as an idea for understanding DIS in the absence of knowledge of an underlying microscopic theory of the strong interaction. It relies on an intuition stated in Sec. 2.3.1 and symbolized in Fig. 2.2.

Feynman proposed that the photon vertex couples to a single constituent of the target hadron, and that it is useful to neglect the strong interactions of the constituents during the collision with the lepton. The word "parton" is a generic term for one of the constituents under the conditions in which it participates in the short-distance part of a collision. In QCD it is therefore often treated as a collective name for quarks and gluons (and antiquarks).

A quantitative formulation is greatly helped by the relativistic kinematics of the process. Consider a parton of momentum k inside its parent hadron of momentum P. To get from the rest frame of the hadron to the frame of Fig. 2.2, we apply a large boost. We use light-front coordinates (App. B) with the positive z axis in the direction of the hadron; we therefore write $k^\mu = (k^+, k^-, \mathbf{k}_\mathrm{T})$, $P^\mu = (P^+, M^2/(2P^+), \mathbf{0}_\mathrm{T})$, where $k^\pm = (k^0 \pm k^z)/\sqrt{2}$. We assume that in the rest frame of the hadron, the components of k are appropriate for a constituent of a bound state whose typical scale is M, i.e., that all components of k are of order M (or smaller) in the hadron rest frame. Then after the large boost, k^+ is by far the biggest: it is of order Q, while k^- and k_T are of order M^2/Q and M. The ratio of the plus momenta is boost invariant, so we define the fractional momentum of the parton by $\xi = k^+/P^+$.

Based on the space-time structure of the reaction, the parton model asserts that we should approximate the inclusive DIS cross section as incoherent scattering of electrons on quasi-free partons. The partons have a probability distribution in fractional momentum ξ and in parton flavor, and the shape of the distribution is determined by the proton's bound state wave function. For the electron-parton interaction, the momentum transfer Q is large, so we approximate the incoming and outgoing partons as massless free particles, and neglect the transverse momentum of the incoming parton. The outgoing parton also has high energy, so the interactions converting it to a hadron final state are also time-dilated, thereby justifying its approximation as a free particle. Most importantly, the *strong* interaction is neglected, and only the lowest-order electromagnetic scattering interaction is used.

Contrary to the impression that might be gained from the literature, the parton model does *not* require that partons are genuinely free massless particles. They are only approximately free, and only for the purposes of estimating a short-distance cross section.

It is by no means obvious, a priori, that the parton model is actually valid. In Ch. 6 and later chapters, we will formulate the parton model in real quantum field theories, and show that modifications are generally needed, because of singularities in the short-distance interactions. Moreover, the concept of a wave function and how to apply Lorentz boosts to it are quite unobvious in relativistic quantum field theories. Nevertheless the parton model has intuitive appeal, so it provides an excellent framework for motivating and organizing a proper treatment. In fact, we will even justify the parton model, in a certain sense, because QCD is asymptotically free; a dimensionless measure of its interactions decreases with distance. The true results are a distorted parton model.

2.4.1 Elementary formulation of parton model; parton densities

We now make a quantitative formulation of the parton model. The logic, as presented here following the original work, involves certain intuitively motivated jumps, the quality of which we can best assess after the more strictly deductive treatment in later chapters.

The hard scattering, i.e., the short-distance scattering of the electron and parton, occurs at a particular time. The proton is in a state consisting of some number of partons, whose fractional plus momenta are $\xi_1, \xi_2, \ldots$, which sum to unity: $\sum \xi_i = 1$. There is a probability distribution over states and the hard scattering samples any one particular parton. So we postulate that there is a number distribution of partons $f_j(\xi)$. Thus $f_j(\xi)\,d\xi$ is the expectation of the number of partons of flavor j with fractional momentum ξ to $\xi + d\xi$. Standard terminology is to call $f_j(\xi)$ a "parton density" or a "parton distribution function". In QCD, the flavor index takes on values for up-quark, anti-up-quark, gluon, etc. If the two u quarks and the d quark in a proton shared its energy roughly equally, we would expect the quark densities to be peaked at around $\xi \sim 1/3$ and the u quark density to be approximately twice the d quark density. We would expect the other quark and antiquark densities to be smaller. In a real QFT, these other densities are in fact non-zero, because of the presence of quark-antiquark pairs from the interaction terms of the Hamiltonian, as can be seen later from the formal operator definitions of parton densities.

We can interpret the initial insight for the parton model in Feynman-graph notation with the aid of Fig. 2.5(a). A parton of momentum k scatters off the virtual photon; it then goes into the final state, undergoing "hadronization" interactions that convert it to observable hadrons. Topologically this diagram is in fact the most general one possible. The parton model consists of an assertion of the typical momenta involved and that the final-state hadronization interactions cancel. In the parton model, the struck quark momentum k has a large plus component, and relatively small minus and transverse components (in the Breit frame), while the outgoing parton $k + q$ has low invariant mass. The final-state interactions rearrange the content of the final state, but time dilation of the outgoing parton suggests that

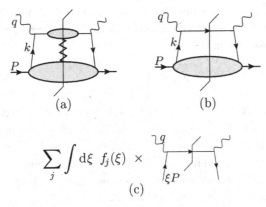

Fig. 2.5. Parton scattering in DIS. (a) Including hadronization and final-state interactions of struck parton. (b) Handbag diagram obtained after cancellation of hadronization and final-state interactions in graph (a). (c) Parton model with parton density and lowest-order DIS on partonic target.

these happen on a long time-scale, and therefore do not greatly affect the probability that a scattering has occurred. That is, the final-state interactions cancel to a first approximation in the *inclusive* cross section. Thus we can approximate graph (a) by the "handbag" graph (b), where the final-state interactions of the quark are ignored.

An analysis can be made from the handbag diagram itself, but that is postponed to Ch. 6. Here we just work with the parton-model assertion of incoherent lowest-order electromagnetic scattering on partons governed by parton densities, as embodied in Fig. 2.5(c).

2.4.2 Quark-parton model calculation

It is convenient to use the Breit frame, and to write the light-front coordinates of q and P as

$$q^\mu = \left(-x_N P^+, \frac{Q^2}{2x_N P^+}, \mathbf{0}_T\right), \qquad P^\mu = \left(P^+, \frac{M^2}{2P^+}, \mathbf{0}_T\right). \tag{2.23}$$

In this equation, x_N is the Nachtmann variable (Nachtmann, 1973)

$$x_N = \frac{2x_{Bj}}{1 + \sqrt{1 + 4M^2 x_{Bj}^2/Q^2}}, \tag{2.24}$$

which differs from the Bjorken variable $x_{Bj} = Q^2/2P \cdot q$ by a power-suppressed correction.

In the partonic scattering we replace k by its plus component: $k \mapsto (\xi P^+, 0, \mathbf{0}_T)$, in accordance with our discussion of the sizes of the components of k. We also approximate the outgoing parton as massless and on-shell. We let $d\sigma_{lj}^{\text{partonic}}$ be the differential cross section for lepton-parton scattering with the following kinematics:

$$l + (\xi P^+, 0, \mathbf{0}_T) \to l' + k', \tag{2.25}$$

where the outgoing parton momentum k' is massless and on-shell. Within the parton model, the partonic cross section is computed at lowest order. Then the parton model asserts that the inclusive DIS cross section is

$$d\sigma = \sum_j \int d\xi \, f_j(\xi) \, d\sigma_{lj}^{\text{partonic}}, \tag{2.26}$$

where the sum is over parton flavors. This formula relates a cross section with a hadron target to a cross section with a calculable partonic cross section. Naturally, these two kinds of cross section should be chosen to be differential in the same variables.

There now follow corresponding formulae for the structure tensor and for the structure functions. Now, in (2.15), we see a factor $1/s$ in converting the hadronic structure tensor to a cross section. But at the partonic level, there is instead a factor $1/\xi s$, because the lepton-parton scattering has a squared center-of-mass energy $(\xi P + q)^2 \simeq 2\xi P \cdot q$, up to power-law corrections. Then the parton model approximation for $W^{\mu\nu}$ is

$$W_{\text{PM}}^{\mu\nu} = \sum_j \int \frac{d\xi}{\xi} f_j(\xi) \, C_{j,\,\text{partonic}}^{\mu\nu}, \tag{2.27}$$

with a factor $1/\xi$ compared with (2.26). Here $C_{j,\,\text{partonic}}^{\mu\nu}$ is like $W^{\mu\nu}$ but computed on a free massless parton of type j and momentum $\hat{k}^\mu = (\xi P^+, 0, \mathbf{0}_T)$, and with neglect of all interactions.

When the partons are quarks of spin $\frac{1}{2}$, we have

$$C_{j,\text{partonic}}^{\mu\nu} = e_j^2 \frac{1}{4\pi} \frac{1}{2} \operatorname{Tr} \hat{k} \gamma^\mu (q + \hat{k}) \gamma^\nu \, 2\pi \delta((q + \hat{k})^2)$$

$$= e_j^2 \left(2\hat{k}^\mu \hat{k}^\nu + q^\mu \hat{k}^\nu + \hat{k}^\mu q^\nu - g^{\mu\nu} q \cdot \hat{k}\right) \frac{x}{Q^2} \delta(\xi - x), \tag{2.28}$$

where e_j is the electric charge of quark j (in units of the positron charge). It immediately follows that

$$F_2^{\text{QPM}} = \sum_j e_j^2 x \, f_j(x), \qquad F_1^{\text{QPM}} = \frac{1}{2x} F_2^{\text{QPM}}, \tag{2.29}$$

where "QPM" means "quark-parton model" (to distinguish these formulae from the correct factorization formulae in QCD). In this calculation, the incoming and outgoing quarks are approximated as massless and on-shell. The on-shell condition for the outgoing parton results in the parton momentum fraction ξ being set to the measurable Bjorken variable x (up to ignored power-law corrections). The measured variable y, defined in (2.13), equals $(1 - \cos\theta)/2$ in the parton model, where θ is the scattering angle of the lepton-parton collision. Thus, a measurement of x and Q in an event immediately gives an estimate of the parton kinematics, Fig. 2.3 providing an illustration of a typical event.

2.4.3 *Bjorken scaling*

A prediction of the parton model embodied in (2.29) is that at fixed x the structure functions are independent of Q (at large Q of course). This is called "Bjorken scaling", and, as we

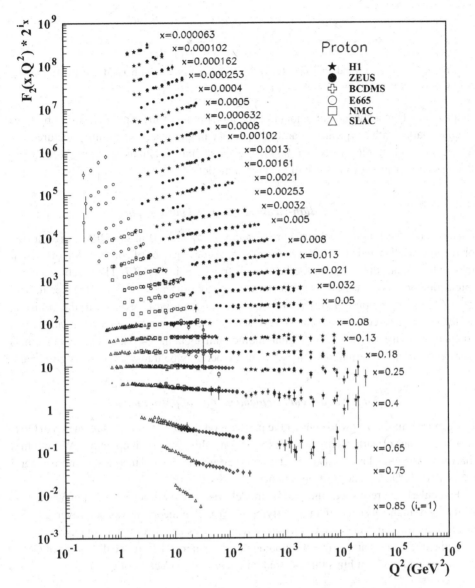

Fig. 2.6. Compilation by the Particle Data Group of data on F_2 on a proton target. For the purpose of separating the different sets of data the values of F_2 have been multiplied by 2^{i_x}, where i_x is the number of the x bin, ranging from $i_x = 1$ for $x = 0.85$ to $i_x = 28$ for $x = 0.000\,063$. Reprinted from Amsler *et al.* (2008), with permission from Elsevier.

will see in Chs. 8 and 11, it is violated after allowing for QCD interactions. We will see that measurements of scaling violation allow a deduction of the strength of the strong interaction. Current data are shown in Fig. 2.6. It can be seen that Bjorken scaling is approximately true at moderate x, for example between 0.1 and 0.5. This region is relevant

Before   After

Fig. 2.7. Initial state and final state for QPM, with conserved right-handed helicity for the
quark. The small arrows indicate the spin.

to a model where a substantial amount of the momentum of the proton is carried by three
similar quarks, with a typical x of around $1/3$. One might expect the intuitive picture to be
less reliable at extreme values of x, so the greater scaling violations as x gets close to 1 or
0 are not in violation of the spirit of the parton model.

2.4.4 Callan-Gross relation and parton spin

Observe that the longitudinal structure function is $F_L^{QPM} = 0$ in the QPM, a result first
obtained by Callan and Gross (1969). It is a simple consequence of conservation of angular
momentum about the z axis in the brick-wall frame, as in Fig. 2.7. The electromagnetic
interaction preserves the helicity of the massless quark (Sterman, 1993, p. 215), i.e., its spin
relative to its *direction of motion*. The quark's 3-momentum is reversed in the collision,
so relative to a fixed axis its spin is reversed. There are no transverse momenta in this
calculation, so there is no orbital angular momentum about the z axis. So one unit of spin is
transferred from the virtual photon, which must therefore be transverse, not longitudinal.

2.4.5 Field theory implementation of parton model

In Ch. 6 we will show how to convert the parton-model idea into formal statements in QFT,
with definitions of parton densities as expectation values of certain operators. We will find
that the parton model is exactly correct only in certain simple field theories. In more general
cases, notably QCD, modifications are needed: Chs. 8 and 11.

Particularly in retrospect the parton model was a natural conjecture, but when first
formulated, in the absence of an underlying microscopic theory, it was controversial. The
need for modifying it in real QCD underscores the basis for the initial scepticism.

Some of the first parts of this development were obtained before the discovery of QCD,
and provided important hints that pointed uniquely to the structure of QCD.

2.5 Asymptotic freedom

A powerful argument by Callan and Gross (1973) used the operator product expansion
and the renormalization group to show that exact Bjorken scaling in DIS requires there to
be an ultra-violet fixed point of the strong-interaction theory at zero coupling. Hence the
observed approximate Bjorken scaling implies that the strong interaction is relatively weak
at short distances.

Since the strong interaction is strong at large distances, this led to a search for theories
that are asymptotically free, i.e., for which the effective coupling goes to zero at zero

distance. One result was the demonstration by Coleman and Gross (1973) that no field theory constructed using only scalar, Dirac, and abelian gauge fields can be asymptotically free. This left only non-abelian gauge theories, which just slightly earlier had been quantized and proved renormalizable. If these theories also failed to give asymptotic freedom, then it would be strong evidence that no quantum field theory could describe strong interactions, a view that was quite popular at the time: there were indeed absolute arguments by Landau and Pomeranchuk based on apparently universally fundamental principles that the effective coupling always had to increase at short distances; see 't Hooft (1999).

Then Gross and Wilczek (1973a, b) and Politzer (1973) calculated the lowest-order renormalization-group β function for the Yang-Mills theory, and demonstrated its asymptotic freedom, even with quark fields present.[7] The previously formulated QCD Lagrangian is therefore able to explain (approximate) Bjorken scaling. The rising coupling in the infrared, even if it does not by itself imply color confinement, is compatible with it and is a precondition that the standard connection between fields and particles can be completely destroyed for quarks and gluons.

The result then is that for the first time there was a unique viable and complete theory of the strong interaction, QCD. Previously mysterious phenomena were direct consequences of the Lagrangian (2.1). From now on we can proceed deductively.

2.6 Justification of QCD

I now summarize the powerful arguments that pick out QCD as the unique field theory of the strong interaction. The following list involves a rearrangement and even a reversal of the historical logic.

1. We can treat any theory of currently known physics as a low-energy effective theory (Weinberg, 1995, p. 499) obtained from some more exact theory. In the normal quantum field theory framework it is a theorem that the low-energy theory is renormalizable. This applies to leading power in the ratio of a large mass scale for the exact theory to currently available energies. To agree with observations, the theory is Poincaré invariant to a very good approximation (Liberati and Maccione, 2009).

2. Bjorken scaling implies either actual asymptotic freedom, or at least a decreasing coupling at currently accessible energies. Hence the theory must be a non-abelian gauge theory with not too many matter fields. See Fig. 3.6 below for a recent plot of measured values of the strong coupling.

3. It must be possible to combine the theory with the known Weinberg-Salam theory of electroweak interactions. Since the couplings are very different, we cannot have anything except a direct product of the SI gauge group and the EW gauge group. Let us call the SI gauge fields "gluon" fields and the SI matter fields "quark" fields (which could be Dirac and/or scalar).

---

[7] In fact, the calculation of this coefficient had already been made slightly earlier by 't Hooft (see 't Hooft, 1999) and in 1969 by Khriplovich (1970). Even earlier, Vanyashin and Terentyev (1965) computed a negative beta function in Yang-Mills theory, but their calculation did not include the not-yet-known ghost contribution. But these authors did not immediately recognize the significance of their results for a theory of strong-interaction physics.

4. Because the electroweak and strong-interaction gauge groups commute, there are no direct gluon couplings to W, Z, and Higgs fields.
5. Thus the strong-interaction theory is the QCD Lagrangian, possibly supplemented only by extra quark fields. It is the original Yang-Mills Lagrangian, but with a different gauge group and with extra fermion fields. No further terms are permitted in the gauge-invariant Lagrangian without violating renormalizability.
6. We now identify the gauge group and the matter fields:
   (a) Asymptotic freedom together with the masslessness of the gluons implies that the effective coupling increases out of the perturbative range for low mass scales or large distances. This allows the connection between fields and directly observable particles to be lost.
   (b) It also indicates that under suitable conditions, quarks and gluons have approximately free-particle behavior for short distances.
   (c) Colored states tend to be unbound or of higher energy.
   (d) The approximation of the quark model indicates that an SU(3) color group together with three light flavors of Dirac quark is needed to explain the observed spectrum of hadrons.
   (e) Extra quarks, in whatever representation of the color group, are a matter for discovery at higher energy, and of obtaining a suitably consistent structure for the electroweak theory. Consistency requirements concern the lack of anomalies in the electroweak theory.
   (f) Certain measurements are key ones in confirming the determination of the color group, and in the measurement of the number of flavors, during and after the discovery of QCD:
      i. the $\pi^0 \to \gamma\gamma$ decay rate, which is obtained from an anomaly in the vacuum matrix element of three currents;
      ii. the total cross section for e^+e^- annihilation to hadrons at high energy gives a measure of the sum of the charges squared of the accessible quarks – see Ch. 4;
      iii. More detailed jet cross sections in e^+e^- give quite direct measurements of the color-group theory coefficients C_A and C_F, etc. – again, see Ch. 4.

These arguments are primarily structural. They do not depend, for the most part, on detailed numerical predictions of the theory. Such predictions are used mainly in determining which gauge group is needed.

Once we have confidence that the theory is a good approximation to reality, we (i.e., people working on the strong interaction) change our attitude. The mathematics is hard, and when useful, we appeal to the real world as a realization of QCD to help us to determine what results are true. A failure of agreement between theory and experiment is expected to indicate that there is an error either in the theoretical methods or their application, or in the experiments, but it does not normally indicate an error in the theory itself. (An extension of the theory, to add another quark, for example, is not regarded as a breakdown in the theory.)

2.7 QCD in the full Standard Model

Many applications of perturbative QCD concern the interaction of hadrons with non-QCD particles, e.g., DIS, and all kinds of production processes for leptons, the Higgs boson, and many hypothesized particles. To put these in context, I now review the definition of the Standard Model (SM). For details, see any standard textbook, such as Halzen and Martin (1984); Peskin and Schroeder (1995); Quigg (1997).

The SM Lagrangian is

$$\mathcal{L}_{SM} = -\frac{1}{4}\sum_{\alpha}\left(G_{\alpha\,\mu\nu}\right)^2 + i\sum_{f}\bar{\psi}_f\,\slashed{D}\psi_f + D\phi^{\dagger}\cdot D\phi + M^2\phi^{\dagger}\phi$$

$$-\frac{\lambda}{4}(\phi^{\dagger}\phi)^2 - \sum_{ij}h_{ij}\bar{\psi}_{i\,R}\phi\psi_{j,L} + \text{gauge-fixing terms, etc.,} \qquad (2.30)$$

with the usual modifications for renormalization. Structurally this is like QCD, except for the addition of a scalar "Higgs" field ϕ, with its self-interaction and its Yukawa couplings to the fermion fields. The main features are as follows.

- In the first line, the sum over α is over the 12 generators of the gauge group $SU(3)\otimes SU(2)\otimes U(1)$. We let the gauge fields for the three commuting components of the gauge group be $A_{\mu}^{\alpha}(x)$, W_{μ}^{j} and B_{μ}. The renormalized couplings of the three commuting component groups are g_s, g and g' respectively, and the $SU(3)$ subgroup is the QCD group.
- When we are working with pure QCD, without any mention of electroweak interactions, we will often replace the notation g_s by g.
- The fermion fields ψ_{paf} carry different representations of the gauge group, unlike the case of simple QCD.
- The covariant derivative is

$$D_{\mu} = \partial_{\mu} + ig_s\sum_{\alpha=1}^{8}T_{col}^{\alpha}A_{\mu}^{\alpha} + ig\sum_{j=1}^{3}W_{\mu}^{j}T_{W}^{j} + ig'B_{\mu}\frac{Y}{2}, \qquad (2.31)$$

where for any given multiplet of fields T_{col} and T_W are the generating matrices for the color $SU(3)$ and the $SU(2)$ groups, while Y is the weak hypercharge of the multiplet.
- The fermion fields are arranged in multiplets of left-handed fields and right-handed fields. "Left-handed" fields are $\frac{1}{2}(1-\gamma_5)$ times the Dirac field, and "right-handed" fields have a $\frac{1}{2}(1+\gamma_5)$ factor.
- All known left-handed fields are doublets under $SU(2)$, and all known right-handed fields are singlets under $SU(2)$.
- There are three generations of fermion, and the assignments of quantum numbers to fields are specified in Table 2.1. Here we have extended the Standard Model slightly beyond its original definition to include right-handed neutrino fields, as needed to accommodate the measured neutrino mixing.
- The vacuum expectation value of the Higgs field is given by $\langle 0|\phi|0\rangle = (0, v/\sqrt{2})^{T}$, with $v = 246\,\text{GeV}$. This breaks three of the electroweak symmetries, with the Z and

Table 2.1 *Quantum numbers of field multiplets in the Standard Model. The symbols for the fields correspond to the particle names.*

	Color singlet	Y	Color triplet	Y
First generation:	$\begin{pmatrix} \nu_{e\,\mathrm{L}} \\ e_{\mathrm{L}} \end{pmatrix}$	-1	$\begin{pmatrix} u_{\mathrm{L}} \\ d_{\mathrm{L}} \end{pmatrix}$	$\frac{1}{3}$
	$\begin{pmatrix} e_{\mathrm{R}} \end{pmatrix}$	-2	$\begin{pmatrix} u_{\mathrm{R}} \end{pmatrix}$	$\frac{4}{3}$
	$\begin{pmatrix} \nu_{e\,\mathrm{R}} \end{pmatrix}$	0	$\begin{pmatrix} d_{\mathrm{R}} \end{pmatrix}$	$-\frac{2}{3}$

The next two generations (ν_μ, μ, s, c) and (ν_τ, τ, b, t) are exactly similar.

	Color singlet	Y
Higgs field:	$\begin{pmatrix} \phi^+ \\ \phi^0 \end{pmatrix}$	1

photon fields being

$$Z_\mu = \cos\theta_{\mathrm{W}}\, W_\mu^3 - \sin\theta_{\mathrm{W}}\, B_\mu, \qquad A_\mu = \sin\theta_{\mathrm{W}}\, W_\mu^3 + \cos\theta_{\mathrm{W}}\, B_\mu, \qquad (2.32)$$

where the measured Weinberg angle obeys $\sin^2\theta_{\mathrm{W}} = 0.22 \pm 0.02$.

- The electroweak couplings are given in terms of the QED coupling and θ_{W} by

$$g = \frac{e}{\sin\theta_{\mathrm{W}}}, \qquad g' = \frac{e}{\cos\theta_{\mathrm{W}}}.$$

- The fermion masses are then obtained from the Yukawa couplings. From global fits to data (Amsler *et al.*, 2008) estimates of the masses of the elementary fields are found (Table 2.2).
- All the formulae for masses, etc., are subject to higher-order electroweak corrections.
- The flavor and mass eigenstates of the two components of the fermion doublets are not aligned, but have mixing given by the CKM and MNS matrices; see, e.g., Amsler *et al.* (2008).

2.8 Beyond the Standard Model

All theories of physics are ultimately approximate, and many possibilities for theories that are better than the SM are under active discussion. To keep agreement with known results, the QFTs considered are generally extensions of the SM, except for the Higgs sector on which there is as yet little direct data. Extensions include both the simple addition of field multiplets and the embedding of the symmetry groups in bigger symmetries, as in Grand Unified Theories and in supersymmetry.

Once gravity enters the picture, space-time becomes dynamical, and so any QFT, including QCD, becomes only an effective low-energy approximation to a radically different kind

Table 2.2 *Standard Model masses for elementary fields, from Amsler* et al. *(2008).*

Leptons and quarks (spin 1/2):

ν	~ 0	d	3.5 to 6 MeV	
e	0.511 MeV	u	1.5 to 3.3 MeV	
		s	$\sim 104 ^{+26}_{-34}$ MeV	
μ	106 MeV	c	$\sim 1.27 ^{+0.07}_{-0.11}$ GeV	
		b	$\sim 4.20 ^{+0.17}_{-0.07}$ GeV	
τ	1.78 GeV	t	171.2 ± 2.1 GeV	

Gauge bosons:

| $W^\pm$ | 80.398 ± 0.025 GeV | Z | 91.1876 ± 0.0021 GeV |

Higgs:

| 100 to 300 GeV (indirect) |

of theory (e.g., string theory), with a very different understanding of space-time. Factorization in QCD remains a vital tool in phenomenological discussions of such theories, because it separates treatment of the ultra-microscopic physics of the new theories from the longer-distance physics which is an integral part of a full scattering process.

For current work in this area, see the proceedings of recent conferences and workshops, e.g., Allanach *et al.* (2006).

2.9 Relation between fields and particles

In a free QFT, there is a direct correspondence between the types of single particle and the fields, and in fact with the normal modes of the corresponding classical field theory. In simple interacting QFTs, this correspondence continues to hold, but it is clear from both QCD and the full Standard Model, that the particle-field correspondence is not general:

- With interactions some of these particles can be become unstable, as exemplified by the muon, with its decay to $e \nu_\mu \bar{\nu}_e$.
- There may be bound states, e.g., atoms. These are not related in a simple way to normal modes of the elementary fields.
- It is also possible that there is no particle, stable or unstable, that corresponds to a particular elementary field of a theory. QCD is an excellent example with its quark and gluon fields. Any corresponding particles are permanently confined, and only behave approximately like particles on short enough distance scales inside collisions. Before the advent of QCD, this possibility was hardly recognized, if at all.
- Moreover, low-energy effective theories approximating a more exact microscopic theory may use fields corresponding to bound states. This is the case for a Schrödinger QFT for atomic physics, which might have fields for atomic nuclei.

Moreover, one must be careful about what is meant by a particle. One standard definition is from the single-particle states that are used to build up the asymptotic in- and out-states of scattering theory. For this purpose completely stable bound states, like the ground state of a hydrogen atom or even of a large macroscopic object like a planet, are particles. But unstable particles, even relatively long-lived ones like the muon and the neutron, are not particles under this definition.

It is clear that the connection between particles and interacting fields is somewhat impressionistic. Even the usage of the word "particle" is quite fuzzy in the real world. Which objects are called particles, which bound states, and which resonances is essentially a linguistic matter: a matter of convention, and usage, and even of context.

Some confusions in the recent literature should be noted. For example, Weinberg in his excellent textbooks on quantum field theory (Weinberg, 1995, p. 110) bases his logic on the concept of a particle in the strict sense of scattering theory. Then his derivation of perturbation theory requires that the set of one-particle states be unchanged after turning on the interaction in a theory. If this were really necessary, it would immediately rule out conventional Feynman perturbation theory for all known interactions.

Weinberg's derivation of perturbation theory is for the S-matrix. Instead, if one bases the logic on perturbation theory for (off-shell) Green functions, one no longer has to assume that the particle spectrum is unchanged under perturbations. The particle spectrum and the S-matrix are derived objects involving examination of poles in the Green functions. Thus, for example, the stability or instability of a particular particle can be an accidental consequence of the particular values of parameters of the theory.

It is evidently important to dispose of this issue at the outset, for otherwise most of our work in *perturbative* QCD would be without a foundation. An account of the logic for perturbation theory that is suitable from our perspective is given in Sterman's textbook (Sterman, 1993).

Exercises

2.1 (a) Show how to compute a particle-particle potential from the non-relativistic limit of a first-order $2 \longrightarrow 2$ scattering amplitude. You might do this by comparing the Born approximation in QED with the Born approximation in non-relativistic potential scattering. Consider both the case of spin-$\frac{1}{2}$ and spin-0 particles.

(b) Apply this method to QCD to find the lowest-order approximation to the quark-antiquark potential with massive quarks. Separately consider the case that the system is a color singlet and a color octet. You should find that the potential is only attractive for a color-singlet bound state.

2.2 *Review problem:* Define the concept of a structure function. Why is it a useful concept?

2.3 In the parton model approximation, compute the electromagnetic structure functions for a scalar quark (i.e., for a spin-0 quark).

2.4 *Formulation of the structure function method for scalar field exchange instead of vector field:* Suppose you wanted to investigate the consequences of a hypothetical theory with an extra neutral scalar field ϕ that has Yukawa couplings to quarks and to leptons:

$$\mathcal{L}_{\text{int}} = \phi \times \left(h_e \bar{e}e + \sum_i h_i \bar{q}_i q_i \right), \qquad (2.33)$$

where h_e and h_i are the couplings to electrons and to quarks of flavor i. (a) What would be an appropriate definition of structure function(s) in this problem? (b) What would be the parton model formula?

Review and revise your answer to problem 2.2 in the light of your answer this problem.

2.5 How do you extend the analysis of problem 2.4 in the presence of interference between scalar and vector exchange?

2.6 Examine the state of the knowledge about current algebra just before the discovery of QCD, e.g., in Treiman, Jackiw, and Gross (1972). How does this compare with the description in this chapter?

The rest of this problem is best done after finishing learning about QCD. During your studies of QCD, determine the extent to which the work in Treiman *et al.* (1972) is (a) true in QCD, (b) needs modification, or (c) still needs proof. How much remains relevant to current research and/or to understanding QCD and the strong interaction?

3

Basics of QCD

In this chapter, we review some basic properties of QCD that directly follow from its definition. This material is completely standard, and will form a foundation for the rest of the book. More details can be found in a standard textbook on quantum field theory, e.g., Peskin and Schroeder (1995), Srednicki (2007), Sterman (1993) and Weinberg (1995, 1996). For specific information on renormalization and the renormalization group see also my book on renormalization (Collins, 1984).

I first review how the theory is quantized and renormalized. Then I discuss the renormalization group (RG) and the calculation of the asymptotic freedom of QCD. A brief review of the flavor symmetries follows. Finally I show some of the complications that arise in perturbative calculations because some of the fields are much more massive than others.

3.1 Quantization

3.1.1 Definition; functional integral

A list of the fields of QCD and the formula for its gauge-invariant Lagrangian density (2.1) are sufficient to specify the theory, with the aid of general principles. Although there are some mathematical issues that have not been solved properly, it is standard to assume that the theory can be constructed (with some complications) through a functional integral. This gives Green functions, i.e., vacuum expectation values of time-ordered products of fields, as

$$\langle 0 | T f[A, \psi, \bar{\psi}] | 0 \rangle = \mathcal{N} \int \mathcal{D}A \, \mathcal{D}\psi \, \mathcal{D}\bar{\psi} \, e^{iS[A, \psi, \bar{\psi}]} \, f[A, \psi, \bar{\psi}]. \tag{3.1}$$

Here $f[A, \psi, \bar{\psi}]$ is a functional of the fields, e.g., a product $G^2(x) \, \bar{\psi}\psi(y)$. On the left-hand side, the fields are the quantum fields of QCD, time-ordered, while $|0\rangle$ is the true vacuum state. But on the right-hand side the fields are corresponding classical fields (Grassmann-valued in the case of the fermion fields ψ and $\bar{\psi}$). The normalization factor $\mathcal{N}$ is set so that $\langle 0 | 0 \rangle = 1$.

From the Green functions can be reconstructed the state space and the operators. This includes an extraction of the particle content of the theory, from an examination of the positions of the poles in propagators and other Green functions. The S-matrix and scattering theory follow by the Lehmann-Symanzik-Zimmermann (LSZ) method. Note that the poles of Green functions need not be the same as in free field theory, and so the particle content can be different from a free field theory of quarks and gluons.

3.1.2 Faddeev-Popov method; Feynman rules

The rules for Feynman perturbation theory are readily derived from the functional integral, with the Faddeev-Popov technique being used for gauge fixing. In this technique, a change of variables is used on sets of field configurations equivalent under gauge transformations. The implementation involves fermion scalar "ghost" fields, η_α and $\bar{\eta}_\alpha$. See most modern textbooks on QFT for details.

In the covariant gauges we will normally use, the gauge-invariant Lagrangian of (2.1) is replaced by

$$\mathcal{L} = \mathcal{L}_{\text{GI from (2.1)}} + \mathcal{L}_{\text{GF}} + \mathcal{L}_{\text{GC}}, \tag{3.2}$$

where the gauge-fixing and "gauge-compensating" terms are

$$\mathcal{L}_{\text{GF}} = -\frac{1}{2\xi_0}(\partial \cdot A_{(0)\alpha})^2, \tag{3.3}$$

$$\mathcal{L}_{\text{GC}} = \partial_\mu \bar{\eta}_{0\alpha} \partial^\mu \eta_{0\alpha} + g_0 \partial^\mu \bar{\eta}_{0\gamma} f_{\alpha\beta\gamma} A^\beta_{(0)\mu} \eta_{0\alpha}, \tag{3.4}$$

in terms of bare quantities. This gives

$$\mathcal{L} = \bar{\psi}_0 (i\slashed{D} - m_0)\psi_0 - \frac{1}{4}(G^\alpha_{(0)\mu\nu})^2 - \frac{1}{2\xi_0}(\partial \cdot A^\alpha_{(0)})^2$$

$$+ \partial_\mu \bar{\eta}_{0\alpha} \partial^\mu \eta_{0\alpha} + g_0 \partial^\mu \bar{\eta}_{0\gamma} f_{\alpha\beta\gamma} A^\beta_{(0)\mu} \eta_{0\alpha}. \tag{3.5}$$

Feynman rules for Green functions are derived in the usual way. In Sec. 3.2, we will formulate Feynman rules for renormalized Green functions with a counterterm method. Rules for elementary perturbation theory in terms of bare quantities can be obtained from those listed in Fig. 3.1 below by replacing each occurrence of $g\mu^\epsilon$ in that figure by the bare coupling g_0, and each renormalized quark mass m_f by the bare mass m_{0f}.

Note that without gauge fixing in the Lagrangian, Green functions of the elementary gauge-variant field operators are zero (Elitzur, 1975).

3.1.3 BRST symmetry

The full Lagrangian (3.2) is not gauge invariant, which considerably complicates the derivation and formulation of generalized Ward identities. The appropriate identities for non-abelian gauge theories were first found by Slavnov (1972) and Taylor (1971). The derivations were greatly simplified by Becchi, Rouet, and Stora (1975, 1976) and by Tyutin (1975), who discovered a new symmetry of the full Lagrangian.

This BRST symmetry is a supersymmetry, i.e., one that relates Bose and Fermi fields. It uses a parameter $\delta\lambda_0$ that takes its value in the fermionic part of some Grassmann algebra. For the gauge and matter fields, the BRST transformations are gauge transformations (2.4) with $\omega_\alpha(x) = \eta_{0\alpha}(x)\delta\lambda_0$. Thus any gauge-invariant operator is also BRST invariant. The linear terms in the variation of the bare fields are

$$\delta_{\text{BRST}}\psi_0 = -ig_0\eta_{0\alpha}\delta\lambda_0 t^\alpha \psi_0 = ig_0\eta_{0\alpha}t^\alpha \psi_0\delta\lambda_0, \tag{3.6a}$$

$$\delta_{\text{BRST}}\bar\psi_0 = ig_0\bar\psi_0 t^\alpha \eta_{0\alpha}\delta\lambda_0, \tag{3.6b}$$

$$\delta_{\text{BRST}}A^\alpha_{(0)\mu} = \left(\partial_\mu\eta_{0\alpha} + g_0 f_{\alpha\beta\gamma}\eta_{0\beta}A^\gamma_{(0)\mu}\right)\delta\lambda_0. \tag{3.6c}$$

The ghost and antighost fields transform as

$$\delta_{\text{BRST}}\eta_{0\alpha} = -\tfrac{1}{2}g_0 f_{\alpha\beta\gamma}\eta_{0\beta}\eta_{0\gamma}\delta\lambda_0, \tag{3.6d}$$

$$\delta_{\text{BRST}}\bar\eta_{0\alpha} = \frac{1}{\xi_0}\partial \cdot A_{0\alpha}\delta\lambda_0. \tag{3.6e}$$

It can readily be checked that the full Lagrangian is BRST invariant, up to a total derivative. With a slight exception, BRST transformations are also nilpotent. That is, applying successive BRST transformations with different anticommuting parameters $\delta\lambda_1$ and $\delta\lambda_2$ gives zero:

$$\left(\frac{\delta_{\text{BRST}}}{\delta\lambda_0}\right)^2 \text{field} = 0. \tag{3.7}$$

The exception is that the second variation of $\bar\eta_0$ only vanishes after using the equation of motion for $A_{(0)}$; a third variation of the field is needed to get zero without use of the equations of motion.

A good formulation of the quantum theory associated with Faddeev-Popov quantization and BRST transformations is given by Nakanishi and Ojima (1990). In particular they give a full formulation of the conditions to be applied to physical quantum-mechanical states.

3.1.4 Relation to Euclidean lattice gauge theory

The functional integral contains an oscillating functional e^{iS}, and it can be defined by analytically continuing to Euclidean space-time, where the time coordinate becomes imaginary, $t = -i\tau$, and by then putting the theory on a lattice in a finite volume of space-time. The functional integral is then an ordinary finite-dimensional convergent integral (with suitable modifications for the fermion integrations). Numerical evaluation of these integrals by Monte-Carlo methods is the core of lattice gauge theory, a key technique for non-perturbative calculations in QCD (DeGrand and Detar, 2006).

The infinite-volume limit is an ordinary thermodynamic limit, but the continuum limit of zero lattice spacing is non-trivial, needing the use of renormalization: Sec. 3.2. However, there is not yet a completely rigorous proof that the limit exists.

The continuation back to real time is potentially problematic. Typical time dependence associated with high-energy states at large times, e^{-iEt}, corresponds to strongly suppressed exponentials $e^{-E\tau}$ in Euclidean time. Small errors in the Euclidean calculation, e.g., due to the neglect of weak-interaction effects or purely numerical errors, do not automatically continue to small errors in the real-time formalism. Further research is clearly needed here. Euclidean lattice methods are not suitable for high-energy scattering problems.

For our purposes, it suffices to assume that some method exists to construct real-time functional integrals, as in (3.1).

3.2 Renormalization

Ultra-violet (UV) divergences appear in QCD (and in most other relativistic quantum field theories) when the continuum limit is taken. These were first found in perturbative calculations, but the divergences are a property of the exact theory, as is shown by a renormalization-group analysis, particularly using Wilsonian methods (Polchinski, 1984). The divergences are from large loop momenta, or, equivalently, from where interaction vertices approach each other in space-time. In renormalizable theories, like QCD, the divergences can be proved to be removed by a modification of the continuum limit, at least in perturbation theory.

1. The theory is first defined with a regulator[1] (or cutoff) of the UV divergences. Standard UV regulators are a non-zero lattice spacing or dimensional regularization.
2. All parameters of the theory consistent with its symmetries are made adjustable as functions of the cutoff. The parameters include the coefficients of terms like $i\bar{\psi}\partial_\mu\psi$.
3. When the limit of no UV cutoff is taken, the cutoff dependence of the parameters is chosen so as to remove the UV divergences and to obtain a non-trivial limiting theory.

Note that an entirely different status is to be given to the infra-red (IR) divergences that appear in perturbation theory for the S-matrix in theories such as QCD and QED that have massless fields. The S-matrix is derived given certain hypotheses about the large-time behavior of Green functions. But in a theory like QED with actual massless particles, these hypotheses are violated, while in QCD the particle content does not even correspond to the elementary fields. In either case, perturbative calculations must be adapted to the correct physics. But IR divergences do not affect the definition of the theory, only the interpretation of its solution, unlike the case of UV divergences.

The general ideas and methods of renormalization are explained in almost any modern QFT textbook, and a more specialized reference is Collins (1984), which is compatible with the presentation here.

3.2.1 Reformulating ℒ: bare parameters

To obtain finite Green functions, we use the freedom not only to change g_0 and m_0 in (3.5), but also to change the normalization of the fields, i.e., to do field strength renormalization.

---

[1] For mathematicians: In much of the mathematical literature, the word "regularization" has a different meaning, equivalent to physicists' "renormalization".

We therefore define the bare fields to be (square roots of) "wave-function renormalization" factors times renormalized fields: $A_{(0)\mu} = Z_3^{1/2} A_\mu$, $\psi_0 = Z_2^{1/2}\psi$, and $\eta_0 = \tilde{Z}^{1/2}\eta$. It is Green functions of the renormalized fields that are to be finite after removal of the UV cutoff. This gives the following formula for $\mathcal{L}$:

$$\mathcal{L} = Z_2\bar{\psi}(i\slashed{\partial} - m_0)\psi - Z_2 Z_3^{1/2} g_0 \bar{\psi} t^\alpha \slashed{A}^\alpha \psi$$

$$- \frac{Z_3}{4}(\partial_\mu A_\nu^\alpha - \partial_\nu A_\mu^\alpha)^2 + \frac{Z_3^{3/2} g_0}{2} f_{\alpha\beta\gamma}(\partial_\mu A_\nu^\alpha - \partial_\nu A_\mu^\alpha)A_\mu^\beta A_\nu^\gamma$$

$$- \frac{Z_3^2 g_0^2}{4}\left(f_{\alpha\beta\gamma}A_\mu^\beta A_\nu^\gamma\right)^2 - \frac{Z_3}{2\xi_0}(\partial\cdot A^\alpha)^2$$

$$+ \tilde{Z}\partial_\mu\bar{\eta}_\alpha\partial^\mu\eta_\alpha + \tilde{Z}Z_3^{1/2} g_0 \partial^\mu\bar{\eta}_\gamma f_{\alpha\beta\gamma}A_\mu^\beta\eta_\alpha. \tag{3.8}$$

Note that Z_2 could be a matrix relating bare and renormalized quark fields, diagonal in quark flavor, but color-independent.

Both of the formulae (3.5) and (3.8) define the same Lagrangian density; they differ only by a change of variables; the physical predictions are the same. Thus, provided that the correct LSZ prescription is used, the S-matrix and cross sections are unchanged under the field redefinitions.

The first form (3.5), with the bare fields, has unit coefficients for the terms $i\bar{\psi}_0\slashed{\partial}\psi_0$, etc., which implies that the bare fields obey canonical (anti)commutation relations. This is a natural standard which then gives an invariant meaning to the normalization of the bare coupling and mass.

We have restricted the change of parameters to those that preserve gauge-invariance properties, admittedly with a renormalization of the definition of the gauge transformations. It is a theorem that this is sufficient to obtain finite Green functions. It can also be proved that ξ_0/Z_3 is finite, so that we can define $\xi_0 = \xi Z_3$ with ξ a finite renormalized gauge-fixing parameter; thus the gauge-fixing term in (3.8) has coefficient $1/\xi$. For proofs, see, for example, Collins (1984).

3.2.2 Renormalized BRST symmetry

The BRST transformations also need renormalization. This is done by a multiplicative renormalization of the parameter $\delta\lambda_0$:

$$\delta\lambda_0 = \delta\lambda Z_3^{1/2}\tilde{Z}^{1/2}. \tag{3.9}$$

In the resulting formulae (Collins, 1984, p. 297) for the renormalized BRST transformations of the renormalized fields, it is convenient to define

$$X = \tilde{Z}Z_3^{1/2} g_0/g_R, \tag{3.10}$$

where g_R is a finite parameter that is a version of the renormalized coupling to be introduced later. (The actual formula is $g_R = g\mu^\epsilon$; see (3.14).) The resulting renormalized BRST transformations are finite:

$$\delta_{\text{BRST, R}}\psi = -ig_R\eta_\alpha\delta\lambda t^\alpha \psi X, \tag{3.11a}$$

$$\delta_{\text{BRST, R}}\bar{\psi} = ig_R\bar{\psi}t^\alpha\eta_\alpha X \delta\lambda, \tag{3.11b}$$

$$\delta_{\text{BRST, R}}A_\mu^\alpha = \left(\partial_\mu\eta_\alpha\tilde{Z} + X g_R f_{\alpha\beta\gamma}\eta_\beta A_\mu^\gamma\right)\delta\lambda. \tag{3.11c}$$

The ghost and antighost fields transform as

$$\delta_{\text{BRST, R}}\eta_\alpha = -\tfrac{1}{2}g_R X f_{\alpha\beta\gamma}\eta_\beta\eta_\gamma\delta\lambda, \tag{3.11d}$$

$$\delta_{\text{BRST, R}}\bar{\eta}_\alpha = \frac{1}{\xi}\partial \cdot A_\alpha\delta\lambda. \tag{3.11e}$$

The finite operators on the right-hand sides of these equations are used in Slavnov-Taylor identities.

3.2.3 Counterterms, renormalized parameters, dimensional regularization

To implement renormalization in perturbation theory, we use a counterterm approach. The Lagrangian is split into three parts:

$$\mathcal{L} = \mathcal{L}_{\text{free}} + \mathcal{L}_{\text{b.i.}} + \mathcal{L}_{\text{c.t.}}. \tag{3.12}$$

Free propagators correspond to the free Lagrangian $\mathcal{L}_{\text{free}}$, which has the standard form in which appear derivative terms with unit coefficient, and mass terms with renormalized masses. The "basic interaction" Lagrangian $\mathcal{L}_{\text{b.i.}}$ contains interaction terms, but with coefficients constructed using only finite renormalized couplings. Graphs constructed with only the basic interaction contain divergences in some of their one-particle-irreducible (1PI) subgraphs. The divergences are canceled by graphs in which divergent subgraphs are replaced by counterterm vertices derived from the counterterm Lagrangian $\mathcal{L}_{\text{c.t.}}$. The rules for perturbation theory ensure that subdivergences in multiloop graphs are correctly canceled, order-by-order in an expansion in powers of the renormalized coupling.

Since the counterterms cancel the divergent contributions to loop graphs from UV momenta, it does not matter how UV divergences are regulated. After removal of the regulator, the same results are obtained for renormalized Green functions expressed in terms of renormalized parameters. The only requirement is a suitable adjustment of the finite parts of the counterterms when the method of UV regulation is changed.

For QCD perturbation theory, the most convenient UV regulator is often dimensional regularization, where the dimension n of space time is a continuous complex parameter, also written[2] as $n = 4 - 2\epsilon$. Although it is not known how to apply dimensional regularization to the exact theory, there are no problems in perturbation theory. A concrete mathematical

---

[2] *Warning*: Although this is the most common definition of ϵ, other definitions also appear in the literature, notably $\epsilon = n - 4$ and $\epsilon = 4 - n$.

definition can be made (Wilson, 1973; Collins, 1984) by using an infinite dimensional space for momenta (and coordinates), and by using pathologies of infinite dimensional spaces to define integration so that it gives the scaling properties appropriate for a non-integer dimension.

3.2.4 Implementation in QCD

The free and basic interaction Lagrangians are defined to be

$$\mathcal{L}_{\text{free}} = \bar{\psi}(i\not{\partial} - m)\psi - \frac{1}{4}\left(\partial_\mu A_\nu^\alpha - \partial_\nu A_\mu^\alpha\right)^2 - \frac{1}{2\xi}(\partial \cdot A_\alpha)^2 + \partial_\mu \bar{\eta}^\alpha \partial^\mu \eta^\alpha, \qquad (3.13)$$

$$\mathcal{L}_{\text{b.i.}} = -g\mu^\epsilon \bar{\psi} t^\alpha \not{A}^\alpha \psi + g\mu^\epsilon f_{\alpha\beta\gamma} A^{\beta\mu} A^{\gamma\nu} \partial_\mu A_\nu^\alpha - \frac{g^2 \mu^{2\epsilon}}{4}\left(f_{\alpha\beta\gamma} A_\mu^\beta A_\nu^\gamma\right)^2$$

$$+ g\mu^\epsilon f_{\alpha\beta\gamma} \partial^\mu \bar{\eta}^\gamma A_\mu^\beta \eta^\alpha. \qquad (3.14)$$

Here is introduced the well-known unit of mass μ for dimensional regularization, so that the renormalized coupling is $g\mu^\epsilon$, with g dimensionless for all ϵ. The Feynman rules that follow from these parts of $\mathcal{L}$ are listed in Fig. 3.1.

The counterterm Lagrangian is everything else in the full Lagrangian (3.8):

$$\mathcal{L}_{\text{c.t.}} = (Z_2 - 1)\bar{\psi} i\not{\partial}\psi - \left(g_0 Z_2 Z_3^{1/2} - g\mu^\epsilon\right)\bar{\psi}\not{A}^\alpha t^\alpha \psi + \ldots \qquad (3.15)$$

In renormalized perturbation theory, the counterterm Lagrangian is treated as part of the interaction. We therefore have an extra set of vertices, the counterterm vertices, listed in Fig. 3.2. These are like those in the basic interaction, Fig. 3.1, but with modified coefficients, together with extra two-point vertices.

3.2.5 Mass-dependence and gauge-invariance relations for counterterms

In renormalization theory (e.g., Collins, 1984) the following is shown:

- The Ward identities that follow from gauge invariance imply that independent renormalization of the different interaction vertices is not needed; a single renormalization factor applied to g_0 is suitable. Thus gauge invariance is preserved.
- No counterterm proportional to the gauge-fixing term is needed. That is, $\xi_0 = Z_3 \xi$, within the class of gauges we are using.
- With the exception of the mass parameters, the renormalization counterterms can be chosen to be independent of the quark masses.
- Renormalization of the bare coupling g_0 and the bare mass m_0 can be chosen to be independent of the gauge-fixing parameter ξ.
- The bare quark mass is linear in the renormalized mass: $m_{(0)f} = Z_m m_f + m_{00}$, with Z_m and m_{00} independent of mass. With dimensional regularization, we can set $m_{00} = 0$, so that $m_{(0)f} = Z_m m_f$.
- Z_2 and Z_m can be chosen to be independent of quark flavor. (But other choices of scheme can be useful in treating heavy quarks: Secs. 3.9 and 3.10.)

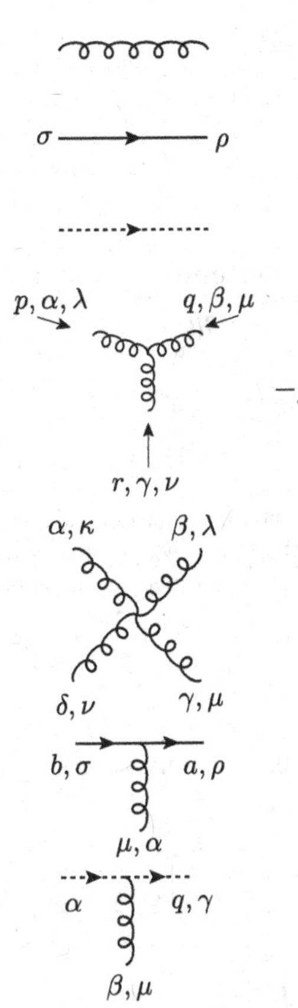

$$\left[-g_{\mu\nu} + (1-\xi)\frac{p_\mu p_\nu}{p^2+i0}\right]\frac{i}{p^2+i0}$$

$$\frac{i(\not{p}+m_f)_{\rho\sigma}}{p^2-m_f^2+i0}$$

$$\frac{i}{p^2+i0}$$

$$-g\mu^\epsilon f_{\alpha\beta\gamma}\left[(p-q)^\nu g^{\lambda\mu} + (q-r)^\lambda g^{\mu\nu} + (r-p)^\mu g^{\nu\lambda}\right]$$

$$\begin{aligned}
&-ig^2\mu^{2\epsilon}f_{\epsilon\alpha\beta}f_{\epsilon\gamma\delta}(g^{\kappa\mu}g^{\lambda\nu}-g^{\kappa\nu}g^{\lambda\mu})\\
&-ig^2\mu^{2\epsilon}f_{\epsilon\alpha\gamma}f_{\epsilon\beta\delta}(g^{\kappa\lambda}g^{\mu\nu}-g^{\kappa\nu}g^{\lambda\mu})\\
&-ig^2\mu^{2\epsilon}f_{\epsilon\alpha\delta}f_{\epsilon\beta\gamma}(g^{\kappa\lambda}g^{\mu\nu}-g^{\kappa\mu}g^{\lambda\nu})
\end{aligned}$$

$$-ig\mu^\epsilon(t^\alpha)_{ab}\gamma^\mu_{\rho\sigma}$$

$$-g\mu^\epsilon f_{\alpha\beta\gamma}q^\mu$$

Fig. 3.1. Basic Feynman rules of QCD. The coupling has been replaced by $g\mu^\epsilon$, according to the standard convention for use in $4-2\epsilon$ dimensions. Propagators and vertices are diagonal in any indices (flavor or color) that are not explicitly indicated. For the renormalization counterterm vertices, see Fig. 3.2.

- Minimal subtraction (Sec. 3.2.6) is among the schemes to which the above statements on the lack of mass, flavor and gauge dependence apply.

3.2.6 Minimal subtraction

In a calculation order-by-order in the renormalized coupling, the requirement that a counterterm cancels its corresponding divergence determines the part of the counterterm that diverges as the UV regulator is removed, but not the finite part. A rule for determining the finite part is called a renormalization prescription. The most common in QCD calculations is minimal subtraction in its modified form, the $\overline{\text{MS}}$ scheme due to Bardeen *et al.* (1978). When dealing with heavy quarks, it is convenient to apply a different scheme for graphs with heavy quark lines: Sec. 3.10.

$$i(Z_3 - 1)\left(-g_{\mu\nu}p^2 + p_\mu p_\nu\right)$$

$$i\left[\not{p}(Z_2 - 1) - (Z_2 Z_m - 1)m_f\right]$$

$$i(\tilde{Z} - 1)p^2$$

Vertex	Basic interaction	Counterterm
3-gluon	$g\mu^\epsilon$	$g_0 Z_3^{3/2} - g\mu^\epsilon$
4-gluon	$g^2\mu^{2\epsilon}$	$g_0^2 Z_3^2 - g^2\mu^{2\epsilon}$
Quark-gluon	$g\mu^\epsilon$	$g_0 Z_2 Z_3^{1/2} - g\mu^\epsilon$
Ghost-gluon	$g\mu^\epsilon$	$g_0 \tilde{Z} Z_3^{1/2} - g\mu^\epsilon$

Fig. 3.2. Counterterm vertices in QCD. The 2-point counterterms have diagonal dependence on all but Dirac indices for quarks and Lorentz indices for gluons. The other counterterm vertices simply correspond to vertices in Fig. 3.1 with the indicated modified coefficients for the coupling factors.

Definition

In the $\overline{\text{MS}}$ scheme, counterterms are pure poles at $\epsilon = 0$, except for unit-of-mass factors and a special factor S_ϵ for each loop:

$$g_0 = g\mu^\epsilon\left[1 + g^2 S_\epsilon\frac{B_{11}}{\epsilon} + g^4 S_\epsilon^2\left(\frac{B_{22}}{\epsilon^2} + \frac{B_{21}}{\epsilon}\right) + \dots\right], \tag{3.16}$$

$$Z_2 = 1 + g^2 S_\epsilon\frac{Z_{2,11}}{\epsilon} + g^4 S_\epsilon^2\left(\frac{Z_{2,22}}{\epsilon^2} + \frac{Z_{2,21}}{\epsilon}\right) + \dots, \tag{3.17}$$

etc. The rationale for the factor S_ϵ and its value are explained below. For normal UV divergences, the strength of the pole is at most $1/\epsilon^L$ in an L-loop counterterm. The only parameter on which the coefficients depend is the gauge-fixing parameter ξ, and this is absent for the bare coupling: the coefficients B_{ij} are pure numbers. In particular, the coefficients are independent of mass and of μ ('t Hooft, 1973; Collins, 1974).

The role played in renormalization by the unit of mass μ is quite central. It is commonly called the "renormalization mass" or "renormalization scale".

The $\overline{\text{MS}}$ scheme differs from the simplest minimal subtraction scheme by inserting a factor S_ϵ for each loop in the counterterms. This was motivated (Bardeen et al., 1978) by the observation that in a one-loop calculation, there is an ϵ-dependent factor that naturally arises from an angular integration in $4 - 2\epsilon$ dimensions, and that would lead to certain universally occurring extra terms in renormalized Green functions. These are eliminated by choosing S_ϵ suitably. I define

$$S_\epsilon = \frac{(4\pi)^\epsilon}{\Gamma(1 - \epsilon)} = 1 + \epsilon[\ln(4\pi) - \gamma_E] + O(\epsilon^2) \simeq 1 + 1.954\epsilon + O(\epsilon^2). \tag{3.18}$$

Here $\gamma_E = 0.5772\cdots$ is the Euler constant, and Γ is the gamma function.

Although there are several ways in which the $\overline{\text{MS}}$ scheme has been defined in the literature, it can be proved (see problem 3.3) that all these definitions lead to identical renormalized Green functions at $\epsilon = 0$. For example, there are different formulae for S_ϵ, but only the order ϵ part of S_ϵ affects renormalized Green functions.[3] The equivalence of the definitions, to all orders of perturbation theory, applies to conventional Green functions, where the UV divergences give at most one pole per loop. But in Chs. 10 and 13, we will define quantities that have a double UV pole per loop. For these, it is the particular definition, (3.18), that gives the maximal simplification.

Advantages

Among the advantages of minimal subtraction is that it automatically preserves simple symmetries. For example, the counterterms for the 4-gluon interaction and for the 3-gluon interaction, etc., will automatically give counterterms with the correct gauge-invariance relations. Counterterms have their minimal mass dependence.

Mathematically, some care is needed in understanding the expansion about $g = \epsilon = 0$. Perturbative renormalization is done by first expanding in g and then analyzing the ϵ dependence. Real physics is defined with $\epsilon \to 0$ taken at fixed g. The direct perturbative calculation of the counterterms is really only valid in a region of g that shrinks to zero as $\epsilon \to 0$. This is enough to obtain the coefficients for renormalized perturbation theory, whose radius of applicability is not expected to shrink with ϵ.

As we will see in Sec. 3.5, renormalization-group methods can be used to calculate the true behavior of the bare parameters when the UV regulator (e.g., dimensional regularization, or a lattice) is removed with the renormalized couplings and mass fixed.

Renormalization group

A change of renormalization scheme, including a change of the unit of mass, can be compensated by a change in the numerical values of the renormalized parameters. All that changes is the parameterization of the set of renormalized theories by coupling(s) and masses. This is the subject of the renormalization group (RG) – Sec. 3.5 – which is a vital technique in perturbative QCD.

Minimal subtraction with other regulators

Although minimal subtraction is normally defined using dimensional regularization, the concept applies to any regularization method. With regularization by a lattice spacing a, one could define the counterterms in each order to be a polynomial in $\ln(a\mu)$ with no constant term. This would define a different scheme, related by a RG transformation to the standard $\overline{\text{MS}}$ scheme, which uses dimensional regularization.

3.3 Renormalization counterterms of QCD

Renormalization plays an essential role in perturbative QCD calculations. Not only does renormalization enable finite results to be obtained, but the counterterms themselves

---

[3] *Warning*: In comparing formulae for S_ϵ, note that some authors use a different convention for ϵ than this book.

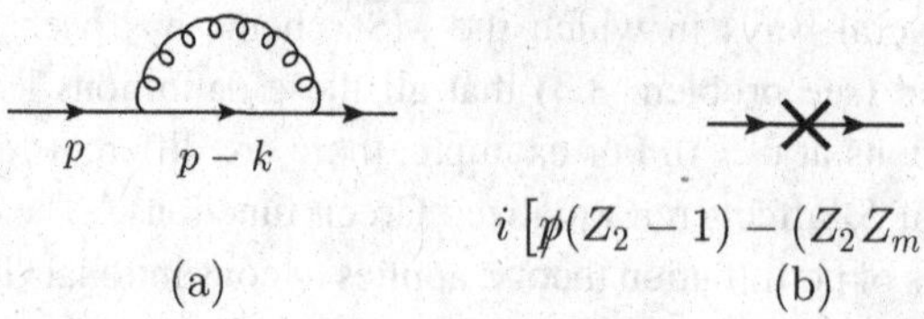

$$i\left[\not{p}(Z_2 - 1) - (Z_2 Z_m - 1)m_f\right]$$

(a)   (b)

Fig. 3.3. (a) Quark self-energy graph. (b) Counterterm.

determine the renormalization-group coefficients that we will see are vital to predicting the scale dependence of measurable quantities. This is useful, since counterterms are much simpler to calculate than the finite parts of graphs.

This section reviews the renormalization of QCD at the one-loop level, giving a complete calculation for some parts and leaving the rest as an exercise. In Sec. 3.5, this will enable us to verify the key result of asymptotic freedom of QCD.

3.3.1 Wave-function renormalization

The wave-function and mass renormalization factors are obtained from propagator corrections, the "self-energy graphs". For the case of the quark, the one-loop graph and its counterterm are shown in Fig. 3.3. The graph's value is

$$\frac{g^2\mu^{2\epsilon}C_F}{(2\pi)^{4-2\epsilon}}\int d^{4-2\epsilon}k \; \frac{\gamma^\mu(\not{p} - \not{k} + m)\gamma^\nu\left[-g_{\mu\nu} + (1-\xi)k_\mu k_\nu/(k^2+i0)\right]}{\left[(p-k)^2 - m^2 + i0\right](k^2+i0)}, \tag{3.19}$$

where the C_F factor is from the color matrices $\sum_\alpha t_\alpha t_\alpha$, which gives 4/3 in QCD. We combine the denominators using the Feynman parameter method (A.55), after which the momentum integral can be shifted so that the denominator loses its linear term in k. The use of standard Dirac algebra gives

$$\frac{g^2\mu^{2\epsilon}C_F}{(2\pi)^{4-2\epsilon}}\int_0^1 dx \int d^{4-2\epsilon}k \left\{ \frac{(2-2\epsilon)\not{p}(1-x) - (4-2\epsilon)m + (1-\xi)(m-\not{p}x)}{\left[-k^2 - p^2 x(1-x) + m^2 x - i0\right]^2} \right.$$

$$\left. - \frac{2(1-x)(1-\xi)(p^2 \not{p} x^2 + \not{k}\not{p}\not{k})}{\left[-k^2 - p^2 x(1-x) + m^2 x - i0\right]^3} + \text{terms odd in } k \right\}. \tag{3.20}$$

A Wick rotation gives a spherically symmetric integral in a Euclidean k variable in $4-2\epsilon$ dimensions, which can be performed analytically by using (A.34) and (A.50) and $\Gamma(1+\epsilon) = \epsilon\Gamma(\epsilon)$ to give

$$\frac{ig^2(4\pi\mu^2)^\epsilon C_F}{16\pi^2}\Gamma(\epsilon)\int_0^1 dx \left[-p^2 x(1-x) + m^2 x - i0\right]^{-\epsilon}$$

$$\times \left\{(2-2\epsilon)\not{p}(1-x) - (4-2\epsilon)m + (1-\xi)\left[m - \not{p}x - \not{p}(1-x)(1-\epsilon)\right]\right.$$

$$\left. - \frac{\epsilon x^2(1-x)(1-\xi)p^2 \not{p}}{\left[-p^2 x(1-x) + m^2 x - i0\right]}\right\}. \tag{3.21}$$

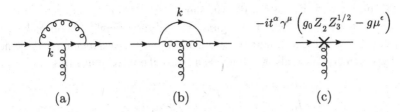

Fig. 3.4. (a) and (b) One-loop graphs for quark-gluon vertex. (c) Counterterm graph.

The pole at $\epsilon = 0$ is easy to extract, since $\Gamma(\epsilon) = 1/\epsilon +$ finite, so that

$$\text{pole part of graph (a)} = \frac{ig^2 C_F}{16\pi^2 \epsilon}\left[-3m + \xi(\not{p} - m)\right]. \tag{3.22}$$

We require the pole part plus the order g^2 part of the counterterm in Fig. 3.3(b) to be finite. This gives

$$Z_2 = 1 - \xi C_F \frac{\alpha_s S_\epsilon}{4\pi \epsilon} + O(\alpha_s^2), \tag{3.23}$$

$$Z_m = 1 - 3C_F \frac{\alpha_s S_\epsilon}{4\pi \epsilon} + O(\alpha_s^2). \tag{3.24}$$

Here, $\alpha_s = g^2/(4\pi)$, a commonly used definition analogous to the fine-structure constant in electromagnetism. The quantity S_ϵ is defined in (3.18), used to define the $\overline{\text{MS}}$ scheme.

Similar calculations for the gluon and ghost give

$$Z_3 = 1 - \frac{\alpha_s S_\epsilon}{4\pi \epsilon}\left[\left(\frac{\xi}{2} - \frac{13}{6}\right)C_A + \frac{4}{3}T_F n_f\right] + O(\alpha_s^2), \tag{3.25}$$

$$\tilde{Z} = 1 + \frac{\alpha_s S_\epsilon}{4\pi \epsilon}C_A\left(\frac{3}{4} - \frac{\xi}{4}\right) + O(\alpha_s^2). \tag{3.26}$$

In QCD, with its SU(3) gauge group and quarks in the triplet representation, the group theory coefficients used here are $C_A = 3$ and $T_F = 1/2$. See Sec. A.11 for more details. The quantity n_f is the number of quark flavors in QCD.

3.3.2 Quark-gluon vertex

To obtain g_0, we need to calculate one of the vertex functions. The simplest is the quark-gluon vertex, because it is only logarithmically divergent. The one-loop graphs and the counterterm are shown in Fig. 3.4. Now the UV divergence is independent of masses and external momenta. So we simplify the calculation by setting these variables to zero, and by ignoring any ϵ dependence that does not affect the pole. From the first graph we need

$$V_a = t_a g \mu^\epsilon S_\epsilon \frac{g^2}{16\pi^4}(C_F - \tfrac{1}{2}C_A)\,\text{PP}\int_{\text{UV}} \frac{d^{4-2\epsilon}k}{(k^2)^3}\left[-\gamma^\nu \not{k}\gamma^\mu \not{k}\gamma_\nu + \frac{1-\xi}{k^2}\not{k}\not{k}\gamma^\mu \not{k}\not{k}\right], \tag{3.27}$$

where "PP" means "pole part at $\epsilon = 0$". The subscript "UV" on the integration means that we restrict the integration to the UV region; we cut out a neighborhood of $k = 0$. The prefactors are those present in the lowest-order vertex. A Wick rotation and elementary spherically symmetric integrals over Euclidean k give the integral in terms of

$$\int_{\text{UV}} \frac{d^{4-2\epsilon}k}{(k^2)^2} = \frac{\pi^{2-\epsilon}}{\Gamma(2-\epsilon)} \int_{\text{finite}}^{\infty} \frac{d|k^2|}{|k^2|^{1+\epsilon}}, \tag{3.28}$$

so that

$$V_a = -i t_\alpha \gamma^\mu g \mu^\epsilon S_\epsilon \frac{g^2}{16\pi^2\epsilon} \xi (C_F - C_A/2). \tag{3.29}$$

Similarly, graph (b) gives

$$V_b = -t_\alpha g \mu^\epsilon S_\epsilon \frac{g^2}{16\pi^4} (\tfrac{1}{2} C_A)$$

$$\times \text{PP} \int_{\text{UV}} d^{4-2\epsilon}k \frac{\gamma^{\kappa'} \slashed{k} \gamma^{\nu'}}{(k^2)^3} (-2k^\mu g^{\kappa\nu} + k^\nu g^{\kappa\mu} + k^\kappa g^{\nu\mu})$$

$$\times \left(-g_{\nu'\nu} + (1-\xi) \frac{k_{\nu'} k_\nu}{k^2} \right) \left(-g_{\kappa'\kappa} + (1-\xi) \frac{k_{\kappa'} k_\kappa}{k^2} \right)$$

$$= -i t_\alpha \gamma^\mu g \mu^\epsilon S_\epsilon \frac{g^2}{16\pi^2\epsilon} C_A \left(\frac{3}{4} + \frac{3}{4}\xi \right). \tag{3.30}$$

From these, we deduce the one-loop counterterm and, hence, from the previously calculated values of Z_2 and Z_3 we get the bare coupling:

$$g_0 = g\mu^\epsilon \left[1 - \frac{\alpha_s S_\epsilon}{4\pi\epsilon} \left(\frac{11}{6} C_A - \frac{2}{3} T_F n_f \right) + O(\alpha_s^2) \right]. \tag{3.31}$$

Note that the manipulations to obtain the coupling are performed with only the first two terms in a strict expansion in powers of g.

The results for the counterterms to higher order, up to four loops, can be deduced from the published values (Tarasov, Vladimirov, and Zharkov, 1980; Larin and Vermaseren, 1993; van Ritbergen, Vermaseren, and Larin, 1997; Czakon, 2005) of the RG coefficients, the primary ones being given in Sec. 3.7. See problem 3.2.

3.4 Meaning of unit of mass, renormalization scale

The unit of mass μ is a rather abstract concept seemingly tied to the use of dimensional regularization. It appears as a renormalization scale in renormalized quantities. We will see later (Sec. 3.5) that the value of the renormalization scale can be freely chosen, provided that the numerical value of the coupling and masses are adjusted in compensation. Perturbative calculations can be optimized in accuracy by a suitable choice of μ.

To understand how to choose μ, I now present a simple example that gives μ an intuitive meaning as approximating a cutoff in the physical dimension at a certain value of transverse momentum.

Now, in many of our calculations for scattering, there will be preferred coordinates determined by the momenta of two of the particles. The Breit frame for DIS is a good example. Let us use these directions to fix a plane of t and z. Then for an integration over a momentum k, we first perform the k^0 and k^z integrals. After that we have a two (or $2-2\epsilon$) dimensional integral over a transverse momentum k_{T}, which is often rotationally symmetric. A generic one-loop integral, relative to a lowest-order calculation, is then

$$
\begin{aligned}
I_0 &= \frac{g^2 \pi \mu^{2\epsilon}}{(4\pi)^{4-2\epsilon}} \int d^{2-2\epsilon} k_{\mathrm{T}} \, \frac{1}{k_{\mathrm{T}}^2 + M^2} \\
&= \frac{g^2}{16\pi^2} \frac{(4\pi\mu^2)^\epsilon}{\Gamma(1-\epsilon)} \int_0^\infty dk_{\mathrm{T}}^2 \, \frac{(k_{\mathrm{T}}^2)^{-\epsilon}}{k_{\mathrm{T}}^2 + M^2}.
\end{aligned}
\tag{3.32}
$$

The factor π in the first line is typical for a two-dimensional integral over the two longitudinal components of k. In an actual application, M would be a function of external longitudinal kinematic variables as well as of masses of particles and fields. For examples, see (9.4) and (10.137).

Using (A.50), we express the integral in terms of Γ functions, and then obtain the pole and finite part using (A.47):

$$
I_0 = \frac{g^2}{16\pi^2} \Gamma(\epsilon) \left(\frac{4\pi\mu^2}{M^2} \right)^\epsilon = \frac{g^2 S_\epsilon}{16\pi^2} \left(\frac{1}{\epsilon} + \ln\frac{\mu^2}{M^2} + O(\epsilon) \right).
\tag{3.33}
$$

Subtraction of the $\overline{\mathrm{MS}}$ pole gives the renormalized value

$$
I_R = \lim_{\epsilon\to 0} \left(I_0 - \frac{g^2 S_\epsilon}{16\pi^2\epsilon} \right) = \frac{g^2}{16\pi^2} \ln\frac{\mu^2}{M^2}.
\tag{3.34}
$$

Without the S_ϵ factor in the definition of the $\overline{\mathrm{MS}}$ counterterm, we would get an extra term containing $\ln(4\pi) - \gamma_{\mathrm{E}}$. The simple logarithmic dependence on the unit of mass μ is a general expectation, but for a more general integral the rest of the result will not be so simple and will not always have a simple analytic form.

To obtain an interpretation, we now rewrite the counterterm as a subtraction at the level of the integrand. Since the divergence is associated with the asymptotic large k_{T} behavior of the integrand, we consider an integral over this asymptotic behavior:

$$
\frac{g^2 \pi \mu^{2\epsilon}}{(4\pi)^{4-2\epsilon}} \int_{k_{\mathrm{T}}^2 > C\mu^2} \frac{d^{2-2\epsilon} k_{\mathrm{T}}}{k_{\mathrm{T}}^2} = \frac{g^2}{16\pi^2} \frac{(4\pi\mu^2)^\epsilon}{\Gamma(1-\epsilon)} \int_{C\mu^2}^\infty \frac{dk_{\mathrm{T}}^2}{(k_{\mathrm{T}}^2)^{1+\epsilon}}.
\tag{3.35}
$$

The integral is of a power of k_{T}, so it is trivial to calculate. Since the extraction of the asymptotic behavior would otherwise expose an IR divergence, we put a lower limit proportional to μ^2 on the integration, with a coefficient C that is to be adjusted to obtain the correct finite part of the counterterm. The integral is

$$
\frac{g^2 S_\epsilon}{16\pi^2\epsilon} \frac{e^{\gamma_{\mathrm{E}}\epsilon}}{\Gamma(1-\epsilon)} C^{-\epsilon}.
\tag{3.36}
$$

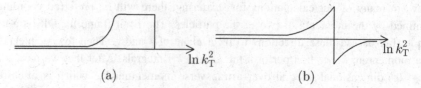

Fig. 3.5. (a) Integrand times k_T^2 of ((3.37) when $\mu \gg M$. (b) Same when μ is close to M.

Since the second factor is $1 + O(\epsilon^2)$, we can reproduce the renormalized graph by using (3.36) in place of the true $\overline{\text{MS}}$ counterterm, provided that we set $C = 1$. Then the renormalized graph is an integral in the physical dimension with a subtracted integrand:

$$I_R = \frac{g^2}{16\pi^3} \int d^2 k_T \left(\frac{1}{k_T^2 + M^2} - \frac{\theta(|k_T| > \mu)}{k_T^2} \right). \tag{3.37}$$

The integrand is plotted in Fig. 3.5. Because of the logarithmic behavior at large k_T, it is convenient to multiply the integrand by k_T^2 and to plot it against $\ln k_T^2$, to correspond to the integrand on the r.h.s. of an integral of the form

$$\int dk_T^2 \, f(k_T) = \int d \ln k_T^2 \, \left[k_T^2 f(k_T) \right]. \tag{3.38}$$

We now interpret (3.37), with a view to generalization.

- The natural expansion parameter for perturbation theory is $g^2/16\pi^2$, which arises as the product of the coupling, the factor $1/(2\pi)^4$ for a loop integral, and π^2 for an angular integration in four dimensions.
- This is multiplied by a group-theory factor and the number of graphs.
- In simple cases, renormalization can be performed by a subtraction of the asymptote of the integrand. The lower bound on k_T in the subtraction is commonly exactly μ.
- The coefficient S_ϵ defining the $\overline{\text{MS}}$ scheme is responsible for the cutoff being μ rather than a factor times μ. This gives a direct connection to the physical scale M in the integrand.
- In a more general graph, finite terms with modest, typically rational, numbers must be added. The need for this can be seen in the quark self-energy calculation, where the ϵ dependence of the numerator algebra enters.
- To get perturbative corrections of a natural size, μ should be close to the scale that is set by the transverse momentum dependence of the integrand, i.e., a scale characterizing the change from $1/k_T^2$ behavior at large k_T to constant behavior at small k_T.
- Although our example integral is exactly zero when $\mu = M$, this is not true in general; also M will generally be a function of external momentum. The best general statement is that for a single graph without a group-theory coefficient, the expected coefficient of $g^2/16\pi^2$ is a modest number of typical size unity if μ is close to a natural scale.
- For large values of μ, μ behaves like a cutoff on k_T in the unsubtracted integral.
- The rationale for these results suggests that they should approximately generalize to higher orders. In a well-behaved L-loop calculation, we can expect the result to be

roughly $(g^2/16\pi^2)^L$ times an effective number of graphs times a typical group-theory factor, provided again that μ is of the order of the physically relevant scale in transverse momentum.

- When we meet badly behaved situations, it is a good idea to search for explanations for large perturbative corrections in terms of the sizes of integrands in relevant kinematic regions.

3.5 Renormalization group

The general idea of renormalization prescribes only that counterterms cancel divergences; thus the finite parts of counterterms can be chosen freely. Within many schemes, like $\overline{\text{MS}}$, there is also a parameter μ that can be chosen freely. At first sight, the choices remove predictive power from the theory since any numerical value can be obtained from a one-loop integral with given external momenta. In reality, as explained more fully in textbook accounts of renormalization, this is not so. Instead we exploit the freedom in choosing μ to optimize the accuracy of finite-order perturbative calculations.

The complete theory is exactly invariant if when changing μ (or, more generally, the renormalization prescription) we also change the numerical values of the renormalized parameters of the theory. This is the renormalization-group (RG) invariance of the theory. An RG transformation amounts to a change in the partitioning of the full Lagrangian $\mathcal{L}$ into the three terms in (3.12). Thus it corresponds to a rearrangement of the perturbation expansion. The most important case for us is the transformation of the renormalized coupling and masses when the renormalization mass μ is changed.

3.5.1 RG evolution

When we perform RG transformations for changes of μ, keeping observable quantities fixed, each numerical value of μ corresponds to particular numerical values of the renormalized parameters $g(\mu)$, $m_f(\mu)$. When we change μ to another value μ', not only do the coupling and masses change, but also the normalization of the renormalized fields. So we write

$$\phi_i(x;\mu) = \zeta_i(\mu,\mu')\phi_i(x;\mu'). \tag{3.39}$$

Here i labels the different types of field (gluon, quark, etc.). A Green function therefore transforms as

$$G(\underline{p};\mu,g(\mu),\underline{m}(\mu)) = \prod_e \zeta_{i_e}(\mu,\mu')G(\underline{p};\mu',g(\mu'),\underline{m}(\mu')), \tag{3.40}$$

where $\underline{p}$ is the collection of external momenta of G, $\underline{m}$ is the set of renormalized masses, and the product is over the external lines, e, of G, with i_e labeling the corresponding types of field.

The S-matrix and hence cross sections are RG invariant. This is because an S-matrix element is obtained by applying to the corresponding off-shell Green function the following operations: (a) divide out a full external propagator; (b) multiply by the square root of the residue of the particle pole; (c) put the external momenta on-shell. In this process there is a cancellation of the ζ factors for each external line. Exactly the same idea applies to Green functions of the composite external fields needed to obtain the S-matrix for composite particles.

We now determine equations for μ dependence of $g(\mu)$. The coefficients in the equations are obtained from the counterterms in the bare parameters, the starting point being RG invariance of the bare parameters, as is necessary to keep the physics unchanged. The normalizations of the bare parameters and the bare fields are fixed because terms like $i\bar{\psi}_0\partial\!\!\!/\psi_0$ have unit coefficients. Our discussion is tailored to the $\overline{MS}$ scheme, but the main principles and methods are general.

3.5.2 Coupling and mass

With a UV cutoff applied ($\epsilon \neq 0$), we hold the bare parameters g_0 and $m_{(0)f}$ fixed and vary μ. For g_0, we get

$$0 = \frac{d}{d\ln\mu}g_0(\mu, g(\mu), \epsilon) = \frac{\partial g_0}{\partial\ln\mu} + \frac{dg}{d\ln\mu}\frac{\partial g_0}{\partial g} = \epsilon g_0 + \frac{dg}{d\ln\mu}\frac{\partial g_0}{\partial g}. \tag{3.41}$$

We distinguish between a total derivative $d/d\mu$, with respect to *all* the μ dependence, and a partial derivative $\partial/\partial\mu$, for which the renormalized parameters $g(\mu)$, etc., are fixed. It is convenient to use a logarithmic derivative, given that renormalized graphs have a μ dependence that is polynomial in $\ln\mu$.

For the masses

$$0 = \frac{d}{d\ln\mu}m_0(m(\mu), g(\mu), \epsilon) = \frac{dg}{d\ln\mu}\frac{\partial m_0}{\partial g} + \frac{dm}{d\ln\mu}\frac{m_0}{m}, \tag{3.42}$$

where we used the lack of explicit μ dependence of m_0 in minimal subtraction.

It is convenient to use as the expansion parameter $\alpha_s/4\pi = g^2/16\pi^2$. Then from (3.41) we find

$$\frac{d\alpha_s/4\pi}{d\ln\mu} = \frac{g}{8\pi^2}\frac{dg}{d\ln\mu} = -\frac{\epsilon g}{8\pi^2}\frac{g_0}{\partial g_0/\partial g}. \qquad \text{(General } \epsilon) \tag{3.43}$$

The left-hand side is finite at $\epsilon = 0$, and therefore the right-hand side is finite also; all poles in ϵ must cancel. In the $\overline{MS}$ scheme each α_s in the counterterms is accompanied by a factor S_ϵ and all the terms in g_0 have negative powers of ϵ. We therefore find that the right-hand side has the form[4]

$$-2\epsilon\frac{\alpha_s}{4\pi} + S_\epsilon^{-1}2\beta(\alpha_s S_\epsilon/(4\pi)), \tag{3.44}$$

where the only ϵ dependence is in the $-\epsilon\alpha_s$ term and in the explicit factors of S_ϵ.

At the physical space-time dimension, i.e., at $\epsilon = 0$, we use the perturbatively calculable β function to give an equation for the scale dependence of the coupling:

$$\frac{d\alpha_s/4\pi}{d\ln\mu} = 2\beta(\alpha_s/4\pi). \qquad (\epsilon = 0) \tag{3.45}$$

---

[4] The factor of 2 multiplying β is to correspond to the definition in Larin and Vermaseren (1993); this arises because these authors use derivatives with respect to $\ln\mu^2$ instead of $\ln\mu$.

The one-loop value of the bare coupling, in (3.31), immediately gives

$$\beta(\alpha_s/4\pi) = -\left(\frac{11}{3}C_A - \frac{4}{3}T_F n_f\right)\frac{\alpha_s^2}{16\pi^2} + O(\alpha_s^3) \quad \text{(for general group)}$$

$$= -\left(11 - \frac{2}{3}n_f\right)\frac{\alpha_s^2}{16\pi^2} + O(\alpha_s^3). \quad \text{(for SU(3))} \quad (3.46)$$

Provided there are at most 16 quark flavors, which is true in currently known strong interactions, the coupling decreases with increasing scale, at least when it is small enough at the outset. The coupling does in fact go to zero as $\mu \to \infty$, as we will see, so that QCD is asymptotically free. The importance of this is clear from the previous chapter.

The results at higher order will be quoted in Sec. 3.7. Here we just note that β can be obtained from the *single* pole terms in g_0. With the conventions of (3.16), we get:

$$\beta(\alpha_s/4\pi) = \sum_{n=1}^{\infty} \frac{g^{2n+2}}{8\pi^2} n B_{n1}. \quad (3.47)$$

The finiteness conditions for $d\alpha_s/d\ln\mu$ enable the higher poles in g_0 to be computed in terms of the single poles.

The RG dependence of the mass is similarly obtained. A dimensionless function is obtained by using logarithmic derivatives:

$$\gamma_m(\alpha_s S_\epsilon/4\pi) \stackrel{\text{def}}{=} \frac{d\ln m}{d\ln\mu}$$

$$= \left(2\epsilon\alpha_s S_\epsilon/4\pi - 2\beta(\alpha_s S_\epsilon/4\pi)\right)\frac{\partial\ln Z_m}{\partial\alpha_s S_\epsilon/4\pi}$$

$$= -6C_F\frac{\alpha_s}{4\pi}S_\epsilon + O(\alpha_s^2). \quad (3.48)$$

Again, the divergences present in Z_m must cancel in this derivative in order that γ_m is finite. This time, it can be shown that the only ϵ dependence is in the S_ϵ multiplying α_s. This RG coefficient is usually less important in practice, since most pQCD calculations are performed with masses set to zero, or with a different scheme for heavy quarks.

The lack of mass dependence in the renormalization group coefficients β and γ_m follows from the mass-independence property of $\overline{\text{MS}}$ counterterms.

3.5.3 Anomalous dimensions and RG equations for Green functions

To unify the treatment of the RG transformation for renormalized fields, let us use the notation ϕ_i for the renormalized fields, with the label i denoting the type of field (gluon, flavor of quark, etc.). We define its anomalous dimension by

$$\gamma_i(\alpha_s S_\epsilon/4\pi, \xi)\phi_i = -\frac{d\phi_i}{d\ln\mu}. \quad (3.49)$$

Given that the corresponding bare field is $\phi_{(0)i} = Z_i^{1/2}\phi_i$, it follows that

$$\gamma_i(\alpha_s S_\epsilon/4\pi, \xi) = \frac{1}{2}\frac{d\ln Z_i}{d\ln\mu}. \quad (3.50)$$

A complication arises in gauge theories, from the gauge dependence of wave-function renormalization. Because of the relation $\xi_0 = \xi Z_3$, the gauge-fixing parameter obeys

$$\frac{d \ln \xi}{d \ln \mu} = -\frac{d \ln Z_3}{d \ln \mu} = -2\gamma_3. \tag{3.51}$$

Then the definition of γ_3 gives

$$\gamma_3 = \left(-\epsilon \frac{\alpha_s S_\epsilon}{4\pi} + \beta(\alpha_s S_\epsilon/4\pi)\right) \frac{d \ln Z_3}{d\alpha_s S_\epsilon/4\pi} - \gamma_3 \frac{\partial \ln Z_3}{\partial \xi}. \tag{3.52}$$

Hence

$$\gamma_3 = \frac{\left(-\epsilon \frac{\alpha_s}{4\pi} S_\epsilon + \beta(\alpha_s S_\epsilon/4\pi)\right) d \ln Z_3/d(\alpha_s S_\epsilon/4\pi)}{1 + \partial \ln Z_3/\partial \xi}. \tag{3.53}$$

For the other anomalous dimensions, we have equations of the form

$$\gamma_2 = \left(-\epsilon \frac{\alpha_s S_\epsilon}{4\pi} + \beta(\alpha_s S_\epsilon/4\pi)\right) \frac{d \ln Z_2}{d\alpha_s S_\epsilon/4\pi} - \gamma_3 \frac{\partial \ln Z_2}{\partial \xi}. \tag{3.54}$$

See Sec. 3.7 for the values of the anomalous dimensions.

From the above results follows the renormalization-group equation (RGE) for a renormalized Green function G. If G has n_2 external quark fields (and the same number of antiquarks) and n_3 external gluons, then

$$\frac{dG}{d \ln \mu} = -(2n_2\gamma_2 + n_3\gamma_3)\, G. \tag{3.55}$$

Exactly similar equations can be derived for other operator matrix elements, where the states can be other than the vacuum and the fields not simple products of the elementary fields of QCD at different space-time points. A simple example is the hadronic tensor $W^{\mu\nu}$ of DIS, (2.18). The electromagnetic current is a symmetry current of QCD and can be shown to have zero anomalous dimension. Hence $W^{\mu\nu}$ is RG invariant:

$$\frac{dW^{\mu\nu}}{d \ln \mu} = 0. \tag{3.56}$$

3.6 Solution of RG equations

3.6.1 General form of solution

The RG equations for the coupling, mass, and Green functions are readily solved to relate these quantities at different values of the $\overline{\text{MS}}$, with the aid of integrals of β, γ_m and the anomalous dimensions:

$$\ln \frac{\mu}{\mu_0} = \int_{\alpha_s(\mu_0)/4\pi}^{\alpha_s(\mu)/4\pi} \frac{d\alpha/4\pi}{2\beta(\alpha/4\pi)}, \tag{3.57}$$

$$\ln \frac{m(\mu)}{m(\mu_0)} = \int_{\mu_0}^{\mu} \frac{d\mu'}{\mu'} \gamma_m(\alpha_s(\mu')/4\pi) = \int_{\alpha_s(\mu_0)/4\pi}^{\alpha_s(\mu)/4\pi} d\alpha/4\pi \frac{\gamma_m(\alpha/4\pi)}{2\beta(\alpha/4\pi)}, \tag{3.58}$$

$$\ln \frac{G(\mu)}{G(\mu_0)} = -\int_{\mu_0}^{\mu} \frac{d\mu'}{\mu'} \gamma_G(\alpha_s(\mu')/4\pi, \xi(\mu')) = -\int_{\alpha_s(\mu_0)/4\pi}^{\alpha_s(\mu)/4\pi} d\alpha/4\pi \frac{\gamma_G(\alpha/4\pi, \xi(\mu'))}{2\beta(\alpha/4\pi)}. \tag{3.59}$$

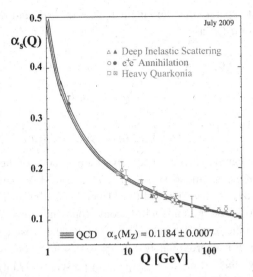

Fig. 3.6. QCD effective coupling. With kind permission from Springer Science+Business Media: Bethke (2009, Fig. 6). The lines represent the solution of the RGE for $\alpha_s(\mu)$ with the $\pm 1\sigma$ limits on the constant of integration. The scheme used is $\overline{\text{MS}}$ with a variable number of active quarks, as in Sec. 3.10. The data are, in increasing order of μ, from fits to the τ width, Υ decays, DIS, e^+e^- event shapes at 22 GeV at JADE, shapes at TRISTAN at 58 GeV, Z width, and e^+e^- event shapes at 91–208 GeV.

Here $\gamma_G = 2n_2\gamma_2 + n_3\gamma_3$ is the anomalous dimension of the Green function G, all of whose momentum and mass arguments we have suppressed.[5]

Since $\beta(\alpha/4\pi)$ is negative and $O(\alpha^2)$ at small coupling, (3.57) shows that $\alpha_s(\mu) \to 0$ as $\mu \to \infty$, i.e., that QCD is asymptotically free.

3.6.2 Effective coupling; scale parameter Λ

The μ dependence of the coupling underlies all other RG calculations in QCD, so a detailed analysis is useful. There is a one-parameter family of solutions of (3.45) for $\alpha_s(\mu)$, and the physical solution is specified, for example, by the value of coupling at a given scale (e.g., "$\alpha_s(M_Z) = 0.1184 \pm 0.0007$ in the $\overline{\text{MS}}$ scheme with five active flavors"). The physical solution is obtained by fitting the one parameter to data, with a result shown in Fig. 3.6.

One often-used procedure is the following, which is particularly useful for assessing the errors due to the limited accuracy with which RG functions are known. It was obtained (Buras *et al.*, 1977) basically by expanding $\alpha_s(\mu)$ in powers of $1/\ln \mu^2$ at large μ.

Let us write the expansion of β as

$$\beta(a_s) \stackrel{\text{def}}{=} \frac{da_s}{d\ln \mu^2} = -\beta_0 a_s^2 - \beta_1 a_s^3 - \beta_2 a_s^4 - \beta_3 a_s^5 + O(a_s^6), \qquad (3.60)$$

where $a_s = \alpha_s/4\pi = g^2/16\pi^2$. (The normalizations of all but β_0 differ from the less systematic conventions of the PDG; Amsler *et al.*, 2008.) In the solution (3.57) the integral

---

[5] Thus $G(\mu_0)$ means $G(\underline{p}; \mu_0, g(\mu_0), \underline{m}(\mu_0), \xi(\mu_0))$.

is of $1/\beta$, so we first separate out the singular parts of $1/\beta$, and represent the general solution of the RGE for $a_s(\mu)$ as

$$\ln \frac{\mu^2}{\Lambda^2} = \frac{1}{\beta_0 a_s} + \frac{\beta_1}{\beta_0^2} \ln(\beta_0 a_s) - f(a_s), \qquad (3.61)$$

where

$$f(a_s) \overset{\text{def}}{=} \int_0^{a_s} da \left(-\frac{1}{\beta(a)} - \frac{1}{\beta_0 a^2} + \frac{\beta_1}{\beta_0^2 a} \right). \qquad (3.62)$$

Here the constant of integration is represented by a parameter Λ, of the dimension of mass; it has an experimentally determined value[6] of around 200 MeV. The constant β_0 in the logarithm in the second term on the r.h.s. of (3.61) merely amounts to a standard convention for the definition of Λ whose rationale will become apparent below. When it is necessary to distinguish Λ from other similar parameters, we will add a subscript, as in Λ_{QCD}.

For small coupling, β is approximately $-\beta_0 a_s^2$, so that $a_s(\mu)$ behaves like $1/(\beta_0 \ln \mu^2)$ at large μ. To improve this estimate, we expand in powers of $1/\ln(\mu^2/\Lambda^2)$ (with some modifications as required). This gives

$$\frac{\alpha_s}{4\pi} = \frac{1}{\beta_0 \ln(\mu^2/\Lambda^2)} - \frac{\beta_1 \ln \ln(\mu^2/\Lambda^2)}{\beta_0^3 \ln^2(\mu^2/\Lambda^2)} + O\left(\frac{\ln^2 \ln(\mu^2/\Lambda^2)}{\ln^3(\mu^2/\Lambda^2)} \right). \qquad (3.63)$$

Normally we would expect a term constant$/\ln^2(\mu^2/\Lambda^2)$, and the absence of this term is effectively the definition of Λ, and is exactly correlated with the use of $\ln(\beta_0 a_s)$ rather than $\ln a_s$ in (3.61). This convention is due to Buras *et al.* (1977). Then Λ can, in principle, be extracted from the large μ asymptote of $a_s(\mu)$:

$$\Lambda^2 = \lim_{\mu \to \infty} \mu^2 \exp\left[-\frac{1}{\beta_0 a_s} - \frac{\beta_1}{\beta_0^2} \ln(\beta_0 a_s) \right]. \qquad (3.64)$$

Notice that this formula requires only the use of the known one- and two-loop terms in β, not any of the higher terms not all of which are known. Of course the higher terms will improve the accuracy of the measurement of Λ since $a_s(\mu)$ is only known at finite μ.

3.6.3 Dimensional transmutation

Suppose we were to approximate quark masses of QCD by zero. Since the masses of the light quarks are considerably smaller than the proton mass, this is in fact a useful approximation, for low-energy processes, if we keep only two (u, d) or three flavors (u, d, s), with the heavier quarks being removed according to the decoupling theorem. Then the mass of any particle, like the proton, would be a function of α_s and μ only. But by dimensional analysis it is μ times a function of α_s:

$$m_p = \mu F(\alpha_s) \qquad \text{in massless QCD}. \qquad (3.65)$$

---

[6] Details depend on a treatment of heavy quark masses which we will present later (Sec. 3.10). The current best value with five active quarks is (Bethke, 2009) $\Lambda = (213 \pm 9)\,\text{MeV}$.

Since m_p is a physical mass, it is RG invariant, which fixes $\alpha_s(\mu)$ up to a multiplicative factor. It follows that m_p equals Λ times a pure number, K_p, which is a property of the solution of massless QCD: $m_p = \Lambda K_p$. The number K_p is non-perturbative and can be obtained from lattice QCD calculations.

Instead of specifying the theory by the numerical value of its dimensionless coupling g, we can instead specify a fixed mass parameter Λ. This is the property of dimensional transmutation (Coleman and Weinberg, 1973).

In fact, there is a certain sense in which even this parameter is illusory. Suppose we consider pure strong interactions with massless quarks. To completely define a measurement of the numerical value of Λ, we must specify a system of units, i.e., specify what a mass of numerical value unity means.[7] But with only the strong interaction under consideration, this can only mean some physical mass like the proton mass, which can be taken as a physical definition of a standard mass. So a measurement of Λ is really a measurement of the dimensionless ratio Λ/m_P, whose value is a unique prediction of the theory.

This is the sense in which *massless* QCD has no parameters. All real predictions of the theory are pure numbers. For example, a cross section as a function of center-of-mass energy $\sigma(E)$ is of the form $m_p^{-2} S(E/m_p)$, where S is a dimensionless function of a dimensionless variable. This function is in principle predicted with no parameters by massless QCD.

Since the masses of the three light quarks are known to give only a relatively small contribution to the nucleon mass, the above statements are approximately true in real QCD. The real intrinsic parameters of QCD are the quark masses, expressed in terms of a suitable chosen unit, e.g., Λ or m_p.

There is a contrast with QED, because of the different physics of its classical long-distance limit. For simplicity consider QED of a photon and electron field only. Then, again by dimensional transmutation, there is only one true parameter m_e/Λ_{QED}. As with its QCD analog, Λ_{QED} is in a region where the coupling is strong. In contrast to the QCD coupling, the QED coupling increases at large scales, and in fact Λ_{QED} is around the Planck scale. At low energies compared to m_e, the electron decouples, giving a free Maxwell field theory which we can solve completely and exactly. It therefore becomes much more sensible than in QCD to use an on-shell renormalization prescription, and to define the expansion parameter of the theory as the usual $\alpha \simeq 1/137$. Within pure single-lepton QED, we can take the unit of mass to be m_e.

Of course, weak coupling methods are very useful and accurate for normal phenomena in QED, including its bound states, in contrast to QCD, where perturbation theory has a more restricted range of applicability.

Although dimensional transmutation has reduced the number of genuine parameters in a quantum field theory by one compared with the apparent number, the parameter is regained when the theory is treated as a component of a more complete theory. For example, we can combine QED and QCD to get a complete theory underlying all nuclear, atomic and

---

[7] The last sentence was carefully worded to avoid confusion between the concept of unit of mass in dimensional regularization and the concept of the unit of mass in a system of units.

molecular phenomena. Then m_e/Λ_{QCD} is a parameter of the combined theory in addition to the intrinsic parameters of the separate theories.

3.6.4 Bare coupling

We used (3.41) to obtain the (finite) β function from the divergent perturbation expansion for the bare coupling. But we can also use it to obtain a formula for the bare coupling as a function of Λ and ϵ. From the ϵ-dependent β function given in (3.44) we get

$$\frac{\partial \ln a_0}{\partial a_s} = \frac{\epsilon}{\epsilon a_s - S_\epsilon^{-1}\beta(a_s S_\epsilon)}, \tag{3.66}$$

where again $a_s = g^2/(16\pi^2)$, while $a_0 = g_0^2/(16\pi^2)$ is the bare equivalent of $a_s\mu^{2\epsilon}$, with mass dimension 2ϵ. The solution is

$$\ln a_0 = \ln(a_s\mu^{2\epsilon}) + \int_0^{a_s S_\epsilon} da \left[\frac{\epsilon}{\epsilon a - \beta(a)} - \frac{1}{a} \right], \tag{3.67}$$

where the boundary condition is set by requiring $a_0/(a_s\mu^{2\epsilon}) \to 1$ as $a_s \to 0$ at fixed ϵ.

An important formula is obtained by expressing this in terms of Λ, and then taking the limit $\epsilon \to 0$ at fixed a_s. This gives

$$\frac{g_0^2}{16\pi^2} = \frac{1}{\beta_0} \epsilon^{1+\epsilon\beta_1/\beta_0^2} \left(\frac{\Lambda^2 e^{\beta_1/\beta_0^2 + \gamma_E}}{4\pi} \right)^\epsilon \left[1 + O(\epsilon^2) \right]. \tag{3.68}$$

When $\epsilon \to 0$, the $O(\epsilon^2)$ fractional correction can be dropped, since it is equivalent to a change in Λ by a fraction of order $O(\epsilon)$: since Λ determines the coupling in the renormalized theory, the correction does not affect renormalized Green functions at $\epsilon = 0$. From the β function, only the scheme-independent coefficients β_0 and β_1 are needed; the scheme choice is manifested in the numerical coefficient multiplying Λ^2. To provide a *full* specification of the renormalization of the theory, only the one- and two-loop renormalization counterterms in the coupling need to be known.

Similar results can be obtained when any regularization scheme is used. With a lattice regulator, we would have

$$\frac{g_0^2}{16\pi^2} = \frac{1}{-\beta_0 \ln(a^2\Lambda^2)} - \frac{\beta_1 \ln(-\ln(a^2\Lambda^2))}{\beta_0^3 \ln^2(a^2\Lambda^2)} - \frac{A}{\ln^2(a^2\Lambda^2)}, \tag{3.69}$$

where a is the lattice spacing, and the coefficient A can be computed (Hasenfratz and Hasenfratz, 1980; Dashen and Gross, 1981) from the perturbative expansion of the bare coupling computed to two-loop order, with the renormalized coupling being in the $\overline{\text{MS}}$ scheme.

Other bare parameters and renormalization factors may be treated similarly.

3.7 Values of RG coefficients

The β function has been calculated in the $\overline{\text{MS}}$ scheme up to three loops by Tarasov, Vladimirov, and Zharkov (1980) and by Larin and Vermaseren (1993), and to four loops by van Ritbergen, Vermaseren, and Larin (1997). The results have been confirmed by Czakon

(2005). The first three coefficients in β are rational numbers. With the notation of (3.60),

$$\beta_0 = \frac{11}{3}C_A - \frac{4}{3}T_F n_f, \tag{3.70a}$$

$$\beta_1 = \frac{34}{3}C_A^2 - 4C_F T_F n_f - \frac{20}{3}C_A T_F n_f, \tag{3.70b}$$

$$\beta_2 = \frac{2857}{54}C_A^3 + 2C_F^2 T_F n_f - \frac{205}{9}C_F C_A T_F n_f - \frac{1415}{27}C_A^2 T_F n_f$$
$$+ \frac{44}{9}C_F T_F^2 n_f^2 + \frac{158}{27}C_A T_F^2 n_f^2. \tag{3.70c}$$

The expression for the four-loop coefficient β_3 is more complicated and includes the irrational number ζ_3; the full expression is given in van Ritbergen, Vermaseren, and Larin (1997). The fact that even the three-loop coefficient is a rational number indicates a fundamental simplicity in the theory and in minimal subtraction that is certainly not apparent in straightforward calculations of Feynman diagrams. In the case of SU(3), i.e., for QCD, the coefficients are

$$\beta_0 = 11 - \frac{2}{3}n_f, \tag{3.71a}$$

$$\beta_1 = 102 - \frac{38}{3}n_f, \tag{3.71b}$$

$$\beta_2 = \frac{2857}{2} - \frac{5033}{18}n_f + \frac{325}{54}n_f^2, \tag{3.71c}$$

$$\beta_3 = \left(\frac{149\,753}{6} + 3564\zeta_3\right) - \left(\frac{1\,078\,361}{162} + \frac{6508}{27}\zeta_3\right)n_f$$
$$+ \left(\frac{50\,065}{162} + \frac{6472}{81}\zeta_3\right)n_f^2 + \frac{1093}{729}n_f^3$$
$$\approx 29\,243.0 - 6946.30n_f + 405.089n_f^2 + 1.499\,31n_f^3. \tag{3.71d}$$

The anomalous dimensions have been computed by Larin and Vermaseren (1993) up to three loops, and by Czakon (2005) to four loops. The full results can be found in these papers.[8] Up to two-loop order, where the coefficients are rational, the values are

$$\gamma_2(\alpha_s/4\pi, \xi) = \frac{\alpha_s}{4\pi}C_F\xi$$
$$+ \left(\frac{\alpha_s}{4\pi}\right)^2\left(-\frac{3}{2}C_F^2 + \frac{25}{4}C_F C_A - C_F n_f + 2\xi C_F C_A + \xi^2\frac{1}{4}C_F C_A\right) + \cdots, \tag{3.72}$$

$$\gamma_3(\alpha_s/4\pi, \xi) = \frac{\alpha_s}{4\pi}\left[-\frac{13}{6}C_A + \frac{2}{3}n_f + \xi\frac{1}{2}C_A\right]$$
$$+ \left(\frac{\alpha_s}{4\pi}\right)^2\left[-\frac{59}{8}C_A^2 + 2C_F n_f + \frac{5}{2}C_A n_f + \xi\frac{11}{8}C_A^2 + \xi^2\frac{1}{4}C_A^2\right] + \cdots \tag{3.73}$$

In these equations, the value $T_F = 1/2$ was used.

---

[8] The definition of γ has different normalization conventions in different books and papers. The conventions of this book agree with those of Larin and Vermaseren (1993).

3.8 Symmetries and approximate symmetries of QCD

In this section, I summarize the standard set of exact and approximate symmetries of QCD. See Narison (2002, Chs. 53 and 54) for a recent account of many of their consequences, especially those that are not further referenced in this section.

3.8.1 Exact symmetries

The QCD Lagrangian is exactly invariant when any one of the quark fields is multiplied by a phase. By Noether's theorem this gives rise to conservation of the number of quarks (minus antiquarks) of each flavor: u-quark number, d-quark number, etc. The sum of all of these, the total quark number, is particularly important because it is not broken by flavor-changing weak interactions. Baryon number is simply one-third of total quark number, and its invariance was established long before QCD.

QCD is also invariant under each of the discrete symmetries of parity, charge conjugation, and time-reversal.

3.8.2 Note on "strong CP problem"

If QCD is specified simply as a renormalizable gauge theory, with an SU(3) gauge group and some set of quark fields in the triplet representation, then one extra term is permitted beyond those in the Lagrangian (2.1). In a standard normalization, the extra term has the form

$$\frac{\theta}{16\pi^2} G^\alpha_{\mu\nu} \tilde{G}^{\alpha\,\mu\nu}, \tag{3.74}$$

where $\tilde{G}^\alpha_{\mu\nu} = \frac{1}{2}\epsilon_{\mu\nu\rho\sigma} G^{\alpha\,\rho\sigma}$. The extra term breaks CP invariance, and there is a stringent observational bound on its coupling, $\theta \ll 10^{-9}$. It is considered problematic as to why θ is so small. This is the strong CP problem, which is reviewed along with possible solutions in Dine (2000).

3.8.3 Isospin and flavor SU(3)

If the up and down quarks were exactly equal in mass, QCD would be invariant under the isospin symmetry of SU(2) transformations on the u- and d-quark fields. This symmetry is quite accurate; we will apply it to the flavor dependence of parton densities and fragmentation functions in Secs. 6.9.7 and 12.4.8.

Rather less accurate is the flavor SU(3) symmetry that would be exact if the masses of the lightest three quarks, u, d, and s, were equal. SU(3) breaking is described by the quark mass terms, which correspond to the Q_3 and Q_8 terms of an operator transforming as an octet under flavor SU(3). Treated to first order in perturbation theory, these give a good description of the mass splittings within the well-known flavor-SU(3) octet and decuplet multiplets of hadrons.

3.8.4 Symmetries at zero mass

The masses of the u and d quarks are quite small. When these masses are neglected, the QCD Lagrangian is further symmetric under separate SU(2) transformations on left- and right-handed quark fields defined by

$$\psi_L = \frac{1}{2}(1 - \gamma_5)\psi, \qquad \psi_R = \frac{1}{2}(1 + \gamma_5)\psi. \qquad (3.75)$$

Then chiral $SU(2)_L \otimes SU(2)_R$ transformations have six parameters ω_L and ω_R for two commuting SU(2) groups, and the quark fields transform as

$$\begin{pmatrix} u_L \\ d_L \end{pmatrix} \mapsto e^{-i\omega_L \cdot \sigma/2} \begin{pmatrix} u_L \\ d_L \end{pmatrix}, \qquad \begin{pmatrix} u_R \\ d_R \end{pmatrix} \mapsto e^{-i\omega_R \cdot \sigma/2} \begin{pmatrix} u_R \\ d_R \end{pmatrix}. \qquad (3.76)$$

The other fields (gluons, other quark flavors) are invariant. This symmetry is in fact spontaneously broken down to isospin SU(2). The low mass of the pions (about 140 MeV) relative to other hadrons is indicative of the expectation that they would be Goldstone bosons for spontaneously broken chiral symmetry in the limit of zero quark mass. Consequences can be successfully derived by the use of Ward identities together with the chiral transformation properties of the quark mass terms. These form much of the subject of current algebra.

3.8.5 Anomalies

When the u and d quarks are massless, the symmetry of their part of the Lagrangian appears also to include separate U(1) transformations on the left- and right-handed fields. (The quark-number symmetry corresponds to the same U(1) to both the left- and the right-handed fields.)

This symmetry is in fact anomalously broken. Thus, unlike the case of $SU(2)_L \otimes SU(2)_R$, there is no approximate Goldstone boson.

3.8.6 Chiral symmetry, hard scattering and factorization

When applying a factorization theorem like (1.1) there is a hard-scattering factor $d\hat{\sigma}(\xi_a, \xi_b, i, j)$. This is normally computed with quark masses set to zero, and thus chiral symmetry applies to it.

Many consequences arise because at the quark-quark-gluon vertex, the coupling is only between quarks of the same helicity, and between quarks and antiquarks of the opposite helicities. That is, only the following transitions are possible:

$$q_L \leftrightarrow q_L + g, \qquad q_R \leftrightarrow q_R + g, \qquad q_L + \bar{q}_R \leftrightarrow g, \qquad q_R + \bar{q}_L \leftrightarrow g. \qquad (3.77)$$

This produces many restrictions on the polarization dependence, as we will see in Secs. 11.6 and 13.16.

3.9 Dealing with quark masses

Our basic technique for exploiting perturbation theory in QCD is to find quantities whose calculation has internal lines of Feynman graphs far off-shell, i.e., with some large virtuality Q^2. In these quantities we set the renormalization scale of order Q, so that the weakness of α_s at large scales allows the use of low-order perturbation theory, and we normally neglect quark masses.

However, there are quarks whose masses are not always negligible in these calculations, so that the general procedure needs modification to deal with heavy quarks. These are defined to be those quarks for which the coupling is small when the renormalization scale is of order the mass: $\alpha_s(m_q) \ll 1$. The known heavy quarks are c, b and t, with the remaining quarks and the gluon being called "light". The charm quark, of mass 1 to 1.5 GeV, is only marginally heavy, but, for robust observables, perturbation theory may be applicable at scales around the charm mass.

Clearly we need improved methods whenever Q, the physical of the process under consideration, is comparable to or smaller than the mass of one or more heavy quarks. First, we should not automatically neglect the mass. Second, the use of a mass-independent scheme, like $\overline{\text{MS}}$, becomes unsuitable whenever the scale is *much* less than one of the quark masses.

The main issues are manifested in a calculation of the one-loop quark contribution to the gauge-field self-energy:

$$
\Pi^{\mu\nu}_{\overline{\text{MS}}} = \sum_j \frac{-g^2 \mu^{2\epsilon} \delta_{\alpha\beta} T_F}{(2\pi)^{4-2\epsilon}} \int d^{4-2\epsilon} k \; \frac{\text{Tr}\,(\slashed{k} + m_j)\gamma^\mu (\slashed{p} + \slashed{k} + m_q)\gamma^\nu}{\left(k^2 - m_j^2 + i0\right)\left[(p+k)^2 - m_j^2 + i0\right]}
$$

$$
+ \text{counterterm, with } \epsilon \to 0
$$

$$
= \sum_j \frac{-2i\alpha_s \delta_{\alpha\beta} T_F}{\pi} (-g^{\mu\nu} p^2 + p^\mu p^\nu) \int_0^1 dx \, x(1-x) \ln \frac{m_j^2 - p^2 x(1-x)}{\mu^2}.
$$

$$\tag{3.78}$$

The following properties apply to this graph and more generally.

- If $|p^2|$ is large compared with m_j^2, then m_j can be neglected, with relative errors of order $m_j^2/|p^2|$.
- Furthermore, in the same situation, $|p^2| \gg m_j^2$, there is logarithmic dependence on p^2. The large logarithm can be removed by taking μ^2 of order $|p^2|$.
- If $|p^2|$ is much less than m_j^2, the integral approaches a constant, $\ln(m_j^2/\mu^2)$. In (3.78), this multiplies a factor quadratic in p, of the same momentum dependence as the UV counterterm.

The last item exemplifies the non-trivial part of the decoupling theorem for heavy particles (Appelquist and Carazzone, 1975). This theorem concerns a situation where we hold fixed the external scales of a Green function and make some internal mass much larger. Then the contributions of *convergent* graphs with the large internal mass are suppressed.

The suppression fails whenever the heavy internal lines are in a divergent loop, but the unsuppressed contributions are equivalent to a contribution to renormalization counterterms. Thus the unsuppressed contributions can be eliminated by a choice of counterterm.

Suppose that we have the real-world situation that the quark masses are widely different. Then we can have a conflict in the choice of μ that eliminates large logarithms, whenever $|p|^2$ lies between two heavy quark masses, e.g., $m_t^2 \gg |p^2| \gg m_b^2$, which is common in practice. Different graphs for the same process involve different heavy quarks.

If we use $\overline{\text{MS}}$ renormalization, then, for the quarks that are heavy on a scale of p^2, we have logarithms $\ln(m_j^2/\mu^2)$, which can be removed by setting $\mu \sim m_j$. For the quarks that are light on a scale of p^2, we have logarithms $\ln(-p^2/\mu^2)$, which are removed by setting $\mu^2 \sim |p^2|$. When the quark masses and $|p^2|$ cover a wide range, we have incompatible conditions on μ.

The original way of using the decoupling theorem was to define a second theory in which all fields are omitted whose masses are much larger than the external scales. This is the low-energy effective theory (LEET) for a given set of heavy quarks. The renormalized parameters of the LEET have numerical values that, in general, differ from those of the full theory. These numerical values can be computed by comparing calculations of Green functions in the two theories and requiring that they give equivalent results.

A LEET removes from calculations quarks whose masses are much larger than the external scales. There can remain quarks with masses comparable to the external scales. For example, in a calculation at $Q \sim 5$ GeV, we would decouple the t quark, but none of the others, so that the LEET has five quark fields. But we could not neglect the mass of the b quark. Depending on the situation and required accuracy, we might be able to neglect the charm quark mass or might need to retain its mass. One normally neglects all three light quark masses in standard perturbative calculations.

For a full set of QCD calculations, we need to successively decouple the top, bottom and charm quarks. This gives us a series of effective theories with three, four and five quarks, with corresponding values of their $\overline{\text{MS}}$ couplings. Non-perturbative calculations at low scales are normally done in the 3-flavor effective theory; these include the well-known lattice Monte-Carlo simulations.

However, the method of LEETs has certain disadvantages, and in the next section I present a better method. The primary disadvantage of a LEET is that it is limited in the ultimate accuracy that it can achieve. For example, consider the 3-flavor effective theory. We could obtain it by sequential decoupling of the three heavy quarks. Now, the decoupling of the charm quark, to get the final 3-flavor LEET, assumes that it is much lighter than the previously decoupled bottom quark; so we have the leading term in an expansion in powers of m_c/m_b. But this ratio is only about 1/3, so the errors could be quite large relative to a desirable accuracy. If instead we decouple both the charm and bottom quarks in one step, then the matching conditions would include logarithmic dependence on m_b/m_c, which would also reduce the accuracy.

A more general approach is to change the renormalization scheme to make decoupling more manifest. The simplest of such schemes is momentum-space subtraction, in which the counterterms are chosen to set certain 1PI Green functions (and/or appropriate derivatives) to zero at a particular point in momentum space. For the quark self-energy, we could choose the renormalization point to be $p^2 = -\mu^2$, obtaining

$$\Pi_{\text{MOM}}^{\mu\nu} = \sum_j \frac{-2i\alpha_s \delta_{\alpha\beta} T_F}{\pi} (-g^{\mu\nu} p^2 + p^\mu p^\nu)$$

$$\times \int_0^1 dx\, x(1-x) \ln \frac{m_j^2 - p^2 x(1-x)}{m_j^2 + \mu^2 x(1-x)}. \tag{3.79}$$

This scheme solves the difficulty of removing all large logarithms; these are eliminated by setting μ^2 of order $|p^2|$, independently of the size of m_j. Thus the scheme satisfies manifest decoupling, which means that we obtain the low-energy effective theory simply by deleting all graphs containing quarks much heavier than the external scale. The errors in doing this are a power of p^2 divided by the square of the mass of the lightest deleted quark.

But the scheme has two technical disadvantages. One is that gauge invariance is not automatically preserved. The defined momentum-space subtractions can only be applied to a limited set of 1PI Green functions, sufficient to determine an independent set of renormalization factors. The counterterms for the remaining 1PI divergent graphs are determined by gauge invariance, and will generally not have an obvious momentum-space definition. Indeed, a separate argument will be necessary to prove decoupling.

The second disadvantage is the practical one that the counterterms are mass dependent, so that the renormalization-group equations for the coupling and mass will be complicated and coupled. So the solution will be much more complicated and more difficult to overview. Moreover, the calculations of counterterms become algorithmically much more complicated: the exact values of off-shell Green functions are needed instead of just the pole part at $\epsilon = 0$. Calculation of on-shell Green functions is generally simpler than when they are off-shell, and calculations of the pole parts are even easier. This was nicely illustrated in our calculation of the quark-quark-gluon vertex graph in Sec. 3.3. This is an important issue, since high-order calculations are extremely expensive in time and effort, which rapidly increases with the order of the calculation. Moreover, for a given desired accuracy in a final phenomenological result, it is generally necessary to compute RG coefficients to one order higher than everything else, because the RG coefficients get integrated over a large range of scales, thereby increasing the effect of an error due to uncalculated higher-order corrections.

3.10 CWZ (ACOT) method for heavy quarks

A method that overcomes these complications was constructed by Collins, Wilczek, and Zee (1978) (CWZ). This method is actually a composite scheme, composed of a sequence of subschemes. The subschemes are parameterized by what is called the number of active quarks, N_{act}. The active quarks are the N_{act} lightest, and the inactive are the remaining, heavier quarks. Since the gluon has zero mass, it is always treated as active. For a 1PI graph containing only active quarks, normal $\overline{\text{MS}}$ counterterms are used. But *zero*-momentum subtractions are used for any 1PI graph that has at least one internal line for an inactive quark.

Normally, zero-momentum counterterms would have undesirable IR divergences in a theory with massless fields, like the gluons in QCD. But the presence of at least one massive line removes these divergences, to all orders of perturbation theory.

The CWZ scheme has the following advantages.

- Each subscheme automatically satisfies gauge invariance. That is, if the counterterms in the Lagrangian are determined by some minimal set of 1PI Green functions, then the remaining 1PI Green functions, with their counterterms determined by gauge invariance of $\mathcal{L}$ also obey the CWZ renormalization condition. No extra finite counterterms are needed.
- Manifest decoupling is satisfied in each scheme. In particular, the numerical value of the coupling in the LEET with N_{act} flavors and pure $\overline{\text{MS}}$ renormalization is the same as in the CWZ subscheme with N_{act} active quarks.
- The RG coefficients in each subscheme are mass independent and in fact exactly identical to those in the theory obtained by deleting the inactive quarks.
- This apparently violates the theorem that we have scheme independence of the one- and two-loop terms in β, and of the one-loop terms in the other RG coefficients. But the theorem only applies if the counterterms are mass independent, which is not the case here, when the number of active quarks changes.
- Normally, calculations of Green functions at zero external momentum are much easier than with a general external momentum.
- No IR divergences are induced by the use of zero-momentum subtractions.

Since there is a sequence of subschemes, relations must be derived between the renormalized parameters in the subschemes. This is quite straightforward, with some results listed below. Moreover, there are no large logarithms in relating the subscheme with N_1 active quarks to the scheme with $N_1 + 1$ active quarks, provided only that μ is of order the mass of the single quark that is making the transition between active and inactive. We will see examples later.

This scheme has become a standard, e.g., Bethke (2009). It extends quite simply to the treatment of parton densities, etc., in which case it is called the ACOT scheme, as expounded by Aivazis *et al.* (1994). It is the one I will use throughout this book, unless otherwise specified.

An important misapprehension needs to be eliminated from the beginning. This is that the $\overline{\text{MS}}$ scheme only applies to massless quarks. It is true that RG coefficients (and their generalizations) do not depend on the quark masses. For this and other reasons, it is often best to do many calculations with massless quarks. But there is no intrinsic reason for it to be restricted to massless quarks. The misapprehension is coupled with some severe conceptual misunderstandings concerning the factorization theorems of QCD, as we will see in later chapters.

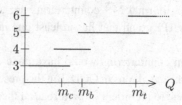

Fig. 3.7. Range of scales for which particular numbers of active flavors are appropriate.

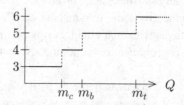

Fig. 3.8. Possible choice of switching points between CWZ subschemes.

3.11 Relating CWZ subschemes with different numbers of active quarks

For a particular CWZ subscheme with a given number, N, of active quarks, the vacuum polarization in (3.78) is replaced by

$$\Pi_{\text{CWZ}}^{\mu\nu} = \frac{-2i\alpha_{s,N}\delta_{\alpha\beta}T_F}{\pi}(-g^{\mu\nu}p^2 + p^\mu p^\nu)\int_0^1 dx\, x(1-x)$$

$$\times \left[\sum_{\text{active } j} \ln\frac{m_j^2 - p^2x(1-x)}{\mu^2} + \sum_{\text{inactive } j} \ln\frac{m_j^2 - p^2x(1-x)}{m_j^2}\right], \qquad (3.80)$$

where $\alpha_{s,N}(\mu)$ is the coupling appropriate to the subscheme. For a particular value of p^2, to eliminate large logarithms, we should (a) take μ^2 of order $|p^2|$, (b) make inactive all quarks with $m_j^2 \gg |p^2|$, and (c) make active all quarks with $m_j^2 \ll |p^2|$. Obviously, for quarks with $m_j^2 \sim |p^2|$ we have a choice of whether to make them active or inactive, as illustrated in Fig. 3.7. In the past, there was a tendency to make a definite switching point between subschemes: quark j was considered active if $\mu > m_j$, and inactive otherwise. But this is now seen as undesirable.

At one-loop, the relations between the subschemes are readily computed from the vacuum polarization graphs, as we will now see. Let us define $Z_{3,N}$ to be the value of Z_3 when the lightest N quarks are active, and similarly for $\tilde{Z}$ and the renormalized masses and coupling. Let $Z_{2,N,j}$ be the field strength renormalization for quark j.

3.11.1 Field-strength renormalization

At one-loop, the self-energies of the first N quarks and the ghost have no inactive quark lines, so $\overline{\text{MS}}$ counterterms apply in both of the subschemes we are relating. Similar considerations apply to the quarks which are inactive in both schemes.

$$\tilde{Z}_N = \tilde{Z}_{N+1} + O(\alpha_s^2), \tag{3.81}$$

$$Z_{2,N,j} = Z_{2,N+1,j} + O(\alpha_s^2), \qquad \text{if } j \leq N \text{ or } j \geq N+2, \tag{3.82}$$

$$Z_{m,N,j} = Z_{m,N+1,j} + O(\alpha_s^2), \qquad \text{if } j \leq N \text{ or } j \geq N+2. \tag{3.83}$$

Here, we use a notation in which the quark label j equals its sequence number in order of mass.

However, the counterterm for the gluon self-energy changes. From the earlier calculations we have

$$Z_{3,N} = Z_{3,\overline{\text{MS}}} + \frac{\alpha_s S_\epsilon}{3\pi} T_F(n_f - N) \sum_{j>N} \left[\frac{\Gamma(\epsilon)}{e^{-\gamma_E \epsilon}} \left(\frac{\mu^2}{m_j^2} \right)^\epsilon - \frac{1}{\epsilon} \right] + O(\alpha_s^2). \tag{3.84}$$

Bare quantities, including fields, are the same in *all* schemes. We therefore obtain the following relations between the fields and masses in the two subschemes:

$$A_N = A_{N+1} \left[1 + \frac{\alpha_s}{6\pi} T_F \ln \frac{\mu^2}{m_{N+1}^2} + O(\alpha_s^2) \right], \tag{3.85a}$$

$$\eta_N = \eta_{N+1}[1 + O(\alpha_s^2)], \tag{3.85b}$$

$$\psi_{j,N} = \psi_{j,N+1}[1 + O(\alpha_s^2)], \qquad \text{if } j \leq N \text{ or } j \geq N+2, \tag{3.85c}$$

$$m_{j,N} = m_{j,N+1}[1 + O(\alpha_s^2)], \qquad \text{if } j \leq N \text{ or } j \geq N+2, \tag{3.85d}$$

3.11.2 Coupling

Now consider the vertex for the ghost to a gluon. Its counterterm is pure $\overline{\text{MS}}$ in both subschemes, and the counterterm is computed from g_0 and the Z factors as proportional to $g_0 \tilde{Z} Z_{3,N} - \mu^\epsilon g_N + O(g^5)$. The bare coupling is the same in both subschemes, so it follows that the renormalized coupling has the relation

$$\alpha_{s,N} = \alpha_{s,N+1} \left[1 - \frac{\alpha_s}{2\pi} T_F \ln \frac{\mu^2}{m_{N+1}^2} + O(\alpha_s^2) \right]. \tag{3.86}$$

Evidently, at the one-loop order, it is sufficient to compute the vacuum polarization.

Higher-order corrections to these relations have been made. For two-loop calculations, see Bernreuther and Wetzel (1982); Bernreuther (1983a, b). For three-loop calculations, see Chetyrkin, Kniehl, and Steinhauser (1997, 1998).

Exercises

3.1 Complete the calculation of the renormalization of QCD at one-loop order. The most economical method is probably to calculate the gluon, quark and ghost self-energies in addition to the quark-gluon vertex.[9] You will have thus verified for yourself the asymptotic freedom of QCD.

3.2 Given the values of the renormalization-group coefficients, reconstruct formulae for the $\overline{\text{MS}}$ renormalization factors for the coupling, and for the fields to at least two-loop order. You may find the results useful if you ever do serious perturbative QCD calculations.

   (One method is to treat (3.41), etc., as differential equations determining renormalization factors from the RG coefficients. Solve these order-by-order in powers of the renormalized coupling. Then apply the boundary conditions that the Z factors and $g_0/g\mu^\epsilon$ go to unity at zero renormalized coupling.)

3.3 (**) There are competing definitions of the $\overline{\text{MS}}$ scheme. Show that these definitions all agree in the values of renormalized Green functions at $\epsilon = 0$, provided that S_ϵ in the different definitions agree to order ϵ.

3.4 Find the next term in the expansion (3.63) of the effective coupling. This will be $1/\ln^3(\mu^2/\Lambda^2)$ times a quadratic polynomial in $\ln\ln(\mu^2/\Lambda^2)$. To check your answer, see (9.5) of Amsler *et al.* (2008), but beware of different conventions for defining the β_j coefficients.

---

[9] The calculation of the three- and four-point gluon gluon functions is substantially more complicated, and should only be attempted if you have much time and wish to verify the general theorems on the renormalizability of non-abelian gauge theories. It is also possible to work with the ghost-gluon coupling, although this is a little more complicated, because it has a derivative coupling.

4

Infra-red safety and non-safety

In this chapter we examine the simplest measurable quantity that can be computed purely perturbatively in QCD: the total cross section for e^+e^- annihilation at high energy Q to hadrons. This is the paradigm of physical single-scale problems: when the renormalization scale μ is of order Q, low-order perturbation theory in $\alpha_s(Q)$ gives a valid estimate of the cross section.

Since the calculation involves quark and gluon final states in a confining theory, we will examine how to justify the use of perturbation theory with apparently incorrect states. There are divergences in individual terms in the calculation. But, in the total cross section, the divergences cancel after a sum over all terms of a given order of α_s. This property is called "infra-red (IR) safety", and in this case is a version of the theorem of Kinoshita (1962) and Lee and Nauenberg (1964) (KLN theorem).

More general situations need a systematic analysis of non-IR-safe situations, and are the primary concern of the rest of this book.

4.1 e^+e^- total cross section

We consider the process $e^+e^- \to$ hadrons, to lowest order in electromagnetism.[1] The amplitude, Fig. 4.1, involves an s-channel exchange of a photon of momentum q^μ, and an incoming electron and positron of momenta l_1 and l_2, with the center-of-mass energy being $Q = \sqrt{q^2}$. The leptonic and hadronic parts of the cross section factorize, as in DIS:

$$\sigma = \frac{e^4}{2Q^6} L_{\mu\nu} W^{\mu\nu}, \tag{4.1}$$

where (with neglect of the electron mass, and with unpolarized beams)

$$L^{\mu\nu} = l_1^\mu l_2^\nu + l_2^\mu l_1^\nu - g^{\mu\nu} l_1 \cdot l_2, \tag{4.2}$$

and

$$W^{\mu\nu}(q) = \int d^4x \, e^{iq \cdot x} \langle 0| \, j^\mu(x) \, j^\nu(0) \, |0\rangle. \tag{4.3}$$

---

[1] There can be large IR-dominated higher-order electromagnetic corrections when the cross section is rapidly varying, e.g., near a narrow resonance. The techniques for unfolding such radiative corrections are standard, and we will not treat them here. A full treatment needs the addition to the amplitude of the Z exchange graph. This does not change the principles, so the reader is referred elsewhere, e.g., Ellis, Stirling, and Webber (1996), for details.

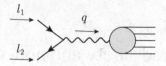

Fig. 4.1. Amplitude for $e^+e^- \longrightarrow$ hadrons.

Conservation of j^μ gives $q_\mu W^{\mu\nu} = 0$, so that we can decompose $W^{\mu\nu}$ in terms of a scalar structure function $R(Q^2)$ as

$$W^{\mu\nu} = \left(-g^{\mu\nu}q^2 + q^\mu q^\nu\right) \frac{1}{6\pi} R(Q^2)\theta(q^0). \tag{4.4}$$

Hence the cross section is

$$\sigma = \frac{4\pi\alpha^2}{3Q^2} R(Q^2). \tag{4.5}$$

The normalization coefficient in (4.4) is chosen so that R is the ratio of σ to the lowest-order cross section $e^+e^- \longrightarrow \mu^+\mu^-$:

$$R = \frac{\sigma(e^+e^- \longrightarrow \text{hadrons})}{\sigma(e^+e^- \longrightarrow \mu^+\mu^-, \text{LO}, \text{em})}. \tag{4.6}$$

Some authors define the denominator to be the complete cross section for $e^+e^- \longrightarrow \mu^+\mu^-$; the definition here is the PDG one.

A compilation of the data is shown in Fig. 4.2. At low energies, there are several large peaks, resonances corresponding to mesons made of light quarks. After that, the cross section generally decreases with energy, approximately as $1/Q^2$ as is generic for processes involving a large virtuality like the photon in Fig. 4.1. The trends are easier to see in the plot of R, whose basically constant value is interrupted at around 4 GeV and 10 GeV by jumps that correspond to the thresholds for production of charm and bottom quarks, preceded by sharp peaks for the bound states of these quarks with their antiquarks. Finally, the addition of a graph with the exchange of a Z instead of a photon in Fig. 4.1 gives rise to the prominent peak at $Q = m_Z \simeq 91$ GeV that interrupts the fall of the cross section.

4.1.1 Short-distance dominance in averaged cross section

When Q is large, the high virtuality of the photon in Fig. 4.1 suggests that it has a short lifetime, of order $1/Q$ in its rest frame, and hence that the process occurs over a short scale in time and distance. This makes it suitable for exploiting asymptotic freedom, so that a first approximation is obtained from the lowest-order graph, Fig. 4.3, for $e^+e^- \to q\bar{q}$.

However, the two currents in (4.3) need not actually have a small space-time separation. Consider a semi-classical approximation in which a quark and antiquark are assigned trajectories after their creation at a particular time and position. Suppose that the quark-antiquark force were such that they repeatedly bounce back to their creation position, as in Fig. 4.4. Now the incoming electron and positron have almost definite momenta

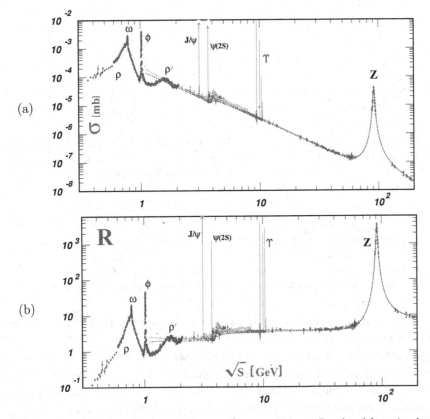

Fig. 4.2. (a) Total cross section and (b) R, for $e^+e^- \longrightarrow$ hadrons. Reprinted from Amsler *et al.* (2008), with permission from Elsevier. The dashed line is the lowest-order "parton-model" prediction, and the solid line is the 3-loop pQCD prediction from equations (1)–(3) of Chetyrkin, Harlander, and Kuhn (2000).

(in a normal experiment), so that their states can be represented by long wave packets, Fig. 4.5. Therefore their collision and the production of quark-antiquark pairs occurs over an extended time.

Now a pair produced late in the collision is in the same spatial position as a pair that is produced early but that has bounced back, and we get interference when we add the quantum-mechanical amplitudes for pair production at different times. At certain energies the phases of the interfering terms could all be the same, giving constructive interference, and a resonance peak. Off-resonance, the phases vary, giving destructive interference. Thus we can get sharp resonances, as seen at certain energies in the data in Fig. 4.2. These correspond to interference between quark-antiquark pairs produced at very different values of space-time positions.

For example, the sharp J/ψ and Υ peaks occur just below the thresholds for the production of c and b quark pairs, respectively; there, the heavy quarks are slowly moving and are easy to bring back to the production point. However, resonances are not present

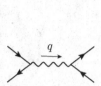

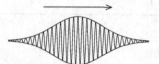

Fig. 4.3. Lowest-order diagram for
$e^+e^- \to q\bar{q}$ (or $e^+e^- \to \mu^+\mu^-$).

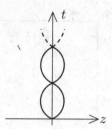

Fig. 4.4. Space-time evolution of semi-classical
trajectory of a $q\bar{q}$ pair created at the origin, if the
quark-antiquark force caused them to bounce back.

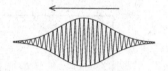

Fig. 4.5. Representation of wave packets for incoming electron and positron.

much above the heavy-quark thresholds, so that we deduce from the data that fast-moving
quarks and antiquarks do not bounce back.

Unfortunately, the relevant long-distance phenomena in QCD are non-perturbative, and
not readily susceptible to a first-principles analysis. So we ask what properties of the cross
section are predicted purely perturbatively without any need to understand long-distance
phenomena. A solution (Poggio, Quinn, and Weinberg, 1976) is to use a local average of
the cross section in energy.

To understand this idea, we investigate the relation between the space-time structure
of the scattering and the momentum spread in a physical initial state. This exemplifies a
general issue that intuition and understanding can be obtained by studying the evolution
of states in coordinate space, even though actual calculations are typically performed in
momentum space.

Now a physical incoming e^+e^- state cannot be exactly a state with particles of definite momenta. We must use a superposition of momentum eigenstates corresponding to
coordinate-space wave packets, as in Fig. 4.5:

$$|\psi\rangle = \sum_{l'_1, l'_2, \lambda_1, \lambda_2} |l'_1, \lambda_1, l'_2, \lambda_2; \text{in}\rangle \, \psi_1(l'_1, \lambda_1) \, \psi_2(l'_2, \lambda_2). \tag{4.7}$$

Here l'_1 and l'_2 are the momenta of the incoming electron and positron, and λ_1 and λ_2
label their spin states. The momentum-space wave functions $\psi_1(l'_1, \lambda_1)$ and $\psi_2(l'_2, \lambda_2)$
are narrowly peaked around central values of momentum l_1 and l_2. We let $q = l_1 + l_2$
be the corresponding central value of total momentum. The notation $\sum_{l'_1, l'_2}$ is the usual
Lorentz-invariant integral over a particle's momentum, (A.15).

As in (4.1) and (4.3), we treat electroweak interactions perturbatively. The initial state
$|\psi\rangle$ evolves to a slightly depleted version of $|\psi\rangle$ plus a hadronic component $|\phi\rangle$, plus

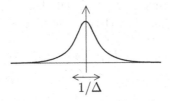

Fig. 4.6. A component of the wave function ϕ_μ for the state in (4.9), as a function of a component of position x in the overall center-of-mass.

components with scattered leptons:

$$|\text{final state}\rangle = |\phi\rangle + |\psi\rangle\,(1 - \ldots) + |\text{leptonic part}\rangle. \tag{4.8}$$

In the graph Fig. 4.1, the hadronic factor is a vacuum-to-X matrix element of the electromagnetic current, $\langle X, \text{out}|\,j^\mu(x)|0\rangle$, for a general hadronic out-state X.

Hence the hadronic final state $|\phi\rangle$ is $j^\mu(x)\,|0\rangle$ integrated with an x-dependent factor to be computed from the Feynman rules for Fig. 4.1, and the wave packet state (4.7). Now the creation of the hadronic final state occurs in the space-time region where the beams collide. So at later times, the QCD part $|\phi\rangle$ of the state in $e^+e^- \to$ hadrons is a time-independent Heisenberg state. We write this as

$$|\phi\rangle \overset{\text{def}}{=} \int d^4x\, j^\mu(x)\,|0\rangle\, e^{-iq\cdot x}\phi_\mu(x), \tag{4.9}$$

where we have extracted a factor $e^{-iq\cdot x}$, anticipating that it is the dominant oscillatory factor in the coefficient, with q being the central value of the total momentum. Lowest-order electromagnetic perturbation theory gives

$$\phi_\mu(x) = \sum_{l_1',l_2',\lambda_1,\lambda_2} \frac{-ie^2\,\bar{v}_{l_2',\lambda_2}\gamma_\mu u_{l_1',\lambda_1}}{(l_1' + l_2')^2 + i0}\,\psi_1(l_1', \lambda_1)\,\psi_2(l_2', \lambda_2)\,e^{i(q - l_1' - l_2')\cdot x}. \tag{4.10}$$

Here the $e^{-i(l_1'+l_2')\cdot x}$ factor arises from Fourier-transforming the leptonic part of the Feynman graph, and the $e^{iq\cdot x}$ factor compensates the corresponding factor in (4.9).

The beams have approximately definite momenta, centered at $l_1 + l_2 = q$, so the oscillatory factor mostly cancels. Let the states be localized to within Δ in momentum. Then each component ϕ_μ is a smooth function with little oscillation, as in Fig. 4.6. Correspondingly the position x in (4.9) is localized to about $1/\Delta$.

Once the hadronic state $|\phi\rangle$ has been created by the current and the current is no longer acting because the coefficients $\phi_\mu(x)$ have become zero, the state cannot be destroyed, to lowest order in electroweak interactions. Thus the probability of the transition $e^+e^- \to$ hadrons is just $\langle\phi|\phi\rangle$. This is genuinely a scattering *probability*, not a cross section. The concept of a cross section arises when one observes that experiments are done with beams of particles which are distributed over an area that is large compared with the scattering region. The relative transverse separation of the beam particles has a broad distribution. The cross section is obtained by displacing one beam transversely with respect to the other, and then integrating over the displacement b_T. Let the hadronic state with a displaced beam be $|\phi_{b_T}\rangle$, with wave function $\phi_\mu(x; b_T)$. The cross section is

$$\sigma = \int d^2 b_T \, \langle\phi_{b_T}|\phi_{b_T}\rangle = \int d^4x \, e^{iq\cdot x} \langle 0| \, j^\mu(x) \, j^\nu(0) \, |0\rangle \, \tilde{t}_{\mu\nu}(x), \qquad (4.11)$$

where

$$\tilde{t}_{\mu\nu}(x) = \int d^2 b_T \int d^4w \, \phi_\mu \left(w + \frac{1}{2}x; b_T \right)^* \phi_\nu \left(w - \frac{1}{2}x; b_T \right), \qquad (4.12)$$

which is localized in x to within $1/\Delta$. After a Fourier transformation

$$\tilde{t}_{\mu\nu}(x) = \int \frac{d^4k}{(2\pi)^4} t_{\mu\nu}(k) e^{i(k-q)\cdot x}, \qquad (4.13)$$

we find that the cross section is a weighted average in momentum space:

$$\sigma = \int \frac{d^4k}{(2\pi)^4} t_{\mu\nu}(k) W^{\mu\nu}(k) = \frac{1}{6\pi} \int_0^\infty dM^2 \, R(M^2) \, f(M^2), \qquad (4.14)$$

where

$$f(M^2) = -\int \frac{d^4k}{(2\pi)^4} t^\mu_\mu(k) \theta(k^0) \delta(k^2 - M^2). \qquad (4.15)$$

We now see one result of the wave-packet construction: that a local average of $R(Q)$ over a range of Q of width Δ corresponds to a localization of the positions of the current operators to $x \sim 1/\Delta$. Of course, real particle beams are very narrow in momenta. But a broader averaging applied to the measured cross section gives a quantity with better localization in position and therefore with better perturbative calculability.

The standard momentum-space analysis gives the cross section in terms of R, from which we deduce the correct normalization of the averaging function f without needing the detailed wave-packet analysis:

$$\int_0^\infty dM^2 \, f(M^2) = \frac{8\pi^2\alpha^2}{Q^2}, \qquad (4.16)$$

up to terms that vanish when $\Delta/Q \to 0$.

It is convenient to consider as a standardized quantity, one particular *normalized* local average of R:

$$\bar{R}(Q^2, \Delta^2) \stackrel{\text{def}}{=} \int ds \, F(s - Q^2, \Delta^2) \, R(s). \qquad (4.17)$$

Here $F(s - Q^2, \Delta^2)$ is one particular averaging function, of unit integral, centered at $s = Q^2$, and of width Δ^2. I choose

$$F(s - Q^2, \Delta^2) = \frac{\Delta^2}{\pi \left[(s - Q^2)^2 + \Delta^4 \right]}. \qquad (4.18)$$

We assume Δ is somewhat less than Q^2, but not enormously so. If R is smooth in a region of Q, as is the case experimentally for most large values of Q, then the local average $\bar{R}$ is almost equal to R; the averaging does nothing. But where R has sharp features, e.g., near the thresholds for c and b quark production, the average smooths out the sharp peaks and the thresholds.

4.1.2 When is perturbation theory good?

To put the concept of perturbative calculability in a general context, current ideas can be summarized in the following assertion:

Consider a situation where all vertices in a perturbative calculation are dominantly separated by small distances, of order $1/M$, and that we set $\mu \sim M$. Then QCD perturbation theory in powers of the weak coupling $\alpha_s(M)$ provides a good approximation.

That is, short-distance-dominated quantities are perturbatively computable. The integration over the positions of vertices is, of course, unrestricted. What matters for the above assertion is whether the vertices are *dominantly* close to some external vertices determined by the problem.

   A similar assertion could be made about momentum-space Green functions, where the premise would be about the lines of the graph being dominated by high virtualities, of order M^2. However, this assertion is not so general. This can be seen from the matrix element (4.3) defining $W^{\mu\nu}$. The currents have fixed ordering and perturbation theory gives final states with on-shell quarks and gluons, so that not all propagators are far off-shell, even when the positions of the current operators are arbitrarily close, a situation that is different for the time-ordered product of operators.

4.2 Explicit calculations

Since the locally averaged quantity $\bar{R}$ is short-distance-dominated, we can use perturbation theory to predict it reliably. Therefore, to the extent we are away from resonances, we predict the unaveraged $R(Q^2)$, in both cases at large Q. The electromagnetic current has zero anomalous dimension within pure QCD.[2] So we change the renormalization scale μ to be of order Q, without changing R, and then expand in powers of the small coupling. We also approximate light-quark masses by zero. Thus:

$$R(Q^2, \mu, g(\mu), \underline{m}(\mu)) = R(Q^2, cQ, g(cQ), \underline{m}(cQ))$$
$$\simeq R(Q^2, cQ, g(cQ), \underline{0})$$
$$= \sum_{n \geq 0} \alpha_s(cQ)^n R^{[n]}(c). \tag{4.19}$$

If we truncate the series at order N, then the error is of order α_s^{N+1}, so that we have an effective method of calculation given that $\alpha_s(cQ)$ is small. From Sec. 3.4, we expect (in the $\overline{\text{MS}}$ scheme) optimal applicability of perturbation theory when μ^2 is of order a

---

[2] This is *not* true beyond QCD (Collins, Manohar, and Wise, 2006), contrary to many statements in the literature.

Fig. 4.7. Lowest-order graph for amplitude used in R.

typical internal virtuality. This could be governed by the width of the smoothing function, so in (4.19) μ is a constant c times Q, and a good value could be $c = \frac{1}{2}$ or $\frac{1}{4}$. RG invariance implies that the value of c is irrelevant in an exact calculation, while in a truncated perturbation calculation the effect of a modest change in c is of order the expected truncation error.

In the remainder of this section, we perform the perturbative calculation of R to order α_s.

4.2.1 Lowest order

The single graph for the lowest-order calculation, Fig. 4.7, is the same as for $\mu^+\mu^-$ production, with the replacement of a muon line by a quark line. So with the neglect of quark particle masses, the lowest-order value of R is

$$R^{[0]} = 3 \sum_f e_f^2. \tag{4.20}$$

The factor 3 is for the sum over quark colors, and the sum is over the accessible flavors of quark, which depends on the value of Q relative to the quark masses.

Some complications now occur because of the non-negligible masses of the c, b and t quarks. Any quark that is not accessible kinematically, i.e., for which $m_f > Q/2$, should certainly be dropped from the sum. The remaining quarks we term "accessible". Provided that Q is much larger than the other quark masses, these masses may be neglected, as in the calculation giving (4.20).

The remaining case is when Q is comparable to $2m_f$ for one of the quarks. As regards perturbation theory, there is a threshold at $Q = 2m_f$ for production of quark f. Since there are sharp resonances just below threshold (Fig. 4.2) we should apply the averaging procedure in Q before using the elementary perturbative prediction.

Hence we deduce that it is a good first approximation to restrict the sum in (4.20) to those quarks with $2m_f < Q$, and to otherwise ignore the effects of quark masses. The known quark charges and masses then give a first prediction of R:

$$R^{[0]} = \begin{cases} 2 & \text{if } Q \lesssim 3 \text{ GeV,} \\ 3\frac{1}{3} & \text{if } 3 \text{ GeV} \lesssim Q \lesssim 10 \text{ GeV,} \\ 3\frac{2}{3} & \text{if } 10 \text{ GeV} \lesssim Q. \end{cases} \tag{4.21}$$

Once Z exchange effects become important, this prediction needs changing, so we do not include a possible last line, to include the t quark. For the inclusion of masses at lowest

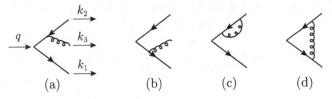

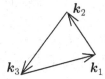

Fig. 4.8. NLO graphs for amplitudes for R.

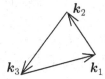

Fig. 4.9. Momentum configuration for 3-body final state.

order, see problem 4.2. For masses and the effects of Z at higher order, see Chetyrkin, Harlander, and Kuhn (2000).

4.2.2 *Next-to-leading order: real gluon*

The next-to-leading order (NLO) terms arise from the graphs of Fig. 4.8. One contribution is from the real-gluon emission graphs, (a) and (b), with a $q\bar{q}g$ final state:

$$\int \text{dfsps} \, |(a) + (b)|^2 \,, \tag{4.22}$$

with dfsps given by (A.17). The other contribution is from the virtual corrections, (c) and (d), with a $q\bar{q}$ final state.

All of the terms individually have divergences which we regulate by using a space-time dimension $4 - 2\epsilon$.

Provided that the integrand involves only Lorentz scalars, the $(5 - 4\epsilon)$-dimensional integral for real-gluon emission can be simplified to a two-dimensional integral, so that angular averages can be performed to give (A.44). So we calculate the trace of $W_{\mu\nu}$:

$$W \stackrel{\text{def}}{=} -g^{\mu\nu} W_{\mu\nu} = (3 - 2\epsilon)\frac{Q^2}{6\pi} R(Q^2). \tag{4.23}$$

In the overall center-of-mass, the 3-momenta of the final state form a triangle – Fig. 4.9, whose perimeter is $\sum_i |\mathbf{k}_i| = Q$, from energy conservation. The integration variables in (A.44) are the relative deficits of the spatial momenta relative to their maximum $Q/2$:

$$y_i = 1 - \frac{2|\mathbf{k}_i|}{Q}. \tag{4.24}$$

The integral is over positive values subject to $\sum_i y_i = 1$. We have

$$(k_1 + k_2)^2 = y_3 Q^2, \qquad (k_2 + k_3)^2 = y_1 Q^2, \qquad (k_3 + k_1)^2 = y_2 Q^2. \tag{4.25}$$

It is also convenient to factor out the lowest-order calculation in $4 - 2\epsilon$ dimensions, derived from (A.43) for the 2-body phase space:

$$W^{[0]} = R_0 \frac{Q^{-2\epsilon}}{2^{4-4\epsilon}\pi^{1/2-\epsilon}\Gamma(\frac{3}{2}-\epsilon)}(-g_{\mu\nu})\,\mathrm{Tr}\,\gamma^\mu\!\not{k}_1\gamma^\nu\!\not{k}_2$$

$$= R_0 \frac{Q^{2-2\epsilon}(1-\epsilon)}{2^{2-4\epsilon}\pi^{1/2-\epsilon}\Gamma(\frac{3}{2}-\epsilon)}. \tag{4.26}$$

After a standard application of Feynman rules, etc., we find that the contribution of graphs (a) and (b) to W is

$$W^{[1]}(q\bar{q}g) = W^{[0]}\frac{\alpha_s C_F}{4\pi\Gamma(1-\epsilon)}\left(\frac{Q^2}{4\pi\mu^2}\right)^{-\epsilon}\int_0^1 dy_1 \int_0^{1-y_1} dy_2\,(y_1 y_2 y_3)^{-\epsilon}$$

$$\times \frac{4(y_3 + y_1 y_2 \epsilon) + 2(1-\epsilon)(y_1^2 + y_2^2)}{y_1 y_2}. \tag{4.27}$$

The sum over flavors and a factor e_f^2 are the same as in lowest order, and are in the factor $W^{[0]}$. Thus, the order-α_s correction factor is the same for all flavors of massless quark.

The integrand is singular when y_1 and/or y_2 is zero, and gives a divergence in the integral for space-time dimension 4 or less, i.e., when $\epsilon > 0$. In Ch. 5, we will analyze the physics of these and other divergences more generally. But for calculational purposes, it suffices that the divergence can be regulated and hence quantified by using a space-time dimension above 4, i.e., $\epsilon < 0$. In the ultimate result, for R, we will find a cancellation against divergences from the virtual-gluon graphs. The configuration of momenta at the singularities is easily deduced from the geometry of Fig. 4.9:

$$\begin{aligned} y_1 &= 0 \quad \text{gluon parallel to } k_2, \\ y_2 &= 0 \quad \text{gluon parallel to } k_1, \\ y_1 &= y_2 = 0 \quad \text{gluon of zero momentum.} \end{aligned} \tag{4.28}$$

In Ch. 5, we will analyze such singularities. The first two give "collinear divergences", where two final-state massless particles are parallel, and the last one gives a "soft divergence", where the gluon has zero momentum.

The integral is readily computed in terms of Γ functions. Its expansion in powers of ϵ exhibits the divergence quantitatively:

$$W^{[1]}(q\bar{q}g) = W^{[0]}\frac{\alpha_s C_F}{\pi}\left(\frac{Q^2}{4\pi\mu^2}\right)^{-\epsilon}\frac{\Gamma(-\epsilon)^2}{\Gamma(2-3\epsilon)}\left[1 - \frac{\epsilon(3-5\epsilon)}{2-3\epsilon}\right]$$

$$= W^{[0]}\frac{\alpha_s C_F}{4\pi}\left(4\pi e^{-\gamma_E}\right)^\epsilon\left[\frac{4}{\epsilon^2} + \frac{1}{\epsilon}\left(-4\ln\frac{Q^2}{\mu^2} + 6\right) + 2\ln^2\frac{Q^2}{\mu^2}\right.$$

$$\left. - 6\ln\frac{Q^2}{\mu^2} + 19 - \frac{7\pi^2}{3} + O(\epsilon)\right]. \tag{4.29}$$

That we obtain a relatively simple analytic result is associated with the masslessness of the quarks and gluons in the calculation.

4.2.3 Next-to-leading order: virtual gluon

For the virtual-gluon corrections, from Fig. 4.8(c) and (d), it is convenient to compute the matrix elements from Green functions with bare fields, rather than using the full counterterm structure of (3.13)–(3.15). One reason is that the electromagnetic current is simplest in terms of bare fields: $j_{em}^\mu = \sum_f e_f \bar\psi_{f,0} \gamma^\mu \psi_{f,0}$. Another is that the implementation of LSZ reduction for massless theories is trivial.

The LSZ reduction formula tells us that to get an on-shell matrix element, we amputate complete external propagators and replace each by the square root of the residue of the particle pole. Let $z_2 = 1 + g^2 z_2^{[1]} + \ldots$ be the residue of the pole in the propagator of a *bare* quark field. The one-loop term is

$$g^2 z_2^{[1]} = \lim_{p^2 \to 0} \frac{i \not p}{p^2} \frac{g^2 C_F (2\pi\mu)^{2\epsilon}}{16\pi^4} \int d^{4-2\epsilon} k \frac{-\gamma^\nu \not k \gamma_\nu}{(k^2 + i0)\,[(p-k)^2 + i0]}$$

$$= \text{coefficient} \times \lim_{p^2 \to 0} (-p^2)^{-\epsilon}$$

$$= 0. \tag{4.30}$$

Since dimensional regularization is used here to regulate *infra-red*-related divergences, we take ϵ negative, which gives the zero result in the last line. No UV counterterm is applied, since we work with bare fields.

The result (4.30) generalizes to all orders: an N-loop calculation gives a factor of $g^{2N}\mu^{2N\epsilon}$, and hence dimensional analysis shows that the power of p^2 is $(-p^2)^{-N\epsilon}$. Thus to all orders in perturbation theory the residue of the pole in the bare propagator is exactly $z_2 = 1$.

The only non-zero one-loop virtual contribution to R is therefore from the vertex graph (d). To get its contribution at order α_s to W, we multiply the graph by the complex conjugated LO graphs, then we add the complex conjugate, we take the trace of $W_{\mu\nu}$ with $-g^{\mu\nu}$ and perform the angular integral in $3 - 2\epsilon$ spatial dimensions. This gives

$$W^{[1]}(q\bar q) = W^{[0]} \Re \frac{i g^2 C_F (2\pi\mu)^{2\epsilon}}{32\pi^4 Q^2 (1-\epsilon)} \int d^{4-2\epsilon} k \frac{\text{Tr}\,\not k_1 \gamma^\kappa (\not k_1 - \not k)\gamma_\mu (\not k_2 + \not k)\gamma_\kappa \not k_2 \gamma^\mu}{(k^2 + i0)\,[(k_1 - k)^2 + i0]\,[(k_2 + k)^2 + i0]}. \tag{4.31}$$

There is an extra factor of $4Q^2(1 - \epsilon)$ in the denominator of the prefactor because of the normalization to $W^{[0]}$. Standard manipulations give

$$W^{[1]}(q\bar q) = W^{[0]} \Re \frac{\alpha_s C_F}{4\pi} \left(\frac{-Q^2 - i0}{4\pi\mu^2} \right)^{-\epsilon} \frac{\Gamma(1+\epsilon)\Gamma(1-\epsilon)^2}{\epsilon^2 \Gamma(1-2\epsilon)} \left[\frac{-4}{1-2\epsilon} + 2\epsilon \right]$$

$$= W^{[0]} \frac{\alpha_s C_F}{4\pi} (4\pi e^{-\gamma_E})^\epsilon \left[-\frac{4}{\epsilon^2} + \frac{1}{\epsilon}\left(4\ln\frac{Q^2}{\mu^2} - 6 \right) - 2\ln^2\frac{Q^2}{\mu^2} \right.$$

$$\left. + 6\ln\frac{Q^2}{\mu^2} - 16 + \frac{7\pi^2}{3} + O(\epsilon) \right]. \tag{4.32}$$

4.2.4 Leading and next-to-leading order: total

In the total for $R^{[1]}$, the divergences cancel, and at $\epsilon = 0$ we find

$$R = R^{[0]}\left[1 + \frac{3\alpha_s(\mu)}{4\pi}C_F + O(\alpha_s^2)\right] = R^{[0]}\left[1 + \frac{\alpha_s(\mu)}{\pi} + O(\alpha_s^2)\right], \qquad (4.33)$$

with the physical value of C_F. Notice that both logarithms of Q/μ have canceled. This follows from the RG invariance of R, which implies that a logarithm of Q/μ first appears in the coefficient of α_s^2 (problem 4.5). We have left the renormalization scale μ arbitrary, but, as explained earlier, a value for μ of order Q should be used to ensure that higher-order calculations do not get large logarithms. Thus $O(\alpha_s^2)$ correctly represents the expected size of the error due to omission of higher-order perturbation theory.

This result is both reassuring and disturbing. It is reassuring that the divergences cancel in a quantity that was supposed to have a valid perturbation expansion. But it is also disturbing: the intermediate steps involve totally unphysical states. In another arena, QED, there are somewhat similar IR divergences because of the masslessness of the photon; but at least electrons and photons are actual identifiable particles. See Sec. 4.3 for a detailed analysis.

The calculation evidently makes important predictions. Among these is that measuring the ratio R gives an estimate of the sum of the squared charges of the accessible quarks. When first obtained, this was a rather dramatic result, and the data (Fig. 4.2) confirm the charge assignments of the quarks. Deviations from this value can be used to measure the strong coupling and to test its evolution with scale.

4.2.5 Full result, and phenomenological implications

The currently most accurate calculations may be traced from Chetyrkin, Harlander, and Kuhn (2000), where the calculation is extended at order α_s^3 to include quartic mass corrections (i.e., of order m^4/Q^4). With massless quarks the current results are

$$\frac{R}{R^{[0]}} = 1 + \frac{\alpha_s}{\pi} + \left(\frac{\alpha_s}{\pi}\right)^2\left[\frac{365}{24} - 11\,\zeta_3 + n_f\left(-\frac{11}{12} + \frac{2}{3}\zeta_3\right)\right]$$

$$+ \left(\frac{\alpha_s}{\pi}\right)^3\left[\frac{87\,029}{288} - \frac{121}{8}\zeta_2 - \frac{1103}{4}\zeta_3 + \frac{275}{6}\zeta_5\right.$$

$$\left. + n_f\left(-\frac{7847}{216} + \frac{11}{6}\zeta_2 + \frac{262}{9}\zeta_3 - \frac{25}{9}\zeta_5\right) + n_f^2\left(\frac{151}{162} - \frac{1}{18}\zeta_2 - \frac{19}{27}\zeta_3\right)\right]$$

$$+ O\left(\frac{\alpha_s}{\pi}\right)^4$$

$$\approx 1 + \frac{\alpha_s}{\pi} + \left(\frac{\alpha_s}{\pi}\right)^2(1.985\,71 - 0.115\,295 n_f)$$

$$+ \left(\frac{\alpha_s}{\pi}\right)^3(-6.636\,94 - 1.200\,13 n_f - 0.005\,178\,36 n_f^2) + O\left(\frac{\alpha_s}{\pi}\right)^4. \qquad (4.34)$$

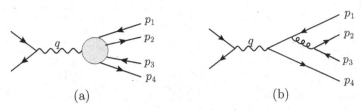

Fig. 4.10. (a) Matrix element of $q\bar{q}q\bar{q}$ fields used to obtain two-pion production. (b) An example of a perturbative graph for the matrix element.

Here, the $\overline{\text{MS}}$ unit of mass was set to $\mu = Q$, so that $\alpha_s = \alpha_s(Q)$. The logarithmic dependence on Q/μ can be restored with the aid of the renormalization group (problem 4.5).

4.3 Evolution of state

Individual terms in the perturbative calculation of R involve quarks and gluons rather than the hadrons that actually appear in the final state. To understand better why we nevertheless obtain a valid prediction of QCD, we examine the evolution of the hadronic final state, between the $j^\mu(x)$ and $j^\nu(0)$ operators in (4.11). The arguments in this section are not intended to be precise and rigorous.

Although the hadronic Heisenberg-picture state, $|\phi\rangle$, is time independent, its interpretation in terms of localized particle content does evolve. It can be analyzed by matrix elements of products of field operators between $|\phi\rangle$ and the vacuum. For example, consider [Fig. 4.10(a)]

$$\langle 0 | T \bar{u}(w_1)d(w_2)\bar{d}(w_3)u(w_4) | \phi \rangle, \tag{4.35}$$

where the fields annihilate $\bar{u}, d, \bar{d}$, and u quarks respectively. Fourier transformation gives a function of momenta p_1, p_2, p_3, p_4. Poles in this function correspond to particles in the asymptotic out-states. For example, a state with a π^- and a π^+ gives a pole in the exact matrix element at $(p_1 + p_2)^2 = m_\pi^2$ and at $(p_3 + p_4)^2 = m_\pi^2$. The poles are related to the coordinate-space asymptotics when the times of all the four fields are taken to $+\infty$; in the $\pi^-\pi^+$ example, the spatial components of w_1 and w_2 are close together and in the direction of the π^-, and similarly for w_3 and w_4 and the π^+.

In finite-order perturbation theory, we have diagrams like Fig. 4.10(b). This has no poles for pions, but only for quarks and gluons, for example at $p_1^2 = p_4^2 = m_u^2$ and $p_2^2 = p_3^2 = m_d^2$. Such poles give a large-time behavior for the individual graph that corresponds to a state that does not exist in a theory with color confinement; fixed-order perturbation theory gives an entirely incorrect approximation to asymptotic large-time Green functions and matrix elements. However, if the times are not too large, perturbation theory should approximate the true results. Thus the poles in fixed-order graphs imply that we do have, but only *approximately*, the propagation of the corresponding quarks and gluons.

Returning to the lowest-order graph for R, Fig. 4.7, we deduce in a rough fashion that at the earliest times we have predominantly an outward-moving q and $\bar{q}$, as at the lower

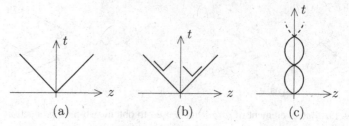

Fig. 4.11. Three semi-classical scenarios for evolution of $q\bar{q}$ system: (a) approximately free, (b) string, (c) spring. In case (b), extra $q\bar{q}$ pairs are produced in the middle.

end of Fig. 4.4. We can reasonably assign them a virtuality of some size M^2 that is much less than Q^2. The Lorentz boost to energy $Q/2$ implies that the lifetime of the $q\bar{q}$ state is of order Q/M^2. Perturbative corrections, like those in Fig. 4.8 or Fig. 4.10(b), alter the state, for example by changing the probability of the $q\bar{q}$ state and by adding a component with a gluon. At late times, QCD perturbation theory is entirely inapplicable, in the domain where, in the real world, the system non-perturbatively hadronizes into a set of isolated color-singlet hadrons.

4.3.1 String model for hadronization

If we ignored any knowledge of the real world we could imagine at least three scenarios for the time development, as illustrated in Fig. 4.11:

- The quarks and gluons continue basically unhindered into the observed final state, as in QED, where there is no confinement. Let us call this the "unconfined" or "free-quark-and-gluon" picture.
- In the gluon field between the quark and antiquark, extra $q\bar{q}$ pairs are made. We call this the "breakable string" picture. The $q\bar{q}$ pairs combine into color-singlet hadrons, mostly pions. Nothing returns to the production point, and the general momentum flow corresponds to the system at short times, which is little deflected. But the space between the ends of the kinematic range is filled in with particles, and the detected particles are hadrons, not quarks and gluons.
- A confining potential exists, which brings the quarks and gluons back. We can call this the "unbreakable elastic spring" picture. The final states form a sequence of bound states. After multiple bounces, it may be that the states decay, perhaps in the style of the string picture, but the directions of the decay products need not be very correlated with the initial $q\bar{q}$ direction.

Purely perturbative calculations in QCD cannot decide between these scenarios. But we can appeal to experiment, semi-classical intuition, modeling, and lattice gauge theory calculations, at least. The unconfined scenario is ruled out experimentally. The increase in α_s in the infra-red is a precondition for a rising potential. But the bound states or resonances in the spring picture do not appear to be relevant except close to quark thresholds, where there is little energy for producing extra particles, and where the initial q and $\bar{q}$ are moving slowly.

It is the string picture that seems to be approximately correct. Embodied quantitatively in the semi-classical Lund string model (Andersson, 1998), it rather successfully describes the hadronization of quarks and gluons. In this model, when the $q\bar{q}$ separation is large enough, the gluon field collapses to a flux tube ("string") with a fixed cross-sectional area, and with a constant energy per unit length. Without $q\bar{q}$ production, this would correspond to a linearly rising potential, which has significant phenomenological support from quark models of hadrons, etc.

The Lund model postulates that creation of light $q\bar{q}$ pairs occurs in the string with a constant rate per unit length and unit time; this is the only Lorentz-invariant possibility. In a strong coupling, strong field situation such as we have here, the string therefore breaks, we have inelastic scattering, and the description in terms of a genuine potential breaks down.

A more detailed investigation shows that the string breaking and the hadronization occur along a hyperbolic region $t^2 - z^2 \sim 1/\Lambda^2$. The fastest particles, with energies of order Q, are generated at the ends of the string in a time of order Q/Λ^2, while the slowest particles are generated in the middle in a time of order $1/\Lambda$.

The Lund model is plausible and natural as a first approximation to real QCD dynamics in situations such as e^+e^- annihilation at high energies. For each outgoing parton, the model leads to the production of a jet of hadrons with approximately the 4-momentum of the parton. This can be seen in event pictures like Fig. 2.3 for a similar situation in DIS.

The validity of the string model depends on specific dynamical properties of real QCD, with its light u and d quarks. In contrast, there is the solvable model of 't Hooft (1974), pedagogically reviewed in Manohar (1998). This model is QCD but in $1 + 1$ space-time dimensions with a gauge group U(N) taken in the limit $N \rightarrow \infty$. This model provides an example of the "elastic spring" scenario;[3] the large N limit suppresses the $q\bar{q}$ production that causes string breaking in the Lund model. In the 't Hooft model, the final states in $e^+e^- \rightarrow$ hadrons form an infinite sequence of meson bound states with no continuum, whereas a simple perturbative calculation gives a continuum. It is a local average of the true cross section that agrees with the perturbative calculation, as we saw earlier. Explicit calculations support the general result, as was particularly clearly shown by Einhorn (1976), where the result was also extended to other cases, like DIS.

4.3.2 Analysis in terms of final states

We decompose the averaged cross section (4.11) in terms of a basis for the hadronic final states:

$$\sigma = \sum_X \int \mathrm{d}^4x \ e^{iq \cdot x} \langle 0| j^\mu(x)|X \rangle \langle X| j^\nu(0)|0 \rangle \tilde{t}_{\mu\nu}(x), \qquad (4.36)$$

---

[3] The 't Hooft model is normally said to give an example of a string model. But I use the name "elastic spring" to emphasize its unbreakability, to contrast with the fragility of the string in real QCD.

and analyze the states in three bases:

- a momentum basis for the true out-states (involving hadrons);
- a spatially localized basis obtained using quark and gluon fields not too long after the creation of $|\phi\rangle$;
- a momentum basis for quark and gluon out-states, as seen in dimensionally regularized weak coupling perturbation theory. Here we must go to a space-time dimension above 4, i.e., to $\epsilon < 0$, so that the IR behavior is mild enough that the ordinary S-matrix exists.

The first basis gives the true distribution of observed final-state particles, but its use in calculations requires an unavailable non-perturbative solution of QCD. The second basis is most fundamentally suited to perturbative calculations, by working only with objects involving short distances. It completely justifies short-distance dominance for averaged cross section, but there is no known formulation explicit enough for actual calculations. The third basis, involving a momentum-space decomposition, is the easiest calculationally, but it involves a basis constructed from the $t \to \infty$ behavior of Green functions when a regulator is applied.

The low-order calculation of the individual terms in the ratio R in the third basis is only appropriate when the coupling is small enough that higher-order terms are not larger than lower-order terms. Given the *double* poles in ϵ that occur per loop, this implies that we should only apply the calculation when $\alpha_s \lesssim \epsilon^2$ (with ϵ *negative*). When the IR regulator is removed, the range of validity of the calculation shrinks to zero. So the cancellation of divergences in R is not sufficient *by itself* to justify the use of the result for R at non-zero $\alpha_s(Q)$.

But the result for R is independent of the basis for the completeness sum $\sum_X |X\rangle \langle X|$, so we can use the short-distance quark-gluon basis to justify the validity of perturbation theory for R for non-zero $\alpha_s(Q)$.

4.4 Dispersion relation and effective virtuality of final-state quarks and gluons

A perturbative calculation of $R(Q)$ involves cut graphs with an on-shell final state. In this section, I show that after a local average in Q, $R(Q)$ is given by an integral over an *uncut* graph, in which the final-state quarks and gluons are effectively off-shell by order Q^2. The derivation provides general principles that we will frequently generalize to other situations.

We consider a Green function $\Pi^{\mu\nu}$ that is defined like $W^{\mu\nu}$, but with *time-ordered* current operators. It has an associated scalar function $\Pi(q^2)$:

$$\Pi^{\mu\nu}(q) = \left(-g^{\mu\nu}q^2 + q^\mu q^\nu\right) \Pi(Q^2) = i \int \mathrm{d}^4x \; e^{iq\cdot x} \langle 0| T \, j^\mu(x) \, j^\nu(0) \, |0\rangle . \qquad (4.37)$$

Note the factor i in the last part. Diagrammatically, $\Pi^{\mu\nu}$ and $W^{\mu\nu}$ are notated in Fig. 4.12 :

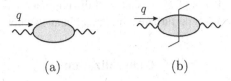

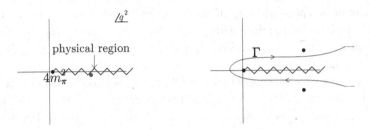

Fig. 4.12. (a) Uncut and (b) cut diagrams for the hadronic part of the photon self-energy, i.e., for $\Pi^{\mu\nu}$ and $W^{\mu\nu}$.

Fig. 4.13. Analyticity in Q^2 for $\Pi(Q^2)$.

Fig. 4.14. Contour to relate $\bar{R}$ to Π.-The off-real-axis singularities are those in the averaging function $F(s - Q^2, \Delta^2)$. The dots represent the singularities of F.

Now, $\Pi(Q^2)$ is an analytic function of Q^2 with a cut and singularities along the positive real axis, as in Fig. 4.13. When Q^2 is below the threshold for physical final states, Π is real, in particular for space-like q^μ. When Q^2 is above threshold, the physical region is on the upper side of the cut. Moreover, the cut amplitude is twice the imaginary part, as is provable from a dispersion relation. Hence

$$R(Q^2) = \frac{2\Pi(Q^2 + i0)}{i} = \frac{\Pi(Q^2 + i0) - \Pi(Q^2 - i0)}{i}. \tag{4.38}$$

Hence we can relate the averaged R to the uncut amplitude Π:

$$\bar{R}(Q^2, \Delta^2) = \frac{6\pi}{i} \int_\Gamma ds \, F(s - Q^2, \Delta^2) \, \Pi(s) \qquad \text{[general } F\text{]}$$

$$= \frac{6\pi}{i} \left[\Pi(Q^2 + i\Delta^2) - \Pi(Q^2 - i\Delta^2) \right] \qquad \text{[standard } F\text{]}. \tag{4.39}$$

In the first line, the contour Γ loops around the positive real axis inside of the singularities of the averaging function F, as in Fig. 4.14. In the second line, F is chosen to have the standard form in (4.18) and the contour is closed on the two poles of F.

Thus we have expressed $\bar{R}$ in terms of Π evaluated at non-physical values of the momentum. If the averaging range Δ is large, then the momentum is correspondingly far from the physical region. The Landau analysis of the singularities of Feynman graphs (Ch. 5) shows that the contour for the loop-momentum integrals in Π can then be chosen to avoid the poles of the internal propagators. If Δ is of order Q, the internal lines have (typically complex) virtualities of order Q^2.

Therefore in the calculation of $\bar{R}$, we can treat the internal quark and gluon lines as far off-shell, thereby justifying its perturbative treatment in an asymptotically free theory.

4.5 Generalizations

One simple generalization of the work in this chapter is to allow for Z exchange as well as photon exchange. This is needed for fits to the high-energy parts of the data in Fig. 4.2. Another generalization (Baikov, Chetyrkin, and Kuhn, 2008) is to the hadronic part of the decay rate of the τ lepton, where the initiating boson is the W.

For these cases, the same principles apply as to the case we treated: There is a cancellation of IR-sensitive regions, leaving a quantity for which perturbation theory is applicable. Such quantities we call "IR-safe".

To analyze more general situations, we use the Libby-Sterman argument to be explained in Ch. 5; this determines both the nature and power-counting of the IR-sensitive regions. The most interesting cases are where the cancellations of divergences fail to occur. For many of these, we will be able to derive factorization theorems, where only part of an amplitude or cross section is IR-safe.

One outcome will be a discussion of IR-safe jet cross sections in Sec. 12.13.4.

Exercises

4.1 Compute the contribution of a scalar quark to R, to lowest order. What is the angular distribution of the $q\bar{q}$ final state in both the spin-0 and spin-$\frac{1}{2}$ cases?

4.2 Compute the value of $R^{[0]}$ with quark masses taken into account. You should get

$$R^{[0]} = \sum_{i:\, m_i < Q/2} 3e_i^2 \left(1 - \frac{2m_i^2}{Q^2}\right) \sqrt{1 - \frac{4m_i^2}{Q^2}}. \tag{4.40}$$

4.3 (**) Compute the order α_s correction to R for scalar quarks.

4.4 (*) I wrote that near a resonance in e^+e^- annihilation, the outgoing state is obtained by quantum-mechanical interference involving sources at a large range of time scales. Despite a large value of Q, the separation x of the currents in (4.3) is not small, of order $1/Q$. Verify these statements explicitly. You could use the following approximation to the cross section near the resonance of mass M and width Γ:

$$\frac{C}{(Q^2 - M^2)^2 + \Gamma^2 M^2}. \tag{4.41}$$

4.5 (*) Using the RG β function for the effective coupling, find the Q/μ dependence of the coefficients in the formula for R, equation (4.34).

5

Libby-Sterman analysis and power-counting

Central assertions in setting up the parton model for DIS (Sec. 2.4) were that hard scattering occurs off a single parton constituent of the target, and that the hard scattering is just the Born approximation for electron-quark scattering. In fact, both assertions fail if taken literally. So in this chapter I show how to derive correct statements about the dominant configurations in DIS and the many other cases of interest. I will interleave a general treatment with a detailed discussion of specific examples.

Key insights were found by Sterman (1978) and Libby and Sterman (1978b), who systematized a correspondence between divergences in massless perturbative calculations and important configurations for high-energy processes. For any suitable process (like DIS) with an energy scale Q much larger than relevant particle masses, the main results are:

1. A one-to-one correspondence between mass divergences[1] in massless perturbation theory and non-UV regions in loop-momentum space that give the large Q asymptote.
2. That mass divergences are at surfaces where the integral over loop momenta cannot be deformed away from singularities of propagators. These surfaces are called pinch-singular surfaces (PSSs).
3. Simple and very general geometrical arguments in four-dimensional momentum space to locate the PSSs for a massless theory. The PSSs are in the typically higher-dimension space of all loop momenta.
4. Simple power-counting results for the strengths of the possible PSSs, and for the power dependence on Q of the contribution of the region associated with each PSS.
5. From the derivation of the power-counting results it is made evident what approximations are appropriate to each region, as needed to derive factorization theorems.
6. Hence error estimates are also obtained for the difference between an exact graph and its approximation in any of the regions.

These results form the logical basis of most further work in perturbative QCD, and in particular for the derivation of factorization theorems. The methods apply not only to QCD but to a general QFT.

---

[1] That is, divergences that appear when fields or particles are made massless, to be distinguished from ultra-violet (UV) divergences, for example.

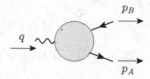

Fig. 5.1. Green function for e^+e^- annihilation to a quark-antiquark pair.

Practical calculations in QCD, as in Sec. 4.2, involve the facile manipulation of mass divergences, so that it is easy to attribute to the divergences an existence in the real world. But this is definitively incorrect: some of the fields have a non-zero mass, so that many of the mass divergences are not actually present. Moreover, even though QCD does have a massless gluon field, color confinement cuts off the divergences and prevents there from being asymptotic quark and gluon states in the exact theory.

The true relation between mass divergences and asymptotic behavior is that the PSSs for the divergences form a *skeleton* for important regions of momentum space. We use PSSs to label the regions, with the regions being *neighborhoods* of the PSSs.

As one gains experience with the methodology, the results gain a reality whose intuitive justification goes far beyond the Feynman-graph domain to which the strict mathematical justification is currently restricted. We have already explored some of these issues in Sec. 4.3, and we will see more in the generalization of the parton model to full QCD. Many of the issues have not been properly formalized. As a symptom, consider the Lund string model (Andersson, 1998), summarized in Sec. 4.3.1. This model gives a useful account of the hadronization of high-energy systems of quarks and gluons. To connect it to the fundamental underlying QCD theory, one needs to formulate the quantum-mechanical evolution of states locally in space-time in highly relativistic situations. A complete appropriate formalism is not yet available. This problem is closely related to important foundational issues in quantum mechanics and QFT.

5.1 High-energy asymptotics and mass singularities

5.1.1 Sudakov form factor, $\gamma^* \to q\bar{q}$

Many of the general principles can be discerned from a paradigmatic example, which is termed the Sudakov form factor, from its discussion by Sudakov (1956). We use the Green function for a quark field, an antiquark field, and a current (Fig. 5.1):

$$\Gamma^\mu \stackrel{\text{def}}{=} \int d^4x \, d^4y \, e^{ip_A \cdot x + ip_B \cdot y} \langle 0 | T \psi(x) \bar{\psi}(y) j^\mu(0) | 0 \rangle$$

$$= G^\mu_{\text{irred}} \times \text{full external quark propagators.}$$

(5.1)

Here j^μ is the electromagnetic current, and ψ and $\bar{\psi}$ are fields for some flavor of quark. The photon momentum is $q = p_A + p_B$, with invariant size $Q \stackrel{\text{def}}{=} \sqrt{q^2}$. Our aim is to understand the asymptotics when Q gets large with p_A^2 and p_B^2 fixed, but not necessarily on-shell. Factoring out external propagators gives the definition of the irreducible amplitude indicated in the last line. The off-shell amplitude appears in high-energy e^+e^- annihilation, as a subgraph of the full amplitude for the process.

In fixed-order perturbation theory, taking p_A and p_B on-shell gives IR divergences because the gluon is massless. Beyond perturbation theory, we expect color confinement in QCD to force on-shell quark amplitudes to be zero, and to cut off the IR divergences. But these issues are quite separate from the association we wish to make between properties of the large Q limit and divergences in a completely massless theory.

In setting up methods for factorization later in this book, a convenient model example is the Sudakov form factor with on-shell quarks treated in an *abelian* gauge theory, normally with a *massive gluon* (Ch. 10).

We work in the overall center-of-mass frame, oriented so that the external 4-momenta in ordinary Cartesian coordinates are

$$p_A = \frac{Q}{2}\left(1, 0, 0, \sqrt{1 - 4p_A^2/Q^2}\right), \tag{5.2a}$$

$$p_B = \frac{Q}{2}\left(1, 0, 0, -\sqrt{1 - 4p_B^2/Q^2}\right), \tag{5.2b}$$

$$q = Q(1, \mathbf{0}). \tag{5.2c}$$

5.1.2 Scaling in units of Q

Consider a particular L-loop graph G for the 1PI factor G_{irred}. Let k denote its loop momenta, and let I denote the integrand, so that

$$G = g^{2L} \int d^{nL}k \; I(k, p_A, p_B; m) + \text{UV counterterms}. \tag{5.3}$$

Imagine first that we were in a situation where all internal momenta have components of order Q, and have virtuality of order Q^2. After using the renormalization group to set the renormalization scale to Q, we could use weak-coupling perturbation theory, and, to the leading power in Q, we could neglect masses. Errors in the massless approximation, from an expansion in powers of m/Q, p_A^2/Q^2, and p_B^2/Q^2, would be asymptotically much less than corrections from higher orders in $\alpha_s(Q)$.

Of course our initial supposition on the sizes of the internal momenta is in general false. Nevertheless, the region of k that it covers forms a useful standard for treating the general situation.

Relatively benign alternative regions are where some or all components of k are much bigger than Q. Since the external momenta are much smaller than these large components, this is the situation handled by renormalization. So let us add renormalization counterterms and then apply an RG transformation to set the renormalization scale μ of order Q. As we saw in Sec. 3.4, this procedure *effectively* cuts off the integration at around Q.

Therefore the interesting regions are where relevant components of momenta are of size Q or smaller, and where some lines have small virtuality, i.e., their momenta l obey

$|l^2| \ll Q^2$. For these lines, a lowest-order Taylor expansion in masses compared with virtuality fails. Such regions form a small part of the whole of loop-momentum space, but they can give large contributions to the integral, because of small propagator denominators.

To systematically locate relevant regions with low virtuality, we use an analysis with momenta and masses scaled in units of Q. Thus we define

$$\tilde{p}_A \overset{\text{def}}{=} \frac{p_A}{Q} \to \tfrac{1}{2}(1, 0, 0, 1), \tag{5.4a}$$

$$\tilde{p}_B \overset{\text{def}}{=} \frac{p_B}{Q} \to \tfrac{1}{2}(1, 0, 0, -1), \tag{5.4b}$$

$$\tilde{q} \overset{\text{def}}{=} \frac{q}{Q} = (1, \mathbf{0}), \tag{5.4c}$$

where the limits apply as $Q \to \infty$. The scaled external quark and antiquark momenta become light-like, while $\tilde{q}$ is a fixed time-like vector. Similarly we have scaled loop momenta, $\tilde{k} \overset{\text{def}}{=} k/Q$, and mass(es), $\tilde{m} \overset{\text{def}}{=} m/Q \to 0$.

Dimensional analysis applied to (5.3) gives

$$G = Q^{D(G)} g^{2L} \int d^{nL}\tilde{k} \; I(\tilde{k}, \tilde{p}_A, \tilde{p}_B; \tilde{m}) + \text{UV counterterms}. \tag{5.5}$$

Here $D(G)$ is the dimension of the integral (in powers of energy), with the coupling excluded. In a space-time dimension $n = 4 - 2\epsilon$, we have

$$D(G) = nL + \dim I = \dim G - 2L \dim g = -2L\epsilon. \tag{5.6}$$

Equations (5.4) and (5.5) show that the infinite Q limit at fixed mass is closely linked to the zero-mass limit at fixed Q, in the scaled integral on the right-hand side of (5.5). As observed earlier, if there were no singularities in the zero-mass limit, we could just set $p_A^2 = p_B^2 = m^2 = 0$ to obtain an elementary RG-controlled calculation of the large Q behavior. Moreover, the Q dependence would just be $Q^{D(G)}$. From (5.6), we see that because of the dimensionless of a gauge theory coupling at the physical space-time dimension, the power of Q is the *same* for all graphs, viz. zero.

5.1.3 Importance of pinch-singular surfaces in massless limit

We now need to locate the situations where the zero-mass limit fails. These situations arise from regions where one or more lines have virtuality much less than Q^2. But often the contour of integration can be deformed away from such regions, and the above scaling arguments work equally well on a deformed contour for k. So our concern is regions where there is an obstacle to any possible deformation to where the lines have virtuality of order Q^2. In fact, as we now show, the only obstacles are those that give a pinch-singular surface (PSS) in the massless limit.

Consider first some region of scaled loop momentum $\tilde{k}$ where certain propagator denominators are not part of a pinch in the massless theory. Then in the scaled integral and on some deformed contour, these denominators have a non-zero minimum size. In the original integral, before scaling, the same denominators have a minimum size proportional to Q^2 in the corresponding region of k. Then the simple massless limit applies for the contribution to the large Q asymptote by these denominators.

Next we consider unscaled momenta in a neighborhood of a PSS of the massless theory. Even with a massless PSS, the minimum virtuality of *unscaled* lines often stays finite as Q gets large, even on a deformed contour. Typically, this virtuality would be of order a mass-squared. But in some cases the minimum virtuality may grow with Q, but less rapidly than Q^2, for example, it might be of order Qm. Even so, in all these cases, the scaled virtuality, i.e., relative to Q^2, goes to zero as $Q \to \infty$. This corresponds to an exact pinch in the massless theory: that is, with masses set to zero, the scaled momenta $\tilde{k}$ in (5.5) have a minimum distance of zero from the lines participating in the PSS.

In the actual case, with non-zero masses and finite Q, the relevant momenta are forced to go close to the PSS, the closeness in units of Q decreasing as Q increases. I summarize this by saying that the PSSs of the massless theory form a skeleton for the important non-UV regions of loop momentum space. This can happen even in a field theory where all the fields have non-zero mass, so that the exact massive theory has no literal PSS.

5.1.4 Location of pinch-singular surfaces: Landau criterion

Therefore we now have to find all possible PSSs in the massless limit and determine their strengths. The general task of locating PSSs is made quite simple by the Landau criteria (e.g., Eden *et al.*, 1966) in the form particularly emphasized by Coleman and Norton (1965): The PSSs (for the physical region, which is all that concerns us) are where the on-shell propagators and momenta correspond to classically allowed scattering processes treated in coordinate space.

Each point on a PSS (in loop momentum space) corresponds to a space-time diagram obtained as follows. First we write a reduced graph by contracting to points all of the lines whose denominators are not pinched. Then we assign space-time points to each vertex of the reduced graph so that the pinched lines and their momenta correspond to classical particles. That is, to each line we assign a particle propagating between the space-time points corresponding to the vertices at its ends. The momentum of the particle is exactly the on-shell momentum carried by the line, correctly oriented to have positive energy. If, for some set of momenta, it is not possible to construct such a reduced graph, then we are free to deform the contour of integration.

Although our argument to this point was presented in the context of the Sudakov form factor, it is in fact a general argument and can be applied to many processes with a large scale Q.

5.2 Reduced graphs and space-time propagation

The construction of the most general reduced graph becomes extremely simple in the zero-mass limit, since at a PSS all pinched lines must carry either a light-like momentum or zero momentum. Moreover, each light-like momentum must be parallel to one of the light-like external lines.

To understand this, we just need to obtain the simple rules for how massless on-shell momenta combine at vertices of a reduced graph.

1. First, adding zero momentum to anything leaves the second momentum unaltered. So a zero-momentum line can attach anywhere.
2. Two non-zero light-like momenta in the same direction are proportional to each other and add to make another parallel light-like momentum, with a special case of giving zero when they are equal and opposite. If we orient the momenta of the lines for a particular light-like direction so that they all have positive energy, then as we follow them forward, the momenta can split and recombine arbitrarily, but the total momentum is fixed.
3. Adding two non-zero light-like momenta with *different* directions produces a non-light-like momentum, necessarily off-shell in a massless theory. Either the non-light-like momentum is external or it is on an internal line. An external non-light-like momentum would be like the virtual photon in the form factor or in DIS. An internal line is off-shell, so it is internal to a reduced vertex, i.e., it does not participate in the pinch under discussion.
4. It is possible for a reduced vertex to correspond to a non-trivial wide-angle scattering of massless particles. But for the classical scattering condition to hold, the other ends of the light-like lines are a long way from the reduced vertex. So further rescattering of the same particles is not possible. See the discussion of Fig. 5.3 below on p. 94 for an example.

The results for massless PSSs can be presented in two forms: (a) the structure of the reduced graphs, with a labeling of lines by momentum type, and (b) the locations of the vertices of the corresponding classical processes in space-time; see the illustrative examples in Sec. 5.3 below.

It is convenient to present the results with the aid of massless but unscaled momenta corresponding to high-energy external lines. For example, in the case of Fig. 5.1, from the limits in (5.4a), we define unscaled massless momenta by

$$p_{A,\infty} \stackrel{\text{def}}{=} \frac{Q}{2}\,(1,\,0,0,\,1)\,, \tag{5.7a}$$

$$p_{B,\infty} \stackrel{\text{def}}{=} \frac{Q}{2}\,(1,\,0,0,\,-1)\,. \tag{5.7b}$$

5.3 Examples of general reduced graphs

5.3.1 Vertex graph

For the vertex graph of Fig. 5.1, a typical reduced graph and the corresponding space-time diagram are shown in Fig. 5.2. In the reduced graph, there is a subgraph H which includes

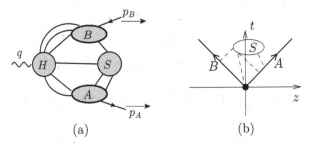

Fig. 5.2. Typical (a) reduced graph, and (b) space-time diagram, for a general PSS for the vertex graph.

the vertex for the current j^μ. This subgraph is intended to be a vertex of the reduced graph, i.e., none of its lines participate in the pinch. Thus, in the space-time diagram all of the lines and Feynman-graph vertices that compose H are contracted to a single point.

From H exit two sets of lines in what we call collinear subgraphs. One collinear subgraph, A, has lines in the $p_{A,\infty}$ direction, and the other, subgraph B, has lines in the $p_{B,\infty}$ direction. Finally the soft subgraph S, not necessarily connected, consists of lines of zero momentum at the PSS, and it can connect to any of the other subgraphs. Notice that we labeled the collinear graphs by the light-like momenta $p_{A,\infty}$ and $p_{B,\infty}$ rather than the actual external momenta p_A and p_B, since we are discussing PSSs in the massless limit.

In the space-time picture the hard subgraph corresponds to a single point at the origin, and the collinear subgraphs A and B correspond to propagation outward along light-like directions. Within each collinear subgraph, there can be arbitrary splitting and recombination of the collinear momenta. Any number of lines can join the A and B subgraphs to the H subgraph. Finally the S subgraph corresponds to zero momentum and so to arbitrarily large separations in space and time. The zero-momentum lines can interact arbitrarily with each other, and any number of lines can connect their subgraph to the other subgraphs.

From the reduced diagram point of view, the collinear and soft subgraphs contain lines of the stated kind, i.e., parallel to $p_{A,\infty}$, $p_{B,\infty}$, or zero. But it should be noted that the reduced-graph vertices that join them within each subgraph may comprise non-trivial (one-particle-irreducible) graphs from the Feynman graph point-of-view.

The collinear lines go outward from the hard vertex and eventually combine to form the momenta $p_{A,\infty}$ and $p_{B,\infty}$ of the outgoing external lines of the vertex, treated as massless momenta. There can be no other massless lines propagating in other directions, or from the past. Any such line would just give a dangling end with no external line(s) to absorb or generate the non-zero momentum.

These conclusions depend not only on the on-shell condition for the lines of the reduced graph, but, critically, also on the condition that they correspond to a physical scattering. As an example, consider the configuration illustrated in Fig. 5.3. Here there are two intermediate massless on-shell lines with 3-momenta not along the z axis:

$$p_{C,\infty} = \frac{Q}{2}(1, \boldsymbol{n}), \qquad p_{D,\infty} = \frac{Q}{2}(1, -\boldsymbol{n}). \tag{5.8}$$

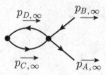

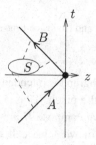

Fig. 5.3. Non-pinched on-shell configuration for Sudakov form factor.

Fig. 5.4. Space-time diagram for PSSs for the vertex graph when the A line is incoming, so that the momentum transfer is space-like.

These rescatter at the right-hand reduced vertex to make the standard external lines. This reduced vertex is for elastic scattering with large momentum transfer. The on-shell configuration obeys momentum conservation, and does contribute in a computation of the imaginary part of the amplitude from on-shell intermediate states. But for the rescattering to be classical, in the sense used for the Landau criterion, the two wide-angle particles have to meet at a single point to rescatter. Thus they would travel only a zero distance from their generation at the electromagnetic vertex, and not the arbitrary non-zero distance needed for classicality. Hence this configuration does not participate in a pinch.

A minor variation can be made by letting the p_A line be incoming rather than outgoing, with the momentum transfer now being space-like. This would be appropriate for a subgraph inside a deeply inelastic scattering amplitude. The general reduced graphs stay the same, except for the orientation of the momenta in the A subgraph. Correspondingly, the space-time structure changes to that shown in Fig. 5.4.

5.3.2 Leading regions for vertex graph

Comparing Fig. 5.2(a) to the structure Fig. 2.5(b) that was used to obtain the parton-model formula for DIS, we see a lot of extra connections between the subgraphs. This endangers the derivation of a factorization theorem. In the parton-model ansatz for DIS, the hard scattering involves only a *single* parton, and the target and outgoing collinear subgraphs are not otherwise coupled. Similar remarks evidently apply to all other processes.

When we derive rules for power-counting, later in this chapter, we will find that for many of the massless PSSs, the corresponding contributions to the actual vertex are in fact suppressed by a power of Q. Generally, we will neglect these power-suppressed

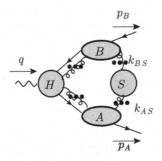

Fig. 5.5. Typical reduced graphs for the vertex graph, but now restricted to those PSSs relevant for the leading power.

contributions. Then we will find that the leading regions for the Sudakov form factor are restricted to those of Fig. 5.5. Compared with the general PSS, Fig. 5.2(a), the changes are that: no lines connect S to H, only gluons connect S to the collinear subgraphs, and exactly one fermion but arbitrarily many gluons connect the collinear subgraphs A and B to the hard subgraph.

The arbitrary number of gluons linking the different subgraphs of a reduced graph still leaves us with an apparent difficulty for proving factorization. A final power-counting result will come to the rescue, concerning the dominant polarization for the extra gluon connections.

Here we only summarize what we will prove later. The relevant polarizations are such as to allow us to use Ward identities to sum over the ways of connecting the extra collinear gluons to the hard subgraph and of connecting the soft gluons to the collinear subgraphs. The end product will be a factorized form, with definitions of parton densities and other non-perturbative quantities as matrix elements of certain non-local operators. Without the extra gluon connections, the operators would not be gauge invariant. Summing the extra gluon connections between the subgraphs converts the operators to a gauge-invariant form.

5.3.3 DIS from uncut amplitude

A very straightforward application of the Landau analysis is to DIS, if we apply the same trick as we used in Sec. 4.4 for the $e^+e^- \longrightarrow$ hadrons cross section.

Instead of the hadronic tensor $W^{\mu\nu}$ defined by (2.18), we use the corresponding uncut amplitude[2] where the current operators are time-ordered:

$$T^{\mu\nu}(q, P) = \frac{1}{4\pi} \int d^4z \; e^{iq \cdot z} \langle P, S| \, T \, J^\mu(z/2) \, J^\nu(-z/2) \, |P, S\rangle . \qquad (5.9)$$

This amplitude is analytic in the plane of $\nu = p \cdot q$, with cuts along the positive and negative real axis starting from $\nu = \pm Q^2/2$ (Fig. 5.6). The ordinary hadronic tensor is the

---

[2] *Warning*: Definitions in the literature disagree on the normalization.

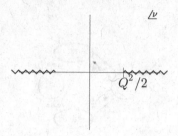

Fig. 5.6. Complex plane in $\nu = P \cdot q$ for $T^{\mu\nu}$, with its cuts.

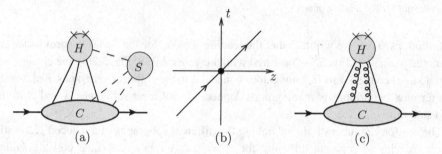

(a)   (b)   (c)

Fig. 5.7. (a) Typical general reduced graph, and (b) space-time diagram, for the most general PSS for the uncut amplitude for DIS. (c) For a leading PSS, there is no soft part, and beyond the main partons, an arbitrary number of gluons connect the collinear and hard subgraphs.

discontinuity

$$W^{\mu\nu}(q, P) = T^{\mu\nu}(\nu + i0) - T^{\mu\nu}(\nu - i0). \tag{5.10}$$

See Ch. 14 of Collins (1984) for more details and an account of earlier work on DIS. There the analyticity properties of $T^{\mu\nu}$ were exploited to allow the use of the short-distance operator product expansion to analyze integer moments of DIS structure functions.

Just in e^+e^- annihilation (Sec. 4.4), a local averaging should be applied, after which we only need to treat $T^{\mu\nu}$ away from its singularities in the complex plane.

The massless PSSs for the amplitude are illustrated by the reduced graph in Fig. 5.7(a). There is a single collinear subgraph C, where the target comes in and undergoes arbitrary collinear splittings and recombinations until the target is reconstituted. The hard scattering H is at the origin in space-time, and there is a soft subgraph S. In a general PSS, there are arbitrarily many lines joining the subgraphs. The graphical structure, Fig. 2.5(b), that we used to formulate the parton model is the simplest example. It corresponds to a minimal PSS where only two lines join the collinear and hard subgraphs, where there is no soft subgraph, and where the hard subgraph is a lowest-order Feynman graph.

The Landau analysis has now indicated, in Fig. 5.7(a), the maximum complication to be considered in the general case. We can again anticipate the power-counting results, in Fig. 5.7(c). At leading power, the soft subgraph is absent. The connections between the collinear and hard subgraphs consist of the primary pair of parton lines, just as in the

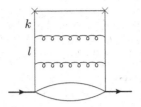

Fig. 5.8. A graph for uncut amplitude for DIS with multiple PSS.

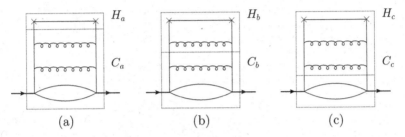

(a)   (b)   (c)

Fig. 5.9. The three leading regions for Fig. 5.8 correspond to these decompositions into hard and collinear subgraphs.

parton model, but they are now accompanied by any number of gluon lines with the special polarization that allows the use of Ward identities to give a factorization theorem.

5.3.4 Higher-order corrections to hard scattering

The following consequence of the general region analysis contains a critical difference between the true results of QCD and the parton model: This is that there are higher-order perturbative corrections to the hard scattering.

Although we will work out the details only in later chapters, it is possible to understand the basic ideas from our analysis so far. First we observe that any particular Feynman graph might have multiple leading PSSs. For example, consider Fig. 5.8, which can appear in a model for DIS in which the target is treated as elementary. This graph, of the form of what is often called a "ladder graph", has three decompositions of the form of Fig. 5.7(c), but, in this particular case, without any of the extra gluonic connections. In one of its PSSs all the quark lines on the sides of the ladder are collinear to the target, i.e., the momenta k and l are target-collinear. This corresponds to the decomposition of Fig. 5.9(a), where the hard subgraph H_a is the smallest possible, and is indeed exactly the same as in the parton model.

A second PSS corresponds to Fig. 5.9(b), where the upper loop momentum k is of high virtuality, while the lower momentum l is still target-collinear. This has a one-loop hard subgraph H_b. Physically it corresponds to production of two jets in the hard scattering, as in the experimental event shown in Fig. 5.10. A third PSS corresponds to Fig. 5.9(c), where both k and l are of high virtuality; this situation corresponds to production of three jets.

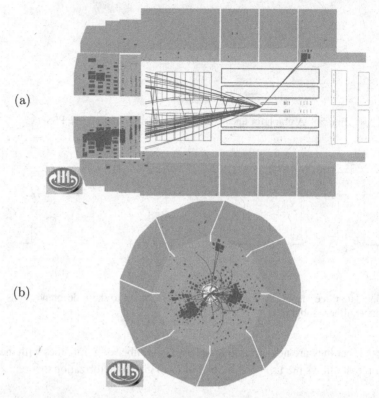

Fig. 5.10. Scattering event with two high-transverse-momentum jets in an *ep* collision in the H1 detector (H1 website, 2010). The final state contains an electron track (to the right in the side view), and two jets of hadrons.

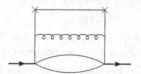

Fig. 5.11. Another graph for uncut amplitude for DIS in which some of the same hard subgraphs occur as for the previous graph.

Each of these hard-scattering subgraphs can occur in other graphs for $T^{\mu\nu}$. For example, the hard subgraphs H_a and H_b also appear in PSSs for Fig. 5.11.

The momentum regions associated with the three PSSs are represented in Fig. 5.12, where the smaller PSSs are boundaries of the bigger ones. Disentangling the contributions associated with different PSSs gives interesting mathematical and technical issues, which occupy much of this book.

We will see that larger hard subgraphs H_b, etc., can be treated as higher-order corrections to the lowest-order subgraph H_a, but with subtractions to compensate for double counting between different contributions.

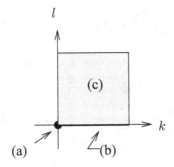

Fig. 5.12. Momentum regions associated with the PSSs in Fig. 5.9. Each axis corresponds to the deviation of the associated momentum from exact collinearity, and the labels "(a)", "(b)" and "(c)" correspond to the PSSs associated with the graphical decompositions in Fig. 5.9.

The idea of higher-order corrections to the hard scattering is readily accommodated by the original space-time motivation for the parton model. This asserted that the cross section was governed by a short-distance scattering of the electron and a single constituent of the target, as in Fig. 2.2. The true hard scattering is the short-distance structure at the origin in the space-time representation, Fig. 5.7(b), but it need not be a lowest-order graph.

A scaling argument of the kind given in (5.5) shows that the power of Q is determined only by the number of external lines of the hard scattering, in any renormalizable theory like QCD, since then the coupling is dimensionless. Thus there is no power-law suppression of higher-order hard scattering. The only suppression is from the smallness of the effective coupling $\alpha_s(Q)$ at large Q. The appropriate scale for the coupling in the hard scattering is of order Q, so that the asymptotic freedom of QCD allows low-order perturbation theory to give useful predictions of the hard scattering.

Physically, the hard subgraph H is not literally at a single point, but is spread over a space-time range of order $1/Q$. Similarly, the collinear subgraph is not exactly on the light-like line indicated in Fig. 5.7(b), but is spread out as appropriate for a highly boosted composite particle. Lorentz contraction indicates that the width of the collinear lines is of order $1/Q$ in the t-z plane, but of order $1/M$ transversely, while time dilation gives a large longitudinal scale to Fig. 5.7(b), of order Q/M^2. This interpretation is another way of explaining the statement that the massless PSSs form a skeleton for the location of the actual physical phenomena. A formal derivation from first principles within QFT of the detailed space-time interpretation would be very useful.

5.3.5 DIS from cut amplitude

To understand how the final states in DIS arise, we now restore the final-state cut. It is evident from our calculations of e^+e^- annihilation that there is a close connection between divergences from virtual gluon emission and those from real gluon emission. Therefore, it is useful to extend our analysis with reduced graphs and space-time diagrams to include the integrals over final states.

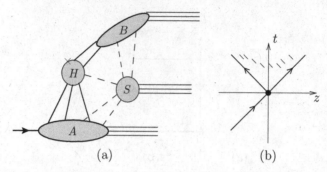

(a)   (b)

Fig. 5.13. (a) Reduced graphs and (b) space-time diagram, for DIS amplitude, in the case that only one jet arises from the hard scattering. The lighter hatching at the top of (b) corresponds to the low momentum or soft particles from the soft subgraph S.

The basic idea is unchanged: taking $Q \to \infty$ at fixed mass is equivalent to a massless limit at fixed Q, and we need to know where propagator denominators fail to have virtuality of order Q^2. Just as before, it is the locations of PSSs in the massless theory that label all the interesting regions. But for final-state lines, we no longer have to appeal to a technical argument as to whether or not a contour deformation is possible. Final-state lines are necessarily on-shell, so they have to be considered always pinched. Since final-state particles can be observed, it is appropriate not to even consider deforming any of the integrals over final-state momenta. Some lines are not part of any loop, as in the real emission graphs considered in Ch. 4; their virtuality is entirely determined by the external lines. At a collinear singularity, it is simply from the topology of the graph plus the simple rules for combining light-like momenta that we get the condition of a classical process. We supplement this by the Landau criterion for lines that are part of a loop.

In the case that we only have one direction for the particles from the hard scattering, the reduced diagrams and space-time picture are shown in Fig. 5.13, for an amplitude $\langle X, \text{out} | j | 0 \rangle$. These correspond quite directly to the picture shown in Fig. 2.2, and the actual scattering event in Fig. 2.3. The collinear subgraph A corresponds to the target hadron, its evolution and its remnants after a quark has been struck out of it. The remnants are around the beam pipe in the actual event. The subgraph B corresponds to collinear evolution of the struck partonic system into an observed jet. Some lines can go out to the final state from the S subgraph; at the exact mass singularity, these have zero momentum. The corresponding actual particles, all of whose momentum components are much less than Q to be close to the PSS, are those that fill in the rapidity gap between the jet and the beam remnant.

Other PSSs arise when there are two or more groups of parallel lines emerging from the hard scattering, as in Fig. 5.14. In experiments one manifestation of momentum configurations near to such singularities are events with extra jets, as in Fig. 5.10.

Naturally, the full DIS cross section has an integral over all accessible final states. This integral includes all intermediate configurations between the extremes given by the reduced diagrams and their associated massless PSSs. Proper factorization theorems, and their proofs, handle the intermediate cases once the extremes are dealt with.

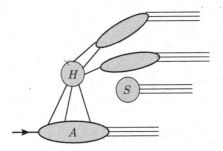

Fig. 5.14. Reduced graph for DIS, in the case that partons in more than one direction arise from the hard scattering. For clarity the connections between the soft subgraph and the other subgraphs have been omitted.

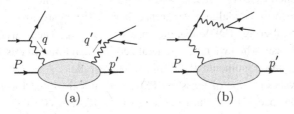

Fig. 5.15. (a) The double deeply virtual Compton scattering process, including the attached leptons. (b) The Bethe-Heitler pair production process that also contributes to the scattering.

5.3.6 Deeply virtual Compton scattering, etc.

So far, we have treated the uncut hadronic tensor $T^{\mu\nu}$ merely as a tool for analyzing DIS, whose true cross section arises from the discontinuity, i.e., from the cut amplitude.

But it is also interesting to examine this quantity in its own right as the hadronic part of an appropriate scattering amplitude. It actually provides the conceptually simplest of all QCD factorization theorems. We therefore take the opportunity to introduce the relevant processes. For this, we attach leptons at the other ends of virtual photon lines. To obtain a realizable scattering, one of the virtual photons is time-like, creating a lepton pair. Thus the relevant process is $lP \to l'p'e^+e^-$ or $lP \to l'p'\mu^+\mu^-$; Fig. 5.15(a). Since one photon has space-like momentum q and the other has time-like momentum q', the hadronic amplitude is not diagonal, unlike the case for DIS. A complication for the analysis of data is that one needs to separate the contribution where the lepton pair arises from a virtual photon attaching to the other leptons: Fig. 5.15(b).

This leads (Müller *et al.*, 1994; Blümlein and Robaschik, 2000) to the study of the process $\gamma^*(q) + P \to \gamma^*(q') + p'$, which corresponds to the off-diagonal hadronic tensor

$$A^{\mu\nu}\left(\gamma^*(q) + p \to \gamma^*(q') + p'\right)$$

$$= \frac{1}{4\pi} \int d^4z \; e^{iz\cdot(q+q')/2} \left\langle p' \left| T \, J^\mu(z/2) \, J^\nu(-z/2) \right| P \right\rangle. \tag{5.11}$$

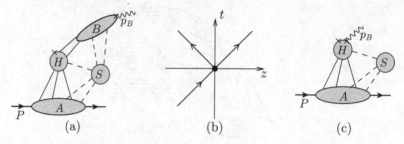

Fig. 5.16. (a) Reduced graphs and (b) space-time diagram, for DVCS and exclusive elec-troproduction of mesons. (c) Extra reduced graphs for DVCS, but *not* exclusive meson electroproduction, with photon directly connected to H.

This was investigated by Berger, Diehl, and Pire (2002), who called it "timelike Compton scattering", and then by Guidal and Vanderhaeghen (2003), who called it "double deeply virtual Compton scattering" (DDVCS), the term we use here. The analysis closely corre-sponds to the DIS case, when we take a generalized Bjorken limit. In this limit q^2, q'^2, etc. are large, and the hadron momenta P and p' become parallel.

Thus the analysis in terms of massless PSSs is identical to that for $T^{\mu\nu}$ for DIS; the reduced graphs and space-time picture are exactly the same. DDVCS has great fundamental importance as the simplest quantity to which factorization methods can be applied. However the cross sections at the leptonic level are high order in electromagnetism and thus very small; see Berger, Diehl, and Pire (2002); Guidal and Vanderhaeghen (2003).

What is studied experimentally at present is the case that the outgoing photon is real. This is deeply virtual Compton scattering (DVCS): Müller *et al.* (1994); Blümlein and Robaschik (2000); Belitsky, Müller, and Kirchner (2002):

$$\gamma^*(q) + P \to \gamma(p_B) + p'. \tag{5.12}$$

The outgoing photon is light-like in what we can choose to be approximately the $-z$ direction. Thus it is convenient to change notation to use p_B for the photon momentum; this corresponds to our notation for other processes with two high-energy particles. Another closely related process has the photon replaced by a meson:

$$\gamma^*(q) + P \to M(p_B) + p', \tag{5.13}$$

the measured meson being typically a ρ. This is actually an exclusive two-body subprocess of DIS, called exclusive electroproduction of mesons. The reduced graphs now acquire a collinear-B subgraph going out from the hard scattering, Fig. 5.16(a), with a corresponding space-time diagram. The power-counting is a bit more subtle, and depends on the polarization of the meson (Brodsky *et al.*, 1994; Collins, Frankfurt, and Strikman, 1997).

For the case of a photon, i.e., DVCS, there are also reduced graphs without the B subgraph, i.e., with the photon connecting directly to the hard subgraph. These are, of course, the same as for a highly virtual photon; it is these reduced graphs that turn out to be the leading ones (Müller *et al.*, 1994; Blümlein and Robaschik, 2000; Belitsky, Müller, and Kirchner, 2002).

5.3.7 Drell-Yan process

Another important process is the Drell-Yan (DY) process, i.e., inclusive production of high-mass lepton pairs in hadron-hadron collisions:

$$P_A + P_B \rightarrow (\gamma^* \rightarrow l^+l^-) + X, \tag{5.14}$$

where we have indicated that in lowest order in electromagnetism, the lepton pair arises from a virtual photon. Essentially all the same theoretical considerations apply to the production of high-mass electroweak bosons, like the W, Z, and Higgs particle, as well as innumerable conjectured particles in extensions of the Standard Model.

In light-front coordinates, we write the momenta as

$$P_A = \left(P_A^+, \, m_A^2/2P_A^+, \, \mathbf{0}_T\right), \tag{5.15a}$$

$$P_B = \left(m_B^2/2P_B^-, \, P_B^-, \, \mathbf{0}_T\right), \tag{5.15b}$$

$$q = \left(x_A P_A^+\sqrt{1+q_T^2/Q^2}, \, x_B P_B^-\sqrt{1+q_T^2/Q^2}, \, \mathbf{q}_T\right). \tag{5.15c}$$

Here the scaling variables are defined by

$$x_A = Qe^y/\sqrt{s}, \quad x_B = Qe^{-y}/\sqrt{s}, \tag{5.16}$$

where $y = \frac{1}{2}\ln\frac{q^+P_B^-}{q^-P_A^+}$ is the center-of-mass rapidity of the lepton pair, and $Q = \sqrt{q^2}$ is its invariant mass. In the center-of-mass, the large components of the hadron momenta are P_A^+ and P_B^-, both equal to $\sqrt{s/2}$ up to power-suppressed corrections. Frequently, the cross section is integrated over q_T, and is presented as $\mathrm{d}^2\sigma/(\mathrm{d}Q^2\,\mathrm{d}y)$.

We first discuss the DY amplitude. Its reduced graphs are constructed by an elementary generalization of the construction for DIS. We now have two collinear subgraphs, A and B, associated with each incoming particle. As in DIS, we classify the reduced graphs by the number of outgoing directions of lines from the hard scattering H. Now H has incoming lines from each of the A and B subgraphs, and has the virtual photon taking out momentum. This allows the minimal situation, illustrated in Fig. 5.17, with no extra collinear groups at all going out from H. The soft subgraph can create particles in the final state that fill in the rapidity gap between the beam remnants.

This is illustrated by the microscopic view of a collision shown in Fig. 5.18 (which corresponds to Fig. 2.2 for DIS). Here we have shown the simplest possibility: a single parton from each parent hadron collides over a short distance scale, of order $1/Q$ at the position indicated by a star, and we have not depicted the possible soft interactions.

One new possibility is that we could have a second hard part, disconnected from the first in which other collinear lines from A and B collide to undergo a wide-angle scattering. Physically, this corresponds to a second partonic collision in Fig. 5.18, typically occurring at about the same time as the one that creates the DY pair, but at a different transverse separation. Later, from the power-counting rules, we will see that this case is power-suppressed.

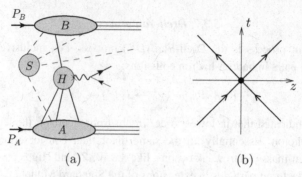

(a)   (b)

Fig. 5.17. (a) An important reduced graph for the amplitude for the Drell-Yan process. (b) Space-time diagram for collinear subgraphs.

Fig. 5.18. Microscopic view of a DY process, corresponding to Fig. 2.2 for DIS.

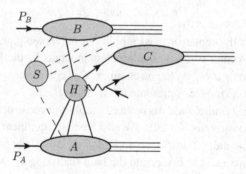

Fig. 5.19. A reduced graph for the amplitude for Drell-Yan process when one extra jet of high transverse momentum is produced.

After this, we will find the usual situation for the leading power that only one main parton from each beam hadron enters a single hard scattering. Each is accompanied only by extra gluons of the longitudinal polarization that can be reorganized by Ward identities into gauge-invariant parton densities. Also the soft subgraph at leading power only connects to the collinear subgraphs and by gluons.

It is possible for the single hard scattering to produce, in addition to the lepton pair, one or more extra partons of high transverse momentum, Fig. 5.19. These manifest themselves as jets in the hadronic final state, just as in the corresponding situation for e^+e^- annihilation or DIS.

If instead we restrict to a minimal reduced graph, and then multiply by the complex conjugate amplitude, we get the cut graph shown in Fig. 5.20. This is the natural

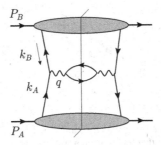

Fig. 5.20. Minimal reduced graph for cross section for the Drell-Yan process.

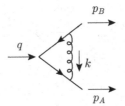

Fig. 5.21. One-loop vertex graph.

generalization of the corresponding structure that led to the parton model in DIS, Fig. 2.5(b). The most elementary treatment of this situation leads to the parton model formula for lepton-pair production, first worked out by Drell and Yan (1970). Here the lepton pair is produced in the lowest-order annihilation of a quark out of one hadron, and an antiquark out of the other, with the same parton densities as in DIS.

We thus see a general pattern: Libby and Sterman's insight leads to the reduced diagram analysis. Approximating the situation by configurations corresponding to the simplest reduced graphs gives us the parton model, with the natural space-time interpretation. The general reduced graph plus the restriction to leading power delimits the maximum way in which we have to distort the parton model to get the results of real QCD.

5.4 One-loop vertex graph

To illustrate the properties of the regions associated with PSSs, we examine the PSSs for the one-loop vertex graph of Fig. 5.21:

$$G_1 = \frac{ig^2}{(2\pi)^n} \int d^n k \, \frac{\text{numerator}}{(k^2 - m_g^2 + i0)\,[(p_A - k)^2 - m_q^2 + i0]\,[(p_B + k)^2 - m_q^2 + i0]}.$$

$$(5.17)$$

The numerator factor is irrelevant for determining the positions of the PSSs. But it is important in computing their strengths, for which different field theory models give interesting characteristic effects. We also allow a gluon mass, which is zero in QCD, but not necessarily

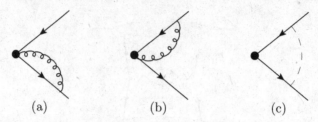

Fig. 5.22. Reduced graphs for PSSs R_A, R_B, and R_S of Fig. 5.21. The dot represents the short-distance reduced graph, the diagonal lines are collinear in the appropriate directions, and the dashed line is soft (zero momentum).

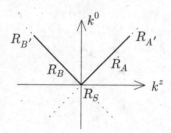

Fig. 5.23. Location of massless PSSs of Fig. 5.21 in the space of the gluon momentum. The singularities are all in the plane of zero transverse momentum, so we just show the plane of k^0 and k^z, with the $2 - 2\epsilon$ transverse dimensions out of the paper.

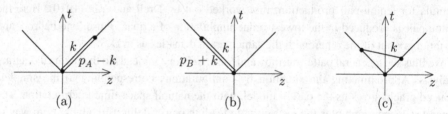

Fig. 5.24. Space-time description of PSSs of Fig. 5.22. For all three plots, the scale for the separation of the vertices is Q/λ^2, where λ is the radial integration variable in (5.29) for a collinear region, but λ^2/Q is the radial variable in (5.49) for the soft region.

in other model theories. Generally, I will assume that the external quark lines are on-shell, equipped with Dirac wave functions as appropriate.

5.4.1 Geometry and topology of PSSs

Useful insights are obtained from each of several ways of examining the PSSs: in terms of reduced graphs (Fig. 5.22), in terms of PSSs' locations in the space of loop momenta (Fig. 5.23), and in terms of the locations of the graph's vertices in space-time (Fig. 5.24).

The criterion, of a classically allowed process in the massless limit, gives the following PSSs, which I label by the nature of the gluon's momentum, (R_A, R_B, etc.[3]):

1. **Gluon collinear to A**: We label this PSS R_A. It has two massless on-shell lines: k and $p_{A,\infty} - k$, each parallel to $p_{A,\infty}$:

$$R_A : \begin{cases} k = zp_{A,\infty}, \\ p_{A,\infty} - k = (1 - z)p_{A,\infty}, \end{cases} \tag{5.18}$$

with z between 0 and 1. The line $p_B + k$ has virtuality of order Q^2: $(p_{B,\infty} + k)^2 = Q^2 z$.

In the reduced graph, Fig. 5.22(a), the far off-shell line $p_B + k$ is contracted with the current vertex to form a composite reduced vertex. Out of this come two massless on-shell momenta in the $p_{A,\infty}$ direction, which later combine to make a single massless on-shell momentum $p_{A,\infty}$.

The momentum fraction variable z must be between 0 and 1, since other values of z do not give a classical scattering configuration. For example, if z is negative, the quark goes out to the future from the current vertex, but the gluon comes in from the past. Thus they are unable to meet at the recombination point if $z < 0$.

2. **Gluon collinear to B**: This PSS, labeled R_B, with reduced graph Fig. 5.22(b), is exactly like the first PSS, but with the roles of the quark lines exchanged:

$$R_B : \begin{cases} -k = zp_{B,\infty}, \\ p_{B,\infty} + k = (1 - z)p_{B,\infty}. \end{cases} \tag{5.19}$$

3. **Soft gluon**: k has zero momentum on this PSS, which we call R_S. Its reduced graph is Fig. 5.22(c), and the quark lines have massless momenta $p_{A,\infty}$ and $p_{B,\infty}$. The quark and antiquark come out of the electromagnetic vertex and a soft gluon is exchanged. This is a rather special case of the Landau-Coleman-Norton criterion.

4. **Soft quark**: Here it is the internal quark instead of the gluon that is soft. Since the gluon now has a maximal collinear momentum $k = p_{A,\infty}$, we label this region $R_{A'}$.

5. **Soft antiquark**: Here the internal *antiquark* is soft, and the gluon has $k = -p_{B,\infty}$. The PSS's label is $R_{B'}$.

The locations of the PSSs in loop-momentum space are shown in Fig. 5.23, from which can be seen some topological relations between the different PSSs. For example, R_S is at the intersection of R_A and R_B, while $R_{A'}$ is an endpoint of R_A. When we derive factorization theorems, we will find contributions and approximations associated with each PSS. The topological relations between different PSSs will determine subtractions that prevent double counting between different contributions. There will also be a contribution from the region R_H where all internal lines are far off-shell. We therefore will speak about regions; intuitively a region connotes a particular part of loop-momentum space. But as a precise mathematical notion we will use the PSSs supplemented by the hard region R_H. The intuitive notion of a region means, roughly, momenta near the corresponding PSS.

---

[3] The subscripts should not be confused with the same symbols used to denote the various subgraphs of a reduced graph.

To formalize the relations between regions we first define a manifold for each PSS:

Name	Manifold	Dimension
R_S	$\{k = 0\}$	0
$R_{A'}$	$\{k = p_{A,\infty}\}$	0
$R_{B'}$	$\{k = -p_{B,\infty}\}$	0
R_A	$\{k = zp_{A,\infty} : 0 < z < 1\}$	1
R_B	$\{k = -zp_{B,\infty} : 0 < z < 1\}$	1
R_H	$\{$all k such that $k \notin R_A, R_B, R_S, R_{A'}, R_{B'}\}$	4

(5.20)

Each manifold *excludes* the manifolds for smaller PSSs. For example, in the regions R_A and R_B we exclude the point $z = 0$, i.e., $k = 0$, since this does not give a collinear gluon momentum.

There is evidently a hierarchy of sizes of region:

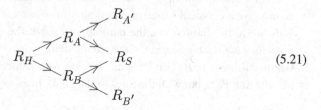

(5.21)

where the biggest region is on the left. A formal definition of the hierarchy is not by simple set-theoretic inclusion, since the manifolds for smaller regions are not part of those for the bigger ones. Instead we define the hierarchy in terms of the topological closures $\bar{R}$ of the manifolds R for the various regions. For example, $\bar{R}_A = \{k = zp_{A,\infty} : 0 \leq z \leq 1\}$, with the endpoints at $z = 0$ and $z = 1$ included. Then we define the statement that a PSS R_1 is bigger than a PSS R_2, $R_1 > R_2$, to mean that $\bar{R}_1 \supset \bar{R}_2$.

For the actual graph with massive propagators, and possibly off-shell external quarks, we have already argued that there are important contributions from momenta close to the PSSs. This suggests a coordinate-space interpretation in terms of the relative positions of the vertices. For example, near the PSS R_A, the upper quark line $p_B + k$ has virtuality of order Q^2, and therefore the vertices at its ends are separated by order $1/Q$. The other two lines, k and $p_A - k$, have low virtuality, so the invariant separation of their ends is much larger than $1/Q$. Moreover, the lines are highly boosted in the $+z$ direction. This gives typical locations for the vertices as shown in Fig. 5.24(a), which corresponds closely to the classical scattering picture given by the Coleman-Norton criterion. Corresponding situations for the PSSs R_B and R_S are also shown in Fig. 5.24(b) and (c). The arguments just given are quite heuristic, and it is left as an exercise to derive them more formally (problem 5.1).

5.4.2 Pinch- and non-pinch-singular surfaces: collinear-to-A

PSS R_A was restricted to $k = zp_{A,\infty}$ with z between 0 and 1. But the massless limit of the integrand in (5.17) is singular for any value of z; it is the criterion of a pinch that restricts z,

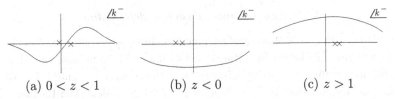

(a) $0 < z < 1$ (b) $z < 0$ (c) $z > 1$

Fig. 5.25. The k^- plane, showing the singularities for the lines k and $p_A - k$ for the three cases $0 < z < 1$, $z < 0$, and $z > 1$, together with appropriate choices of contour. The scale of the diagram is roughly $(k_T^2 + m^2)/p_A^+$; the pole for the $p_B + k$ line is far off-scale, at $k^- = -O(Q^2/p_A^+)$.

as we now verify explicitly. We use light-front coordinates, as is natural for collinear PSSs, to give

$$G_1 = \frac{ig^2}{(2\pi)^n} \int d^n k \frac{\text{numerator}}{[2(p_B^- + k^-)(zp_A^+ + p_B^+) - k_T^2 - m_q^2 + i0]}$$

$$\times \frac{1}{(2zp_A^+ k^- - k_T^2 - m_g^2 + i0)\,[2(1-z)p_A^+(p_A^- - k^-) - k_T^2 - m_q^2 + i0]}. \quad (5.22)$$

Here we wrote $k^+ = zp_A^+$, so that $d^n k = dz\, p_A^+\, dk^-\, d^{n-2}k_T$.

In the following discussion, there are order-of-magnitude estimates for denominators, and it is convenient to use the symbol m as a generic size for all masses in the problem.

To understand the R_A region, we choose k_T to be much less than Q, and we examine the contour integral for k^-. In the center-of-mass frame, the large components of external momenta are p_A^+ and p_B^-, of order Q, while the small components, p_A^- and p_B^+, are of order m^2/Q. The poles on the collinear lines k and $p_A - k$ are at small values of $|k^-|$, of order $(k_T^2 + m^2)/Q$, and, when $0 < z < 1$, they are on opposite sides of the real axis, trapping the contour, as in Fig. 5.25(a). In contrast, the remaining pole, from the $p_B + k$ line, is much further away, at $k^- \simeq -p_B^- = -O(Q)$, corresponding to the line's large virtuality in the R_A region.

Naturally, when z approaches 0 or 1, the accuracy of this argument degrades. For example, the separation of the poles in k^- is of order

$$\frac{k_T^2 + m^2}{p_A^+}\left(\frac{1}{z} + \frac{1}{1-z}\right), \quad (5.23)$$

and this gets large close to the endpoints of R_A, i.e., near the R_S and $R_{A'}$ regions. This formula also exhibits the exact pinch in the massless limit. That is, when $m = 0$, the minimum distance between the poles is zero, obtained at $k_T = 0$.

Outside the PSS region, i.e., for z below 0 or above 1, the two collinear denominators are on the same side of the real axis: Fig. 5.25(b) and (c). Then we can deform k^- to be of order Q, so that all the denominators are of order Q^2, i.e., the momenta are in the hard region. Note that we cannot deform the contour all the way to infinity, to give a zero integral, because of the singularity on the $p_B + k$ line.

5.4.3 *Multidimensional contour deformation*

For one variable, like k^-, the analysis of the pinch condition is straightforward, because the contour deformation is visualizable. But the actual integral is multidimensional, and thus hard to visualize. Is there a cunning deformation of the contour in z and/or k_T that would allow the four-complex-dimensional[4] contour to avoid the poles? The Landau criterion asserts in complete generality that this cannot be done.

A devil's advocate would search for a proof in the literature that the Landau equations are both necessary and sufficient for a PSS, and would be rewarded by *not* finding a published explicit and complete proof. Textbook treatments, when examined closely, are incomplete. For example, in the authoritative book on analyticity properties in QFT, by Eden *et al.* (1966), we read (p. 48): "A proper proof needs the use of topology; ... We shall be content with plausibility arguments." The reference given for a real proof is an *unpublished* paper, by Fotiadi, Froissart, Lascoux, and Pham; the paper, as far as I can find out, is still unpublished forty years later. Devil's advocates are recommended to investigate further (problem 5.3); there is something in this subject that is not fully understood.

I now present some techniques to help formalize issues about contour deformation in the general case, with the momentum integral for L loops having nL dimensions. The aim is to make very transparent the concepts that relate exact PSSs in the massless theory to properties of actual integrals with non-zero masses but large Q.

First we write the loop momentum in terms of real and imaginary parts:

$$k = k_R + i\kappa \, k_I(k_R). \tag{5.24}$$

Here a contour deformation is characterized by increasing the real parameter κ from 0 to 1, with each point on the contour labeled by its (nL-dimensional) real part k_R. The imaginary part is some function of the real part, and naturally $d^{nL}k$ includes a Jacobian for the transformation between k and k_R. An allowed contour deformation is one for which no poles are crossed in going from $\kappa = 0$ to $\kappa = 1$. We also require a uniform upper bound on the derivatives $\partial k_{Ia}/\partial k_{Rb}$, so that the Jacobian stays finite; otherwise, an arbitrarily large size for Jacobian would ruin our derivation of power-counting. Thus in a one-dimensional contour integral we might require the deformed contour to have an angle of at most $45°$ to the real axis. The precise bound does not matter, but having an angle close to $90°$ would give a very big Jacobian.

Next consider a denominator $D(k) + i0$ at a zero of $D(k)$. Our aim is to determine whether this denominator participates in a pinch at this value of momentum, or whether the contour of k can be deformed away. We avoid the corresponding pole if D acquires a positive imaginary part when κ becomes slightly positive, i.e., if

$$k_I \cdot \frac{\partial D}{\partial k} > 0 \quad \text{pole avoidance criterion} \tag{5.25}$$

at the zero of D. We have an exact pinch if, no matter what choice we make for k_I, (5.25) fails for at least one of the on-shell lines.

---

[4] Or $4 - 2\epsilon$-dimensional contour.

The criterion just stated applies to determining whether there is an exact pinch. In our context, the PSSs we are cataloging are those of the massless theory. But our use of these PSSs is also in the massive theory, where we are concerned not with whether or not there is an exact pinch, but with whether or not the integration contour is forced to be close to particular propagator poles. So we now ask: What are the appropriate criteria for avoiding or not avoiding poles in the massive theory?

We do not consider a particular pole to be avoided unless the minimum value of $|D(k)|$ on the deformed contour is of order Q^2 in a whole neighborhood of some candidate for a PSS. The neighborhood should be of a size of order Q in the components of loop momentum k_R. Now all the momentum components of interest are at most of order Q, and similarly for the derivatives $\partial D/\partial k$. For the denominator to be of order Q^2 when the real part of k is at a zero of $D(k)$, it must be true that the imaginary part has a component of order Q.

It also follows that the first-order term in an expansion in powers of k_I, i.e., the l.h.s. of (5.25), must itself be of order Q^2. Otherwise the first derivative would change sign near our initially chosen k_R, since the second derivative is of order unity, and then we would find places where, as we deform the contour, the denominator gets a negative imaginary part. Because of the limit on the gradient of k_I with respect to k_R, the pole avoidance condition (5.25) is obeyed, not just exactly at the PSS, but in a neighborhood. It also follows that the component of $\partial D/\partial k$ in the direction k_I is of order Q.

In the example of the two collinear denominators for region R_A of the vertex graph, the derivatives are

$$\frac{\partial(k^2 - m_g^2)}{\partial k} = 2k \simeq 2zp_{A,\infty}, \quad \frac{\partial((p_A - k)^2 - m^2)}{\partial k} \simeq -2(1 - z)p_{A,\infty}. \tag{5.26}$$

On PSS R_A, these two vectors are opposite in direction, so that the pole avoidance criterion (5.25) cannot be simultaneously satisfied by both denominators. The exact PSS is in the massless theory, but small changes in the pole positions, to allow for masses, do not break this argument. As just explained, any contour deformation that successfully avoids a singularity has to work over a large neighborhood of the propagator poles. If we tried deforming another component of k than k^-, its imaginary part would multiply a small derivative on the l.h.s. of (5.25), and would not make this l.h.s. of order Q^2.

In contrast, when we extrapolate the PSS to $z < 0$ or to $z > 1$, the two derivative vectors have the same direction. Therefore if we choose k_I to give one denominator a large positive imaginary part, then the other denominator also gets an imaginary part of the same sign. Thus we can avoid the pole. Since $k_I \cdot p_{A,\infty} = k_I^- p_{A,\infty}^+$, it is the minus component of k_I that needs to be made large to avoid the pole; this again justifies our choice to examine contour integration only over k^-. Therefore the singular surfaces at $z < 0$ and $z > 1$ are not PSSs.

5.5 Power-counting for vertex graph

I next use the one-loop vertex graph to motivate the primary tools for power-counting. In addition, we will encounter the so-called Glauber region of gluon momenta. Glauber momenta form a subset of soft momenta, but require a different treatment than generic soft momenta; in particular standard factorization is only obtained after a contour deformation away from momenta in the Glauber region.

For the power-counting, I will usually set the space-time dimension to $n = 4$. But to discuss properties of regulated integrals, I will sometimes change to $n = 4 - 2\epsilon$.

Characteristic differences between QFTs are controlled by the numerator factor in (5.17), and we can see the spectrum of possibilities from specific examples:

- A ϕ^3-type theory where both the quarks and gluons are scalar fields, and the vertex for the electromagnetic current is replaced by one for a ϕ^2 operator. It gives a numerator factor of unity.
- A Yukawa theory with a scalar "gluon" and fermionic quarks. It gives a numerator $\bar{u}_A[\gamma \cdot (p_A - k) + m_q] \gamma^\mu [-\gamma \cdot (p_B + k) + m_q] v_B$, where u_A and v_B are Dirac wave functions.
- A gauge theory with fermion quarks. The numerator factor is

$$\bar{u}_A \gamma^\kappa [\gamma \cdot (p_A - k) + m_q] \gamma^\mu [-\gamma \cdot (p_B + k) + m_q] \gamma^\lambda v_B N_{\kappa\lambda}. \tag{5.27}$$

In Feynman gauge, the gluon part of numerator is $N_{\kappa\lambda} = -g_{\kappa\lambda}$.

(Further cases are left as an exercise; problem 5.6.) In addition, we will examine how the power laws change with the dimension of space-time.

Our main interest is in the size and power law of the loop graph *relative to the lowest-order* graph. For the ϕ^3 theory, the lowest-order graph is unity, but for the other two theories, the lowest-order graph is of order Q, since the largest component of a Dirac wave function grows like $Q^{1/2}$.

5.5.1 Hard region R_H: power corresponds to UV divergence

In region R_H, all momentum components are of order Q and all virtualities are of order Q^2. As we found around (5.6), the power of Q is given by dimensional analysis, and is the same as for UV divergences. Thus in ϕ^3 theory at $n = 4$, region R_H's contribution to the vertex graph is of order $1/Q^2$. In Yukawa and gauge theories, which are renormalizable, the numerators provide factors of Q^2 times Dirac wave functions, so the contribution is of the same power as the lowest-order vertex, and we call R_H a leading region. Of course, if we increased the space-time dimension to 6 in ϕ^3 theory we also get leading behavior. These arguments apply after UV renormalization, provided we apply an RG transformation to set the renormalization scale μ of order Q.

In any of the renormalizable theories, we therefore write the contribution of region R_H as

$$G_1 \text{ in } R_H = O(1) \times \text{LO}. \tag{5.28}$$

This simply means that we have a bound. That is, for large Q/m, the size of this contribution is less than some constant number times the lowest-order graph. In QCD (for example), a useful bound is the product of

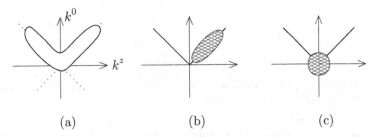

Fig. 5.26. (a) Integration domain used for region H; it excludes the blanked-out area around the PSSs. The size of the regions shown is a modest factor less than Q. This diagram should be treated as having two more dimensions perpendicular to the ones shown. (b) Integration domain used for region A, the cross-hatched area. (c) Integration domain used for region S.

- a factor of a few, from the approximations on the denominators and from the multiple terms in the Dirac algebra;
- a factor $g^2/(16\pi^4)$ explicitly in the Feynman rules; and
- π^2 from an angular integral in four space-time dimensions.

This gives a modest factor times $g^2/16\pi^2$. In principle there could be cancellations, since the sign and complex phase of the integrand are not fixed. But, in general, if such cancellations occur frequently and are strong, we should expect this to have a specific cause.

The integration domain for an actual numerical estimate should be like that in Fig. 5.26(a). Here we cut out pieces surrounding each of the (smaller) PSSs, perhaps of size $Q/2$. The precise positions of the borders will not bother us. But we must insist that a contour deformation is applied to stay away from all propagator poles where there is not a PSS. For example, suppose k is close to a *negative* number times $p_{A,\infty}$. Without the contour deformation, we would have two low-virtuality denominators, which falsifies the derivation of the estimate. A convenient way of interpreting Fig. 5.26(a) is to treat the variables plotted there as the real parts k_R. Imaginary parts, as in (5.24), give denominators of order Q^2, for example from the contour deformation in Fig. 5.25(b).

5.5.2 Basic treatment of collinear region R_A

Next we integrate around the PSS for region R_A, Fig. 5.26(b), excluding neighborhoods of the smaller PSSs, R_S and $R_{A'}$. The dimensionless variable z parameterizes the PSS; we call it an intrinsic variable for the PSS. At fixed z, consider the integral over k^- and k_T, which parameterize the deviation from the PSS, and which we therefore term normal variables for the PSS. Near the PSS the momentum $p_B + k$ is off-shell by approximately zQ^2. On the other hand, the momenta k and $p_A - k$, which we call collinear, are approximately parallel to p_A.

To understand the integral's behavior near the PSS as an example of a general case, we change to a set of dimensionless variables $\bar{k}$ parameterizing a surface surrounding the PSS, together with a radial variable λ with the units of mass that scales this surface and is chosen

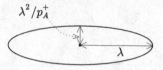

Fig. 5.27. Surface of fixed λ surrounding a collinear PSS. The surface is drawn asymmetrically, to correspond to the scalings defined in (5.29).

to lie in the range $0 \le \lambda \le Q$. Observe that the collinear denominators are quadratic in k_T but linear in k^-. So we choose different scalings for k^- and k_T:

$$\text{Collinear to } A: \qquad k^- = \lambda^2 \overline{k}^- / p_A^+, \quad k_T = \lambda \overline{k_T}, \tag{5.29}$$

as illustrated in Fig. 5.27. These variables should be thought of as generalized polar coordinates, with $\overline{k}$ being treated as two-dimensional angular variables. The definition is non-unique, and we can specify it by giving λ as a function of k^- and k_T:

$$\lambda = f(|k^-|, |k_T|, p_A^+), \tag{5.30}$$

of a form consistent with the scaling law (5.29). I choose

$$f(|k^-|, |k_T|, p_A^+) = \sqrt{|p_A^+ k^-| + |k_T|^2}. \tag{5.31}$$

Such a definition is not Lorentz invariant, but is intended to be applied in a natural frame for the process, which is the center of mass. I have arranged for the definition to be invariant under z boosts, and for the angular variables $\overline{k}^-$ and $\overline{k}_T$ to be dimensionless. Given (5.29) and (5.31), the angular variables satisfy the normalization condition $|\overline{k}^-| + |\overline{k_T}|^2 = 1$.

To understand the size of the integrand, and the consequent power-counting, we examine the dependence on λ. In each collinear denominator there are terms of order λ^2 and of order m^2, e.g., $-\lambda^2 \times 2(1-z)\overline{k}^- - \lambda^2 \overline{k}_T^2 + p_A^2(1-z) - m^2$ for $p_A - k$. Since the angular variables parameterize a (two-dimensional) surface surrounding a point on the PSS, they cover over a finite range independent of λ, and only one of $p_A^+ \overline{k}^-$ and $\overline{k}_T$ can go to zero simultaneously. Thus in estimating sizes, we write the collinear denominators as $\lambda^2 O(1) + m^2 O(1)$, where "$O(1)$" denotes a quantity that goes over a finite range, never approaching infinity.

However, this is not sufficient to obtain a result for the integral. The problem is that the argument so far only gives us an upper bound on the denominators, and the denominators can and do get arbitrarily small. Thus for the integral itself we cannot directly deduce an upper bound. But we can limit the closest approach to the poles by applying a contour deformation like that in Fig. 5.25(a), where the separation of the poles is given by (5.23). On the deformed contour there is a minimum size for each denominator, and a minimum size for k^-, for a given value of k_T.

Now the definition of λ in (5.30) was deliberately written with absolute values of the momentum components. Thus it can be applied on the deformed contour, and the integration over the purely real-valued radial variable λ can be regarded as a slicing of the k integral. We now find that on the deformed contour we can always treat the denominator as being of order $\lambda^2 + m^2$, but in a much stricter sense. The size of each collinear denominator obeys $C_1(\lambda^2 + m^2) < |\text{denom.}| < C_2(\lambda^2 + m^2)$, where C_1 and C_2 are two constants with C_1 strictly non-zero and C_2 finite. These bounds apply uniformly for all values of $\overline{k}$ on the contour and for all relevant values of λ. We could use separate bounds for the λ^2 and m^2 terms, but we would not gain anything useful.

There is in fact a notation for this which has become standard in some areas, and which is defined in App. A.17:

$$|\text{collinear denominator}| = \Theta(\lambda^2 + m^2). \tag{5.32}$$

The use of $\Theta(\lambda^2 + m^2)$ instead of $O(\lambda^2 + m^2)$ indicates that we have a lower as well as an upper bound, so that we can deduce a similar result also for the inverse

$$\left| \frac{1}{\text{collinear denominator}} \right| = \Theta\left(\frac{1}{\lambda^2 + m^2} \right). \tag{5.33}$$

This lets us obtain the power law associated with the R_A region. We have the following sizes, in the sense of the Θ notation:

- $1/Q^2$ for the far off-shell denominator;
- $d\lambda\, \lambda^3$ for the radial integration;
- $1/(\lambda^2 + m^2)$ for each of the two collinear denominators;
- unity for the integral over the angular variables $\overline{k}$;
- a numerator factor.

First, we ignore the numerator, and provide an estimate for the ϕ^3 theory:

$$R_A \text{ region} = \frac{g^2}{Q^2} \int_0^Q \frac{d\lambda\, \lambda^3}{\Theta(\lambda^2 + m^2)^2} \tag{5.34}$$

$$= O\left(\frac{g^2 \ln(Q^2/m^2)}{Q^2} \right). \tag{5.35}$$

Since the integrand has a variable complex phase, there is a possibility of a cancellation, so that we must use the symbol $O(\ldots)$ rather than $\Theta(\ldots)$ for our estimate of the integral.

From (5.34), we see that for *large* λ, of order Q, the estimate matches our result $1/Q^2$ for the hard region R_H in ϕ^3 theory. For small λ, when m is set to zero, we get a logarithmic (collinear) divergence at $\lambda = 0$, i.e., the degree of collinear divergence is zero. This symptomizes two properties of the actual massive integral: (a) for λ of order m, we get the same size as in the hard region R_H; (b) there there is a logarithmic enhancement from the region $m \ll \lambda \ll Q$. This is an example of a general result, that if the two regions have the same power law, then there is a logarithmic enhancement from the integral between the extremes, with the exponent of the power being unchanged.

If we change the space-time dimension from 4 to n, the power for $\lambda \sim m$ is changed to $g^2 m^{n-4}/Q^2$. Thus in ϕ^3 theory, i.e., without the numerator factor, the collinear region always has a $1/Q^2$ suppression independent of space-time dimension; i.e., this region is never leading. There is a contribution from the hard region of order $g^2 Q^{n-6}$.

5.5.3 *Where are the vertices in space-time?*

We did not associate the space-time picture of a classical space-time process at a PSS with any specific distance scale. We now remedy this defect. The argument is sketchy, and making a more detailed argument is left for problem 5.1.

It is reasonable that the typical time between two ends of a line of a Feynman graph is the inverse of the deviation of its energy from being on-shell:

$$\Delta t \sim \frac{1}{|k^0 - E_k|} = \frac{|k^0 + E_k|}{|k^2 - m^2|} \sim \frac{E_k}{|k^2 - m^2|}. \tag{5.36}$$

Naturally, we assume that the integration contour has been deformed as far away as possible from propagator poles.

For a collinear line, with its momentum scaled as in (5.29), we find a time of order Q/λ^2. This can be interpreted as a time $1/\lambda$ in the rest-frame of the collinear system multiplied by a time-dilation factor Q/λ. The boost argument shows that this is also the separation of the vertices in x^+, and that the separation in the other light-front coordinate x^- is of order $1/Q$, the same as the size of the hard scattering. The separation in transverse position is invariant under a boost in the z direction, and is therefore $1/\lambda$.

This therefore gives a scale for the drawings in Figs. 5.24(a) and (b), and for their generalizations to higher-order graphs.

One caveat is needed. When λ becomes less than the quark and gluon masses, the virtuality of the lines remains of order m^2 instead of scaling down like λ^2, so we should really equip the estimate with a minimum:

$$\Delta t \sim \frac{Q}{\max(\lambda^2, m^2)}, \tag{5.37}$$

from the pole separation value given in (5.23). Naturally, if both the quark and gluon have zero mass, then the time scale goes to infinity as λ goes to zero; this corresponds to the actual collinear divergence in the massless case.

5.5.4 *Collinear region boosted from rest frame*

We now consider a general case of a collinear subgraph, and more generally a non-perturbative amplitude for a collinear subgraph, as in the lower bubble in Fig. 2.5(b) for the parton model for DIS. We can regard a collinear subgraph or amplitude as being obtained by a boost from its rest frame. We always define a collinear subgraph to include all its attached collinear lines and the integral over all the small components of the collinear momenta.

For scalar fields, a collinear subgraph is boost invariant. Thus the collinear subgraph counts as Q^0, and the power law for the whole graph is just that for the hard part of the graph, i.e., $1/Q^2$ in our one-loop example, independent of the space-time dimension.

For a field with spin s, the biggest component of a matrix element of its field grows like $(Q/m)^s$ under a boost to energy of order Q from a rest frame associated with mass m. This gives enhancements that we now investigate. We will find them to be particularly notable for the exchange of a field of the highest spin, i.e., for the gluon.

5.5.5 *Yukawa theory, region* R_A

First we examine the on-shell electromagnetic vertex of a fermion in Yukawa theory. The Dirac wave functions for a spin-$\frac{1}{2}$ fermion grows like $Q^{1/2}$ in the center-of-mass frame, so the tree-graph amplitude grows like Q.

For the one-loop graph in the Yukawa theory in a collinear region, the boost argument of Sec. 5.5.4 shows that the same $Q^{1/2}$ growth applies to the whole collinear subgraph (lines k and $p_A - k$) as to the Dirac wave function. Thus the power of Q for the whole graph in the R_A region is given by the off-shell propagator $i[-\gamma \cdot (p_B + k) + m]/[(p_B + k)^2 - m^2]$. This now has dimension -1, so it contributes $1/Q$, and we get a power suppression.

From the overall numerator factor, $[\gamma \cdot (p_A - k) + m_q] \gamma^\mu [-\gamma \cdot (p_B + k) + m_q]$, this is not quite so obvious, since it contains two factors with momentum components of order Q. These might compensate the $1/Q^2$ suppression from the $p_B + k$ denominator. But the large part of the $p_A - k$ numerator can be eliminated by the equations of motion for a Dirac spinor:

$$\bar{u}_A \gamma^- (p_A^+ - k^+) = (1 - z)\bar{u}_A \gamma^- p_A^+ = (1 - z)\bar{u}_A(m - \gamma^+ p_A^-). \tag{5.38}$$

The boost argument shows that this is part of a general result, not an accident of a one-loop calculation.

5.5.6 *Gauge theory, region* R_A

The situation changes when the exchanged line is for a vector field, as in QCD. The collinear part of the graph is proportional to

$$\int d^4k \; \frac{\bar{u}_A \gamma^\kappa [\gamma \cdot (p_A - k) + m_q]\gamma^\mu}{(k^2 - m_g^2 + i0)[(p_A - k)^2 - m_q^2 + i0]}. \tag{5.39}$$

Under a boost, the $\kappa = +$ component gains a factor of order Q relative to the size in a Yukawa theory; this removes the $1/Q$ suppression from the off-shell $p_B + k$ line. The gluon collinear region is therefore leading, independently of the space-time dimension. The same leading power applies to any graph in which arbitrarily many gluons go from a collinear subgraph to a hard subgraph. This immediately implies that substantial modifications are needed to the derivation of even the elementary parton model. Instead of considering graphs like Fig. 2.5(b), we must allow extra gluon exchanges to the hard subgraph, as in Fig. 5.7(c).

The resulting complications are tamed, as we will see in later chapters, by noticing that the enhancement is associated with the one component, $\kappa = +$, of the gluon field that scales like Q/m under the boost to the collinear-to-A direction. In (5.27), the dominant part of the gluon numerator is N_{+-}. This dominance can be eliminated by a gauge transformation, e.g., by a suitable choice of axial gauge $n \cdot A = 0$, for which the gluon numerator is

$$N_{\kappa\lambda}^{\text{axial-gauge}} = -g_{\kappa\lambda} + \frac{k_\kappa n_\lambda + n_\kappa k_\lambda}{k \cdot n} - \frac{k_\kappa k_\lambda n^2}{(k \cdot n)^2}. \tag{5.40}$$

If we choose n at rest in the center-of-mass, say $n \propto p_A + p_B$, the all-important N_{+-} component is $-k^- k^+ n^2/(k \cdot n)^2$. It is readily checked that when k is a collinear momentum, this is of order m^2/Q^2; the contribution of region R_A in this gauge is thus suppressed by two powers of Q.

Another common choice is the light-front gauge: $n \cdot A = A^+ = 0$; in that case N_{+-} is exactly zero. However, this gauge is not symmetric between p_A and p_B, so that it causes difficulties in a general treatment. Even the non-light-like case, with $n^2 \neq 0$, is not adequate for our later work, because the singularity at $k \cdot n = 0$ breaks standard analyticity rules for propagators that are needed in proofs of factorization; see Ch. 14.

Therefore we will generally stay in the Feynman gauge, with the implication that regions with collinear gluon exchange, such as region R_A, will be leading. However, the fact that these regions can be made non-leading by a certain choice of gauge, implies that important simplifications can be made by the use of Ward identities.

We can see the basic idea of the argument by the following chain of approximations for the numerator. We consider a general situation in which one gluon connects a collinear-to-A subgraph to a hard subgraph:

$$\text{collinear-}A^\kappa \, N_{\kappa\lambda} \, \text{hard}^\lambda \simeq \text{collinear-}A^\kappa \, N_{\kappa -} \, \text{hard}^-$$

$$= \text{collinear-}A^\kappa \, N_{\kappa -} \, \frac{1}{k^+} \, k^+ \text{hard}^-$$

$$\simeq \text{collinear-}A^\kappa \, N_{\kappa -} \, \frac{1}{k^+} \, k \cdot \text{hard}. \tag{5.41}$$

All the approximations are accurate at the leading power of Q. In the first line, we replaced the hard subgraph by its minus component, that dominates in the contraction with the collinear-to-A subgraph. Then we multiplied and divided by k^+, which allows us in the last line to replace $k^+ \ldots$ by $k \cdot \ldots$ for the gluon connecting to the hard scattering, accurate to the leading power of Q. Having k contracted with the hard subgraph is exactly of the form to which a Ward identity applies. This method was obtained by generalizing the argument of Grammer and Yennie (1973) that was devised for treating IR divergences in QED.

5.5.7 *Effect of different degree of divergence*

The above calculations exhibit some quite general phenomena in the estimation of the sizes of the contributions of different regions. For each PSS, we parameterize the approach to the PSS by a radial variable λ. The general structure of the momentum-space integrands for Feynman graphs is of products of very simple rational functions. This generally gives a power-law behavior in λ as $\lambda \to 0$, with a cutoff provided by masses.

Because the power-law dependence gives useful order-of-magnitude estimates all the way from $\lambda = 0$ to $\lambda = Q$, we can now obtain some interesting relations between the power laws for different regions. The basic general form of the size of the contribution from a region is

$$\frac{1}{Q^\alpha} \int_0^Q \frac{d\lambda \, \lambda^\beta}{\Theta(\lambda^2 + m^2)^\gamma}, \tag{5.42}$$

where we now allow general exponents. Situations with nested leading regions often require us to modify these estimates by logarithmic corrections from integrals over the angular variables, but this will not change the basic power laws and exponents. See p. 115 for an explanation of the Θ notation.

For order-of-magnitude estimates we have a power-law integral $d\lambda \, \lambda^{\beta-2\gamma}$ cutoff at the lower end by mass effects. Let us define the infra-red degree of divergence in the massless limit by $\Delta = 2\gamma - \beta - 1$.

A common, but not universal, situation in QCD and other theories in four-dimensional space-time is that we have a logarithmic divergence, $\Delta = 0$. Then, as we have seen, the total contribution has the same power $1/Q^\alpha$ as the contribution from the hard region, i.e., from $\lambda \sim Q$, but there is a logarithmic enhancement. If the integrand is modified by a logarithm, then the number of logarithms increases by one after integration, e.g.,

$$\frac{1}{Q^\alpha} \int_0^Q \frac{d\lambda \, \lambda^{2\gamma-1}}{\Theta(\lambda^2 + m^2)^\gamma} \ln^\delta\left(\frac{Q}{\lambda}\right) = O\left(\frac{\ln^{\delta+1}(Q/m)}{Q^\alpha}\right). \tag{5.43}$$

In this situation all scales between m and Q are important.

In contrast, if we have a power-law divergence $\Delta > 0$, then the lower end of the integral, $\lambda \sim m$, dominates, and the power there is $1/Q^\alpha$ (times $O(1/m^\Delta)$). The power from the hard region $\lambda \sim Q$ is weaker: $1/Q^{\alpha+\Delta}$. From a UV-centric point of view, we can say that in this situation there are power-law enhancements as we go from large to small momenta. Alternatively we can take an IR-centric view: momenta near the IR scale dominate, and there is a convergent extrapolation of the integral to infinite λ. This situation is typical in a model super-renormalizable QFT in a space-time of dimension less than 4.

The reverse holds if Δ is negative. In that case the hard region $\lambda \sim Q$ dominates and we can legitimately neglect masses.

In all cases, the power law for the region for the PSS at $\lambda = 0$ is $1/Q^\alpha$ and the power for the hard region is $1/Q^{\alpha+\Delta}$, with the proviso that we may have logarithmic enhancement(s) associated with IR degree of divergence zero.

5.5.8 Soft-gluon region R_S

For the soft-gluon region R_S we integrate over a domain like that in Fig. 5.26(c) that surrounds R_S (a single point in this case). To parameterize the approach to R_S, we use the same scaling for all components of k:

$$k^\mu = \lambda_S \bar{k}^\mu. \tag{5.44}$$

Again the radial variable λ_S has the dimensions of mass and is specified by a (non-Lorentz-covariant) function

$$\lambda_S = f_S(k^\mu), \tag{5.45}$$

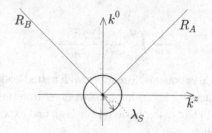

Fig. 5.28. Surface of fixed λ_S surrounding the soft PSS R_S. In contrast to the collinear case, Fig. 5.27, we have the same scaling on all components of k. The diagonal lines are the *soft ends* of the PSSs R_A and R_B.

with an appropriate scaling property. We choose

$$f_S(k^\mu) = \sum_\mu |k^\mu|. \tag{5.46}$$

(Three-dimensional) surfaces of fixed λ_S surround the point $k = 0$, which is the PSS for R_S (Fig. 5.28). From (5.46), the angular variables are normalized to $\sum_\mu |\bar{k}^\mu| = 1$.

Many interesting complications in perturbative QCD arise from soft gluons and their couplings to collinear subgraphs. One is simply that soft gluon connected to collinear subgraphs give leading-power contributions. As we will see in Sec. 5.5.10, another complication arises because soft loop momenta circulate through collinear subgraphs, so that the power-counting for λ_S depends non-trivially on properties of the collinear subgraphs and the relative sizes of the components of soft momenta.

We first derive a basic scaling argument for the integral near the PSS R_S for the one-loop vertex graph. It applies for generic values of the angular variables $\bar{k}$, i.e., when any considered combination of the components of $\bar{k}$ is of order unity. Later, in Sec. 5.5.10, we will consider the relatively small Glauber region, where the argument needs to be changed. For the generic case:

1. The integration measure is $d\lambda_S \, \lambda_S^{n-1} \, d^{n-1}\bar{k}$, which gives a power λ_S^n, where n is the dimension of space-time.
2. The gluon denominator $k^2 - m_g^2$ is $\lambda_S^2 \bar{k}^2 - m_g^2$, i.e., its size is $O(\lambda_S^2 + m^2)$. In the massless limit, or when m_g is negligible, this is simply $O(\lambda_S^2)$. The gluon mass becomes important when λ_S is around m_g.
3. The lower quark denominator is

$$(p_A - k)^2 - m_q^2 = p_A^2 - m_q^2 - 2p_A \cdot \bar{k}\lambda_S + \lambda_S^2 \bar{k}^2. \tag{5.47}$$

Since we treat all the components of k as comparable, the biggest k-dependent term is $-2p_A^+ \bar{k}^- \lambda_S$, so that the denominator is $O(\lambda_S Q + m^2)$. In the massless limit, the dominant term is $-2p_A^+ \bar{k}^- \lambda_S$, i.e., $O(\lambda_S Q)$.
4. The upper antiquark denominator is treated similarly, with its dominant part in the massless limit being $2p_B^- \bar{k}^+ \lambda_S$, also $O(\lambda_S Q)$.

As regards the massless limit, this gives an overall result of order

$$\int d\lambda_S \, \lambda_S^{n-1} \frac{1}{\lambda_S^2} \frac{1}{(\lambda_S Q)^2} \times \text{numerator} = \int_0^{\sim Q} d\lambda_S \, \lambda_S^{n-5} Q^{-2} \times \text{numerator}. \quad (5.48)$$

When we set $n = 4$, the physical space-time dimension, and restore the mass cutoff, we find a logarithmic enhancement multiplying the explicit power $1/Q^2$.

The dependence on the spin of the soft line is rather interesting. The boost argument of Sec. 5.5.4 shows that the numerator gains a factor of Q^{2s}, where s is the spin of the exchanged soft line. This is an enhancement relative to the power obtained for coupling the collinear graphs to the hard subgraphs. Hence:

- For the case of both ϕ^3 and Yukawa theory, the exchanged gluon is a scalar. Therefore the explicit power $1/Q^2$ in (5.48) shows that the region R_S gives a non-leading power.
- For a vector gluon, the boost argument shows that the case $\kappa = +$ and $\lambda = -$ gives an extra factor of Q^2. So the soft region R_S is leading, independently of the space-time dimension n.

We have now seen that in a gauge theory all of the regions R_A, R_B, and R_S for the vertex graph are of leading power. In contrast, none is leading in theories without a gauge field. The remaining regions $R_{A'}$ and $R_{B'}$ are always non-leading. In the absence of a vector field, only the hard region R_H could be leading. Hence a large number of complications in the parton physics of QCD result from QCD being a gauge theory.

5.5.9 Where are the vertices in space-time for the soft region?

Although the virtualities are different for soft and collinear lines (λ_S^2 and $\lambda_S Q$ respectively), both kinds of line give the same time scale $1/\lambda_S$ in the center-of-mass frame. This arises from time dilation of the collinear lines, and can be deduced from (5.36).

When we work with more complicated regions, it is useful for the time scale to match the one in (5.37) for the collinear region. So we define λ by $\lambda_S = \lambda^2/Q$, so that

$$\text{Soft:} \qquad k^\mu = \frac{\lambda^2}{Q} \bar{k}^\mu. \qquad (5.49)$$

Then the time scale is the same as for the collinear region, i.e., Q/λ^2, to the extent that we neglect masses. It is naturally appropriate to use the λ as a redefined radial variable for the soft region.

The effect of masses is different for collinear and soft momenta. For the collinear case, masses give a lower cutoff of m on λ. For the soft region, this also applies to the *quark* mass. But the gluon mass implies a more stringent cutoff, at $\lambda \sim \sqrt{m_g Q}$. So for the soft region we replace (5.37) by

$$\Delta t \sim \frac{Q}{\max(\lambda^2, m_g Q, m^2)}. \qquad (5.50)$$

Of course this makes no difference if the gluon is massless. But in real QCD there is some kind of non-perturbative infra-red cutoff due to confinement, so in real QCD physics m_g in the above equation should be replaced by Λ.

Even so, the two widely different scales of cutoff indicate that when we go to higher-order diagrams there can be complications. It will turn out that most of these will be avoided after we use Ward identities to sum over different ways of attaching soft lines to collinear subgraphs. Moreover the non-perturbative cutoff does not apply directly to Feynman graphs, so there will be some interesting issues in the leading regions and their interpretation in Feynman graphs with massless gluons and massive quarks, that will involve us with regions that are not really physical.

5.5.10 "Glauber" region

Just as with the collinear regions, there are certain parts of the integration over angular variables $\bar{k}$ where denominators get much smaller than the estimates used above. Again we need to investigate to what extent contour deformation can rescue them, but the conclusions will now be less trivial. The necessary contour deformations work for some situations like the vertex graph, but fail for others.

This issue does not concern only the determination of the power law associated with the soft region. More importantly, it gives a danger of violating the Grammer-Yennie approximation that is essential in deriving factorization, by allowing us to apply Ward identities to the sum over soft gluon connections to collinear subgraphs. The approximation is a simple generalization of (5.41):

$$(\text{coll. } A)^\kappa N_{\kappa\lambda} (\text{coll. } B)^\lambda \simeq (\text{coll. } A)^+ N^{-+} (\text{coll. } B)^-$$

$$\simeq (\text{coll. } A) \cdot k \, \frac{N^{-+}}{k^+ k^-} \, k \cdot (\text{coll. } B). \tag{5.51}$$

Here, our aim is a formula in which the gluon momentum k is contracted with each collinear factor, so that we can apply Ward identities. The critical step is in the second line, where we use the following approximations that are valid to the leading power of Q if the components of k are not too much different: $k \cdot (\text{coll. } A) \simeq k^-(\text{coll. } A)^+$ and $(\text{coll. } B) \cdot k \simeq (\text{coll. } B)^- k^+$.

When these approximations are valid, we will find that in our actual applications further approximations of k in the collinear factors are useful and valid: to replace k inside the collinear-B part by its plus component and to replace k inside the collinear-A part by its minus component.

These approximations rely on all components of k being comparable. Thus one or more of the approximations fails when k^- and/or k^+ gets too small with respect to the other components. By examining the relative sizes of components of collinear momenta, we find that the approximations are accurate under the following conditions:

$$\frac{m^2}{(p_B^-)^2} \ll \left| \frac{k^+}{k^-} \right| \ll \frac{(p_A^+)^2}{m^2}, \tag{5.52}$$

$$\left| \frac{k^+}{k_T} \right| \gg \frac{m}{p_B^-}, \qquad \left| \frac{k^-}{k_T} \right| \gg \frac{m}{p_A^+}. \tag{5.53}$$

The first line simply states that the rapidity of the gluon must be well inside the range between the collinear rapidities, which is essentially the simplest definition of the soft region. The conditions on the second line are that the longitudinal components of k should not be too much smaller than the transverse momentum. Where the approximations and standard power-counting hold for the soft region, we deduce that

$$|k^- k^+| \gg k_T^2 \frac{m^2}{Q^2}. \tag{5.54}$$

We now ask when the conditions fail. If the failure is only of the conditions on the rapidity of k, that simply takes us to one of the collinear regions; this does not concern us here since we treat the collinear regions separately. However, a failure of (5.54) is problematic. When this condition fails, we have $|k^+ k^-| \ll k_T^2$. This puts k in a region called the Glauber region (Bodwin, Brodsky, and Lepage, 1981) in view of its importance in final-state interactions in high-energy scattering. The same region was also termed the "Coulomb region" in Collins and Sterman (1981).

In the case we are currently treating, the vertex function, we can perform a contour deformation on either or both of k^+ and k^- to get out of the Glauber region. Consider first the k^+ integral in the Glauber region. We can neglect k^+ compared with p_A^+ in the $p_A - k$ denominator; this is generally true when k is soft. We can also neglect k^+ in the gluon denominator, specifically because of the Glauber-region condition $|k^+ k^-| \ll k_T^2$. This leaves the denominator $(p_B + k)^2 - m^2 + i0 \simeq 2p_B^- k^+ - k_T^2 + p_B^2 - m^2 + i0$, and we can therefore deform k^+ into the upper half plane. Similarly we can deform k^- into the lower half plane.

The limits of the deformation on $k^\pm$ are given by other poles, notably that of the gluon. The deformed contour no longer goes through the Glauber region. So on the deformed contour in the soft region, the standard power-counting and the Grammer-Yennie approximation are valid. However, the denominators in the Grammer-Yennie approximation give extra singularities at $k^+ = 0$ and $k^- = 0$, i.e., in the Glauber region close to the poles on the quark propagators. Thus the denominators must be equipped with $i0$ prescriptions that do not block the contour deformation:

$$(\text{coll. } A)^\kappa \, N_{\kappa\lambda} \, (\text{coll. } B)^\lambda \simeq (\text{coll. } A) \cdot k \, \frac{1}{k^- - i0} \, N^{-+} \, \frac{1}{k^+ + i0} \, k \cdot (\text{coll. } B). \tag{5.55}$$

In the previous paragraphs, there is a change of the kind of pole avoidance under discussion compared with the earlier part of this chapter. Initially, we viewed momenta relative to the large scale Q, and determined whether or not momentum components were forced go through regions where they are much smaller than Q. Now we are examining a soft momentum, of size $\lambda_S \ll Q$, and are determining whether or not its plus and/or minus components are forced to go through regions where they are much smaller than λ_S.

Although we derived it only for the one-loop graph, the contour deformation applies very generally to avoid the Glauber region in our process. Consider a general reduced graph (Fig. 5.29) for the vertex, and let k be a momentum flowing down on a soft line from the upper collinear graph B. We know that the flow of minus momentum in the B

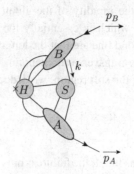

Fig. 5.29. Reduced graph for vertex.

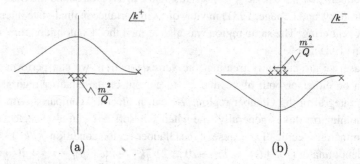

(a)   (b)

Fig. 5.30. Contour deformations out of Glauber region for (a) k^+, (b) k^-. The crosses near the origin are final-state Glauber-region poles in collinear subgraphs. The crosses near the edges are other poles that limit the contour deformation.

subgraph is all towards the future, from the hard subgraph H to the final-state particle p_B. There must be a sequence of lines in B that gets to the vertex with line k from H by going forward with the flow of the minus component of momentum. We can choose to set up k as a loop momentum that goes along these lines, and completes its loop through H and A.

If k^+ is small enough for k to be in the Glauber region, then the only important dependence on k^+ is in B. Since k goes with the flow of collinear minus momentum, all the nearby poles are in the lower half plane, as in

$$\frac{1}{(k_B + k)^2 - m^2 + i0} \simeq \frac{1}{2k_B^- k^+ - D + i0}. \tag{5.56}$$

Here k_B is a generic collinear momentum on a line of subgraph B, and D does not depend on k^+. Thus the same contour deformation into the upper half plane works as for the one-loop graph. A similar argument applies to a Glauber momentum attaching to the A subgraph.

This situation is illustrated in Fig. 5.30, and we characterize it by saying that all singularities in subgraphs A and B are in the final state; the lines in A and B all go out to the future from the hard scattering. To see a direct relation to a space-time picture, we simply Fourier-transform (5.56) into coordinate space, with the soft-gluon approximation that only the k^+ dependence of the propagator is retained:

$$f(x) = \int \frac{d^4 k}{(2\pi)^4} \frac{e^{-ik \cdot x}}{2k_B^- k^+ - D + i0} \simeq \frac{-i}{2k_B^-} \delta(x^+) \delta^{(2)}(x_T) \, \theta(x^-) e^{-ix^- D/(2k_B^-)}. \quad (5.57)$$

When $x^- < 0$, we get zero, because the integrand decreases rapidly to zero in the upper half plane of k^+, so that we can close the k^+ contour in the upper half plane. But when $x^- > 0$ we close the contour in the lower half plane and pick up the residue of the pole. More generally, for any function whose singularities in k^+ are only in the lower half plane, its Fourier transform is non-zero only for positive values of x^-. This is so general a property that the contour-deformation result applies beyond perturbation theory.

The delta functions in (5.57) show that, from the point of view of a soft gluon, the collinear subgraph is a line going to the future in a light-like direction from the hard scattering, so that the soft gluon does not resolve any internal structure of the collinear system.

5.5.11 Soft-quark regions, $R_{A'}$ and $R_{B'}$

The remaining PSSs for the one-loop vertex graph are $R_{A'}$ and $R_{B'}$ where one of the fermion lines is soft. Power-counting like that for the soft-gluon case, R_S, gives a suppression by at least one power of Q. Our general treatment, Sec. 5.8, will show that this happens because one end of a soft-quark line is at the hard subgraph instead of a collinear subgraph.

5.6 Which reactions have a pinch in the Glauber region?

For the vertex graph, the ability to deform out of the Glauber region is tied to the collinear lines all being final-state lines. We now ask for situations in which we cannot perform this deformation. This requires reactions in which both initial-state and final-state collinear lines are present. See Ch. 14 for some of the resulting complications. The reduced-diagram technique enables us to diagnose these cases very readily, and in fact we already have a supply of interesting examples.

Reactions for hadron production in $e^+ e^-$ annihilation via a single virtual photon will always have the hadrons in the final state. Hence these reactions are always safe from the Glauber region.

For DIS (Figs. 5.13 and 5.14) the jets are always outgoing, so contour deformation out of the Glauber region is possible for k^-. Target-collinear lines can be in both the initial and final state (Fig. 5.13(b)) so k^+ is trapped. But to avoid the Glauber region, it turns out to be sufficient that a deformation can be made on k^- (Collins, 1998b; Collins and Metz, 2004). This applies equally to variations on DIS, like deeply virtual Compton scattering and exclusive meson production in DIS.

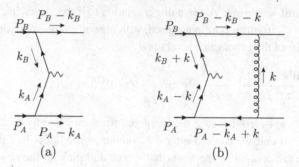

Fig. 5.31. (a) Simple Feynman graph for DY process. (b) The same with addition of a gluon exchanged between the spectator lines; the gluon's momentum is trapped in the Glauber region.

5.6.1 Remnant-remnant interactions in Drell-Yan

The situation changes for the Drell-Yan (DY) process,[5] since the initial state has two oppositely moving hadrons, and the final state contains the beam remnants (Fig. 5.17).

Physically, what happens can be seen in the microscopic view of a scattering reaction in Fig. 5.18. One parton out of each hadron collides at the short-distance hard interaction indicated by the star. The transverse separation of these two active partons is of order $1/Q$, corresponding to the scale of the hard collision. Inside the hadrons, partons are spread out over a transverse area proportional to r^2, where $r \simeq 1$ fm is the size of a hadron. The transverse area is not changed under a boost. The probability that a pair of partons is within $1/Q$ of each other in the transverse direction is therefore proportional to $1/(Qr)^2$, which corresponds to a hard-scattering cross section decreasing with $1/Q^2$ at large Q.

But when the active partons collide, the remnants of the two hadrons overlap, and can therefore interact. Remnant-remnant interactions of small momentum transfer occur with high probability, since such hadronic interactions are strong. One direct manifestation is that the total hadron-hadron cross section is of order r^2 (Amsler *et al.*, 2008). Thus we know experimentally that interactions happen with high probability whenever the impact parameter of a pair of hadrons is less than about r. The strong remnant interactions involve momentum exchanges in the Glauber region.

5.6.2 Glauber pinch in momentum space

We now verify from an example that spectator-spectator interactions are trapped in the Glauber region for the Drell-Yan process, and that they give a leading power. In Fig. 5.31 are shown two graphs for the Drell-Yan amplitude when the beam particles are modeled by elementary particles. In both graphs, each beam particle splits into a quark-antiquark pair. A quark out of one beam annihilates with an antiquark out of the other to make a high-mass

---

[5] And generally for hard processes in hadron-hadron collisions.

virtual photon. Graph (a) gives an example of pure parton-model physics, but graph (b) has a gluon exchanged between the beam remnants, and I will show that the gluon is trapped in the Glauber region.

The value of graph (a) is

$$-\bar{u}_B \, \Gamma_B \, \frac{-\not{k}_B + m}{k_B^2 - m^2 + i0} \, \gamma^\mu \, \frac{\not{k}_A + m}{k_A - m^2 + i0} \, \Gamma_A \, v_A. \tag{5.58}$$

Here u_B and v_A are the Dirac wave functions for the final-state fermions. The matrices Γ_B and Γ_A give the coupling between the beam particles and quarks. We choose the kinematic region where the fermions are prototypically collinear, with transverse momenta of order m, as is appropriate for the parton model. The large components of k_B and k_A are determined by the virtual photon momentum (5.15), so that

$$k_A = (x_A P_A^+, 0, \mathbf{0}_{\mathrm{T}}) + \big(O(m^2/Q), O(m^2/Q), O(m)\big), \tag{5.59}$$

$$k_B = (0, x_B P_B^-, \mathbf{0}_{\mathrm{T}}) + \big(O(m^2/Q), O(m^2/Q), O(m)\big). \tag{5.60}$$

Graph (b) gives

$$ig^2 \int \frac{d^4 k}{(2\pi)^4} \, \frac{1}{k^2 - m_g^2 + i0} \, \bar{u}_B \, \gamma^- \, \frac{\not{P}_B - \not{k}_B - \not{k} + m}{(P_B - k_B - k)^2 - m^2 + i0}$$

$$\times \Gamma_B \, \frac{-\not{k}_B - \not{k} + m}{(k_B + k)^2 - m^2 + i0} \, \gamma^\mu \, \frac{\not{k}_A - \not{k} + m}{(k_A - k)^2 - m^2 + i0}$$

$$\times \Gamma_A \, \frac{-\not{P}_A + \not{k}_A - \not{k} + m}{(P_A - k_A + k)^2 - m^2 + i0} \, \gamma^+ \, v_A, \tag{5.61}$$

where the gluon couplings are replaced by their dominant minus and plus components. The gluon has transverse momentum of order the usual radial variable λ_S for the soft PSS, and the most characteristic value to model non-perturbative hadronic interactions is $\lambda_S \sim m$.

We first make approximations that are always valid when the gluon is soft, independently of whether it is in the Glauber subregion. So we neglect k^- with respect to k_B^- in the collinear-to-B denominators, and similarly for k^+ in the collinear-to-A denominators. Thus

$$(k_B + k)^2 - m^2 + i0 \simeq 2(k^+ + k_B^+)k_B^- - (k_{\mathrm{T}} + k_{B\mathrm{T}})^2 - m^2 + i0$$

$$= 2k^+ k_B^- + k_B^2 - m^2 - 2k_{\mathrm{T}} \cdot k_{B\mathrm{T}} - k_{\mathrm{T}}^2 + i0$$

$$= 2k^+ k_B^- + O(m^2, m\lambda_S, \lambda_S^2) + i0. \tag{5.62}$$

This approximation needs the assumption that all components of k are much less than Q, but it needs no assumption on the relative sizes of the components.

If k were in the generic part of the soft region we could further approximate by noting that $k^+ k_B^-$ would be of order $\lambda_S Q$, so that

$$(k_B + k)^2 - m^2 + i0 \simeq 2k^+ k_B^- + k_B^2 - m^2 + i0. \quad (k \text{ not Glauber}) \tag{5.63}$$

But this further approximation fails in the Glauber region, $|k^+ k^-| \ll k_{\mathrm{T}}^2$.

The relevant part of the integral (5.61) now becomes

$$
\int \frac{dk^+ \, dk^-}{(2\pi)^2} \frac{\text{numerator}}{2k^+k^- - k_T^2 - m_g^2 + i0}
$$

$$
\times \frac{1}{[-2k^+(P_B^- - k_B^-) + \ldots + i0]\,[2k^+k_B^- + \ldots + i0]} \tag{5.64}
$$

$$
\times \frac{1}{[-2k^-k_A^+ + \ldots + i0]\,[2k^-(P_A^+ - k_A^+) + \ldots + i0]},
$$

where the terms indicated by "..." are independent of k^+ and k^-, and are of order $m^2, m\lambda_S, \lambda_S^2$. In the Glauber region, $|k^+k^-| \ll |k_T|^2$, only the poles on the collinear lines are relevant. We see immediately that k^+ and k^- are trapped there, with $k^\pm = O(m^2, m\lambda_S, \lambda_S^2)/Q$, to be compared to $k_T = O(\lambda_S)$.

The dominant contribution is in fact where $\lambda_S = O(m)$. Smaller values are cut off by the gluon mass, while there are enough powers of k_T^2 in the denominators to suppress larger values, given our assumption about the collinear kinematics of k_A and k_B.

The asymmetric sizes, $k^\pm = O(m^2/Q)$ and $k_T = O(m)$, correspond to the momentum exchanged in small-angle elastic scattering. They are therefore natural values for spectator-spectator interactions. The sizes of $k^\pm$ correspond to the small components of collinear momenta.

To obtain the power law in Q, we compute the size of the graph compared with the basic graph, Fig. 5.31(a). The extra Glauber gluon brings in the following powers:

- integration measure: m^6/Q^2, from the sizes of $k^\pm$ and k_T;
- three denominators each of order $1/m^2$;
- a numerator of order Q^2 because the gluon is a vector particle.

This is independent of Q, with the numerator canceling the small range of $k^\pm$. If the space-time dimension is changed from $n = 4$, we still get the same power of Q. The basic graph, Fig. 5.31(a), has the power-counting of the parton model, which we use to define the leading power for the process. Therefore, there is an unsuppressed contribution from Glauber corrections. This result is unchanged if we make the collinear subgraphs arbitrarily complicated.

5.6.3 Generalized Landau-equation analysis for Glauber region

The actual integrals for Feynman graphs are in a high dimension. So, as in the elementary association between regions and massless PSSs, one can ask whether there is a possibility of an unforeseen exotic deformation in the high-dimensional complex space, and one can ask for a general characterization of Glauber regions. In a one-loop example, it was sufficient to visualize the relevant one-dimensional contour integrals. I now give an appropriate argument, generalized from the Libby-Sterman method.

In the first part of this chapter, we scaled all momentum components with Q. From this, we showed that integration momenta are trapped at small virtualities in the vicinity of exact

PSSs in the massless limit. The Landau method determined the locations of the PSSs quite generally.

To determine the existence or non-existence of a Glauber pinch, we generalize this strategy. We devise a scaling such that a trapping in a Glauber region corresponds to an exact pinch in a certain limit. Then we use a variation of the Landau analysis to locate the exact pinches systematically.

First I show that an exact Glauber pinch occurs when we replace the collinear denominators by just the terms of the form that are the non-dotted terms in (5.64). These terms are given by taking the asymptotics of large Q while holding the overall size of the soft momentum fixed at order λ_S, and treating the collinear scaling factor as λ_S. (Thus the transverse parts of collinear momenta are treated as order λ_S.) For this limit we also require that λ_S is of order m or bigger. Asymptotically, the propagator of the soft line remains unaltered, but the collinear denominators are simplified, so that they are just a factor of k^+ or k^- times a large component of the collinear momentum, e.g.,

$$\frac{1}{k^2 - m_g^2 + i0} \times \frac{1}{[-2k^+(P_B^- - k_B^-) + i0]\,[2k^+k_B^- + i0]}$$

$$\times \frac{1}{[-2k^-k_A^+ + i0]\,[2k^-(P_A^+ - k_A^+) + i0]}. \tag{5.65}$$

The trapping of $k^\pm$ at $k^\pm \ll \lambda_S$ has now become an exact pinch at $k^\pm = 0$. The on-shell condition for the collinear-to-B propagators is $k^+ = 0$, and for the collinear-to-A propagators is $k^- = 0$. In the chosen scaling limit, the on-shell conditions apply independently of $\mathbf{k}_T$, which represents a significant change from the standard Landau analysis. At the singularities, at $k^+ = 0$ and/or $k^- = 0$, the gluon denominator is non-zero, so the gluon line counts as part of a vertex of a reduced graph for this analysis: it is a hard subgraph relative to the collinear propagators.

To determine allowed directions of contour deformation, we need derivatives of the collinear propagators, as in (5.25). The derivatives of the collinear denominators are now exactly light-like directions. In space-time, these correspond to propagation along a light-like line, as in (5.57). For example, the collinear-to-B lines give

$$\frac{\partial D(P_B - k_B - k)}{\partial k^\mu} \longrightarrow \begin{pmatrix} -2(P_B^- - k_B^-) \\ 0 \\ \mathbf{0}_T \end{pmatrix}, \tag{5.66}$$

$$\frac{\partial D(k_B + k)}{\partial k^\mu} \longrightarrow \begin{pmatrix} 2k_B^- \\ 0 \\ \mathbf{0}_T \end{pmatrix}. \tag{5.67}$$

We have used column vectors for the derivatives, to distinguish them from the row vectors we use for normal contravariant momentum vectors. In the asymptotic limit these vectors are opposite in direction, so that when we apply a contour deformation, as in (5.24), the imaginary parts generated by the deformation are opposite; the deformation fails.

Applying this analysis in general shows that the general Glauber-pinch configuration is like having one or more extra hard scatterings (of the spectator collinear lines). The condition for a classical scattering applies, and the only change with respect to the standard hard-scattering case is the much lower momentum transfer.

5.7 Coordinates for a PSS

We now resume our general analysis. So far, we have used the Landau-equation/reduced-diagram method to locate PSSs; this led to a catalog of important momentum regions. We next formalize and systematize the variables we use for a general treatment, after giving a general characterization of the class of problems we address.

For each PSS R we will define "intrinsic coordinates", which parameterize location on the PSS itself, and normal coordinates, which parameterize deviations off the PSS. The normal coordinates are required to be zero on the PSS. From the normal coordinates, we will define a radial coordinate λ_R, with the dimensions of mass, to give a notion of distance from the PSS. Then we will define what we term angular coordinates to parameterize surfaces of fixed λ_R surrounding the PSS.

This gives us a language, which lets us perform power-counting in Sec. 5.8, to determine which PSSs are leading. These results then support all the later work in this book.

For any of the reactions that we discuss, there is an intimidating multiplicity of regions, and this comes from a genuine complexity: there are infinitely many graphs, and high-order graphs have high-dimensional loop integrations, with a large number of leading regions. In QCD, unadorned low-order perturbative calculations are not adequate for estimating cross sections, except in very few cases, as in Ch. 4. So, to get a useful and productive analysis of the behavior of some amplitude or cross section, we need general methods that do not require detailed analysis of individual graphs.[6]

The general strategy is essentially a recursive divide-and-conquer. We discuss each leading region separately, and arrange to analyze it in terms of diagrammatic decompositions such as Fig. 5.17. By our choice of coordinates, the analysis of a general region can be visualized by a diagram that appeared in one of our examples, Fig. 5.28. At the end, it will (perhaps) be evident that there are structures here that go beyond the perturbatively based situations in which we derive them.

5.7.1 Relations between regions

The key elements of a general discussion are the geometrical and topological relations between different regions, as in (5.21) and in Fig. 5.28. We take a particular point on some PSS R for a graph, and examine a neighborhood, parameterized by a radial variable λ_R.

- Some propagators are off-shell at the PSS. For these, the effect of varying λ_R is suppressed by a power of λ_R/Q, and the denominators have a fixed order of magnitude.
- Denominators of the other propagators go to zero when λ_R and masses go to zero.

---

[6] But motivations can be obtained by analyzing suitable low-order graphs.

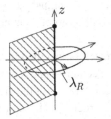

Fig. 5.32. Representation of line/surface of constant λ surrounding a PSS R at a particular value of the intrinsic coordinate(s), together with the relation to bigger and smaller PSSs. See the text for details.

- At a generic point around a surface of fixed λ_R, we perform elementary power-counting for the order of magnitude of the graph at R.
- But close to certain submanifolds of a fixed λ_R surface, some denominators get much smaller than the power-counting estimate. The location of these submanifolds will be obtained in Sec. 5.10 by iterating the Libby-Sterman analysis. With certain exceptions, each such submanifold corresponds to the intersection of the surface of fixed λ with the PSS for another region R' larger than the first one.

The general situation is illustrated in Fig. 5.32. The thick vertical line represents the PSS R, and there may be smaller PSSs, represented by the dots at the ends of R. There may be one or more larger PSSs, exemplified by the shaded plane at the left of the figure. Surrounding R, at a fixed value of the intrinsic coordinate(s), is a line of constant λ_R. The integration contour, and therefore Fig. 5.32, is deformed in the space of complex momenta to avoid non-pinch singularities.

In the figure the dimension of R is one, while the dimensions of the smaller and larger PSSs are zero and two. But in general, R may have any dimension from zero for a soft-gluon region in a one-loop vertex to a very high dimension in a multiloop graph, with appropriate ranges for the smaller and bigger PSSs.

- There are exceptions to the rule that, in the integration over angular variables, intersections with larger PSSs determine the locations where the integrand gets much smaller than the standard for R. These are typified by the Glauber region we met in Secs. 5.5.10 and 5.6. In processes without a Glauber pinch, we do not have to worry about the exceptions.
- After the intrinsic coordinates for R are integrated over, the integration includes smaller PSSs, and we need to mesh the analysis of R with the analysis of the smaller PSSs.
- Factorization theorems generalizing the parton model are obtained by expanding in powers of λ_R about a PSS, and then (typically) taking the leading power. The previous items will tell us how to modify this analysis to deal with multiple regions.

5.7.2 *Formulation of problem*

We denote by $G(p_1, \ldots p_n; q_1, \ldots; m, \mu, a_s(\mu))$ the Green function, amplitude or cross section to be treated. It depends on external momenta $p_1, p_2, \ldots; q_1, \ldots$ We divide these into two classes, to be defined below, distinguished by the letters p and q. This generalizes the usage in DIS, where p is the target momentum and q is the virtual photon momentum.

The asymptotic behavior to be treated is specified by a scalar variable Q that gets large. We work in a particular frame like the center-of-mass frame for the DY process, or the Breit frame for DIS. We call this the reference frame for the process. In this frame all the external momenta have some components of order Q. The p_j momenta have fixed masses, while the q_j momenta have invariant sizes of order Q^2. [Here we mean $\Theta(Q^2)$ in the notation of App. A.17.] The q_js are typically fixed vectors proportional to Q, or are obtained from such a vector by at most a finite, bounded boost from the reference frame. We also rule out the trivial but irrelevant case of giving a common large boost to a set of fixed momenta p_j. A Lorentz-invariant characterization is as follows.

1. Define scaled momenta by $\tilde{p}_i^\mu = p_i^\mu / Q$, $\tilde{q}_i^\mu = q_i^\mu / Q$.
2. We take Q large (i.e., much larger than particle masses) with each of the scaled external momenta smoothly approaching a fixed limit as $Q \to \infty$.
3. The limit of each $\tilde{p}_i$ is a light-like vector, and the limit of each $\tilde{q}_i$ is a non-light-like vector.
4. From the light-like limit vectors, we construct a set of unscaled light-like momenta $p_{A,\infty}$, $p_{B,\infty}$, etc., as in our examples, e.g., $p_{A,\infty} = Q \lim_{Q \to \infty} \tilde{p}_A$. Associated with each collinear subgraph is one such light-like momentum, which we will call the reference momentum for the subgraph. At the PSS, the momenta of the lines of the collinear subgraph are proportional to its reference momentum.
5. At least one of the Lorentz invariants $q_i \cdot q_j$, $q_i \cdot p_j$, and $p_i \cdot p_j$ increases like Q^2 as $Q \to \infty$; none increases more rapidly.

Since this is intended to be a universal characterization, the following caveats apply.

- Some of the limiting light-like vectors may be proportional to each other. This is the case, for example, for the momenta p and p' in the DVCS process. So we just pick one of these to be in the set of $p_{A,\infty}$, etc.
- Certain minor variations on the theme are also covered; for example:
  - In the Drell-Yan cross section, the transverse momentum may range from very small to order Q; it may also be integrated over. The key point for the asymptotic analysis is that the invariants q^2, $p_A \cdot q$, $p_B \cdot q$, and $p_A \cdot p_B$ are all of order Q^2.
  - Some quark and hadron masses may be large, of order Q or bigger.
- There may be no need for the q_j momenta. This is the case for high-energy elastic scattering at wide angle, where the momenta of the external particles are sufficient to specify the process. The previously stated principles tell us to define $Q = \sqrt{s}$, up to some constant factor.
- We take G to be connected. A disconnected amplitude can always be discussed in terms of its connected components.

A more serious complication is when the invariants have a range of sizes. A typical and important case is DIS at small x, when $p \cdot q \propto Q^2/x \gg Q^2$. Another case would be high-energy elastic scattering at small angle, where $|t| \ll s$. A complete discussion of such situations requires a generalization of our analysis.

5.7.3 Intrinsic and normal coordinates

We now show how to define intrinsic and normal coordinates for a PSS. These generalize our earlier examples.

For the collinear-to-A PSS of the vertex graph, we used $z = k^+/p_A^+$ as the sole intrinsic coordinate, in (5.18), with k^- and k_T as the normal coordinates. The smallest PSSs, R_S, $R_{A'}$ and $R_{B'}$, were just points. Thus they had no intrinsic coordinates, while suitable normal coordinates were 4-momentum deviations from the PSS, e.g., k for R_S.

Naturally, the choice for these coordinates is non-unique. But certain general guidelines apply. Each coordinate system is particularly useful in a neighborhood of its own PSS. But it must apply to the whole of loop-momentum space, or at the very least to a large region of size of order Q including the PSS. The transformation from the local coordinates around the PSS to ordinary momentum variables must be analytic, certainly near the PSS *and its smaller PSSs*. The intrinsic coordinates extend uniquely beyond the boundaries of their PSS. Without this requirement, artificial coordinate singularities would complicate all our discussions.

Each line in a *reduced* diagram for a PSS in a massless theory has a momentum parallel to one of the light-like limit momenta, or is zero. For the collinear lines we choose intrinsic coordinates as fractional momenta, each with respect to the light-like limit momentum, e.g., $p_{A,\infty}$, of its collinear subgraph. The remaining intrinsic variables are the hard loop momenta. Now each PSS is a segment of a flat hyperplane in loop-momentum space. So with the definitions just given, the intrinsic coordinates of a PSS extend simply and naturally to the whole of the hyperplanes, beyond the boundaries of the regions where there is a pinch. Similarly we take the normal coordinates to be ordinary linear coordinates in momentum space. Thus there is a unique natural extension of the coordinates to the whole of loop-momentum space. (Our treatment of collinear regions for the vertex graph illustrated this.)

5.7.4 Radial coordinates

We obtain the power-counting for a PSS from the integral over a radial coordinate λ, for which we now present a suitable definition. We choose λ to have the dimensions of mass.

To make the definition, we split the normal coordinates into two sets. One set consists of soft loop momenta circulating through soft and possibly some collinear and hard subgraphs. The other set consists of collinear loop momenta each circulating through a particular collinear subgraph and possibly through hard subgraph(s).

We will write each individual normal component as a power of λ times a dimensionless angular variable and a possible Q-dependent normalizing factor, as in (5.29) and (5.49), with a chosen normalization condition on the angular variables.

General collinear momenta

We specify the scalings for a collinear momentum exactly as for a collinear region for the vertex graph, but we need to define a light-front coordinate system separately for each collinear subgraph.

Let k be a collinear momentum in a particular collinear subgraph, and let p_∞ be the light-like reference momentum for the subgraph. We define time and spatial parts of vectors in the reference frame (e.g., center-of-mass) of the process as a whole. Plus and minus coordinates relative to p_∞ are

$$k^\pm \overset{\text{def}}{=} \frac{1}{\sqrt{2}}(k^0 \pm n \cdot k). \tag{5.68}$$

Here n is a unit vector for the spatial direction of p_∞. Then the transverse momentum is

$$k_{\mathrm{T}} \overset{\text{def}}{=} k - \frac{k^+}{\sqrt{2}}(1, n) - \frac{k^-}{\sqrt{2}}(1, -n), \tag{5.69}$$

where the representations of vectors are in normal time-space coordinates, in the reference frame.

Then the scalings of k are defined by exactly (5.29), with p_A^+ replaced by $p_\infty^+ \propto Q$; that is, a scaling with λ^2 for k^- and a scaling λ for k_{T}.

Note that a covariant specification of the plus and minus coordinates needs *two* 4-vectors. In effect, we have taken these as the light-like reference vector for the collinear subgraph, and the rest vector of the overall reference frame for the process.

Soft momenta

As in Sec. 5.5.9, we define the scaling for soft momenta by

$$k_S = \frac{\lambda^2}{Q} \overline{k}_S. \tag{5.70}$$

Thus the power-counting of a soft momentum flowing through a collinear subgraph is the same as the smallest component of a collinear momentum, and the time scales of the soft and collinear lines are the same.

Normalization condition

A possible normalization condition on the angular variables is

$$\sum_{\text{collinear } k} \left(|\overline{k^-}| + |\overline{k_{\mathrm{T}}}|^2 \right) + \sum_{\text{soft } k} \sum_\mu |\overline{k}^\mu|, \tag{5.71}$$

which generalizes (5.31) and (5.46), with suitable homogeneity properties under rescaling of λ.

5.8 Power-counting

A basic issue in analyzing processes of the kind described in Sec. 5.7.2 is to understand the general size of the cross section or amplitude. The primary complication is that propagator denominators vary widely in size in the integration over loop momenta. To handle this issue, we use the language of PSSs in the massless theory.

In this section, for each PSS R, we categorize by a power of Q the contribution from integration over a neighborhood of R. We will identify those PSSs that give the leading power for the processes we consider. For deriving factorization, we normally only retain the leading power, e.g., for DIS the power Q^0, which corresponds to Bjorken scaling. The restriction to the leading power is important because PSSs with a non-leading power often have a much more complicated structure than those for the leading power.

In deriving the power laws, we will see that logarithmic enhancements arise from integrations between different nested regions. But logarithms do not affect the utility of dropping non-leading power terms.

The derivation involves estimates of the sizes of propagator denominators near a PSS. In later chapters a consequence will be the construction of an appropriate approximation for each leading PSS. These approximations enable the derivation of useful factorization theorems.

We will vary terminology between "PSS" and "region". Precise formulations use PSSs in the massless theory. But we talk about a region, rather than the associated PSS, when we wish to emphasize, for example, that the associated power of Q concerns the contribution from a neighborhood of the PSS in the real theory.

The present formulation originates in the work of Sterman (1996), but with improvements, closely following the treatment of Collins, Frankfurt, and Strikman (1997). This treatment relies on general properties of dimensional analysis and of Lorentz transformations rather than on a detailed analysis of the numbers of loops, lines and vertices of graphs and subgraphs. Using such general properties, in particular the transformation of collinear subgraphs under large boosts, gives the results a validity beyond strict perturbation theory. Although much of the treatment concerns Feynman graphs, the collinear and soft factors should really be non-perturbative.

Much earlier work used an axial gauge (e.g., $A^0 = 0$, $A^3 = 0$, or $A^+ = 0$) or the Coulomb gauge. However, the unphysical singularities in the gluon propagator for such "physical gauges" prevent us from using contour deformation arguments. Thus we prefer to work in a covariant gauge – see the discussion of the Glauber region in Sec. 5.5.10, where unphysical singularities in physical gauges would have obstructed a contour deformation out of the Glauber region.

Therefore we normally use a covariant gauge, like the Feynman gauge. The price is that leading regions (e.g., Fig. 5.7(c) for DIS) have arbitrarily many extra gluons joining the collinear and hard subgraphs. But these gluons have a particular "scalar" polarization for which Ward identities apply to convert the sum over all possibilities to a factorized form.

5.8.1 Comments on power of Q and dimensions

A danger in formulating general results is that one misses nuances of particular cases. Consider the simplest general statement of the leading power of Q, that it corresponds

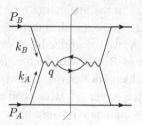

Fig. 5.33. Elementary contribution to Drell-Yan with single spectator.

to the dimension of the cross section or amplitude under consideration, as in (5.5). This rule is indeed correct for the DIS structure functions F_1 or F_2, with their power Q^0 that corresponds to Bjorken scaling. But modifications are needed for certain other cases. For example, in Sec. 5.8.2 we will find that a different power is needed for the Drell-Yan cross section, in the case that the transverse momentum q_T is much less than Q.

The culprit is a delta function for transverse-momentum conservation. Essentially the derivations of power-counting are at their most straightforward when applied to ordinary functions, not to delta functions.

In general, a reliable strategy for dealing with such issues is to start by analyzing very simple graphs for the process under consideration, e.g., graphs such as Fig. 5.20 that gives the parton model for the Drell-Yan process, or, better still, just the lowest-order case, Fig. 5.33. A general region for the process can have more-complicated region subgraphs, and can have more lines joining the subgraphs. The changes *relative* to the simplest graph are robustly handled by our general derivation of the power-counting rules, in later sections. It is just the most basic situation that needs to be treated in a more process-specific fashion.

5.8.2 *Power-counting for DY*

To see these issues concretely, consider the fully differential cross section for the Drell-Yan process (5.14). This can be written as $d\sigma / d^4q \, d\Omega$, where q is the momentum of the lepton pair and $d\Omega$ is for the polar angles θ and ϕ that give the directions of the individual leptons. Since the lepton pair results from a single virtual photon, the angular distribution is a second-order polynomial in the sine and cosine of the polar angles; thus no special issues arise that depend on different regions of θ and ϕ. The cross section has an energy dimension of -6, and the natural power law is Q^{-6}, where Q is the mass of the lepton pair $Q = \sqrt{q^2}$, assumed to be comparable with the center-of-mass energy.

This power law is in fact correct when the transverse momentum q_T of the lepton pair is comparable with Q. But I will now illustrate, by examining graphs of the form of Fig. 5.33, that when q_T is much smaller than Q, the power law must be changed to the much bigger value $1/(Q^4 q_T^2)$. This power is cut off by the effects of hadronic masses when q_T is of order a hadronic mass.

Let us write the graph as

$$\int d^4k_A\, d^4k_B\; B(k_B, P_B)\, A(k_A, P_A)\, H(k_A, k_B, q)\, \delta^{(4)}(q - k_A - k_B), \qquad (5.72)$$

where A and B represent the upper and lower bubbles, while H represents the product of the amplitudes for the $q\bar{q} \to \mu^+\mu^-$ amplitude. Initially, we assume that the initial particles P_A and P_B are elementary, and for the purposes of understanding the power-counting it is sufficient to take the subgraphs A and B to be the simplest tree graphs, as in Fig. 5.33.

We must now investigate the following regions for Fig. 5.33, which can be distinguished by the values of the transverse components of the loop momenta:

- the purely hard region, where the whole graph forms the hard subgraph, so that both k_{AT} and k_{BT} are of order Q;
- the single-collinear regions where only one transverse momentum is of order Q; the other is power-counted as order $\lambda \ll Q$;
- the double-collinear region, where both k_{AT} and k_{BT} are much less than Q; they are both counted as order λ.

In the purely hard region, dimensional analysis applies unambiguously to give the basic $1/Q^6$ power. There are delta functions in A and B to put the spectator particles on-shell. But these set the sizes of momentum components that are of order Q, and the dimensional analysis argument still works. In this region of loop momenta, the $1/Q^6$ power law applies independently of the value of the external transverse momentum q_T. The hadronic final state contains two jets of high transverse momentum, corresponding to the remnant partons, which have large transverse momentum.

Next, consider a single-collinear region; for definiteness let us choose k_A to be collinear to P_A, so that k_{AT} is of order λ. The other transverse momentum k_{BT} is large and must therefore flow out into the virtual photon. Hence this region only exists when the lepton pair has transverse momentum of order Q. This large transverse momentum is approximately balanced by a final-state jet formed by a remnant parton on the B side.

We can think of the collinear subgraph as having an approximate rest frame in which all components of its momenta are of order λ. Given that parton A is a quark, the collinear subgraph has dimension -3; the measure of the k_A integral has dimension $+4$, for a total dimension of $+1$. This corresponds to a power λ, instead of a power Q which we would obtain for the same subgraph in the purely hard region. But the subgraph is boosted from its rest frame. Each of the lines connecting it to the hard subgraph has spin-$\frac{1}{2}$, so that largest components of the spinors on each line gain a factor of $(Q/\lambda)^{1/2}$ from the boost, for a total of Q/λ. Thus the complete power law is $(\lambda/Q)^0$ relative to the purely hard case; that is, the overall power remains unchanged. Thus we still get the overall power Q^{-6} for a single-collinear region, but this region only exists for the large-q_T region.

It is worth noting that although the detailed argument depends on the spin of the quark, the power law does not. If we were to use a model with a scalar quark, then there would be no Q/λ enhancement from the boost, but the collinear subgraph, complete with its integral, has dimension 0. So the overall power is unchanged. The same argument applies to gluons with transverse polarization. But a virtual gluon can also have a polarization in the direction of the large momentum, and boosting each of two gluons gives an enhancement by a factor $(Q/\lambda)^2$. But a Ward-identity argument will show that this part cancels after a sum over all possible hard subgraphs: see Ch. 11 for the simplest such derivation.

Finally, we examine the double-collinear region. The virtual photon has transverse momentum $k_{AT} + k_{BT}$, which is of order λ. The separate collinear subgraphs give an unchanged power of Q just as in the single-collinear region. However, in the delta function for transverse momentum conservation, $\delta^{(2)}(q_T - k_{AT} - k_{BT})$, all the momenta are of order λ. So the delta function power counts as $1/\lambda^2$ instead of $1/Q^2$. As previously announced, the result is an enhanced overall power of $1/(Q^4 q_T^2)$ instead of $1/Q^6$.

The phenomenological result is a strong enhancement at small q_T, as we will see in the data in Fig. 14.13. Evolution effects, to be treated in Chs. 13 and 14, will strongly modify the actual power law, so the power law just derived has exact applicability only for individual graphs. Of course the decrease is cut off at small enough λ that mass effects need to be taken into account.

We made an initial assumption, for simplicity, that the initial-state particles are elementary. But the dimensional-analysis argument applies to the collinear subgraphs even when the initial particles are composite. So we get the same power laws when the initial-state particles are normal hadrons, which entails the use of bound-state wave functions.

It is also useful to examine the cross section integrated over the transverse momentum q_T of the pair and also over the angle of the leptons, to give $d\sigma / dQ^2 dy$, where y is the rapidity of the lepton pair relative to the center-of-mass. In the small-q_T region, a factor λ^2 arises from the integration measure $d^2 q_T$, which compensates the $1/\lambda^2$ factor in the differential cross section. Hence the integrated cross section power counts as Q^{-4} in all regions. Naturally there is a logarithmic enhancement from the integral to small transverse momentum, which leaves the power law itself unchanged.

We can summarize the source of the enhancement in the differential cross section at small q_T as being in the creation of virtual photon from two oppositely moving collinear partons, without production of extra jets. Technically the enhancement is associated with the transverse-momentum delta function in this situation, so that the collinear transverse-momentum integrals are linked. In regions with production of jets of high transverse momentum, as in Fig. 5.19, there is no enhancement. We therefore see a simplification of the leading regions relative to the case that q_T is of order Q. In compensation, the linking of the collinear transverse-momentum integrals introduces some very interesting extra features in the derivation and formulation of factorization, as we will see in Chs. 13 and 14.

5.8.3 *Powers of Q and λ*

We consider a generic point in the intrinsic variable(s) z, and examine the integral over the radial variable λ. There is an angular integral, represented by the ellipse in Fig. 5.32. Over most of the angular integration, the sizes of the denominators (with masses neglected for now) obey the standard power-counting, Sec. 5.5.8:

$$
\begin{aligned}
\text{Hard:} &\quad Q^2, \\
\text{Collinear:} &\quad \lambda^2, \\
\text{Soft:} &\quad \lambda^4/Q^2 = \lambda_S^2.
\end{aligned}
\tag{5.73}
$$

These sizes are not exceeded. Much smaller values may be obtained, but only close to certain submanifolds of the intrinsic variables or of the angular integration. These are where z gets close to the PSS for a smaller region than or where the angular variables get close to a larger region: Sec. 5.10.

Our basic strategy is to use dimensional analysis to convert these estimates for single lines to estimates for the whole hard, collinear and soft subgraphs. This is supplemented by factors implementing boosts of collinear subgraphs from their rest frame: these produce enhancements that increase with the spin of the lines connecting the collinear to hard and soft subgraphs.

For small λ all the denominators participating in the PSS R get small. For large $\lambda \sim Q$, they all become hard, i.e., they all have virtualities of order Q^2. (Of course, this is holds except for neighborhoods of PSSs R' that are bigger than R. In neighborhoods of these, some denominators remain small. But this happens only in a small part of the angular integration.)

To obtain the power of Q for a region, we start from an estimate of the λ-dependent part of the integral in the form

$$
Q^{p_1} \int_0^{O(Q)} \frac{\mathrm{d}\lambda}{\lambda} \lambda^{p_2},
\tag{5.74}
$$

or some variation thereof, with the exponents p_1 and p_2 to be determined. At small λ, we should cut off the integral by the effects of masses, and at large λ we get to a purely hard region when $\lambda \sim Q$. We distinguish three different cases:

- The power of λ is zero: $p_2 = 0$. Then the integral is logarithmic and each order of magnitude in λ contributes equally. The resulting Q dependence is Q^{p_1} modified by logarithms, a very typical situation in QCD.
- The power of λ is negative. Then the integral would have a power-law divergence at $\lambda = 0$ were it not for mass effects. The physical result is therefore dominated by small λ, and we must examine the cutoff provided by masses. If the dominant cutoff is on collinear lines, then it is at $\lambda \sim m$, and the power law is still Q^{p_1}. If the dominant cutoff is on soft lines, then the cutoff on λ is $\sqrt{mQ}$, and the power of Q is $Q^{p_1+p_2/2}$.
- The power of λ is positive. Then the integral is dominated by its upper end, $\lambda \sim Q$, i.e., by a hard region rather than R. The power of Q for this hard region is $Q^{p_1+p_2}$. The contribution of the region R for a particular size of λ is of order $Q^{p_1}\lambda^{p_2}$, which is a power of Q less than the contribution of the hard region. Thus the region R itself is non-leading.

5.8.4 Overall form of power law

We now derive the power law for a general PSS R in the form

$$Q^p \left(\frac{\lambda}{Q} \right)^{\alpha(R)} \left(\frac{\lambda_S}{Q} \right)^{\beta(R)} \left(\frac{m}{Q} \right)^{\text{s.r.}(H)} \left(\frac{m}{\lambda} \right)^{\text{s.r.}(C)} \left(\frac{m}{\lambda_S} \right)^{\text{s.r.}(S)}. \tag{5.75}$$

The first factor is the characteristic power of Q for the process, e.g., the dimensional-analysis power for the Sudakov form factor or for DIS. As such, it is independent of the particular PSS R. The exponents in the second and third factors indicate how the power is modified by collinear and soft subgraphs. We will obtain formulae for the exponents in terms of the numbers of external lines of the various subgraphs defining the process and the regions.

The last three factors arise if there are super-renormalizable couplings in the theory. Although super-renormalizable couplings do not exist in QCD, it is useful to work with models with extra couplings. First, they allow us to see the result for a general QFT. Second, they do arise when we dimensionally regulate QCD. Finally, they help to give insight into the physical phenomena associated with the power-counting theorems. Furthermore, some equivalents of super-renormalizable couplings occur when external particles are bound states and collinear subgraphs contain their wave functions.

The above power law is intended to apply when we integrate over λ of some order of magnitude. Similarly, we assume that we have integrated over a range of the intrinsic variables that is of order their typical size, all the while staying away from smaller PSSs. Notice that, to let us easily read off the different effects of masses in the collinear and soft subgraphs, we wrote some factors in terms of λ and some in terms of the soft scaling variable $\lambda_S = \lambda^2/Q$. Factors involving λ_S are associated with the soft subgraph.

5.8.5 Basic power Q^p

Subject to the caveat in Sec. 5.8.1, the first factor Q^p in (5.75) is the dimensional analysis power for the amplitude or cross section under discussion. For a connected amplitude, dimensional analysis gives

$$p = 4 - \#(\text{ext. lines}), \tag{5.76}$$

where $\#(\text{ext. lines})$ is the number of external particles and external hard currents. In this estimate are included Dirac wave functions for external spin-$\frac{1}{2}$ fermions, which grow with energy like $Q^{1/2}$; the exponent is independent of the types of the external particles.

For example, for the current-quark-antiquark vertex of Fig. 5.1, we have three external lines, and therefore the power is Q^1. In the case of a scalar quark, at lowest order the power is from the factor of momentum at the photon-quark-quark vertex. In the case of an ordinary Dirac quark, the vertex is a Q-independent Dirac matrix γ^μ; the two external Dirac wave functions give the overall power Q^1.

Another example is the DIS structure tensor $W^{\mu\nu}$, for which there are four external lines. This gives Q^0, i.e., Bjorken scaling.

5.8.6 Formulae for the other exponents

I first state the formulae for the exponents $\alpha(R)$ and $\beta(R)$:

$$\alpha(R) = \#(CH) - \#(\text{scalar pol. glue } CH) - \#(\text{ext. lines}), \qquad (5.77)$$

$$\beta(R) = \#(0 \text{ or } 1, SH) + \frac{3}{2}\#(\tfrac{1}{2}, SH) + \frac{1}{2}\#(0 \text{ or } \tfrac{1}{2}, SC), \qquad (5.78)$$

and explain the meanings of the terms on the r.h.s.; these give the exponents in terms of the numbers of external lines of the different subgraphs for the PSS. Then I will state the formulae for the remaining exponents. In later sections I will derive the formulae.

In (5.77), $\#(CH)$ is the number of lines joining the collinear and hard subgraphs. When $\alpha(R)$ is used in (5.75) there is therefore a power suppression as the number of collinear lines joining the subgraphs is increased. We will see in Sec. 5.8.8 that, in the polarization sum for a gluon connecting a collinear to the hard subgraph, there is no power suppression when gluon has what we call scalar polarization. The second term on the right, $\#(\text{scalar pol. glue } CH)$, provides the necessary compensation to the first term. Finally $\#(\text{ext. lines})$ is the number of external particles of collinear subgraphs (e.g., a total of two for the two collinear subgraphs in a PSS for the Sudakov form factor).

The power $\beta(R)$ in (5.78) depends on the numbers of lines connecting the soft subgraph to the other subgraphs. The value depends on the spin of the lines, so we write, for example, $\#(0 \text{ or } \tfrac{1}{2}, SC)$ for the number of lines of spin 0 or $\tfrac{1}{2}$ connecting the soft to collinear subgraphs, and similarly for the other terms.

Notice that the formula for $\beta(R)$ implies that there is generally a suppression by a power of Q whenever lines join the soft to the collinear or hard subgraphs, the suppression increasing with the number of lines. But there is an exception, that there is no penalty for gluons joining soft to collinear subgraphs. Thus $\beta(R)$ is zero when the connections of the soft subgraph consist only of gluons to the collinear subgraphs. In all other cases $\beta(R) > 0$.

Finally, the other exponents in (5.75), s.r.(H), s.r.(C), and s.r.(S), are the dimensions of the super-renormalizable couplings in the hard, collinear and soft subgraphs. In the corresponding factors, we use m to denote a mass scale for the typical size for these couplings.

5.8.7 Exponent for hard subgraph

Let the hard subgraph H have N_F external fermionic (Dirac) lines and N_B external boson lines. In normal QCD processes, this means that N_F is the number of quark plus antiquark external lines, while N_B is the number of external gluon lines, plus the number of external photon, W, Z, and Higgs lines. We always take the hard part to be one-particle irreducible in its external lines, so the dimension of H is $d_H = 4 - \frac{3}{2}N_F - N_B$. In the usual case that all the couplings are dimensionless, the power associated with the hard subgraph is just the usual UV power from dimensional counting with all momenta of order Q:

$$Q^{d_H} = Q^{4 - \frac{3}{2}N_F - N_B}. \qquad (5.79)$$

The use of dimensional analysis shows that this power depends only on the external lines of the subgraph, not on the internal details. We will combine the above exponent with the results for collinear and soft subgraphs to give (5.75), (5.77), and (5.78).

If there are super-renormalizable couplings with combined mass-dimension D, they count as m^D instead of Q^D. This gives a correction factor relative to (5.79) of $(m/Q)^D$, i.e., the fourth factor in (5.75).

5.8.8 *Exponents for collinear subgraph*

Rest frame

The region may have one or more collinear subgraphs C, as in Figs. 5.7 and 5.16. For each collinear subgraph, we express the momenta of its lines in the light-front coordinates defined in (5.68) and (5.69). Our definition of the radial variable λ in Sec. 5.7.4 gives exactly the same power-counting as in the one-loop example in Sec. 5.5.2, so that both the integration measure for a generic collinear momentum k and denominators for collinear propagators count according to their dimensions, λ^4 and λ^2 respectively. In a collinear subgraph we include any collinear loop momenta that circulate through the hard subgraph.

In the collinear subgraphs, we also include the wave functions for external particles of the relevant collinearity class, and numerator factors. Their effect is assessed by boosting from (an approximate) rest frame.

Now in the *rest frame* of the collinear momenta, the power of λ is just given by its dimension: $\lambda^{\text{dim of subgraph}}$, apart from super-renormalizable couplings. The dimension of a connected collinear subgraph, including external Dirac wave functions, is

$$\#(C \text{ to } H) + \frac{1}{2}\#\left(\tfrac{1}{2}: C \text{ to } H\right) - \#(S \text{ to } C) - \frac{1}{2}\#\left(\tfrac{1}{2}: S \text{ to } C\right) - \#(C \text{ to ext.}), \qquad (5.80)$$

with a notation like that in (5.77), and with $\#(C \text{ to ext.})$ representing the number of external lines connecting to the collinear subgraph. The different signs of the terms in (5.80) arises from the differences between amputated and unamputated lines at the edge of the subgraph, and from the loop integrals coupling the graph to the hard subgraph. If there are super-renormalizable couplings, they give a correction factor which is the fifth factor in (5.75), similarly to the case of the hard subgraph.

We sum (5.80) over all the connected collinear subgraphs, and obtain the same formula, with the terms like $\#(C \text{ to } H)$ now denoting the number of lines connecting all collinear subgraphs to the hard subgraph.

Boost of collinear subgraph

Next we boost each collinear subgraph to the overall center-of-mass frame. The result depends on the spins of the lines connecting the subgraph to the hard and soft subgraphs. For a field of spin-s, standard properties of representations of the Lorentz group show that its biggest component increases under the boost like $(Q/\lambda)^s$. For a Dirac field we have a power $(Q/\lambda)^{1/2}$, while for a gluon[7] we have Q/λ. For a whole collinear subgraph, we need

---

[7] Any result for a gluon applies also to any other spin-1 field, e.g., for the photon.

the product of one such power for each line joining it to the hard subgraph, and for each line joining it to the soft subgraph. (We have included external Dirac wave functions in the collinear subgraph(s), so they do not need to be allowed for separately.) Combining all the powers so far gives (5.75) except for soft-subgraph associated factors:

$$\text{Result in (5.75)} \times \lambda_S^{-\#(S)-\frac{1}{2}\#(\frac{1}{2}:S)}. \tag{5.81}$$

With one exception, the exponents, p, $\alpha(R)$, and $\beta(R)$ in the referenced formula (5.75) are given by (5.76)–(5.78), while $\#(S)$ is the number of all external lines of the soft subgraph, and $\#(\frac{1}{2}:S)$ is its number of spin-$\frac{1}{2}$ external lines [which also count in $\#(S)$]. We wrote the second factor in (5.81) in terms of $\lambda_S = \lambda^2/Q$, since that is the natural variable for the soft factor.

The exception about (5.81) concerns the gluons, where the derivation so far gives

$$\text{Preliminary:} \qquad \alpha(R) = \#(CH) - \#(\text{spin } 1: CH) - \#(\text{ext. lines}). \tag{5.82}$$

This is the exponent of λ/Q, so it implies that we have a penalty for every extra line joining the collinear and hard subgraphs, except for the gluons. The non-suppression of gluons arises from the plus component of the gluon polarization (in the direction of the collinear group it belongs to), because of the corresponding boost factor Q/λ.

But we will also use the transverse components, which do not undergo this boost. We now examine how to separate the contributions.

Collinear gluon polarization

We have already seen this phenomenon in examples. So let us examine a general decomposition of a connection of a gluon of momentum k from a collinear subgraph C to the hard subgraph H. We have a factor $C(k) \cdot H(k)$, where there is a contraction of the Lorentz index at the H end of the gluon. The gluon is collinear, so we define the collinear factor C to include the gluon's propagator. We decompose $C \cdot H$ with respect to the light-front components for C:

$$C \cdot H = C^+ H^- + C^- H^+ - C_T \cdot H_T. \tag{5.83}$$

After the boost from the rest frame for the collinear subgraph, the largest component of C^μ is the C^+ component, which increases like Q/λ. Next is the transverse component C_T, which is boost invariant, and finally C^-, which decreases like λ/Q.

The largest term is therefore $C^+ H^-$, and this gives the power derived above, in (5.81) and (5.82). So we define a Grammer-Yennie decomposition:

$$H \cdot C = H \cdot k \frac{C^+}{k^+} + H_\mu \left(C^\mu - k^\mu \frac{C^+}{k^+} \right). \tag{5.84}$$

The highest power Q/λ for C^μ is in the first term alone, which we call the scalar polarization term, since it has a polarization vector proportional to the momentum of the gluon. It is of a form suitable for applying a Ward identity. The second term, a transverse polarization term, has the highest power removed: the quantity in parentheses is exactly zero when $\mu = +$. Therefore this term power counts as 1 instead of Q/λ.

We now apply this decomposition to every gluon joining the collinear subgraphs and H. Each gluon line gives a scalar polarization term and a transverse polarization term. This converts the exponent $\alpha(R)$ from the one in (5.82) to the one in (5.77).

The importance of this operation is as follows. We start with a case like the parton model for DIS where the hard scattering is induced by fermion lines, and to get a leading power, we use the minimum possible number of such lines, which is two for the structure function in DIS. Replacing the fermions by scalar polarized gluons increases the power of Q to Q^2, giving a super-leading contribution. The super-leading contribution in fact cancels, as shown by the use of Ward identities (Labastida and Sterman, 1985). The remaining term is leading, and involves transversely polarized gluons.

Similar decompositions can be applied on fermion lines, but we will not need them here, because we will not have the same cancellation of the highest power.

5.8.9 Derivation of exponent for soft subgraph

We now bring in the soft subgraph S. All its external lines attach to the collinear and hard subgraphs. We include in S the integrals over loop momenta that circulate from S through the hard and collinear subgraphs, since these loop momenta are necessarily soft. The soft subgraph S may have one or more connected components.

A complication we have already noticed is that of choosing an appropriate scaling of the momenta. We let λ_S be the scaling factor for all the components of soft momenta. We have seen that to match the time scales of soft and collinear graphs, we need to take $\lambda_S = \lambda^2/Q$, where λ is the overall radial variable for the region under discussion. This contrasts with the treatment in Sterman (1996) where λ_S and λ were taken to be the same.

Without super-renormalizable couplings, our usual dimensional analysis argument applies in terms of λ_S to give a power

$$\lambda_S^{\#(S)+\frac{1}{2}\#(\frac{1}{2}:S)}, \tag{5.85}$$

where the exponent is the dimension of the soft subgraph, including its loop integrals to the collinear and hard subgraph. This power applies independently of the number of connected components of the soft subgraph. This power evidently cancels the second factor in (5.81), so the final power law is (5.75), with the exponents defined in (5.76), (5.77), and (5.78). If there are super-renormalizable couplings, they give the last factor in (5.75), by the same reasoning as for the other subgraphs.

5.8.10 Other scalings

The derivation of the power law assumed what we can call the canonical scaling of momenta for a region R – (5.29), (5.49), which led to (5.73) for the denominators. Could other cases matter? We have cataloged all pinch-singular surfaces of massless graphs for our process. The scalings parameterize a neighborhood of each region by a radial variable. To the extent that the estimates of the denominators in (5.73) are correct, our derivations are correct.

Where the denominators are much smaller than the estimates, our derivation is incomplete. We will see in detail later that these situations occur in three ways. One is around an intersection of a surface of constant λ with the PSS for a bigger region than R, as in Fig. 5.32. The second is where the intrinsic variables of R approach a smaller PSS. The final possibility is where there is a trap of the integration region in a Glauber-type region.

We will show that the power laws remain correct in the first two cases, but if there is logarithmic behavior in λ, i.e., a power λ^0, then logarithmic enhancements in the Q dependence occur relative to the basic power. This is quite common.

For many processes of interest, the Glauber region does not contribute or cancels after a sum over allowed final-state cuts.

One complication arises when some particle masses are actually zero, and we have an actual infra-red or collinear divergence at $\lambda = 0$. In a theory of confined quarks and gluons these divergences are not genuinely physical, but they do appear in Feynman graphs. They are handled by a sufficiently careful treatment of the soft region as we have defined it.

5.8.11 Power of Q

From (5.75), we derive the power of Q associated with a region after integration over λ. An important case is that all the exponents $\alpha(R), \ldots,$ s.r.(S) are zero, which corresponds to leading regions for processes like DIS and Drell-Yan. Then we simply get Q^p, which is the power corresponding to the dimension of the amplitude or cross section under consideration. There is no λ dependence in (5.75), so the integral (5.74) gives a logarithm of Q divided by a mass scale. When we discuss nested regions, in Sec. 5.10, we will find an extra logarithm for every level of nesting where power-counting gives a logarithmic radial integral. The actual result is then

$$\text{Standard leading power:} \qquad Q^p \times \text{logarithms.} \tag{5.86}$$

When one or more of the exponents is non-zero, the precise power of Q will depend on how masses cut off the integral at small λ. If there is no soft subgraph, then the cutoff is dominated by masses on collinear lines, so that the power of Q is determined by setting $\lambda \sim m$ and we get

$$\text{Coll. cutoff:} \qquad Q^{p-\alpha(R)-\text{s.r.}(H)} m^{\alpha(R)+\text{s.r.}(H)} \times \text{logarithms.} \tag{5.87}$$

If there is a soft subgraph, then the cutoff is at $\lambda_S \sim m$, i.e., $\lambda \sim \sqrt{mQ}$, and we get

$$\text{Soft cutoff:} \qquad Q^{p-\frac{1}{2}\alpha(R)-\beta(R)-\text{s.r.}(H)-\frac{1}{2}\text{s.r.}(C)}$$

$$\times\ m^{\frac{1}{2}\alpha(R)+\beta(R)+\text{s.r.}(H)+\frac{1}{2}\text{s.r.}(C)} \times \text{logarithms.} \tag{5.88}$$

If there are both collinear and soft loops, the cutoffs can be different on the collinear and soft loops. This will result in an important contribution where the $\overline{k}$ variables [see (5.29)] are particularly small on collinear lines. This will refer to a small part of the angular integral. In our discussion of nested regions, we will assign this part to another region.

5.9 Catalog of leading regions

We now obtain general rules for determining the leading regions for a process.

5.9.1 General principles

The general power law was given in (5.75). Rather than presenting a power of Q alone, we have included powers of λ and λ_S. Thus we can read off the effects of masses that cut off the integrals at their lower ends.

For each process there is a minimum number of collinear lines entering the hard scattering if the process is to occur kinematically. For example, this is one on each side of the final-state cut in DIS, and two on each side of the final-state cut for Drell-Yan. In all these cases this is the same as the number of external hadrons for the process as a whole. With this minimum number of lines, we get $\alpha(R) = 0$. This is provided that we exclude gluons of scalar polarization in the minimally connected graphs; as will be proved later, we get zero after summing over graphs when all the lines joining a collinear subgraph to the hard scattering are zero.

Thus with the minimal number of connections between the collinear and hard subgraphs, the power of Q is the same as the pure UV power, $Q^p = Q^{4-\#(\text{ext. lines})}$, which we define to be the leading power for the process, e.g., Q^0 for DIS.

After this we read off from (5.75)–(5.78) that we get a power suppression, when we do any of the following:

- attach extra collinear lines to the hard scattering, except for scalar polarized gluons;
- attach any soft lines to the hard scattering;
- attach the soft subgraph to the collinear subgraphs by anything but gluons.

But there is no penalty for extra scalar-polarized collinear gluons attaching to the hard scattering, and there is no penalty for soft subgraphs that attach to collinear subgraphs by gluon lines only.

As to super-renormalizable couplings, they always give a penalty in the hard scattering. But in the collinear and soft subgraphs, there is no penalty as long as the momenta are at the lower end of their range, near the mass cutoff. Note that in the limit of zero mass, super-renormalizable couplings convert otherwise logarithmic IR singularities to power-law singularities.

It is worth observing that our rules give no penalty for having quark loops *inside* the soft subgraph. This is a fact that is sometimes forgotten, because in the corresponding IR-divergence problem in QED, no loops of massive fermions need to be considered.

One complication that sometimes arises is that when one actually does a particular calculation, the coefficient of the leading power might be zero. Typically this arises because of some symmetry. A simple example is the polarization dependence of DIS. The power-counting argument permits a Q^0 behavior in $W^{\mu\nu}$ for the dependence on both longitudinal and transverse polarization. In fact only longitudinal polarization gives this behavior, in the structure function g_1 – see (2.20). But for transverse polarization, there is a power suppression – see Sec. 6.1.4 for the parton-model case – which results from the chiral symmetry of QCD and QED perturbation theory for hard scattering.

5.9.2 Prescription for leading regions

From the results just derived, the leading regions in the examples earlier in this chapter are indeed those stated: e.g., Fig. 5.5 for the quark-quark-current vertex, Fig. 5.7(c) for DIS and DVCS. The general principles are:

1. The soft subgraph connects only to collinear subgraphs and only by gluons.
2. The collinear subgraph(s) each connect to the hard subgraph(s) by the minimum number of lines consistent with the desired process or reaction occurring at all.
3. In addition, arbitrarily many gluons of scalar polarization may connect a collinear subgraph to a hard subgraph.

Thus in DIS, two quarks, one on each side of the final-state cut, can join the target-collinear subgraph to the hard subgraph. This exactly corresponds to the idea that motivated the parton model. But the rules just stated show that it is possible to replace the quark lines by transversely polarized gluon lines. This corresponds to a short-distance scattering off a gluon constituent in the target, compatible with the basic short-distance scattering idea. However, the minimal hard scattering is the reaction $\gamma^* + g \rightarrow q\bar{q}$, for which the amplitude is one order higher in QCD perturbation theory than for scattering off a quark.

5.9.3 Possibility of multiple hard scatterings

A particularly non-trivial example is elastic scattering of protons at wide angle. The reaction is $P_1 + P_2 \rightarrow p_3 + p_4$. The incoming protons are in opposite directions, and the outgoing protons are in very different (and again opposite) directions. Thus there are four collinear directions, two in the initial state and two in the final state.

If we restrict our attention to reduced graphs with collinear and hard subgraphs, then one possibility is a single hard subgraph, as in Fig. 5.34(a). Now a single quark has baryon number $\frac{1}{3}$, so a minimum of three quarks out of the collinear subgraph for each proton must attach to the hard scattering; otherwise, for example, remnants of the incoming protons would be left in the final state, approximately parallel to the incoming hadrons.

The connected hard scattering subgraph has 12 quark lines, which, from (5.76), corresponds to a power $1/Q^8$ in the amplitude, or equivalently $1/s^4$. Converting to a cross section gives $d\sigma / dt \propto 1/s^{10}$, as first found by Brodsky and Farrar (1973).

But it is also possible to have three separate quark-quark hard scatterings: Fig. 5.34(b). As shown by Landshoff (1974), this results in less of a suppression, giving $d\sigma / dt \propto 1/s^8$. The derivation needs a generalization of the results earlier in this section, both because the hard scattering is disconnected, and because of the associated momentum-conservation delta functions.

There are also a number of other possibilities that need to be examined, including a single quark-quark hard scattering, with the other quarks being soft. Soft quarks normally give

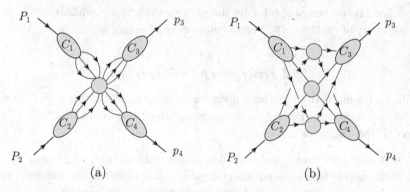

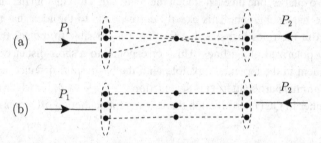

Fig. 5.34. Possible reduced graphs for wide-angle elastic proton-proton scattering: (a) with connected hard subgraph, (b) with three separate hard subgraphs. The elliptical blobs labeled C_j are collinear, and the unlabeled circular blobs are hard.

Fig. 5.35. Side views of the spatial structure corresponding to the reduced graphs of Fig. 5.34.

a power-suppression, but here this is compensated by not needing so many lines entering the hard-scattering subgraph(s). A correct analysis also needs to account for the Sudakov suppression of the hard scattering in the Landshoff graph, because each subgraph involves isolated color.

The difference between the mechanisms can be understood in space-time. With a single hard scattering, all the quarks in each proton must come down to within a transverse distance $1/Q$ of each other: Fig. 5.35(a). This gives a strongly power-suppressed probability, since the normal transverse separation of the quarks is of the order of 1 fm.

For the Landshoff process, it is merely necessary that each quark in one hadron comes within $1/Q$ of *one* of the quarks in the other hadron, Fig. 5.35(b), which is more probable. In order for this to match the same picture for the outgoing protons, the three intersections must line on a line transverse to the scattering plane, which gives a further suppression in the final result in Landshoff (1974).

5.10 Power-counting with multiple regions

The power-counting scheme of the preceding section arose from estimates of the sizes of propagator denominators around any given region R. We call this the canonical power estimate. It not only gives us the power of Q associated with the region; it also indicates what kind of approximator is appropriate, where we neglect certain components of momentum on a line. Such approximators are critical to deriving factorization theorems.

So we must ask where the estimates fail. As we will now show, with a certain exception, the failures occur in two situations: (a) where the particular values of the intrinsic variables for R approach a smaller PSS; and (b) where the angular variables take us to the vicinity of a larger PSS. The true results in these cases are essentially obtained from the canonical power-counting for these other regions. The canonical power law of Q will be modified by logarithmic corrections. The one exception to the above statements concerns regions of the Glauber type, which are avoided by contour deformation in many cases, or otherwise need special discussion.

For high-order graphs there are many possible PSSs which intersect in many ways. An important feature of the following discussion will be to reduce the general case to a collection of a very few generic situations.

5.10.1 Locations of failures of power-counting

Consider a region R, with its radial variable λ. We use (5.29), (5.49) for the scaling of collinear and soft momenta, which gives the canonical sizes (5.73) for propagator denominators. Because of the normalization condition on the angular variables, the size of each momentum component is limited by its canonical scaling value, apart from a constant factor. Numerators are all bounded by their canonical values. Thus the only possibility of a failure of the power-counting is for one (or more) denominators to be much less than their canonical values.

To determine where this happens, we use a variation on the Libby-Sterman scaling argument. It involves the ratios of the propagator denominators to their canonical values:

$$r_l = \left| \frac{\text{denominator}_l}{\text{canonical}_l} \right|, \tag{5.89}$$

where l labels the line. Our concern is the minimum value of these ratios. First, suppose there is a non-zero lower bound to all the ratios: $r_l \geq r_{\min} \neq 0$, that applies uniformly over all propagators, over all the angular variables, over λ from zero to order Q, and for all large enough Q. Then the canonical value of the denominator is unambiguously correct for our power-counting.

Next we locate failures of such a bound by integrating around a surface of constant λ (Fig. 5.36) with the intrinsic coordinate(s) fixed. We call this surface $\Sigma(\lambda, R)$. Often the minimum value of the ratio is set by mass effects, so that the ratio is very small when λ is increased, thereby wrecking the power-counting. We therefore set masses to zero to give an appropriate diagnostic. If the minimum value of one or more ratios is zero in the massless theory, then the power-counting has failed, and we must examine a neighborhood of the subsurface where the minimum is zero. In this situation we have a singularity in the integrand in the massless theory.

Naturally, as in all our arguments, if it is possible to deform the contour of integration away from a singularity, we do so. Thus we only need treat cases where one or more of

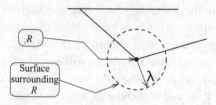

Fig. 5.36. The dashed line represents a surface $\Sigma(\lambda, R)$ of constant λ surrounding the PSS for R. The dot in the center represents the PSS R, and the three solid lines represent other PSSs. Although the surface $\Sigma(\lambda, R)$ is *diagrammed* as having radius λ, some momentum components may scale differently, e.g., as λ^2/Q.

the ratios r_l is pinched at zero. This is exactly the condition for a PSS, and in fact that the surface of constant λ intersects another PSS R'. There are now two cases, depending on whether or not the second PSS intersects the first.

If R' is like the upper solid line in Fig. 5.36, it does not intersect the original PSS R. In this case we reduce the maximum value of λ under consideration to avoid R'. The maximum value of λ is still of order Q, leaving our methodology unaltered. The region around R' can be treated by power-counting methods adapted to R' without the need to consider R. Any leftover gaps involve purely hard momenta.

The other case is that R' intersects the original PSS R, as for the lower two solid lines in Fig. 5.36. Then we must examine a neighborhood of R', and use its power-counting to modify our original estimate – which we will do in Sec. 5.10.2. We can treat each such R' separately. In angular sectors not near these PSSs, the original estimate applies unchanged.

So one possible failure of the simple power-counting occurs at the intersection of $\Sigma(\lambda, R)$ with a PSS R' bigger than R.

But there are other possibilities for the intersection of the new surface R' with R. One is that the intersection $R' \cap R$ is a lower-dimension surface. In that case, we reorient the discussion. The intersection is itself a PSS, which we will call R_1. Our power-counting applies for a fixed value of the intrinsic coordinates of R, in which case we treat R' and R as non-intersecting. We will separately treat the situation the intrinsic coordinates approach the position of a sub-PSS, of which R_1 will be a typical example.

A final possibility is that the intersection of R' with R has the same dimension as R, but is not the whole of R. There are possibly several such intersections. In that case we consider each of the intersections as a separate PSS. That is, we replace R by a set of PSSs which combine to form R. The edges of these small PSSs, particularly where they abut, are themselves lower-dimension PSSs.

It is also possible that the minimum value of one or more of the r_l ratios is non-zero on $\Sigma(\lambda, R)$ when λ is fixed, but that the minimum decreases to zero as $\lambda \to 0$. In other words the non-zero lower bound is not uniform in λ.[8] This is behavior that we term Glauber-like, whose general criteria we will determine in Sec. 5.11.

---

[8] We take for granted that if it is possible to deform the contour of integration to avoid such a situation, then we do so.

To see that the name is appropriate, we examine the quark and gluon propagators in Fig. 5.31(b). For a normal soft region, a soft denominator is of order λ^4/Q^2, while a collinear denominator is of order λ^2, for a large ratio

$$\frac{\text{collinear denom.}}{\text{soft denom.}} \sim \frac{Q^2}{\lambda^2}. \tag{5.90}$$

In our discussion of the Glauber pinch for Fig. 5.31(b), we used a soft transverse momentum of order m, which we now translate to $\lambda \sim \sqrt{mQ}$. In the Glauber region $k^\pm \sim m^2/Q$, so collinear denominators are of order m^2, i.e., λ^4/Q^2 instead of λ^2. Thus for a given gluon virtuality the collinear denominators are much smaller in the Glauber region than in the normal soft region.

The above discussion covers the case of fixed intrinsic coordinate(s) for the PSS R. A further issue occurs when we integrate over the intrinsic coordinate(s) of R, and approach a smaller PSS R_1. This case is handled by observing that it involves the treatment of power-counting for the smaller region R_1. If we change to the viewpoint of integrating around R_1, we have treated that case already.

In summary, there are just two situations we need to cover: (a) the intersection of R with a bigger PSS R' at a generic point on R, which by a change of point-of-view also includes the approach on R to a smaller PSS; and (b) a Glauber-type situation.

5.10.2 Intersection of $\Sigma(\lambda, R)$ with PSS bigger than R

Relations between regions and subgraphs

Let R' be a PSS bigger than R, like one of the lower solid lines in Fig. 5.36, or the shaded surface in Fig. 5.32. We consider the integral over the constant λ surface $\Sigma(\lambda, R)$ near its intersection with R', and we let λ' be the radial variable for R'.

Some of the propagators are not trapped at R'. Their denominators retain their powers from the first region, i.e., Q^2 for a hard line, λ^2 for a collinear line, and λ^4/Q^2 for a soft line. As we have already seen, the time scale for these lines is Q/λ^2 for the soft and collinear lines, or $1/Q$ for the hard lines; in all cases this is at most Q/λ^2.

Since these lines are not pinched at R', they constitute the hard subgraph H' for R'. When $\lambda \to 0$, the intersection of $\Sigma(\lambda, R)$ and R' approaches the original PSS R, which we can think of as an endpoint of R'. Thus in the situation we consider, i.e., $\lambda \ll Q$, the virtualities of some lines of H' are much smaller than the standard value Q^2 for a hard subgraph, the smallness being controlled by λ.

In contrast, the denominators of those lines that are pinched at R' have arbitrarily much smaller denominators, governed by λ' rather than λ. The time scale for these lines is Q/λ'^2, much longer than that for the non-pinched lines. In the case of a graph for the Sudakov form factor, this is illustrated in Fig. 5.37. There, the placement of the collinear and soft subgraphs is meant to be like the space-time diagram Fig. 5.2(b).

With respect to each PSS, each line of the graph can be assigned a category: soft, collinear with respect to an external line, or hard. There are corresponding subgraphs: e.g., for the vertex graph we have subgraphs S, A, B, H with respect to R, and subgraphs S',

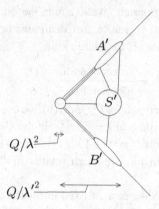

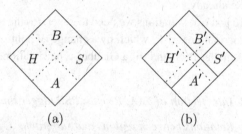

Fig. 5.37. Reduced graph for region R' near a smaller region R. The time scales for the different subgraphs are indicated.

Fig. 5.38. Decomposition of graph into subgraphs for momentum classes, according to PSS (a) R and (b) $R' > R$. The subgraphs for R are delimited by the dotted lines and those for R' by solid lines.

A', B', H' with respect to R'. Now lines with energy of order Q on R retain approximately this energy near R', while lines that are far off-shell at R are also far off-shell near R'. Thus we have the following possible transitions relating the categories of a line with respect to the different regions:

$$S \to S', \ A', \ B', \ H';$$

$$A \to A', \ H';$$

$$B \to B', \ H';$$ (5.91)

$$H \to H';$$

as illustrated in Fig. 5.38.

As we integrate around $\Sigma(\lambda, R)$, λ' varies from zero to a maximum. We need to know the order of magnitude of the maximum value of λ', which is in fact λ. To see this, we assign to the momentum components in S', A' and B' their canonical power-counting with respect to R' and match with the powers with respect to region R. The powers agree when $\lambda' \sim \lambda$. The only exception concerns the minus components for momenta in $S \cap B'$ and similarly for $S \cap A'$. These components would be of order Q for a fully collinear region,

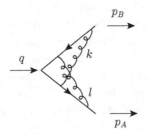

Fig. 5.39. Two-loop vertex graph.

but are now of order λ: the smallness of λ causes these momenta to be close to an endpoint of a collinear region by the standards of R'.

Example

A convenient illustration uses the two-loop graph in Fig. 5.39, with its two gluons of momenta k and l. Let region R be where k is soft and l is collinear to A. Thus the canonical sizes of $(+, -, \mathrm{T})$ components are

$$R: \quad k \sim \left(\frac{\lambda^2}{Q}, \frac{\lambda^2}{Q}, \frac{\lambda^2}{Q}\right), \quad l \sim \left(Q, \frac{\lambda^2}{Q}, \lambda\right). \tag{5.92}$$

To avoid issues with IR problems, we assume the gluon is massive. Then the effective cutoff on λ is $\sqrt{mQ}$. We let the region R' be where k is collinear to B and l is hard:

$$R': \quad k \sim \left(\frac{\lambda'^2}{Q}, Q, \lambda'\right), \quad l \sim (Q, Q, Q). \tag{5.93}$$

Now consider the particular orders of magnitude:

$$k \sim \left(\frac{m^{3/2}}{Q^{1/2}}, Q^{1/2}m^{1/2}, m\right), \quad l \sim \left(Q, Q^{1/2}m^{1/2}, Q^{3/4}m^{1/4}\right). \tag{5.94}$$

We could consider this as near to PSS R with $\lambda \sim Q^{3/4}m^{1/4}$: the components of l have exactly the standard sizes $(Q, \lambda^2/Q, \lambda)$ for a collinear-to-A momentum. All components of k are much less than Q, with a maximum size λ^2/Q, so k is soft. But notice that the plus and transverse components of k are much smaller than the standard λ^2/Q for a soft momentum.

But we can also consider the configuration as near to R' with $\lambda' \sim m^{3/4}Q^{1/4}$: we can treat k as collinear-to-B, since it has large negative rapidity: $y_k \sim -\frac{1}{2}\ln(Q/m)$, although l^- is much less than Q. We can consider l hard since its virtuality is much bigger than λ'^2.

It can be checked that the time scales of the lines are

$$p_B - k - l : \frac{1}{Q},$$

$$l, \; p_A + k + l, \; p_A + l : \frac{Q}{\lambda^2} \sim \frac{1}{Q^{1/2}m^{1/2}}, \tag{5.95}$$

$$k, \; p_B - k : \frac{Q}{\lambda'^2} \sim \frac{Q^{1/2}}{m^{3/2}}.$$

Thus we have a clear separation of scales, and the configuration has characteristics of both the regions R and R'. It can be verified that (5.94) gives a leading-power contribution. We must ensure that our treatment of factorization correctly handles it and the obvious myriad of similar possibilities in this and higher-order graphs.

In the above case, we assumed the gluon was massive, so the lower cutoff on transverse momentum was of order m, a typical mass. To keep k of low energy with respect to Q we were forced to keep its rapidity well short of that of p_B, and to put the lower quark lines far off-shell.

But if the gluon is massless, a rather more extreme situation arises. For example, try

$$k \sim \left(\frac{m^4}{Q^3}, \frac{m^2}{Q}, \frac{m^3}{Q^2} \right), \quad l \sim \left(Q, \frac{m^2}{Q}, m \right). \tag{5.96}$$

In the sense of rapidities, k is fully collinear to B: $y_k \sim -\ln(Q/m)$, and l is fully collinear to A: $y_l \sim \ln(Q/m)$. But k is also very soft by having its maximum component much less than Q. This configuration has $\lambda' \sim m^2/Q$, $\lambda \sim m$.

Obviously, we should not treat all such configurations separately, if at all possible, otherwise we could easily have much too complicated a problem to solve systematically. In fact we will be able to treat all such situations by a combination of methods that directly deal with the canonical scalings only.

But when we derive factorization, we will need to apply approximators suitable for neighborhoods of the different regions. Awareness of situations such as we have examined will inform our choice of approximators.

The physical property that will keep the situation under control is that the time scales associated with different lines are widely different, unlike the canonical case for the soft and collinear lines: we can treat one scale at a time and examine directly only the relations to neighboring time scales. Thus we only need to treat the relation between pairs of regions, each treated quite generically.

Effect on power-counting

To get the correct power-counting near the intersection of the constant λ surface $\Sigma(\lambda, R)$ and the PSS R', we integrate over a range of λ of some particular order of magnitude, and then we decompose the result by the variable λ', which measures the approach to R'. There will be powers of Q, λ and λ':

$$Q^\alpha \lambda^\beta \lambda'^\gamma \tag{5.97}$$

appropriate to the strongly ordered situation $\lambda \ll \lambda' \ll Q$. To obtain the exponents we match to the canonical power-counting for the regions R and R'. The canonical power for region R' applies to the case that $\lambda \sim Q$ with $\lambda' \lesssim Q$. Thus we have

$$\text{power for } R' = Q^{\alpha+\beta} \lambda'^\gamma. \tag{5.98}$$

The canonical power for R applies when λ' has it maximum value, i.e., of order λ, so

$$\text{power for } R = Q^{\alpha} \lambda^{\beta + \gamma}. \tag{5.99}$$

This determines the powers in (5.97) from the canonical ones for R and R'.[9]

As in (5.75), the powers are those for the situation that we integrate over a range of a radial variable comparable to its size. Thus they are the sizes of integrands to be used in integrals with respect to $\ln \lambda$ and $\ln \lambda'$:

$$\int^{\ln Q} d \ln \lambda \int^{\ln \lambda} d \ln \lambda' \; Q^{\alpha} \lambda^{\beta} \lambda'^{\gamma}. \tag{5.100}$$

The lower limits of the integrals are either $\ln m$ or $\ln \sqrt{mQ}$, depending on whether the cutoff is governed by masses on collinear lines or masses on soft lines.

We now read off the results for the Q dependence of the integration over $\Sigma(\lambda, R)$.

The most common case we use is for the leading regions in QCD, for which $\beta = \gamma = 0$. Then the leading power of Q is Q^{α}, and the integrals over λ and λ' give logarithmic enhancements. Naturally we can have multiply nested regions. Iterating our argument gives the general rule that there is one logarithm of Q for each nesting. Thus for the one-loop Sudakov form factor, we have the nestings of leading regions: $H > A > S, H > B > S$. In (5.21), which depicts the hierarchy of regions, these nestings give ordered paths of length two and hence two logarithms. When we make the decomposition around the soft region, the two collinear regions A and B occur at distinct places in the angular integral. Thus the logarithms associated with the two different ordered paths add, rather than giving a more complicated situation.

In the other cases, one end or the other wins, which greatly simplifies the extraction of the leading power. There are several cases:

- If $\beta > 0$, then the top end $\lambda \sim Q$ of the λ integral wins. Then for the highest power of Q, the situation is the same as for region R'.
- If $\gamma > 0$, then the top end $\lambda' \sim \lambda$ of the λ' integral wins. Then for the highest power of Q, the situation is the same as for region R.

  Note that if both $\beta > 0$ and $\gamma > 0$, the integral is dominated by $\lambda \sim \lambda' \sim Q$, i.e., by the hard region; both R and R' are non-leading by a power of Q.
- If both $\beta < 0$ and $\gamma < 0$, then the integral is dominated by the lower ends of both integrals. If they both have the same lower cutoff, then at the cutoff we have $\lambda' \sim \lambda$, which is just reproduces the generic situation for region R.

  It is possible that the lower cutoffs are different: m for λ' and $\sqrt{mQ}$ for λ. This needs special discussion.
- If $\beta < 0$ and $\gamma = 0$, then the lower end of the λ integral wins and there is at most a logarithm from the λ' integral. The power for R remains correct.

---

[9] Situations where there is an apparent mismatch of power laws between regions were found in Bacchetta *et al.* (2008). These situations concern certain spin-dependent cross sections, and they can be handled by a generalization of our argument by allowing for powers of quark mass as well as of Q, λ, and λ'.

- If $\beta = 0$ and $\gamma < 0$, then the lower end of the λ' integral wins and there is a logarithm from the λ integral. The power for R' remains correct.

Aside from the case $\beta < 0$ and $\gamma < 0$, the general rule is that the overall power of Q is the highest power of Q as determined from the pure canonical powers for the individual regions.

5.11 Determination of Glauber-like regions

For each PSS, we found a canonical scaling law, and we saw that modifications to the canonical values of propagators were generally associated with canonical scaling for other intersecting PSSs. The only exception was what we called Glauber-like. This is where at some locations on the surface $\Sigma(\lambda, R)$ surrounding a PSS R, some denominators get much smaller than their canonical sizes, but that the ratio r_l on these lines only goes to zero at $\lambda = 0$.

We now show how to determine where the Glauber-like situation arises. We use another variation on the Libby-Sterman scaling argument, after first showing in an example how the Glauber region can be obtained from the standard scaling for a region by taking some of the angular coordinates to be very small.

5.11.1 Example

Consider Fig. 5.31(b) for the Drell-Yan process in the region where the quarks are collinear and the gluon is soft. With the canonical scalings, we parameterize the momenta of the gluon and the collinear momenta by

$$k = (S^+\lambda^2/Q, \ S^-\lambda^2/Q, \ S^T\lambda^2/Q), \tag{5.101a}$$

$$k_A = (z_A p_A^+, \ A^-\lambda^2/Q, \ A^T\lambda), \tag{5.101b}$$

$$k_B = (B^+\lambda^2/Q, \ z_B p_B^-, \ B^T\lambda). \tag{5.101c}$$

Here S^μ, A^μ, and B^μ give the angular coordinates for the soft and collinear momenta. Our usual normalization conditions show that the angular coordinates are at most about unity, and that the biggest is of order unity.

The canonical power-counting for this region applies when all the angular coordinates are of order unity. Note that in the interesting case that the transverse momentum of the Drell-Yan pair is of order m, a leading power is obtained only for $\lambda \sim m$, not for higher λ. When the gluon has a non-zero mass, the lowest effective value of λ is $O(\sqrt{mQ})$, and we get a power-suppression.

But we can also have a different scaling, the Glauber scaling, for which

$$k \sim (\lambda'^2/Q, \ \lambda'^2/Q, \ \lambda'), \tag{5.102a}$$

$$k_A \sim (Q, \ \lambda'^2/Q, \ \lambda'), \tag{5.102b}$$

$$k_B \sim (\lambda'^2/Q, \ Q, \ \lambda'), \tag{5.102c}$$

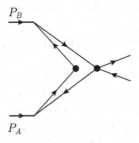

P_B

P_A

Fig. 5.40. Reduced diagram for Fig. 5.31(b) in Glauber region. The dots are the reduced vertices, and the lines are collinear to either p_A (in the bottom half of the diagram) or p_B (in the top half).

where we take all the coefficients of order unity. This can be obtained from the standard soft parameterization by making all the angular coefficients sufficiently small except for S^T:

$$S^{\pm}, A^-, B^+ \sim \lambda^2/Q^2, \qquad A^T, B^T \sim \lambda/Q, \qquad S^T \sim 1. \qquad (5.103)$$

We follow this by the change of variable $\lambda' = \lambda^2/Q$.

From the point of view of the canonical soft scaling, this is a region where the soft denominator retains it canonical size, $\lambda^4/Q^2 = \lambda'^2$, but the collinear denominators are also of this size instead of their canonical value $\lambda^2 = \lambda'Q$. This is actually the minimum possible for the ratio of the collinear denominators to their canonical values, and approaches zero as $\lambda \to 0$.

We have seen that the integration contour is trapped in this region, unlike the case of DIS and e^+e^- annihilation

5.11.2 Application of Libby-Sterman argument

In the general case, with many loop momenta, there appears to be an explosion of the number of possible cases for different scalings of the momentum components, with a corresponding difficulty in determining the cases that are relevant. We overcome this problem by the Libby-Sterman method.

For some alternative scaling, we define a reduced diagram in which the vertices are obtained from those denominators with the canonical scaling. The lines of the reduced diagram are those with denominators that are much smaller than canonical. For the Glauber region of Fig. 5.31(b), the reduced diagram is obtained by shrinking the gluon to a point, to give Fig. 5.40.

We now apply the Landau criterion for a pinch in the massless version of the reduced diagram. This works just as in the standard Libby-Sterman argument. The only difference is in the interpretation of the vertices of the reduced graph: in the original argument, the vertices corresponded to subgraphs whose internal momenta are hard, with virtuality Q^2. It is now possible to have vertices with much smaller internal virtualities. The common feature is that the time and distance scales of the vertices of the reduced graph are much smaller than those for the lines.

The result is of the form of a possible PSS for the original graph. But the power-counting may have changed.

For a general starting region, some of the new PSSs are the same as leading PSSs of the original massless graph, so we can cover them by the original argument.

In Fig. 5.40 we have acquired a second hard scattering. The generic case would be to have multiple extra hard scatterings. These would be non-leading if all the hard scatterings had large virtuality. They are all covered by the original space-time diagram, Fig. 5.17(b), where the diagonal lines correspond to the on-shell lines in the reduced graph. What has changed with respect to the standard regions is that at the origin we have multiple colliding lines. Since each extra hard collision needs a minimum of two incoming and two outgoing on-shell lines, such a situation cannot arise in e^+e^- annihilation and DIS; in the hadronic part of these processes there is zero or one incoming hadron (respectively).

Viewing the space-time structure of the collision, Fig. 5.18, gives further intuition. Each incoming hadron contains multiple constituents which are located at a longitudinal distance $1/Q$ of each other, but with a transverse separation $1/M$. The single genuine hard collision has a quark out of one hadron getting within a transverse distance $1/Q$ of each other. The remaining constituents undergo soft collisions over a transverse range $1/M$; since these are soft collisions, the momentum transfer is restricted to small values, and the partons remain approximately collinear to their parent beams.

These situations are exactly of the kind that corresponds to spectator-spectator inter-actions with exchanged Glauber momentum. Therefore the Glauber region represents the general alternative scaling that we need to consider. The power-counting used for the Drell-Yan example readily generalizes to show that these situations contribute at leading power. Part of the factorization proof for the Drell-Yan process, in Ch. 14, will be to show a cancellation of the Glauber region.

Naturally, interesting variations on this theme can arise, e.g., if the transverse radius for the scattering differs substantially from the size of the hadron. This happens for nuclei. Similar adjustments to the picture are needed if the hard collision is at very large or small x, so that the size $1/Q$ of the hard collisions substantially differs from the longitudinal size of the fast-moving beam hadrons.

Exercises

5.1 (***) From the coordinate-space representation of Feynman graphs (or otherwise), determine the regions in coordinate space that correspond to the regions R_H, R_A, R_B, R_S, $R_{A'}$, and $R_{B'}$ for the vertex graph. As far as possible determine the locations quantitatively.

There are some non-trivial complications in this problem because the final answer involves integrals over oscillating functions, with a lot of cancellation. A good answer probably involves significant original research.

If possible, verify the validity of estimates such as those given in Secs. 5.5.3 and 5.5.9, and that were used in the caption of Fig. 5.24.

5.2 (***) The standard Landau-type analysis of singularities of Feynman graphs and of associated asymptotic problems is in momentum space. Reformulate it in coordinate space. The Coleman-Norton (Coleman and Norton, 1965) paper shows how the internal momentum configurations correspond to classical scattering processes. Show that this is literally true in a coordinate-space analysis.

Extend this result to treat asymptotics governed by nearby pinch singularities to show what regions of coordinate space dominate. Be as quantitative as possible. You should, for example, be able to recover the intuitive picture of the parton model, with its hard scattering on a short time scale on a constituent of a Lorentz-contracted, time-dilated hadron.

Are any corrections to this picture needed?

5.3 (***) Find in the published literature, or construct for yourself, a proof that the Landau equations are actually necessary and sufficient for a PSS of a Feynman graph. To see that this is a non-trivial exercise, critically examine the accounts given in a standard textbook, e.g., Bogoliubov and Shirkov (1959); Eden *et al.* (1966); Itzykson and Zuber (1980); Peskin and Schroeder (1995); Sterman (1993). Are full proofs actually given? Do they actually work, and cover both necessity and sufficiency? Do they apply to the massless case, or do they make implicit assumptions only valid in the massive case?

You should also find or devise a proof that extends to certain modified integrals that occur in perturbative QCD. Such cases include graphs with eikonal propagators for Wilson lines: Ch. 10. These do not mesh particularly well with the Feynman-parameter representations often used in the treatments of the Landau equations.

For applications to pQCD, as we will see, it is important not merely to know that there is a PSS, but also to know exactly which lines participate in a particular pinch and which not, and to know exactly which loop-momentum variables actually participate in the pinch. Extend results in the literature to cover these issues explicitly.

Preferably any proof should be comprehensible by ordinary students of QFT.

5.4 Catalog the most general leading regions for graphs for the following processes. Describe the corresponding space-time structure.
  (a) $q(P_A) + \gamma^*(q) \to q(p_B)$, i.e., the space-like version of the process treated in Sec. 5.3.1, with the state of momentum P_A in the initial state instead of the final state.
  (b) $H(P_A) + H(P_B) \to H(p_C) + X$, i.e., inclusive production of hadrons of large transverse momentum in hadron-hadron collisions.

5.5 For elastic hadron-hadron scattering, derive the power law given in Sec. 5.9.3 when there are multiple hard scatterings. Pay careful attention to the effects of momentum conservation at the hard scattering on the collinear loop integrals.

5.6 Extend the power-counting analysis given in Sec. 5.5 to the following cases:

(a) other vertices replace the external electromagnetic current, e.g., a $\bar{\psi}\psi H$ vertex that might be for the interaction of a fermionic quark with a Higgs field;

(b) scalar quark in gauge theory.

These represent possible variations on the basic ideas that might occur in applications of the Standard Model, or in extensions of it (e.g., scalar quarks appear in supersymmetric extensions).

5.7 Verify that the general rules given for power-counting apply in these specific cases. If not, improve the rules.

5.8 (**) Prove that the PSSs for a massless Feynman graph are flat surfaces.

6

Parton model to parton theory: simple model theories

Basic ideas on the space-time structure of deeply inelastic scattering (DIS), symbolized in Figs. 2.2 and 2.5, led us to the parton model in Sec. 2.4. However, as we saw in Ch. 5, the leading regions can be more general than those that give the parton model. Indeed, the properties needed for the literal truth of the parton model are violated in any QFT that needs renormalization or that is a gauge theory, or both, like QCD.

Even so, the ideas that led to the parton model (the distance scales, time dilation and Lorentz contraction) are such basic properties that one should expect the parton model to be some kind of approximation to real QCD.

Because of the complications inherent to a sound treatment in QCD, it is useful to build up methodologies step by step. In this chapter, we treat situations where the parton model is correct, which happens in suitable model field theories. For these we will construct a strict field-theoretic implementation of the parton model.

One key result will be operator definitions of the parton distribution functions (parton densities or pdfs). Another result will be light-front quantization, whereby a probability interpretation of a pdf can be completely justified, in those model theories where the parton model is exact.

6.1 Field theory formulation of parton model

DIS concerns electron scattering off a hadronic target, $e + P \rightarrow e + X$, to lowest order in electromagnetism, with kinematic variables and structure functions defined in Sec. 2.3. Our aim is to understand the asymptotics when the momentum transfer Q is much larger than a typical hadronic scale, with the Bjorken variable x held fixed, away from 0 and 1.

In the parton model (Sec. 2.4), the process is treated as being caused by a short-distance scattering of an electron off a parton, i.e., a quasi-free constituent of the target, with the electron-quark scattering taken to lowest order.

We implement the parton-model idea field-theoretically by an assertion that the dominant contribution arises from cut graphs of the form of the "handbag diagram" of Fig. 6.1, with the virtualities of the explicitly drawn quark lines being much less than Q^2. The methods of Ch. 5 tell us that this is equivalent to the statement that the only leading regions are those also symbolized by Fig. 6.1, where now the lower subgraph consists of lines collinear to the target, and the upper subgraph consists of lines collinear in another direction.

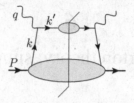

Fig. 6.1. Parton model in field theory starts from "handbag graphs" of this form. The assertion that the parton model is exactly valid is that all leading regions correspond to reduced graphs of the handbag form with the two bubbles being collinear subgraphs.

From the power-counting results in Sec. 5.8 and especially Sec. 5.8.11, we find the conditions that Fig. 6.1 gives all the leading regions: (a) there are no gauge fields, so that no extra gluons connect the hard scattering and the collinear subgraphs, and (b) the theory is super-renormalizable, so that higher-order terms in the hard scattering are power-suppressed.

Evidently, these conditions do not hold in QCD. It is nevertheless useful to investigate the consequences of assuming that Fig. 6.1 gives the whole leading-power behavior of the structure functions.

Even with this restriction, the power-counting results show that leading regions include those with non-trivial corrections on the struck quark line, i.e., that we should use Fig. 6.1 rather than Fig. 2.5(b), where we omitted the upper bubble. The final state for quark k' must therefore be considered a jet of hadrons, in agreement with experiment. The quark k' does not need to give a single particle in the final state; it can only be treated as a single particle over distance scales of order $1/Q$. Of course, if Fig. 6.1 were the whole story, then we would have particles in the final state with fractional electric charge. But Fig. 6.1 is not the whole story, because there are other leading regions in QCD.

6.1.1 Analysis of parton kinematics

We now analyze regions of the form of Fig. 6.1 on the hypothesis that they are the only leading regions. Our aim is to make a formal derivation of the parton model, and to obtain a definition of the parton densities.

It is convenient to use light-front coordinates (App. B) in the Breit frame, as described in Sec. 2.4. (A finite boost will not greatly affect the derivation.) In the Breit frame, the (space-like) photon has zero energy and its large 3-momentum is in the $-z$ direction. Then, as we saw in Sec. 2.4, the big light-front component of the target's momentum P is the plus component. We define the fractional plus momentum of the incoming quark to be ξ relative to the target, and write

$$q^\mu \doteq \left(-xP^+, \frac{Q^2}{2xP^+}, \mathbf{0}_\mathrm{T} \right), \tag{6.1}$$

$$P^\mu = \left(P^+, \frac{M^2}{2P^+}, \mathbf{0}_\mathrm{T} \right), \tag{6.2}$$

$$k^\mu = \left(\xi P^+, k^-, \mathbf{k}_\mathrm{T}\right), \tag{6.3}$$

$$k'^\mu = k^\mu + q^\mu = \left((\xi - x)P^+, \frac{Q^2}{2x P^+} + k^-, \mathbf{k}_\mathrm{T}\right), \tag{6.4}$$

with $x P^+ = Q/\sqrt{2}$ in the Breit frame. The collinear property of the momenta is that the only large component of k is its plus component and the only large component of k' is its minus component. From the analysis in Secs. 5.7 and 5.8, we find that the leading contribution is from where the transverse momentum $\mathbf{k}_\mathrm{T}$ is of order m, and the small components of longitudinal momentum, k^- and k'^+, are of order m^2/Q, where m characterizes the particle masses of the theory. Thus $\xi - x$ is of order $x m^2/Q^2$.

The contribution of Fig. 6.1 to $W^{\mu\nu}$ is

$$W^{\mu\nu} = \sum_j \frac{e_j^2}{4\pi} \int \frac{\mathrm{d}^4 k}{(2\pi)^4} \,\mathrm{Tr}\, \gamma^\mu \, U_j(k+q) \, \gamma^\nu \, L_j(k, P), \tag{6.5}$$

where $U_j(k')$ and $L_j(k, P)$ are the upper and lower bubbles, which are color- and Dirac-matrix-valued functions of their external momentum and quark flavor. The sum over j is over quark flavors and antiflavors, with e_j being the charge of the struck quark. The trace is over color as well as Dirac indices, and the factor $1/(4\pi)$ is from the definition of $W^{\mu\nu}$.

For the leading power in m/Q, a suitable approximation is to neglect the small components of momentum, $k^-, \mathbf{k}_\mathrm{T}$, and $(\xi - x)P^+$, with respect to Q where possible. A convenient way to do this is:

1. Apply a Lorentz transformation to U so that its quark k' has zero transverse momentum, and then neglect k^- with respect to q^-:

$$\left(k^+ + q^+ - \frac{k_\mathrm{T}^2}{2(q^- + k^-)}, q^- + k^-, \mathbf{0}_\mathrm{T}\right) \simeq \left(k^+ + q^+ - \frac{k_\mathrm{T}^2}{2q^-}, q^-, \mathbf{0}_\mathrm{T}\right). \tag{6.6}$$

   The matrix for the Lorentz transformation approaches unity as $k_\mathrm{T}/Q \to 0$.
2. Change the integration variable for the plus component of momentum from k^+ to $l^+ = k^+ + q^+ - k_\mathrm{T}^2/2q^-$, so that $k^+ = -q^+ + l^+ + k_\mathrm{T}^2/2q^-$. In the region of interest $k^+ \simeq -q^+ = x P^+$, up to a small fractional correction.
3. Therefore, in the lower part of the graph, L, we approximate k^+ by the fixed value $x P^+$. For this we need to assume that L is a smooth function of k^+/P^+, which is normally true in QCD, as evidenced by the smooth dependence of structure functions on x in Fig. 2.6. When the smoothness assumption is false, we can instead apply the derivation to a local average of the x dependence of a structure function, as a generalization of Secs. 4.1.1 and 4.4.
4. Project out the leading part of the Dirac matrix trace.

After the first three steps, we find

$$W^{\mu\nu} \simeq \sum_j \frac{e_j^2}{4\pi} \, \mathrm{Tr}\, \gamma^\mu \left[\int \frac{\mathrm{d}l^+}{2\pi} U_j(l^+, q^-, \mathbf{0}_\mathrm{T}) \right] \gamma^\nu \left[\int \frac{\mathrm{d}k^- \, \mathrm{d}^2 k_\mathrm{T}}{(2\pi)^3} L_j((xP^+, k^-, k_\mathrm{T}), P) \right].$$

$$(6.7)$$

The leading-power approximation has short-circuited the integrations, so that the integrations over k^- and k_T are restricted to L, and the l^+ integration is restricted to U. So we have two factors coupled by a trace in Dirac spinor space, and a trivial trace over color indices.

6.1.2 Projection of Dirac matrix structure

Projectors on matrix space

To project out the leading part of the Dirac trace, we apply (A.23) to write L in terms of numerical basis matrices:

$$L = A + \gamma_5 B + \gamma_\mu C^\mu + \gamma_\mu \gamma_5 D^\mu + \sigma_{\mu\nu} E^{\mu\nu}, \tag{6.8}$$

where we temporarily drop the flavor index j. Now L is highly boosted from the target rest frame, and we know the transformation properties of the coefficients, which are a Lorentz scalar A, pseudo-scalar B, vector C, etc. In the target rest frame, each of the coefficients $A, \ldots, E^{\mu\nu}$ has a fixed order of magnitude. Boosting to the Breit frame increases plus components and decreases minus components by a large factor. The large terms are C^+, D^+ and E^{+i}, which multiply γ^- factors. Only these can give leading-power contributions to (6.5). They may be obtained from L by, for example, $C^+ = \frac{1}{4} \mathrm{Tr}\, \gamma^+ L$. Note that the antisymmetry of $\sigma_{\mu\nu}$ removes the possibility of an otherwise dominant term with E^{++}. A similar decomposition applies to U_j, for which the coefficients of γ^+ are biggest.

Projectors on spinor space

The above method works for the quantities L and U as a whole. We now show an alternative method that works more locally in the Feynman graphs: to extract the large Q asymptote, it applies projectors on the individual lines joining the electromagnetic vertices to L and U. This method will show that the hard scattering is computed with Dirac wave functions for on-shell massless quarks, exactly as in the parton model.

Now each of L and U is obtained by a large boost from a rest frame. Since Dirac spinors are in the $(\frac{1}{2}, 0) \oplus (0, \frac{1}{2})$ representation of the Lorentz group, spinors in one two-dimensional subspace increase like $Q^{1/2}$, and those in the other subspace decrease like $Q^{-1/2}$. The first subspace is the part that gives the leading power as $Q/m \to \infty$. The same subspace is also obtained by taking the zero mass limit, and is the space of Dirac wave functions for the appropriate massless momentum, in the plus direction for L and the minus direction for U.

To project the leading power in the Dirac trace, we therefore use a matrix that projects onto the space of massless wave functions. Let $u_s(p_\infty)$ be a Dirac wave function for a massless particle of momentum p_∞ and spin label s. A covariantly defined projection onto their space is

$$\mathcal{P}(p_\infty) \overset{\text{def}}{=} \frac{\sum_s u_s(p_\infty)\bar{u}_s(p_\infty)\gamma \cdot n}{\bar{u}(p_\infty)\gamma \cdot nu(p_\infty)} = \frac{\not{p}_\infty \gamma \cdot n}{2p_\infty \cdot n}. \tag{6.9}$$

Here n is any vector such that $p_\infty \cdot n \neq 0$.

How do we resolve the ambiguity from the choice of n? Notice first that $\mathcal{P}(p_\infty)$ is invariant when n is simply scaled by a factor. We actually need a projection matrix for each external line of the hard scattering. The primary constraint on the vector n in each projector is that the projection matrix should not upset the power-counting. Thus if in the center-of-mass frame the largest components of p and n are of order E and $n_{\max}$, then $p_\infty \cdot n$ is at most approximately $En_{\max}$. Preserving the power-counting requires that $p_\infty \cdot n$ should *not* be a large factor smaller than $En_{\max}$. Since the largest component of a on-shell momentum is the energy, it is easiest to satisfy the requirement by setting n to be the rest vector of the center-of-mass.

In the case of DIS, we need two projectors, onto the wave functions for target and jet sides in (6.7). We can choose the n vectors in the $(0, z)$ plane, e.g., the rest vector of the Breit frame. The results are then unique, and the two projectors are

$$\mathcal{P}_A \overset{\text{def}}{=} \mathcal{P}(k^+, 0, \mathbf{0}_T) = \frac{\gamma^-\gamma^+}{2}, \qquad \mathcal{P}_B \overset{\text{def}}{=} \mathcal{P}(0, q^-, \mathbf{0}_T) = \frac{\gamma^+\gamma^-}{2}. \tag{6.10}$$

For projections onto the conjugate spinors $\bar{u}$ we use

$$\overline{\mathcal{P}}(p_\infty) \overset{\text{def}}{=} \frac{n \cdot \gamma \sum_s u_s(p_\infty)\bar{u}_s(p_\infty)}{\bar{u}(p_\infty)n \cdot \gamma u(p_\infty)} = \frac{n \cdot \gamma \not{p}_\infty}{2n \cdot p_\infty} = 1 - \mathcal{P}(p_\infty), \tag{6.11}$$

so that

$$\overline{\mathcal{P}}_A = \mathcal{P}_B, \qquad \overline{\mathcal{P}}_B = \mathcal{P}_A. \tag{6.12}$$

Using these in (6.7) to project the leading-power terms gives

$$W^{\mu\nu} \simeq \sum_j \frac{e_j^2}{4\pi} \operatorname{Tr} \gamma^\mu \left[\int \frac{dl^+}{2\pi} \mathcal{P}_B U_j(l^+, q^-, \mathbf{0}_T)\mathcal{P}_A \right]$$
$$\times \gamma^\nu \left[\int \frac{dk^- d^2k_T}{(2\pi)^3} \mathcal{P}_A L_j\big((xP^+, k^-, k_T), P\big) \mathcal{P}_B \right]. \tag{6.13}$$

Notice that $\mathcal{P}_A L \mathcal{P}_B$ projects out exactly the terms in L involving C^+, D^+ and E^{+i}. Thus the projection-matrix technique reproduces the results of the first method in this section.

6.1.3 *Parton densities: unpolarized and polarized*

We now show how to organize (6.13) into a form involving parton densities and what we will call hard-scattering coefficients. The hard scattering corresponds, as we will see, to

DIS on a free quark target, i.e., the process $\gamma^* + q \to q$ with the quarks on-shell and of zero transverse momentum.

Definitions

First we define quantities f_j, λ_j and $b^i_{j\mathrm{T}}$ by[1]

$$f_j(\xi) \stackrel{\text{prelim}}{=} \int \frac{\mathrm{d}k^- \, \mathrm{d}^2 k_\mathrm{T}}{(2\pi)^4} \operatorname{Tr} \frac{\gamma^+}{2} L_j(k, P), \qquad (6.14)$$

$$\lambda_j f_j(\xi) \stackrel{\text{prelim}}{=} \int \frac{\mathrm{d}k^- \, \mathrm{d}^2 k_\mathrm{T}}{(2\pi)^4} \operatorname{Tr} \frac{\gamma^+}{2} \gamma_5 L_j(k, P), \qquad (6.15)$$

$$b^i_{\mathrm{T}j} f_j(\xi) \stackrel{\text{prelim}}{=} \int \frac{\mathrm{d}k^- \, \mathrm{d}^2 k_\mathrm{T}}{(2\pi)^4} \operatorname{Tr} \frac{\gamma^+}{2} \gamma^i \gamma_5 L_j(k, P), \qquad (6.16)$$

with the traces being over both color and Dirac indices. We have a sum over quark colors, and it is not useful to define separate quark densities for different colors. The variable ξ is k^+/P^+, and is equal to x in the use of these definitions in the parton-model approximation for $W^{\mu\nu}$. We keep the more general variable ξ to emphasize that it is not in the first instance to be identified with the Bjorken x variable of DIS.

These definitions correspond to the leading terms C^+, D^+ and E^{+i} in (6.8). But there is a change in normalization that lets $f_j(\xi)$ etc. have simple interpretations when we use light-front quantization. We will find that $f_j(\xi)$ is the number density in ξ of quarks of flavor j. The terminology "parton density", "parton distribution" or "parton distribution function" (pdf) is therefore appropriate – all three names are in common use.

We will also find that λ_j is the longitudinal quark polarization and $b_{j\mathrm{T}}$ is the transverse quark polarization, both normalized to maximum values of unity. For a spin-$\frac{1}{2}$ parton these variables suffice to specify the most general spin state, pure or mixed; see Sec. 6.5. We will also see that the quark polarizations are functions of ξ times the corresponding variables specifying target polarization. We therefore define the polarized parton densities $\Delta f_j(\xi)$ and $\delta_\mathrm{T} f_j(\xi)$ as the coefficients of proportionality:

$$\lambda_{\text{targ}} \Delta f_j(\xi) = \lambda_j f_j(\xi), \qquad (6.17)$$

$$b_{\mathrm{T}\text{targ}} \delta_\mathrm{T} f_j(\xi) = b_{\mathrm{T}j} f_j(\xi). \qquad (6.18)$$

An interpretation will be that Δf_j is the number density of parallel-helicity quarks minus that of antiparallel-helicity quarks of flavor j in a target of maximal right-handed helicity, i.e., it is the helicity asymmetry. Similarly, $\delta_\mathrm{T} f_j(\xi)$ is an asymmetry in transverse spin.

Notation and terminology The transverse spin density is also called the transversity density and the symbols δf, $\Delta_\mathrm{T} f$, h_T, $\Delta_1 f$ and h_1 are also used.

---

[1] The notation $\stackrel{\text{prelim}}{=}$ indicates that these definitions are preliminary. In full QCD, modified definitions will be necessary. The definitions given here are exactly correct only when all of the leading regions in a theory are of the kind depicted in Fig. 6.1.

Parton-model factorization

We now write (6.13) in terms of quark densities and polarization:

$$W^{\mu\nu} \simeq \sum_j \frac{e_j^2}{4\pi} f_j(x) \int \frac{dl^+}{\hat{k}^+} \underset{\mathrm{D}}{\mathrm{Tr}} \left[\gamma^\mu \, \mathcal{P}_B U_j(l^+, q^-, \mathbf{0}_{\mathrm{T}}) \mathcal{P}_A \, \gamma^\nu \frac{\slashed{\hat{k}}}{2} \left(1 - \gamma_5 \lambda_j - \gamma_5 b^i_{j\mathrm{T}} \gamma^i \right) \right].$$

(6.19)

Here $\hat{k}$ is an approximate version of k,

$$\hat{k} = (xP^+, 0, \mathbf{0}_{\mathrm{T}}),$$

(6.20)

which is massless and of zero transverse momentum. In (6.19), we choose the trace with U to be only over Dirac indices (subscript "D"); a color average is assumed, a triviality since U is a unit matrix in color space. This formula is of the form of a parton density times a structure tensor for DIS on a massless quark target of momentum $\hat{k}$. It still has an integral over the jet factor U_j, which we will convert to the Dirac matrix for a spin sum for a final-state quark in Sec. 6.1.4.

6.1.4 *Result for structure functions; including polarization*

We now analyze the jet factor, obtained from the upper part U_j of the graphs. The result will be a cancellation of all but the lowest-order graph, after which we will get exactly the standard parton model result, complete with its generalization to polarized scattering.

To do this, we use an argument from our discussion of e^+e^- annihilation, around Figs. 4.13 and 4.14, applied to the integral over l^+ of

$$\mathcal{P}_B U_j(l^+, q^-, \mathbf{0}_{\mathrm{T}}) \mathcal{P}_A = q^- \gamma^+ \tilde{U}_j(2l^+q^-),$$

(6.21)

which is a cut 2-point function and therefore a discontinuity of an ordinary uncut propagator. In this equation, we have noted that the projectors pick out the coefficient of γ^+ in U, and have observed that its coefficient is q^- times a function $\tilde{U}$ of the virtuality of the quark. Terms proportional to $\gamma^+\gamma_5$ or to $\gamma^+\boldsymbol{\gamma}_{\mathrm{T}}$ are absent because of parity invariance and because of rotational invariance of the integral over final states at zero transverse momentum.

Initially we have a contour integral in l^+ around the cut of the propagator. We deform the contour out into complex plane, to where the quark has virtuality $2l^+q^-$, i.e., of order Q^2. Here we may correctly approximate all masses in the propagator by zero. Moreover, as usual, the decrease of the projected U (or of the uncut projected propagator) at large l^+ is the decrease of $\tilde{U}$ in (6.21) at large virtuality, which is governed by dimensional analysis of Feynman-graph integrands.

For the moment, we are working under the hypothesis that the parton model is exact, in which case our theory is super-renormalizable. Then all graphs for U beyond lowest order decrease by a power faster than $1/l^+$, and thus they provide a contribution to the integral suppressed by a power of Q. This leaves the lowest-order propagator, which decreases only as $1/l^+$. Therefore, we replace U by the lowest-order cut massless propagator

$\mathcal{P}_B U \mathcal{P}_A = \acute{q}^- \gamma^+ (2\pi)\delta(2l^+ q^-)$ to obtain

$$\int \frac{dl^+}{\hat{k}^+} \underset{D}{\mathrm{Tr}} \left[\gamma^\mu \, \mathcal{P}_B U_j(l^+, q^-) \mathcal{P}_A \gamma^\nu \frac{\hat{k}}{2} \left(1 - \gamma_5 \lambda_j - \gamma_5 s^i_{jT} \gamma^i\right) \right]$$

$$= \frac{2\pi}{Q^2} \mathrm{Tr} \, \gamma^\mu \, (\hat{k} + \acute{q}) \gamma^\nu \frac{\hat{k}}{2} \left(1 - \gamma_5 \lambda_j - \gamma_5 s^i_{jT} \gamma^i\right). \tag{6.22}$$

Then the parton-model approximation to $W^{\mu\nu}$ is

$$W^{\mu\nu} = \sum_j e_j^2 f_j(x) \left[\frac{1}{2} \left(-g^{\mu\nu} + q^\mu q^\nu / q^2\right) + \frac{(\hat{k}^\mu - q^\mu \hat{k} \cdot q/q^2)(\hat{k}^\nu - q^\nu \hat{k} \cdot q/q^2)}{\hat{k} \cdot q} \right.$$

$$\left. + \frac{1}{2} i \epsilon^{\mu\nu\alpha\beta} \frac{q_\alpha \lambda_j \hat{k}_\beta}{\hat{k} \cdot q} \right]. \tag{6.23}$$

To relate this to our original statement of the parton model, we first recognize the last factor in (6.22) as the numerator factor for DIS on a free massless quark target, i.e., for the process $\gamma^* + q_j(\hat{k}) \to q_j(\hat{k} + q)$. Next we observe that if we assign the incoming quark a fractional momentum ξ, i.e., if we replace $\hat{k}$ by $(\xi P^+, 0, \mathbf{0}_\mathrm{T})$, then the final-state cut propagator gives a factor

$$2\pi \delta\big((\hat{k} + q)^2\big) = \frac{2\pi}{Q^2} x \delta(\xi - x). \tag{6.24}$$

The first factor appears on the right of (6.22), and the delta function sets the parton momentum fraction equal to x.

Comparison of (6.23) with the definitions of the structure functions in (2.20) gives the parton-model results for all four structure functions:

$$F_2^{\mathrm{QPM}} = \sum_j e_j^2 x \, f_j(x), \qquad\qquad F_1^{\mathrm{QPM}} = \frac{1}{2} \sum_j e_j^2 f_j(x), \tag{6.25a}$$

$$g_1^{\mathrm{QPM}} = \frac{1}{2} \sum_j e_j^2 \Delta f_j(x), \qquad\qquad g_2^{\mathrm{QPM}} = 0. \tag{6.25b}$$

The first two agree with the previous results, Bjorken scaling being a prediction. But now we have a concrete derivation, which is susceptible to improvement. We also have a definite definition of the parton densities, and an extension to polarized DIS.

6.1.5 *Parton transverse momentum and virtuality*

The quark lines entering and leaving the hard scattering have momenta that we approximated as being of zero transverse momentum, massless and on-shell. However, it is important that this is an approximation applied only in a certain part of the diagrams. The actual quarks have non-zero transverse momentum, are off-shell, and have non-zero masses. Thus, in the definition, (6.14) etc., of the parton densities, the parton transverse momentum and virtuality are *non*-negligible and are actually *integrated* over. Failure to recognize this important distinction can lead to all kinds of unphysical paradoxes.

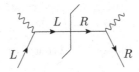

Fig. 6.2. Interference between left-handed and right-handed initial quark in DIS is prevented by helicity conservation at the electromagnetic vertex.

6.1.6 Parton densities vs structure functions

The parton density for transverse spin drops out of the result for $W^{\mu\nu}$, so that the g_2 structure function is zero in the parton-model approximation. This is associated with helicity conservation at the electromagnetic vertex in massless electron-quark scattering, in (6.22). To see this, observe that a transversely polarized state is a linear combination of states of left-handed and right-handed helicity, with a relative phase dependent on the azimuthal angle ϕ of the transverse spin vector around the direction of motion of the particle:

$$|\phi\rangle = \frac{1}{\sqrt{2}} \left(e^{i\phi/2} |L\rangle + e^{-i\phi/2} |R\rangle \right). \tag{6.26}$$

Getting a transverse-spin dependence of a cross section, i.e., a dependence on ϕ, requires interference between amplitudes for a left-handed and a right-handed initial state that produce some common final state. But helicity conservation at the electromagnetic coupling of massless particle implies that the final-state quark has the same helicity as the initial state, so that there is no interference (Fig. 6.2).

Because the unpolarized and the longitudinal-polarization quark densities have simple relations to structure functions in the parton model, one often sees a confusion between the concepts of parton density and structure function, with parton densities sometimes being called structure functions. The error of confusing the concepts must be strongly avoided. The structure functions are properties of cross sections, needing only elementary properties of electroweak interactions for their definition. But parton densities are more abstract theoretical constructs in QCD, with definite definitions; they are only related to experiment because factorization theorems can be derived to relate structure functions and other cross sections to parton densities in certain approximations. An excellent example of confusion between parton distributions and structure functions is in Jaffe and Ji (1991), where even the same notation is used for some structure functions and their corresponding parton densities.

The issue becomes particularly noticeable in the case of transverse polarization (Barone, Drago, and Ratcliffe, 2002), since transverse spin dependence drops out of $W^{\mu\nu}$ (at leading power). While the formalism clearly allows for a possible transverse spin dependence, it is the dynamics of a particular theory that determine whether or not there is a non-zero transverse spin dependence for a particular reaction. A reaction other than fully inclusive DIS is needed for a non-zero effect. This has been a topic of intense study in recent years – see Secs. 13.16 and 14.5.4 for examples.

Confusion has arisen from incorrect results in the older literature which apparently indicate that transverse-polarization effects are universally suppressed in hard collisions, contrary to reality. One example is in Feynman (1972), where on p. 157 an incorrect derivation related a combination of the g_1 and g_2 structure functions to the transverse spin densities. Another example is in Wandzura and Wilczek (1977), where we read (p. 196):

For a highly relativistic quark, the quark spin is, of course, nearly always parallel to its momentum.

and (p. 197):

In the parton model combination $g_1(x) + g_2(x)$ is equal to the difference $k_+(x) - k_-(x)$ of distribution functions for a parton with momentum fraction x in the infinite momentum frame to be spinning up ($k_+(x)$) or down ($k_-(x)$) in a nucleon spinning up (perpendicular to the infinite momentum). Now, again, if the parton is moving rapidly we expect that with overwhelming probability it is spinning along its direction of motion, and therefore

$$k_+(x) \approx k_-(x) \ldots$$

Their notation follows that of Feynman (1972), and $k_+(x) - k_-(x)$ is to be identified with $\delta_T f(x)$. The problem is that the large size of the longitudinal component of a boosted spin vector is entirely misleading.

This can be seen in the formula (A.26) for the expression of a spin state in Dirac spinor state. The spin vector appears in the combination $\$/M$, whose biggest component is of order E/M for a particle of high energy. However the effect of the big component disappears, because it is multiplied by $\not{p} + M$.

This can be seen from the non-singular massless limit (A.27). Thus for our purposes, it is generally preferable to use a helicity density matrix to parameterize the spin state of a particle or a parton (Sec. 6.5). The helicity variable λ is invariant under boosts along the direction of motion. It is true that DIS structure functions on a spin-$\frac{1}{2}$ target are defined, (2.20), in terms of the spin vector; but in a more general situation, the density matrix gives a better route to correct power-counting.

6.2 When is the parton model valid?

The word "valid" in the title of this section means "correct to the leading power of Q".

6.2.1 Properties needed to derive parton model

To understand the generalization of the parton model to QCD, it is useful to pinpoint the assumptions used to derive the parton model. Then we can determine QFTs in which the assumptions are derivable or easily repairable. The inter-related assumptions are as follows.

1. The dominant contributions have the structure of Fig. 6.1, i.e., the hard scattering occurs off a single parton, with no final-state interactions between the outgoing parton and the spectator part of the target.

   Note that final-state *collinear* interactions of the struck quark are explicitly allowed for, and they cancel, as we showed, so that the final-state quark can be treated as if it were free.

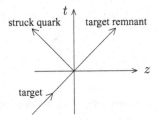

Fig. 6.3. The Libby-Sterman analysis associates these world lines of massless particles in the Breit frame with the leading region that gives the parton model.

2. The hadronic amplitude L falls off sufficiently rapidly at large k_T that the integrals defining the parton densities are convergent.
3. The corrections to U at large virtuality of k' fall more rapidly than the free-field term. Thus when we integrate over the virtuality of k', as in Sec. 6.1.4, all but the free-field term drop out. This leaves us with an effectively free final-state quark: we can replace Fig. 6.1 by Fig. 2.5(b).
4. The parton density is smooth and slowly varying on a scale of x.

6.2.2 When are they true?

In Secs. 5.8 and 5.9, we found rules that determine all the regions that contribute at the leading power of Q. If all the leading regions are those represented in Fig. 6.1, then we need a super-renormalizable model theory without gauge fields. The lack of gauge fields removes the possibility of a soft subgraph, and of extra gluons connecting the collinear subgraphs to the hard subgraph. Super-renormalizability implies that higher-order corrections to the hard scattering are power-suppressed.

In such a theory (e.g., Yukawa theory in three space-time dimensions) it is also true that the decrease of L at large k_T and of U at large virtuality is sufficient to give convergence of the integrals on the right of (6.13). To see this, we observe that if the integrals did not converge, there would be an unsuppressed contribution from large values of the integration momenta. Then there would be extra leading regions beyond those of Fig. 6.1.

Related to the Libby-Sterman analysis is that the trajectories of the target and its constituents, including the struck quark and the target remnant, are in the vicinity of the light-like world line from bottom left to top right in Fig. 6.3. At the origin, the virtual photon injects negative momentum to make the struck quark go to the left. Near its world line are the collinear interactions that convert the outgoing quark into a jet.

6.2.3 Smoothness or otherwise of parton density

It was known, even in the earliest days, that to derive exactly the parton model from QFT one needs a sufficiently fast decrease of U and L, and that this assumption is violated in typical QFTs in a four-dimensional space-time. However, a less obvious assumption is that the L factor and hence the parton densities are smooth functions of ξ, so that one

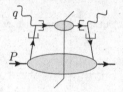

Fig. 6.4. Notation for parton-model approximation to the graph in Fig. 6.1.

can replace k^+ by xP^+, given that $|k^+ - xP^+| = O(xm^2/Q^2) \ll xP^+$. The necessary quantitative property is that the x derivative of a parton density should obey

$$\left| x \frac{\partial f(x)}{\partial x} \right| \lesssim f(x). \tag{6.27}$$

If this condition is badly violated, the relative errors in the parton-model approximation are much bigger than m^2/Q^2. When we generalize the parton model to the *standard* factorization theorems of QCD, the same smoothness property is needed.

From experimental measurements, the smoothness property in fact holds at moderate and small x for the real strong interaction, and hence for QCD. This is seen from the plots of the F_2 structure function in Fig. 2.6, or from many successful fits of factorization formulae to data that give measured values for parton densities.

However, the smoothness assumption is not universally true. In the first place, parton densities decrease to zero at $x = 1$ roughly like a power: $f \sim (1 - x)^n$, where the exponent is around 3 to 6, depending on the flavor of parton. Then (6.27) is violated as $x \to 1$:

$$\left| \frac{x \partial (1 - x)^n \partial x}{(1 - x)^n} \right| = \frac{nx}{1 - x} \sim \frac{n}{1 - x}. \tag{6.28}$$

In the second place, we can apply parton-model methods to other theories. For example in electromagnetic interactions at high energies it can be useful to apply parton methods (and the associated factorization theorems). In that case we need parton densities for electrons and photons in on-shell electron and photon states. As is readily seen in model calculations, these have delta-function terms at $x = 1$. This is the epitome of non-smoothness.

6.2.4 Notation for parton-model approximation

A diagrammatic notation for the approximations used in (6.19) is useful. For an unapprox-imated graph, Fig. 6.1, we represent the approximation by Fig. 6.4. Crossing the quark lines entering and leaving the hard scattering are thin bent lines ("hooks") denoting where approximations are applied. The approximations are as follows. On the hard-scattering side, i.e., the concave sides of the hooks, the momenta k and $k' = k + q$ are replaced by $(k^+, 0, \mathbf{0}_{\mathrm{T}})$ and $(0, k^- + q^-, \mathbf{0}_{\mathrm{T}})$ respectively, and masses are set to zero (which for this graph is a triviality). Momentum conservation then requires the approximated momenta to equal $(-q^+, 0, \mathbf{0}_{\mathrm{T}})$ and $(0, q^-, \mathbf{0}_{\mathrm{T}})$. The approximation also includes the insertion of Dirac projection matrices, $\mathcal{P}_A$ or $\mathcal{P}_B$, as appropriate.

These operations are all applied on the concave, hard-scattering sides of the hooks. Further operations are applied outside the hard scattering, to change the momentum of the quark in the target bubble from k^+ to $x P^+ = -q^+$, and to change the momentum of the final-state quark so that it has no transverse momentum.

One way of implementing the approximations on momenta is as a replacement of the hard vertex and the associated momentum conservation delta function. Let us use $T_{\text{PM, L}}$ and $T_{\text{PM, R}}$ to denote the application of the approximator on, respectively, the left-hand and right-hand sides of the final-state cut. Because of the Dirac projection matrices, these have slightly different formulae:

$$T_{\text{PM, L}} \, \gamma^\nu \delta^{(4)}(q + k - k') = \mathcal{P}_A \gamma^\nu \mathcal{P}_A \, \delta(q^+ + k^+) \, \delta(q^- - k'^-) \, \delta^{(2)}(k'_{\text{T}}), \qquad (6.29\text{a})$$

$$T_{\text{PM, R}} \, \gamma^\mu \delta^{(4)}(q + k - k') = \mathcal{P}_B \gamma^\mu \mathcal{P}_B \, \delta(q^+ + k^+) \, \delta(q^- - k'^-) \, \delta^{(2)}(k'_{\text{T}}). \qquad (6.29\text{b})$$

Thus we formulate the approximations locally at the places indicated by the hooks, rather than as global operations on a complete Feynman graph.

6.2.5 Shift of final-state momentum

Our parton-model approximation employed a shift of the plus component of k and the minus component of k'. This implies a shift of the momenta of both parts of the final state, i.e., the target remnant and the struck quark's jet. The approximation is certainly valid under the conditions we consider, i.e., in the parton-model kinematic region, when the parton density is a smooth function of x, and for the fully inclusive structure function, i.e., integrated over hadronic final states.

However, there are more general situations. For example, Monte-Carlo event generators generate complete simulated events for processes like DIS. When they are based on the usual partonic methods, the standard kinematic approximations result in events that violate momentum conservation. Thus it is necessary to adjust (Bengtsson and Sjöstrand, 1988) the parton kinematics so that generated events obey 4-momentum conservation.

In this and similar cases, if one wishes to obtain a more systematic treatment, there is a conflict between the need to maintain exact kinematics and the kinematic approximations used in standard factorization. This has been particularly emphasized by Watt, Martin, and Ryskin (2003, 2004); Collins and Zu (2005); Collins and Jung (2005); Collins, Rogers, and Staśto (2008). These authors show that more general methods are needed. One case, to be treated in this book in Chs. 13 and 14, concerns cross sections sensitive to partonic transverse momentum.

For our immediate purposes, of treating inclusive cross sections, the standard kinematic approximations are appropriate. But it is important to be aware of the flexibility of adjusting the approximations to the actual situations under discussion. Thus it is useful to make very explicit the form of the approximations, with an aim of recognizing situations where changes are needed. The form of the kinematic approximations is closely tied to the detailed structure of the corresponding factorization theorem, and to the definitions of the parton densities (or their generalizations used with different approximations).

6.3 Parton densities as operator matrix elements

6.3.1 Unpolarized quark density

The parton density defined in (6.14) is an integral over the lower bubble in Fig. 6.4, together with a trace with $\gamma^+/2$:

$$f_{j/h}(\xi) \stackrel{\text{prelim}}{=} \text{Tr} \frac{\gamma^+}{2} \int \frac{dk^- \, d^2 k_T}{(2\pi)^4} \quad \text{(6.30)}$$

where h denotes the type of the target hadron, and $k^+ = \xi P^+$. The diagram is a certain amplitude times its conjugate, with the amplitude involving one off-shell quark, the target state, and a final state. When the quark line on the left of the final-state cut is directed away from the lower bubble, then its top end corresponds to annihilation of a quark by the field ψ_j. It is left as an exercise (problem 6.2) to derive an explicit formula for the quark density as a matrix element of a bilocal operator:

$$f_{j/h}(\xi) \stackrel{\text{prelim}}{=} \int \frac{dw^-}{2\pi} \, e^{-i\xi P^+ w^-} \left\langle P | \overline{\psi}_j(0, w^-, \mathbf{0}_T) \frac{\gamma^+}{2} \psi_j(0) | P \right\rangle_c . \quad \text{(6.31)}$$

With standard conventions, it is the *right*-hand part of the matrix element, with the ψ_j field, that corresponds to the part of the diagram to the *left* of the final-state cut, and the left-hand part of the matrix element corresponds to the complex conjugated amplitude on the right of the cut. Only the contribution with the quark fields connected to the target state $|P\rangle$ are to be included, and this is indicated by the subscript "c".

The field $\psi_j(0)$ represents the extraction of a quark by the hard scattering. Because we integrate over all momentum in the minus and transverse directions, the antiquark field in the complex conjugate amplitude has zero relative position in w^+ and $\mathbf{w}_T$; note that w^+ is Fourier conjugate to the opposite light-front component k^- in momentum space. The average position of the quark and antiquark fields is irrelevant, since the definition is actually applied to a momentum eigenstate, i.e., a target state uniformly spread out over all space. The space-time locations of the fields are shown in Fig. 6.5.

We have again tagged the definitions as preliminary, in view of the adjustments that will be needed in QCD.

The restriction to connected amplitudes can be implemented by subtracting disconnected graphs, Fig. 6.6, i.e., as subtraction of the vacuum expectation value (VEV) of the operator. This can be written as

$$\left\langle P' | \overline{\psi}_j(y) \gamma^+ \psi_j(0) | P \right\rangle_c \stackrel{\text{def}}{=} \left\langle P' | \overline{\psi}_j(y) \gamma^+ \psi_j(0) | P \right\rangle - \left\langle P' | P \right\rangle \left\langle 0 | \overline{\psi}_j(y) \gamma^+ \psi_j(0) | 0 \right\rangle . \quad \text{(6.32)}$$

An off-diagonal matrix is used here, since momentum eigenstates are non-normalizable. After the subtraction, the diagonal matrix element can be taken: i.e., with $P' = P$. Without this manoeuvre, we would subtract an unquantified infinity proportional to $\langle P | P \rangle$.

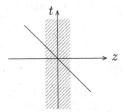

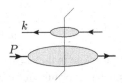

Fig. 6.5. The space-time location of the fields is along a light-like line. The shaded region represents the approximate location of the target hadron in its rest frame.

Fig. 6.6. Disconnected graphs, of this form, must be removed from the definition of the parton density when extended to negative ξ.

6.3.2 Antiquark density

For the density of an antiquark, whose flavor we denote by $\bar{j}$, we have similarly

$$f_{\bar{j}/h}(\xi) \stackrel{\text{prelim}}{=} \int \frac{dw^-}{2\pi} e^{-i\xi P^+ w^-} \operatorname{Tr} \frac{\gamma^+}{2} \left\langle P \left| \psi_j(0, w^-, \mathbf{0}_T) \overline{\psi}_j(0) \right| P \right\rangle_c, \tag{6.33}$$

where the trace is over the Dirac and color indices of the fields. In the parton model, the antiquark density appears in contributions to the structure function where the direction of the quark line in Figs. 6.1 and 6.4 is reversed.

6.3.3 Lorentz-covariant definition

The definitions of parton densities are not Lorentz invariant, but they have Lorentz-covariant expressions in terms of a single light-like vector $n^\mu \propto (0, 1, \mathbf{0}_T) = \delta^\mu_-$, so that $\xi = k \cdot n/P \cdot n$. Thus:

$$f_{j/h}(\xi) \stackrel{\text{prelim}}{=} \int \frac{d\lambda}{2\pi} e^{-i\xi n \cdot P \lambda} \left\langle P \left| \overline{\psi}_j(\lambda n) \frac{n \cdot \gamma}{2} \psi_j(0) \right| P \right\rangle_c. \tag{6.34}$$

Here the right-hand side is a scalar, so it is a function of Lorentz invariants only, i.e., of $k \cdot n$ and $P \cdot n$, with n^2 fixed at zero. The formula is invariant under scaling of n by a positive factor, so that only the combination $k \cdot n/P \cdot n$, i.e., ξ, is allowed. Hence, as a function of ξ, the numerical values of the quark density are independent of n, provided only that it is light-like and future-pointing. But for deriving factorization a suitable choice of n is needed, which is determined by the directions of the external momenta p and q.

6.3.4 Relation to wave function?

A parton density can be thought of as some property of the target. But since it is an integral along the light-like line in Fig. 6.5, there can be no simple relation to an ordinary wave function as used in non-relativistic physics, which corresponds to properties of the target at a fixed time in the target rest frame. The transformation to a light-like line involves the interactions of the theory.

We will need to use light-front quantization to interpret parton densities in terms of wave functions (Secs. 6.6 and 6.7).

6.3.5 Support properties

The intermediate state in the parton density, between the two fields, has momentum $P - k$. For it to be physical, it must have non-negative energy, so that $P^+ - k^+ \geq 0$, i.e., $\xi \leq 1$. Thus the parton density is zero if $\xi > 1$.

In the parton-model factorization formula, (6.19) and Fig. 6.4, the final state in the *upper* part of the graph has plus momentum $\xi P^+ + q^+ = (\xi - x)P^+$. This must be positive, in order that the state have positive energy, so that $\xi \geq x > 0$. This restriction applies quite generally in standard factorization formulae for cross sections. Thus we will use parton densities only in the range $0 < \xi \leq 1$.

However, the matrix element for the parton density is generally non-vanishing for negative ξ. We will see later that we can relate $f_{j/h}(\xi)$ for negative ξ to the antiquark density with the opposite sign of ξ: $f_{j/h}(\xi) = -f_{\bar{j}/h}(-\xi)$. This will be critical to the derivation of sum rules. But to make it work, it will be important that we have removed disconnected graphs, Fig. 6.6, from the definition; the disconnected graphs are non-zero for negative values of ξ.

6.3.6 Polarized quark densities

We defined polarized quark densities in (6.15) and (6.16). By the same methods as we used for the unpolarized densities, these can be re-expressed as expectation values in a target state of the operators $\overline{\psi}(z)\gamma^+\gamma_5\psi(0)$ and $\overline{\psi}(z)\gamma^+\gamma^i_\perp\gamma_5\psi(0)$:

$$\lambda_{\text{targ}}\Delta f_j(\xi) = \int \frac{dw^-}{2\pi} e^{-i\xi P^+ w^-} \left\langle P, S \left| \overline{\psi}_j(0, w^-, \mathbf{0}_T) \frac{\gamma^+\gamma_5}{2} \psi_j(0) \right| P, S \right\rangle_c, \quad (6.35)$$

$$\boldsymbol{b}_{\text{Ttarg}}\delta_T f_j(\xi) = \int \frac{dw^-}{2\pi} e^{-i\xi P^+ w^-} \left\langle P, S \left| \overline{\psi}_j(0, w^-, \mathbf{0}_T) \frac{\gamma^+\gamma^i_\perp\gamma_5}{2} \psi_j(0) \right| P, S \right\rangle_c. \quad (6.36)$$

Here $|P, S\rangle$ denotes a state with normalized helicity λ_{targ} and normalized transverse spin $\boldsymbol{b}_{\text{Ttarg}}$. These definitions presuppose proportionalities between quark and target spin variables, to be proved in Sec. 6.4. Then the quantities $\Delta f_j(\xi)$ and $\delta_T f_j(\xi)$ are independent of target polarization, i.e., they are parton densities par excellence.

6.3.7 Polarized antiquark densities

Similarly, definitions of polarized antiquark densities are

$$\lambda_{\text{targ}}\Delta f_{\bar{j}}(\xi) = -\int \frac{dw^-}{2\pi} e^{-i\xi P^+ w^-} \left\langle P, S \left| \psi_j(0, w^-, \mathbf{0}_T)\overline{\psi}_j(0) \frac{\gamma^+\gamma_5}{2} \right| P, S \right\rangle_c, \quad (6.37)$$

$$\boldsymbol{b}_{\text{Ttarg}}\delta_T f_{\bar{j}}(\xi) = \int \frac{dw^-}{2\pi} e^{-i\xi P^+ w^-} \left\langle P, S \left| \psi_j(0, w^-, \mathbf{0}_T)\overline{\psi}_j(0) \frac{\gamma^+\gamma^i_\perp\gamma_5}{2} \right| P, S \right\rangle_c. \quad (6.38)$$

Note the extra minus sign in (6.37) compared with the other antiquark densities. For the moment, we can regard this as purely a strange convention. Later we will see that the signs are those needed to give a number-density interpretation.

6.4 Consequences of rotation and parity invariance: polarization dependence

In this section, we examine how parton densities depend on the polarization of the particles and the quarks. This will introduce us to techniques for analyzing the consequences for parton physics of symmetries of QCD, and will justify the definitions given in Secs. 6.3.6 and 6.3.7.

Mental health warning: There are no fixed conventions for the normalizations of many of the objects discussed in this section. The objects concerned range from the definitions even of basic mathematical quantities, like $\epsilon_{\kappa\lambda\mu\nu}$, through the definitions of various kinds of spin vector, to the definitions of structure functions and parton densities. Quantities of the same name and symbol change their normalizations between different papers, even by the same authors. If one needs to make numerical results, it is important to check all conventions very carefully.

The conventions used in this book are defined in Apps. A.7 and A.10.

6.4.1 Polarization state

The target can be polarized, and in the most general case a spin density matrix is needed to specify the polarization state. So the target state $|P, S\rangle$ has an extra argument specifying the polarization. For the general case, this argument can be the density matrix, with respect to some basis. But for massive spin-$\frac{1}{2}$ hadron, like a proton, we can use the covariant spin vector S^μ, as defined in App. A.7. Although our notation, as in (6.35) etc., is as if we are working with pure states, there is actually an implicit trace with a helicity density matrix, as defined in (A.8) and (A.13), to allow the target to be in a mixed state.

A helicity basis is rather natural when we work with high-energy particles or with a massless limit. Helicity states are obtained in the theory of irreducible representations of the Poincaré group for massless particles. Moreover, the chiral symmetry ubiquitous in the massless limit of QCD perturbation theory effectively tells us to treat left-handed and right-handed quarks as if they were separate particles. Even so, transversely polarized quarks, i.e., states that are linear combinations of equal amounts of left- and right-handed components, are allowed physically, and have interesting properties.

When we obtain the most general dependence on target spin, it is important that expectation values as defined by (A.13) have a spin dependence that is linear in the spin vector S, Equally, it is linear in the normalized helicity λ and normalized transverse spin b_T, as defined in App. A.7. Helicity and transverse spin are well behaved in the massless limit unlike the covariant spin vector, and they apply also to the spin state of a quark parton.

The formula (A.12) for S in terms of b_T and λ exhibits some oddities in the zero-mass limit. In the rest frame of a massive spin-$\frac{1}{2}$ particle, the spin vector has only a spatial component, which unproblematically corresponds to standard usage in non-relativistic physics. Boosting along the z axis does not change the transverse component of S, but increases its longitudinal components. In contrast, λ and $\boldsymbol{b}_{\mathrm{T}}$ are invariant under the boost (except for the obvious change of sign when the direction of motion is reversed!).

But a massless particle has no rest frame. Instead one works either with the helicity density matrix or with its decomposition (A.9) in terms of $(\boldsymbol{b}_{\mathrm{T}}, \lambda)$. The spin vector is useful for a proton but not for an on-shell massless quark such as we use in hard scattering. In contrast, the density matrix and the Bloch vector formalisms work for both quarks and protons.

6.4.2 *Rotations about z axis*

The definition of what we called the unpolarized quark density $f_j(\xi)$ makes no reference to any azimuthal direction, in the (x, y) plane. Therefore we expect this parton density to be independent of the direction of the transverse spin vector of the target. Since matrix elements are linear in the target's spin vector, this implies that there is no dependence even on the size of the transverse spin vector.

To derive this and similar properties formally, we define an operator $U(\phi)$ on state space that corresponds to a rotation by an angle ϕ around the z axis; its action on a helicity eigenstate is

$$U(\phi)|P, \alpha\rangle = e^{-i\alpha\phi}|P, \alpha\rangle .\tag{6.39}$$

Hence the matrix element of an operator between helicity eigenstates obeys

$$\langle P, \alpha'|\text{ op }|P, \alpha\rangle = e^{i(\alpha-\alpha')\phi}\langle P, \alpha'|\, U(\phi)^{\dagger}\text{ op }U(\phi)\,|P, \alpha\rangle .\tag{6.40}$$

The combination $U(\phi)^{\dagger}\text{ op }U(\phi)$ is the rotated operator. Of the operators defining the parton densities the following two are rotation invariant: $\overline{\psi}(0, w^-, \boldsymbol{0}_{\mathrm{T}})\gamma^+\psi(0)$ and $\overline{\psi}(0, w^-, \boldsymbol{0}_{\mathrm{T}})\gamma^+\gamma_5\psi(0)$. From (6.40) follows that their matrix elements are diagonal in helicity eigenstates of the target.

For the case of a spin-$\frac{1}{2}$ target, we can apply the rotation to the spin vector S of a general spin state, (A.13), to get

$$\langle P, \text{rotated } S|\text{ op }|P, \text{rotated } S\rangle = \langle P, S|\text{ rotated op }|P, S\rangle .\tag{6.41}$$

6.4.3 *Implications for unpolarized quark density*

Spin-$\frac{1}{2}$ target

We now show that the unpolarized quark density $f_j(\xi)$ defined by (6.31) is independent of the polarization state of a spin-$\frac{1}{2}$ target.

We already proved that the matrix elements of the operator defining the unpolarized parton density are diagonal in helicity. But the transverse part of the spin vector only results in off-diagonal terms in the density matrix, so the unpolarized density is independent of transverse spin.

Now a parity transformation reverses the helicity of a state, but also reverses the 3-momentum. A rotation can then be applied to bring the momentum of the state back to its original value, and makes no change to the already reversed helicity. Thus we can apply the same method as above, but with $U(\phi)$ replaced by a unitary operator U_P that reverses helicity while preserving P:

$$U_P |P, j\rangle = |P, -j\rangle . \tag{6.42}$$

Since the operator in (6.31) is invariant under U_P, it follows that the unpolarized parton density is the same in states of opposite helicity, in a parity-invariant theory.

For a spin-$\frac{1}{2}$ target there are only two helicity states, so we now have shown that the unpolarized parton density $f_j(\xi)$ is independent of the polarization state.

Higher spin

When the target has spin higher than $\frac{1}{2}$, there is a wider range of possibilities. For example, in a spin-1 target, there is one unpolarized quark density for targets of helicity ± 1 and one for targets of helicity zero. There are corresponding generalizations for the DIS structure functions. See problem 6.8 for an exercise to fill in the details, and see Hoodbhoy, Jaffe, and Manohar (1989) for results on DIS, including several new structure functions.

6.5 Polarization and polarized parton densities in spin-$\frac{1}{2}$ target

To treat the polarized densities in a spin-$\frac{1}{2}$ target, we use its helicity density matrix $\rho_{\alpha\alpha'}(S)$ from (A.9), now written as

$$\rho(S) = \frac{1}{2} \left(I + \lambda_{\text{targ}} \sigma_z + \boldsymbol{b}_{\text{T targ}} \cdot \boldsymbol{\sigma} \right) , \tag{6.43}$$

where the label "targ" is used to distinguish the spin variables for the target from those for the quark. Expectation values of operators are linear in λ_{targ} and $\boldsymbol{b}_{\text{T targ}}$.

The operator defining the polarized parton density $\Delta f_j(\xi)$ in (6.35) is invariant under rotations around the z axis. Therefore its matrix elements are diagonal in helicity and hence independent of transverse spin, just like the unpolarized density. But unlike that case, the operator has the reversed sign under a parity transformation, as does normalized helicity λ_{targ}. So the analog of (6.41) used with the operator U_P of (6.42) shows that the matrix element of the operator is linear in λ_{targ}, as asserted in the matrix element representation (6.35). Thus all the dependence on target polarization is in the explicit factor of λ_{targ}, so that $\Delta f_j(\xi)$ is polarization independent.

Finally, the remaining parton density $\delta_{\text{T}} f_j(\xi)$ in (6.36) is obtained from the matrix element of an operator $\overline{\psi}_j(0, w^-, \boldsymbol{0}_{\text{T}}) \frac{\gamma^+ \gamma^i_\perp \gamma_5}{2} \psi_j(0)$ that transforms as a (two-dimensional transverse) vector. Because of the γ_5 factor, it actually transforms as a pseudo-vector, i.e., under a parity transformation it acquires a minus sign relative to the transformation of an ordinary momentum. The transverse spin vector is also a pseudo-vector.

To get the correct rotation properties, the matrix element of the operator must be a coefficient times the transverse spin vector, but possibly with the application of a rotation of some angle around the z axis. This rotation, as a function of ξ, would be a property of the target; it would represent some analog of optical rotation phenomena in a chiral medium. But let us apply a parity transformation followed by a 180° rotation about the x axis (say). This preserves the momentum and the x component of spin of the target, but it reverses the y component of spin. The same transformation applies to the operator. Thus a spin in the x direction gives only a non-zero x component to the matrix element, and similarly for the y component. Thus parity invariance requires there to be exactly no rotation between the spin vector and the matrix elements. (Actually a 180° rotation is also allowed, but for the two transverse directions in question, this is equivalent to a reversal of sign, i.e., to an overall coefficient.) Thus all the dependence on target polarization is in the explicit factor of b_{Ttarg}, so that $\delta_{\text{T}} f_j(\xi)$ is polarization independent.

6.6 Light-front quantization

A standard method of formulating quantum field theory uses the usual canonical quantization rules for a quantum theory: equal-time commutation (or anticommutation) relations are obtained from the Lagrangian density, and then the Heisenberg equations of motion determine the fields at all times from their values at one particular time. An alternative, first proposed by Dirac (1949), is to use a light-like surface $t + z = 0$ as the initial surface on which (anti)commutation relations are fixed. This is called light-front quantization, with the terms "light-cone quantization" and "null-plane quantization" being synonyms.

Light-front quantization is useful for DIS and other processes where a target system is probed along an almost light-like surface. Inspired by initial approaches using the so-called "infinite momentum frame", Bardakci and Halpern (1968) developed light-front quantization in field theories. Then Kogut and Soper (1970) made a very clear fundamental treatment. See Brodsky, Pauli, and Pinsky (1998) and Heinzl (2001) for reviews. See also Heinzl and Werner (1994) for a careful treatment of the issue that in solving the equations of motion, it is not sufficient to specify initial conditions on a light-like surface.

As we will now show, light-front quantization gives a direct probability interpretation of parton densities and yields a convenient decomposition of hadronic states in terms of partonic states. Further advantages are explained in the literature just quoted. In extending the method to theories like QCD that are renormalizable or have gauge fields, there are a number of complications that imply that light-front quantization must be used with care. Nevertheless, it provides important insights.

6.6.1 Formulation

To understand the general principles of light-front quantization, we examine the simple case of a Yukawa field theory with Lagrangian density

$$\mathcal{L} = \frac{i}{2}[\overline{\psi}\gamma^\mu \partial_\mu \psi - (\partial_\mu \overline{\psi})\gamma^\mu \psi] - M\overline{\psi}\psi + \frac{1}{2}(\partial\phi)^2 - \frac{m^2}{2}\phi^2 - g\overline{\psi}\psi\phi - \frac{h}{3!}\phi^3 - \frac{\lambda}{4!}\phi^4.$$

$$(6.44)$$

This theory is renormalizable at space-time dimension $n = 4$, and is super-renormalizable when $n < 4$. We use the scalar rather than the pseudo-scalar coupling for the Yukawa interaction, to avoid complications with γ_5 in dimensional regularization.

We use light-front coordinates $(x^+, x^-, \boldsymbol{x}_T)$ as defined in App. B. Then the equations of motion are

$$0 = i\partial\!\!\!/\psi - M\psi - g\psi\phi, \tag{6.45}$$

$$0 = 2\partial_+\partial_-\phi - \nabla_T^2\phi + m^2\phi + g\bar{\psi}\psi + \frac{h}{2}\phi^2 + \frac{\lambda}{3!}\phi^3. \tag{6.46}$$

In light-front quantization we treat these equations as giving evolution in x^+ from fields on the initial surface $x^+ = 0$.

Now the term with an x^+ derivative of the Dirac field is $i\gamma^+\partial_+\psi$, which only affects two independent components of ψ. Therefore we project onto what are called "good" and "bad" components of ψ by the matrices

$$\mathcal{P}_G = \frac{1}{2}\gamma^-\gamma^+, \qquad \mathcal{P}_B = \frac{1}{2}\gamma^+\gamma^-. \tag{6.47}$$

These are exactly the same as we used in projecting out the leading power of Q in the Dirac trace in the parton-model approximation to DIS, but now they appear with a more fundamental significance. In view of the jargon of this part of the subject, I replaced the subscript "A" by "G" for good: $\mathcal{P}_G = \mathcal{P}_A$. These matrices obey the usual properties of projectors ($\mathcal{P}_G + \mathcal{P}_B = 1$, $\mathcal{P}_G^2 = \mathcal{P}_G$, etc., and especially $\mathcal{P}_B\mathcal{P}_G = \mathcal{P}_G\mathcal{P}_B = 0$). Then we define the good and bad parts of the fermion field by

$$\psi_G = \mathcal{P}_G\psi, \qquad \psi_B = \mathcal{P}_B\psi, \tag{6.48}$$

so that $\overline{\psi_G} = \bar{\psi}\mathcal{P}_B$.

The equation of motion for ψ then separates into two separate two-dimensional pieces:

$$0 = 2i\,\partial_+\psi_G + \gamma^-\left(i\gamma^j\nabla_j - M - g\phi\right)\psi_B, \tag{6.49a}$$

$$0 = 2i\,\partial_-\psi_B + \gamma^+\left(i\gamma^j\nabla_j - M - g\phi\right)\psi_G, \tag{6.49b}$$

where the sums over j are over transverse components. The first equation gives the evolution of ψ_G in x^+, while we treat the second equation as a constraint: it fixes ψ_B at a given value of x^+ in terms of ψ_G, up to boundary conditions. So we treat ψ_G as the independent set of components. The solution of the constraint equation is

$$\psi_B(x) = \gamma^+\frac{i}{2\partial_-}\left(i\gamma^j\nabla_j - M - g\phi\right)\psi_G$$

$$= \frac{i\gamma^+}{4}\int_{-\infty}^{\infty} dy^-\,\mathrm{sign}(x^- - y^-)\left(i\gamma^j\nabla_j - M - g\phi\right)\psi_G(x^+, y^-, \boldsymbol{x}_T) + C_\psi(x^+, \boldsymbol{x}_T).$$

$$\tag{6.50}$$

There is a term C_ψ independent of x^- that is not determined by the equation of motion. When a Fourier transform over x^- and x_T is made, to momentum variables k^+ and k_T, the C_ψ term is proportional to a delta function at $k^+ = 0$. It is therefore characterized as contributing to the zero mode only. A similar zero mode arises in using the equation of motion (6.46) for the ϕ field to determine $\partial\phi/\partial x^+$ in terms of the fields on a surface of fixed x^+.

The zero-mode issue is quite important to the vacuum structure, and it is not clear to me that it has been properly treated in the literature. But much of what we do will not need a professional treatment of the zero modes. The primary issue is that the equations of motion alone are not sufficient to determine the evolution in x^+. Extra boundary conditions must be imposed. In contrast, for equal-time quantization, the Euler-Lagrange equations are sufficient to determine the time derivatives of the fields in terms of the independent fields and canonical momentum fields. A related complication concerns the $1/k^+$ singularity in mode sums like (6.59). For treatments of these and related issues, see Nakanishi and Yamawaki (1977), Yamawaki (1998), Heinzl (2003), Heinzl and Ilderton (2007, Sec. 4), and Steinhardt (1980).

We now arrange to form the quantum mechanics of our system by using Hamilton methods, but with evolution in the variable x^+ instead of conventional time.[2] For this we need commutation relations on surfaces of constant x^+, and a Hamilton to control the evolution by the standard Heisenberg equation

$$i\frac{\partial A(x)}{\partial x^+} = [A(x), P_+], \tag{6.51}$$

which applies to any field operator $A(x)$. Now the Lagrangian is linear in derivatives with respect to x^+, so the standard elementary rules of quantization need generalization, for which we use the simple formulation given by Faddeev and Jackiw (1988).

The appropriate Hamilton is just the Noether charge for translations in the x^+ direction, i.e., the appropriate component of momentum:

$$P_+ = \int dx^- \, d^2x_T \left[\overline{\psi}\left(-i\gamma^-\partial_- \gamma - i\gamma^j \nabla_j + M + g\phi\right)\psi \right.$$

$$\left. + \frac{1}{2}(\nabla_T\phi)^2 + \frac{m^2}{2}\phi^2 + \frac{h}{3!}\phi^3 + \frac{\lambda}{4!}\phi^4 \right]. \tag{6.52}$$

As with conventional equal-time quantization, for which the founding papers are Born and Jordan (1925); Dirac (1926) and Born, Heisenberg, and Jordan (1926), the equal-x^+ commutation/anticommutation are to be such that the equation of motion in the Heisenberg form (6.51) and in the Euler-Lagrange form (6.46) and (6.49) are equivalent. Thus we have

$$\frac{\partial}{\partial x^-}\left[\phi(x^+, x^-, x_T), \, \phi(x^+, w^-, w_T)\right] = \frac{-i}{2}\delta(x^- - w^-)\delta^{(2)}(x_T - w_T), \tag{6.53}$$

$$\left[\psi_G(x^+, x^-, x_T), \, \overline{\psi}_G(x^+, w^-, w_T)\right]_+ = \frac{\gamma^-}{2}\delta(x^- - w^-)\delta^{(2)}(x_T - w_T), \tag{6.54}$$

[2] For this reason, x^+ and the evolution operator P_+ are sometimes called "light-front time" and "light-front Hamiltonian".

with the other commutators involving ϕ, ψ_G and $\overline{\psi}_G$ being zero. The subscript $+$ in $\left[\psi, \overline{\psi}\right]_+$ etc. denotes the anticommutator appropriate for fermionic fields. Now the first of the above equations is the derivative of the commutator of the scalar field. From it we obtain the commutator of the field with itself:

$$\left[\phi(x^+, x^-, \mathbf{x}_T), \phi(x^+, w^-, \mathbf{w}_T)\right] = \frac{-i}{4}\,\text{sign}(x^- - w^-)\delta^{(2)}(\mathbf{x}_T - \mathbf{w}_T), \qquad (6.55)$$

with the boundary condition for inverting $\partial/\partial x^-$ being determined by the antisymmetry of the commutator of two ϕ fields under exchange of the position arguments.

To verify the correctness of this setup, one applies the (anti)commutation relations (6.53) and (6.54) to the the right-hand side of the Heisenberg equations of motion (6.51), for the fields ϕ and ψ_G. In this calculation we do not need the (anti)commutators of ϕ and ψ_G with ψ_B. For example, the term involving $[\phi(x), \psi_B(x^+, y^-, \mathbf{y}_T)]$ is

$$\int dy^- d^2 \mathbf{y}_T \left[\phi(x), \psi_B(x^+, y^-, \mathbf{y}_T)\right] \frac{\delta P_+}{\delta \psi_B(x^+, y^-, \mathbf{y}_T)}, \qquad (6.56)$$

where $\delta P_+/\delta \psi_B(y)$ denotes a functional derivative. This functional derivative is zero by the constraint equation of motion. It follows from elementary algebra that the Heisenberg equations are also valid for sums and products of fields. Unitary evolution implies that the (anti)commutation relations are true at all x^+ when they are true at $x^+ = 0$.

Since ψ_B is determined from the other fields by the interaction-dependent (6.50), the commutators and anticommutators of ψ_B are interaction dependent. Therefore, because the right-hand side of (6.50) is non-linear in fields, the equal-x^+ (anti)commutators of ψ_B are field dependent. That is, they are not simply numerical-valued functions times the unit operator. This is the primary reason for the jargon of calling ψ_B the bad components of the fermion field. For example, in current algebra one deals with operators constructed out of the elementary fields of a theory. Only for operators constructed solely out of good components at a given value of x^+ can one obtain their commutators directly from the canonical (anti)commutators of the elementary fields, without investigating how to solve the theory.

A similar issue arises with the quark densities. Because of the factor of γ^+ in their defining operators (see (6.31) etc.) only the good components are used:

$$\overline{\psi}(0, w^-, \mathbf{0}_T)\gamma^+\psi(0) = \overline{\psi}_G(0, w^-, \mathbf{0}_T)\gamma^+\psi_G(0). \qquad (6.57)$$

In Sec. 6.7, we will show that this operator can be represented in terms of light-front annihilation and creation operators for the quark, and this directly gives an interpretation of the quark density as a number density, i.e., as a probability density. This interpretation requires commutation relations for the annihilation and creation operators, which in turn arise from the anticommutation relation (6.54).

It is possible to treat quark correlators constructed from bad components of fields. But the resulting (anti)commutation relations for the Fourier-transformed quantities would be interaction dependent, and hence would be not those of conventional creation and annihilation operators. Therefore we do not expect any simple interpretation as number densities for parton-density-like quantities constructed using the bad components of the fields.

6.6.2 Light-front annihilation and creation operators

We now obtain annihilation and creation operators in terms of light-front fields (Kogut and Soper, 1970), and derive their commutators.

The annihilation and creation operators are defined by Fourier-transforming the scalar field and the *good* components of the fermion field:

$$\phi(x) = \sum_k \left(a_k(x^+) e^{-ik^+x^- + ik_T \cdot x_T} + a_k(x^+)^\dagger e^{ik^+x^- - ik_T \cdot x_T} \right), \tag{6.58a}$$

$$\psi_G(x) = \sum_{k,\alpha} \left(b_{k,\alpha}(x^+) u_{k,\alpha} e^{-ik^+x^- + ik_T \cdot x_T} + d_{k,\alpha}(x^+)^\dagger u_{k,-\alpha} e^{ik^+x^- - ik_T \cdot x_T} \right). \tag{6.58b}$$

The sum over α is over the two possible values $\alpha = \pm\frac{1}{2}$ for the "light-front helicity" for the fermion, as defined below. The integral over momentum modes is denoted by $\sum_k$, and is restricted to $k^+ > 0$:

$$\sum_k \ldots \overset{\text{def}}{=} \frac{1}{(2\pi)^3} \int_0^\infty \frac{dk^+}{2k^+} \int d^2k_T \ldots \tag{6.59}$$

This is just the normal *Lorentz-invariant* form for the integral over a single particle momentum:

$$\frac{1}{(2\pi)^3} \int_0^\infty \frac{dk^+}{2k^+} \int d^2k_T \ldots = \frac{1}{(2\pi)^3} \int d^4k \, \delta(k^2 - m^2)\theta(k^0) \ldots, \tag{6.60}$$

but without the need to specify the value of the mass. This is an advantage since the physical mass is an interaction-dependent quantity, not known before solving the theory, and moreover the formula applies to quarks and other confined particles that do not have a definite physical mass.

Although the integral is restricted to positive k^+, Fourier modes with the opposite sign of k^+ are allowed for by using terms with a complex-conjugated exponential in (6.58). The distinction between annihilation operators a_k etc. and creation operators $a_k^\dagger$ etc. is made by the sign of the exponential of x^-. (This contrasts with the situation in the Fourier decomposition of fields in equal-time quantization.)

The Dirac wave functions $u_{k,\alpha}$ are defined to be wave functions for massless particles with zero transverse momentum, which span the space of good components (because $\gamma^- u_{k,\alpha} = 0$). They are normalized to obey

$$\bar{u}_{k,\alpha}\gamma^+ u_{k,\alpha'} = 2k^+ \delta_{\alpha\alpha'}, \tag{6.61}$$

and hence

$$\sum_\alpha u_{k,\alpha} \bar{u}_{k,\alpha} = k^+ \gamma^-. \tag{6.62}$$

The label α corresponds to "light-front helicity" in the sense that

$$\sigma^{xy} u_{k,\alpha} = 2\alpha u_{k,\alpha}, \tag{6.63}$$

which is exactly normal helicity for particles of zero transverse momentum. (Note that $2\alpha = \pm 1$, that $\sigma^{xy} = \frac{i}{2}[\gamma^x, \gamma^y]$, and that the wave function for an antiquark of helicity α is $u_{k,-\alpha}$, with argument $-\alpha$.)

In (6.58), the x^+ dependence is in the annihilation and creation operators, not in the exponential factor, since the x^+ dependence depends on solution of the interacting theory, which is not a simple linear problem.

Unlike the case of the corresponding decomposition at equal time, the annihilation and creation operators correspond to different Fourier components. Thus we obtain these operators simply by inverting the Fourier transform:

$$a_k(x^+) = 2k^+ \int dx^- d^2x_T \, e^{ik^+ x^- - ik_T \cdot x_T} \phi(x), \tag{6.64a}$$

$$b_{k,\alpha}(x^+) = \int dx^- d^2x_T \, e^{ik^+ x^- - ik_T \cdot x_T} \bar{u}_{k,\alpha} \gamma^+ \psi(x), \tag{6.64b}$$

$$d_{k,\alpha}(x^+) = \int dx^- d^2x_T \, e^{ik^+ x^- - ik_T \cdot x_T} \bar{\psi}(x) \gamma^+ u_{k,-\alpha}. \tag{6.64c}$$

Values of masses do not appear in these formulae, in contrast to the corresponding formulae in equal-time quantization, which involve $E_k = \sqrt{k^2 + m^2}$. Which value of a mass to use would be unobvious and ambiguous. The possibilities include: the physical mass, the bare mass, and the $\overline{\text{MS}}$ renormalized mass, none of which are equal, with the relationships only known after the theory is solved. But we are formulating the Fourier transform before solving the theory.

From (6.64) follow the (anti)commutation relations appropriate for annihilation and creation operators:

$$[a_k, a_l^\dagger] = \delta_{kl}, \qquad [b_{k\alpha}, b_{l\alpha'}^\dagger]_+ = [d_{k\alpha}, d_{l\alpha'}^\dagger]_+ = \delta_{kl}\delta_{\alpha\alpha'}, \tag{6.65}$$

where δ_{kl} means $(2\pi)^3 2k^+ \delta(k^+ - l^+)\delta^{(2)}(k_T - l_T)$. The other (anti)commutators are zero.

6.7 Parton densities as number densities

From the operator definitions (6.31) etc., we now derive the interpretation of parton densities as number densities, as found by Bouchiat, Fayet, and Meyer (1971) and by Soper (1977). See problem 6.6 for corresponding results for the parton density for a scalar field.

6.7.1 Statement of result

Our field-theoretic analysis of DIS structure functions led us to the formal definition of a parton density by (6.31). But previously, in Sec. 2.4, we had introduced the concept of a parton density rather intuitively as a number density. We now complete the picture by showing that the abstract field-theoretic definition is exactly a number density, defined with the aid of light-front annihilation and creation operators:

$$f_{j/h}(\xi) \overset{\text{prelim}}{=} \frac{1}{2\xi(2\pi)^3} \sum_\alpha \int d^2k_T \frac{\langle P, h | b^\dagger_{k,\alpha,j} b_{k,\alpha,j} | P, h \rangle}{\langle P, h | P, h \rangle}. \tag{6.66}$$

Here we have inserted labels j and h for the quark and target type. The prefactor $1/[2\xi(2\pi)^3]$ is present merely to correspond to our chosen continuum normalization of b and $b^\dagger$ operators: The (anti)commutation relations in (6.65) imply that the right-hand side of (6.66) is exactly the number density in ξ of quarks of flavor j in hadron h; its unweighted integral over ξ is a number of quarks.

In the previous section, we explained light-front quantization in the context of a simple model, whereas in the present section our notation is intended to cover more general theories with more than one flavor of quark. We use the terminology "hadron" for the target state, as is appropriate in QCD. In a general field theory, the target state $|P, h\rangle$ can be any stable single-particle state of definite momentum P, the label h serving to distinguish different stable particles. Similarly the parton label j just refers to any particular field in the theory's Lagrangian.

We explicitly flag (6.66) as preliminary because of important modifications needed in QCD. Even within a super-renormalizable non-gauge model QFT, where the unmodified parton model is valid, there are two important complications:

- Momentum eigenstates have infinite normalization, so the quotient in (6.66) needs interpretation, in terms of an expectation value in a wave packet state, in the limit of a state of definite momentum – see below.
- Our original operator definition had a subtraction of the VEV of the operator, as indicated by the subscript "c" in (6.31). This will not be relevant for the normal situation of positive ξ.

The number density interpretation immediately suggests several sum rules that we will derive. Simple generalizations of the derivation of (6.66) will give corresponding interpretations for the polarized parton densities, and for the parton densities for antiquarks and for scalar (spin-0) partons.

Finally, this result shows that a parton density is an integral of a number density over parton transverse momentum. It is natural to define an unintegrated density, a density in ξ and k_T, by simply deleting the integral over k_T. This we will do in Sec. 6.8. Unintegrated densities are important to the treatment of reactions with sensitivity to partonic transverse momentum – see Chs. 13 and 14. The original kind of parton density naturally gets called an "integrated parton density" whenever the distinction with unintegrated densities is needed.

6.7.2 Wave-packet state

Now we return to the derivation of the number density formula (6.66). We first replace the non-normalizable momentum eigenstate $|P, h\rangle$ by a wave-packet state $|P, h; \Delta\rangle$ whose central value of momentum is P and whose momentum-space width, Δ, we will eventually take to zero. The state is a linear combination of momentum eigenstates:

$$|P, h; \Delta\rangle = \sum_{P'} |P', h\rangle F(P'; P, \Delta), \tag{6.67}$$

which we assume to be normalized:

$$\langle P, h; \Delta | P, h; \Delta\rangle = \sum_{P'} |F(P'; P, \Delta)|^2 = 1. \tag{6.68}$$

A suitable form for the wave function is a Gaussian in rapidity and transverse momentum

$$F(P'; P, \Delta) = \frac{4M^{1/2}(2\pi)^{3/4}}{\Delta^{3/2}} \exp\left[-\frac{(y'-y)^2 M^2}{\Delta^2} - \frac{P_{\mathrm{T}}'^2}{\Delta^2}\right], \tag{6.69}$$

where M is the mass of the target, and we choose the central value P of momentum to have zero transverse component, as usual. To give the wave function a trivial transformation under boosts in the z direction, it is written as a function of rapidity $y = \frac{1}{2}\ln(P^+/P^-)$. The exact form of the wave function will be irrelevant for our work; all that will matter is the peak value and the width. The theorem to be proved is:

$$f_{j/h}(\xi) \stackrel{\text{prelim}}{=} \lim_{\Delta \to 0} \sum_{\alpha} \int d^2 k_{\mathrm{T}} \frac{1}{2\xi(2\pi)^3} \langle P, h; \Delta | b_{k,\alpha,j}^\dagger b_{k,\alpha,j} | P, h; \Delta\rangle, \tag{6.70}$$

with $f_{j/h}(\xi)$ defined by (6.31).

6.7.3 Derivation

First we verify that the right-hand side is indeed correctly normalized for a number density in ξ and k_{T}. To do this, we integrate the operator $b_{k,\alpha,j}^\dagger b_{k,\alpha,j}/[2\xi(2\pi)^3]$ with a smooth function $t(\xi, k_{\mathrm{T}})$ and then check its commutation relation with the $b_{k,\alpha,j}^\dagger$. So we define

$$N_t \stackrel{\text{def}}{=} \int d\xi\, d^2 k_{\mathrm{T}}\, t(\xi, k_{\mathrm{T}}) \frac{1}{2\xi(2\pi)^3} b_{k,\alpha,j}^\dagger b_{k,\alpha,j}. \tag{6.71}$$

Then

$$[N_t, b_{k,\alpha,j}^\dagger] = \int d\xi\, d^2k_T\, t(\xi, k_T)\, \frac{1}{2\xi(2\pi)^3} b_{k,\alpha,j}^\dagger \delta_{kl}$$

$$= t(k^+/P^+, k_T)\, b_{k,\alpha,j}^\dagger. \tag{6.72}$$

One implication is that when we set the function t to be unity everywhere, the resulting operator N_1 counts the total number of partons of type j. To see this, we apply $b_{k,\alpha,j}^\dagger$ to an eigenstate of N_1. The commutation relation (6.72) shows that the resulting state is an eigenstate of N_1, with an eigenvalue increased by unity.

For the main proof, we first use (6.64b) to express the right-hand side of (6.70) in terms of field operators. Before the integral over quark transverse momentum this gives

$$\sum_\alpha \frac{\langle P, \Delta | b_{k,\alpha,j}^\dagger b_{k,\alpha,j} | P, \Delta \rangle}{2\xi(2\pi)^3}$$

$$= \sum_{P'',P'} \frac{2k^+}{2\xi(2\pi)^3} F(P'')^* F(P') \int dw^-\, dz^-\, d^2w_T\, d^2z_T$$

$$\times e^{-ik^+(w^- - z^-) + ik_T \cdot (w_T - z_T)} \left\langle P'' \Big| \overline{\psi}_j(0, w^-, w_T)\, \gamma^+\, \psi_j(0, z^-, z_T) \Big| P' \right\rangle$$

$$= \sum_{P'',P'} \frac{P^+}{(2\pi)^3} F(P'')^* F(P') \int dw^-\, dz^-\, d^2w_T\, d^2z_T$$

$$\times e^{-ik^+(w^- - z^-) + ik_T \cdot (w_T - z_T) + i(P'' - P') \cdot z} \left\langle P'' \Big| \overline{\psi}_j(w - z)\, \gamma^+\, \psi_j(0) \Big| P' \right\rangle$$

$$= \sum_{P'} \frac{P^+}{2P'^+(2\pi)^3} |F(P')|^2 \int dw^-\, d^2w_T$$

$$\times e^{-ik^+(w^- - z^-) + ik_T \cdot (w_T - z_T)} \left\langle P' \Big| \overline{\psi}_j(w - z)\, \gamma^+\, \psi_j(0) \Big| P' \right\rangle. \tag{6.73}$$

In the first step, we used $\sum_\alpha \gamma^+ u_{k,\alpha} \bar{u}_{k,\alpha} \gamma^+ = 2k^+ \gamma^+$. In the third step we performed the integrals over z^- and z_T with $w - z$ held fixed; the resulting delta function between P' and P'' removed the P'' integral except for a factor $1/(2P''^+)$ implicit in $\sum_{P''}$. In the above manipulations observe the different kinds of momentum label for the target state. The fixed central value is P and this is used to define $\xi = k^+/P^+$. The other variables P' and P'' are dummy variables of integration.

Taking the limit that the wave function is very narrow gives

$$\lim_{\Delta \to 0} \sum_\alpha \frac{\langle P, \Delta | b_{k,\alpha,j}^\dagger b_{k,\alpha,j} | P, \Delta \rangle}{2\xi(2\pi)^3}$$

$$= \int \frac{dw^-\, d^2w_T}{(2\pi)^3} e^{-i\xi P^+ w^- + ik_T \cdot w_T} \left\langle P \Big| \overline{\psi}_j(0, w^-, w_T)\, \frac{\gamma^+}{2}\, \psi_j(0) | P \right\rangle, \tag{6.74}$$

whose right-hand side we will take as the definition, (6.79) below, of a quantity $f_{j/h}(\xi, \mathbf{k}_T)$ that we call the unintegrated quark density, or the transverse-momentum-dependent (TMD) quark density.

Integrating the TMD quark density over $\mathbf{k}_T$ reproduces the definition (6.31) of the integrated density. Thus we obtain both the desired theorem, (6.70), and the natural relation that the integrated density is the integral over $\mathbf{k}_T$ of the unintegrated density:

$$f_{j/h}(\xi) \overset{\text{prelim}}{=} \int d^2\mathbf{k}_T \, f_{j/h}(\xi, \mathbf{k}_T). \qquad (6.75)$$

Our derivation does not result in the restriction to connected graphs that was implied by the subscript c in (6.31). We will repair this omission when we discuss support properties of parton densities in Sec. 6.9.3.

In view of the particularly significant complications that arise in QCD in the relation between integrated and unintegrated parton densities, please note that assuming any typical naive generalization of (6.75) to QCD will result in conceptually and phenomenologically wrong results. The literature is rife with such results. See Ch. 13, where we will show how the above derivations are to be generalized.

6.7.4 Interpretation of polarized parton densities

The above derivations can readily be generalized to the polarized quark and antiquark densities. The results are as follows.

The quantity $\Delta f_{j/h}$ is the helicity asymmetry of quarks of flavor j. That is, in a target spin-$\frac{1}{2}$ state of definite helicity,

$$\Delta f_{j/h}(x) = \text{density of quark } j \text{ of helicity parallel to target}$$

$$- \text{density of quark } j \text{ of helicity antiparallel to target.} \qquad (6.76)$$

This applies also to the antiquark helicity density defined by (6.37). The minus sign in (6.37) compensates the reversed sign for the helicity dependence in the matrix elements of $\gamma^+\gamma_5$:

$$\bar{u}_{k,\alpha}\gamma^+\gamma_5 u_{k,\alpha'} = 4\alpha k^+ \delta_{\alpha\alpha'}, \quad \bar{v}_{k,\alpha}\gamma^+\gamma_5 v_{k,\alpha'} = -4\alpha k^+ \delta_{\alpha\alpha'}. \qquad (6.77)$$

For transverse-spin dependence, there is no such minus sign in the matrix elements of $\gamma^+\gamma^i\gamma_5$, and therefore no minus sign is needed in the transverse-spin asymmetry of the antiquarks, (6.38). Again it can be checked that

$$\delta_T f_{j/h}(x) = \text{density of quark } j \text{ of spin parallel to target}$$

$$- \text{density of quark } j \text{ of spin antiparallel to target,} \qquad (6.78)$$

where the spin-$\frac{1}{2}$ target is now chosen to be fully polarized transversely to its direction of motion.

6.8 Unintegrated parton densities

Equations (6.74) and (6.75) show that it is natural to define an unintegrated quark density by

$$f_{j/h}(\xi, \boldsymbol{k}_{\mathrm{T}}) = \int \frac{\mathrm{d}w^- \, \mathrm{d}^2 \boldsymbol{w}_{\mathrm{T}}}{(2\pi)^3} \, e^{-i\xi P^+ w^- + i\boldsymbol{k}_{\mathrm{T}} \cdot \boldsymbol{w}_{\mathrm{T}}} \left\langle P \left| \overline{\psi}_j(0, w^-, \boldsymbol{w}_{\mathrm{T}}) \frac{\gamma^+}{2} \psi_j(0) \right| P \right\rangle_{\mathrm{c}}$$

$$= \int \frac{\mathrm{d}k^-}{(2\pi)^4} \, \mathrm{Tr} \, \frac{\gamma^+}{2} \qquad \qquad \tag{6.79}$$

to be interpreted as a TMD number density $\mathrm{d}N/(\mathrm{d}\xi\,\mathrm{d}^2\boldsymbol{k}_{\mathrm{T}})$. This has a Fourier transform on the relative transverse position of the two fields as well as on w^-, to give a two-argument function of a longitudinal momentum fraction ξ and a quark transverse momentum $\boldsymbol{k}_{\mathrm{T}}$. The last line of this formula is an expression in terms of momentum-space matrix elements from which Feynman rules immediately follow – see Sec. 6.10.

Particularly non-trivial modifications to (6.79) will be needed in QCD: Ch. 13. But in a simple theory – which means a super-renormalizable non-gauge theory – the modifications are absent. In this case it is trivial that an unintegrated density gives the integrated density by an integral over all $\boldsymbol{k}_{\mathrm{T}}$, as in (6.75).

There are natural generalizations for polarized densities and other kinds of parton. But because of the presence of an extra vector in the problem, $\boldsymbol{k}_{\mathrm{T}}$, the polarization dependence of the unintegrated parton densities is more complicated and interesting than that of integrated parton densities. No longer are the quark transverse and longitudinal polarizations simply proportional to those of the target (in the spin-$\frac{1}{2}$ case). See Secs. 13.16 and 14.5.4 for details.

6.9 Properties of parton densities

In this section we derive some basic properties of the pdfs. The proofs are non-perturbative, and many of the results apply, with only small changes, to the correctly defined parton densities of QCD. See Collins and Soper (1982b) and Jaffe (1983) for the original treatments.

6.9.1 Positivity

The number operator formulae (6.66) and (6.74) show that, up to normalization, the matrix element in a parton number density is of the form

$$\langle P | a^\dagger a | P \rangle = \left| a | P \rangle \right|^2, \tag{6.80}$$

i.e., the square of the length of a state vector. So all parton densities are non-negative:

$$f_j(\xi), \ f_j(\xi, \boldsymbol{k}_{\mathrm{T}}) \geq 0. \tag{6.81}$$

Note that this particular result *will not hold exactly in renormalizable theories, because of the need for renormalization of the parton densities*; see Sec. 8.3.

6.9.2 Lorentz invariance/covariance

The definitions of the pdfs depend on a choice of a coordinate system, where the axes are determined by the scattering process being treated. As we saw in (6.34), integrated parton densities can be given explicitly Lorentz-covariant definitions, by use of an auxiliary light-like vector n.

Unintegrated densities need a second vector for a covariant definition. For this, we let n_B be a future-pointing light-like vector with $n_B \cdot n \neq 0$. Up to irrelevant factors, we interpret n and n_B as defining light-front coordinates: $k^+ = n \cdot k$ and $k^- = n_B \cdot k$. Thus n and n_B point in the minus and plus directions respectively. Then we define longitudinal momentum fraction and covariant transverse momentum by

$$\xi = \frac{k \cdot n}{P \cdot n}, \quad k_T^\mu = k^\mu - n_B^\mu \frac{k \cdot n}{n_B \cdot n} - n^\mu \frac{k \cdot n_B}{n_B \cdot n}, \tag{6.82}$$

so that

$$k_T^2 = -k^2 + \frac{2k \cdot n_B k \cdot n}{n_B \cdot n}. \tag{6.83}$$

The unintegrated density (6.79) is

$$f_{j/h}(\xi, k_T) = \int \frac{d^4 w}{(2\pi)^3} \delta(w \cdot n) e^{-iw \cdot k} \left\langle P \left| \overline{\psi}_j(w) \frac{\gamma \cdot n}{2} \psi_j(0) \right| P \right\rangle_c. \tag{6.84}$$

This is invariant when k is shifted in the n direction: $k \mapsto k + cn$.

It is interesting that n_B does not enter this definition, but only in the definition of the variables in (6.82). This situation changes in a gauge theory, where, as we will see in Ch. 13, the definition of unintegrated densities needs Wilson lines in the operators. (Wilson lines are exponentials of integrals of the gauge field along particular lines.)

6.9.3 Support properties, negative ξ

Between the fields in the definition of a parton density, there is a sum over final states, notated by the cut in (6.30). The states have momentum $P - k$, and physical eigenvalues of the plus momentum are positive, so that $P^+ \geq k^+$. Thus pdfs vanish for $\xi > 1$, no matter whether they are integrated pdfs $f(\xi)$ or unintegrated pdfs $f(\xi, k_T)$.

This argument, by itself, provides no restriction for negative ξ. However, we can (anti)commute the two fields in the definition of the pdfs. Since they are at light-like or space-like separation, their (anti)commutator is just the unit operator times a coefficient (localized at $w^- = \boldsymbol{w}_T = 0$). Since we subtract the vacuum expectation value to get the connected matrix element for the pdf, the unit operator from the (anti)commutator gives no contribution. Thus we get a relation between the quark densities at negative x and the antiquark densities at positive x.

The actual relation has an extra minus sign:

$$f_{j/h}(\xi) = -f_{\bar{j}/h}(-\xi), \quad f_{j/h}(\xi, k_T) = -f_{\bar{j}/h}(-\xi, -k_T). \tag{6.85}$$

When the parton is a fermion (e.g., a normal quark), the minus sign arises because we applied an anticommutator. When the parton corresponds to a scalar, the minus sign arises from an explicit factor of ξ in the definition of the scalar-parton density; see problem 6.6. As an example of a derivation, here is the one for the unintegrated densities of a *charged* scalar parton:

$$
\begin{aligned}
f_s(-\xi, -\boldsymbol{k}_T) &= -\xi P^+ \int_{-\infty}^{\infty} \frac{dw^- \, d^2\boldsymbol{w}_T}{(2\pi)^3} \, e^{i\xi P^+ w^- - i k_T \cdot \boldsymbol{w}_T} \, \langle P | \phi^\dagger(0, w^-, \boldsymbol{w}_T) \phi(0) | P \rangle_c \\
&= -\xi P^+ \int_{-\infty}^{\infty} \frac{dw^- \, d^2\boldsymbol{w}_T}{(2\pi)^3} \, e^{i\xi P^+ w^- - i k_T \cdot \boldsymbol{w}_T} \, \langle P | \phi(0) \phi^\dagger(0, w^-, \boldsymbol{0}_T) | P \rangle_c \\
&= -\xi P^+ \int_{-\infty}^{\infty} \frac{dw^- \, d^2\boldsymbol{w}_T}{(2\pi)^3} \, e^{i\xi P^+ w^- - i k_T \cdot \boldsymbol{w}_T} \, \langle P | \phi(0, -w^-, -\boldsymbol{w}_T) \phi^\dagger(0) | P \rangle_c \\
&= -f_{\bar{s}}(\xi, \boldsymbol{k}_T).
\end{aligned}
\tag{6.86}
$$

Since antiparton densities vanish for $\xi > 1$, it immediately follows that all parton densities also vanish for $\xi < -1$.

When the scalar field is a hermitian scalar field, the relation is between the parton density and itself, e.g.,

$$
f_{\phi/q}(\xi) = -f_{\phi/q}(-\xi) \qquad \text{when } \phi \text{ is hermitian.}
\tag{6.87}
$$

A further insight is from the derivation of the probability interpretation. Let us reverse the order of the steps in (6.73), and apply them for negative ξ. Then in place of an annihilation operator $b_{k,\alpha,j}$ we get a creation operator $d^\dagger_{-k,-\alpha,j}$ at the opposite momentum and helicity and for the opposite quark. But we get the operators in the order $d \, d^\dagger$. To get them in the standard order for a number operator, we must anticommute them, leaving the matrix element of the operator for the number of antiquarks (apart from a sign). To this is added the expectation value of the anticommutator, which is a c number, and therefore removed by subtraction of the vacuum expectation value.

6.9.4 Time-ordered bilocal operators

The definitions given so far for the parton densities involved a fixed ordering of the operators. In Feynman-graph calculations, there is a sum and integral over the final states between two operators, as indicated by the vertical line in the cut-graph notation. Now ordinary Green functions and Feynman-graph calculations involve a matrix element between an in-state and an out-state. So with the final states made explicit, as in

$$
f_{j/h}(\xi, \boldsymbol{k}_T) = \sum_X \int \frac{dw^- \, d^2\boldsymbol{w}_T}{(2\pi)^3} \, e^{-i\xi P^+ w^- + i k_T \cdot \boldsymbol{w}_T}
$$

$$
\times \langle P; \text{in} | \overline{\psi}_j(0, w^-, \boldsymbol{w}_T) | X; \text{out} \rangle \, \frac{\gamma^+}{2} \, \langle X; \text{out} | \psi_j(0) | P; \text{in} \rangle_c
\tag{6.88}
$$

for the TMD quark density, we see the density as an integral over amplitude times complex-conjugated amplitude (on its left in the formula, on the right in a cut diagram).

However, the two fields may be (anti)commutated through each other without changing the value of the parton density. Hence we can replace the fixed-order operator product by a time-ordered product:

$$
f_{j/h}(\xi, \boldsymbol{k}_{\mathrm{T}}) = -f_{\bar{j}/h}(-\xi, -\boldsymbol{k}_{\mathrm{T}})
$$

$$
= \int \frac{\mathrm{d}w^- \, \mathrm{d}^2 \boldsymbol{w}_{\mathrm{T}}}{(2\pi)^3} e^{-i\xi P^+ w^- + i\boldsymbol{k}_{\mathrm{T}} \cdot \boldsymbol{w}_{\mathrm{T}}} \left\langle P \left| T \overline{\psi}_j(0, w^-, \boldsymbol{w}_{\mathrm{T}}) \frac{\gamma^+}{2} \psi_j(0) \right| P \right\rangle_c
$$

$$
= \int \frac{\mathrm{d}k^-}{(2\pi)^4} \operatorname{Tr} \frac{\gamma^+}{2} \quad P \overset{k}{\Longrightarrow} \qquad (6.89)
$$

with similar formulae for the integrated densities and for unpolarized densities. Feynman-graph calculations then involve uncut amplitudes, and use exactly the same Feynman graphs for a quark density as for an antiquark density (except for the labeling of the momentum direction). As we will see in explicit calculations, in Sec. 6.11, application of contour integration to the k^- integral gives relations between the two methods of calculation, between the uncut and the cut Feynman graphs. In particular, when a particular graph gives a zero contribution in the cut-graph method for a certain range of ξ, we will find that the poles in k^- in the uncut graph will either all be in the upper half plane or the lower half plane of k^-. Thus the uncut-graph method also gives zero, by use of contour integration for k^-.

Normal Feynman-graph methods apply when the states $\langle P|$ and $|P\rangle$ in (6.89) are, respectively, out- and in-states. But because stable single-particle states are the same for both in- and out-states, this change makes no difference. But it could affect potential generalizations to use hadronic resonances instead of stable single-particle states.

To show that the cut-graph and uncut-graph methods give the same result, we used the fact that the (anti)commutators of the relevant fields are proportional to the unit operator. This applies only to the good components of fields. In contrast, the bad components of the fields have non-trivial (anti)commutators. Thus if we imagined generalizing the definitions of parton densities to correlators of other components of quark fields, the equality between definitions with fixed ordering and with time-ordering will no longer hold. Thus it is a good idea to transform such definitions by use of the equations of motion to write them in terms of the good components of fields.

One use of the definition using time-ordered operator products and uncut graphs is to relate ordinary parton densities to limits of what are called generalized parton densities (GPDs). GPDs are used to analyze the *amplitudes* for certain exclusive reactions; for a review, see Diehl (2003). The definitions of GPDs generalize those of parton densities, by having off-diagonal matrix elements but with the same operators. Since GPDs are applied to amplitudes, the operators are naturally time-ordered:

$$
\int \frac{\mathrm{d}w^-}{2\pi} e^{-i\xi P^+ w^-} \left\langle P' \left| T \overline{\psi}_j \left(0, \frac{1}{2}w^-, \boldsymbol{0}_{\mathrm{T}}\right) \frac{\gamma^+}{2} \psi_j \left(0, -\frac{1}{2}w^-, \boldsymbol{0}_{\mathrm{T}}\right) \right| P \right\rangle_c, \qquad (6.90)
$$

where the position arguments of the fields are in the symmetric form used in Diehl (2003).

6.9.5 Number sum rules

Suppose there is a conserved quark number, as is the case for each flavor (u, d, etc.) in QCD. Then the total number of quarks minus the number of antiquarks of that flavor should equal the value determined by the flavor content of the target state. In QCD we therefore expect the following sum rules for a proton target:

$$\int_0^1 d\xi \left[f_{u/p}(\xi) - f_{\bar{u}/p}(\xi) \right] = 2, \tag{6.91a}$$

$$\int_0^1 d\xi \left[f_{d/p}(\xi) - f_{\bar{d}/p}(\xi) \right] = 1, \tag{6.91b}$$

$$\int_0^1 d\xi \left[f_{j/p}(\xi) - f_{\bar{j}/p}(\xi) \right] = 0 \quad \text{(other flavors);} \tag{6.91c}$$

and of course a baryon number sum rule:

$$\sum_j \int_0^1 d\xi \left[f_{j/p}(\xi) - f_{\bar{j}/p}(\xi) \right] = 3. \tag{6.91d}$$

Obvious changes apply for other target states (e.g., a neutron or a particular nucleus). We now show how these rules (and similar ones in model QFTs) are derived when the parton-model hypotheses are obeyed. The full proof in QCD will involve using the correct definitions and treating renormalization effects, but the final answer is the same.

The basic observation is that when we integrate over all ξ in the definition of a pdf, we get a delta function that sets $w^- = 0$, and the operator becomes a component of the Noether current for quark number. Then we use the fact that parton densities vanish for $|\xi| > 1$ and the relation between parton and antiparton densities to get the sum rule

$$\int_0^1 d\xi \left[f_j(\xi) - f_{\bar{j}}(\xi) \right] = \int_{-\infty}^{\infty} d\xi \int \frac{dw^-}{2\pi} e^{-i\xi P^+ w^-} \left\langle P \left| \bar{\psi}_j(0, w^-, \mathbf{0}_{\mathrm{T}}) \frac{\gamma^+}{2} \psi_j(0) \right| P \right\rangle_c$$

$$= \frac{1}{2P^+} \langle P | \bar{\psi}_j(0) \gamma^+ \psi_j(0) | P \rangle_c \,. \tag{6.92}$$

We now have the expectation value of the plus component of the Noether current for the number of quarks of flavor j. From standard properties of currents, this expectation value is the charge of the state times a factor of twice the momentum of the state, which is canceled in the last line. From this result all the above-listed sum rules follow. The subtraction of the VEV implies that the number density is relative to the vacuum.

6.9.6 Momentum sum rule

A very similar argument gives the momentum sum rule:

$$\sum_{\text{all } j} \int_0^1 d\xi \, \xi f_j(\xi) = 1. \tag{6.93}$$

Here we weight the number densities of partons by ξ, to give a density of fractional momentum. So the sum rule says that the total fractional momentum carried by partons is unity. Note that the sum is over all flavors of parton, including separate terms for antipartons as well as partons. In our Yukawa model this means fermion, antifermion and scalar partons.

The proof is left as an exercise (problem 6.15). It simply involves converting the sum and integral over parton densities to an expectation value of a certain component of the energy-momentum tensor (relative to the vacuum).

6.9.7 Isospin and charge conjugation relations

Consider a theory with an SU(2) isospin symmetry and quarks, like QCD, where we have u and d quarks, which form an isodoublet, and s and heavier quarks, which are all isosinglet.

In real QCD, isospin symmetry is slightly broken by the different masses of the u and d quarks. By neglecting this breaking, we can obtain relations between parton densities in different targets, which hold to the accuracy that isospin symmetry holds. Unlike the sum rules, these relations are valid point-by-point in x.

We will illustrate this for the important cases of the proton and neutron and for the pions. (Scattering experiments are done with all of these particles.) We will obtain a further set of relations for pions using charge conjugation invariance. The general form of all of these arguments is to insert a symmetry transformation operator U times its inverse next to the target state in the definition of a parton density:

$$\langle P, h | U^\dagger U \, A \, U^\dagger U | P, h \rangle = \langle P, h' | A' | P, h' \rangle. \tag{6.94}$$

Here h' labels the state obtained by transforming the target, label h, by transformation U, A is the operator whose matrix element is the parton density, and A' denotes the transformed operator.

Since only the transformation properties under simple symmetries are involved in our derivation, the results apply equally to unintegrated parton densities, as well as the more usual integrated parton densities. As explained in Sec. 6.9.8, the results apply equally to the correct QCD definitions of parton densities, so they are presented in their QCD applications.

Proton and neutron

Physical targets are always eigenstates of I_z. So let us take U to be an operator that exchanges the $I_z = \pm \frac{1}{2}$ elements of an isodoublet. We then get the following relations between parton densities on a neutron and a proton:

$$f_{u/p}(x) = f_{d/n}(x), \quad f_{d/p}(x) = f_{u/n}(x), \tag{6.95a}$$

$$f_{\bar{u}/p}(x) = f_{\bar{d}/n}(x), \quad f_{\bar{d}/p}(x) = f_{\bar{u}/n}(x), \tag{6.95b}$$

$$f_{j/p}(x) = f_{j/n}(x), \quad f_{\bar{j}/p}(x) = f_{\bar{j}/n}(x) \quad (j \text{ is } s, c, \text{etc.}). \tag{6.95c}$$

In electromagnetic DIS, the structure functions are dominated by the density of the u quark, since it has the larger charge. The above relations allow the use of scattering on a nuclear target to gain information on $f_{u/n}$ and hence on $f_{d/p}$, the density of the lower-charge quark.

Antiproton

One standard beam particle is the antiproton. Parton densities in the antiproton are related to those in the proton by letting U be the charge conjugation operator. This gives $f_{j/\bar{p}}(x) = f_{\bar{j}/p}(x)$ for all species of parton. Particular cases are

$$f_{u/p}(x) = f_{\bar{u}/\bar{p}}(x), \quad f_{d/p}(x) = f_{\bar{d}/\bar{p}}(x). \tag{6.96}$$

These relations are very important for the phenomenology of data from the Tevatron, which uses proton-antiproton collisions.

Gluon in proton, neutron and antiproton

Since the gluon is its own antiparticle as well as being isosinglet, the gluon density is the same in all the targets we have mentioned:

$$f_{g/p}(x) = f_{g/n}(x) = f_{g/\bar{p}}(x). \tag{6.97}$$

Proton target is default

The combination of all the above results means that we can express results for all kinds of nucleon target in terms of parton densities in the proton. So for real QCD applications, when we write a parton density without a hadron label, e.g., $f_u(x)$, it is to be understood that a proton target is intended.

Densities of definite isospin

It is sometimes convenient to use combinations of parton densities that correspond to isotriplet and isosinglet operators, e.g.,

$$f_{I=0}(x) = f_u(x) + f_d(x), \tag{6.98}$$

$$f_{I=1}(x) = f_u(x) - f_d(x), \tag{6.99}$$

with a proton target understood.

Nuclear targets

Data on non-trivial larger nuclei are often analyzed in terms of parton densities in the constituent proton and neutron; this needs a compensation for nuclear-physics effects in nuclear binding. But it is also possible to treat parton densities on the nucleus as a whole. It is often possible to treat nuclei as approximately or exactly isosinglet, notably for the deuteron. In that case isospin relates u and d quark densities, e.g.,

$$f_{u/D}(x) = f_{d/D}(x), \quad f_{\bar{u}/D}(x) = f_{\bar{d}/D}(x). \tag{6.100}$$

(See Schienbein *et al.*, 2009; Eskola, Paukkunen, and Salgado, 2009.)

Pion

The three pions, π^+, π^-, and π^0, are related by both isospin and charge conjugation. We leave as an exercise to derive

$$f_{u/\pi^+}(x) = f_{d/\pi^-}(x) = f_{\bar{d}/\pi^+}(x) = f_{\bar{u}/\pi^-}(x), \tag{6.101a}$$

$$f_{d/\pi^+}(x) = f_{u/\pi^-}(x) = f_{\bar{u}/\pi^+}(x) = f_{\bar{d}/\pi^-}(x), \tag{6.101b}$$

$$f_{g/\pi^+}(x) = f_{g/\pi^-}(x), \tag{6.101c}$$

$$f_{s/\pi^+}(x) = f_{s/\pi^-}(x) = f_{\bar{s}/\pi^+}(x) = f_{\bar{s}/\pi^-}(x). \tag{6.101d}$$

It can be seen that there are very few independent densities, which considerably assists the analysis of data with pion beams. The parton densities in the π^0 are determined in terms of the above:

$$f_{u/\pi^0}(x) = f_{d/\pi^0}(x) = f_{\bar{d}/\pi^0}(x) = f_{\bar{u}/\pi^0}(x) = \frac{1}{2}\left(f_{u/\pi^+}(x) + f_{d/\pi^+}(x)\right), \tag{6.102a}$$

$$f_{g/\pi^0}(x) = f_{g/\pi^+}(x), \tag{6.102b}$$

$$f_{s/\pi^0}(x) = f_{\bar{s}/\pi^0}(x) = f_{s/\pi^+}(x). \tag{6.102c}$$

These last relations are of relatively little use, since we do not normally deal with beams of neutral pions.

6.9.8 *Are the sum rules etc. valid in QCD?*

The derivations just presented apply as they stand to a theory which is super-renormalizable and contains only fields of spin zero and spin half. Evidently, QCD violates both prerequisites, and later in the book we will make the necessary improvements. But here it is possible to assess the difficulties and to state the extent to which the results presented continue to apply in QCD.

Our specific model field theory was a very simple Yukawa theory with one field of each type, but the principles immediately generalize when there are multiple fields. Thus we were able to conceive of a theory with the same flavor symmetries as QCD, and to prove certain sum rules.

Isospin relations preserved

In Sec. 6.9.7, we derived relations between parton densities for different flavors of parton and hadron. The only properties that were used of the operators defining parton densities were their transformations under charge conjugation and isospin. These properties are entirely unaffected by the changes needed to accommodate renormalization and the use of gauge fields. This will become fully evident when we construct the definitions of parton densities in QCD.

Renormalization

A renormalizable theory, as opposed to a super-renormalizable theory, is exemplified by the Yukawa theory in four space-time dimensions, $n = 4$. All of the above derivations apply when a UV cutoff is applied, for example dimensional regularization with $n = 4 - 2\epsilon$. The fields in the derivations should be bare fields, i.e., the ones with canonical commutation relations. The bare fields are those for which the coefficients of the first term in each line for the right-hand side of (6.44) is exactly as given. We then remove the UV cutoff after applying renormalization.

To implement renormalization, we first relabel all the fields and parameters in (6.44) with a subscript 0, to denote bare quantities, e.g., g_0. Then we write the bare fields as renormalized fields times "wave-function-renormalization factors", e.g., $\psi_0 = \psi \sqrt{Z_2}$ with a conventional notation. Thus the Lagrangian density defining the theory becomes

$$\mathcal{L} = \frac{iZ_2}{2} \left[\bar{\psi} \gamma^\mu \partial_\mu \psi - (\partial_\mu \bar{\psi}) \gamma^\mu \psi \right] - M_0 Z_2 \bar{\psi} \psi$$

$$+ \frac{Z_3}{2} (\partial \phi)^2 - \frac{m_0^2 Z_3}{2} \phi^2 - g_0 Z_2 Z_3^{1/2} \bar{\psi} \psi \phi - \frac{h_0 Z_3^{3/2}}{3!} \phi^3 - \frac{\lambda_0 Z_3^2}{4!} \phi^4. \tag{6.103}$$

Finally we adjust the bare parameters, g_0, Z_2, etc., in an ϵ-dependent way to remove the divergences. In perturbation theory, this is implemented by using renormalized couplings and masses, g_R, M_R, etc., and using an expansion of the bare parameters in powers of the renormalized coupling, with coefficients adjusted to cancel the divergences order-by-order.

It is Green functions of the renormalized fields ψ and ϕ that are finite rather than those of the bare fields. So we should define the light-front annihilation and creation operators in terms of the renormalized fields. Then the (anti)commutation relations of these operators are changed by wave function renormalization, as in

$$[a_k, a_l^\dagger] = \delta_{kl} Z_3^{-1}, \qquad [b_{k\alpha}, b_{l\alpha'}^\dagger]_+ = [d_{k\alpha}, d_{l\alpha'}^\dagger]_+ = \delta_{kl} \delta_{\alpha\alpha'} Z_2^{-1}, \tag{6.104}$$

since it is the bare fields that obey the canonical (anti)commutation relations. An RG analysis can be used to investigate/compute the true value of the renormalization coefficients when the UV cutoff is removed. Generally, the coefficients in (6.104) diverge to $+\infty$ in this limit, with the (rare) exceptions being if the anomalous dimension of a field vanishes strongly enough at the UV fixed point of the theory. The Källen-Lehmann representation of the propagator tells us that $0 \leq Z_i \leq 1$ when an on-shell renormalization prescription is used, so we expect Z_i^{-1} to go to infinity rather than zero in the UV limit.

As we will see later, there are further UV divergences in the integrated parton densities, beyond those removed by wave-function renormalization. We will also see that renormalized integrated parton densities can be defined by a further kind of renormalization, which is completely analogous to what is done for local composite operators.

Since the finite operators no longer have the standard generalized-harmonic-oscillator (anti)commutation relations, and since renormalization of the integrated parton densities is needed, the strict probability interpretation of the parton densities is lost.

Nevertheless, we will show in Sec. 8.6 that the UV divergences cancel in the sum rules, which remain true in a renormalizable theory.

Gauge theories

We will examine the light-front quantization of gauge theories in Sec. 7.4.

Extending the definitions of parton densities to QCD will require significant modifications to the definitions. These involve insertion of what are called Wilson lines to make them gauge invariant: Sec. 7.5. These will further complicate the probability interpretation of parton densities and their renormalization. Nevertheless, the derivation of the sum rules will still work.

6.9.9 *Axial currents; Bjorken sum rule*

We derived sum rules that related certain integrals over unpolarized parton densities to expectation values of conserved vector currents. Axial currents are also of interest in QCD, so we now discuss the associated sum rules. Even though our discussion of QCD is only later in this book, we can explain the sum rules without this discussion. We simply assume that the definitions given for parton densities can still be used, and then apply them in a theory with the same flavor symmetries as QCD.

The use of axial currents is rather more tricky than vector currents. One reason is that for the $SU(2) \otimes SU(2)$ symmetry of QCD (broken in the Lagrangian only by light quark masses) there is spontaneous symmetry breaking of the axial part of the symmetry. So the expectation values of the axial currents and hence the right-hand sides of the equivalents of (6.91) are determined by the dynamics of QCD, not by the charges of the target. Some of the currents appear in the coupling of quarks to weak gauge bosons, and the matrix elements can be measured, for example, in semi-leptonic decays of hadrons. A second complication is that the isosinglet axial current has an anomaly and is not prone to easy measurement or prediction. A third complication is that whereas there are conserved vector currents in QCD for each of the heavy quarks, resulting in (6.91d), the conservation laws for the axial currents for heavy quarks are badly broken by quark masses.

An elementary generalization of (6.92) leads to the following result for each quark flavor:

$$\int_0^1 d\xi \left[\Delta f_j(\xi) + \Delta f_{\bar{j}}(\xi) \right] = \frac{1}{2P^+} \langle P | \overline{\psi}_j(0) \gamma^+ \gamma_5 \psi_j(0) | P \rangle_c . \qquad (6.105)$$

Note that the antiquark term now has a plus sign instead of the minus sign in the number sum rules. *In some sense* the left-hand side measures the total contribution of quarks and antiquarks of flavor j to the spin of the target. Unlike the case of the quark number currents, the current does not correspond to a conserved charge. So there is no direct determination of the right-hand side (although one can well imagine calculating it non-perturbatively by lattice QCD methods).

For the non-singlet combination, we get

$$\int_0^1 d\xi \left[\Delta u(\xi) + \Delta \bar{u}(\xi) - \Delta d(\xi) - \Delta \bar{d}(\xi)\right] = \frac{1}{2P+} \langle P|\overline{\psi}(0)\gamma^+\gamma_5\tau_3\psi(0)|P\rangle_c , \quad (6.106)$$

where τ_3 is a Pauli matrix acting on the doublet of fields for the u and d quarks. We used the quark symbols to denote their parton densities. The current on the right-hand side is one of the generators of the approximate chiral $SU(2) \otimes SU(2)$ symmetry of QCD. It is also related by an isospin transformation for the axial part of the current which couples the W boson to u and d quarks. The matrix element can therefore be deduced from the rate and angular distribution of neutron decay (to $p + e\bar{\nu}$), presented as a value conventionally denoted by G_A/G_V, whose measured value (Amsler *et al.*, 2008) is 1.2695 ± 0.0029.

Roughly speaking, the sum rule can be probed in the difference between g_1 structure function on the proton and neutron, for which recent data and an analysis related to the sum rule can be found in Airapetian *et al.* (2007). To indicate the idea, we observe that the parton model approximation to g_1 is

$$g_1(x, Q) = \frac{1}{2} \sum_q e_q^2 [\Delta q(x) + \Delta \bar{q}(x)]. \quad (6.107)$$

Using the isospin relations between the polarized parton densities in the neutron and proton, which are immediate generalizations of (6.95), and then using the sum rule (6.106) we get

$$\int_0^1 dx[g_1^p(x, Q) - g_1^n(x, Q)] = \frac{G_A}{6G_V} \simeq 0.21 \quad \text{(parton model)}. \quad (6.108)$$

This is one of two results due to Bjorken that are both called Bjorken sum rules.

6.9.10 Moments

The derivation of (6.92) can be readily extended to general integer moments of parton densities by inserting a factor of ξ^{n-1} on the left-hand side and a suitable sign with the antiquark density. The factor of ξ^{n-1} gives $n - 1$ derivatives with respect to the position w^- and we obtain a matrix element of a local operator:

$$\int_0^1 d\xi \, \xi^{n-1} \left[f_j(\xi) + (-1)^n f_j(\xi)\right] = \frac{i^{n-1}}{2(P+)^n} \langle P|\overline{\psi}_j(0)\gamma^+(\partial^+)^{n-1}\psi_j(0)|P\rangle_c . \quad (6.109)$$

In the early days of the study of DIS, the operator product expansion was used to express moments of the DIS in terms of perturbative coefficients times expectation values of local operators, exactly like those on the right-hand side of the above equation; see Ch. 14 of Collins (1984). (Of course, in QCD we need renormalized, gauge-invariant versions of the operators.)

Equation (6.109) shows how these operators are related to parton densities. The expectation values of local operators are susceptible to calculation by Euclidean lattice Monte-Carlo methods, unlike parton densities, whose operators are strictly Minkowski-space objects.

Thus the equation also provides a way that lattice Monte-Carlo methods can be used to give predictions for properties of parton densities.

6.10 Feynman rules for pdfs

In this section, I show how the definitions of parton densities are to be applied in Feynman-graph calculations, by defining special rules for vertices corresponding to the operators in the definitions of the parton densities. Motivated by applications in QCD, I use the word "quark" to refer to the fermion field in our Yukawa model theory, and to its associated particle.

In (6.30), we saw that a quark density can be expressed as an integral over a cut amplitude. A convenient notation is to write

$$f(\xi) = \int \frac{dk^- \, d^{2-2\epsilon} k_T}{(2\pi)^{4-2\epsilon}} \frac{\gamma^+}{2}$$

$$(6.110)$$

$$= \int \frac{d^{4-2\epsilon} k}{(2\pi)^{4-2\epsilon}} \frac{\gamma^+}{2} \delta(k^+ - \xi P^+)$$

which gives the Feynman rule for the operator vertices in an *integrated* unpolarized quark density, in $4 - 2\epsilon$ space-time dimensions. The crosses in the first part indicate the operations that are to be applied to the quark fields to obtain the actual pdf. They denote the integrals over k^- and k_T and the trace with $\gamma^+/2$. The plus component of the momentum at the quark vertices is fixed to be ξP^+. In view of the extensive use that is made of dimensional regularization, the vertex is given for a general space-time dimension. Were there a color degree of freedom, there would be an unweighted sum over the colors of the field. The bubble indicates the basic matrix element of the quark fields in an on-shell target state of momentum P.

Generalizations to the polarized parton densities are simply made by changing the factor $\gamma^+/2$ to the appropriate Dirac matrix in the definition of the parton density. Similarly, the definitions for the antiquark densities are made simply by changing the direction of the arrow on the quark line. These are all illustrated in Fig. 6.7. Note that the minus sign in the definition of the helicity density of an antiquark requires a corresponding minus sign in the Feynman rule for the antiquark helicity density.

Further generalizations to TMD densities, e.g., Fig. 6.8, are trivially obtained by deleting the integral over transverse momentum. Generally the context will indicate whether we are using integrated or unintegrated densities, so we make no distinction in the graphical

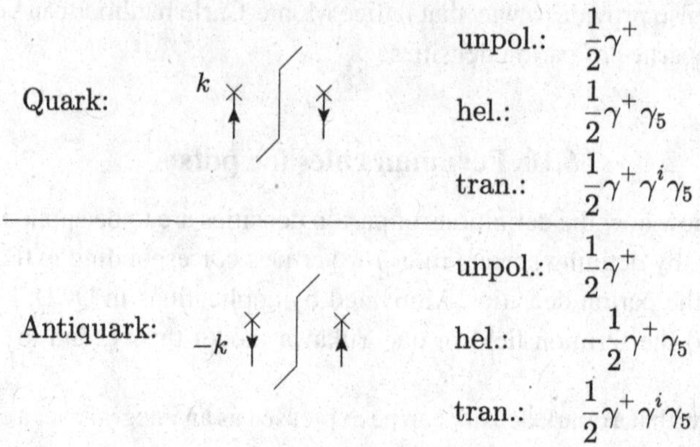

Quark:

unpol.: $\dfrac{1}{2}\gamma^+$

hel.: $\dfrac{1}{2}\gamma^+\gamma_5$

tran.: $\dfrac{1}{2}\gamma^+\gamma^i\gamma_5$

Antiquark:

unpol.: $\dfrac{1}{2}\gamma^+$

hel.: $-\dfrac{1}{2}\gamma^+\gamma_5$

tran.: $\dfrac{1}{2}\gamma^+\gamma^i\gamma_5$

Fig. 6.7. Gamma matrix factors for all the unpolarized and polarized quark and antiquark densities. For the helicity densities, the target should be in a state of maximum right-handed polarization. For the transversity densities, the target should be in a state of maximum transverse spin, and the rules listed above will give the transversity densities times a unit vector in the direction of the transverse spin of the target. *Note the minus sign in the definition of the helicity density of an antiquark.* Note also that the quark momentum is assumed to be in the direction of the arrow of the quark line. Thus the momentum for the line at the antiquark density is written as $-k$.

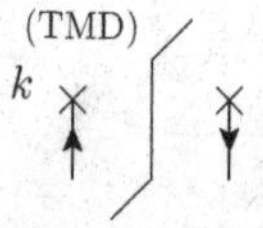

$$\text{(TMD)} \qquad = \int \frac{\mathrm{d}k^-}{(2\pi)^{4-2\epsilon}}\frac{\gamma^+}{2}\cdots$$

Fig. 6.8. The rule for the vertex, as in (6.110), but for a TMD, or unintegrated, quark density. Note that we have not made any notational distinction between the vertices for integrated and unintegrated densities; generally the distinction can be determined from the context.

notation. The common feature of all the definitions is the unweighted integral over all k^-, so that the field operators in the parton density definition are at equal values of x^+.

The change to the definition with time-ordered products can be made simply by deleting the symbol for the final-state cut.

6.11 Calculational examples

In QCD, parton densities with hadronic targets are strictly non-perturbative objects. But it is useful to examine low-order Feynman-graph calculations of parton densities with the target being an elementary particle of a theory.

So in this section, I present some calculations in the model Yukawa theory used in our treatment of light-front quantization. The calculations introduce the methods in their simplest form, and they enable us to see basic principles without being confused by many of the complications – one might almost say pathologies – that arise in QCD. Moreover, such calculations can be used as self-consistent models for interesting effects in QCD – e.g., Brodsky, Hwang, and Schmidt (2002); Collins (2002). In our model calculations, we

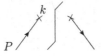

Fig. 6.9. Lowest-order quark density in quark.

will be introduced to the UV divergences of parton densities in renormalizable theories. Perturbative calculations of parton densities also appear as components of perturbative calculations of hard-scattering coefficients.

In the calculations, the target state is a physical on-shell elementary-particle state corresponding to one of the basic field of the theory like the quark. Our calculations in the Yukawa theory of (6.44) are of the density of a quark in a quark, $f_{q/q}(\xi)$, and of a scalar in a quark $f_{\phi/q}(\xi)$.

The concept of the "density of a quark in a quark" is confusing, initially: Why should this not be a trivial delta function at $\xi = 1$? In fact, the word "quark" in that phrase has two meanings. One is for the target state, which is an on-shell physical state. The second meaning is for a state created by the corresponding light-front creation operator. Thus the different instances of the word "quark", as well as the two instances of the symbol "q" in $f_{q/q}(\xi)$, refer to different bases of theory's state space. In an interacting QFT, on-shell single-particle states, as used in scattering theory, are normally non-trivial combinations of multiparticle states when expressed in the basis given by the creation operators.

6.11.1 Tree approximation

In an expansion in powers of the coupling(s) for $f_{q/q}(\xi)$, the first term is of zeroth order (Fig. 6.9). This is deceptively similar to the representation of just the vertices for the parton density. It is intended to denote the combination of those vertices with the lowest-order amplitude for the bubble in (6.110). The lowest-order bubble consists of $(2\pi)^{4-2\epsilon}\delta^{(4-2\epsilon)}(k - P)$ for momentum conservation in a disconnected graph, and a factor of the on-shell wave function for the target. We allow the most general polarization state for the target, which can be specified by a spin vector S, as in (A.26). We therefore obtain

$$f_{q/q}^{[0]}(\xi) = \int \frac{dk^- \, d^{2-2\epsilon}k_{\mathrm{T}}}{(2\pi)^{4-2\epsilon}} (2\pi)^{4-2\epsilon}\delta^{(4-2\epsilon)}(k - P) \, \mathrm{Tr}(\slashed{P} + M)\frac{1}{2}\left(1 + \gamma_5\frac{\slashed{S}}{M}\right)\frac{\gamma^+}{2}$$

$$= \delta(\xi - 1). \tag{6.111}$$

Here we use the superscript "[0]" to denote the lowest-order value with zero loops. This calculation provides a basic verification of the normalization of our definition. Without interactions the single on-shell quark is also a single particle in the light-front creation operator basis, and it carries the whole momentum of the target, i.e., it has $\xi = 1$.

6.11.2 One-loop quark in quark

At one-loop order, there are two kinds of graph for $f_{q/q}$ (Fig. 6.10): (a) self-energy corrections on the external line, and (b) a graph with a scalar particle emitted into the final state.

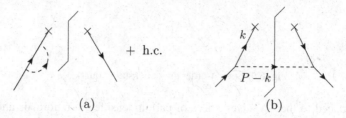

Fig. 6.10. One-loop graphs for density of quark in quark.

(We consider graph (b) a loop graph since there is a momentum integral through the vertex for the parton density.)

Self-energy graph

The full effects of self-energy corrections for external on-shell lines are given by the LSZ method. This tells us that for each external particle we need a factor of the *square root* of the residue of the pole of the propagator. To calculate this, we start from the one-loop self-energy of the quark:

$$\frac{g^2}{16\pi^2}\Sigma^{[1]} = ig^2\mu^{2\epsilon}\int\frac{d^{4-2\epsilon}k}{(2\pi)^{4-2\epsilon}}\frac{\not{P}-\not{k}+M}{[(P-k)^2-M^2+i0](k^2-m^2+i0)}. \tag{6.112}$$

The superscript "[1]" denotes the coefficient of the one-loop approximation. As usual, the coupling is written as $g\mu^\epsilon$, where g is dimensionless and μ is the unit of mass for dimensional regularization.

Now the full quark propagator is $i/(\not{p}-M-\Sigma)$. So the one-loop contribution to the residue is given by differentiating $\Sigma^{[1]}$ with respect to $\not{p}$ and by then setting p on-shell. After performing the k integral by the Feynman parameter method, we find that to one-loop order, the residue is

$$1+\frac{g^2}{16\pi^2}\text{residue}^{[1]} = 1 - \frac{g^2}{16\pi^2}\Gamma(\epsilon)\int_0^1 dx\left[\frac{4\pi\mu^2}{m^2x+M^2(1-x)^2}\right]^\epsilon$$

$$\times\left[x+\frac{2\epsilon M^2x(1-x^2)}{m^2x+M^2(1-x)^2}\right]. \tag{6.113}$$

We have a factor of the square root of the residue for both external quark lines, so that the resulting one-loop contribution to the quark density is

$$\frac{g^2}{16\pi^2}f_{q/q}^{[1,V]}(\xi) = \delta(\xi-1)\times\frac{g^2}{16\pi^2}\text{residue}^{[1]}. \tag{6.114}$$

The "V" in the superscript denotes "virtual correction". Equation (6.113) shows that this contribution is negative. This reduces the size of the one-light-front-particle component in the normalized target state, leaving room for a multiparton component.

Of course, when we go to four space-time dimensions, $\epsilon=0$, this term is UV divergent. We will explain what happens for the parton density, when we discuss its renormalization.

Real emission

The integral for the real-emission term (Fig. 6.10(b)) is readily written down from the Feynman rules:

$$\frac{g^2}{16\pi^2} f_{q/q}^{[1,R]}(\xi) = g^2 \mu^{2\epsilon} \int \frac{dk^- \, d^{2-2\epsilon} k_T}{(2\pi)^{4-2\epsilon}} \frac{2\pi \delta\big((P-k)^2 - m^2\big)}{(k^2 - M^2)^2}$$

$$\times \theta(P^+ - k^+) \operatorname{Tr} \frac{\gamma^+}{2}(\slashed{k} + M)(\slashed{P} + M)\frac{1}{2}\left(1 + \gamma_5 \frac{\slashed{S}}{M}\right)(\slashed{k} + M). \quad (6.115)$$

We set $k^+ = \xi P^+$, and then use the delta function to perform the k^- integral, whereby

$$-2P^+ k^- = \frac{k_T^2 + m^2 - M^2(1-\xi)}{1-\xi}. \quad (6.116)$$

This gives

$$\frac{g^2}{16\pi^2} f_{q/q}^{[1,R]}(\xi) = \frac{g^2 (4\pi\mu^2)^\epsilon}{16\pi^2 \, \Gamma(1-\epsilon)} \int_0^\infty dk_T^2 (k_T^2)^{-\epsilon} \frac{(1-\xi)\,[k_T^2 + (1+\xi)^2 M^2]}{[k_T^2 + \xi m^2 + (1-\xi)^2 M^2]^2}$$

$$= \frac{g^2 \Gamma(\epsilon)}{16\pi^2} \left[\frac{4\pi\mu^2}{\xi m^2 + (1-\xi)^2 M^2}\right]^\epsilon \left[1 - \xi + \frac{\epsilon\xi(1-\xi)(4M^2 - m^2)}{\xi m^2 + (1-\xi)^2 M^2}\right]. \quad (6.117)$$

Here, we have used a standard result, (A.34), to perform the angular part of the transverse-momentum integral. The restriction of the final-state momentum $P - k$ to physical positive energy implies that the above formula should have an implicit theta function that restricts it to $\xi \leq 1$. In addition, for negative ξ, as we will see, the calculation is not the complete one; a correct calculation (Sec. 6.11.6) for $\xi < 0$ gives zero. Thus there should also be a restriction to positive ξ. Then in the physical range, we have a non-singular function.

Notice that the denominator is identical to the one in the self-energy. This is related to a cancellation needed to verify sum rules.

Naturally the real-emission contribution is positive, since parton densities are positive, and for the situation that ξ is not equal to unity, our calculation gives the lowest-order contribution.

When the theory is super-renormalizable, in less than four space-time dimensions, i.e., for $\epsilon > 0$, the k_T integral is convergent. But at the physical space-time dimension, with $\epsilon = 0$, there arises a logarithmic divergence at $k_T \to \infty$. This in fact should be considered a conventional UV divergence, since the virtual line k goes far off-shell, and masses become negligible in the region that gives the divergence. We will discuss the UV divergences later in Sec. 8.3.

6.11.3 One-loop scalar in quark

The remaining one-loop contribution to parton densities in an on-shell quark is the density of the scalar. For this, we need the Feynman rule for the density of a scalar parton (Fig. 6.11).

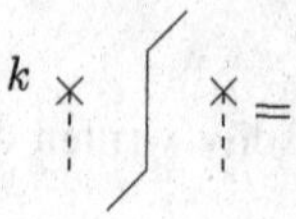

$$k \underset{\vdots}{\overset{\times}{\times}} \int \underset{\vdots}{\overset{\times}{\vdots}} = \xi P^+ \int \frac{dk^- \, d^{2-2\epsilon} \mathbf{k}_T}{(2\pi)^{4-2\epsilon}} \text{with } k^+ = \xi P^+ \dots$$

$$= \int \frac{d^{4-2\epsilon} k}{(2\pi)^{4-2\epsilon}} \delta(k^+/\xi P^+ - 1) \dots$$

Fig. 6.11. Feynman rule for operator for the density of a scalar parton.

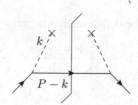

Fig. 6.12. Scalar density in quark.

It has a factor ξP^+ in place of the $\gamma^+/2$ for the quark density. The derivation is left as an exercise (problem 6.6), and it results in the definition in (6.124) below.

Then we readily find the one-loop scalar density from Fig. 6.12:

$$\frac{g^2}{16\pi^2} f_{\phi/q}^{[1]}(\xi) = g^2 \mu^{2\epsilon} \int \frac{dk^- \, d^{2-2\epsilon} \mathbf{k}_T}{(2\pi)^{4-2\epsilon}} \frac{2\pi \delta((P-k)^2 - M^2)}{(k^2 - m^2)^2}$$

$$\times \xi P^+ \operatorname{Tr}(\not{P} - \not{k} + M)(\not{P} + M) \frac{1}{2} \left(1 + \gamma_5 \not{S}/M\right)$$

$$= \frac{g^2 (4\pi \mu^2)^\epsilon}{16\pi^2 \, \Gamma(1-\epsilon)} \int_0^\infty dk_T^2 (k_T^2)^{-\epsilon} \frac{\xi \, [k_T^2 + (2-\xi)^2 M^2]}{[k_T^2 + (1-\xi)m^2 + \xi^2 M^2]^2}$$

$$= \frac{g^2 \Gamma(\epsilon)}{16\pi^2} \left[\frac{4\pi \mu^2}{(1-\xi)m^2 + \xi^2 M^2} \right]^\epsilon \left[\xi + \frac{\epsilon \xi (1-\xi)(4M^2 - m^2)}{(1-\xi)m^2 + \xi^2 M^2} \right]. \quad (6.118)$$

Notice that the denominator is obtained from the denominator in the quark density by changing ξ to $1-\xi$, as is appropriate now that the scalar line has its plus component of momentum equal to k^+ instead of $P^+ - k^+$. Again, we have a positive contribution, with a UV divergence when $\epsilon = 0$.

The above calculation is valid when $0 < \xi < 1$. As usual, the positive-energy condition on $P - k$ ensures that parton densities are zero if $\xi > 1$. For negative ξ, a more elaborate argument, with extra graphs, is needed, and is given in Sec. 6.11.6.

6.11.4 Sum rules

We now check that the number and momentum sum rules are obeyed by our calculation. Naturally the lowest-order term contributes unity to both the quark number and to the

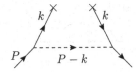

Fig. 6.13. Real-emission contribution to one-loop quark density in quark when the definition with time-ordered operators is used.

momentum sum rules. So to confirm the sum rules at order g^2, we must show that the one-loop contributions to each sum rule are zero.

For the number sum rule we have

$$\int d\xi\, f_{q/q}^{[1,V]}(\xi) + \int d\xi\, f_{q/q}^{[1,R]}(\xi)$$

$$= \text{residue}^{[1]} + \int d\xi\, f_{q/q}^{[1,R]}(\xi)$$

$$= \Gamma(\epsilon) \int_0^1 dx \left[\frac{4\pi\mu^2}{m^2 x + M^2(1-x)^2}\right]^\epsilon \left[1 - 2x + \frac{\epsilon M^2 x(1-x)[2(1-x)M^2 - m^2]}{m^2 x + M^2(1-x)^2}\right]$$

$$= 0. \tag{6.119}$$

The zero in the last line can be easily calculated by using the fact that the integrand in the previous line is proportional to the derivative with respect to x of $x(1-x)[m^2 x + M^2(1-x)^2]^{-\epsilon}$.

The momentum sum rule is checked similarly.

6.11.5 Uncut graphs

We saw in Sec. 6.9.4 that because the fields in the definition of a parton density commute or anticommute, except for an irrelevant "c-number" term, the operator product in the definition of a parton density can be replaced by a time-ordered product, as in (6.89). So we now examine how this alternative definition can be used and verify in an example that it gives the same results as when the original definition is used.

When a time-ordered product is used the Feynman rules for parton densities are simply given by deletion of the final-state cut in (6.110) and all its relatives. For the case of the one-loop calculation of the density of a quark in a quark that we examined earlier, this results in the replacement of Fig. 6.10(b) by Fig. 6.13. Applying the Feynman rules gives

$$\frac{g^2}{16\pi^2} f_{q/q}^{[1,R]}(\xi) = ig^2\mu^{2\epsilon} \int \frac{dk^- \, d^{2-2\epsilon} k_T}{(2\pi)^{4-2\epsilon}} \frac{\text{Tr}\,\frac{\gamma^+}{2}(\not{k} + M)(\not{P} + M)\frac{1}{2}\left(1 + \gamma_5 \not{S}/M\right)(\not{k} + M)}{\left(k^2 - M^2 + i0\right)^2 \left[(P-k)^2 - m^2 + i0\right]}.$$

$$\tag{6.120}$$

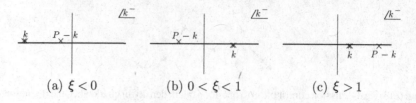

(a) $\xi < 0$ (b) $0 < \xi < 1$ (c) $\xi > 1$

Fig. 6.14. Singularities in k^- plane for (6.121).

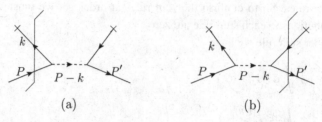

(a)   (b)

Fig. 6.15. Extra cuts of the one-loop graph for the quark density in a quark. These contribute only for negative ξ, and then cancel the contribution of the standard term Fig. 6.10(b). To avoid a division by zero in the uncut quark propagator, the matrix element is temporarily made off-diagonal in the target momentum.

All the lines now have regular propagators. Notice the overall factor of i compared with (6.115). In terms of light-front coordinates, the denominator factor is

$$\frac{1}{\left(2\xi P^+k^- - k_{\mathrm{T}}^2 - M^2 + i0\right)^2 \left[2(1-\xi)P^+(P^- - k^-) - k_{\mathrm{T}}^2 - m^2 + i0\right]}. \tag{6.121}$$

We now perform the integral over k^- by the residue theorem. This works in almost exactly the same way as in Sec. 5.4.2 for the collinear-to-A contribution to the Sudakov form factor. As illustrated in Fig. 6.14, when $\xi < 0$ and when $\xi > 1$ all the poles are in either the upper or lower half plane, so that we can deform the contour to infinity away from the poles and get zero.

The only non-zero contribution is when $0 < \xi < 1$. Closing on the single pole at $(P - k)^2 = m^2$ sets this line on-shell, and exactly reproduces the previous result, (6.117).

6.11.6 Negative ξ

One additional feature of the calculation in the previous section is that a vanishing value is obtained when ξ is negative. From the relation (6.85), this corresponds to a vanishing density of antiquarks in the quark at this order of perturbation theory.

In contrast, in the formalism with fixed ordering for the operators. the cut graph (Fig. 6.10(b)) gives a non-zero value. This appears paradoxical until we observe that there are two further cuts of the same graph, as shown in Fig. 6.15, where the quark propagator is cut, to give a final state consisting of the target and an antiquark of momentum

$-k$. When ξ is positive, the cut lines in Fig. 6.15 do not obey the positive-energy condition for physical particles, and therefore these diagrams give zero. But for negative ξ the
positive-energy condition is satisfied, and we get a non-zero contribution from the extra
cuts.

A further problem now arises: when we set $k^2 = M^2$ in one quark propagator, the other
quark propagator is exactly at its pole and gives infinity. How is one to show in a principled
way that the infinities cancel between the two cut graphs in Fig. 6.15 and that the finite
part cancels against Fig. 6.10(b)? We could solve this by using a wave-packet state as
we did in finding the probability interpretation of parton densities. An alternative, which
we will use here, is to start with the matrix element defining the parton density being
off-diagonal in target momentum: $\langle P | \dots | P \rangle \mapsto \langle P' | \dots | P \rangle$. We only take the diagonal
limit $P' \to P$ after summing over cuts. The off-diagonal matrix element shifts one of the
quark propagators from momentum k to $k + P' - P$, thereby taking the uncut propagator
slightly away from its pole. As a function of k^-, the pole and delta function structure for
the three cuts is of the form

$$\delta(k^- - A)\,(-ig)\,\frac{-i}{k^- - B - i0}\,(-ig)\,\frac{-i}{k^- - A' - i0}$$

$$+ \frac{i}{k^- - A + i0}\,(ig)\,\delta(k^- - B)\,(-ig)\,\frac{-i}{k^- - A' - i0} \qquad (6.122)$$

$$+ \frac{i}{k^- - A + i0}\,(ig)\,\frac{i}{k^- - B + i0}\,(ig)\,\delta(k^- - A'),$$

up to a common overall factor. The quantities A, B and A' are functions of masses, of ξ
and the difference between P' and P. The diagonal-matrix-element limit $P' \to P$ gives
$A' \to A$. Integrating over k^- gives

$$g^2 \frac{1}{A - B}\frac{1}{A - A'} + g^2 \frac{1}{B - A}\frac{1}{B - A'} + g^2 \frac{1}{A' - A}\frac{1}{A' - B}, \qquad (6.123)$$

which sums to zero, even before taking the limit $A' \to A$.

This calculation is a verification in an example of a general result that we proved using
operator (anti)commutation relations. The cancellation corresponds to the fact that in the
time-ordered-operator formalism, all the poles in the propagators are on one side of the real
axis, as in Fig. 6.14(a).

An interesting variant of this problem occurs when we try computing the density of a
scalar parton in the fermion target. Exactly the argument we have just given shows that the
graph in Fig. 6.12 has two extra cuts and that the sum vanishes for negative ξ. However, we
have also shown that, since the scalar particle is its own antiparticle, its density at negative
ξ is the negative of the density at positive ξ, (6.87), and therefore is non-zero.

To recover this result, we observe that there are other possible graphs, Fig. 6.16, in
which the vertices of the scalar line on the fermion line are reversed. For *positive* ξ, these
graphs are zero, and so do not affect the calculation we have already done. But when ξ is in
the range $-1 < \xi < 0$, similar arguments to those we gave earlier in this section show that

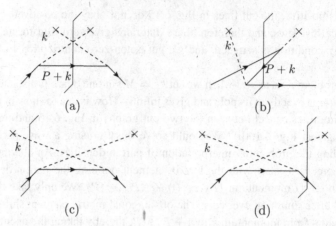

Fig. 6.16. Cut graphs at one-loop order when $\xi < 0$ for the density of scalar partons. Graph (a) is in fact zero, because the coupling of the three on-shell particles violates 4-momentum conservation. Graph (b) only contributes when $\xi < -1$, by the positive-energy condition on the particles on the cut.

the sum of these extra graphs is non-zero, and in fact they result in (6.87). When $\xi < -1$ the graphs sum to zero. Verification of these statements is left as an exercise.

Exercises

6.1 Find a/the k_T-dependent Lorentz transformation that converts k to k' in (6.6).

6.2 Derive (6.31) from (6.14).

6.3 Similarly derive (6.33).

6.4 (a) Derive the corresponding results for polarized antiquark densities. Pay careful attention to signs.

(b) Fill in any other missing details in Sec. 6.5.

6.5 What would happen if the theory were parity violating?

6.6 (a) Using the methods of this chapter, derive the parton model when the quarks have spin 0. Then derive a formula for the corresponding parton density:

$$f_s(\xi) = \xi P^+ \int_{-\infty}^{\infty} \frac{\mathrm{d}w^-}{2\pi} e^{-i\xi P^+ w^-} \langle P | \phi^\dagger(0, w^-, \mathbf{0}_T) \phi(0) | P \rangle_c , \qquad (6.124)$$

including the, perhaps unexpected, factor ξP^+. [Note: A scalar quark might appear in a model field theory or an extension to QCD, notably a super-symmetric extension.]

(b) Obtain the corresponding formulae for the unintegrated density.

6.7 Carefully derive the signs in the exponents in (6.26).

6.8 Generalize whatever needs to be generalized in this chapter to deal with DIS on a spin-1 target like the deuteron. [See Hoodbhoy, Jaffe, and Manohar (1989) for an account of some of the theory, one of the features of which is a new structure function b_1. See Airapetian *et al.* (2005) for the first measurement of b_1.]

6.9 Check the statement given in the text that, in light-front quantization in the theory specified by (6.44), the standard field equations (6.45) and (6.46) do indeed follow from the canonical (anti)commutation relations (6.53) and (6.54) and the Heisenberg equations of motion (6.51).

6.10 Check that the other equations in the sections on light-front quantization and their relations to parton densities are correctly derived, notably (6.65).

6.11 Verify the results (6.76) and (6.78) for the interpretation of the polarized parton densities. Do this for both quarks and antiquarks. [Note: There are some subtleties in discussing the spin states needed in the wave-packet derivation that may impinge on this discussion. See Bakker, Leader, and Trueman (2004).]

6.12 Generalize the relation between quark for negative ξ and antiquark densities with positive ξ to the polarized case.

6.13 Derive the relations (6.101) and (6.102) for parton densities in pions.

6.14 Extend these results to kaons.

6.15 Generalize the proof in Sec. 6.9.5 to derive the momentum sum rule (6.93). You will need to convert the left-hand side of the sum rule to a matrix element of the energy-momentum tensor.

6.16 At one-loop order verify the momentum sum rule (6.93) for a quark target in the Yukawa model theory. The sum over j is over the fermion, the antifermion, and the scalar.

6.17 Perform the one-loop calculation of the parton densities for a target that corresponds to the scalar field in our Yukawa field theory. Again verify the momentum sum rule. (The number sum rule is trivially satisfied, since, as you can verify, $f_{\bar{q}/\phi}(\xi) = f_{q/\phi}(\xi)$.)

6.18 Verify by explicit calculations the statements at the end of Sec. 6.11.6.

6.19 (**) (This problem is quite hard, probably very difficult, and might even deserve three stars.) Suppose we take field theory to be defined by Feynman graphs for Green functions. Derive equal-time and equal-x^+ commutation relations. Thus Feynman perturbation theory does in fact correctly solve the operator formulation of the theory, despite any doubts one might have about the rigor of the derivation of perturbation theory.

Note that there is quite a bit of literature on obtaining commutation relations from time-ordered Green functions, but that most of this dates from the heyday of current algebra and therefore pre-dates QCD. These techniques have not propagated to modern textbooks. I refer here to the Bjorken-Johnson-Low (BJL) method (Bjorken, 1966; Johnson and Low, 1966).

6.20 (***) What happens in the previous problem if you apply it in the presence of renormalization and/or of gauge fields? [Note: Either or both of these conditions is liable to need techniques from the later part of this book, but probably in their simpler forms.]

7

Parton theory: further developments

In the previous chapter, we formalized the parton model in a simple quantum field theory. A number of further developments follow fairly simply, and this chapter's purpose is to give an account of them, before we go on to the full QCD treatment.

We first extend the parton model for DIS to the very important case of charged-current weak-interaction processes. Then we examine a particularly influential form of perturbation theory: light-front (or x^+-ordered) perturbation theory. After that I present the light-front quantization of gauge theories, a natural extension of what we did earlier for non-gauge theories. We will thereby be able to introduce appropriate definitions for parton densities in a gauge theory, and to convert them to a gauge invariant form with the aid of what are known as Wilson-line operators.

7.1 DIS with weak interactions, neutrino scattering, etc.

We have extensively discussed DIS for the case of virtual photon exchange. The same principles apply equally to all lepto-production processes $l + N \longrightarrow l' + X$, and thus they apply whether the exchanged electroweak boson is a W, Z or photon. There are a large number of different cases, and, as far as the theory by itself is concerned, all are a minor variation on the purely electromagnetic case, both at the parton-model level, and with all the QCD modifications. The structure-function review in Amsler *et al.* (2008, Ch. 16) is an authoritative source for the relevant results including corrections of errors in the literature and commonly used standards for notations, the bulk of which we follow. See also Hobbs and Melnitchouk (2008) for a recent treatment of the role of γ–Z interference in the parity-violating part of neutral current DIS.

7.1.1 Structure functions

In view of its particular importance to the determination of the flavor-separated quark densities, we restrict our attention to the charged-current processes in neutrino scattering on unpolarized nucleons. These are the processes $\nu + N \longrightarrow \mu + X$ and $\nu + N \longrightarrow e + X$, with the exchanged boson being the W^+. The hadronic tensor is

$$W^{\mu\nu}(q, P) = \frac{1}{4\pi} \int d^4z \; e^{iq \cdot z} \langle P, S| J^\mu(z/2)^\dagger J^\nu(-z/2) |P, S\rangle , \qquad (7.1)$$

where J is now the non-hermitian hadronic current coupling to the W boson. We normalize this charge-changing current to

$$J^\mu = \bar{u}\gamma^\mu(1-\gamma_5)d' + \bar{c}\gamma^\mu(1-\gamma_5)s' + \bar{t}\gamma^\mu(1-\gamma_5)b'$$

$$= \begin{pmatrix} \bar{u} & \bar{c} & \bar{t} \end{pmatrix} \gamma^\mu(1-\gamma_5)\, U_{\text{CKM}} \begin{pmatrix} d \\ s \\ b \end{pmatrix}. \tag{7.2}$$

Here u, c, and t are the fields for the corresponding quarks, and d', s', and b' are for the down-type quarks that are associated with them in multiplets of weak isospin. The fields for *mass-eigenstate* quarks, d, etc., are obtained by a CKM rotation, reviewed in Amsler *et al.* (2008, Ch. 11), and implemented in the above equation by the matrix U_{CKM}, which acts on quark flavor indices.

We next decompose $W^{\mu\nu}$ in scalar structure functions. There are two differences compared with the pure electromagnetic case, both of which increase the number of structure functions. First is that parity is not conserved and second is that the currents are not conserved because the quark masses are non-zero. A thorough analysis was given by Ji (1993), who found 14 structure functions on a spin-$\frac{1}{2}$ hadron target, of which 5 appear when the hadron is unpolarized and 9 concern hadron-polarization dependence.

For most purposes we can neglect the structure functions allowed by non-conservation of the current. Normally, we neglect quark masses compared with Q within the hard scattering, so that the extra structure functions are suppressed by a power of m_q/Q. This of course does not always work for heavy quarks, notably the b and t.

Thus the extra structure functions could be significant when there is a heavy quark in the hard scattering. However, the associated tensors almost all have a factor of q^μ or q^ν, the one exception being in polarized scattering. Now a factor q^μ or q^ν times the leptonic tensor is non-zero only because a lepton mass is non-zero, and therefore we obtain a suppression by a power of a small *lepton* mass divided by Q.

The result is that for neutrino scattering on an unpolarized target, we have one extra relevant structure function F_3:

$$W^{\mu\nu} = \left(-g^{\mu\nu} + q^\mu q^\nu/q^2\right) F_1(x, Q^2) + \frac{(P^\mu - q^\mu P\cdot q/q^2)(P^\nu - q^\nu P\cdot q/q^2)}{P\cdot q} F_2(x, Q^2)$$

$$- i\epsilon^{\mu\nu\alpha\beta}\frac{q_\alpha P_\beta}{2P\cdot q} F_3(x, Q^2) + \text{irrelevant}. \tag{7.3}$$

See Amsler *et al.* (2008, Ch. 16) for a definition that includes structure functions for polarized scattering.

7.1.2 Parton model with low-mass quarks

The parton model and its derivation work equally well with neutrino scattering at large Q. As before, the parton-model approximation to the hadronic tensor is just a sum over parton densities times the tensor computed to lowest order on an on-shell quark target, as

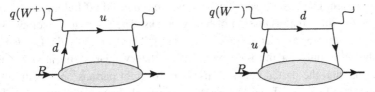

Fig. 7.1. Examples of lowest-order parton-model processes for charged-current DIS with a W^+ or a W^- exchanged.

in (2.27). Let $\hat{k}^\mu = (x P^+, 0, \mathbf{0}_T)$ be the approximated quark momentum. Then the partonic tensor (2.28) is simply replaced by

$$C_j^{\mu\nu} = \frac{1}{4\pi}\frac{1}{2}\operatorname{Tr}\hat{k}\gamma^\mu(1-\gamma_5)(\not q+\hat{\not k})\gamma^\nu(1-\gamma_5)\,2\pi\delta((q+\hat{k})^2),\qquad(7.4)$$

with the restriction to the allowed partonic subprocesses, e.g., $d \to u$, $\bar{u} \to \bar{d}$, etc. for W^+ exchange. Note that the formula must be slightly changed on an antiquark. It is readily deduced that the parton-model structure functions are

$$F_2^{\mathrm{QPM},W^+} = 2x\,[d(x)+\bar{u}(x)+s(x)+\bar{c}(x)+\ldots],\qquad(7.5\mathrm{a})$$

$$F_1^{\mathrm{QPM},W^+} = \frac{1}{2x}F_2,\qquad(7.5\mathrm{b})$$

$$F_3^{\mathrm{QPM},W^+} = 2\,[d(x)-\bar{u}(x)+s(x)-\bar{c}(x)+\ldots].\qquad(7.5\mathrm{c})$$

For processes like $\bar{\nu}p \to e^+X$ with W^- exchange, the roles of the quarks in each isospin doublet are exchanged, to give

$$F_2^{\mathrm{QPM},W^-} = 2x\left[u(x)+\bar{d}(x)+\bar{s}(x)+c(x)+\ldots\right],\qquad(7.6\mathrm{a})$$

$$F_1^{\mathrm{QPM},W^-} = \frac{1}{2x}F_2,\qquad(7.6\mathrm{b})$$

$$F_3^{\mathrm{QPM},W^-} = 2\left[u(x)-\bar{d}(x)-\bar{s}(x)+c(x)+\ldots\right].\qquad(7.6\mathrm{c})$$

Only those heavy quarks whose mass is low enough to participate in the process should be included. Notice the restricted set of quark flavors allowed in each structure function (Fig. 7.1). Notice also the reversal of sign for the antiquark terms in the F_3 structure function. These properties indicate how important charged-current scattering is for the flavor separation of quark and antiquark densities from data.

7.1.3 Quark masses

So far our arguments have relied on Q being much larger than all the particle masses of a theory; the reduced diagram analysis concerned the zero-mass limit $m/Q \to 0$. But the wide range of quark masses in QCD shows that there is an interesting region where Q is much larger than the lightest masses, but less than or comparable to some of the heavy quark masses. A full and systematic treatment in QCD will appear in Sec. 11.7.

Charged-current DIS, where heavy quarks can be produced off light quarks, provides a useful place to initiate the discussion of heavy quarks. The basic methodology is to treat the heavy quark masses as a large scale, just like Q. Then we apply the Landau analysis to locate the PSSs only for the light partons. The heavy quarks appear only inside the hard scattering, and the parton densities used are for light partons only. Naturally, when Q is increased sufficiently above the mass of a particular quark, the status of the quark changes.

For neutral-current processes, heavy quarks are made in pairs, and the hard scattering analysis is closely tied to the higher-order corrections to the hard scattering, to be studied later. But for charged-current processes, the production of a heavy quark can occur at lowest order, e.g., $W^- + \bar{s} \to \bar{c}$, $W^+ + s \to c$. Then the parton-model approximation for the hard scattering retains the massless approximation for the incoming quark, but we insert the mass m_h for the outgoing heavy quark. Thus we replace the parton-level structure function (7.4) by

$$C_j^{\mu\nu} = \frac{1}{4\pi} \frac{1}{2} \operatorname{Tr} \hat{k} \gamma^\mu (1 - \gamma_5)(\slashed{q} + \hat{k} + m_h)\gamma^\nu (1 - \gamma_5) \, 2\pi \delta((q + \hat{k})^2 - m_h^2), \qquad (7.7)$$

applicable to a process $W + q_j \to q_h$, with a transition from a light quark of flavor j to a heavy quark h. The mass shell condition now sets the parton momentum fraction to

$$\xi = x(1 + m_h^2/Q^2) \qquad (7.8)$$

rather than simply x. It can readily be checked that the contributions to the hadronic structure functions are

$$F_1^{j \to h} = f_j(\xi), \qquad F_2^{j \to h} = 2\xi f_j(\xi), \qquad F_3^{j \to h} = 2 f_j(\xi). \qquad (7.9)$$

This should be used to replace the relevant terms in (7.5) and (7.6).

In addition there are terms in (7.7) proportional to $q^\mu q^\nu / Q^2$ and $(P^\mu q^\nu + q^\mu P^\nu)/P \cdot q$, which are allowed because of non-conservation of the currents when quark masses are non-zero. As already stated, the factors of q^μ or q^ν multiply the leptonic tensor, and give a suppression by a power of *lepton* mass divided by Q.

There is in principle a sharp structure in the structure functions at the threshold for production of a heavy quark, at

$$x_{\text{exact threshold}} = \frac{1}{1 + (M_{\min}^2 - M^2)/Q^2}, \qquad (7.10)$$

where $M_{\min}$ is the lowest-mass final state in $W^\pm + P \to X$ that includes a hadron containing a particular heavy quark. For a b quark this might be the lightest B-flavored baryon, Λ_b^0. In contrast the partonic calculation gives a threshold in x given by setting $\xi = 1$ in (7.8), i.e., at

$$x_{\text{parton threshold}} = \frac{1}{1 + m_h^2/Q^2}. \qquad (7.11)$$

This differs from the exact threshold because the heavy baryon's mass is not exactly equal to the heavy quark's mass and because there is an effect due to the proton mass, both effects being neglected in the partonic calculation. For practical purposes the difference is not important because the parton densities vanish strongly at $\xi = 1$ and the main contributions arise from small ξ. Thus the main numerical contributions to the structure functions from heavy quark occur for values of x well beyond the threshold for heavy quark production.

Even so, the disagreement between the thresholds in x illustrates a principle mentioned in our derivation of the parton model in Sec. 6.1.1. This is that the approximations only apply to a local average of the structure functions, which would smear out sharp structures. Observe that the exact and parton-model thresholds in x differ by an amount proportional to hadronic-mass-squared divided by Q^2, a power-suppressed quantity. This is an important principle to remember whenever thresholds appear in partonic calculations. Improvements can only be made by treating parton kinematics better.

7.2 Light-front perturbation theory

In analyzing a collinear region, a number of interesting simplifications arise when we integrate over the minus components of loop momenta. A simple example was in Sec. 6.11.5 for a one-loop calculation of a parton density, where it gave the restriction that fractional momenta for internal lines correspond to forward-moving particles: the target splits into two forward-moving partons, both with positive plus momentum, and one of the partons initiates the hard scattering.

In the example, and as we will now see quite generally for any Feynman graph, integrating over the minus momenta led to restrictions on plus momenta that correspond to to the restrictions imposed by the reduced graph analysis of Ch. 5. This leads to an interesting generalization of the method of time-ordered perturbation theory (Sterman, 1993, Sec. 9.5). In this older method, in contrast to Feynman perturbation theory, the effect of interactions is to cause transitions from one state to another, the interactions occur as a sequence in time, and there are energy denominators corresponding to the intermediate states. Time-ordered perturbation theory gives a useful intuition as to the time evolution of the system's state. But in relativistic theories time-ordered perturbation theory is inefficient (Heinzl, 2007), because a Feynman graph with n vertices has $n!$ time orderings. If time ordering is replaced by ordering in x^+, it turns out that many of the orderings of the vertices give zero. This formulation corresponds to a natural version of perturbation theory within light-front quantization, when the role of time in ordinary quantum-mechanical evolution equations is replaced by evolution in x^+.

The method arose first – see Brodsky, Pauli, and Pinsky (1998) for a review – in the use of what was called the "infinite-momentum" frame for understanding the parton model. The systematization in terms of light-front variables and then x^+-ordered perturbation theory was made by Chang and Ma (1969) and Kogut and Soper (1970). Chang and Ma also showed how the rules arise by performing the k^- integrals in Feynman graphs.

Naturally, if one wishes to discuss collinear regions with the high-energy particle(s) moving in some direction other than the $+z$ direction, a different definition of light-front coordinates is appropriate.

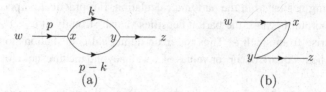

Fig. 7.2: (a) Self-energy graph. (b) An x^+ ordering that gives zero.

7.2.1 Example

The basic principles are illustrated by the simple example of the propagator correction graph in ϕ^3 theory, shown in Fig. 7.2(a). We first examine the case that the external momentum obeys $p^+ > 0$. Since we need to consider also the graphs in coordinate space, each vertex is labeled with a position variable. The value of the graph is

$$\Gamma(p) = \frac{-g^2}{(p^2 - m^2)^2 \, 2(2\pi)^n} \int d^n k \, \frac{1}{2k^+ k^- - k_T^2 - m^2 + i0}$$

$$\times \frac{1}{2(p^+ - k^+)(p^- - k^-) - (\boldsymbol{p}_T - \boldsymbol{k}_T)^2 - m^2 + i0}. \tag{7.12}$$

We perform the integral over k^- by closing the contour in the upper or lower half plane. This gives zero except when k^+ is between 0 and p^+, with the result

$$\Gamma(p) = \frac{(-ig)^2}{2(2\pi)^{n-1}} \int_0^{p^+} dk^+ \int d^{n-2} k_T \, \frac{1}{(2p^+)^2 \, 2k^+ \, 2(p^+ - k^+)}$$

$$\times \frac{i^3}{\left[p^- - \frac{p_T^2 + m^2}{2p^+} + i0 \right]^2 \left[p^- - \frac{k_T^2 + m^2}{2k^+} - \frac{(\boldsymbol{p}_T - \boldsymbol{k}_T)^2 + m^2}{2(p^+ - k^+)} + i0 \right]}. \tag{7.13}$$

This has been organized so as to correspond to the general form we will find in x^+-ordered perturbation theory. The relevant ordering is given in Fig. 7.2(a). This has three intermediate states, between the vertices w and x, between x and y, and between y and z. The last line of (7.13) is a product of an energy denominator factor for each intermediate state. Each of these factors is $i/(p^- - \text{on-shell} + i0)$, where "on-shell" denotes the value of minus momentum the state would have if the particles were on-shell. Note that two of the denominators, for the first and last intermediate states, are equal. In common with the Feynman-graph method, there are the factors of $-ig$ for each vertex and a symmetry factor $\frac{1}{2}$. The integration is only over the plus and transverse momenta, and for each line there is a factor of one divided by twice its plus momentum. This corresponds to the denominator in the light-front version of the "Lorentz-invariant phase-space" measure, (6.59). At each internal vertex are conserved the independent components of momentum needed to specify physical states, i.e., the plus and transverse components.

In advance of their general proof, we can use this graph as an illustration of the results that yield the main simplifications given by light-front perturbation theory. Each line has to carry a physical positive value of plus momentum, when considered flowing from left to right. Because $p^+ > 0$, the x vertex is to the right of the w vertex, and similarly for z relative to y, i.e., $x^+ > w^+$, $z^+ > y^+$. This still leaves one other possible ordering, shown in Fig. 7.2(b), where y^+ is earlier than x^+. However, this ordering does not give an allowed situation, since at the x vertex we have three positive plus momenta coming in from the left and none going out to the right. This is how the simplification compared with time-ordered perturbation theory occurs. With time-ordered perturbation theory, the ordinary energy k^0 would have been integrated over, and the independent variables for each line would be the ordinary spatial momentum, none of whose components has any constraint on its sign.

For this one Feynman graph we have one x^+ ordering that gives a non-zero result. In contrast, time-ordered perturbation theory would have $4! = 24$ time orderings for the same Feynman graph.

One can readily verify that the above calculation reproduces the standard result for the Feynman graph by performing the k_T integral. After a change of variable to $x = k^+/p^+$, an integral is obtained that is the same as is obtained when the momentum integral in (7.12) is performed by the conventional Feynman parameter method.

7.2.2 Paradox at $p^+ = 0$

We obtained (7.13) for the case that p^+ was positive, and observed that it gives the correct value. Similarly when p^+ is negative, we also get the correct value, but with a reversed ordering for the vertices.

But if p^+ is exactly zero, the poles in k^- in the original integral (7.12) are either both in the lower half plane (if $k^+ > 0$) or both in the upper half plane (if $k^+ < 0$). Completing the contour of k^- away from the poles then gives zero for all values of k^+. This disagrees with the definite non-zero limit of (7.13) as $p^+ \to 0$, and hence with the non-zero value of the ordinary Feynman graph.

This issue is rather important, because the same method can be used to show that disconnected vacuum bubbles are apparently all zero in light-front perturbation theory e.g., Weinberg (1966) and the *introduction* (but not the later sections) of Chang and Ma (1969). This has contributed to a general impression (e.g., Brodsky, Pauli, and Pinsky, 1998) that the vacuum is trivial in light-front quantization, unlike the case for equal-time quantization; i.e., the interactions do not change the vacuum state. Now, although vacuum bubbles are normally discarded, they physically give an energy density to the vacuum, which can be related to the energy-momentum tensor in the vacuum. Vacuum energy-momentum is equivalent (Weinberg, 1989) to a contribution to the cosmological constant in general relativity, i.e., it has observable consequences. (There is, of course, an infinite renormalization of the cosmological constant to cancel UV divergences in vacuum bubbles.) Evidently, when different results are obtained for the same graph by different methods of calculation in the same theory, at least one method is wrong.

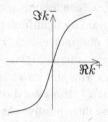

Fig. 7.3. Partially deformed contour for evaluation of integral at zero external p^+. Note that the axes are the real part of k^+ and the imaginary part of the *other* variable k^-, and that there are therefore two other dimensions not shown, for $\Re k^-$ and $\Im k^+$.

Our propagator calculation indicates that there must be a problem with the derivation of a zero value for the propagator correction at zero external plus momentum, since that disagrees with the limit from non-zero p^+, and propagators are analytic functions of external momentum. This problem was recognized and solved, at least in examples, by Chang and Ma (1969). A more general solution was provided by Yan (1973) with generalizations by Heinzl (2003).

At zero external p^+ we need the integral over k^- and k^+ of

$$\frac{1}{[2k^+k^- - k_{\mathrm{T}}^2 - m^2 + i0]\,[2k^+(k^- - p^-) - (\boldsymbol{p}_{\mathrm{T}} - \boldsymbol{k}_{\mathrm{T}})^2 - m^2 + i0]}. \tag{7.14}$$

We deform the (*two*-real-dimensional) contour of integration so that the imaginary part of k^- is infinite and positive when $k^+ > 0$, but infinite and negative when $k^+ < 0$. As illustrated in Fig. 7.3, the contour of integration is a connected manifold and so at k^+ close to zero, the deformed contour has to pass through small values of $\Im k^-$. This leaves the possibility of a non-zero contribution, where k^+ is very small and k^- is very large, leaving k^+k^- of a fixed size. (Such a contribution does not arise when the external plus momentum is non-zero, because there is then a large denominator containing a p^+k^- term.)

Contour integration shows that the integral over k^- of (7.14) is zero whenever k^+ is non-zero. On the other hand the integral of (7.14) over both k^+ and k^- is definitely non-zero. This indicates that the integral over the one variable k^- must be treated as a generalized function of the other variable k^+, e.g., a coefficient times $\delta(k^+)$: it is zero everywhere except at one point, but with a non-zero integral. In fact, Yan (1973) and Heinzl (2003) showed that

$$\int dk^- \frac{1}{(M^2 - 2k^+k^- - i0)^\nu} = \pi i \frac{\delta(k^+)}{(\nu - 1)(M^2)^{\nu - 1}}. \tag{7.15}$$

This formula can be used to calculate the integral of (7.14) with the aid of a Feynman parameter combination for the denominators, and results in agreement with the Feynman-graph calculation and with the limit from $p^+ \neq 0$. The paradox is now resolved.

For disconnected vacuum diagrams, this solves the disagreement between light-front perturbation theory and regular Feynman perturbation theory; Feynman perturbation theory is correct, and there is in this case little notable advantage to use of light-front perturbation theory.

But for graphs with non-zero external momenta, we can choose the external momenta to avoid the problematic situations, as was shown quite generally by Chang and Ma (1969). We avoid the problems if no subgraph is forced by the configuration of external momenta to have exactly zero for its external plus momentum.

Certain other complications arise when there are numerator factors with dependence on k^-. These can affect the convergence of the k^- integrals, and a naive application of light-front methods to a Feynman graph can give a wrong result. This is particularly the case for calculations in a massless on-shell approximation.

7.2.3 General rules

Statement

The general rules for perturbation theory in the x^+-ordered form can be found in Chang and Ma (1969), Kogut and Soper (1970), and Yan (1973), but with a different normalization. There are some complications associated with momentum dependence in vertices and in propagator numerators, so we first state and derive the rules for scalar theories.

1. The graphs are like Feynman graphs except that the vertices are assigned an ordering in x^+, which in drawing diagrams we will take to increase from left to right. All possible graphs and orderings are to be used.
2. Coupling factors at vertices and symmetry factors are the same as in Feynman graphs.
3. Each line l is assigned a plus and transverse momentum: k_l^+, $k_{l\mathrm{T}}$, and these components of momentum are subject to conservation at the vertices.
4. The sign of each line momentum is chosen to correspond to propagation from lower to higher x^+, and then k_l^+ is always physical, i.e., positive.
5. For each loop there is an integral of a loop momentum, but only over its plus and transverse components: $\int \dfrac{\mathrm{d}k^+ \, \mathrm{d}^{n-2}k_{\mathrm{T}}}{(2\pi)^{n-1}}$.
6. For each line l there is a factor $\dfrac{1}{2k_l^+}$.
7. For each intermediate state α there is a factor

$$\frac{i}{P_\alpha^- - P_{\alpha \text{ on-shell}}^- + i0}. \tag{7.16}$$

Here P_α^- is the total *external* minus momentum entering the graph to the left (earlier x^+) than the intermediate state α, while $P_{\alpha \text{ on-shell}}^-$ is the value of the minus momentum of the particles contained in the state when they are on-shell. That is,

$$P_{\alpha \text{ on-shell}}^- = \sum_{l \in \alpha} \frac{k_{l\mathrm{T}}^2 + m_l^2}{2k_l^+}. \tag{7.17}$$

These rules can be derived by normal time-dependent perturbation theory, with the change that light-front quantization is used and the evolution variable is x^+ instead of ordinary time (Kogut and Soper, 1970). What we will do here instead is to derive them from Feynman perturbation theory in the coordinate-space representation, with the integrals over the positions of the vertices split up according to their ordering in x^+. This second method directly shows the equivalence with Feynman perturbation theory; it will also provide techniques for analyzing Feynman graphs in terms of particles propagating in space-time.

Derivation

See also Ligterink and Bakker (1995) for a derivation.

We start with the momentum-space representation for Feynman graphs and perform appropriate Fourier transforms to obtain the coordinate-space representation, but only as regards plus components of vertex positions. In all of the following, we will use x_j to represent the position of a vertex j in a graph.

First we Fourier-transform a free propagator to plus position, and decompose according to the ordering of its vertices:

$$
\tilde{G} \equiv \int \frac{dk^-}{2\pi} e^{ik^-(x_j^+ - x_k^+)} \frac{i}{2k^+k^- - k_T^2 - m^2 + i0}
$$

$$
= \theta(x_k^+ - x_j^+) \frac{\theta(k^+)}{2k^+} e^{ik_{\text{on-shell}}^-(x_j^+ - x_k^-)} + \theta(x_j^+ - x_k^+) \frac{\theta(-k^+)}{-2k^+} e^{i(-k)_{\text{on-shell}}^-(x_k^+ - x_j^-)}. \quad (7.18)
$$

Here k is regarded as flowing from x_j to x_k, and the explicit minus signs in the second term serve to indicate that we always have physical (positive) plus momentum flowing from the earlier vertex to the later vertex. The above formula is readily derived by contour integration.

So we decompose the free momentum-space propagator as

$$
G = \frac{\theta(k^+)}{2k^+} \frac{i}{k^- - (k_T^2 - m^2)/(2k^+) + i0} + \frac{\theta(-k^+)}{-2k^+} \frac{i}{-k^- - (k_T^2 - m^2)/(-2k^+) + i0},
$$

$$
(7.19)
$$

with each term being associated with one of the two possible x^+ orderings of the ends of the line.

A common textbook derivation of Feynman rules starts from the coordinate representation (from Wick's theorem or a similar result from the functional integral), and then writes the result in terms of Fourier transforms into momentum space. Here we first partially reverse the derivation by writing the delta function for conservation of minus momentum at a vertex j as

$$
2\pi\delta\left(p_j^- + \sum k_{l\,\text{in}}^- - \sum k_{l\,\text{out}}^-\right) = \int dx_j^+ \, e^{ix_j^+(-p_j^- - \sum k_{l\,\text{in}}^- + \sum k_{l\,\text{out}}^-)}, \quad (7.20)
$$

where p_j^- is the external momentum entering at the vertex, the $k_{l\,\text{in}}^-$ are the momenta on lines coming to the vertex from earlier vertices, and $k_{l\,\text{out}}^-$ are the momenta on lines leaving the vertex on lines to later vertices. In this formulation we have an integral over the minus momentum of each and every line, with explicit delta functions at the vertices.

We next obtain the contribution from a particular ordering of the x_j^+. We choose the vertex labels to correspond to the ordering $x_1^+ < x_2^+ < x_3^+ \ldots$, and we implement this by

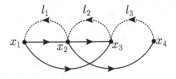

Fig. 7.4. Momentum-space representation of an x^+ ordering of a Feynman graph. See the text for explanation.

multiplying the original Feynman graph by theta functions: $\prod \theta(x_{j+1}^+ - x_j^+)$. Then we write a momentum-space representation for the theta functions:

$$\theta(x_{j+1}^+ - x_j^+) = \int \frac{dl_j^-}{2\pi} e^{il_j^-(x_j^+ - x_{j+1}^+)} \frac{i}{l_j^- + i0}. \tag{7.21}$$

We represent this in Fig. 7.4, where the dotted lines represent minus momenta flowing from each vertex to the next, together with the factor $i/(l_j^- + i0)$ in the above equation. Next to each vertex is a label for its position.

The integral over the vertex positions now gives us back conservation of minus momentum at each vertex, but with the momenta on the dotted theta-function lines included. We treat the minus momenta on the regular (non-dotted) lines as the independent variables of integration, with the vertex delta functions determining the l_j^- variables. If a line of momentum k goes from vertex j to a later vertex j', then k^- gets routed back along all the dotted lines from j' to j.

Finally, we apply contour integration on each k^- integral, closing in the lower half plane on the poles of the regular propagators. This then sets k^- in each of the dotted-line factors between j and j' to be the on-shell value. Repeating this for every line then results in the dotted line joining two vertices having a contribution of the on-shell k^- in the corresponding intermediate state. The momentum-conservation delta functions also route the external momenta along the dotted lines. Thus the final result is to turn the dotted-line factors into exactly the energy denominators we announced in the rules for x^+-ordered perturbation theory.

This completes the derivation.

Fermion lines

The numerator factor for a free fermion propagator G_f has k^- dependence, and this entails an extension to the decomposition (7.19) that we wrote for a scalar field propagator. We write the k^- dependence of the numerator as $i\gamma^+ k^- = i\gamma^+ k_{\text{on-shell}}^- + i\gamma^+(k^- - k_{\text{on-shell}}^-)$, where the first term works just as in the scalar propagator, but the second term cancels the on-shell pole, to give

$$G_f = \frac{\theta(k^+)}{2k^+} \frac{i(\slashed{k}_{\text{on-shell}} + m)}{k^- - (k_T^2 - m^2)/(2k^+) + i0} + \frac{\theta(-k^+)}{-2k^+} \frac{i(\slashed{k}_{\text{on-shell}} + m)}{-k^- - (k_T^2 - m^2)/(-2k^+) + i0}$$

$$+ \frac{i\gamma^+}{2k^+}. \tag{7.22}$$

Fig. 7.5. Instantaneous interaction for fermion, denoted by the line with a bar across it. It has the value of $i\gamma^+/(2k^+)$ times the attached interaction vertices.

After the x^- integrals are performed, the last term Fourier-transforms to a delta function in x^+, giving an instantaneous interaction (Kogut and Soper, 1970) denoted diagrammatically by a line with a bar across it, as in Fig. 7.5. Naturally there is no associated intermediate state.

Similar issues arise with momentum-dependent vertices, as for the 3-gluon vertex in QCD and with couplings of gauge bosons to scalar fields.

7.2.4 Interpretation for pdf; time scales

Originally, the methods of x^+-ordered perturbation theory and its predecessor, the infinite-momentum technique, were applied to scattering processes at high energy. But as factorization theorems became systematized, the applications of x^+-ordered perturbation theory shifted more to treating the properties of a collinear region; more specifically, they became of use in analyzing the state of a fast-moving particle, e.g., the target in DIS.

An example is the calculation of a parton density, e.g., from Fig. 6.13. The two vertices defining the parton density are at equal x^+, and thus there is no intermediate state between them. In x^+-ordered perturbation theory the target splits into two particles. There is an intermediate state of the partons k and $P - k$ which propagates until one of the partons gets to the parton-density vertex. In the application of a pdf to DIS, this corresponds to where the virtual photon knocks out the parton, over a short time scale. Then the amplitude is squared to make a probability density. In the space-time picture of DIS, Fig. 6.3, the outgoing struck quark goes almost exactly in the x^- direction.

In the paradigmatic parton-model region, the incoming quark has transverse momentum of order a normal hadronic mass M. Then the denominator for the intermediate state in light-front perturbation theory is of order M^2/P^+, i.e., of order M^2x/Q in the Breit frame. We expect the typical lifetime of the state to be the inverse of this. This gives a typical intrinsic hadronic time scale $1/M$ times a time-dilation factor P^+/M. Thus x^+-ordered perturbation theory nicely and quantitatively implements the parton-model intuition. We can summarize the parton-model approximation as neglecting the duration of the hard collision compared with this long time scale P^+/M^2.

DIS exhibits the situation that in the interesting cases one always has at least two different directions of motion for the high-energy particles. While x^+-ordered perturbation theory is very natural for discussing the target state and its evolution, including that of the target remnant, a corresponding discussion of the outgoing struck quark is more naturally made with ordering with respect to the other light-front variable x^-. Naturally in more complicated situations one has even more relevant directions for collinear sets of particles, and a correspondingly appropriate light-front variable for each set. Observe that discussion of the struck quark jet in DIS involves lines with plus momentum of order M^2/Q at large Q. Thus in the version of light-front perturbation theory appropriate to the *target*, the lines of the outgoing struck quark have close to zero plus momentum, i.e., they are close to the zero modes.

A unified description of the whole process is best made using ordinary Feynman perturbation theory, with light-front methods being applied separately to each collinear group (e.g., the target, or the outgoing struck quark together with its associated jet).

7.2.5 Frame dependence of ordering of x^+ and x^- in DIS

Consider the struck quark in DIS before it collides with the virtual photon in the parton-model region of low transverse momentum. In a Feynman graph its longitudinal momentum components have opposite signs: $k^+ > 0$, $k^- < 0$. This has the following interesting consequence.

In x^+-ordered perturbation theory, the parton travels forward from its last interaction inside the target, precisely because $k^+ > 0$. The value of k^- was integrated over to obtain this form of perturbation theory. Viewed in the Breit frame, this shows that the last interaction in the target happens earlier than the hard collision with the virtual photon.

Suppose instead we used x^--ordered perturbation theory. This would appropriate for discussing physics in the target rest frame, in which case the virtual photon is moving with large momentum in the *negative* z direction. The longitudinal variable parameterizing the parton state is now k^-. Since this is negative, the propagation is from the photon vertex to an interaction with the target. Thus the ordering of the events is reversed. This is illustrated in Fig. 7.6. We have thus found that the time-ordering of the ends of the line of momentum k gets reversed in different frames. This requires the separation of the ends to be space-like, and is specifically associated with the opposite signs of k^+ and k^-, and thus with the fact that the momentum of the line is space-like.

7.3 Light-front wave functions

7.3.1 Definitions

The treatment in this section is based on Brodsky and Lepage (1989) and Brodsky, Pauli, and Pinsky (1998), but the normalizations of the states and wave functions are adjusted to be Lorentz invariant.

In any quantum field theory, the states of the theory can be obtained by applying products of fields to the true vacuum and then taking linear combinations. A convenient basis with a Fock-space structure is made by using the creation operators obtained in light-front quantization.

Let us define basis states by applying bare creation operators to the true vacuum. We label the states by the particle type, their plus and transverse momenta, and helicities. For

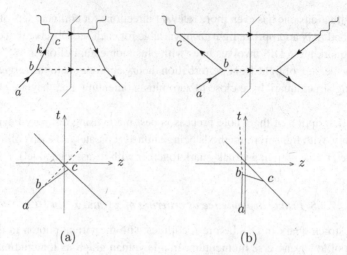

Fig. 7.6. x^+ and x^- ordering and DIS viewed in (a) the Breit frame, and (b) the target rest frame. The top line shows Feynman graphs organized for x^+- and x^--ordered perturbation theory, and the bottom line shows the positions of the vertices in space-time.

example, in the Yukawa theory treated in the previous chapter, the list of basis states would start as

$$|0\rangle,$$
$$|f : k, \alpha\rangle \qquad = b_{k,\alpha}^{\dagger} |0\rangle,$$
$$|s : k\rangle \qquad = a_k^{\dagger} |0\rangle,$$
$$|f\bar{f} : k_1, \alpha_1; k_2, \alpha_2\rangle = b_{k_1,\alpha_1}^{\dagger} d_{k_2,\alpha_2}^{\dagger} |0\rangle,$$
$$|fs : k_1, \alpha_1; k_2\rangle \qquad = b_{k_1,\alpha_1}^{\dagger} a_{k_2}^{\dagger} |0\rangle,$$
$$\cdots \qquad\qquad\qquad (7.23)$$

Here f and s denote the fermion and the scalar particles, and α is for fermion helicity. The momentum label for each particle is of the form $k_j = (k_j^+, \boldsymbol{k}_{jT})$. Naturally, this generalizes to any theory, simply by the use of suitable particle labels. In a theory with color confinement, like QCD, it is necessary to restrict to color singlet states. (At this point we gloss over complications that happen in real QCD.)

We use bare creation operators, i.e., those obtained from the bare fields, so that they obey the standard (anti)commutation relations (6.65), and the states have standard orthonormality conditions, e.g.,

$$\langle f : k', \alpha' | f : k, \alpha\rangle = (2\pi)^{n-1} 2k^+ \delta_{\alpha\alpha'} \delta(k'^+ - k^+) \delta^{(n-2)}(\boldsymbol{k}_T' - \boldsymbol{k}_T). \qquad (7.24)$$

A general single-particle state $|h : P\rangle$ of momentum P, with $\boldsymbol{P}_T = 0$, is expanded as

$$|h : P\rangle = \sum_{F,\{\alpha_j\}} \int d[\{x, k_T\}] |F : \{x_j P^+, \boldsymbol{k}_{jT}, \alpha_j\}\rangle \, \psi_{F/h}(\{x_j, \boldsymbol{k}_{jT}, \alpha_j\}), \qquad (7.25)$$

where the sum is over the numbers of particles, and their types and helicities. The notation $\{\dots\}$ denotes an array of single-particle quantities. The measure for the integral is

$$d[\{x_j, k_{jT}\}] \stackrel{\text{def}}{=} \frac{1}{\#(f)! \#(\bar{f})! \#(s)!} \prod_j \left(\frac{dx_j \, d^{n-2} k_{jT}}{2x_j (2\pi)^{n-1}} \right)$$

$$\times \delta \left(1 - \sum_j x_j \right) 2(2\pi)^{n-1} \delta^{(n-2)} \left(\sum_j k_{jT} \right), \quad (7.26)$$

with the factorials in the prefactor chosen to compensate the multiple counting of configurations of identical partons.

The decomposition (7.25) of a state $|h : P\rangle$ was defined to apply at $P_T = 0$. It is left as an exercise (problem 7.4) to show that to obtain the state with non-zero P_T one makes the replacement

$$\psi_{F/h}(\{x_j, k_{jT}, \alpha_j\}) \mapsto \psi_{F/h}(\{x_j, k_{jT} - x_j P_T, \alpha_j\}). \quad (7.27)$$

The coefficients $\psi_{F/h}(\{x_j, k_{jT}, \alpha_j\})$ are called the light-front wave functions,[1] and they obey the normalization condition (to be proved in problem 7.4)

$$\sum_{F, \{\alpha_j\}} \int d[\{x, k_T\}] \left| \psi_{F/h}(\{x_j, k_{jT}, \alpha_j\}) \right|^2 = 1. \quad (7.28)$$

A projection onto basis states gives the wave functions

$$\langle F : \{x_j P^+, k_{jT}, \alpha_j\} | h : P\rangle = \psi_{F/h}(\{x_j, k_{jT}, \alpha_j\})$$

$$\times 2(2\pi)^{n-1} \delta \left(1 - \sum_j x_j \right) \delta^{(n-2)} \left(\sum_j k_{jT} \right), \quad (7.29)$$

where we now assume $P_T = 0$ again.

7.3.2 Uses

Light-front wave functions are directly used in factorization theorems for exclusive scattering. The parton densities can be expressed in terms of light-front wave functions (problem 7.6).

7.4 Light-front quantization in gauge theories

We have seen the value of light-front quantization in gaining understanding and intuition for the parton model. So in this section we examine its application to QCD. At first sight, if we use the light-cone gauge $A^+ = 0$, all the same considerations as we used above seem to apply. Notably, the same results about the number density interpretation of parton densities

---

[1] In (7.25) the measure was normalized to match the covariant normalization (7.24) for partonic states. Thus the normalization of the wave functions differs from those in Brodsky, Pauli, and Pinsky (1998) and Heinzl (2001).

appear to apply. However, a number of complications are caused by the use of light-cone gauge, symptomized by important divergences, as we will see strongly in later chapters.

Nevertheless, it is useful to make a start by ignoring the complications and divergences. Such an approach has been enormously influential. Among other things one gains candidate definitions of parton densities, of light-front wave functions, and of related quantities, not to mention substantial intuition and insight. The true results will be distortions of those presented here.

7.4.1 Light-cone gauge

For treating light-front quantization, on a null plane of constant x^+, it is convenient (Kogut and Soper, 1970; Srivastava and Brodsky, 2001) to use the gauge-fixing condition $A^+ = 0$: only transverse degrees of freedom propagate, and there are no Faddeev-Popov ghosts. This is the "light-cone gauge" or "light-like axial gauge".

The determining issue for using this gauge to treat parton physics is that the leading regions for DIS are then the same as in non-gauge theories. In contrast, in a general gauge, there are extra gluon lines attaching to the hard subgraph H in Fig. 5.7(c), and the leading part involves the plus component of gluon polarization, which vanishes in $A^+ = 0$ gauge.

We first examine this gauge (Bassetto *et al.*, 1985; Leibbrandt, 1987) independently of the issues of light-front quantization and of parton physics. This can be done in a coordinate-independent and Lorentz covariant fashion by introducing a future-pointing light-like vector $n^\mu = \delta^\mu_-$, so that for any vector V we have $V^+ = n \cdot V$. The gauge condition is $n \cdot A = 0$, and a fractional longitudinal momentum is $\xi = k^+/P^+ = n \cdot k/n \cdot P$. Results for Green functions etc. are invariant under scaling of n by a positive real number.

There are no ghost fields in this gauge. The Feynman rules (Bassetto *et al.*, 1985; Leibbrandt, 1987) are obtained from those in covariant gauges (Fig. 3.1) by making two changes: the Faddeev-Popov fields are removed, and the free gluon propagator is changed to

$$\frac{i\delta_{\alpha\beta} N_{\mu\nu}}{k^2 + i0}, \tag{7.30}$$

where the numerator is

$$N_{\mu\nu} = -g_{\mu\nu} + \frac{k_\mu n_\nu + n_\mu k_\nu}{k \cdot n}. \tag{7.31}$$

The singularity at $k^+ = k \cdot n = 0$ causes problems. It is (Bassetto *et al.*, 1985; Leibbrandt, 1987) to be defined as a principal value in loop integrals. In many cases this works and gives physical results equivalent to those in covariant gauge, despite some complications in renormalization (Bassetto, Dalbosco, and Soldati, 1987).

However, the gauge gives some non-trivial divergences in TMD parton densities, etc. See Ch. 13 for the non-trivial details and how this is related to physically observable effects.

$$\frac{}{} = \frac{in_\mu n_\nu}{(k \cdot n)^2}$$

Fig. 7.7. Instantaneous interaction for gluon, when x^+-ordered perturbation theory in light-cone gauge is used. This barred gluon connects two regular interaction vertices at equal values of x^+. The ends of the line are connected to any normal gluon-containing vertices.

7.4.2 Light-front perturbation theory for gauge theory

For the derivation of the x^+-ordered rules for perturbation theory, the k^- dependence of the numerator of the gluon propagator causes a complication. Just as with the fermion propagator, we will find we need an extra interaction, now with instantaneous gluon exchange. We derive it by extracting from the gluon numerator a term that contains the k^- dependence and that is proportional to the denominator, i.e., k^2. Thus we obtain a modified gluon numerator:

$$N_{\mu\nu}^{\text{lf pert.}} = -g_{\mu\nu} + \frac{k_\mu n_\nu + n_\mu k_\nu}{k \cdot n} - \frac{k^2 n_\mu n_\nu}{(k \cdot n)^2}$$

$$= -g_{\mu\nu} + \frac{k_\mu n_\nu + n_\mu k_\nu}{k \cdot n}\bigg|_{k^- \mapsto k_T^2/2k^+}. \tag{7.32}$$

This does not affect the $k^2 = 0$ pole of the free gluon propagator, and is the appropriate form for making the transition to light-front perturbation theory (Srivastava and Brodsky, 2001). To keep the physical predictions of the theory the same, the extra term in the gluon propagator is compensated by an extra instantaneous interaction (Kogut and Soper, 1970). The new element in the Feynman rules, Fig. 7.7, corresponds to extra terms in the light-front Hamiltonian (Srivastava and Brodsky, 2001). See the quoted references for details. Note that in ordinary Feynman perturbation, the correct numerator is *not* (7.32), but is (7.31), with the ordinary interactions.

7.5 Parton densities in gauge theories

Initially we defined the parton density for a fermion as an expectation value of a certain bilocal operator, (6.31). This was motivated by the derivation of the parton model in a model field theory. We then saw that this parton density is an expectation value of a light-front number operator, (6.66), which is a natural implementation of the intuition embodied in the picture of scattering off constituents of the target.

For QCD, we could apply this same operator definition in $A^+ = 0$ gauge, because of the already mentioned simplification of the leading regions in this gauge. We simply modify the definition to include a sum over the three quark colors.

But we wish also to be able to use a general gauge. For this we need to find a gauge-invariant definition that agrees with (6.31) in $A^+ = 0$ gauge. As I now explain, this is done (Collins and Soper, 1982b) by inserting between the $\bar{\psi}$ and ψ operators a suitable path-ordered exponential of the gluon field. Such an exponential is called a Wilson line.

7.5.1 Wilson lines

A general Wilson line[2] is defined as a path-ordered exponential of the integral of the gluon field (times generating matrix) along a line (or path) joining two points. If we parameterize a path C by a function $x^\mu(s)$ where s goes from 0 to 1, then the associated Wilson line is

$$W(C) \overset{\text{def}}{=} P\left\{\exp\left[-ig_0 \int_0^1 ds \, \frac{dx^\mu(s)}{ds} A^\alpha_{(0)\mu}(x(s)) \, t_\alpha\right]\right\}, \tag{7.33}$$

which is more compactly written as

$$W(C) = P\left\{\exp\left[-ig_0 \int_C dx^\mu \, A^\alpha_{(0)\mu}(x) \, t_\alpha\right]\right\}. \tag{7.34}$$

Here t_α are generating matrices of the gauge group, in the fundamental representation. The path-ordering symbol P means that when the exponential is expanded, the fields with higher values of s are to the left. The Wilson line is invariant if the path is reparameterized, but it does change if the location of the path is changed even with fixed endpoints.

Under a gauge transformation, the Wilson line transforms as

$$W(C) \mapsto e^{-ig_0 t^\alpha \omega_\alpha(x(1))} \, W(C) \, e^{ig_0 t^\alpha \omega_\alpha(x(0))}, \tag{7.35}$$

which involves only the transformations at the ends of the path. Note that it is the bare field and coupling that appear in the formula for $W(C)$, since the transformations giving invariance of the Lagrangian are those in (2.4) with bare gauge fields and couplings.

From the transformation of $W(C)$ it follows that if C is a path from v to w then the combination $\bar{\psi}(w)W(C)\psi(v)$ is gauge invariant.

A simple generalization is to replace t_α by the generating matrices in another representation. We use this for the gluon density, where the fields at the ends of the Wilson line are gluon field-strength tensors, and the Wilson line uses the adjoint representation.

7.5.2 Path dependence of Wilson line

In general $\bar{\psi}(w)W(C)\psi(v)$ depends not only on the endpoints v and w of the Wilson line, but also on the exact path used to join them.

However, for the case of the standard parton densities, a simplification occurs, because we use a light-like separation in the minus direction: $v = 0$ and $w = (0, w^-, \mathbf{0}_T)$, and it is appropriate to take the Wilson line along the x^- axis. In that case, we now show that the Wilson line depends only on the endpoints. This will enable us to obtain a useful simplification in the Feynman rules for the Wilson line.

---

[2] Another commonly used name is a "gauge link".

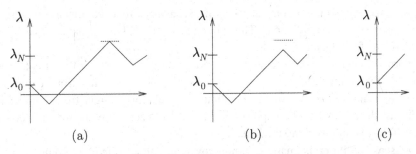

Fig. 7.8. (a) Example of possible path for Wilson line along a single line, as in an ordinary pdf, but with possible backtracking. The vertical axis denotes the coordinate λ along the line. (b) The path altered by changing one of the extreme points. The original coordinate of the altered point is marked by the dotted line. (c) The path after removal of all backtracking. The corresponding Wilson-line factor is unchanged.

We now prove the following general result:

Let C be a path restricted to a line in a fixed direction n, so that points on the line can be written $w^\mu = \lambda n^\mu$. The path is a sequence of N segments joined by direct lines:

$$W(C) = \prod_{j=1}^{N} W(\lambda_j, \lambda_{j-1}), \tag{7.36}$$

where

$$W(\lambda_j, \lambda_{j-1}) \overset{\text{def}}{=} P\left\{ \exp\left[-i g_0 \int_{\lambda_{j-1}}^{\lambda_j} d\lambda \, n^\mu A^\alpha_{(0)\mu}(\lambda n) \, t_\alpha \right] \right\}. \tag{7.37}$$

Then $W(C)$ depends only on the endpoints:

$$W(C) = W(\lambda_N, \lambda_0). \tag{7.38}$$

We illustrate the proof in Fig. 7.8, where the path oscillates on its way from the start to the end, and does some backtracking.

The proof is made by differentiating $W(\lambda_{j+1}, \lambda_j) W(\lambda_j, \lambda_{j-1})$ with respect to λ_j, which gives zero. Thus the product is independent of λ_j, so that we can replace λ_j by λ_{j-1}, and we can remove the $W(\lambda_j, \lambda_{j-1})$ factor. Repeating this $N - 2$ times gives the desired result.

Notice that this proof would not work if we tried to deform the path off the chosen line. For example, moving one of the break points $\lambda_j n$ off the line would shift positions of the gauge fields in the neighboring segments, and the differentiation would involve more than the endpoints of the factors $W(\lambda_j, \lambda_{j-1})$.

A particular case used in the parton densities, is to replace the direct line joining the endpoints by a trip to infinity and back:

$$W(\lambda_N, \lambda_0) = W(\lambda_N, +\infty) W(+\infty, \lambda_0) = [W(+\infty, \lambda_N)]^\dagger W(+\infty, \lambda_0). \tag{7.39}$$

7.5.3 Time ordering v. path ordering

Feynman rules apply to time-ordered Green functions, so conflicts can arise between the path ordering defining Wilson lines and the time ordering used for Green functions. In a covariant gauge, the fields commute at space-like separation, so no conflict arises if we use Wilson lines in space-like directions. This will be the case for TMD densities (Ch. 13) and for the Sudakov form factor (Ch. 10).

For normal integrated parton densities, one uses light-like lines in the direction $n = (0, 1, \mathbf{0}_T)$. If any serious difficulty arises, we take the line as the limit from a space-like direction. We can also use the canonical commutation relations in light-front quantization, in which case the relevant field component $n \cdot A = A^+$ has zero commutator with the same field at different positions along the line.

Such issues can be problematic in a non-covariant gauge, where the commutators of elementary fields may be non-vanishing at space-like separation.

7.5.4 Gauge-invariant quark density in QCD

To define a quark density gauge-invariantly, we use (Collins and Soper, 1982b) a Wilson line exactly along the light-like line joining the quark and antiquark fields. Then the Wilson line uses only A^+ component of the gauge field, and is unity in $A^+ = 0$ gauge; thus the gauge-invariant definition reduces to the basic definition (6.31) in this gauge. The gauge-invariant definition is

$$f_{(0)\,j/h}(\xi) = \int \frac{dw^-}{2\pi}\, e^{-i\xi P^+ w^-} \left\langle P \left| \overline{\psi}_j^{(0)}(0, w^-, \mathbf{0}_T) W(w^-, 0) \frac{\gamma^+}{2} \psi_j^{(0)}(0) \right| P \right\rangle_c, \quad (7.40)$$

where

$$W(w^-, 0) = P\left\{ e^{-ig_0 \int_0^{w^-} dy^- A^+_{(0)\alpha}(0, y^-, \mathbf{0}_T) t_\alpha} \right\}. \quad (7.41)$$

Here we have written the bare parton density, in which all the fields are bare fields, since this is the object to which the probability interpretation applies. In real QCD, in four space-time dimensions, there are UV divergences, so a complete definition requires us to apply renormalization to obtain our final and correct definition of the parton densities. The same applies in more elementary theories, as we will discuss later in Sec. 8.3.

The gluon operators in the Wilson line commute, so a time ordering can be applied to the definition without changing the value of the quark density, just as in Sec. 6.9.4. If we use a fixed ordering for the quark operators, with a final-state cut, then it is better to use a path that goes out to infinity on the left of the final-state cut and back to $(0, w^-, \mathbf{0}_T)$ on the right, as in (7.39). This does not change the value of the quark density, as shown in Sec. 7.5.2.

Antiquark densities are defined by exchanging the roles of the ψ and $\overline{\psi}$ fields, as in (6.33), or equivalently by going to negative ξ in the quark density and using (6.85). Gauge-invariant polarized quark densities are naturally defined by replacing γ^+ by the appropriate Dirac matrix, exactly unchanged from (6.35) and (6.36).

7.5.5 Gluon density

In light-cone gauge, A^+ is zero, while A^- is a field expressed in terms of other fields by a constraint equation. Therefore the independent components of the gluon field are its transverse components A^j. Their free-field action is $\frac{1}{2}\sum_j g^{\mu\nu}\partial_\mu A^j \partial_\nu A^j$, the same as for two independent scalar fields. Thus the operator quantization conditions are the same as for scalar fields. In particular, the expression relating A^j to the light-front creation and annihilation operators is the same, as are the commutation relations. So the gluon density is the same as for a scalar field, (6.124), with a sum over colors and transverse indices:

$$f_{(0)g}(\xi) = \sum_{j,\alpha} \int \xi P^+ \frac{dw^-}{2\pi} e^{-i\xi P^+ w^-} \langle P| A^j_{(0)\alpha}(0, w^-, \mathbf{0}_T) A^j_{(0)\alpha}(0)|P\rangle_{\text{lcg}}. \qquad (7.42)$$

See below for the polarized densities

To convert this to a gauge-invariant expression that has the same value in light-cone gauge, it is not enough just to insert a Wilson-line factor, because of the derivative term in the gauge transformation of the gluon field. Instead we observe that the bare field-strength tensor $G^{\mu\nu}_{(0)}$ transforms without a derivative, so that a gauge-invariant operator can be constructed by joining two field-strength tensors by a Wilson line. Naturally, the representation matrices in the Wilson line must be those for the adjoint representation. Next we observe that in light-cone gauge $G^{+j}_{(0)} = \partial^+ A^j_{(0)}$. In momentum space, this is $A^j_{(0)}$ times a factor of a plus component of momentum (up to a phase). Thus the gauge-invariant form of the bare gluon density is (Collins and Soper, 1982b)

$$f_{(0)g}(\xi) = \sum_{j,\alpha} \int \frac{dw^-}{2\pi\xi P^+} e^{-i\xi P^+ w^-} \langle P| G^{+j}_{(0)\alpha}(0, w^-, \mathbf{0}_T) W_A(w^-, 0)_{\alpha\beta} G^{+j}_{(0)\beta}(0)|P\rangle_c,$$

$$\qquad (7.43)$$

where the subscript A on W_A denotes that the Wilson line is in the adjoint representation.

Just as with a quark, the gluon has a polarization state described by a 2×2 density matrix. But because the gluon has spin 1 instead of spin $\frac{1}{2}$, the decomposition in terms of a Bloch vector is not appropriate, because of the different transformation properties under rotations. Proofs of the many unproved statements in the following discussion are left as an exercise (problem 7.10).

A convenient method starts by modifying (7.42) and (7.43) to provide the gluon density matrix $\rho_{g,j'j}$:

$$\rho_{g,j'j}(\xi, S) f_{(0)g}(\xi) = \sum_\alpha \xi P^+ \int \frac{dw^-}{2\pi} e^{-i\xi P^+ w^-}$$

$$\times \langle P, S| G^{+j}_{(0)\alpha}(0, w^-, \mathbf{0}_T) W_A(w^-, 0)_{\alpha\beta} G^{+j'}_{(0)\beta}(0)|P, S\rangle_c. \qquad (7.44)$$

Here we have simply removed the sum over the transverse spin index of the gluon field and allowed the two fields to have independent indices. Naturally we now allow a polarization specified by S for the target state. Notice the reversal of the order of the indices j and j' between the left- and right-hand sides of the equation. The factor $f_{(0)g}$ on the left-hand side ensures that ρ has the unit trace appropriate to a density matrix. The density matrix is a function of the longitudinal momentum fraction of the gluon and of the spin state of the target. But the gluon density $f_{(0)g}$ is independent of the spin state of the target, because it is the expectation value of an azimuthally symmetric operator, just like a quark density.

We next note that a gluon with a polarization vector ϵ has a density matrix $\epsilon_{j'}^* \epsilon_j$, and in $A^+ = 0$ gauge, ϵ is a 2-component transverse vector. Important pure states can be made with linear polarization, where both components are relatively real, and with circular polarization. A convenient decomposition of a general density matrix is in terms of a helicity α and a linear polarization L:

$$
\rho = \frac{1}{2}\begin{pmatrix} 1 + L_x^2 - L_y^2 & 2L_x L_y - i\alpha \\ 2L_x L_y + i\alpha & 1 - L_x^2 + L_y^2 \end{pmatrix}
$$
$$
= \frac{1}{2}\begin{pmatrix} 1 + |L|^2 \cos 2\phi & |L|^2 \sin 2\phi - i\alpha \\ |L|^2 \sin 2\phi + i\alpha & 1 - |L|^2 \cos 2\phi \end{pmatrix}, \tag{7.45}
$$

where ϕ is the azimuthal angle of the linear polarization relative to the x axis, and there is a positivity restriction $|L|^4 + \alpha^2 \le 1$. The helicity terms give the imaginary part of the off-diagonal elements of ρ, and their sign arises from the polarization vectors $(\epsilon_x, \epsilon_y) \propto (1, i)/\sqrt{2}$ for helicity $+1$ and $(\epsilon_x, \epsilon_y) \propto (-1, i)/\sqrt{2}$ for helicity -1.

We can project the helicity part of ρ by using the matrix[3]

$$
P_{11}^{\text{hel}} = P_{11}^{\text{hel}} = 0, \qquad P_{12}^{\text{hel}} = -i, \qquad P_{21}^{\text{hel}} = i, \tag{7.46}
$$

to give

$$
\alpha_g f_{(0)g}(\xi) = \sum_{j,j'=1}^{2} P_{jj'}^{\text{hel}} \int \frac{dw^-}{2\pi \xi P^+} e^{-i\xi P^+ w^-}
$$
$$
\times \langle P, S | G_{(0)}^{+j}(0, w^-, \mathbf{0}_{\text{T}}) W_A(w^-, 0) G_{(0)}^{+j'}(0) | P, S \rangle. \tag{7.47}
$$

We can use parity invariance (actually parity and a $180°$ rotation in the (x, y) plane) to relate the parton densities in target states of opposite helicity. As with a quark, it follows that in a spin-$\frac{1}{2}$ target, like a proton, the gluon helicity is proportional to the target helicity, so that we can define the bare gluon helicity density $\Delta f_{(0)g}$ by

$$
\alpha_g f_{(0)g}(\xi) = \alpha_{\text{target}} \Delta f_{(0)g}(\xi). \tag{7.48}
$$

Then in a target state of maximal helicity, $\Delta f_{(0)g}$ has the interpretation of a helicity asymmetry: the number density of gluons polarized parallel to the target minus the number polarized antiparallel.

The linear polarization of a gluon can also be defined, but there is no standard definition of a corresponding parton density. It would have little practical use, because the linear polarization of a gluon is zero in the most important case of a spin-$\frac{1}{2}$ hadron, as follows from conservation of angular momentum about the z axis (Artru and Mekhfi, 1990). (Linear polarization is measured by an operator that flips helicity by two units. Since no helicity

---

[3] Note that the formula for this matrix in Brock *et al.* (1995) is incorrect.

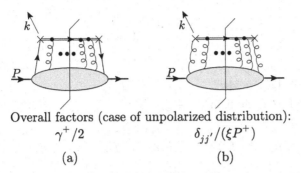

Overall factors (case of unpolarized distribution):

$$\gamma^+/2 \qquad\qquad \delta_{jj'}/(\xi P^+)$$

$$\text{(a)} \qquad\qquad\qquad \text{(b)}$$

Fig. 7.9. Feynman-graph notation for gauge-invariant (a) quark density, (b) gluon density. The double lines are for the Wilson lines, whose rules are in Figs. 7.10, 7.11, and 7.12. The short line at the top merely represents the flow of external momentum at the parton density vertex. The overall factors in the case of a quark are the same as in Fig. 6.7.

is absorbed by the azimuthally symmetric space-time part of the definition of the parton densities, the helicity flip in the operator equals the helicity flip in the density matrix for the hadron.)

In the occasionally used case of a spin-0 target (pion), the gluon is unpolarized, as follows by combining the above two arguments.

7.6 Feynman rules for gauge-invariant parton densities

To represent gauge-invariant parton densities in Feynman graphs, we notate the Wilson line by a double line joining the fields at the ends of the Wilson line, as in Fig. 7.9. Any number of gluons (zero or more) connect the Wilson line to the rest of the graph. An overall trace with a Dirac matrix in a quark density is independent of the presence of the Wilson line. To derive Feynman rules for the Wilson lines we expand the exponential of the field in powers of its argument. Each term gives a target matrix element of several gluon fields (and the fields at the ends of the Wilson line), integrated over certain positions. The factors of $-ig_0 A^+_{(0)\alpha} t_\alpha = -ig_0 n \cdot A_{(0)\alpha} t_\alpha$ result in the rules for vertices on the Wilson lines shown in Fig. 7.10. The rules are first written for bare fields. If we calculate with renormalized fields, the factor $Z_3^{1/2}$ in the relation between bare and renormalized gluon fields requires that we associate a factor $Z_3^{1/2}$ with each of the gluon fields in the rules for the parton density and the Wilson line, as shown in Fig. 7.10. The generating matrices t_α are those for the color representation of the quark or gluon whose density is being used, and these are multiplied along the Wilson line.

Next we write the Wilson line in the form of exponentials going to infinity, (7.39), we expand each exponential in a power series in its argument, and write the necessary coordinate-space Green function in terms of momentum-space Green function. The order g_0^n term has an integral over n coordinates. We express each integral as an integral over ordered variables, which cancels the factor $1/n!$ in the series expansion, and then we have

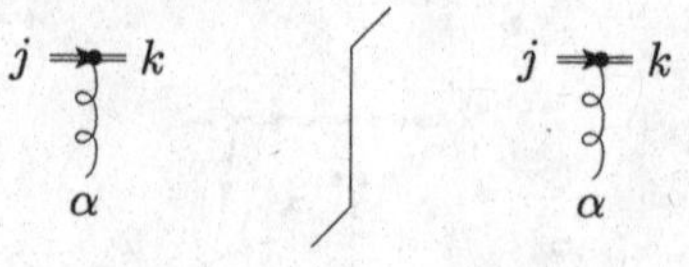

Quark pdf, bare fields:	$-ig_0 n^\mu(t_\alpha)_{kj}$	$ig_0 n^\mu(t_\alpha)_{kj}$
Quark pdf, renorm. fields:	$-ig_0 Z_3^{1/2} n^\mu(t_\alpha)_{kj}$	$ig_0 Z_3^{1/2} n^\mu(t_\alpha)_{kj}$
Gluon pdf, bare fields:	$g_0 n^\mu f_{k\alpha j}$	$g_0 n^\mu f_{k\alpha j}$
Gluon pdf, renorm. fields:	$g_0 Z_3^{1/2} n^\mu f_{k\alpha j}$	$g_0 Z_3^{1/2} n^\mu f_{k\alpha j}$

Fig. 7.10. Feynman rules for vertex on Wilson lines in parton densities. Here, $n^\mu = \delta^\mu_- = (0, 1, \mathbf{0}_T)$. In the Wilson line for a gluon pdf, the generating matrix for the adjoint representation was used: $(T_\alpha)_{kj} = if_{k\alpha j}$. The sign of the vertex is reversed compared with Collins and Soper (1982b), and corresponds to the sign of the coupling in our Lagrangian, whose Feynman rules are in Fig. 3.1.

an integral of the form

$$\prod_j \int \frac{\mathrm{d}^{4-2\epsilon}k_j}{(2\pi)^{4-2\epsilon}} G(k_1, \mu_1, \alpha_1; \ldots) \prod_j (-ig_0 t_{\alpha_j} n^{\mu_j}) \int_0^\infty \mathrm{d}y_1^- \int_{y_1^-}^\infty \mathrm{d}y_2^- \ldots \int_{y_{n-1}^-}^\infty \mathrm{d}y_n^- \prod_j e^{ik_j^+ y_j^-},$$

(7.49)

where G represents the rest of the graph, i.e., a shaded bubble in Fig. 7.9, including the lines connecting it to the Wilson line. This particular formula applies on the left of the final-state cut, with the gluon momenta k_j directed *down*, into the bubble. Applying the standard result

$$\int_z^\infty \mathrm{d}y\, e^{iky} = \frac{i}{k + i0} e^{ikz},$$

(7.50)

gives a value for each double line segment shown in the left part of Fig. 7.11. Thus for the Wilson line on the left of the cut

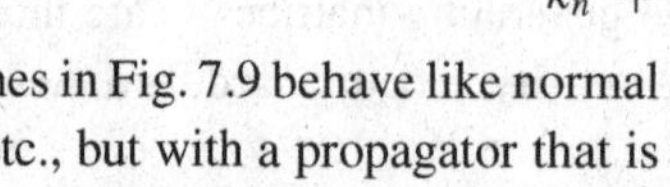

$$= (-ig_0 t_{\alpha_n} n^{\mu_n}) \frac{i}{k_n^+ + i0} (-ig_0 t_{\alpha_{n-1}} n^{\mu_{n-1}}) \frac{i}{k_n^+ + k_{n-1}^+ + i0}$$

$$\times \cdots \times (-ig_0 t_{\alpha_1} n^{\mu_1}) \frac{i}{k_n^+ + \cdots + k_1^+ + i0}.$$

(7.51)

So the double lines in Fig. 7.9 behave like normal lines in a Feynman graph, with circulating loop momenta etc., but with a propagator that is the Fourier transform of a theta function. Of course the whole Wilson-line structure occurs once in the parton density and therefore once in the Feynman graph. There is naturally a hermitian conjugation of the above rules in the part of graphs to the *right* of the final-state cut, as usual, and as indicated in the figures.

In the definition of a parton density there is an integral over the external k^- and k_T. Since the Wilson-line propagator is independent of these momentum components, the integral

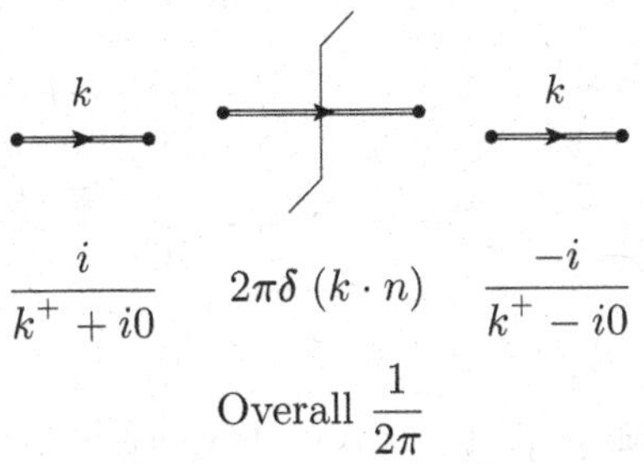

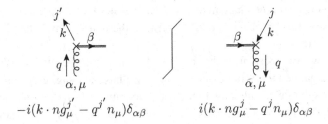

$$\frac{i}{k^+ + i0} \qquad 2\pi\delta\,(k \cdot n) \qquad \frac{-i}{k^+ - i0}$$

$$\text{Overall } \frac{1}{2\pi}$$

Fig. 7.11. Feynman rules for the line part of a Wilson line.

$$-i(k \cdot n g_\mu^{j'} - q^{j'} n_\mu)\delta_{\alpha\beta} \qquad\qquad i(k \cdot n g_\mu^{j} - q^{j} n_\mu)\delta_{\alpha\beta}$$

Fig. 7.12. Feynman rules for attachment of gluon at end of Wilson line in gluon density. The indices α and β are for color, μ is the Lorentz index of the gluon, and j and j' are as in (7.44).

over them can be conveniently notated by routing them along the Wilson line and across the final-state cut. We give the cut line the natural delta function $2\pi\,\delta(k^+)$ for a cut propagator, and then we simply have to extract plus momentum ξP^+ at the end of the Wilson line. We also have to cancel the 2π in the cut-line propagator, as indicated in Fig. 7.11. This in fact results from the explicit factor $1/(2\pi)$ in the definitions of the parton density, e.g., (7.40).

The above completes the definition of the quark density. For the gluon density, we also need the vertex with $G_{(0)\alpha}^{+j} = \partial^+ A_{(0)\alpha}^{j} - \partial^j A_{(0)\alpha}^{+} - g_0 f_{\alpha\beta\gamma} A_{(0)\beta}^{+} A_{(0)\gamma}^{j}$, shown in Fig. 7.12. The derivatives give factors of $-iq^+$ and $-iq^j$, with q being the momentum of the gluon line. We apparently also need a two-gluon coupling at the end of the Wilson line. But we remove it (Collins and Soper, 1982b) by using the identity

$$\partial^+\left(A_{(0)\alpha}^{j}(w^-)W_A(w^-)\right) = \left(\partial^+ A_{(0)\alpha}^{j}(w^-) - g_0 f_{\alpha\beta\gamma} A_{(0)\beta}^{+} A_{(0)\gamma}^{j}\right) W_A(w^-), \qquad (7.52)$$

which accounts for the appearance of $k \cdot n$ rather than $q \cdot n$ in Fig. 7.12.

The application of the above rules will be illustrated by calculational examples in Sec. 9.4.

7.7 Interpretation of Wilson lines within parton model

Our first definition of a quark density was without a Wilson line and it arose from examining a theory in which DIS structure functions are dominated by the handbag diagram,

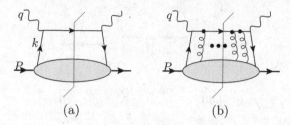

Fig. 7.13. Handbag graph: (a) standard, (b) with extra gluon exchanges.

Fig. 7.13(a), with the exchanged quark collinear to the target. Making suitable approximations converted the top of the diagram to a coefficient times the vertex for the quark density.
In a gauge theory, this procedure gives the term in the quark density that has no gluons
attached to the Wilson line.

We now show how there arise the terms with gluons attached to the Wilson line, as
in Fig. 7.9. In a gauge theory in Feynman gauge, we have leading regions in which extra
gluons couple the collinear subgraph to the hard subgraph. So we examine the generalized
handbag diagrams shown in Fig. 7.13(b), where arbitrarily many gluons are exchanged
between the top rung and the lower bubble. This provides a gauge-invariant extension
of the parton model. In real QCD, we will also need more complicated hard-scattering
graphs.

The extra gluons are to be collinear to the target, just like the exchanged quark, and we
now show that to the leading power of Q, each of these graphs gives a corresponding term in
the quark density, Fig. 7.9(a), times the same coefficient as with the handbag diagram. That
this result is expected, since in $A^+ = 0$ gauge, the gluon-exchange graphs in the structure
function are suppressed, and the gluon couplings to the Wilson line are zero.

To formalize the result, let $F_{[N]}(x, Q)$ be the contribution to a structure function from
graphs of the form of Fig. 7.13(b) with N gluons attached to the upper line. Similarly, let
$f_{[N],j}(x)$ be contribution to the parton density for a quark of flavor j with N gluons attached
to the Wilson line. Then the result to be proved is that

$$F_{[N]}(x, Q) = \sum_j C_j f_{[N],j}(x) + \text{p.s.c.} \tag{7.53}$$

The important property is that the coefficient C_j is the same no matter how many gluons
are exchanged. By "p.s.c." are denoted power-suppressed corrections, i.e., corrections
suppressed by a power of Q. When we sum over N, on the left-hand side we get the
full structure function $\sum_{N=0}^{\infty} F_{[N]} = F$. The sum of the right-hand side gives the full
gauge-invariant parton density: $\sum_{N=0}^{\infty} f_{[N],j} = f_j$, multiplied by C_j. Thus we recover the
standard parton-model formulae for the structure functions (6.25). The independence of
the coefficient from N implies that it is correctly calculated from the case $N = 0$, and that
it is the same as in the simple parton model, e.g., $C_j = e_j^2 x$ for the F_2 structure function in
electromagnetic DIS.

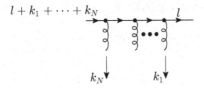

Fig. 7.14. Attachment of collinear gluons to hard quark.

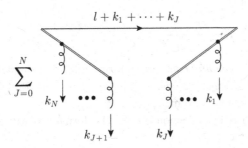

Fig. 7.15. Result of applying collinear approximation to Fig. 7.14. The double lines represent Feynman rules for a Wilson line on the left and for a conjugate Wilson line on the right.

The proof of (7.53) uses a result of Collins and Soper (1981) illustrated in Figs. 7.14 and 7.15. We write an upper quark line, Fig. 7.14, for the generalized handbag graph, as

$$U_N = \prod_j \left(-ig_0 t_{\alpha_j}\right) \frac{i}{\slashed{l} - m + i0} \gamma^{\mu_1} \frac{i}{\slashed{l} + \slashed{k}_1 - m + i0} \cdots \gamma^{\mu_n} \frac{i}{\slashed{l} + \slashed{k}_1 + \ldots \slashed{k}_N - m + i0}.$$

(7.54)

Since all the gluons are target-collinear, we can replace each gluon momentum by its plus component: $k_j \mapsto \hat{k}_j \overset{\text{def}}{=} (k_j^+, 0, \mathbf{0}_T)$, and we can restrict the Dirac matrices at the vertices to their minus components: $\gamma^{\mu_j} \mapsto n^{\mu_j} \gamma^-$, where $n = (0, 1, \mathbf{0}_T)$, as defined earlier. The resulting approximation to the quark line is

$$\hat{U}_N = \prod_j \left(-ig_0 n^{\mu_j} t_{\alpha_j}\right) W_N,$$

(7.55)

where

$$W_N = \frac{i}{\slashed{l} - m + i0} \gamma^- \frac{i}{\slashed{l} + \hat{\slashed{k}}_1 - m + i0} \gamma^- \cdots \gamma^- \frac{i}{\slashed{l} + \hat{\slashed{k}}_1 + \cdots + \hat{\slashed{k}}_N - m + i0}.$$

(7.56)

With a proof summarized below, this can be rewritten as

$$W_N = \sum_{J=0}^{N} R_J M_J L_{N,J},$$

(7.57)

which we write as a diagram in Fig. 7.15, where the left-side factor is

$$L_{N,J} = \frac{i}{k_{J+1}^+ + i0} \cdots \frac{i}{k_{J+1}^+ + \cdots + k_N^+ + i0},$$

(7.58)

the middle factor is

$$M_J = \frac{i}{1 + \hat{k}_1 + \cdots + \hat{k}_J - m + i0},$$

(7.59)

and the right-side factor is

$$R_J = \frac{-i}{k_J^+ + i0} \cdots \frac{-i}{k_1^+ + \cdots + k_J^+ + i0},$$

(7.60)

Note that because of the standard conventions for lines for Dirac particles, the ordering of the objects is reversed between the equation and the diagram. "Right" and "left" refer to the sides of the diagram, not the formula.

The proof of (7.57) is by induction on N. The formula is trivially true for $N = 0$. Suppose that (7.57) is true for W_{N-1}. Then

$$W_N = W_{N-1}\gamma^- M_N$$

$$= \sum_{J=0}^{N-1} R_J M_J L_{N-1,J} \left(\frac{1}{M_N} - \frac{1}{M_J} \right) \frac{i}{k_{J+1}^+ + \cdots + k_N^+ + i0} M_N$$

$$= \sum_{J=0}^{N-1} R_J M_J L_{N,J} - \sum_{J=0}^{N-1} R_J L_{N,J} M_N.$$

(7.61)

In the second line, we replaced γ^- by $(\hat{k}_{J+1} + \cdots + \hat{k}_N)/(k_{J+1}^+ + \cdots + k_N^+)$, and then wrote $\hat{k}_{J+1} + \cdots + \hat{k}_N$ as the difference of two inverse propagators. To complete the proof of (7.57), we use the result

$$\sum_{J=0}^{N} R_J L_{N,J} = 0 \qquad \text{if } N \geq 1,$$

(7.62)

also proved by induction.

The double lines in Fig. 7.15 have the Feynman rules for a Wilson line on the left and a conjugate Wilson line on the right, with the Wilson lines going in the minus direction out to infinity.

We now apply (7.57) to the upper quark line on the left of the final-state cut in Fig. 7.13(b). Since the quark at the final-state end is on-shell, the only surviving term in Fig. 7.15 is where the Wilson-line factor is at the left, next to the current vertex. Similarly, applying (7.57) to the upper quark line on the right of the final-state cut gives a Wilson-line factor at the right of the line (again, next to the current vertex).

The result is to give a factor of the lowest-order hard scattering times a factor corresponding to the rules for the gauge-invariant quark density defined in (7.40), with the application of (7.39) to write the Wilson line as one that goes out to infinity and comes back. The $i0$

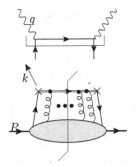

Fig. 7.16. Gauge-invariant form of the parton model.

prescription is chosen to be compatible with a deformation out of the Glauber region away from final-state poles. This is the appropriate direction, as we will see in Ch. 13.

We have now completed the derivation of the parton-model approximation in its gauge-invariant form, illustrated in Fig. 7.16. The coefficient function is the same as without the gluon exchanges.

Exercises

7.1 Verify that performing the k_T integral in (7.13) does reproduce the result of applying the Feynman parameter method to (7.12).

7.2 Find in the literature or derive the full rules for x^+-ordered perturbation theory in a general renormalizable gauge theory, including a proper treatment of the 3-gauge field vertex and the coupling of gauge fields to scalar fields.

7.3 Verify that (7.28) follows from (7.25), the (anti)commutation relations (6.65), and the standard covariant normalization of a single particle state $|h : P\rangle$. Suggestion: investigate $\langle h : P'|h : P\rangle$.

7.4 (a) Find the general form of Lorentz transformations that preserve the plane $x^+ = 0$. Use one such transformation to transform the state $|h : P\rangle$ in (7.25) with $\boldsymbol{P}_T = 0$ to a general value of P with non-zero $\boldsymbol{P}_T$.

(b) The wave-function decomposition (7.25) of a state $|h : P\rangle$ was intended to apply at $\boldsymbol{P}_T = 0$. Show that it also applies at non-zero $\boldsymbol{P}_T$ if the replacement (7.27) is made.

(c) Obtain $\langle h, P'|h, P\rangle$, and deduce the normalization condition (7.28) from the Lorentz-invariant normalization (A.14) for single-particle states.

7.5 Express the left-hand side of (7.29) in terms of field operators in momentum space, integrated over k_j^-.

7.6 Derive an expression for the unintegrated parton densities $f_{j/h}(\xi, k_T)$ in terms of the light-front wave functions in (7.25). The result should be of the form of an integral over $\left|\psi_{F/h}(\{x_j, \mathbf{k}_{jT}, \alpha_j\})\right|^2$, with the values of one of the x_j, $\mathbf{k}_{jT}$ pairs set to ξ, $\mathbf{k}_T$.

7.7 Obtain Feynman rules for computing light-front wave functions in perturbation theory; these will generalize the rules we constructed for parton densities in Sec. 6.10.

7.8 Apply them to the first non-trivial order in the Yukawa field theory we have used for examples. Verify the normalization condition (7.28). (Warning: Use dimensional regularization, so that the calculations can be done in the UV-regulated bare theory.)

7.9 In the parton model for CC processes with production of a heavy quark, in Sec. 7.1.3, we effectively assumed that the quark flavor and mass eigenstates coincided. In other words we assumed that the CKM matrix is unity. Correct the calculation to use a non-trivial CKM matrix.

7.10 Derive all the statements about polarized gluon densities in Sec. 7.5.5. *Check carefully the signs in the polarization vectors for gluons of definite light-front helicity.* You should be able to verify that there is a sign error in the formula for P^{hel} in Brock *et al.* (1995), and hence in the formula in that paper for Δf_g. [Thanks are due to Markus Diehl (private communication) for pointing out the error.]

7.11 As mentioned in Sec. 7.5.5, a linear polarization is possible for the gluon (although not in a spin-$\frac{1}{2}$ target). Work out the appropriate generalization of the work in this chapter to deal with this. An alternative formulation is in a helicity-density-matrix formalism, where linear polarization corresponds to a term with a gluon helicity flip of 2 units. If you get stuck, consult Artru and Mekhfi (1990).

7.12 Complete the derivations of (7.57) and (7.62).

8

Factorization for DIS, mostly in simple field theories

In this chapter, I treat the complications caused by renormalizability of the underlying field theory when one analyzes the asymptotics of processes like DIS. There are four inter-related issues:

- The leading regions include hard-scattering subgraphs that can be of arbitrarily high order in the coupling.
- There are logarithmic unsuppressed contributions from momenta that interpolate between the different regions for a graph.
- The definitions of the parton densities are modified to remove their UV divergences. This we do by renormalization.
- The parton densities acquire a scale argument μ, the dependence on which is governed by renormalization-group (RG) equations, the famous Dokshitzer-Gribov-Lipatov-Altarelli-Parisi (DGLAP) equations. In applications, we set μ of order Q, the large scale in the hard scattering.

I will give a derivation of factorization that in the absence of gauge fields is complete and satisfactory, and is also reasonably elementary. In QCD, the same factorization theorem is also valid for simple processes, like DIS, but its derivation needs enhancement, to be given in later chapters.

8.1 Factorization: overall view

To motivate the factorization idea, we still use the ideas about the space-time structure of DIS that motivated the parton model. As illustrated in the spatial diagram of Fig. 8.1, an electron undergoes a wide-angle hard scattering off a single parton in a high-energy target hadron. In the center-of-mass frame, the target is time-dilated and Lorentz contracted. Thus over the short time and distance scale $1/Q$ of the hard scattering, the struck parton's interactions with the rest of the target can be neglected; in the hard scattering, the incoming parton can be approximated as a free particle. A single struck parton dominates, because the other partons are separated from it by a hadronic scale of ~ 1 fm, large compared with $1/Q$.

Relative to the parton model, an important change in a renormalizable theory is that the dimensionlessness of the coupling allows multiple particles to be created in the hard

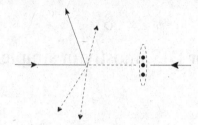

Fig. 8.1. Deeply inelastic scattering of an electron on a hadron. This is like Fig. 2.2, but with more partons exiting the short-distance hard scattering. The struck parton and the partons resulting from the hard scattering are indicated by dashed lines.

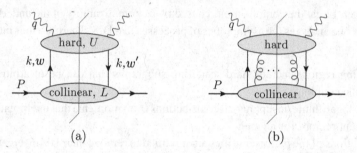

Fig. 8.2. Most general leading regions for DIS. The lines in the lower bubble are collinear to the target hadron, and the lines in the upper bubble have large transverse momentum, of order Q. (a) In a theory without gauge fields, exactly one line on each side of the final-state cut joins the two bubbles. The labels ω and ω' are for the flavor, color and spin of the intermediate parton lines. (b) In a theory with gauge fields, arbitrarily many extra gauge-field lines may join the bubbles. The solid lines may be quarks or transversely polarized gluons. In a gauge theory there may also be a soft subgraph at leading power.

scattering without a power-law suppression. This is manifested experimentally in events like that in Fig. 5.10. Naturally, an appropriate coupling for the short-distance scattering is $\alpha_s(Q)$, whose smallness in QCD allows the use of perturbation theory.

Our calculations in Sec. 6.11 showed another consequence of a dimensionless coupling, that the number density of partons only falls off in transverse momentum roughly as $1/k_{\mathrm{T}}^2$. Therefore the number of partons, integrated over k_{T}, and naively interpreted, diverges. The picture of limited transverse momentum for the constituents, implicit in Fig. 8.1, therefore needs to be distorted.

The formalization of these ideas starts from the Libby-Sterman analysis in Ch. 5, which determines that the leading regions for DIS are those illustrated in Fig. 8.2.

8.1.1 Leading-power regions without gauge fields

In a model field theory without gauge fields, all the leading regions are of the form of Fig. 8.2(a). The lower bubble consists of lines whose momenta are collinear to the target. The upper bubble consists of lines with very different directions than the target or that are far off-shell. On each side of final-state cut, one line connects the collinear subgraph to the hard subgraph. This corresponds to the single struck parton in Fig. 8.1. Scattering off multiple partons would correspond to extra lines connecting the upper and lower bubbles in Fig. 8.2(a), and is power-suppressed, by Libby and Sterman's power-counting.

While keeping the restriction to this single pair of connecting lines, the upper subgraph can be arbitrarily complicated. This gives the possibility of multijet production, as seen experimentally in Fig. 5.10. Associated with this is an essential complication, that a single graph for DIS can have multiple decompositions of the form of Fig. 8.2(a).

The upper bubble, the "hard subgraph", has on-shell final-state lines, but we will nevertheless treat it as if it is a short-distance object, with all internal lines off-shell by order Q^2. The demonstration uses arguments given in Secs. 4.1.1, 4.4, and 5.3.3, where the short-distance property applies to a local average in the cross section (e.g., an average in x). Further details are found later in Secs. 11.2 and 12.7.

8.1.2 *Leading-power regions with gauge fields*

In a gauge theory, like QCD, leading regions can also have extra target-collinear gluons attaching to the hard scattering, as in Fig. 8.2(b). In the methodology where we treat the upper bubble as a pure hard scattering, this exhausts the leading regions; this applies, for example, to the uncut hadronic tensor and the structure functions averaged in x, as in Sec. 5.3.3. But it is also possible to consider the actual on-shell final states in the upper bubble; in that case there are final-state jet subgraphs, and a soft subgraph that connects any or all of the collinear subgraphs.

We now use the first methodology. The leading part of each extra gluon exchange involves the product of the minus component of the vertex at the upper end of the gluon line and a plus component at the lower end, schematically $U^- L^+$. Thus the extra gluons can be eliminated by the use of the light-cone gauge, $A^+ = 0$: in light-cone gauge, the leading regions have the same form, Fig. 8.2(a), as in a non-gauge theory.

Therefore once we have proved factorization in a non-gauge theory, which is done in an elementary fashion in this chapter, we can copy the proof in light-cone-gauge QCD. To take it literally, one must be concerned about problems with the $1/k^+$ singularities in the light-cone-gauge gluon propagator, (7.30) and (7.31). These problems will become particularly apparent when we work with TMD distributions in Ch. 13. Nevertheless, divergences due to the $1/k^+$ singularities cancel in the treatment of DIS, although giving a full satisfactory proof is non-trivial.

For a fully satisfactory treatment, it will be better to return to Feynman gauge. We have already seen, in Sec. 7.7, that at the level of the parton model, the extra gluons can be extracted from the hard scattering to give the Wilson lines in gauge-invariant definitions of the parton densities. This is a result that generalizes, but I postpone a treatment to Ch. 11.

8.2 Elementary treatment of factorization

Before going to a strict derivation of factorization in non-gauge theories, it is useful to give an approximate proof. Its inspiration is a naive interpretation of the diagram Fig. 8.2(a) for the leading regions. This is that the momenta of lines can be unambiguously split into two classes, corresponding to the two subgraphs in the figure. Hard momenta, in the upper subgraph U, have virtualities of order Q^2. Collinear momenta, in the lower subgraph L, have orders of magnitude typical for target momenta, i.e., $(k^+, k^-, k_T) \sim (Q, m^2/Q, m)$, where m is a typical light hadron scale; their virtualities stay fixed when Q becomes large.

This supposition enables a simple proof to be given, and gives a mental picture linking the leading-region diagram Fig. 8.2(a) with the factorization formula. We will take the opportunity to introduce notation that will be useful more generally.

But a clear division between the regions of momenta does not exist; there are important contributions from intermediate momenta. We will overcome this problem by the use of a subtractive formalism, in Sec. 8.9.

8.2.1 Decomposition by regions

We now start from an assumption that there is a clear decomposition of momenta by regions. Then we can decompose each graph into a sum of terms of the form of Fig. 8.2(a), each term corresponding to a particular assignment of momentum types to subgraphs. Let F denote a structure function or the hadronic tensor. Then we have

$$F = \sum_{\text{2PR graphs } \Gamma} \Gamma + \text{non-leading power}$$

$$= \sum_{\text{2PR graphs } \Gamma} \sum_{\text{leading regions } R} U(R) L(R) + \text{non-leading power}, \qquad (8.1)$$

where the summation over Γ is restricted to those graphs that are two-particle reducible in the t channel and that therefore have at least one decomposition of the form of Fig. 8.2(a). A region R corresponds to assignments of momentum types to subgraphs, and is determined by the subgraphs: $U(R)$ for the upper bubble, restricted to hard momenta, and $L(R)$ for the lower bubble, restricted to collinear momenta. We define $L(R)$ to include the full propagators for the two lines that join the L and U subgraphs, since these lines carry collinear momenta.

The product $U(R) L(R)$ is defined as a convolution product, with an integral over the momentum k flowing between U and L, and with summations over the color, spin and flavor indices for the fields. So we write $U = U(k, \omega, \omega'; q)$ and $L = L(k, \omega, \omega'; P, S)$, where ω and ω' are composite indices for the flavor, color and spin of the fields, while P and S are the momentum and spin vector of the target state. Then

$$UL = \sum_{\omega, \omega'} \int \frac{d^{4-2\epsilon} k}{(2\pi)^{4-2\epsilon}} U(k, \omega, \omega'; q) L(k, \omega, \omega'; P, S). \qquad (8.2)$$

A region is completely specified by its hard and target subgraphs, so we replace the sum over graphs and regions by independent sums over graphs for $U(R)$ and $L(R)$. So we write

$$F = UL + \text{non-leading power}, \qquad (8.3)$$

where U and L, without a region specifier, are the sum over all possibilities for the hard and target-collinear subgraphs of Fig. 8.2(a), with the momenta being restricted to the appropriate regions.

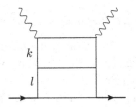

Fig. 8.3. Graph with three decompositions of the form of Fig. 8.2(a).

In (8.3) we have a *multiplicative* structure: the structure function is a product of a hard part and a collinear part. In contrast, at the level of individual graphs for the structure function, we have an *additive* structure: in the second line of (8.1) there is a sum over regions for a given graph. An illustration is given by Fig. 8.3. The possible regions are: where the top rung alone is hard, where the top two rungs are hard, and where all three rungs are hard.

Diagonality in flavor and color

In the convolution of U and L, there is a sum over indices ω and ω', which we now simplify. For QCD, each index has 88 independent values: There are 6 flavors of quark, of antiquark, and a gluon. The quarks and antiquarks each have 3 colors, the gluons have 8, and each flavor-color combination has 2 spin values. We can separate the parts of the index ω as j, c and α, for flavor, color and spin. Here we refer to the QCD version even though in this chapter we will only present proofs in a non-gauge theory: the ideas are general.

In principle, there are separate sums over the indices ω and ω' for the two parton lines connecting U and L. I now show that the sums over flavor and color indices are diagonal in the cases of interest; i.e., the flavor and color parts of ω and ω' are equal.

We choose the flavor label to correspond to the different types of mass eigenstate for the partons (e.g., u, d, etc.). Normal targets (nucleons and pions) are flavor eigenstates, so the lower subgraph L is flavor-diagonal. An exception would be DIS on a K_L^0 or K_S^0, which is not a likely experiment. Note that, for charged-current weak-interaction processes, the upper subgraph U can be flavor changing. Thus, in neutrino DIS, we can have the sequence of quark flavor transitions $d \mapsto u \mapsto s$. But diagonality of L implies that off-diagonal terms in U do not contribute.

As for color, all electroweak currents are color-singlet. Therefore U is diagonal in color, and all the diagonal color components of U are equal.

In contrast the spin-sum need not be diagonal. So we rewrite (8.2) as

$$F = \sum_{j,\alpha,\alpha'} \int \frac{d^{4-2\epsilon}k}{(2\pi)^{4-2\epsilon}}\, U\left(k, j, \alpha, \alpha'; q\right) \sum_c L(k, j, c, \alpha, \alpha'; P, S)$$

$$+ \text{ non-leading power.} \tag{8.4}$$

Here we have left a single flavor label j on U and L, and a single color label c on L. The remaining sum, over α and α', is for Dirac spin indices. The U part can be considered a color average, as will fit its later interpretation in terms of a parton-level cross section.

8.2.2 *Parton approximator*

To get the factorization theorem, we use exactly the same method we applied in Sec. 6.1 for the parton model. (a) In U we neglect the small components of momenta, k^- and k_T, entering it from L, and we also neglect particle masses. (b) In the sum over Dirac indices, we project onto those parts that give the leading power. This operation, which we call the parton approximator, results in an error that is suppressed by one or two powers of Q.

We notate the result as

$$U \overleftarrow{T} | VL = C_{\text{region}} \otimes f_{\text{region}} \overset{\text{def}}{=} \sum_j \int \frac{d\xi}{\xi} C_{j,\,\text{region}}(x, \xi)\, f_{j,\,\text{region}}(\xi), \qquad (8.5)$$

which is a factorized form for the cross section. Here we have defined a fractional momentum variable $\xi = k^+/P^+$. The tripartite symbol $\overleftarrow{T} | V$ denotes the parton approximator, for which we will give a precise definition below. The arrow in $\overleftarrow{T}$ implies that the kinematic part of the approximation is applied to the object to its left, i.e., to U. The quantity V symbolizes the vertex for a parton density that is a factor in the approximator. We separate these symbols by a vertical bar, which will be a useful notation in treating renormalization of the parton densities. Although the above formula makes it appear that the parton approximator is a linear operator, certain features of the approximator, notably that it sets to zero the parton masses in U, take us beyond ordinary linear algebra. Even so, many of the rules of linear algebra still apply.

The parton approximator will give a factor that has a vertex for a parton density integrated with the L factor. Therefore on the right-hand side of (8.5), we have used a notation to express this. The resulting object has the standard definition of a parton density, except that the momenta inside L are restricted to be collinear. So f is equipped with a subscript "region", to label this variation in the definition. A parton density is a function of just one kinematic argument ξ, so we represent the corresponding kind of convolution by the symbol $\otimes$, which is defined as on the rightmost part of the equation. The quantity C is the approximated U, but with a particular normalization. It goes by several names: coefficient function, short-distance partonic scattering, Wilson coefficient. To save extra notational complication, only the unpolarized terms are written explicitly.

Kinematic approximation

The first, kinematic, part of the approximator gives

$$F = \sum_{j,\alpha,\alpha'} \int_{x-}^{1+} \frac{d\xi}{\xi}\, U\big(\hat{k}, j, \alpha, \alpha'; q, m = 0\big)$$

$$\times \sum_c \int \frac{dk^-\, d^{2-2\epsilon} k_T}{(2\pi)^{4-2\epsilon}}\, \xi P^+ L(k, j, c, \alpha, \alpha'; P, S) + \text{non-leading power}. \qquad (8.6)$$

Here, we have changed variable from k^+ to ξ, and we have defined $\hat{k} = (\xi P^+, 0, \mathbf{0}_T)$, for the approximated parton momentum in U. The integral over k^- and k_T is now confined to the L factor, as in a parton density, and we included with it a factor of ξP^+ for the sake of boost invariance. The upper limit on ξ is imposed by the parton density, for positivity of the energy in its final state, $P^+ - \xi P^+ \geq 0$. The lower limit is set from positivity of the energy in the hard part of the graph, $q^+ + \xi P^+ \geq 0$. In general, the integrand can be a generalized function (distribution) with singularities at the endpoints. For example, there can be delta functions at $\xi = 1$ in L, and at $\xi = x$ in U (after approximation). The singularities are properly treated if we take the range of integration over ξ to extend beyond the kinematic limits, so I notate the limits as $x-$ and $1+$.

Approximator for scalar parton

When j denotes a scalar parton, there are no spin labels, so (8.4) gives the definition of $U \overleftarrow{T} | V L$ for a scalar quark:

$$
(U \overleftarrow{T} | V L)_{\text{scalar } j} \overset{\text{def}}{=} \int_{x-}^{1+} \frac{d\xi}{\xi} \, U(q; \hat{k}, j; m = 0)
$$

$$
\times \sum_c \int \frac{dk^- \, d^{2-2\epsilon} k_{\text{T}}}{(2\pi)^{4-2\epsilon}} \, \xi P^+ L(k, j, c; P, S). \tag{8.7}
$$

The second line, including the color sum and the factor ξP^+, reproduces exactly the definition of the density of a scalar quark, (6.124). The first factor, the approximated U, has the normalization appropriate to DIS on an on-shell massless parton target, but with internal momenta restricted to being in the hard region. The integral joining the two factors is a convolution with measure $d\xi/\xi$, which we choose as its standard form.

Approximator for spin-$\frac{1}{2}$ parton

When j denotes a fermion quark, we have two formulations. One involves projection matrices $\mathcal{P}_A$ and $\mathcal{P}_B$ on each line, as in (6.13). The other reorganizes this, as in (6.19), into terms involving different kinds of spin-projected parton density. Thus we have

$$
(U \overleftarrow{T} | V L)_{\text{Dirac } j}
$$

$$
\overset{\text{def}}{=} \sum_{\alpha, \beta, \alpha', \beta'} \int dk^+ \, U(q; \hat{k}, j, \alpha, \alpha'; m = 0)
$$

$$
\times (\mathcal{P}_A)_{\alpha\beta} (\mathcal{P}_B)_{\beta'\alpha'} \sum_c \int \frac{dk^- \, d^{2-2\epsilon} k_{\text{T}}}{(2\pi)^{4-2\epsilon}} L(k, j, c, \beta, \beta'; P, S)
$$

$$
= \int \frac{d\xi}{\xi} \underset{\text{D}}{\text{Tr}} \left[U(q; \hat{k}, j; m = 0) \frac{\hat{k}}{2} \right] \sum_c \int \frac{dk^- \, d^{2-2\epsilon} k_{\text{T}}}{(2\pi)^{4-2\epsilon}} \underset{\text{D}}{\text{Tr}} \frac{\gamma^+}{2} L(k, j, c; P, S)
$$

$$
+ \text{ terms with polarized parton densities.} \tag{8.8}
$$

The factor $\hat{k}/2$ is exactly the external line factor for U that corresponds to a spin-averaged on-shell Dirac particle. See (6.19) and the preceding definitions (6.17) and (6.18) for the form of the polarized terms. They can be allowed for by replacing the factor $\hat{k}/2$ by the form (A.27) with polarization for the quark.

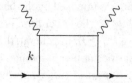

Fig. 8.4. One-loop graph for DIS in a model theory. The lines may represent any kind of field.

8.2.3 Factorization

We have now completed the definition of the parton approximator, and the result is a factorization of the form shown in (8.5).

8.2.4 Why the simple derivation does not work

The above derivation of the factorization theorem would be valid if one could use a fixed decomposition of momentum space into regions appropriate for U and L, at least up to power-suppressed terms. But in renormalizable theories, no clear separation of scales can be made. The issue is quite generic, so I illustrate it by examining a one-loop graph related to the calculations, in Sec. 6.11, of UV divergences in parton densities.

Consider a one-loop graph for DIS with an elementary-particle target, Fig. 8.4. We perform the $k^{\pm}$ integrals by the mass-shell delta functions for the two final-state particles, to leave only an integral over k_{T}^2. By the Libby-Sterman analysis, we obviously have leading-power contributions when k_{T}^2 is comparable to m^2 and when it is comparable to Q^2; these correspond, respectively, to regions where only the top rung is the hard subgraph, and where the whole graph is the hard subgraph.

But, as we now show, there is also a leading contribution from intermediate momenta, i.e., where $m \ll k_{\mathrm{T}} \ll Q$. Since $k_{\mathrm{T}} \ll Q$, we can apply the parton-model approximation to the top rung, and replace the calculation by the calculation of a parton density, as in Sec. 6.11. Then because, $m \ll k_{\mathrm{T}}$, we can neglect m, thereby obtaining a logarithmic integral:

$$\text{constant} \times \int_{\sim m^2}^{\sim Q^2} \frac{\mathrm{d}k_{\mathrm{T}}^2}{k_{\mathrm{T}}^2}. \tag{8.9}$$

That this is a logarithmic integral follows from the fact that the couplings are dimensionless. The whole graph has the same dimension as a lowest-order graph. Hence the momentum integral is dimensionless. Corrections to this formula are suppressed by powers of k_{T}/Q and of m/k_{T}.

Each range of a factor of 2 (say) in k_{T}^2 gives the same contribution. This contribution is also comparable in size to that from the hard range, $k_{\mathrm{T}} \sim Q$, and from the collinear range, $k_{\mathrm{T}} \sim m$. There is therefore no power-suppression (in m/Q) of the intermediate region. Indeed the intermediate region is slightly enhanced, i.e., logarithmically, by a factor $\ln(Q/m)$.

The elementary proof in Secs. 8.2.1–8.2.3 relied on a strict separation of scales: some momenta have $k_T \sim m$ and some have high virtuality, $O(Q^2)$, with unimportant contributions from intermediate momenta. When this is valid, errors of order m/Q result from neglecting k_T relative to Q. But the logarithmic contribution from the intermediate region violates the initial assumption.

One could try rescuing the argument by using an intermediate scale μ to separate collinear and hard momenta. In a one-loop graph this would result in errors of order μ/Q and of order m/μ: the first is from neglecting collinear transverse momenta relative to Q, and the second error is from neglecting masses with respect to hard momenta. The minimum error is of order $\sqrt{m/Q}$, obtained when $\mu \sim \sqrt{mQ}$. This is very non-optimal compared with the m/Q error (modified by logarithms) that is obtained from a better derivation of factorization.

Moreover in higher-order graphs, like Fig. 8.3, the errors from using a simple cutoff to separate the regions are actually unsuppressed. To see this consider a configuration in which the transverse momentum l_T in the lower loop of Fig. 8.3 is slightly below the cutoff, while k_T in the upper loop is slightly above the cutoff. Then l is target-collinear while k is hard. The elementary derivation tells us to neglect k_T with respect to l_T, producing a 100% error.

So we need a more powerful method, which we will come to in due course.

8.3 Renormalization of parton densities

We saw in calculations, Sec. 6.11, that parton densities have UV divergences at or above the space-time dimension $n = 4$ where the theory has a dimensionless coupling. This is one symptom that the parton model is not strictly correct. The Feynman graphs and momentum region that give the parton model still exist in such theories, but there are additional contributions.

In such a situation parton densities continue to be useful, but we have to adjust the definitions to make the parton densities finite. Motivated by what happens with the operator product expansion (OPE), reviewed in Collins (1984), we now construct such a definition by applying conventional UV renormalization. This gives renormalized parton densities as theoretical constructs, which can be studied in and of themselves, without regard to applications. Of course, it is the applications that provide *post hoc* motivation for studying parton densities.

8.3.1 Cutoff or renormalization?

An alternative to renormalization is to impose a cutoff in transverse momentum, e.g., to modify (6.75) to

$$f_{j/h}(\xi) \stackrel{\text{def}}{=} \int_{k_T < \mu} d^2 k_T \, f_{j/h}(\xi, k_T). \tag{8.10}$$

This definition has been particularly advocated by Brodsky and his collaborators (e.g., Lepage and Brodsky, 1980; Brodsky *et al.*, 2001) and clearly has certain advantages. Both kinds of definition, by a cutoff and by renormalization, are legitimate, and there is a choice between them influenced by practicalities and by actual practice, not by absolute necessity. Recall the calculation in Sec. 3.4, where we showed that renormalization with a scale μ is similar to a cutoff at approximately the same scale. Thus the two kinds of definition of finite parton densities have similar properties and intuitive meanings. But one must not take the equivalence between renormalization and a cutoff as a strict mathematical property.

Serious work beyond leading order, or beyond leading-logarithm approximation, requires us to take the definitions rather literally. Here certain disadvantages of the cutoff method appear that lead us to use the renormalization method. One is simply that although the cutoff method lends itself very nicely to getting an overall view, detailed calculations can be harder. A second rather severe disadvantage is that the definition with a cutoff relies on the definition of the unintegrated or TMD density. Now, in a gauge theory, the basic definition of the unintegrated parton density entails the use of light-front gauge $A^+ = 0$. But this results in further divergences even before the k_T integral, and therefore requires even more complicated redefinitions (Ch. 13). This problem is often hidden in elementary discussions, but comes to the forefront once higher-order corrections are considered correctly and is a continuing topic of research and debate.

8.3.2 Statement of renormalization of parton densities

In the theory of renormalization (e.g., Collins, 1984) there are two ways of viewing the renormalization of composite operators. One is the multiplicative view, where renormalized operators are factors times the bare operators. The other view is the counterterm view, where for each Feynman graph a series of counterterms is subtracted to remove the divergences. It is very useful to switch between the views as the occasion demands; we will see their equivalence.

For the parton densities, the multiplicative view will result in the following formula:

$$f_{j/H}(\xi) = \sum_{j'} \int_{\xi-}^{1+} \frac{dz}{z} Z_{jj'}(z, g, \epsilon) f_{(0)\, j'/H}(\xi/z). \qquad (8.11)$$

On the right-hand side is a bare parton density for a parton of flavor j'. Here a bare parton density is defined directly by whichever of operator formulae like (6.31) is appropriate, with the convention that the field operators are bare fields (i.e., that have canonical (anti)commutation relations). The theorem of renormalization is that one can obtain UV-finite parton densities $f_{j/H}(\xi)$ by a proper choice of the renormalization factor $Z_{jj'}$ in (8.11). The multiplication is in the sense of a convolution in the longitudinal momentum fraction and of matrix multiplication on the flavor indices. In the $\overline{\text{MS}}$ scheme, the renormalization factor is a function only of the ratio of the momentum fractions, the renormalized coupling and the dimension of space-time.

We have written limits $\xi-$ and $1+$ in the integral over z in (8.11), with the same meaning as in factorization formulae, such as (8.7). The upper limit is set by the renormalization kernel (which includes a delta function at $z = 1$). The lower limit is set by the bare parton density, which is non-zero only for $\xi/z \le 1$.

8.3.3 Polarization dependence

Formula (8.11) applies both to the unpolarized densities and to the various kinds of polarized densities (helicity, transversity, etc.). The transformations of the densities under rotations and parity imply that there is no mixing between the different kinds of polarized density. That is, one copy of (8.11) applies to the unpolarized densities, a second copy, with different renormalization factors, applies to the helicity densities, and a third copy, with yet different renormalization factors, applies to the transversity densities.

8.3.4 Regions giving UV divergences

An ordinary UV divergence (such as is canceled by renormalization of the Lagrangian) comes from regions where all the components of momenta in a subgraph get large. It might appear that the divergences in parton densities are different because they involve large values only for the minus and transverse components of loop momentum, as we saw in a calculational example. The momentum components are power-counted as $(k^+, k^-, k_T) \sim (P^+, \Lambda^2/P^+, \Lambda)$ where $\Lambda \to \infty$. However, this appearance that the divergence is of a new kind is misleading. We see this in light-front perturbation theory. Plus momenta are restricted to fractions of the external momenta, and even ordinary UV divergences also arise from large minus and transverse momenta, again with the power-counting $(P^+, \Lambda^2/P^+, \Lambda)$. An example is given by the self-energy graph that we calculated at (7.13).

The apparent difference arises because of a different choice of contour deformation: a Wick rotation of energy integrals in the usual case, and a contour integral in k^- for the light-front case. Of course, in a parton density with its integral over k^-, the light-front view is natural.

So the large k_T divergences in parton densities are actually genuine UV divergences to which we can apply normal methods of renormalization.

Further analysis proceeds by examining the momentum regions that give the UV divergence. We use the formalism in which the operators defining the parton density are time-ordered and the graphs are uncut. We take it for granted that renormalization has been applied in the Lagrangian, so that all UV divergences in self-energies and vertex corrections, etc., are canceled by counterterms. The remaining divergences involve loop momentum integrals that include the vertices that define the parton densities. Thus we represent the regions giving divergences by diagrams such as Fig. 8.5(a). In the upper part, labeled "UV", the minus and transverse components of all momenta get large, with plus momenta obeying their normal restrictions (in particular not to be bigger than P^+). In the lower part, labeled "collinear", the momenta stay finite. The collinear part includes the connecting lines of momentum l, while the UV part includes the lines of momentum k that go to the parton density vertices. In addition to being far off-shell, the momenta in the UV part have large negative rapidity relative to the target.

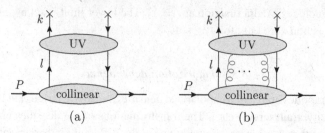

Fig. 8.5. (a) Regions giving UV divergence for pdf in renormalizable non-gauge theory or in a gauge theory (in light-cone gauge, $A^+ = 0$). The lines joining the UV and collinear subgraphs can be of any type, e.g., flavor of quark, antiquark or (transverse) gluon. (b) In a gauge theory arbitrary gluon connections between the collinear and UV subgraphs also give divergences.

We now do power-counting to determine the strength of the divergence and to determine what external lines are allowed for the UV part. For this we use the appropriate general-ization of the rules for ordinary UV divergences given that we treat the components of UV momenta as having sizes $(P^+, \Lambda^2/P^+, \Lambda)$. By observing that such a momentum configu-ration can be obtained by a boost from a frame in which all the components of UV momenta are of order Λ, we readily see that the power-counting works just like the power-counting for hard scattering in DIS: Ch. 5. In the rest frame of a UV momentum, the collinear lines are indeed collinear to the fast-moving target. The basic degree of divergence for a graph with two lines connecting the UV and collinear subgraphs is logarithmic. We saw this in an example, and the property extends to higher-order graphs for the UV subgraph. The reason is that this subgraph is dimensionless, and so the power-counting of a UV divergence follows dimensional analysis in a renormalizable theory.

The estimate of the power can equally well be done in a fixed frame. In that case the key point in relating dimensional analysis to the size of the divergence of the integral is that in Lorentz-invariant quantities, a minus momentum k^-, which has two powers of Λ, always appears multiplied by a plus momentum k^+, which has zero powers of Λ; thus the power of Λ is the dimension of k^+k^-.

Therefore adding external collinear lines to the UV subgraph generally reduces its degree of divergence, and therefore gives convergence. The one exception, just as in our discussion of hard scattering in Ch. 5, is in a gauge theory when there are gluon lines with a minus index in the UV subgraph and a plus index at the attached gluon line. Thus in addition to regions of the form Fig. 8.5(a), we also have divergences with extra gluons joining the UV and collinear subgraphs, Fig. 8.5(b).

For this chapter we restrict our attention to a non-gauge theory, for which the catalog of divergent regions is Fig. 8.5(a). (This set of leading regions also applies to a gauge theory in $A^+ = 0$ gauge. But this chapter's treatment of renormalization does not genuinely apply, because of problems with divergences associated with the $1/k^+$ singularities in the gluon propagator.)

The details of constructing a renormalized parton density follow very closely the con-struction of matrix elements of renormalized local operators in conventional renormalization theory (e.g., Collins, 1984).

8.3.5 Momentum dependence of counterterms

There is one new feature, which concerns the dependence of the counterterms on external momenta. In conventional operator renormalization, when there is a logarithmic divergence, the counterterm can be chosen to be independent of momentum and mass. One general method of proof is to differentiate graphs with respect to the external momenta and/or masses. This reduces the degree of divergence, and thus for a logarithmic divergence shows that after differentiation there is no overall divergence, and therefore no counterterm is needed. There can be subdivergences in multiloop graphs, but these are canceled by their own counterterms; the overall divergence is what determines the need for a counterterm for a whole graph. In general, the counterterms are polynomials in momentum and mass with the degree of the polynomial equal to the degree of divergence.

We now apply the differentiation argument to the UV divergences in parton densities. The examples in Sec. 6.11 provide illustrations of the general principles. We will now show that differentiating with respect to a mass, or an external minus or transverse momentum, does reduce the degree of divergence. But differentiating with respect to an external plus momentum leaves the degree of divergence unchanged. Thus the divergence is allowed to be a function of plus momenta. This gives the convolution form in (8.11) for the renormalization of parton densities, rather than the multiplicative form that applies to local operators.

Differentiating a graph with respect to an external momentum gives a sum over terms where particular propagator (or numerator) factors are differentiated. So we consider a generic propagator, carrying an internal momentum k and an external momentum P:

$$\frac{1}{(P-k)^2 - M^2} = \frac{1}{2(P^+ - k^+)(P^- - k^-) - (\boldsymbol{P}_{\mathrm{T}} - \boldsymbol{k}_{\mathrm{T}})^2 - M^2}. \tag{8.12}$$

The external momentum may be off-shell and may have non-zero transverse momentum. The UV divergence concerns the situation where $\boldsymbol{k}_{\mathrm{T}}$ and k^- go to infinity with k^+ fixed. There are three cases:

1. Differentiation of (8.12) with respect to P^+ reduces the dimension by one, but introduces a factor of a minus momentum:

$$\frac{\mathrm{d}}{\mathrm{d}P^+} \frac{1}{(P-k)^2 - M^2} = \frac{-2(P^- - k^-)}{[2(P^+ - k^+)(P^- - k^-) - (\boldsymbol{P}_{\mathrm{T}} - \boldsymbol{k}_{\mathrm{T}})^2 - M^2]^2}. \tag{8.13}$$

   In power-counting for the degree of divergence, the factor k^- in the numerator is treated as k_{T}^2 rather than as the single power k_{T} that matches its dimension. Thus the degree of divergence is unaffected by differentiating with respect to P^+.

   This is a general result: in Lorentz-invariant quantities, a plus momentum always appears multiplied by a minus momentum. Thus the unchanged degree of divergences is effectively a consequence of invariance under boosts in the z direction.

2. Differentiation with respect to a mass M or transverse momentum $\boldsymbol{P}_{\mathrm{T}}$ brings no extra factor; this reduces the degree of divergence by one unit, just as with local operators.

3. Differentiation with respect to an external minus momentum P^- gives an extra reduction of the degree of divergence, by two units instead of one unit.

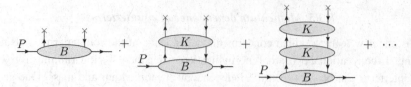

Fig. 8.6. Ladder decomposition of graphs for a bare parton density in terms of two-particle irreducible subgraphs.

When we use $\overline{\text{MS}}$ renormalization, the counterterms are just the divergent pole parts, so the coefficients of the poles obey the above rules for lack of dependence on minus momenta, transverse momenta and masses. In a general renormalization scheme, it is permitted to perform a further *finite* renormalization which does depend on these momentum components and masses. We choose not to.

We now summarize the form of the divergence in a parton density as

$$\int \frac{\mathrm{d}l^+}{l^+} H(l^+, k^+) \int \mathrm{d}l^-\, \mathrm{d}^{2-2\epsilon} l_{\mathrm{T}}\, L(l, P), \tag{8.14}$$

where H denotes the divergence of a UV subgraph. Since the divergence is independent of l^- and l_{T}, the integral over these variables can be confined to the collinear subgraph, corresponding to the rules for a parton density, in fact. But the UV and collinear parts are linked by an integral over l^+.

Now parton densities are invariant under boosts in the z direction. Generally we will arrange for the factors in formulae such as (8.14) to be boost invariant. Notice that this is the case for the measure $\mathrm{d}l^+ / l^+$ of the convolution. Then the UV divergence factor H must be a function, not of k^+ and l^+ separately, but only of their ratio. This gives the kinematic dependence of the renormalization factor $Z_{jj'}$: it is a function of the ratio between the fractional momentum ξ of the renormalized parton density and the fractional momentum of the bare parton density.

8.3.6 Ladder graphs and renormalization

In this section, we will prove the renormalization theorem for parton densities, and we will see how the subtractive counterterm formalism is set up. The methodology (Collins, 1998a) is inspired by Curci, Furmanski, and Petronzio (1980).

The issue that makes the discussion quite non-trivial is that the characterization of UV divergences just given is somewhat incomplete. It assumed that we could assign the estimate $(P^+, \Lambda^2/P^+, \Lambda)$ uniformly to all the different momenta in the UV subgraph. But in fact there can be a variety of sizes.

Notation

A given graph for a bare parton density can have many decompositions of the form of Fig. 8.5(a). Given that they all have the two subgraphs connected by two lines, a convenient way to enumerate all possibilities is to perform a ladder decomposition, as in Fig. 8.6.

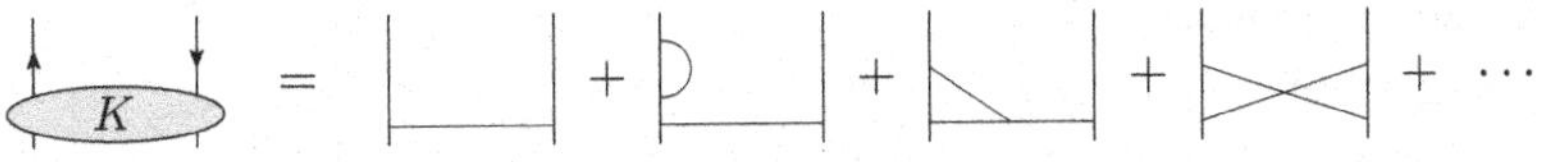

Fig. 8.7. Examples of topologies of graphs for the ladder rung K in Fig. 8.6. The lines and vertices are of any type allowed by the theory. The shortness of the lines at the lower end indicates that these propagators are amputated.

Each of the objects B and K is a sum over two-particle irreducible (2PI) graphs multiplied by full propagators for the upper two lines. Typical examples of graphs for K are shown in Fig. 8.7. They are connected and have two upper external lines and two lower external lines. Propagators with all possible corrections are used for the upper lines, but the lower lines are amputated. The two-particle irreducibility of the core part of K means that its top cannot be disconnected from the bottom by cutting only two lines; at least three lines must be cut. The base of the ladder, B, similarly has two propagators on its upper side times a 2PI amplitude, but it is now connected to the target state, including a bound state wave function if needed. The types of the lines can be any that is allowed in the theory.

We therefore represent the bare parton density for a parton of type j as a sum over ladder graphs with different numbers of rungs:

$$f_{(0)\,j} = Z_j V(j) \sum_{n=0}^{\infty} K^n B$$

$$= Z_j V(j) \frac{1}{1-K} B. \tag{8.15}$$

The products are in the sense of a convolution, i.e., an integral over the momentum of the loop joining the factors, and sums over the flavor, color and spin indices, just as in (8.2).

At the top of the ladder we have the vertex defining the parton density, and in (8.15) we denote it by the factor V. A complete notation is cumbersome:

$$V(\xi P^+, j, s; k, j_1, c, c', \alpha, \alpha') = \delta(\xi P^+ - k^+) s_{\alpha,\alpha'} \delta_{jj_1} \delta_{cc'}, \tag{8.16}$$

which is set up to be used in the convolution notation, as in (8.2). The color and spin indices on the two attached parton lines are (c, α) and (c', α'). There is a common flavor index j_1 for the two lines. We let $s_{\alpha,\alpha'}$ be the matrix with which the vertex couples the spin indices, e.g., $\gamma^+/2$ for an unpolarized quark density.

We require that both of K and B are Green functions of renormalized fields, so that they are UV finite. Since we define the bare parton density by an expectation value of bare operators, we inserted in (8.15) a factor of the wave function renormalization Z_j for the field for parton j.

The rung factor is

$$K(k_1, j_1, c_1, c_1', \alpha_1, \alpha_1'; k_2, j_2, c_2, c_2', \alpha_2, \alpha_2'). \tag{8.17}$$

Here, k_1 is the momentum of each of the upper lines, and j_1 is the flavor, while (c_1, α_1) and (c_1', α_1') are color and spin indices for the upper lines. The other variables are for the lower lines. There is also dependence on the coupling etc which is not indicated. Similarly for the base of the ladder we write

$$B(k, j_1, c_1, c'_1, \alpha_1, \alpha'_1; P, S). \tag{8.18}$$

Note that if the target is an elementary particle, as in our calculational examples in Sec. 6.11, then the base factor B will be just a delta function, e.g.,

$$B(k, j_1, c, c', \alpha, \alpha'; P, S) = (2\pi)^{4-2\epsilon} \delta^{(4-2\epsilon)}(l - P) \delta_{j_1, \text{target}}$$
$$\times \text{ spin density matrix.} \qquad \text{(Elementary target)} \tag{8.19}$$

Just as in (8.4), the two lines joining neighboring rungs (V, K, or B) have equal flavors. But we have allowed unequal values for the spin and color indices. But then we observe that V times any number of Ks is color singlet, and so gives a coefficient times a unit matrix in color space. Hence we only use K and B in combinations with a diagonal sum over colors at their upper end, and so we write, for example,

$$\sum_{c_1} K(k_1, j_1, c_1, c_1, \alpha_1, \alpha'_1; k_2, j_2, c_2, c'_2, \alpha_2, \alpha'_2) = K(k_1, j_1, \alpha_1, \alpha'_1; k_2, j_2, \alpha_2, \alpha'_2) \delta_{c_2, c'_2}.$$
$$\tag{8.20}$$

In the mathematical manipulations that follow, K is to be thought of as a matrix, with two composite indices, V as a row vector, and B as a column vector.

Divergences, subtractions, renormalization

We now define a renormalized parton density by the standard procedure of subtracting counterterms for each subgraph with an overall UV divergence. We first remove the wave-function renormalization factor Z_j. Then we consider the UV divergences in one term $V K^n B$. Each possible divergent subgraph in Fig. 8.5(a) is associated with a subgraph consisting of V and some number $N_1 > 0$ of the nearest rungs.

A zero-rung graph $V B$ therefore has no UV divergences. A one-rung graph $V K B$ has one divergence, in the $V K$ subgraph. We can cancel the divergence in $V K$ by subtracting, for example, its pole part at $\epsilon = 0$, $V K \overleftarrow{\mathcal{P}}$, to give a finite result $V K (1 - \overleftarrow{\mathcal{P}}) B$. The left arrow in $\overleftarrow{\mathcal{P}}$ signifies that the pole part is taken of everything to its left. The significance of the pole part is that it is independent of the external l^- and l_T of $V K$, since this is a property of the elementary UV divergence derived above. Thus $V K \overleftarrow{\mathcal{P}}$ is of the form of a vertex for a parton density at momentum fraction l^+/P^+ times a function of ξP^+ and l^+. This will enable us to obtain multiplicative renormalization after we sum over all graphs and UV-divergent subgraphs. Naturally, the pole part may be replaced by any other operation that achieves the same effect, of canceling the divergence with a counterterm that is a coefficient times a vertex for a parton density.

From now on we will define $\overleftarrow{\mathcal{P}}$ to denote whatever such definition we choose to use, and the choice defines the renormalization scheme for the parton density. The standard choice is the $\overline{\text{MS}}$ scheme, Sec. 3.2.6, with its extra factor S_ϵ for each loop in a counterterm; see (3.16) and (3.18).

For a two-rung ladder, $V K K B$, we first cancel the divergence in the $V K$ subgraph, to get $V K (1 - \overleftarrow{\mathcal{P}}) K B$. The remaining divergence is in the two-rung part, and to cancel it we can subtract $V K(1 - \overleftarrow{\mathcal{P}}) K \overleftarrow{\mathcal{P}} B$. Here the second pole part is the pole part of everything to its left, i.e., the pole part of $V K(1 - \overleftarrow{\mathcal{P}}) K$. After these subtractions the UV-finite result is $V K(1 - \overleftarrow{\mathcal{P}}) K(1 - \overleftarrow{\mathcal{P}}) B$. It is straightforward to extend this result to bigger ladders: we simply insert a factor $1 - \overleftarrow{\mathcal{P}}$ to the right of every factor of K.

We now convert this into a form that we use to demonstrate multiplicative renormalizability:

$$f_j = V(j) \left[1 + K(1 - \overleftarrow{\mathcal{P}}) + K(1 - \overleftarrow{\mathcal{P}}) K(1 - \overleftarrow{\mathcal{P}}) + \ldots \right] B$$

$$= V(j) \sum_{n=0}^{\infty} \left[K(1 - \overleftarrow{\mathcal{P}}) \right]^n B$$

$$= V(j) \sum_{n=0}^{\infty} K^n B - V(j) \sum_{n=1}^{\infty} \sum_{n_1=1}^{n} \left[K(1 - \overleftarrow{\mathcal{P}}) \right]^{n_1 - 1} K \overleftarrow{\mathcal{P}} K^{n - n_1} B$$

$$= V(j) \sum_{n=0}^{\infty} K^n B - V(j) \sum_{n_1=1}^{\infty} \left[K(1 - \overleftarrow{\mathcal{P}}) \right]^{n_1 - 1} K \overleftarrow{\mathcal{P}} \sum_{n_2=0}^{\infty} K^{n_2} B$$

$$= V(j) \frac{1}{1 - K} B - V(j) \sum_{n=0}^{\infty} \left[K(1 - \overleftarrow{\mathcal{P}}) \right]^n K \overleftarrow{\mathcal{P}} \frac{1}{1 - K} B. \tag{8.21}$$

To get from the second to the third line, we expanded all the products and classified the result by where the rightmost $\overleftarrow{\mathcal{P}}$ is. There is one term with no $\overleftarrow{\mathcal{P}}$ factors at all. The last line is in fact of the form of a coefficient convoluted with the bare parton density, (8.11). To see this, we first observe that the term $V(j) \frac{1}{1 - K} B$ is of the desired form, giving a contribution

$$Z_j^{-1} \delta_{jj'} \delta(z - 1) \tag{8.22}$$

to $Z_{jj'}$. Now a renormalization pole part is a coefficient times a vertex for a parton density. So the last term in (8.21) is a pole part times the $\frac{1}{1 - K} B$ factor in the bare parton density. Thus we also get something of the form of the right-hand side of (8.11). In fact we can write the renormalization coefficient as

$$Z_{jj'} V(j') = \frac{1}{Z_{j'}} \left[\delta_{jj'} \delta(z - 1) - V(j) \sum_{n=0}^{\infty} [K(1 - \overleftarrow{\mathcal{P}})]^n K \overleftarrow{\mathcal{P}}(j') \right], \tag{8.23}$$

where the (j') argument of the last $\overleftarrow{\mathcal{P}}$ indicates that we restrict to graphs whose rightmost line pair has flavor j'.

This completes the proof of the renormalization theorem for parton densities, at least when the theory has no gauge fields. The proof also applies in a gauge theory (e.g., QCD) in $A^+ = 0$ gauge, if we assume that the non-trivial complications in this gauge do not matter.

For performing calculations, it is useful that the proof also applies to off-shell Green functions of the parton vertex operator, with the actual on-shell parton densities being obtained by applying LSZ reduction.

Nature of subtractive approach

The starting point of (8.21) was a modification of the definition of a parton density where all UV divergences were subtracted out. Then this was converted to a form that exhibited multiplicative renormalization of bare parton densities.

Now methods using subtractions are fundamental to all aspects of perturbative QCD, as we will see. So in the next few paragraphs I give further insights into the subtractive approach, with renormalization of parton densities giving an example of a general methodology.

Let us focus attention on the third line in (8.21). It starts with a sum over all graphs for the parton density partitioned by the number n of rungs; a generic term is $V K^n B$. Note that K and B themselves are sums of graphs of the appropriate irreducibility properties. Possible ways of getting UV divergences are enumerated by partitioning the product of rungs into two factors:

$$[V K^{n_1}] [K^{n-n_1} B], \tag{8.24}$$

where n_1 can range from 1 to n. Applying this to a single graphical structure, we have n ways of doing the partition. For each partition, there is a divergence where the momenta in the left part of (8.24) get large while the momenta in the right-hand part stay finite. The left factor corresponds to the upper part of Fig. 8.5(a).

An initial idea for removing the divergence is simply to subtract the UV pole part of the subdiagram $V K^{n_1}$. We can notate the subtraction diagrammatically as

$$- \qquad \qquad , \tag{8.25}$$

where the box denotes the taking of the pole part, as in $\overline{\mathrm{MS}}$ renormalization.[1] Such subtractions do not actually remove the divergences correctly, for two related reasons. The first is the possibility of subdivergences: if $n_1 > 1$, the K^{n_1-1} factor has a pole from subdivergences, where only some of the rungs inside the box are in a UV region. The second is that of double counting: there can be further UV divergences when not only the momenta inside the box are UV, but also some momenta further down are also UV, which situation occurs if $n_1 < n$.

Both problems are solved by applying the pole-part operation only after subtractions have been made for subdivergences. In the third line of (8.21), this is done by the $(1 - \overleftarrow{\mathcal{P}})$ factors inside the $V K^{n_1}$ part.

To see this as a prevention of double counting, we imagine constructing the counterterms one by one, starting with the smallest, $n_1 = 1$. Let $C_{n_1}(V K^n B)$ be the counterterm for the n_1-rung graph. It is made by applying minus the pole part to the original graph together

---

[1] Or the corresponding operation in some other renormalization scheme.

with the counterterms for smaller numbers of rungs:

$$C_{n_1}(VK^n B) = - \left[VK^{n_1} + \sum_{n_1'=1}^{n_1-1} C_{n_1'}(VK^{n_1}) \right] \overleftarrow{\mathcal{P}} \; K^{n-n_1} B. \tag{8.26}$$

The internal counterterms $C_{n_1'}$ remove subdivergences. As for double counting, consider the sum over n_1: $\sum_{n_1=1}^{n} C_{n_1}(VK^n B)$. For the overall UV divergence in a particular VK^{n_1}, there will be contributions both from the original graph and from the counterterms for subdivergences in the set of terms $VK^{n_1} + \sum_{n_1'=1}^{n_1-1} C_{n_1'}(VK^{n_1})$. The use of (8.26) to define C_{n_1} deals with this problem.

Equation (8.26) is an example of the Bogoliubov operation in renormalization theory, and it provides a recursive definition of the counterterm. The recursion starts at $n_1 = 1$ where there are no subdivergences:

$$C_1(VK^n B) = - VK \overleftarrow{\mathcal{P}} \; K^{n-1} B. \tag{8.27}$$

It is not too hard to prove by induction that

$$C_{n_1}(VK^n B) = -[VK(1 - \overleftarrow{\mathcal{P}})]^{n_1-1} K \overleftarrow{\mathcal{P}} K^{n-n_1} B, \tag{8.28}$$

which gives the counterterms in the third line of (8.21).

An illustration of the box notation for counterterms is the case $n = 2$:

$$= VK^2 B - VK\overleftarrow{\mathcal{P}} \; KB - \left(VK^2 - VK\overleftarrow{\mathcal{P}}K \right) \overleftarrow{\mathcal{P}} \; B$$
$$= VK(1 - \overleftarrow{\mathcal{P}}) \; K(1 - \overleftarrow{\mathcal{P}}) \; B.$$

8.4 Renormalization group, and DGLAP equation

Renormalized quantities depend on the renormalization scale μ. When we apply the factorization theorem we will enable the effective use of perturbative theory in the hard scattering by setting μ to be of order Q. Therefore to make predictions, we need to transform parton densities between different values of μ, for which we need their renormalization-group (RG) equations.

These are obtained by applying $d/d \ln \mu$ to (8.11) and using the RG invariance of the unrenormalized parton density. The resulting equation are known as the Dokshitzer-Gribov-Lipatov-Altarelli-Parisi (DGLAP) equations[2] (Altarelli and Parisi, 1977; Gribov and Lipatov, 1972; Dokshitzer, 1977). They have the form

$$\frac{d}{d \ln \mu} f_{j/H}(\xi; \mu) = 2 \sum_{j'} \int \frac{dz}{z} P_{jj'}(z, g) f_{j'/H}(\xi/z; \mu), \qquad (8.30)$$

where, with its standard normalization, the (finite at $\epsilon = 0$) DGLAP evolution kernel $P_{jj'}$ obeys

$$\frac{d}{d \ln \mu} Z_{jk}(z, g, \epsilon) = 2 \sum_{j'} \int \frac{dz'}{z'} P_{jj'}(z', g, \epsilon) Z_{j'k}(z/z', g, \epsilon), \qquad (8.31)$$

i.e., essentially

$$P = \frac{1}{2} \frac{d}{d \ln \mu} \ln Z, \qquad (8.32)$$

with algebra (multiplication in particular) for Z being interpreted in the sense of convolutions on z, and in the sense of matrices on the partonic indices. Recall that the RG derivative when applied to such counterterms is just the beta function for a coupling times a derivative with respect to the coupling, and then summed over couplings. In the model Yukawa theory, this is

$$\frac{1}{2} \frac{d}{d \ln \mu} Z_{jk} = \left(-\epsilon \frac{g^2}{16\pi^2} + S_\epsilon^{-1} \beta_{g^2} \right) \frac{\partial Z_{jk}}{\partial g^2/(16\pi^2)} + \left(-\epsilon \frac{\lambda}{16\pi^2} + S_\epsilon^{-1} \beta_\lambda \right) \frac{\partial Z_{jk}}{\partial \lambda/(16\pi^2)}. \qquad (8.33)$$

Here $\beta_{g^2} \overset{\text{def}}{=} \frac{1}{2} dg^2 / d \ln \mu$, etc., with the normalizations like those of Sec. 3.5.2. Each β is a function of $S_\epsilon \lambda$ and $S_\epsilon g^2$, but not of ϵ separately (in the $\overline{\text{MS}}$ scheme). In QCD there would only be the β_{g^2} term.

8.5 Moments and Mellin transform

The connection to the renormalization of local operators can be exhibited by taking an integral with a power of ξ. We define

$$\tilde{f}_{j/H}(J) = \int_0^{1+} d\xi \, \xi^{J-1} f_{j/H}(\xi), \qquad (8.34)$$

$$\tilde{Z}_{jj'}(J) = \int_0^{1+} dz \, z^{J-1} Z_{jj'}(z), \qquad (8.35)$$

---

[2] The original derivations were rather different to the strict RG one presented here.

and similarly for the unrenormalized parton densities and the DGLAP kernels. Then (8.11) gives a matrix-multiplication form for the moments:

$$\tilde{f}_{j/H}(J) = \sum_{j'} \tilde{Z}_{jj'}(J)\tilde{f}_{(0)\,j'/H}(J). \tag{8.36}$$

The DGLAP equation similarly becomes

$$\frac{d}{d\ln\mu}\tilde{f}_{j/H}(J;\mu) = \sum_{j'} 2\tilde{P}_{jj'}(J;g)\tilde{f}_{j'/H}(J;\mu). \tag{8.37}$$

If J is allowed to range over general (complex) values, then we have constructed the Mellin transform of the parton density and shown that renormalization looks particularly simple for the Mellin transform. The Mellin transformation can be inverted to recover the parton densities in ξ space. In numerical calculations, it can be an advantage of the Mellin-transformed formulation that equations like (8.37) involve matrix multiplication rather than convolutions.

If J is restricted to non-negative integer value, and the combinations of parton and antiparton densities are used that correspond to local operators, as in (6.109), then we have the formula for renormalization of the local operators used in the OPE for DIS.

8.6 Sum rules for parton densities and DGLAP kernels, including in QCD

In Secs. 6.9.5 and 6.9.6, we derived number and momentum sum rules in a theory where no renormalization of parton densities was needed. We now extend the treatment to a renormalizable theory. The derivation will also apply to QCD, but only after we show that the renormalization theorems also apply to QCD.

Before renormalization we have bare parton densities in the UV-regulated theory. For a bare quark density, we derived a number sum rule in (6.92); the derivation applies also in QCD, since the Wilson line now needed between the quark and antiquark fields becomes unity when the fields are at the same position. The derivation must be applied to the bare parton densities in order to get the correctly normalized Noether current. In contrast, for the derivation of the momentum sum rule in QCD, the Wilson line requires a slight change in the derivation. Because of the extra factor ξ in the integrand of the sum rule (6.93), a derivative is needed with respect to the position of one of the fields in the quark density definitions. The derivative applies to both the field and the Wilson line, and the result is to give a covariant derivative of the quark field, and so to give the correct quark term in the energy momentum tensor. The gluon term also comes out correctly. After that the derivation is as before.

Each of these derivations applies to a particular moment of parton densities and results in a target matrix element of a Noether current, whose value we know exactly and which is finite. We now need to show that the sum rules also apply to *renormalized* densities and to obtain corresponding constraints on the renormalization coefficients. We first take the inverse transformation to (8.36):

$$\tilde{f}_{(0)i/H}(J) = \sum_j \tilde{Z}_{ij}^{-1}(J)\tilde{f}_{j/H}(J), \tag{8.38}$$

where Z^{-1} is the matrix inverse of Z. The number sum rule for a quark q is that $\tilde{f}_{(0)q/H}(1) - \tilde{f}_{(0)\bar{q}/H}(1)$ is the number of this type of quark in the target H. Since this is finite, the corresponding renormalization coefficients are also finite: $\tilde{Z}_{qj}^{-1}(1) - \tilde{Z}_{\bar{q}j}^{-1}(1)$. Let us use the $\overline{\text{MS}}$ scheme, in which case finiteness only happens if the counterterms are zero, leaving the lowest-order terms. Thus we get the following sum rules for the first moments of the renormalization coefficients:

$$\tilde{Z}_{qj}^{-1}(1) - \tilde{Z}_{\bar{q}j}^{-1}(1) = \delta_{qj} - \delta_{\bar{q}j}, \tag{8.39a}$$

$$\tilde{Z}_{qj}(1) - \tilde{Z}_{\bar{q}j}(1) = \delta_{qj} - \delta_{\bar{q}j}, \tag{8.39b}$$

where the second line follows using the definition of the inverse matrix $Z^{-1}Z = I$. From (8.39b) and the sum rule for the bare parton densities follows the corresponding sum rule for the renormalized densities. Hence (6.91) applies to both bare and renormalized densities, provided the $\overline{\text{MS}}$ scheme is used.

The same argument applies to the momentum sum rule. It also leads to a sum rule for the renormalization coefficients:

$$\sum_j \tilde{Z}_{jj'}^{-1}(2) = \sum_j \tilde{Z}_{jj'}(2) = 1, \tag{8.40}$$

where the sum is over all flavors of parton: quarks, antiquarks, and gluon.

Combining the sum rules for Z with the definition of the DGLAP kernels (8.31) gives sum rules for the kernels:

$$\tilde{P}_{qj}(1) - \tilde{P}_{\bar{q}j}(1) = 0, \tag{8.41}$$

$$\sum_j \tilde{P}_{jj'}(2) = 0. \tag{8.42}$$

These sum rules have important testable consequences for the evolution of parton densities; they also provide useful checks on calculations.

8.7 Renormalization calculations: model theory

In this section we show how to calculate the renormalization of parton densities in the model Yukawa theory used earlier, to illustrate the principles without any confusion by the complications that arise in QCD.

8.7.1 Renormalization of the theory

The Lagrangian of the theory with renormalization for the interactions was given in (6.103). We use dimensional regularization and the $\overline{\text{MS}}$ scheme. We will express all quantities in terms of renormalized couplings g, etc. As usual, to keep the dimension of the coupling fixed, we write the bare couplings in terms of the renormalized couplings with the unit of mass as $g_0 = \mu^\epsilon g(1 + \text{counterterms})$, etc. We will use a counterterm approach, as in Sec. 3.2. Thus we write the Lagrangian as the sum of a free Lagrangian that gives the free propagators, a basic set of interactions, with renormalized couplings, and a counterterm Lagrangian.

Of the renormalization factors in the Lagrangian, the ones that we will need in our calculations are for the self-energy, for which completely standard calculations give

$$Z_2 - 1 = -\frac{g^2 S_\epsilon}{32\pi^2 \epsilon} + \dots, \qquad Z_2 M_0 - M = \frac{g^2 S_\epsilon M}{16\pi^2 \epsilon} + \dots, \qquad (8.43)$$

where the dots indicate terms of yet higher order.

8.7.2 Unintegrated density

First we examine unintegrated, i.e., transverse-momentum-dependent, parton densities. The bare densities in the UV-regulated theory are, e.g.,

$$f_{(0)q/H}(\xi, k_\mathrm{T}) = \int \frac{dw^- d^2 w_\mathrm{T}}{(2\pi)^3} e^{-i\xi P^+ w^- + i k_\mathrm{T} \cdot w_\mathrm{T}} \langle P| \overline{\psi}_0(0, w^-, w_\mathrm{T}) \frac{\gamma^+}{2} \psi_0(0) |P\rangle. \quad (8.44)$$

These have an immediate probability interpretation.

Since there are no extra divergences beyond those renormalized in the Lagrangian, the renormalized unintegrated quark density is obtained simply by using renormalized fields:

$$f_{q/H}(\xi, k_\mathrm{T}; \mu) = Z_2^{-1} f_{(0)q/H}(\xi, k_\mathrm{T}). \qquad (8.45)$$

To get its RG equation, we observe that the bare parton density is a matrix element of bare fields with physical states, and hence is RG invariant. Taking a total derivative of the renormalized density with respect to the renormalization scale μ gives the RG equation of the renormalized density:

$$\frac{d}{d \ln \mu} f_{q/H}(\xi, k_\mathrm{T}; \mu) = -2\gamma_2 f_{q/H}(\xi, k_\mathrm{T}; \mu). \qquad (8.46)$$

Here γ_2 is the anomalous dimension associated with the fermion field:

$$\gamma_2 = \frac{1}{2} \frac{d \ln Z_2}{d \ln \mu} = -\frac{g^2 S_\epsilon}{32\pi^2} + \dots, \qquad (8.47)$$

which has a finite limit at $\epsilon = 0$.

8.7.3 Integrated density

For renormalization of the integrated densities, we use a counterterm approach with subtractions applied in Green functions of renormalized fields. Therefore we first write

(8.11) as

$$f_{j/H}(\xi) = \sum_j \int \frac{dz}{z} [Z_{j'}(g, \epsilon) Z_{jj'}(z, g, \epsilon)] \, [Z_{j'}^{-1} f_{(0) \, j'/H}(\xi/z)]. \qquad (8.48)$$

Here the factor $Z_{j'}^{-1} f_{(0) \, j'/H}$ is the parton density with *renormalized* rather than bare fields used in its definition. Thus it is calculated using the standard Feynman rules for the theory and for the parton density; counterterms from the Lagrangian are used as needed. In compensation for the $Z_{j'}^{-1}$ factor the $Z_{jj'}$ factor is combined with a factor of $Z_{j'}$.

The renormalization factor gives UV-finite parton densities independently of the target state H. For calculations of $Z_{jj'}$, it is therefore convenient to choose the state to correspond to any of the elementary fields of the theory (as opposed to a bound state). To obtain the perturbation expansion of $Z_{jj'}$ from Feynman graphs, we expand (8.48) in powers of the renormalized couplings, and identify the necessary counterterms. We use the following expansions:

$$f_{j/H}(\xi) = \sum_{n=0}^{\infty} \left(\frac{g^2}{16\pi^2} \right)^n f_{j/H}^{[n]}(\xi) + \cdots, \qquad (8.49a)$$

$$f_{(0) \, j/H}(\xi) = \sum_{n=0}^{\infty} \left(\frac{g^2}{16\pi^2} \right)^n f_{(0) \, j/H}^{[n]}(\xi) + \cdots, \qquad (8.49b)$$

$$Z_{jj'}(z, g, \epsilon) = \sum_{n=0}^{\infty} \left(\frac{g^2}{16\pi^2} \right)^n Z_{jj'}^{[n]}(z, g, \epsilon) + \cdots. \qquad (8.49c)$$

To avoid complicated formulae, we have written only the terms with the Yukawa coupling g, and the *dots indicate terms involving the other couplings*. The lowest-order term in Z is unity in the sense of a matrix in parton type and of a convolution in z:

$$Z_{jj'}^{[0]}(z) = \delta_{jj'} \delta(z - 1). \qquad (8.50)$$

When the target is elementary, the lowest-order renormalized and bare parton densities are simply

$$f_{j/j'}^{[0]}(\xi) = f_{(0) \, j/j'}^{[0]}(\xi) = \delta(\xi - 1)\delta_{jj'}. \qquad (8.51)$$

Note the notational distinction between "[0]" in a superscript to denote "lowest order", and "(0)" to denote "bare" (normally in a subscript). Note also a *shift of notation* from Sec. 6.11: there we did not treat renormalization, so the expansion parameter was actually the bare coupling; now the expansion parameter is strictly the finite renormalized coupling.

The key equation for calculations of the renormalization factor is the n-loop expansion of the renormalization equation (8.48):

$$f_{j/k}^{[n]}(\xi) = \sum_{n'=0}^{n} \sum_{j'} \int \frac{dz}{z} \, Z_{jj'}^{[n']}(z, g, \epsilon) \, f_{(0) \, j'/k}^{[n-n']}(\xi/z). \qquad (8.52)$$

8.7.4 One-loop renormalization calculations in model theory

Quark in quark

The one-loop case of (8.52) for the density of a quark in a quark is

$$f_{q/q}^{[1]}(\xi) = \sum_j \int \frac{dz}{z} \left[(Z_2 Z_{qj})^{[0]}(z, g, \epsilon) \, (Z_2^{-1} f_{(0)\,j/q})^{[1]}(\xi/z) \right.$$

$$+ (Z_2 Z_{qj})^{[1]}(z, g, \epsilon)(Z_2^{-1} f_{(0)\,j/q})^{[0]}(\xi/z) \Big]$$

$$= (Z_2^{-1} f_{(0)\,q/q})^{[1]}(\xi) + (Z_2 Z_{qq})^{[1]}(\xi, g, \epsilon). \tag{8.53}$$

We carried out the calculations of the bare version of $f^{[1]}$ in Sec. 6.11, and we now read off the necessary modifications to renormalize the parton densities.

Virtual correction to quark in quark

The one-loop virtual correction to the parton density Fig. 6.10(a) is to be modified by adding wave function and mass renormalization counterterms to the self-energy, so that we replace (6.114) by

$$\frac{g^2}{16\pi^2} f_{q/q}^{[1,V]}(\xi) = -\delta(\xi - 1) \frac{g^2}{16\pi^2}$$

$$\times \int_0^1 dx \left\{ x \ln \left[\frac{\mu^2}{m^2 x + M^2(1-x)^2} \right] + \frac{2M^2 x(1-x^2)}{m^2 x + M^2(1-x)^2} \right\}, \tag{8.54}$$

in the limit that the UV regulator is removed, $\epsilon = 0$. Since this is finite by itself, no delta function contribution to $Z_2 Z_{qq}$ is needed: the UV divergence in the self-energy is removed by a counterterm from the interaction, and so does not affect renormalization of the parton density.

Real correction to quark in quark

For the real emission term, we need

$$\frac{g^2}{16\pi^2}(Z_2 Z)_{qq}^{[1]}(z, g, \epsilon) = -\frac{g^2 S_\epsilon}{16\pi^2 \epsilon}(1 - z) \tag{8.55}$$

to cancel the UV divergence in (6.117), with the result that the real-emission contribution for the renormalized density $\epsilon = 0$ is

$$\frac{g^2}{16\pi^2} f_{q/q}^{[1,R]}(\xi) = \frac{g^2}{16\pi^2} \left\{ (1-\xi) \ln \left[\frac{\mu^2}{\xi m^2 + (1-\xi)^2 M^2} \right] + \frac{\xi(1-\xi)(4M^2 - m^2)}{\xi m^2 + (1-\xi)^2 M^2} \right\}. \tag{8.56}$$

Renormalization of quark in quark

The renormalization coefficient times Z_2 is therefore

$$(Z_2 Z)_{qq}(z, g, \epsilon) = \delta(z - 1) - \frac{g^2 S_\epsilon}{16\pi^2 \epsilon}(1 - z) + \ldots, \tag{8.57}$$

so that

$$Z_{qq}(z, g, \epsilon) = \delta(z - 1) + \frac{g^2 S_\epsilon}{16\pi^2 \epsilon}\left[\tfrac{1}{2}\delta(1 - z) - 1 + z\right] + \ldots \tag{8.58}$$

It is easily verified at order g^2 that this obeys the sum rule $\int_0^{1+} dz\ Z_{qq}(z) = 1$, as is necessary so that the number sum rule is obeyed. From (8.31) and (8.33) then follows the one-loop qq term in the DGLAP kernel:

$$P_{qq}(z) = \frac{g^2}{16\pi^2}\left[-\tfrac{1}{2}\delta(1 - z) + 1 - z\right] + \ldots \tag{8.59}$$

Scalar in quark

Similarly we can renormalize the first off-diagonal term, in the distribution of a scalar parton in a quark from (6.118). The renormalization coefficient and the DGLAP kernel are

$$Z_{\phi q}(z) = -\frac{g^2 S_\epsilon}{16\pi^2 \epsilon}z + \ldots, \tag{8.60}$$

$$P_{\phi q}(z) = \frac{g^2}{16\pi^2}z + \ldots, \tag{8.61}$$

with a corresponding renormalized value for $f_{\phi/q}$.

Verification of sum rules

It is readily checked that the quark number and momentum sum rules are obeyed at this order:

$$\int_0^{1+} dz\left[P_{qq}(z) - P_{\bar{q}q}(z)\right] = 0, \tag{8.62}$$

$$\int_0^{1+} dz\, z\left[P_{qq}(z) + P_{\phi q}(z) + P_{\bar{q}q}(z)\right] = 0. \tag{8.63}$$

Note that these sum rules are written in their complete form, including a term for evolution of a quark to an antiquark $P_{\bar{q}q}$. Of course this last term is zero at one-loop order; the lowest order in which the $q \to \bar{q}$ occurs is order g^4, from the graphs of Fig. 8.8.

Support properties

The continuum terms in all the above calculations of $Z_{jj'}$ and $P_{jj'}$ should be considered to have an implicit theta function to restrict z to lie between zero and one: $\theta(0 \leq z \leq 1)$.

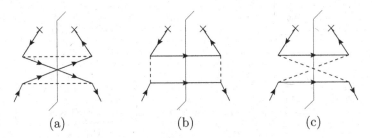

Fig. 8.8. Lowest-order graphs, order g^4, for evolution of quark to antiquark.

8.8 Successive approximation method

I now outline an approach that creates a factorization formula like (8.5) as a series of successive approximations, with the parton model as the first term. This will motivate the technical proof, and will suggest a route for generalization in more complicated situations.

The parton model for the hadronic tensor $W^{\mu\nu}$ for electromagnetic DIS was derived from the handbag diagram as an approximation valid in the momentum region where the struck quark is collinear to the target. We call this the leading-order (LO) approximation to the $W^{\mu\nu}$, notated in Fig. 6.4(b). The graph and region continue to exist in the complete theory. Of course, the approximation breaks down when the transverse momentum or virtuality of the struck quark gets large, and there are graphs other than the handbag diagram. Let us regard the complete $W^{\mu\nu}$ as the LO approximation plus a remainder:

$$W^{\mu\nu} = W^{\mu\nu}_{(LO)} + \left(W^{\mu\nu} - W^{\mu\nu}_{(LO)} \right)$$

$$= \; \vcenter{\hbox{\includegraphics{fig1}}} \; + \; \left(\vcenter{\hbox{\includegraphics{fig2}}} - \vcenter{\hbox{\includegraphics{fig3}}} \right). \qquad (8.64)$$

The hooks on the quark line of momentum k in the first term denote a parton-model approximator. This means that k^- and $\mathbf{k}_T$ are replaced by zero in the part of the diagram above the hook, and that projectors onto the leading power of the Dirac algebra are inserted. The result is a good approximation in the collinear region. We define the approximator to include an integral over all k, thereby obtaining a parton density, exactly as we defined it. Although not explicitly notated, we define the parton density to be renormalized, so that the LO approximation is finite. The unrestricted integral over k and the associated renormalization are the only changes from the parton approximator defined in Sec. 8.2.2.

We now analyze the remainder term, in parentheses. The most general leading-power contributions still have the form summarized in Fig. 8.2(a). However, if we take the hard-scattering subgraph to be lowest order, i.e., to be the top rung only, then in the parenthesized term in (8.64) this lowest-order case no longer gives a leading-power contribution, precisely because the subtraction cancels the relevant region.

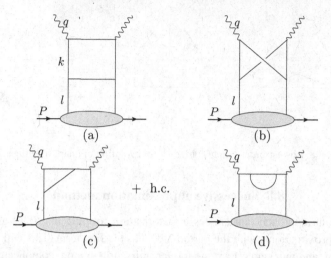

Fig. 8.9. Topologies of graphs needed for NLO approximation. The hermitian conjugate of graph (c) is also needed. UV counterterms are added to (c) and (d), as appropriate for the interaction and the current.

Not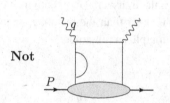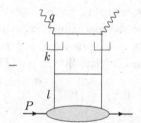

Fig. 8.10. Graphs like this with self-energy corrections are in in the handbag category, and are not used in Fig. 8.9.

Fig. 8.11. Subtraction graph.

For the leading approximation to the remainder term, we examine graphs of the form of Fig. 8.9. At the bottom, we have a complete parton-target amplitude, and at the top, we have a one-loop quantity. We are concerned with the case that the top loop is hard and the lower bubble is target-collinear. There is a sum over the flavors of the lines of the graphs. Notice that graph (a) is also among those included in the basic handbag diagram. Since the lower bubble represents an infinite sum over all graphs with the given external lines, it continues to represent the same quantity as in the handbag diagram. We do not include the case that there is a self-energy on the vertical parton lines, as in Fig. 8.10: these are included in the handbag category, for this part of the argument. To obtain the contribution to the parenthesized term in (8.64), we must subtract the parton-model approximation to graph (a), as symbolized in Fig. 8.11.

The graphs of Fig. 8.9 all have leading-power contributions when the momentum l of the line from the lower bubble to the upper one-loop subgraph, is collinear to the target. Contributions when l is larger will be dealt with in even higher-order corrections to the hard scattering. The first graph (a) also has a leading-power contribution when the line k is target-collinear. But the subtraction, Fig. 8.11, cancels this contribution (to the leading power of k_T/Q). Thus the upper one-loop subgraph in all cases is dominated by large loop momenta.

We therefore apply a parton-model approximation on the line l, and obtain the following form for the NLO contribution to the structure tensor:

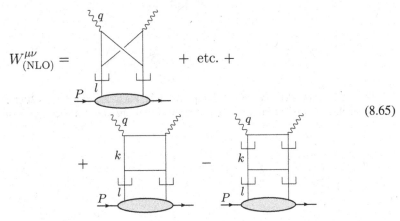

$$(8.65)$$

The lower part again is a parton density, which we define to be renormalized. The definition of the approximator is that, in the upper part of the graph, l is replaced by just its plus component: $l \mapsto (l^+, 0, \mathbf{0}_T)$, with appropriate Dirac-algebra projectors. Thus the upper factor is essentially the one-loop approximation for DIS on an on-shell parton of longitudinal momentum l^+. But there is a subtraction, to remove whatever was already taken care of at LO.

Further improvements can be made simply by iterating the procedure. In place of (8.64) we use

$$W^{\mu\nu} = W^{\mu\nu}_{(\text{LO})} + W^{\mu\nu}_{(\text{NLO})} + \left(W^{\mu\nu} - W^{\mu\nu}_{(\text{LO})} - W^{\mu\nu}_{(\text{NLO})} \right), \tag{8.66}$$

from which we obtain a further parton-model-like correction by analyzing the parenthesized term. This is the next-to-next-to-leading order (NNLO) approximation to DIS. Repeating the above procedure leads to a series of successive approximations that in fact correspond to an expansion in powers of $\alpha_s(Q)$.

8.9 Derivation of factorization by ladder method

We now make a complete derivation (Collins, 1998a) of factorization by using a decomposition in terms of 2PI subgraphs just as we did in Sec. 8.3.6 to discuss renormalization of parton densities.

8.9.1 Ladder expansion

The ladder decomposition is shown in Fig. 8.12, where B at the base of the ladders and K for the rungs are the same as in Fig. 8.6. There are two new features. The first is that

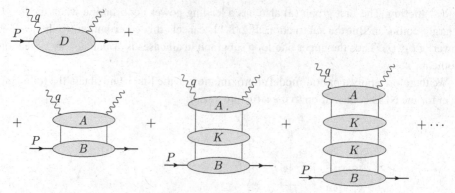

Fig. 8.12. Ladder decomposition of graphs for DIS. Each shaded bubble is 2PI in the vertical channel, except that K and B include the two full propagators on their upper side.

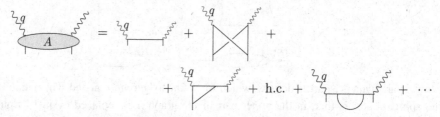

Fig. 8.13. Examples of topologies of graphs for the top A of ladder graphs for DIS in Fig. 8.12. The lines and vertices are of any type allowed by the theory. The shortness of the lines at the lower end indicates that these propagators are defined to be amputated.

because each current has two partonic lines we can have completely 2PI graphs. Their sum we denote by D, and these graphs are power-suppressed in Q because they have no decomposition of the generalized ladder form. The second new feature is that at the upper end of the ladder graphs we have, not a vertex for a parton density, but the sum A of 2PI graphs with two currents. Its expansion up to one-loop order is shown in Fig. 8.13.

Therefore we write a structure function (or the hadronic tensor $W^{\mu\nu}$) as

$$W = A \frac{1}{1-K} B + D, \qquad (8.67)$$

with exactly the same notation as in (8.15). The factor connected to the current has the functional dependence $A = A(q; k, j, c, c', \alpha, \alpha')$, where k is the momentum of the parton on the lower side of A, j is its flavor, and c, c', α and α' are indices for the color and spin of the parton, c and α on the left, and c' and α' on the right of the final-state cut.

8.9.2 Application of parton-model approximator

The proof of factorization generalizes to all orders the method of successive approximation of Sec. 8.8. Its implementation is by an algebraic method using the parton approximator

$\overleftarrow{T}$ defined in (8.7) and (8.8), and the pole-part extractor $\overleftarrow{P}$ used in Sec. 8.3 in the renormalization of parton densities.

To explain the algebraic method, I first apply it to low-order terms in the method of successive approximation, but applying it to the ladder sum (8.67). The first term is obtained by applying the parton approximator at the lower end of A:

$$W_{\text{ELO}} = A\overleftarrow{T}\,|V\frac{1}{1 - K(1 - \overleftarrow{P})}B. \tag{8.68}$$

The parton approximator is applied to the complete top rung of the ladder, i.e., to A, rather than just to the lowest-order rung. So to label the resulting approximation, I use "ELO" for "extended leading order" rather than just "LO". Unlike the use of the parton approximator in Sec. 8.2, there are no longer any restrictions on the internal momentum of any of the factors. But the parton densities are renormalized. This is accomplished by replacing the $1/(1 - K)$ factor by $1/(1 - K(1 - \overleftarrow{P}))$ as already derived for the renormalization of parton densities. To use this definition, we require that the pole-part operation is only applied within the parton density, i.e., only between the $|$ symbol and the $\overleftarrow{P}$ symbol. The reason for the emphasizing this is that the hard part $A\overleftarrow{T}\,|$ can have (finite) dependence on the UV regulator, which should not affect the pole-part operation; the pole-part operation is concerned only with defining the parton density, i.e., only with the objects to the right of the $|$ symbol.

The W_{ELO} term correctly treats the region where the parton below the A bubble is collinear. So in the remainder $W - W_{\text{ELO}}$, this region is suppressed. Therefore the region giving the first leading contribution to $W - W_{\text{ELO}}$ is where the hard subgraph consists of both A and one neighboring rung K. To obtain the associated contribution, we exhibit this first rung by writing the $1/(1 - K)$ factor as

$$\frac{1}{1 - K} = 1 + K\frac{1}{1 - K}. \tag{8.69}$$

Then the contribution in question is

$$W_{\text{ENLO}} = \left[AK - A\overleftarrow{T}\,|VK(1 - \overleftarrow{P})\right]\overleftarrow{T}\,|V\frac{1}{1 - K(1 - \overleftarrow{P})}B$$

$$= \left[A(1 - \overleftarrow{T}\,|V)K + \text{c.t.}\right]\overleftarrow{T}\,|V\frac{1}{1 - K(1 - \overleftarrow{P})}B. \tag{8.70}$$

This is of the form of an ENLO coefficient convoluted with a complete renormalized parton density. The factor of $1 - \overleftarrow{T}\,|V$ between A and the first rung K suppresses the collinear region for the connecting momentum. A UV counterterm removes the UV divergence that is thereby introduced.

8.9.3 General case

The organization of the full proof is first to construct what we call the remainder, in which all leading behavior is subtracted out, and then to show that this remainder is the difference between the exact hadronic tensor W and a factorized form.

Remainder

The remainder is defined by the insertion of a $1 - \overleftarrow{T}\,|V$ factor between each rung in (8.67):

$$r = \sum_{n=0}^{\infty} A\,(1 - \overleftarrow{T}\,|V)\left[K(1 - \overleftarrow{T}\,|V)\right]^n B + D$$

$$= A\,\frac{1}{1 - (1 - \overleftarrow{T}\,|V)K}\,(1 - \overleftarrow{T}\,|V)\,B + D$$

$$= A\,(1 - \overleftarrow{T}\,|V)\,\frac{1}{1 - K(1 - \overleftarrow{T}\,|V)}\,B + D. \tag{8.71}$$

We now show that this is power suppressed. We also show that there are no extra UV divergences, unlike the case in (8.68) and (8.70), so that no UV subtractions need to be applied.

Before inserting $1 - \overleftarrow{T}\,|V$, we recall that leading-power contributions come from regions symbolized in Fig. 8.2(a). Thus inserting $\overleftarrow{T}\,|V$ between the hard and collinear subgraphs gives a good approximation in this region. Hence, inserting a factor $1 - \overleftarrow{T}\,|V$ gives a power suppression. In the general case, where we extend the loop-momentum integrations out of the core of the region, the factor $1 - \overleftarrow{T}\,|V$ gives a suppression which we can represent as

$$\left(\frac{\text{highest virtuality in collinear}}{\text{lowest virtuality in hard}}\right)^p. \tag{8.72}$$

Furthermore, in the rung A, closest to the virtual photon, we have virtualities of order Q^2, while in the rung B, closest to the target, we have virtualities of order M^2. Within a given rung, the leading-power contribution comes where all the lines have comparable virtualities, since leading-power contributions with regions of very different virtualities involve the structure of Fig. 8.2(a), with subgraphs connected by just two lines. Given that in (8.71) we have a factor $1 - \overleftarrow{T}\,|V$ between every 2PI rung, there is a suppression whenever there is a strong decrease of virtuality in going from one rung to its neighbor to the right. Thus we find the ladder part of (8.71) has an overall suppression of order

$$\left(\frac{M}{Q}\right)^p, \tag{8.73}$$

when it is compared to the structure function itself. The 2PI term D is power-suppressed by itself, and thus the whole of r is power-suppressed, as appropriate for what we wish to consider as a remainder.

This suppression of course gets degraded as one goes to higher order for the rungs, since the lines within K can have somewhat different virtualities. The larger a graph we have for K, the wider the range of virtualities we can have without meeting a significant suppression.

A potential problem arises because $\overleftarrow{T}$ removes kinematic restrictions and thereby allows UV divergences to be induced, just as in the lowest-order approximation, (8.68). However, the UV divergences arise from the same kind of two-particle reducible structures as the leading regions, and the $1 - \overleftarrow{T}|V$ factors in r are just as effective at canceling the UV-divergence regions as they are at canceling leading-power contributions. Thus in fact r is finite and power suppressed. The UV divergences, with their attendant renormalization, only need to be treated when we expand the products.

Factorized form for $W - r$

I now show that $W - r$ factorizes. To present the algebra cleanly, I will first present the proof without renormalization of the parton densities, in the UV-regulated theory.

From (8.67) and (8.71), we find

$$
W - r = A \left[\frac{1}{1 - K} - \frac{1}{1 - (1 - \overleftarrow{T}|V)K} (1 - \overleftarrow{T}|V) \right] B
$$

$$
= A \frac{1}{1 - (1 - \overleftarrow{T}|V)K} \left[1 - (1 - \overleftarrow{T}|V)K - (1 - \overleftarrow{T}|V)(1 - K) \right] \frac{1}{1 - K} B
$$

$$
= A \frac{1}{1 - (1 - \overleftarrow{T}|V)K} \overleftarrow{T} | V \frac{1}{1 - K} B. \tag{8.74}
$$

This proof looks like straightforward linear algebra. In fact, there is a subtlety that $\overleftarrow{T}$ is defined to set masses to zero on its left. The quotient $1/[1 - (1 - \overleftarrow{T}|V)K]$ is fundamentally defined as the infinite sum $\sum_{n=0}^{\infty}[(1 - \overleftarrow{T}|V)K]^n$, and the manipulations in (8.74) apply to this definition just as they do in ordinary linear algebra.

The last factor on the last line, $V[1/(1 - K)]B$ is exactly a bare parton density, so we see that $W - r$ is of the form of some coefficient convoluted with a parton density. This is a form of factorization, so we write

$$
W = \sum_j \int_{x-}^{1+} \frac{d\xi}{\xi} C_{B,j}(Q/\mu, \xi/x) f_{B,j}(\xi; \mu)
$$

$$
+ \text{ terms with polarized parton densities} + \text{power-suppressed}
$$

$$
= C_B \otimes f_B + \text{polarized terms} + \text{p.s.c.} \tag{8.75}
$$

Here, "p.s.c." denotes "power-suppressed correction", and we have defined a parton density by

$$
f_{B,j}(\xi) = V \frac{1}{1 - K} B, \tag{8.76}
$$

when the parton at V has flavor j and $k^+ = \xi P^+$. For simplicity, we only indicate explicitly the term with unpolarized densities; the polarized terms are similar in structure. The coefficient function is

$$C_{B,j}(Q/\mu, \xi/x) = A \frac{1}{1 - (1 - \overleftarrow{T}|V)K} \overleftarrow{T}. \tag{8.77}$$

We use the "$\otimes$" notation to indicate a convolution in ξ and a sum over parton flavor, defined by the structure on the first line of (8.75).

We have one remaining complication, that of UV divergences. There are divergences in the parton density factor and in the coefficient function. Of course, these divergences cancel, since the left-hand side of (8.74) is finite, as we have already proved. As a first step, let us apply a UV regulator, e.g., dimensional regularization. We have defined all the rung factors as Green functions with renormalized fields. Thus the parton density $f_{B,j}(\xi)$ used in the above equations is a factor $1/Z_j$ times the bare parton density defined in terms of bare fields.

We now reorganize the (8.74) in terms of UV-finite quantities. From earlier work we know that the renormalized parton density is the convolution of a renormalization factor with the parton density f_B

$$f = G \otimes f_B. \tag{8.78}$$

So we simply define the renormalized coefficient function to be

$$C = C_B \otimes G^{-1}, \tag{8.79}$$

where the inverse in G^{-1} is in the sense of convolutions over ξ and matrix multiplication for parton flavor. Then, trivially $C_B \otimes f_B = C \otimes f$, and the factorization theorem becomes

$$W^{\mu\nu} = C^{\mu\nu} \otimes f + \text{p.s.c.}$$

$$= \sum_j \int_{x-}^{1+} \frac{d\xi}{\xi} C_j^{\mu\nu}(Q/\mu, \xi, x) f_j(\xi; \mu) + \text{polarized terms} + \text{p.s.c.} \tag{8.80}$$

8.10 Factorization formula for structure functions

In this section, we will convert the general structure of factorization, (8.80), into several forms directly suitable for practical calculations, to be carried out in Ch. 9. The formulae are also true in QCD, although their proof needs the enhancements to be given in Ch. 11. So the treatment will be presented with reference to its QCD applications.

8.10.1 Factorization for hadronic tensor

Polarization dependence appears in the trace over spin indices between the parton density and the hard-scattering factor. Exactly as in the parton model, Sec. 6.1, polarization can be allowed for by introducing a helicity density matrix $\rho_j(\xi)$ for the parton initiating the hard scattering. Then factorization of the hadronic tensor has the form:

$$W^{\mu\nu} = \sum_j \int_{x-}^{1+} \frac{d\xi}{\xi} \, \mathrm{Tr}\, C_j^{\mu\nu}(q, \xi P; \alpha_s, \mu) \, \rho_j(\xi; \mu) \, f_j(\xi; \mu) + \text{p.s.c.}$$

$$= C^{\mu\nu} \otimes \rho f + \text{p.s.c.} \tag{8.81}$$

With a slight change of notation, the hard-scattering coefficient, $C_j^{\mu\nu}$, has acquired helicity indices, and is traced with the partonic helicity density matrix. It is to be thought of as giving DIS on a parton target of flavor j and fractional longitudinal momentum ξ. There is a sum over all parton flavors j and an integral over all kinematically accessible ξ. A convenient notation for the integral over ξ, the sum over j and the trace with ρ is the convolution symbol $\otimes$ in the last line.

As explained in Sec. 6.5, the combination of $\rho_j f_j$ can be written in terms of the unpolarized densities f_j and asymmetry densities Δf_j and $\delta_T f_j$ for helicity and transversity, for the case of a spin-$\frac{1}{2}$ target. (A generalization is needed for higher spin targets like the deuteron.)

We express $C_j^{\mu\nu}$ in terms of scalar coefficient functions $\hat{F}_{ij}$ by relations like those for the regular structure functions, (2.20), except for the use of the momentum of the struck (massless) parton instead of the momentum of the target hadron:

$$\mathrm{Tr}\, C_j^{\mu\nu} \rho_j = \left(-g^{\mu\nu} + q^\mu q^\nu / q^2\right) \hat{F}_{1j}(x/\xi, Q^2)$$

$$+ \frac{(\xi \hat{P}^\mu - q^\mu \xi \hat{P} \cdot q/q^2)(\xi \hat{P}^\nu - q^\nu \xi \hat{P} \cdot q/q^2)}{\xi \hat{P} \cdot q} \hat{F}_{2j}(x/\xi, Q^2)$$

$$+ i\epsilon^{\mu\nu\alpha\beta} \frac{q_\alpha S_{j,\beta}}{\hat{P} \cdot q} \hat{g}_{1j}(x, Q^2) + F_3 \text{ term} + \text{extra gluon term.} \tag{8.82}$$

Here $\hat{P} = (P^+, 0, \mathbf{0}_T)$ is a massless projection of the target momentum, so that $\hat{k} \overset{\text{def}}{=} \xi \hat{P}$ is the momentum of the struck parton, in the approximation that is used in the hard scattering. An exact transcription of (2.20) would also include a $\hat{g}_2$ structure function associated with transverse quark spin. We omit it since $\hat{g}_2$ is zero to all orders of perturbation theory (Sec. 8.10.5). Therefore we need only the longitudinal polarization of the parton, and we assign it a spin vector $S_{j,\mu} = \lambda_j \hat{k}_\mu$, where λ_j is the parton's helicity. This is used with the $\hat{g}_1$ structure function.

In QCD, the gluon has spin 1, and when the hadronic target has spin greater than $\frac{1}{2}$, there is a possible term in the gluon's density matrix that flips helicity by two units: see Artru and Mekhfi (1990) and problem 7.11. This results in the "extra gluon term" in (8.82). I have left it as a (probably academic) exercise, to sort out the details (problem 8.3).

8.10.2 Factorization for structure functions

To get factorization formulae for the structure functions, we insert (8.82) in the factorization formula (8.81). Then we use the results from Sec. 6.5 that a parton in an unpolarized target is itself unpolarized and that its helicity is proportional to the target helicity. These results were derived in a simple model theory, but they depend only on symmetry properties of the theory, and are therefore generally true. Hence

$$F_1 = \sum_j \int_{x-}^{1+} \frac{d\xi}{\xi} \hat{F}_{1j}(Q/\mu, x/\xi; \alpha_s) f_j(\xi; \mu) + \text{p.s.c.}, \tag{8.83a}$$

$$F_2 = \sum_j \int_{x-}^{1+} d\xi \, \hat{F}_{2j}(Q/\mu, x/\xi; \alpha_s) f_j(\xi; \mu) + \text{p.s.c.}, \tag{8.83b}$$

$$g_1 = \sum_j \int_{x-}^{1+} \frac{d\xi}{\xi} \hat{g}_{1j}(Q/\mu, x/\xi; \alpha_s) \Delta f_j(\xi; \mu) + \text{p.s.c.} \tag{8.83c}$$

The second formula also applies to the longitudinal structure function $F_L \overset{\text{def}}{=} F_2 - 2xF_1$. Notice that:

- F_1 and F_2 only involve the unpolarized number densities;
- g_1 only involves the helicity asymmetry density;
- in the formula for F_2 the integration measure is $d\xi$ instead of $d\xi/\xi$;
- the coefficients are functions of x/ξ, rather than ξ and x separately;
- the transversity density $\delta_T f_j$ does not appear;
- the structure function g_2 does not have a formula. As we will see, its contribution to $W^{\mu\nu}$ is power suppressed, and therefore its leading-power approximation is zero.

The first two items depend on the parity invariance of the theory. In a parity non-invariant theory, it would be possible, for example, for partons to be polarized even when the parent hadron is unpolarized. We now give derivations of the other items.

8.10.3 Integration measure for F_2

The changed integration measure for F_2 is associated with its transformation under boosts of the target momentum. In the hadronic tensor (2.20), it multiplies the tensor $(P^\mu - q^\mu P \cdot q/q^2)(P^\nu - q^\nu P \cdot q/q^2)/P \cdot q$, which is linear in P. Now the coefficient function depends only on the momenta $\xi \hat{P}$ and q, but not on $\hat{P}$ or ξ separately. Then in the part associated with the F_2 structure function, there appears the tensor $(\xi \hat{P}^\mu - q^\mu \xi \hat{P} \cdot q/q^2)(\xi \hat{P}^\nu - q^\nu \xi \hat{P} \cdot q/q^2)/\xi \hat{P} \cdot q$, which scales linearly with ξ. To obtain the correctly normalized structure function F_2, we extract the factor ξ, which cancels the $1/\xi$ in the integration measure in (8.80). (There is further slight mismatch between the tensors, by a factor $1 + x^2 M^2/Q^2$, which is irrelevant to leading power in Q.)

8.10.4 Functional dependence of partonic structure functions

Both the hadronic tensor $W^{\mu\nu}$ and its hard-scattering counterpart $C_j^{\mu\nu}$ are dimensionless. Each of the partonic structure functions in (8.82) is also dimensionless, and the tensors multiplying them are independent of Q^2. Power-counting in a renormalizable theory therefore shows that order-by-order in perturbation theory all these quantities behave like Q^0 times logarithms of Q.

Each of the partonic structure functions in (8.82) is a Lorentz scalar, so the only kinematic variables it depends on are the invariants constructed out of its external momenta, i.e., Q^2 and $\xi \hat{P} \cdot q = \xi Q^2/(2x)$. The structure functions are also dimensionless. Therefore their independent arguments can be taken as Q/μ and x/ξ.

8.10.5 Transverse polarization

For the polarized structure functions, we first examine their scaling properties. In the Breit frame, the proton is highly boosted, so we count its momentum P as of order Q. When it has a longitudinal polarization λ, the spin vector S scales approximately as P, so that S is of order Q also. The tensor $i\epsilon^{\mu\nu\alpha\beta} q_\alpha S_\beta / P \cdot q$ associated with g_1 therefore scales as the zeroth power of Q, just like the tensors associated with F_1 and F_2. But the tensor multiplying g_2 has the longitudinal part subtracted, in the $S_\beta - P_\beta S \cdot q/P \cdot q$ factor; it is suppressed in fact by order M^2/Q^2 for longitudinal polarization. Thus to leading power, for longitudinal polarization, we have a contribution to g_1 times its tensor, and this is proportional to the longitudinal polarization λ of the target. Correspondingly, the factorization formula (8.83c) for g_1 uses the helicity parton density Δf.

There remains the case of transverse spin, and associated with it the transversity distributions $\delta_T f$. First we observe that the transverse components of the spin vector are invariant under boosts in the z direction. For this case, the tensors multiplying both of g_1 and g_2 are of order M/Q.

Now the only way transverse-spin dependence enters into the factorization (8.80) is through the transversity density, and thus through a transverse polarization for quarks entering the hard scattering and the coefficient function. But we set masses to zero in the hard scattering, and as we now show, there is then exactly zero contribution from transverse quark polarization. (As shown in Sec. 7.5.5, rotation invariance prohibits a gluon distribution that is transverse-spin dependent.)

In the case of the lowest-order calculation, in Sec. 6.1.4, the reason for the zero contribution of transverse spin is quite elementary. In the parton-model hard scattering (6.19), spin dependence arises from the factor $\not{k}(1 - \gamma_5\lambda_j - \gamma_5 b^i_{jT}\gamma^i)$. The transverse-spin dependent term, with b_{jT}, gives a trace of an odd number of elementary Dirac matrices which is always zero. (Recall that $\gamma_5 = i\gamma^0\gamma^1\gamma^2\gamma^3$ so that it counts as four elementary Dirac matrices.)

The same property generalizes to higher order. This is particularly clear in QCD. Let us go around the quark loop in which the struck quark is involved. There is an equal number of propagator numerators and vertices for gluons and photons. Except for the external line factor, each vertex and propagator numerator contains one Dirac matrix, giving a total number that is even. (This is where the masslessness of the calculation enters.) This is modified only on the external line factor with its extra odd number of Dirac matrices. Thus we get zero for the transverse spin dependence, as claimed.

The presence of subtractions in the hard scattering [see (8.77)] does not affect this argument. The subtractions involve kinematic approximants and the insertion of spinor projection matrices $\mathcal{P}_A$ and $\mathcal{P}_B$. The spinor projections each have two elementary Dirac matrices, so that they leave unchanged the evenness or oddness of the number of Dirac matrices.

With couplings to a scalar field, as in a Yukawa theory, there is no Dirac matrix at the scalar vertex. Thus we can get an even number of Dirac matrices in the trace with a transversely polarized quark provided that we have an odd number of scalar vertices on the quark line. But for the leading power in Q, we must keep only those interactions with a dimensionless coupling. All such couplings (in a four-dimensional theory) involve an even number of scalar fields, as in a ϕ^4 coupling or an interaction between a scalar field and a gauge field. If there is an odd number of scalar vertices on the quark loop including the external line, then some other quark loop also has an odd number of scalar vertices. This other loop has no transverse polarization matrix, and therefore an odd number of Dirac matrices, and therefore its Dirac trace vanishes.

The result is that in all cases the coefficient function with the transversity distribution is zero at the leading power of Q. Now transverse-spin dependence of the hard scattering arises from off-diagonal terms in the helicity density matrix. So the result on g_2 can be expressed by saying that in the hard scattering there is helicity conservation, i.e., there is no interference between a left-handed quark and a right-handed quark:

$$= 0. \tag{8.84}$$

Note that helicity is defined in only at space-time dimension 4. But our derivation used only the evenness or oddness of the number of Dirac matrices along quark lines, so the derivation applies without an anomaly when we use dimensional regularization in calculations.

Discussion of g_2 and of transverse-spin dependence in fully inclusive DIS therefore requires us to go beyond the leading power of Q, in fact to twist-3 operator contributions in the jargon of the subject. This is beyond the subject matter of this book. Unlike the case for the unpolarized and helicity parton densities, DIS is not a good place to measure the transversity density.

The whole of the above discussion assumed the target had spin $\frac{1}{2}$, in which case the target's spin state is completely specified by the spin vector S^μ. More general cases, notably spin 1, as for a deuteron target, can be discussed. But the results are of mostly lesser interest.

8.11 Transverse-spin dependence at leading power?

An interesting line of research over the past two decades has found useful observables that depend on transverse spin at the leading power. In this section, we give a general

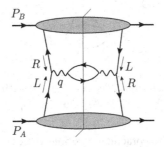

Fig. 8.14. At leading power, LO Drell-Yan has a double-transverse spin asymmetry from amplitudes such as this. Both hadrons are transversely polarized.

characterization of these observables. See, e.g., Boer (2008) for a detailed review, and see Sec. 13.16 for examples.

The whole discussion is conditioned by chirality conservation in the massless limit and hence in the hard scattering. Chirality conservation is the correct generalization of helicity conservation when we include antiquarks; it means the helicity of a quark and the negative of helicity for an antiquark. Thus the vertices of gauge bosons couple a left-handed quark to a left-handed quark or to a right-handed antiquark, but not to a right-handed quark or a left-handed antiquark.

There are two ways of getting dependence on transverse spin. One is to find a more general hard scattering that has off-diagonal helicity dependence. The other is to find parton-density-like objects with more general spin dependence than ordinary parton densities.

8.11.1 Hard scattering with transverse spin

Transverse spin gives one unit of helicity flip in a parton density, and this must be matched in the hard scattering to get a leading-power effect. To avoid violating chirality conservation, we need a hard scattering with (at least) another pair of external quark lines, so that we have two compensating helicity flips (Artru and Mekhfi, 1990). Such processes are needed to measure transversity densities.

One possibility is in hadron-hadron collisions, where the hard scattering is initiated by two partons, one out of each hadron (to be treated in detail in Ch. 14). A classic example is the Drell-Yan process, Sec. 5.3.7, where the lowest-order hard scattering is quark-antiquark annihilation to a virtual photon. If both initial-state hadrons are transversely polarized, then (Ralston and Soper, 1979) we can have a leading-power double-spin asymmetry, as shown in Fig. 8.14.

Another similar possibility is in semi-inclusive DIS, where the cross section is differential in a final-state hadron. In Ch. 12, we will generalize factorization to include a fragmentation function that parameterizes the conversion of an outgoing quark to a jet containing the detected hadron(s). Then the interference diagram Fig. 6.2, which gave zero in ordinary DIS, gets a fragmentation function inserted into it, Fig. 8.15. The fragmentation function needs to be off-diagonal in helicity for our purposes. It could be that the outgoing hadron

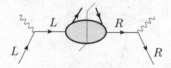

Fig. 8.15. Interference between left-handed and right-handed initial quark in DIS with the fragmentation providing the necessary helicity flip.

has its polarization measured; a practical example (Efremov, 1978; Artru and Mekhfi, 1990) is production of the Λ^0, whose decay allows its polarization to be measured. In addition, since fragmentation is non-perturbative, the chiral symmetry breaking of full QCD allows the fragmentation function to break chirality conservation while keeping leading-power behavior (Collins, Heppelmann, and Ladinsky, 1994), provided a suitable final-state distribution is measured.

8.11.2 Transverse-momentum-dependent densities, etc.

Finally, some reactions require the use of transverse-momentum-dependent (TMD) parton densities (and/or fragmentation functions). As we will see in Ch. 13, a TMD number density can have a correlation between the azimuthal angle of a parton and transverse spin of the target. Thus at leading power, we can have dependence on the transverse spin of a target hadron without needing transverse-spin dependence in the hard scattering.

A considerable number of variations on this idea exist, especially when fragmentation functions are included (Boer, 2008).

Exercises

8.1 (****) In a renormalizable theory, it is natural to define the light-front creation and annihilation operators by Fourier transformation of the renormalized fields instead of bare fields, since it is the renormalized fields that have finite Green functions. For a field with wave function renormalization factor Z, the commutation relations of the creation and annihilation operators are enhanced by a factor $1/Z$, which is infinite unless the anomalous dimension of the field is zero at the UV fixed point. This messes up the normalizations of the basis states (7.23) by an infinite amount, in the limit that the UV cutoff is removed.

Find a good way of specifying basis states in the renormalized theory in the limit that the UV cutoff is removed. What is the relation between these states and the standard basis states in the cutoff theory? [Conjectures and suggestions: 1. Some of the techniques used in treating factorization later in this book may be useful. 2. Fourier-transforming at fixed x^+ corresponds to maximal uncertainty on k^-. It may help to perform a local average over x^+. 3. Useful references include: Yamawaki (1998); Nakanishi and Yamawaki (1977); Heinzl (2003); Sec. 4 of Heinzl and Ilderton (2007); Nakanishi and Yabuki (1977); Steinhardt (1980).]

8.2 (****) Find the relation between parton densities and the basis found in problem 8.1.

8.3 (***) *Extension of problem 6.8 to full QCD:* Generalize the work in this chapter to deal with DIS on a polarized spin-1 target like the deuteron. What is the form of the extra gluonic term indicated in (8.82)? What is the corresponding NLO hard-scattering coefficient corresponding to this extra term? Notes:

- Much of the necessary work on defining structure functions has been done by Hoodbhoy, Jaffe, and Manohar (1989). But it is good to check their results. Note that they used the OPE rather than factorization for their QCD analysis. But they restricted their attention to the quark operators, and did not indicate what to do with gluon operators.

- Since the gluon has spin 1, their analysis definitely needs generalization to deal with a gluon-induced hard scattering. You will need to work out a version of their analysis to the hard-scattering coefficient for a gluon $C_g^{\mu\nu}$. This will result in significant changes, since there are no gluons of helicity zero. Hoodbhoy, Jaffe, and Manohar (1989) also normalized the polarization vector E^μ of a spin-1 particle of mass M to $E^2 = -M^2$, which is clearly a bad idea for a massless particle.

- You should find another polarized gluon density related to linear gluon polarization (so that its operator gives a helicity flip of 2 units); see Artru and Mekhfi (1990).

- You should match the results of this problem with your solution of problem 7.11 and the results in Artru and Mekhfi (1990).

- In the light of the above, you may find better characterizations of the structure functions on a spin-1 target.

- I do not guarantee the phenomenological importance of the results of solving this problem.

9

Corrections to the parton model in QCD

In Ch. 8, factorization was formulated for DIS. The proofs were, however, restricted to non-gauge theory. But the results remain true in QCD, with some complications to be treated in Ch. 11.

So in this chapter we will simply assume factorization holds in QCD, and on that basis introduce methods of applying it phenomenologically. In QCD, with an unpolarized target, we will calculate: (a) the first correction terms to the hard scattering for DIS, and (b) the leading term in the kernel for DGLAP evolution of quark and gluon densities. These are the primary phenomenological tools for quantitatively analyzing DIS in QCD.

The calculations also provide an opportunity to introduce some of the complications that arise in QCD and that must be taken into account in a correct proof of factorization.

The results on which this chapter depends are: factorization for the hadronic tensor, (8.81); factorization for the structure functions (8.83); the decomposition of the partonic hard scattering tensor in terms of parton structure functions (8.82); the definition of parton densities in QCD in Sec. 7.5; the structure of their renormalization (8.11); the corresponding DGLAP evolution equations, from Sec. 8.4.

9.1 Lowest order

The parton-model calculation in (2.28) gives the first terms in the expansion of the partonic structure functions in powers of α_s:

$$\hat{F}_{1j}(Q^2, x/\xi; \alpha_s, \mu) = \frac{e_j^2}{2} \delta(x/\xi - 1) + O(\alpha_s), \qquad (9.1a)$$

$$\hat{F}_{2j}(Q^2, x/\xi; \alpha_s, \mu) = e_j^2 \delta(x/\xi - 1) + O(\alpha_s), \qquad (9.1b)$$

and of course $\hat{F}_{jL} = 0 + O(\alpha_s)$. These are the lowest-order (LO) terms, and they apply to quarks; the gluonic coefficients start at order α_s.

9.2 Projections onto structure functions

In Feynman-graph calculations we will use projectors of a hadronic or partonic tensor onto corresponding structure functions. In the partonic case these follow simply from (8.82). It is convenient to use the longitudinal structure function:

$$\hat{F}_{Lj} \overset{\text{def}}{=} \hat{F}_{2j} - 2\frac{x}{\xi}\hat{F}_{1j} = \frac{8(x/\xi)^3}{Q^2}\hat{k}_\mu\frac{1}{2}\operatorname{Tr}C_j^{\mu\nu}\hat{k}_\nu, \tag{9.2a}$$

$$\hat{F}_{2j} = \frac{x/\xi}{1-\epsilon}\left(-g_{\mu\nu}\frac{1}{2}\operatorname{Tr}C_j^{\mu\nu}\right) + \frac{3-2\epsilon}{2-2\epsilon}\hat{F}_{jL}, \tag{9.2b}$$

where we give the result for a general space-time dimension $4 - 2\epsilon$, as needed later. The factor $\frac{1}{2}$ Tr projects onto the partonic tensor for an unpolarized parton.

9.3 Complications in QCD

9.3.1 Use of on-shell quarks and gluons

It would be possible to obtain hard-scattering coefficients and DGLAP kernels from direct use of the subtractive methods of Ch. 8. Instead we use a method where we start from calculations of structure functions and parton densities with massless quarks and gluons used as the target states.

Now starting from calculations of structure functions and parton densities on some set of target states, we can use the factorization and renormalization formulae to deduce the hard-scattering coefficient functions and the renormalization factors (of parton densities). From the renormalization factors, we deduce the DGLAP kernels. It is the coefficient functions and the DGLAP kernels that are of actual phenomenological interest, since they are perturbative.

Because these quantities are independent of the target state, we are entitled to use whatever targets are convenient for calculations. This leads us to use single on-shell quarks and gluons as the target states, with all calculations done in low-order perturbation theory. Moreover, the quantities to be calculated are independent of mass, so we also set masses to zero everywhere, since this considerably simplifies calculations of Feynman graphs.

Thus a noteworthy feature of many QCD calculations is that they use on-shell quarks and gluons as the target state. This is in striking contrast to the fact that (as far as is currently known) all true particle states in QCD are composites, i.e., bound states like the proton. Moreover there are IR and collinear divergences in perturbative calculations with on-shell massless target states. These can be regulated satisfactorily and cancel in the calculations of the coefficients, which are all short-distance dominated.

9.3.2 Choice of gauge

Another complication in QCD concerns the choice of gauge. We could use $A^+ = 0$ gauge, in which case the structure of the leading regions, for renormalization and for factorization, appears to be simplified to be the same as in a non-gauge theory (Ch. 8). However, calculations are plagued by divergences associated with the $1/k^+$ singularity in the gluon propagator. The divergences cancel, but in a non-trivial manner. This of course indicates that extensions are needed for the proofs of factorization and renormalization that we gave in Secs. 8.3.6 and 8.9.

The alternative, which we will adopt here, is to use Feynman gauge (or a standard covariant gauge). The necessary proofs will come later. For the purposes of calculations, we simply rely on the full statement of renormalization (and factorization) applied with gauge-invariant parton densities. We will in fact still find extra divergences, characterized as rapidity divergences. We will see that the rapidity divergences cancel, non-trivially. The Feynman gauge lends itself better to good derivations of renormalization and factorization than the $A^+ = 0$ gauge.

It is interesting that there was a long-standing disagreement for calculations at two-loop order for the DGLAP kernels. This was between a calculation in light-cone gauge (Furmanski and Petronzio, 1980), and ones in Feynman gauge (Floratos, Ross, and Sachrajda, 1979; Gonzalez-Arroyo and Lopez, 1980; Floratos, Lacaze, and Kounnas, 1981). It turned out that the light-cone gauge calculation is the correct one. The actual calculations are done with massless quarks and gluons; one has a choice between on-shell calculations and off-shell calculations. As we will see, on-shell calculations are much easier algorithmically, but suffer from various kinds of IR and collinear divergence that need to be disentangled from the UV divergences of interest. Off-shell, there are extra parton-density-like objects defined by operators other than the gauge-invariant ones needed in physical matrix elements. A subtle interaction between the IR problems and the non-gauge-invariant operators needed to be sorted out (Hamberg and van Neerven, 1992; Collins and Scalise, 1994), over a decade later than the original calculations. See Sec. 11.4 for some more details.

These problems will not affect our one-loop calculations.

9.4 One-loop renormalization calculations in QCD

In this section, we calculate the one-loop renormalization of the parton densities in QCD, starting from the definitions (7.40) and (7.43) for the bare parton densities. Then we will deduce one-loop values for the DGLAP kernels, which are phenomenologically very important in determining the evolution of parton densities with scale. The results are also essential to calculations of the hard-scattering coefficient functions.

9.4.1 General principles of calculation

Just as in our calculations in Yukawa theory, Sec. 8.7, we work with target states that are in turn a gluon or any flavor of quark. The primary new feature is that each parton density has a Wilson line, for which the Feynman rules were given in Figs. 7.10–7.12. The renormalization coefficients are adjusted so that the renormalized parton densities defined by (8.11) have no UV divergences. The general notation for the expansions in α_s was given in (8.49), and the relation between the n-loop expansion of the bare and renormalized parton densities was given in (8.52).

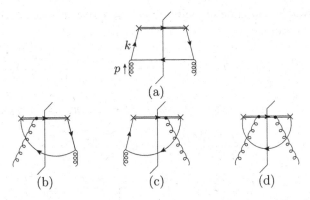

Fig. 9.1. (a) One-loop graph for renormalization of density of quark in gluon. (b)–(d) Graphs that are zero when the gluon polarization is chosen to obey $e_p^+ = 0$.

At one loop this is simple, because of the trivial zero-loop terms (8.50) and (8.51) for the renormalization and the parton densities. The factorized form for renormalization thus shows that the one-loop renormalized parton density in a quark or gluon target is the sum of the one-loop bare parton density and the one-loop renormalization coefficient:

$$f_{j/k}^{[1]}(\xi) = (Z_{2j}^{-1} f)_{(0)j/k}^{[1]}(\xi) + (Z_{2j} Z)_{jk}^{[1]}(\xi, g, \epsilon)$$

$$= (Z_{2j}^{-1} f)_{(0)j/k}^{[1]}(\xi) + Z_{2j}^{[1]} \delta_{jk} \delta(\xi - 1) + Z_{jk}^{[1]}(\xi, g, \epsilon). \qquad (9.3)$$

To obtain this, we wrote the bare parton density as $Z_{2j}(Z_{2j}^{-1} f_{(0)j/k})$, where Z_{2j} is the wave function renormalization for the field for parton j. Then we separated out the one-loop terms for the Z_{2j} and for $(Z_{2j}^{-1} f_{(0)j/k})$. The reason is that $(Z_{2j}^{-1} f_{(0)j/k})$ is the parton density defined with renormalized fields instead of bare fields, so that it is a natural object to compute in perturbation theory.

We now apply the above formula to each possibility for j and k.

9.4.2 Quark in gluon

The simplest calculation is for the order g^2 off-diagonal gluon-to-quark term, i.e., in (9.3) we set k to a gluon and j to any quark flavor. The target state is a on-shell gluon with a physical polarization vector e_p^μ that has zero plus and minus components. The single graph we need is shown in Fig. 9.1(a). Since $e_p^+ = 0$, graphs (b)–(d), in which the gluon attaches to the Wilson line, are zero. [Generally the polarization vector of a on-shell gluon (or photon) of momentum p must obey $p \cdot e_p = 0$, and $e_p \cdot e_p^* = -1$. It is arbitrary up to a gauge transformation, i.e., up to the addition of a multiple of p. The choice of a gauge condition on the polarization vector may be made separately for each on-shell gluon. We have chosen the condition $e_p^+ = 0$.]

A straightforward application of the Feynman rules gives the value of the bare graph (before renormalization):

$$
\frac{g^2}{16\pi^2} f^{[1]}_{(0),q/g}(\xi) = -T_F g^2 \mu^{2\epsilon} \int \frac{dk^- \, d^{2-2\epsilon} k_{\mathrm{T}}}{(2\pi)^{4-2\epsilon}} \frac{2\pi \delta\big((p-k)^2 - m_q^2\big)}{(k^2 - m_q^2)^2}
$$

$$
\times \operatorname{Tr} \frac{\gamma^+}{2} (\slashed{k} + m_q)\slashed{\epsilon}_p (\slashed{k} - \slashed{p} + m_q)\slashed{\epsilon}_p^* (\slashed{k} + m_q).
$$

$$
= \frac{g^2 T_F (4\pi \mu^2)^\epsilon}{8\pi^2 \Gamma(1-\epsilon)} \int_0^\infty dk_{\mathrm{T}}^2 \frac{k_{\mathrm{T}}^{-2\epsilon}}{(k_{\mathrm{T}}^2 + m_q^2)^2}
$$

$$
\times \left\{ (k_{\mathrm{T}}^2 + m_q^2)\left[1 - \frac{2\xi(1-\xi)}{1-\epsilon} \right] + m_q^2 \frac{2\xi(1-\xi)}{1-\epsilon} \right\}
$$

$$
= \frac{g^2 T_F}{8\pi^2} \left(\frac{m_q^2}{4\pi \mu^2} \right)^{-\epsilon} \Gamma(\epsilon)\left[(1-\xi)^2 + \xi^2 \right]. \tag{9.4}
$$

The overall minus sign in the first line arises because of the fermion loop. For information about T_F and other group theory coefficients, see Sec. A.11. The dependence on the direction of the polarization vector has dropped out because of invariance under rotations around the z axis. Unlike the case of our later calculations we have kept a non-zero mass.

The renormalization counterterm $Z^{[1]}_{qg}$ in (9.3) is added to give a finite result at $\epsilon = 0$. In the $\overline{\mathrm{MS}}$ scheme

$$
\frac{g^2}{16\pi^2} Z^{[1]}_{qg}(z) = -\frac{g^2 T_F}{8\pi^2} \frac{S_\epsilon}{\epsilon} \left[(1-z)^2 + z^2 \right]. \tag{9.5}
$$

From the QCD version of (8.33), the corresponding term in the DGLAP kernel is

$$
\frac{g^2}{16\pi^2} P^{[1]}_{qg}(z) = \frac{g^2 T_F}{8\pi^2} \left[(1-z)^2 + z^2 \right]. \tag{9.6}
$$

To this order the finite renormalized density of a quark in a gluon is

$$
\frac{g^2}{16\pi^2} f^{[1]}_{q/g}(\xi) = \frac{g^2 T_F}{8\pi^2} \left[(1-\xi)^2 + \xi^2 \right] \ln \frac{\mu^2}{m_q^2}. \tag{9.7}
$$

This calculation, with its non-zero quark mass, will appear as a subtraction component in calculations of hard-scattering coefficients for heavy quark production. But the $\overline{\mathrm{MS}}$ renormalization coefficient is independent of mass, so its calculation can equally well be performed with a zero quark mass. Moreover hard-scattering calculations, which we will examine later, are considerably simplified when masses are neglected with respect to the hard scale Q. So we now examine what happens when we set $m_q = 0$. The bare graph's integral is now

$$
\frac{g^2 T_F (4\pi \mu^2)^\epsilon}{8\pi^2 \Gamma(1-\epsilon)} \int_0^\infty dk_{\mathrm{T}}^2 \frac{k_{\mathrm{T}}^{-2\epsilon}}{k_{\mathrm{T}}^2} \left[1 - \frac{2\xi(1-\xi)}{1-\epsilon} \right]. \tag{9.8}
$$

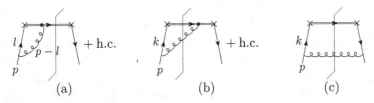

Fig. 9.2. One-loop graphs for renormalization of density of quark in quark. Hermitian conjugates of (a) and (b) should be added. As explained in the text, graphs with a quark self-energy graph need not be considered explicitly, and graphs where the gluon connects the Wilson line to itself are zero.

The integral is of a simple power of k_T, which is elementary compared to (9.4), with its beta function. However, the integral has an extra divergence at $k_T = 0$. This is a collinear divergence, since it happens when the quark and antiquark are parallel to the gluon. Dimensional regularization regulates both the UV and the collinear divergence, but only by going in opposite directions in ϵ. Even so, such integrals can be consistently defined (e.g., Collins, 1984, Ch. 4) and it is a theorem that integrals of a power of the integration variable are zero in dimensional regularization. Thus the collinear and UV divergences are equal and opposite. The UV pole can be obtained by examining the part of the integral in (9.8) from a non-zero value of k_T to infinity. Then the renormalized value of the graph is the negative of the UV pole:

$$\frac{g^2}{16\pi^2} f_{q/g}^{[1]}(\xi; m = 0) = \frac{g^2}{16\pi^2} f_{(0)q/g}^{[1]}(\xi; m = 0) + \frac{g^2}{16\pi^2} Z_{qg}^{[1]}(z)$$

$$= 0 - \frac{g^2 T_F}{8\pi^2} \frac{S_\epsilon}{\epsilon} \left[(1 - \xi)^2 + \xi^2 \right]. \tag{9.9}$$

That the renormalized value is collinear divergent reflects the masslessness of both the quark and the gluon, and that the asymptotic scattering states do not obey the standard rules. Of course, neither the massless limit (for quarks) nor the existence of an isolated gluon (or quark) is a feature of real QCD. As already stated, such massless calculations are useful as components of calculations of hard-scattering coefficients, for which the massless limit does exist, as we will verify explicitly. Thus the existence of a collinear (or other kind of IR) divergence in a renormalized partonic matrix element is not a fundamental problem.

9.4.3 Quark in quark

We next apply the same principles to the density of a quark in a quark, for which the one-loop graphs are shown in Fig. 9.2, with virtual gluon emission in graph (a) and real gluon emission in graphs (b) and (c). There is, in principle, a term where both ends of the gluon attach to the Wilson line. But as we will review below, this term is effectively zero.

We do not include a self-energy correction for the incoming quark, since its renormalization is done by a counterterm in the Lagrangian. Indirectly its effects will appear, in the renormalization factor of the parton density, because of the Z_2 term in (9.3).

Gluon polarization sum

In the graphs with real gluon emission, we use a physical gluonic final state, so that the sum over gluon polarizations, is a sum over physical (transverse) polarizations for the gluon. However, very generally, the sum over physical final states can be extended to a sum over all final states (including when necessary ghost-antighost pairs, which will not concern us here). This is shown in field theory textbooks (e.g., Ch. 11 of Sterman, 1993) under the heading of "Unitarity of the S-matrix". Thus we may replace the sum over transverse gluon polarizations in Figs. 9.2(b) and (c) by the same numerator $-g^{\alpha\beta}$ that appears in the Feynman-gauge gluon propagator. Since $g^{++} = 0$, graphs where both ends of the gluon attach to the Wilson line are zero, so we omit these graphs.

The proof at the level of the emission of one gluon of momentum l goes as follows. Representatives of physical polarizations obey $l \cdot e = 0$, and it is easy to check that the polarization sum obeys

$$\sum_{\text{phys. pols.}} e^{\alpha}(e^{\beta})^* = -g^{\alpha\beta} + l^{\alpha}b^{\beta} + b^{\alpha}l^{\beta} = -g^{\alpha\beta} + \text{terms giving zero by WI}, \quad (9.10)$$

where b is some vector. The terms with a factor l give zero by a Ward identity, after a sum over graphs.

Virtual correction

The virtual gluon correction in Fig. 9.2(a) (*with its hermitian conjugate*) gives

$$\frac{g^2}{16\pi^2} f^{(a+a^\dagger)}_{(0),q/q}(\xi) = 2\delta(p^+ - \xi p^+)\frac{-ig^2 C_F \mu^{2\epsilon}}{(2\pi)^{4-2\epsilon}}$$

$$\times \int d^{4-2\epsilon}l \, \frac{\text{Tr} \, \frac{\gamma^+}{2}\slashed{l}\gamma^+\frac{\slashed{p}}{2}}{(l^2+i0)[(p-l)^2+i0](-p^+ + l^+ + i0)}$$

$$= -\delta(1-\xi)\frac{g^2 C_F(4\pi\mu^2)^\epsilon}{4\pi^2\Gamma(1-\epsilon)}\int_0^1 d\alpha \, \frac{\alpha}{1-\alpha}\int_0^\infty dl_{\mathrm{T}}^2 \, \frac{l_{\mathrm{T}}^{-2\epsilon}}{l_{\mathrm{T}}^2}, \quad (9.11)$$

where $\alpha = l^+/p^+$. The missing steps are to express the integral in light-front coordinates, and then to perform the l^- integral by contour methods. We have chosen to do the calculation with all masses set to zero. As before, the transverse-momentum integral is of the scale-free kind that gives zero. The negative of the UV divergence gives the graph's contribution to the renormalization:

$$\frac{g^2}{16\pi^2}(Z_2 Z)^{(a+a^\dagger)}_{qq}(z, g, \epsilon) = \frac{g^2 C_F S_\epsilon}{4\pi^2\epsilon}\delta(1-z)\int_0^1 d\alpha \, \frac{\alpha}{1-\alpha}. \quad (9.12)$$

Notice that we have now explicitly needed to show the factor of Z_2 in the renormalization factor.

An important new feature is that there is an *un*regulated divergence in the integral over α at $\alpha = 1$. We will see that the divergence cancels against a similar divergence in graph (b), but it is first worth examining the source of the divergence. There are multiple sources of divergence in the integral in the last line of (9.11), and they each have a different status for our ultimate phenomenological uses of the results of our calculations. So we need to make their nature apparent. We first insert non-zero quark and gluon masses, m_q and m_g, in the calculation to regulate with the IR problems. It is readily checked that the effect is to replace the $1/l_T^2$ factor in (9.11) by

$$\frac{1}{l_T^2 + m_g^2\alpha + m_q^2(1 - \alpha)^2}. \tag{9.13}$$

Now, when the gauge symmetry is non-abelian, as in QCD, a non-zero gluon mass is not allowed. However, to understand the divergences we temporarily consider the same calculation in an abelian theory, where a non-zero gauge boson mass can be used.

With the non-zero masses, there is no longer a divergence at $l_T = 0$, but we still have a divergence at $\alpha \to 1$. Relative to the simpler parton densities which we calculated earlier, the $1/(1 - \alpha)$ singularity arises from the Wilson-line denominator. After a contour deformation, the divergence occurs when the $(+, -, \text{T})$ components of the gluon momentum are of order $((1 - \alpha)p^+, l_T^2/((1 - \alpha)p^+), l_T)$, for fixed l_T. The rapidity of the gluon goes to $-\infty$; the gluon can in fact be regarded as collinear to the Wilson line, which has rapidity $y = \frac{1}{2}\ln(n^+/n^-) = -\infty$. The quark goes far off-shell here.

So we call the divergence at $\alpha = 1$ a rapidity divergence. The region evidently has nothing to do with the parton-model physics that a parton density is supposed to capture. When we investigate transverse-momentum-dependent parton densities, we will need to use a Wilson line with a finite rapidity to get an appropriate definition with no rapidity divergence. But for an integrated density we will see a cancellation.

Notice from the denominator in (9.13) that if the gluon mass is zero, there is in addition a divergence at $l_T = 0$ and $\alpha = 1$. This is just like the IR divergence in QED. Finally, if also the quark mass is zero, there is also a divergence when the gluon is collinear to the initial state (at $l_T = 0$ and $\alpha \neq 0, 1$).

Real correction, first part

Figure 9.2(b) plus its hermitian conjugate give

$$\frac{g^2}{16\pi^2} f_{(0),q/q}^{(b+b^\dagger)}(\xi) = 2\frac{-g^2 C_F \mu^{2\epsilon}}{(2\pi)^{4-2\epsilon}} \int d^{2-2\epsilon} k_T \, dk^- \, 2\pi \delta((p - k)^2)\frac{\text{Tr} \frac{\gamma^+}{2}\slashed{k}\gamma^+\frac{\slashed{p}}{2}}{k^2(p^+ - k^+)}$$

$$= \frac{g^2 C_F(4\pi\mu^2)^\epsilon}{4\pi^2\Gamma(1 - \epsilon)}\frac{\xi}{1 - \xi}\int_0^\infty dk_T^2 \frac{k_T^{-2\epsilon}}{k_T^2}, \tag{9.14}$$

The minus sign in the first line arises from the gluon numerator, which is $-g^{\alpha\beta}$ in accordance with the discussion around (9.10). Notice that this formula is almost the same as the integrand for the virtual correction, which comes from a graph related by moving the final-state cut. In fact, we can get the virtual term from the above formula by: (1) changing ξ to α and integrating over it; (2) changing the label of the transverse momentum; (3) inserting a delta function; (4) reversing the sign. If we integrated over ξ (from 0 to 1 of course), there would be a perfect cancellation.

The corresponding contribution to the renormalization is

$$\frac{g^2}{16\pi^2} Z_{qq}^{(b+b^\dagger)}(z, g, \epsilon) = -\frac{g^2 C_F S_\epsilon}{4\pi^2 \epsilon} \frac{z}{1-z}. \tag{9.15}$$

9.4.4 Cancellation of divergence: the plus distribution

All of the quantities involved – parton densities, renormalization factors, DGLAP kernels – have rapidity divergences in individual graphs. For a systematic treatment, we must regard all of these quantities not as ordinary functions, but as a generalized functions. That is, they only have numerical values when integrated with a smooth test function. After this, we will see a cancellation of the rapidity divergences.

So we integrate the sum of graphs (a) and (b) (plus conjugates) with a smooth function $T(\xi)$, to obtain

$$\frac{g^2}{16\pi^2} f_{(0),q/q}^{(a+b+h.c.)}[T] \stackrel{\text{def}}{=} \frac{g^2}{16\pi^2} \int d\xi \, f_{(0),q/q}^{(a+b+h.c.)}(\xi) \, T(\xi)$$

$$= \frac{g^2 C_F (4\pi \mu^2)^\epsilon}{4\pi^2 \Gamma(1-\epsilon)} \int_0^1 d\xi \, \frac{\xi[T(\xi) - T(1)]}{1-\xi} \int_0^\infty dk_T^2 \, \frac{k_T^{-2\epsilon}}{k_T^2}. \tag{9.16}$$

To obtain the contribution from the virtual graph, we used the $\delta(\xi - 1)$ factor to perform the ξ integral, and then changed the name of the variable α to ξ. The divergence at $\xi \to 1$ has now canceled.

To express these graphs directly in ξ space, it is convenient to define the so-called plus distribution:

$$\int_0^1 dx \left(\frac{1}{1-x}\right)_+ T(x) \stackrel{\text{def}}{=} \int_0^1 dx \, \frac{T(x) - T(1)}{1-x}. \tag{9.17}$$

We will often meet this distribution multiplied by polynomials in ξ, in which case we will put the $+$ subscript on the denominator:

$$\int_0^1 dx \, \frac{A(x)}{(1-x)_+} T(x) \stackrel{\text{def}}{=} \int_0^1 dx \, \frac{A(x)T(x) - A(1)T(1)}{1-x}. \tag{9.18}$$

Then the combination we need in the sum of graphs is

$$\int_0^1 d\xi \, \frac{[T(\xi) - T(1)]\xi}{1-\xi} = \int_0^1 d\xi \left[\frac{\xi T(\xi) - T(1)}{1-\xi} + T(1)\right]$$

$$= \int_0^{1+} d\xi \left[\frac{\xi}{(1-\xi)_+} + \delta(\xi - 1)\right] T(\xi), \tag{9.19}$$

so that the sum of graphs (a) and (b) is

$$\frac{g^2}{16\pi^2} f_{(0),q/q}^{(a+b+h.c.)}(\xi) = \frac{g^2 C_F (4\pi \mu^2)^\epsilon}{4\pi^2 \Gamma(1-\epsilon)} \left[\frac{\xi}{(1-\xi)_+} + \delta(\xi - 1)\right] \int_0^\infty dk_T^2 \, \frac{k_T^{-2\epsilon}}{k_T^2}. \tag{9.20}$$

Real correction, second part

Figure 9.2(c) gives no such complications. Its value is

$$\frac{g^2}{16\pi^2} f^{(c)}_{(0),q/q}(\xi) = \frac{-g^2 C_F \mu^{2\epsilon}}{(2\pi)^{4-2\epsilon}} \int d^{2-2\epsilon} k_T \, dk^- \, 2\pi \delta((p-k)^2) \frac{\text{Tr} \frac{\gamma^+}{2} \slashed{k} \gamma^\mu \frac{\slashed{p}}{2} \gamma_\mu \slashed{k}}{(k^2)^2}$$

$$= \frac{g^2 C_F (4\pi \mu^2)^\epsilon}{8\pi^2 \Gamma(1-\epsilon)} (1-\xi)(1-\epsilon) \int_0^\infty dk_T^2 \frac{k_T^{-2\epsilon}}{k_T^2}. \tag{9.21}$$

Total one-loop value for renormalization and DGLAP kernel

We can now combine the UV divergences from the various graphs with the Z_2 term in (9.3), whose value is in (3.23). Then the one-loop renormalization of the quark density is

$$\frac{g^2}{16\pi^2} Z^{[1]}_{jk}(z; \text{quark}) = -\frac{g^2 C_F \delta_{jk}}{8\pi^2} \frac{S_\epsilon}{\epsilon} \left[\frac{1+z^2}{(1-z)_+} + \frac{3}{2} \delta(z-1) \right]. \tag{9.22}$$

From (8.31) and (8.33), the resulting DGLAP kernel is

$$\frac{g^2}{16\pi^2} P^{[1]}_{jk}(z; \text{quark}) = \frac{g^2 C_F \delta_{jk}}{8\pi^2} \left[\frac{1+z^2}{(1-z)_+} + \frac{3}{2} \delta(z-1) \right]$$

$$= \frac{g^2 C_F \delta_{jk}}{8\pi^2} \left[\frac{2}{(1-z)_+} - 1 - z + \frac{3}{2} \delta(z-1) \right]. \tag{9.23}$$

9.4.5 Gluon-in-gluon and gluon-in-quark

Similar calculations can be done for the case of a gluon in a gluon, and for a gluon in a quark. The actual calculations we leave as an exercise, with the results being (Altarelli and Parisi, 1977)

$$\frac{g^2}{16\pi^2} P^{[1]}_{gg}(z) = \frac{g^2}{8\pi^2} \left\{ 2C_A \left[\frac{z}{(1-z)_+} + \frac{1-z}{z} + z(1-z) \right] + \delta(z-1) \frac{11C_A - 4n_f T_R}{6} \right\}, \tag{9.24}$$

$$\frac{g^2}{16\pi^2} P^{[1]}_{gq}(z) = \frac{g^2 C_F}{8\pi^2} \left[\frac{1+(1-z)^2}{z} \right]. \tag{9.25}$$

9.5 One-loop renormalization by subtraction of asymptote

We saw in Sec. 3.4 that UV renormalization, at least at one-loop order, could be implemented by subtraction of the asymptotic large transverse-momentum asymptote of a Feynman graph. This enabled us to give a strictly four-dimensional interpretation of minimal subtraction.

In this section we show how to apply this method to the renormalization of parton densities. This will serve two aims. One is to show how to make a physically appropriate choice of the renormalization scale μ. The second aim concerns calculations of hard-scattering coefficients, which normally employ massless quarks and gluons. At intermediate stages of the calculations, collinear and soft divergences appear, which cancel in the final result. Generally dimensional regularization is used to regulate the divergences, but it is useful to show how to work with a purely four-dimensional integral. One virtue of this method is to allow the immediate use of the compendium of purely four-dimensional amplitudes in Gastmans and Wu (1990).

It is important that our results have extra finite counterterms compared with the illustrative example in Sec. 3.4.

9.5.1 *Quark in gluon*

The unsubtracted one-loop integral for the density of a quark in a gluon is (9.4). The renormalized value is given by adding an $\overline{\text{MS}}$ counterterm, obtained from the renormalization term (9.5) by substituting $z \mapsto \xi$. We write the counterterm as the integral over the asymptote of the original integrand plus a finite correction $R_{q/g}$, to be determined:

$$-\frac{g^2 T_F}{8\pi^2} \frac{S_\epsilon}{\epsilon}[1 - 2(1 - \xi)\xi]$$

$$= -\frac{(4\pi\mu^2)^\epsilon}{\Gamma(1 - \epsilon)} \frac{g^2 T_F}{8\pi^2} \int_{\mu^2}^\infty \frac{dk_{\mathrm{T}}^2}{(k_{\mathrm{T}}^2)^{1+\epsilon}} \left[1 - \frac{2(1 - \xi)\xi}{1 - \epsilon}\right] + R_{q/g}$$

$$= -\frac{g^2 T_F}{8\pi^2} \frac{(4\pi)^\epsilon}{\epsilon\Gamma(1 - \epsilon)} \left[1 - \frac{2(1 - \xi)\xi}{1 - \epsilon}\right] + R_{q/g}, \tag{9.26}$$

where S_ϵ is given in (A.41). Hence

$$R_{q/g} = -\frac{g^2 T_F}{8\pi^2} \frac{S_\epsilon}{1 - \epsilon} 2(1 - \xi)\xi \stackrel{\epsilon \to 0}{\to} -\frac{g^2 T_F}{8\pi^2} 2(1 - \xi)\xi. \tag{9.27}$$

Only the value of $R_{q/g}$ at $\epsilon = 0$ is needed in a purely four-dimensional formula.

With this method the renormalized density at $\epsilon = 0$ is

$$\frac{g^2}{16\pi^2} f_{q/g}^{[1]}(\xi) = \frac{g^2 T_F}{8\pi^2} \left\{ \int_0^\infty dk_{\mathrm{T}}^2 \left[\frac{1 - 2\xi(1 - \xi)}{k_{\mathrm{T}}^2 + m_q^2} + \frac{m_q^2 2\xi(1 - \xi)}{(k_{\mathrm{T}}^2 + m_q^2)^2} \right. \right.$$

$$\left. \left. - \theta(k_{\mathrm{T}} - \mu) \frac{1 - 2\xi(1 - \xi)}{k_{\mathrm{T}}^2} \right] - 2(1 - \xi)\xi \right\}. \tag{9.28}$$

It can be checked that this is the same as the previously calculated value (9.7), but the integrals are algorithmically simpler, because they do not involve the beta functions that arise with the dimensionally regulated integrals. Because of the extra term $2(1 - \xi)\xi$, it cannot be literally said that the integrated parton density is the integral of the unintegrated density with a cutoff at $k_{\mathrm{T}} = \mu$, even for large μ. This is contrary to statements that appear in the literature (e.g., Watt, Martin, and Ryskin, 2003).

9.5.2 *Other cases*

The remaining cases are left as an exercise (problem 9.3) with the results:

$$R_{g/q}(\epsilon = 0) = -\frac{g^2 C_F}{8\pi^2} 4\xi, \tag{9.29}$$

$$R_{q/q}(\epsilon = 0) = -\frac{g^2 C_F}{8\pi^2} 4(1 - \xi), \tag{9.30}$$

$$R_{g/g}(\epsilon = 0) = 0. \tag{9.31}$$

9.6 DIS on partonic target

To calculate the hard-scattering coefficients for DIS, we observe that the factorization theorem applies to any target state, while the coefficient functions $C^{\mu\nu}$ are target independent. Therefore we apply the factorization theorem in perturbation theory with targets that are on-shell quark or gluon states. Computing both the structure functions and the parton densities on partonic targets up to some order in perturbation theory enables us to deduce the hard-scattering coefficients to the same order. Moreover, since the coefficient functions are independent of masses, we will set masses to zero everywhere.

We organize perturbation expansions as we did for the renormalization of parton densities in Sec. 8.7.3. Define $W_j^{\mu\nu}$ to be the hadronic tensor for DIS with a massless on-shell partonic target of flavor j. We write perturbation expansions of $W_j^{\mu\nu}$ and $C_j^{\mu\nu}$ as

$$W_j^{\mu\nu}(x, Q) = \sum_{n=0}^{\infty} \left(\frac{g^2}{16\pi^2}\right)^n W_j^{[n], \mu\nu}(x, Q), \tag{9.32a}$$

$$C_j^{\mu\nu}(x, Q) = \sum_{n=0}^{\infty} \left(\frac{g^2}{16\pi^2}\right)^n C_j^{[n], \mu\nu}(x, Q). \tag{9.32b}$$

The nth order term in the factorization theorem (8.81) is

$$W_j^{[n], \mu\nu}(x, Q) = \sum_{n'=0}^{n} \sum_{j'} \int_{x-}^{1+} \frac{d\xi}{\xi} C_{j'}^{[n'], \mu\nu}(x/\xi, Q) \otimes f_{j'/j}^{[n-n']}(\xi). \tag{9.33}$$

Since masses are set to zero, the power-suppressed corrections in (8.81) are not present. Throughout our calculations we will work with the unpolarized case, so the partonic density matrix ρ is dropped.

We deduce a formula for the nth order hard-scattering coefficient:

$$C_j^{[n], \mu\nu}(z, Q) = W_j^{[n], \mu\nu}(z, Q) - \sum_{n'=0}^{n-1} \sum_{j'} \int_{z-}^{1+} \frac{d\zeta}{\zeta} C_{j'}^{[n'], \mu\nu}(z/\zeta, Q) f_{j'/j}^{[n-n']}(\zeta). \tag{9.34}$$

Here, to avoid confusion with symbols used when the coefficient function is substituted in the factorization formula (8.81) for a hadronic target, the names of partonic variables were changed to z and ζ. In the factorization formula, z would be replaced by x/ξ.

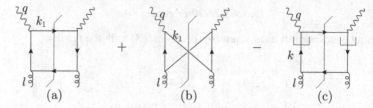

Fig. 9.3. Graphs for NLO gluon coefficient function for DIS. There are, in addition, three other graphs with the direction of the arrow on the quark loop reversed. The hooks on the quark lines in the subtraction graph (c) indicate where a parton-model approximation is made.

Equation (9.34) provides an effective recursive procedure for calculating the nth order term in C starting from the case $n = 0$, for which the result was given in (2.28), with corresponding structure functions in (9.1). At next-to-leading order (NLO) we have

$$C_j^{[1],\mu\nu}(z, Q) = W_j^{[1],\mu\nu}(z, Q) - \sum_{j'} \int_{z-}^{1+} \frac{d\zeta}{\zeta} C_{j'}^{[0],\mu\nu}(z/\zeta, Q) f_{j'/j}^{[1]}(\zeta). \qquad (9.35)$$

Our calculations in Sec. 9.4 of renormalized one-loop parton densities gave the values of $f_{j'/j}^{[1]}(\zeta)$.

Perturbation theory for W and f in massless QCD suffers from IR and collinear divergences. So the radius of convergence[1] in g for these quantities goes to zero as the IR regulator ϵ goes to zero. But this is sufficient to obtain the perturbation expansion of the hard-scattering coefficients C. Since divergences cancel in the coefficient functions, their radius of convergence remains non-zero as $\epsilon \to 0$.

9.7 Computation of NLO gluon coefficient function

Applied to the NLO gluon coefficient, (9.35) requires us to compute the graphs of Fig. 9.3. The external gluons are massless and on-shell, with zero transverse momentum, and the internal quarks are massless and have a sum over flavors. Figure 9.3(c) implements the subtraction in (9.35), and we will call it a double-counting-subtraction graph, since it cancels the contribution in the first two graphs that is taken into account in the lowest-order parton model.

9.7.1 Kinematics

Let k_1 and k_2 be the momenta of the final-state quark and antiquark, and let l be the momentum $(l^+, 0, \mathbf{0}_T)$ of the gluon, so that $k_2 = q + l - k_1$. The scalar kinematic variables

---

[1] Strictly speaking, perturbation series are expected to be asymptotic series but not convergent, so the term "radius of convergence" should be replaced by some better terminology concerning the region of coupling where perturbation theory has some chosen accuracy.

relevant to the problem are Q and

$$z = \frac{Q^2}{2l \cdot q} = \frac{-q^+}{l^+}, \tag{9.36a}$$

$$\hat{s} = (k_1 + k_2)^2 = \frac{Q^2(1-z)}{z}, \tag{9.36b}$$

$$\hat{t} = (l - k_2)^2 = -\frac{Q^2(1 + \cos\theta)}{2z}, \tag{9.36c}$$

$$\hat{u} = (l - k_1)^2 = -\frac{Q^2(1 - \cos\theta)}{2z}, \tag{9.36d}$$

where θ is the scattering angle in the photon-gluon center of mass. Of these variables, only three are independent, of course.

9.7.2 *Calculation of unsubtracted graphs*

Graph (a) of Fig. 9.3 gives

$$-\sum_j \frac{g^2 e_j^2 T_F}{32\pi^2} \left(\frac{16\pi^2 \mu^2}{\hat{s}}\right)^\epsilon \int \frac{d\Omega}{4\pi} \frac{\text{Tr} \, \not{k}_1 \gamma^\nu (\not{l} - \not{k}_2) \not{e}(-\not{k}_2) \not{e}^*(\not{l} - \not{k}_2) \gamma^\mu}{\left[(l - k_2)^2\right]^2}, \tag{9.37}$$

where $d\Omega$ represents the integration over the angle of the quarks in the photon-gluon center of mass, and e^μ is the (transverse) polarization vector of the gluon. The overall minus sign is for a fermion loop, and the normalization arises from the $1/(4\pi)$ in the definition of $W^{\mu\nu}$, and from two-body phase space (A.43). We choose the sum over j to be *over flavors of quark only (not over antiquarks)*. Then we must add, to this and the terms for the other graphs, the contribution with the quark line reversed; this is obtained simply by exchanging k_1 and k_2.

Similarly graph (b) gives

$$-\sum_j \frac{g^2 e_j^2 T_F}{32\pi^2} \left(\frac{16\pi^2 \mu^2}{\hat{s}}\right)^\epsilon \int \frac{d\Omega}{4\pi} \frac{\text{Tr} \, \not{k}_1 \gamma^\nu (\not{l} - \not{k}_2) \not{e}(-\not{k}_2) \gamma^\mu (\not{k}_1 - \not{l}) \not{e}^*}{(l - k_2)^2 (l - k_1)^2}. \tag{9.38}$$

We are only treating unpolarized processes, so we average over gluon polarizations:

$$\frac{1}{2 - 2\epsilon} \sum e^i (e^j)^* = \frac{\delta^{ij}}{2 - 2\epsilon}, \tag{9.39}$$

with a Kronecker delta in the transverse dimensions. Then we use standard Dirac algebra, and use (9.2) to project the sum of the terms for the two graphs onto the tensor structures for $\hat{F}_{Lg}$ and $\hat{F}_{2g}$. The integrands are now independent of the azimuthal direction of the quark momenta, so we use (A.36) and (A.37) to give

$$\hat{F}_{Lg} = \sum_j \frac{g^2 e_j^2 T_F}{4\pi^2} \left(\frac{16\pi \mu^2 z}{Q^2(1-z)}\right)^\epsilon \frac{2z^2(1-z)}{(1-\epsilon)\Gamma(1-\epsilon)} \int_{-1}^1 d\cos\theta \, (\sin\theta)^{-2\epsilon}$$

$$\stackrel{\epsilon \to 0}{\longrightarrow} \sum_j \frac{g^2 e_j^2 T_F}{4\pi^2} 4z^2(1-z), \tag{9.40}$$

$$\hat{F}_{2g} = \sum_j \frac{g^2 e_j^2 T_F}{4\pi^2} \left(\frac{16\pi\mu^2 z}{Q^2(1-z)}\right)^\epsilon \frac{z}{\Gamma(1-\epsilon)} \int_{-1}^{1} d\cos\theta \ (\sin\theta)^{-2\epsilon}$$

$$\times \left\{\frac{1}{\sin^2\theta}\left[1 - \frac{2z(1-z)}{1-\epsilon}\right] + \frac{-2+5\epsilon}{4(1-\epsilon)^2} + \frac{3-2\epsilon}{(1-\epsilon)^2}z(1-z)\right\}$$

$$- \text{term from graph (c)}, \tag{9.41}$$

up to higher-order corrections ($O(g^4)$). In $\hat{F}_{Lg}$, we have omitted the subtraction from graph (c), since that involves the lowest-order parton-model hard scattering, for which there is no contribution to F_L, with fermion quarks.

9.7.3 Double-counting-subtraction graph

The subtraction graph (c) is obtained from the rules for the quark density and the LO hard scattering, which contributes only to F_2. Using the integral from (9.4) at $m_q = 0$, we get

$$\hat{F}_{2g}(\text{graph (c)}) = -\sum_j \frac{g^2 e_j^2 T_F}{4\pi^2} \frac{(4\pi\mu^2)^\epsilon z}{\Gamma(1-\epsilon)} \int_0^\infty dk_T^2 \frac{k_T^{-2\epsilon}}{k_T^2}\left[1 - \frac{2z(1-z)}{1-\epsilon}\right]$$

$$+ \sum_j \frac{g^2 e_j^2 T_F S_\epsilon z}{4\pi^2\epsilon}[1 - 2z(1-z)], \tag{9.42}$$

where the second line is the $\overline{\text{MS}}$ counterterm for the UV divergence. As announced earlier, both of (9.41) and (9.42) are collinear divergent, at $\theta = 0$ and $\theta = \pi$, and at $k_T = 0$. Dimensional regularization with ϵ *negative* regulates the divergence. By making the change of variable $k_T^2 = (\hat{s}/4)\sin^2\theta$, we can see that the collinear singularities in the integrands are equal and opposite, and that the cancellation includes the explicit ϵ dependence. The cancellation is guaranteed by the construction of the subtraction term (c) to cancel the collinear contribution in the other graphs, to prevent double counting with the parton-model term. [When checking the cancellation, note that two values of θ correspond to a single value of k_T. Note also that the maximum value of k_T^2 for graphs (a) and (b) is $\hat{s}/4$, whereas the integral for graph (c) extends to $k_T = \infty$.]

9.7.4 Total

The $\cos\theta$ integral in (9.41) gives a beta function, with a pole at $\epsilon = 0$ caused by the collinear divergence. The k_T integral in (9.42) gives zero, leaving the UV counterterm. So we get the NLO gluonic coefficient function

$$\hat{F}_{2g}(Q^2, x/\xi; \alpha_s, \mu)$$

$$\overset{\epsilon=0}{=} \sum_j \frac{g^2 T_F e_j^2}{4\pi^2} z\left\{[(1-z)^2 + z^2]\ln\left[\frac{Q^2(1-z)}{\mu^2 z}\right] - 1 + 8z(1-z)\right\} + O(g^4). \tag{9.43}$$

There is a somewhat complicated pattern of divergences at $\epsilon = 0$, which can be summarized as follows:

Graph	Collinear	UV	total
(a)	−1	0	−1
(b)	0	0	0
(c) graph	+1	−1	0
(c) counterterm	0	+1	+1

where the coefficients apply to the factor $\sum_j [1 - 2z(1-z)] g^2 T_F e_j^2 / (4\pi^2 \epsilon)$. Since the transverse momentum integral in the subtraction term is exactly zero, it could be said that the $\overline{\text{MS}}$ counterterm cancels the collinear divergence. It is, in fact, a common misconception that this represents the true state of affairs. However, it is also profoundly misleading.

For example, suppose one retained the quark mass in the calculation, as might be appropriate for a quark of large mass. Then the collinear region would no longer give an actual divergence. Instead, graph (a) would be finite, but with a logarithmic enhancement from the region of small transverse momentum. Graph (c) (without its counterterm) would now be non-zero, with a UV divergence. The counterterm cancels the UV divergence. For the dominant part of the collinear contributions (that give divergences at $m_q = 0$) there is a cancellation between graphs (a) and (c). The collinear cancellation is guaranteed by the nature of the subtraction term: (c) is to prevent double counting of the parton-model contribution.

9.7.5 *Use of subtraction of asymptote for UV divergence*

We can also use the method of subtraction of the asymptote for the renormalization of the UV divergence, from Sec. 9.5. This gives

$$\hat{F}_{2g}(\text{NLO}) \overset{?}{=} \sum_j \frac{g^2 T_F e_j^2}{4\pi^2} z \int_{-1}^{1} d\cos\theta \left[\frac{1 - 2z + 2z^2}{\sin^2\theta} - \frac{1}{2} + 3z(1-z) \right]$$

$$+ \sum_j \frac{g^2 T_F e_j^2}{4\pi^2} z \left[2z(1-z) - \int_0^{\mu^2} \frac{dk_T^2}{k_T^2}(1 - 2z + 2z^2) \right], \qquad (9.44)$$

where the $2z(1-z)$ on the second line is from $R_{q/g}(z)$ in (9.27). Each integral is separately divergent, hence the query on the equality sign. To make the integrals correspond, we convert them to use a common variable $k_T^2 = (\hat{s}/4)\sin^2\theta$. Then

$$\hat{F}_{2g}(\text{NLO}) = \sum_j \frac{g^2 T_F e_j^2}{4\pi^2} z \left\{ (1 - 2z + 2z^2) \int_0^{\infty} \frac{dk_T^2}{k_T^2} \left[\frac{\theta(k_{T,\text{max}}^2 - k_T^2)}{\sqrt{1 - k_T^2/k_{T,\text{max}}^2}} - \theta(\mu^2 - k_T^2) \right] \right.$$

$$\left. - 1 + 8z(1-z) \right\}, \qquad (9.45)$$

where $k_{\mathrm{T,max}}^2 = Q^2(1-z)/(4z)$. It can be checked that this agrees with the previous result, (9.43). The advantage of this integral is that it is a fundamentally an integral in the physical space-time dimension. It also enables us to gauge the general order of magnitude of the coefficient.

9.8 Choice of renormalization scale μ

It is necessary to choose the renormalization scale μ when applying a factorization theorem. As can be seen from an example calculation, e.g., (9.43), hard-scattering coefficients depend logarithmically on Q/μ. The general situation follows from the DGLAP equation for the μ dependence of parton densities. Since structure functions are RG invariant, the hard-scattering coefficients obey an inverse DGLAP equation. It follows that at order α_s^n the hard-scattering coefficients have dependence on $\ln(Q/\mu)$ that is polynomial with a highest term $\ln^n(Q/\mu)$.

The effective expansion parameter of the hard scattering is therefore $\alpha_s(\mu)\ln(Q/\mu)$, and to make optimal use of perturbative calculations one should choose μ of order Q. Then the expansion parameter is $\alpha_s(Q)$.

However, we need more precise information about an appropriate value for the ratio μ/Q. To see that this is a non-trivial problem, consider a change of scheme for renormalizing QCD and the parton densities. A concrete example is to replace S_ϵ in the $\overline{\mathrm{MS}}$ scheme by $S_\epsilon e^{2c\epsilon}$ for some constant c. Call this the c scheme. It is related to the $\overline{\mathrm{MS}}$ scheme by a simple substitution: $\mu_{\overline{\mathrm{MS}}} = \mu_c e^c$, so that $\ln(\mu_{\overline{\mathrm{MS}}}/Q) = \ln(\mu_c/Q) + c$. Then if we set $\mu_c = Q$, the coefficients of the perturbative expansion are made arbitrarily large simply by making c large.

Evidently we can remove these large coefficients by setting μ_c to a suitable factor times Q, e.g., $\mu_c = Qe^{-c}$. But this provokes the question of what is so special about the $\overline{\mathrm{MS}}$ scheme that in this scheme one should choose equality of μ and Q (a common choice in practice).

An answer is suggested by the method of renormalization subtraction of the asymptote given in Sec. 9.5. We found that $\mu_{\overline{\mathrm{MS}}}$ is like a cutoff at $k_{\mathrm{T}} = \mu_{\overline{\mathrm{MS}}}$, rather than some factor times this.

The method was applied to a coefficient function in (9.45), where there is a subtraction of the collinear region (e.g., by Fig. 9.3(c)), and then a renormalization of the UV divergence in the subtraction. After that there remains only a contribution from transverse momenta of some natural scale associated with Q, provided that z is not close to 0 or 1, and provided that μ is at this same scale. So the integral is of order unity, and is multiplied by the standard prefactor $g^2/4\pi^2$, and a group theory factor. This justifies the choice that $\mu_{\overline{\mathrm{MS}}}$ is within a modest factor of Q.

If instead we used the c scheme, then Sec. 9.5 shows that an appropriate choice would now be $\mu_c = Qe^{-c}$. Naturally, there is no need to require exactly one particular value of μ. The exact value of a structure function (or cross section) is independent of μ. Changing μ by a factor of 2 (for example) in a finite-order calculation of the hard scattering changes the numerical value of a computed structure function by an amount corresponding to the expected truncation error of the perturbative calculation. Thus the effect of a modest change in μ is within the expected errors.

The simplest version of subtraction of the asymptote applies if there is no extra ϵ dependence in the integrand. If there is extra ϵ dependence, then it results in an extra finite term, as in the last line of (9.45). This can be regarded as being of a natural size for the quantity under consideration, so it does not affect arguments about large logarithms.

The idea that the cutoff should be of the natural size of the transverse momentum for a hard scattering (after subtraction of collinear and UV divergences) suggests that problems can occur when z is close to 0 or 1. This is visible in the logarithm of $(1 - z)/z$. An obvious choice of scale would then be $\mu^2 = Q^2(1 - z)/z$, corresponding to the range of the transverse-momentum integral.

However, in this case there are (at least) two very different physical scales in the hard scattering. Besides Q^2 there is the (square of) the photon-parton center-of-mass energy, $Q^2(1 - z)/z$. Even if we removed the large logarithm in this particular calculation, because it is dominated by the second scale, there would be other graphs with a natural scale Q. An example is the virtual vertex correction Fig. 9.4(d), in whose calculation the range of final-state energies is irrelevant. When different graphs need very different scales, a single choice of μ cannot eliminate all large logarithms. Instead improved factorization theorems are needed, for a genuinely fundamental solution of the problem.

When does this situation arise? Since $z = x/\xi$ and actual parton densities decrease with increasing ξ, one should not expect the case that z is small to be a concern. But when x gets large, the maximum $q\bar{q}$ mass is restricted: the kinematic limits on z are $x < z < 1$. This phenomenon is enhanced by the fact that typical parton densities fall rather rapidly with ξ above about a half, which disfavors the larger masses and keeps z close to unity.

This subject has been under active investigation, with improved factorization methods and resummation techniques being discovered. In any case the outcome is that when the typical value of z gets too close to unity, simple factorization is not an optimal technique.

9.9 NLO quark coefficient

To compute the NLO quark coefficient, we again use (9.35), but now with a quark target. The necessary graphs, including subtractions, are shown in Fig. 9.4. In all the calculations, we use (9.10) to replace the gluon polarization sum in the real-emission graphs by $-g^{\alpha\beta}$. Kinematics and normalization factors are the same as for the gluon-induced graphs (e.g., (9.36)) except for the replacement of the group theory factor T_F by C_F. We take the quark to be unpolarized, and perform the integral over azimuthal angles, using (A.36) and (A.37).

9.9.1 NLO quark coefficient for $\hat{F}_{Lj}$

The contribution to the longitudinal structure function is particularly simple. Because of the factors of l in the projection (9.2a) onto $\hat{F}_{Lj}$, graphs (b)–(e) all have a factor of l next

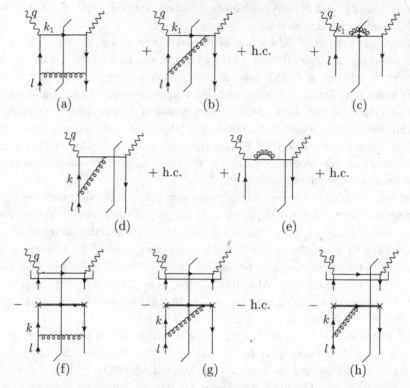

Fig. 9.4. Graphs for NLO quark coefficient function for DIS. Graphs with quark self-energies on the *incoming* quark line are not needed, since they cancel completely and exactly between the graphs for DIS on a quark target and the subtraction terms.

to the l factor for the incoming quark. Thus all these graphs give zero: $(l)^2 = l^2 = 0$. The subtraction graphs are also zero, because $\hat{F}_{Lj}$ vanishes in the parton model. All that remains is graph (a), which gives

$$\hat{F}_{Lj}((a)) = -\frac{g^2 e_j^2 C_F}{64\pi^2}\left(\frac{16\pi\mu^2}{\hat{s}}\right)^\epsilon \frac{1}{\Gamma(1-\epsilon)}\int_{-1}^1 d\cos\theta \, (\sin\theta)^{-2\epsilon}$$

$$\times \frac{8z^3}{Q^2}\frac{\frac{1}{2}\mathrm{Tr}\, l\gamma^\alpha(l - k_2)l k_1 l(l - k_2)\gamma_\alpha}{[(l - k_2)^2]^2}$$

$$\overset{\epsilon\to 0}{=} \frac{g^2 e_j^2 C_F z^2}{8\pi^2}\int_{-1}^1 d\cos\theta\,(1-\cos\theta). \tag{9.46}$$

This has no divergences, so the limit $\epsilon \to 0$ is safe, and we get

$$\hat{F}_{Lj} = \frac{g^2 e_j^2 C_F z^2}{4\pi^2} + O(g^4). \tag{9.47}$$

9.9.2 Real-gluon graphs for $\hat{F}_{2j}$

We apply (9.2b) to the real-gluon graphs for $\hat{F}_{2j}$. For graph (a):

$$\hat{F}_{2j}(a) = \frac{g^2 e_j^2 C_F}{8\pi^2} \left(\frac{16\pi\mu^2}{\hat{s}}\right)^\epsilon \frac{1}{\Gamma(1-\epsilon)} \int_{-1}^{1} d\cos\theta \ (\sin\theta)^{-2\epsilon}$$

$$\times \left\{\frac{z(1-z)(1-\epsilon)}{1+\cos\theta} + 4(3-2\epsilon)z^2(1-\cos\theta)\right\}. \qquad (9.48a)$$

The second part of the factor in braces arises from the $\hat{F}_L$ term in (9.2b). For graph (b), we have

$$\hat{F}_{2j}(b + \text{h.c.}) = \frac{g^2 e_j^2 C_F}{8\pi^2} \left(\frac{16\pi\mu^2}{\hat{s}}\right)^\epsilon \frac{z}{\Gamma(1-\epsilon)} \int_{-1}^{1} d\cos\theta \ (\sin\theta)^{-2\epsilon}$$

$$\times \left\{\frac{z}{1-z}\frac{1-\cos\theta}{1+\cos\theta} + \epsilon\right\}, \qquad (9.48b)$$

where we include a factor 2 to allow for the hermitian conjugate graph. For graph (c)

$$\hat{F}_{2j}(c) = \frac{g^2 e_j^2 C_F}{8\pi^2} \left(\frac{16\pi\mu^2}{\hat{s}}\right)^\epsilon \frac{1}{\Gamma(1-\epsilon)} \int_{-1}^{1} d\cos\theta \ (\sin\theta)^{-2\epsilon} \frac{z(1-\epsilon)}{1-z}(1+\cos\theta).$$

$$(9.48c)$$

Positions of the divergences

Graph (a) simply has a divergence at $\theta = \pi$, i.e., $\cos\theta = -1$. With the conventions by which the momentum k_2 is defined, this is where the gluon is collinear to the initial-state quark. Accordingly it will cancel against the same collinear divergence in the subtraction graph (f).

The other graphs have a more complicated pattern of divergences, involving soft gluons and gluons collinear to the outgoing quark, as is evidenced by the divergence in both graphs at $z \to 1$. Naturally, the divergence only fully manifests itself when we integrate over z. To analyze this quantitatively, we use the principles explained in Sec. 9.4.4, where we needed to treat parton densities as generalized functions. We now do the same for structure functions and the coefficient functions. The existence of the extra divergence(s) indicates, of course, that we will need to improve the proof of factorization. For the moment we just examine the phenomena.

Since both the extra kinds of divergence occur at $z = 1$, some care is needed to identify their kinematics correctly. The general nature of the divergences can be extracted, as always, from the Libby-Sterman analysis. For this analysis, it is convenient to boost to the Breit frame, where $q^+ = -Q/\sqrt{2}, q^- = Q/\sqrt{2}$, and $q_T = 0_T$. Then:

- An initial-state collinear divergence is at $\theta \to \pi$ (i.e., $\cos\theta \to -1$) with z fixed and not equal to unity.
- A final-state collinear divergence is at $z \to 1$, with θ fixed and away from π. Each final-state particle is in the minus direction with momentum fractions $k_1^-/q^- = (1-\cos\theta)/2$

and $k_2^-/q^- = (1 + \cos\theta)/2$. Notice that the quark and gluon form an outgoing system, and that θ is the polar angle of each particle in the Breit frame.

• A soft-gluon divergence is at $\theta \to \pi$ and $z \to 1$.

It is misleadingly tempting to identify all of the $z \to 1$ divergences as soft.

Graph (b) has all three types of divergence, evidenced by its singularities at both $z \to 1$ and $\theta \to \pi$. But graph (a) has only an initial-state collinear divergence, and graph (c) only a final-state collinear divergence. As can be seen from (9.48), dimensional regularization with $\epsilon < 0$ regulates all the divergences.

After integral

We know that after we average over x (or z), the final-state lines become effectively off-shell. This will entail cancellation of final-state collinear and soft divergences between real and virtual graphs. The initial-state collinear divergences cancel against the subtraction graphs.

We could exhibit the cancellation at the level of the integrands. Instead we will evaluate the graphs separately, with dimensional regularization, and see the cancellations of the resulting poles at $\epsilon = 0$. The graphs give the following values, all multiplied by $g^2 e_j^2 C_F/(8\pi^2)$:

$$\text{(a):} \quad -\frac{z(1-z)}{\epsilon} + z(1-z)\left[T + \ln\frac{1-z}{z} + 1\right] + 3z^2, \tag{9.49a}$$

$$\text{(b):} \quad \frac{2}{\epsilon^2}\delta(z-1) + \frac{2}{\epsilon}\left[\delta(z-1)(-T+1) - \frac{z^2}{(1-z)_+}\right]$$

$$+ \delta(z-1)\left(T^2 - 2T + 4 - \frac{\pi^2}{2}\right) +$$

$$+ 2z^2\left[\frac{1}{(1-z)_+}(T-1) - \frac{\ln z}{1-z} + \left(\frac{\ln(1-z)}{1-z}\right)_+\right], \tag{9.49b}$$

$$\text{(c):} \quad -\frac{1}{2\epsilon}\delta(z-1) + \frac{1}{2}\delta(z-1)(T-1) + \frac{z}{2(1-z)_+}. \tag{9.49c}$$

where we have dropped terms of order ϵ and beyond, and we have defined

$$T = \ln\frac{Q^2}{\mu^2} + \gamma - \ln(4\pi). \tag{9.50}$$

The integrals over $\cos\theta$ were performed using (A.49). Then an expansion in powers of ϵ was made using (A.47), (A.48), and (A.54). We again see the appearance of plus distributions, which is very characteristic of QCD calculations.

The double pole in graph (b) is a result of the nesting between the soft and collinear divergences.

9.9.3 *Virtual-gluon graphs for* $\hat{F}_{2j}$

We already calculated the on-shell vertex subgraph used in Fig. 9.4(d); see Sec. 4.2.3. But now: (a) we have space-like instead of time-like q; (b) the trace with the external currents is slightly different. We add to the graph a counterterm for its UV divergence, which is the lowest-order graph times $-[g^2 e_j^2 C_F/(16\pi^2)]S_\epsilon/\epsilon$, times a factor of 2 to allow for the hermitian conjugate graph. The result for graph (d) and its conjugate is

$$(d + h.c.): \quad -\frac{2}{\epsilon^2}\delta(z-1) + \frac{2}{\epsilon}\delta(z-1)(T-2)$$

$$+ \delta(z-1)\left(-T^2 + 4T - \ln\frac{Q^2}{\mu^2} - 8 + \frac{\pi^2}{6}\right), \quad (9.51)$$

again times $g^2 e_j^2 C_F/(8\pi^2)$. This has a double pole, a logarithm in the single pole, and a double logarithm in the ϵ-independent term, all due to the combination of soft and collinear divergences. All of these terms cancel against the corresponding terms for graph (b), which is the only graph related by moving the final-state cut.

Graph (e) just involves a self-energy times the lowest-order hard scattering. As we saw in e^+e^- total cross section, in Sec. 4.1, we apply the LSZ prescription. The dimensionally regulated massless self-energy gives a zero contribution. There remains the UV wave-function renormalization counterterm, which gives

$$(e + h.c.): \quad \frac{S_\epsilon}{2\epsilon}\delta(z-1) = \frac{1}{2\epsilon}\delta(z-1) + \delta(z-1)(\ln(4\pi - \gamma) + O(\epsilon). \quad (9.52)$$

9.9.4 Subtraction graphs for $\hat{F}_{2j}$

The subtraction graphs (f)–(h) are simply a factor of $e_j^2 z$, for the parton-model coefficient function, times the one-loop quark-in-quark density, with the external self-energies omitted, all times a factor -1 because they are subtracted. As usual, the graphs themselves vanish in the massless limit, by the use of dimensional regularization. So we just need the UV counterterm, which is for $Z_2 Z_{jj}$, the factor Z_2 arising because we use the counterterm that allows the use of renormalized fields. With the same conventions as before we get

$$(f-h): \quad \frac{S_\epsilon}{\epsilon}\left[\frac{z(1+z^2)}{(1-z)_+} + \frac{5}{2}\delta(z-1)\right]. \quad (9.53)$$

9.9.5 Total

Adding the contributions of all the graphs and taking the $\epsilon \to 0$ limit gives the quark coefficient function. With the LO term, we have

$$\hat{F}_{2j}(Q^2, z; \alpha_s, \mu)$$

$$= e_j^2\delta(z-1) + \frac{g^2 e_j^2 C_F}{16\pi^2}z\left[4\left(\frac{\ln(1-z)}{1-z}\right)_+ - 3\left(\frac{1}{1-z}\right)_+ - 2(1+z)\ln(1-z)\right.$$

$$\left. - 2\frac{1+z^2}{1-z}\ln z + 6 + 4z - \left(\frac{2\pi^2}{3} + 9\right)\delta(1-z)\right] + O(g^4). \quad (9.54)$$

9.10 Hard scattering with quark masses

In the calculations so far, we have set quark masses to zero, and some of the methods relied on the property of dimensional regularization that scale-free integrals are zero. It is useful to see how to bring in non-zero quark masses. One purpose is to allow the effects of quark masses to be computed, although we will not give a detailed treatment of the effects of quark masses here. A second purpose is to show that calculations of hard-scattering coefficients are not tied to properties of the dimensional regularization scheme with massless particles.

A convenient method to allow for heavy quarks in the hard scattering is to always set to to zero the masses of external particles of the hard scattering, but to allow heavy particles to circulate inside the hard scattering (Collins, 1998a). We will not try to justify this prescription here.

We will restrict our attention to the simplest case of the gluon-induced NLO coefficient functions. The structure of the calculation is unchanged from that with massless quarks; i.e., we use (9.35) to determine the one-loop coefficient function, with a projection onto individual structure functions by (9.2). The actual graphs are Fig. 9.3, just as before.

Analytic calculations of one-loop graphs with masses are harder than with zero masses. We first quote the results for the unsubtracted graphs (a) and (b), which can be deduced from Aivazis *et al.* (1994). First for F_L:

$$\hat{F}_{Lg} = \sum_j \frac{g^2 e_j^2 T_F z}{4\pi^2} \theta(\hat{s} - 4m_j^2) \left\{ \frac{4Q^2 \Delta}{(Q^2 + \hat{s})^2} - L \frac{8m_j^2 Q^2}{(Q^2 + \hat{s})^2} \right\}, \qquad (9.55)$$

where

$$L = 2 \log \left[\frac{\sqrt{\hat{s}} + \sqrt{\hat{s} - 4m_j^2}}{2m_j} \right], \qquad (9.56)$$

$$\Delta = \sqrt{\hat{s}(\hat{s} - 4m_j^2)}, \qquad (9.57)$$

and $\hat{s} = Q^2(1 - z)/z$, as usual. There is a theta function implementing the quark-flavor-dependent threshold in $\hat{s}$. In the general factorization formulae, like (8.83), the threshold restricts ξ to the range $x(1 + 4m_j^2/Q^2) < \xi < 1$.

Note that there are some differences in conventions for defining structure functions in Aivazis *et al.* (1994), and that there appears to be a factor of T_F missing from their formulae. The result for $\hat{F}_{Lg}$ reduces to the previous one, (9.40), in the limit that the quark masses are zero.

As for F_2, we get

$$\hat{F}_{2g} = \sum_j \frac{g^2 e_j^2 T_F z}{4\pi^2} \left\{ \theta(\hat{s} - 4m_j^2) \left[L\frac{Q^4 + \hat{s}^2}{(Q^2 + \hat{s})^2} + \frac{[4Q^2\hat{s} - (\hat{s} - Q^2)^2]\Delta}{\hat{s}(Q^2 + \hat{s})^2} \right. \right.$$

$$\left. + L\frac{4m_j^2(\hat{s} - 2Q^2 - 2m_j^2)}{(Q^2 + \hat{s})^2} - \frac{4m_j^2 \Delta}{(Q^2 + \hat{s})^2} \right]$$

$$\left. - [1 - 2z(1 - z)] \ln \frac{\mu^2}{m_j^2} \right\} + O(g^4), \tag{9.58}$$

where the logarithmic term in the last line is for the subtraction graph (c), calculated at (9.7), here multiplied by 2 to include both the quark and antiquark contributions. The remaining terms are for graphs (a) and (b), and were obtained from Aivazis *et al.* (1994). In the massless limit, the logarithmic divergences cancel, and the limit reproduces the previous calculation (9.43).

Observe the mismatch between the allowed ranges of z in the integrand. The term from graphs (a) and (b) obeys a threshold condition, but the subtraction term allows z to go up to unity, where $\hat{s} = 0$, i.e., to an unphysical value. The parton-model approximation applied to a quark line is responsible for the mismatch. The approximation changes final-state momenta, so that the approximated final state violates conservation of 4-momentum. The same violation is present in the integrand for the parton-model formula, i.e., the LO cross section.

Strictly speaking our formalism was derived for the inclusive cross section, integrated over hadronic final states, and the results correctly apply to that situation. But if one wishes to extend the formalism to observables more differential in the final state, the violation of momentum conservation can have important consequences. Genuinely solving this issue requires the avoidance of approximations on parton momenta when they are related to final-state momenta. As seen in recent work (Collins and Jung, 2005; Collins, Rogers, and Staśto 2008), one must rethink the whole formalism; new methods do not use parton densities, but more general quantities, parton correlation functions, which do not have the integral over k^- and k_T in their definition.

Note that the above calculation applies when the $\overline{\text{MS}}$ scheme is used. This is appropriate for quarks whose mass is at most of order Q. For heavier quarks, a change in scheme is appropriate. There are various ways proposed to do this. A method I prefer is a generalization of the CWZ scheme of Sec. 3.10 to deal with parton densities and factorization; this is the ACOT scheme of Aivazis *et al.* (1994), which is probably best used in a modified version as given in Kretzer *et al.* (2004); Krämer, Olness, and Soper (2000). See Thorne and Tung (2008) for a wider ranging review.

9.11 Critique of conventional treatments

Compared with our presentation so far, a very different approach to factorization is found in much of the literature (e.g., Dissertori, Knowles, and Schmelling, 2003; Ellis, Stirling, and

Webber, 1996). It involves a strong emphasis on the mass divergences in massless on-shell partonic reactions, and it asserts that factorization is a method of absorbing mass divergences into a redefinition of parton densities. In contrast, in our presentation the divergences were canceled by subtraction terms that were needed to avoid double counting between, for example, NLO contributions to hard-scattering coefficients and LO contributions.

In this section, we assess the other approach and see that it is physically misleading, if not actually wrong. As such, it is a profound obstacle to further progress in applying perturbative methods to more complicated situations in QCD. Luckily from a practical point of view, the two approaches give the same results for hard-scattering coefficients when parton masses are set to zero. Thus the physical errors do not propagate to numerical results in phenomenology, at least for the simplest reactions.

The approach can be traced back to certain of the early literature on factorization, notably Ellis *et al.* (1979) and Curci, Furmanski, and Petronzio (1980), and it can be summarized as follows:

1. *Assert* that the structure function (or cross section) under consideration is a convolution of a partonic structure function and parton densities:

$$W = \text{partonic struct fn.} \otimes \text{bare parton density}$$
$$= W^{\text{parton}} \otimes f^{\text{bare}}. \tag{9.59}$$

   The convolution is defined in (8.81). In view of later steps in the presentation, the parton densities are called "bare parton densities".
2. All parton masses in the partonic structure function are set to zero. The parton(s) entering it from the parton density are set *on-shell* and massless, with zero transverse momentum.
3. There are IR/collinear divergences in the parton cross section. It was shown (Ellis *et al.*, 1979; Curci *et al.*, 1980) that the partonic cross sections are a convolution of a divergence factor and a finite cross section.

$$W^{\text{parton}} = C \otimes D. \tag{9.60}$$

4. The final factorization formula is obtained by use of the associativity of convolution to allow the divergences to be absorbed into a redefinition of the parton densities.

$$W = (C \otimes D) \otimes f^{\text{bare}} = C \otimes \left(D \otimes f^{\text{bare}} \right) = C \otimes f^{\text{ren}}, \tag{9.61}$$

   where $f^{\text{ren}} = D \otimes f^{\text{bare}}$.

The final result is of the same form as the factorization formula in (8.81). Moreover, if the collinear divergences are quantified by poles in dimensional regularization, their removal is by the same formula as in our approach. This can be obtained from the remarks at the end of Sec. 9.7.4. The factorization of collinear divergences in massless parton scattering, (9.60), can in fact be obtained from factorization applied to a massless parton target, assisted by the observation that loop graphs for massless parton densities in partonic targets are exactly zero in dimensional regularization.

However, the identity of the results should not obscure the profound problems with the argument just presented.

The first problem is that the starting point, (9.59), is not given a proof. In Ellis *et al.* (1979) a reference is given to the classic book on the parton model by Feynman (1972), which very much predates knowledge of the complications caused by QCD. The bare parton densities are also not defined; they cannot coincide with any of the parton densities we have defined.

A serious physics issue is that the partonic structure function in (9.59) is exactly a structure function initiated by an on-shell parton with zero transverse momentum. For example, the first gluonic term has the form

$$\left(\begin{array}{c} \text{(diagram)} \end{array} + \begin{array}{c} \text{(diagram)} \end{array} \right) \otimes \text{Bare gluon density} \tag{9.62}$$

Here, the gluon is set on-shell, just as in our calculations in Sec. 9.7. There the justification was that there was a subtraction in the coefficient function and therefore it is dominated by wide-angle scattering. We could therefore neglect small components of l with respect to large components. But in (9.59) and (9.62) this is no longer justified, since there is no subtraction. Indeed a gluon confined inside a hadron is not exactly on-shell, and therefore the collinear divergence is cut off.

Similarly in a model theory where all the fields have mass, there are no true collinear divergences. An approximation in which partons are made massless in unsubtracted NLO graphs therefore introduces spurious divergences. In such a theory, parton densities defined by the standard operator formulae have no collinear divergences, before or after renormalization, so the idea of absorbing collinear divergences into a redefinition is not tenable.

Note carefully that there is terminological ambiguity between the two approaches. In our approach "bare parton density" refers to a parton density before renormalization; renormalization is then strictly an issue of eliminating UV divergences by a suitable redefinition, commonly with the $\overline{\text{MS}}$ scheme. In the other approach, "bare parton density" refers to the undefined quantities in (9.59). The renormalization-like procedure applied in (9.61) is a different procedure, even when the $\overline{\text{MS}}$ scheme is said to be used.

We conclude that it is entirely unphysical to describe the basis of factorization in terms of moving collinear divergences from partonic structure functions or cross sections into redefined parton densities. Naturally, attempting to extend an incorrect method to more general situations leads to a conceptual morass. It is more by luck than good physics that the same hard-scattering coefficients are obtained for standard reactions.

9.12 Summary of known higher-order corrections

Here I summarize the available information on the higher-order terms in the DGLAP kernels and the coefficient functions for DIS. They are both known to order α_s^3.

The non-singlet part of DGLAP kernels was calculated to this order by Moch, Vermaseren, and Vogt (2004), and the singlet part by Vogt, Moch, and Vermaseren (2004). The order α_s^2 kernel was found by Furmanski and Petronzio (1980). See also Hamberg and van Neerven (1992) for some issues concerning the gauge invariance of the calculation. We have already given the order α_s kernels in (9.6), (9.23), (9.24), and (9.25).

The DIS coefficient functions were calculated by Vermaseren, Vogt, and Moch (2005) to α_s^3. The order α_s^2 calculation was by Zijlstra and van Neerven (1992) and by Moch and Vermaseren (2000). We have already given the order α_s coefficients in (9.40), (9.43), (9.47), and (9.54), with the parton model (α_s^0) at (9.1).

It is also worth mentioning the results at order α_s^2 for the Drell-Yan process, in Anastasiou *et al.* (2003, 2004), which are relevant to the same kind of precision phenomenology.

9.13 Phenomenology

Much of the predictive power of QCD is from factorization properties, both for inclusive DIS and for many other reactions. The equations used are for factorization of structure functions and cross sections, and for DGLAP evolution:

$$\sigma = \hat{\sigma} \otimes f, \quad \sigma = f_1 \otimes \hat{\sigma} \otimes f_2, \tag{9.63}$$

$$\frac{\mathrm{d}f}{\mathrm{d}\ln\mu} = 2P \otimes f. \tag{9.64}$$

Here σ is a measurable cross section or structure function, $\hat{\sigma}$ is a corresponding hard-scattering coefficient, while f, f_1 and f_2 are parton densities. Factorization is accurate up to power-law corrections in a hard scale Q. The second form of factorization applies to hard reactions in hadron-hadron collisions, where there is a parton density in each hadron.

The hard-scattering coefficients and the DGLAP kernel P are perturbative calculable in powers of the small coupling $\alpha_s(Q)$, and so we regard them as approximately calculable from first principles. The non-perturbative information is contained in the parton densities at some chosen fixed large scale, since the evolution to other large scales is perturbatively controlled. However, at present there is little ability to estimate or model the non-perturbative parton densities from first principles.

The predictive power lies in the universality of the parton densities. Parton densities are the same in all reactions, and, apart from the perturbative DGLAP evolution, they are the same at all values of Q. Thus essentially the following scheme works:

- Fit parton densities for some value of the scale μ from data on a limited set of experiments at one energy, using perturbatively calculated hard-scattering coefficients and DGLAP kernels.
- Evolve the parton densities to other scales.
- Predict cross sections at other energies and for other reactions.

In reality, data is of limited precision, and data on each individual reaction is only useful in determining some particular flavor combinations of parton densities. Therefore global

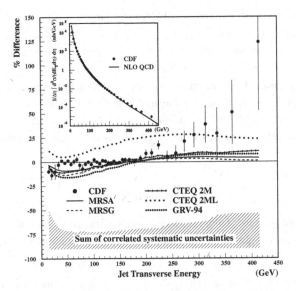

Fig. 9.5. Jet cross section and QCD *predictions* at CDF experiment (Abe *et al.*, 1996). The figure is copyright (1996) by The American Physical Society, and reproduced by courtesy of the CDF collaboration.

analyses are made to a wide variety of data, chosen for situations where the likely errors on both theory and experiment are judged to be sufficiently small. Thus the global analyses simultaneously fit parton densities and test QCD through measures of the goodness of fit. The amount of data is large, so this is a non-trivial undertaking.

Currently the main global analyses are:

- by the members of the CTEQ collaboration (Tung *et al.*, 2007);
- by a group in the UK going under the acronyms MRST and recently MSTW (Martin *et al.*, 2007);
- by Alekhin and collaborators (Alekhin, Melnikov, and Petriello, 2006).

In addition, the two *ep* collider experiments at DESY have made fits to their own data: ZEUS (Chekanov *et al.*, 2005) and H1 (Adloff *et al.*, 2003). They have taken advantage of the availability of charged-current processes to gain flavor separation of the parton densities.

Another group (Del Debbio *et al.*, 2007) is working towards a global fit using rather different calculational technique based on neural-network methods.

An example of the predictive power is shown in Fig. 9.5. Here a measurement (Abe *et al.*, 1996) by the CDF collaboration is shown for the production of jets of high transverse energy, E_T, in proton-antiproton collisions, and it is compared with QCD predictions. Although this is now a rather old comparison, its importance is that there is a genuine prediction. Parton densities at that period were measured in other processes and the perturbative hard-scattering calculations are, of course, from QCD first principles.

The agreement is good, except possibly at the largest values of E_T, but even there not outside the rather large errors. Since then it has been realized that this reaction is a most sensitive one for measuring the gluon density at large parton ξ. Therefore later work has frequently used jet data from hadron-hadron collisions in making global fits for parton densities. Thus the QCD calculations presented with the latest data can no longer be considered pure predictions. Results are available from both CDF (Abulencia *et al.*, 2007) and D0 (Abazov *et al.*, 2008) collaborations.

There are many other processes where QCD predictions have been made, by and large with success.

Exercises

9.1 Finish the calculations of the one-loop renormalization of parton densities by doing the calculations for gluon-in-quark and gluon-in-gluon, thereby verifying (9.24) and (9.25).

9.2 Verify the sum rules (8.41) and (8.42) for quark number and for momentum at one-loop order.

9.3 Verify the results in Sec. 9.5.2.

9.4 Find the gluon-induced NLO correction in a version of QCD where quarks are scalars.

9.5 (***) Using pdfs from some standard fit, obtain some estimates of the typical value of z in integrals of parton densities and hard scattering like those in (9.43), etc. You can probably do this by obtaining diagnostics from a numerical quadrature, although it should also be possible to obtain some order-of-magnitude results more analytically. Draw some conclusions about the reliability of standard perturbative QCD calculations under various kinematic conditions.

9.6 Consider the graph of Fig. 9.3(a) for the photon-gluon process, and suppose that the quarks are given a mass m_q. Show that the minimum fractional plus momentum of the intermediate quark line is $\chi = x(1 + 4m_q^2/Q^2)$. Fractional plus momentum of the intermediate quark means $(k_1^+ - q^+)/P^+$. [See the definition given in Tung, Kretzer, and Schmidt (2002) for the ACOT(χ) scheme for treating heavy quarks in factorization.]

9.7 Generalize the result of problem 9.6 to the case that the current is flavor changing between quarks of different masses, m_1 and m_2.

9.8 Verify the calculations giving the NLO quark contribution to F_2, i.e., (9.48).

10

Factorization and subtractions

In Sec. 9.13 we saw how factorization theorems give a lot of predictive power to QCD. They are essential in the analysis of data at high-energy colliders, not just for understanding the QCD aspects but also in searches for new physics, for example.

So far we have seen a genuine proof (Sec. 8.9) only for inclusive DIS, and only in a model theory without gauge fields. In this chapter we will formulate the principles that apply very generally, to other reactions, and when dealing with the full complications of a gauge theory.

The general class of problem concerns the extraction of the asymptotic behavior of amplitudes and cross sections as some external parameter, like a momentum, gets large. In general discussions, we denote the large parameter by Q. As well as factorization theorems in their broadest sense, such asymptotic problems also encompass simpler situations like renormalization, the operator product expansion (OPE), and the IR divergence issue[1] in QED.

There is a common and general mathematical structure in these different problems that could undoubtedly use further codification. Perhaps methods based on Hopf algebras, or some generalization, would provide an appropriate mathematical structure. So far these methods have been applied to renormalization (e.g., Connes and Kreimer, 2000, 2002).

In this chapter, I interleave a general formal treatment with its application to the Sudakov form factor, including explicit calculations at one-loop order. The general treatment will underlie all further work in this book. The Sudakov form factor illustrates the issues that are characteristic of asymptotic problems in Minkowski space, especially in a gauge theory. Factorization for the Sudakov form factor is a prototype for many important applications.

First I will give an overview of the method, which is a general subtractive procedure generalizing Bogoliubov's procedure for renormalization. The Libby-Sterman analysis is used to determine the leading regions R for a graph Γ for the process under consideration. For each region R of a graph Γ there is defined an approximator T_R. From T_R, with the aid of subtractions to cancel double counting between regions, is constructed the contribution $C_R\Gamma$ associated with the region.

Then I will define an implementation of these ideas for the Sudakov form factor, complete with a specific calculation for a one-loop graph. After that will be a proof that the general

---

[1] Which concerns a small photon mass instead of a large scale Q.

subtraction method works. This will require that the region approximators T_R obey certain conditions, that will be especially critical in QCD. The one-loop example will help to explain the rationale for these conditions and to show how to satisfy them in general.

Then I will derive factorization and evolution equations for the Sudakov form factor.

Many elements of the proofs given here can be found in the literature. However, the presentation as a whole represents a new treatment, which is intended to be a substantial improvement on previous work.

Although the methods presented here apply to perturbation theory, it should be evident, just as in Sec. 8.9, that much structure is seen that has a reality beyond perturbation theory. But exactly how to capture this structure in a strict deductive framework is not so clear, and there are some important open problems.

10.1 Subtraction method

To understand the rationale for a subtraction procedure, recall the successive approximation method outlined in Sec. 8.8. This starts from the smallest region for a graph for some process, for which we find a useful approximation. The approximation typically corresponds to a product or convolution of a lowest-order partonic subgraph and a matrix element of some operator. The operator in the matrix element determines the definition of, for example, a parton density.

We then sequentially construct approximations suitable for successively larger regions. When constructing the contribution C_R associated with some region R, subtractions must be applied to compensate double counting of the contributions $C_{R'}$ from smaller regions R', contributions that have already been constructed. Finally, we sum over the regions for each graph Γ, and over graphs. This results in factorization, by an argument with the pattern given in Sec. 8.2.

A simple example was given by the derivation of leading-twist factorization for DIS in a non-gauge theory in Sec. 8.9. It is a useful exercise to show how the formulae in that section, like (8.70) and (8.74), give particular cases of the more general formulae in the present chapter.

In a gauge theory like QCD, the basic argument will need to be supplemented, notably by an application of Ward identities to extract gluons of scalar polarization from the hard scattering, to convert them to attachments to Wilson lines. Further issues concern the exact nature of the leading regions and the accuracy of the approximators T_R. These are much harder than for relatively simple Euclidean asymptotic problems like the OPE.

10.1.1 Overall view

We let Q denote the large scale for the process under consideration. Each graph Γ has a set of leading regions, and up to power-suppressed terms, we aim to write Γ as a sum over terms for its leading regions:

$$\Gamma = \sum_{R \text{ of } \Gamma} C_R \Gamma + \text{power-suppressed}. \tag{10.1}$$

For the processes of interest, the regions and the associated powers of Q are determined by the Libby-Sterman analysis (Ch. 5). Normally we treat only the leading power. As explained in Ch. 5, each region is specified by a skeleton in loop-momentum space, i.e., the position of the associated pinch-singular surface (PSS) in a massless theory. Each region also corresponds to a decomposition of the whole graph Γ into subgraphs (e.g., Fig. 5.17) where each subgraph has momenta of a particular kind: hard, collinear in some direction, or soft. There can be finer decompositions needed under some circumstances, but that does not affect the principles.

The general definition of the contribution $C_R\Gamma$ associated with a region R of a graph Γ will be made in (10.4) in terms of an "approximator" T_R, together with subtractions to eliminate double counting between regions. A key element in applying (10.1) and in enabling factorization to be derived is the construction of suitable approximators T_R.

10.1.2 Regions: terminology

We review some terminology and definitions from Ch. 5.

- A region R of a graph Γ is specified by a PSS in the massless theory, as determined by the Libby-Sterman method.
- A region is called leading if Libby-Sterman power-counting gives it a leading power, usually defined by dimensional analysis, e.g., Q^0 for a DIS structure function.
- Some regions occur with a super-leading power in individual graphs, when all the gluons exchanged between hard and collinear subgraphs are of scalar polarization. Since such super-leading contributions cancel very generally after a sum over graphs, we choose the definition of the leading power accordingly.
- Our factorization arguments will be applied to regions which give at least a certain chosen power of Q. The term "power-suppressed" in (10.1) means with respect to the chosen power of Q.

  Typically, this is the power of Q we call leading. But extensions of our methods to non-leading powers are possible. Since T_R is essentially a truncation of a Taylor series expansion about a PSS, keeping more terms in the Taylor series corresponds to keeping more non-leading powers of Q.

  When we use dimensional regularization, with $4 - 2\epsilon$ dimensions, some exponents in power laws have ϵ dependence. In categorizing powers as leading or non-leading, we generally work close to $\epsilon = 0$ and ignore changes in exponents that are of order ϵ.
- At each PSS R we choose a set of intrinsic coordinates labeling points within the PSS, and there is a set of normal coordinates labeling deviations off the surface (Sec. 5.7).
- We can convert the normal coordinates for a region R into a radial coordinate λ_R and a set of angle-like coordinates specifying direction. We saw a number of examples in Ch. 5. Power-counting is conveniently done using the one-dimensional integral over λ_R. We require λ_R to have the dimensions of mass.
- Ordering between the regions is defined by set-theoretic inclusion on the skeletons defined technically in Sec. 5.4.1, and reviewed in the next section, 10.1.3.

10.1.3 Regions: properties

Relations between regions

In simple cases, all the leading regions for a graph are nested. A typical example is DIS in a non-gauge theory (Sec. 8.9). For that case, the leading regions are where some number of rungs at the top of a ladder graph, Fig. 8.12, form the hard subgraph, and the rest of the graph is target-collinear. The hard subgraph corresponds to a graphical factor AK^j in Fig. 8.12. If we use R_j to denote the corresponding region, then the ordering of leading regions can be represented along a line:

$$R_0 < R_1 < R_2 < \cdots < R_N. \tag{10.2}$$

This situation is called a total ordering, i.e., any two leading regions, R_1 and R_2, obey exactly one of $R_1 < R_2$, $R_2 < R_1$ or $R_1 = R_2$.

But in general, the ordering is only a partial ordering. That is, between any two regions R_1 and R_2, exactly one of the following holds:

- $R_1 < R_2$: R_1 is smaller than R_2.
- $R_1 > R_2$: R_1 is bigger than R_2.
- $R_1 = R_2$: they are the same region.
- They overlap. That is, the intersection of their skeletons is non-empty, $R_1 \cap R_2 \neq \emptyset$, but none of the preceding three cases hold.[2] Thus $R_1 \cap R_2$ is non-empty and strictly smaller than both of R_1 and R_2. An example is given by R_A and R_B in (5.21). We denote this situation by R_1 ovrlp R_2.
- R_1 and R_2 do not intersect at all: $R_1 \cap R_2 = \emptyset$. An example is given by $R_{A'}$ and $R_{B'}$ in (5.21).

Separation of non-intersecting regions

Suppose two regions R_1 and R_2 do not intersect. Then there is a non-zero separation between them, because the (empty) intersection is of their skeletons, which are *closed* sets. Thus if λ_1 and λ_2 are radial variables for the two regions, then there is a non-zero range $0 \leq \lambda_j \leq L_j$ for which points around each PSS do not intersect the other. Since the PSS are defined from the massless theory, each of these ranges in λ_j is of order Q.

Minimal region(s)

We define a region R_0 to be minimal if it has no smaller regions, i.e., if there is no R' for which $R' < R_0$. One example is for a handbag diagram for DIS. Its minimal region gives the parton model. A non-trivial example is for the one-loop vertex graph treated in Sec. 5.4. It has three minimal regions $R_{A'}$, $R_{B'}$ and R_S. (But only R_S is leading.)

Note that a minimal region R_0 cannot overlap with any region. For every other region, either R_0 is contained in it or does not intersect it.

---

[2] $\emptyset$ denotes the empty set.

Hierarchy

Ordering between the different regions of a graph allows them to be organized in a hierarchy which can be diagrammed as in (5.21).

10.1.4 Definition of region term $C_R \Gamma$

C_R for minimal region

For a minimal region R_0, its contribution is simply defined to be the action of its approximator on the unapproximated graph:

$$C_{R_0} \Gamma \stackrel{\text{def}}{=} T_{R_0} \Gamma. \tag{10.3}$$

In DIS in a non-gauge theory in Ch. 8, a suitable approximator for a minimal leading region was given in (8.68).

As that equation illustrates, a natural definition of the approximator can lead to extra UV divergences, which are to be removed by renormalization of parton densities (and of similar objects, in the general case). Therefore we define the approximator to include such renormalization.

Alternatively, the approximator can be defined to include a suitable cutoff. The comparative advantages and disadvantages of the renormalization and cutoff approaches were discussed in Sec. 8.3.1.

C_R for larger regions

In the contributions from larger regions, we use subtractions to avoid double counting of the contributions from smaller regions. So we define

$$C_R \Gamma \stackrel{\text{def}}{=} T_R \left(\Gamma - \sum_{R' < R} C_{R'} \Gamma \right). \tag{10.4}$$

For a minimal region, (10.4) reduces to (10.3). Thus (10.4) gives a valid recursive definition of $C_R \Gamma$, starting from the minimal region(s).

The factor in parentheses is the original graph minus subtractions for regions smaller than R. For the case treated in Ch. 8, this factor was found in (8.74); it is $A \left[1 - (1 - \overleftarrow{T} | V) K \right]^{-1}$ on the last line of that equation.[3] In that situation, it was evident that the factor is power-suppressed in regions smaller than R. Thus the smallest region where $C_R \Gamma$ is leading is actually R.

But in more general cases, like the Sudakov form factor, such statements will need some modifications.

It is also possible to start from an approximation for a maximal region, and then work to smaller regions, as in Tkachov (1994). But starting from the smaller regions, as we have done, gives a more direct relation to the parton model and makes clearer the relation to a non-perturbative definition of the parton densities.

---

[3] That formula does not explicitly include the needed parton-density renormalization.

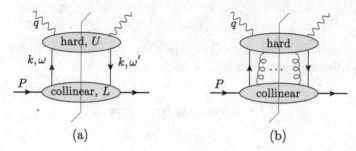

Fig. 10.1. Leading regions for DIS: (a) in a theory without gauge fields, (b) in a gauge theory.

10.1.5 Remainder

We define the remainder of a graph to be

$$r(\Gamma) \overset{\text{def}}{=} \Gamma - \sum_{R \text{ of } \Gamma} C_R \Gamma. \qquad (10.5)$$

It is essential to prove that this is actually power-suppressed, given a particular implementation of the region approximators $T_R \Gamma$.

10.1.6 Relation to factorization

The above formalism focuses on an additive structure for a particular graph. To get a factorized form, we sum over graphs. As observed in Sec. 8.2, the sum over regions and graphs corresponds to independent sums over subgraphs associated with the regions, e.g., independent sums over the hard and collinear subgraphs for DIS in Fig. 10.1.

In the simplest cases, exemplified by Fig. 10.1(a), we have a fixed number of lines joining the subgraphs, and the graphical structure directly corresponds to a factorization formula. Then to prove factorization we need to prove that (1) the approximators T_R respect the factorized structure, (2) UV renormalization needed on the parton densities respects the factorized structure, and (3) the subtractions in (10.4) actually have their intended effect of removing double counting between the terms for different regions.

But as illustrated in Fig. 10.1(b), the situation is more complicated in a gauge theory, because arbitrarily many gauge-field lines can connect the collinear and hard subgraphs,[4] without any power-suppression.

Therefore the graphical representation of the regions does not directly correspond to factorization.

An example of the necessary argument was given in Sec. 7.7 for a gauge-theory version of the parton model. We applied Ward identities to convert the extra gluons into attachments to the Wilson line in the definition of a gauge-invariant quark density. To do this requires an appropriate choice of the approximators T_R, together with a demonstration that the Wilson

---

[4] And also soft and collinear subgraphs in a general case

lines are actually obtained. Only after this work do we find that

$$\sum_{R,\Gamma} C_R \Gamma = \text{factorized form}. \tag{10.6}$$

We could conceive that Fig. 10.1(b) itself represents a generalized factorization structure. But the structure would involve an infinite collection of parton-density-like objects, each with a different number of gluon lines, and each with a different hard-scattering factor. Without further information, such a factorization would not be useful for phenomenology.

10.1.7 Which formulation for calculations?

Often in realistic QCD calculations, there are many graphs to consider. The decomposition (10.1) produces multiple terms for each graph, resulting in an apparently even more elaborate structure. Is it actually necessary to use it?

An alternative calculational approach was described in Sec. 9.6, and corresponds to many practical calculations. The aim is to compute the hard-scattering coefficient in a factorization formula, and the method uses the observation that the hard scattering does not depend on the type of particle used for the target. One first makes a direct computation of Feynman graphs for the process under consideration, but with a partonic target. Then one computes the densities of partons in partons to the relevant order, and then applies factorization on a partonic target to deduce the hard-scattering coefficients. Because factorization has taken account of simplifications due to the use of Ward identities, there are generally fewer terms to calculate than by a direct use of (10.1), which requires a listing of all the leading regions for every graph computed.

This would appear to relegate the subtraction formalism to a key tool in a careful derivation of factorization.

However, direct calculation of partonic Feynman graphs involves the cancellation of various kinds of collinear and soft divergences between different graphs; it thereby entails the use of a regulator. This is satisfactory if calculations are done analytically rather than numerically. But if numerical calculations are used, the cancellation of divergences between graphs is tricky to implement; it is a classic situation where rounding errors can dominate a numerical calculation. To set up a numerical integral for the hard scattering one can apply subtractions directly to the integrand of a hard-scattering subgraph. All necessary cancellations of divergences are then in the integrand, and the integral can be evaluated directly in four dimensions, without a regulator. We saw a very simple example in Sec. 9.7.5.

There is much recent work in implementing subtractions numerically, e.g., Binoth *et al.* (2008); Dittmaier, Kabelschacht, and Kasprzik (2008); Frederix, Gehrmann, and Greiner (2008); Hasegawa, Moch, and Uwer (2008); Seymour and Tevlin (2008).

Since there are many regions involved in high-order graphs, practical application of a subtraction procedure must be automated. If the subtractions are not formulated correctly, there can remain divergences, which manifest themselves in badly behaved numerical integrals over a high-dimensional space.

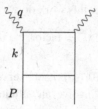

Fig. 10.2. One-loop graph for DIS on elementary target.

10.2 Simple example of subtraction method

With suitable definitions of the region approximators T_R, we will derive factorization for many processes of interest from the structure of the sum over regions and graphs, $\sum_{\Gamma,R} C_R \Gamma$. So to prove factorization is accurate up to a power-law error, we need to prove that for each individual graph the sum over regions, $\sum_R C_R \Gamma$, itself approximates Γ, up to a power-law error, i.e., that the remainder $r(\Gamma)$ is power-suppressed.

Now the approximator T_R is always designed so that $T_R \Gamma$ gives an accurate approximation when the momentum configuration is both close to the PSS defining the region R, and away from the intersections with the PSSs for regions that are smaller than or overlap with R. The complications in making a satisfactory proof that $r(\Gamma)$ is power-suppressed arise from the combination of multiple regions, with the possibility of double counting, and from the fact that there are intermediate configurations of momenta where the individual approximations degrade in accuracy.

The simplest proof is when all the relevant regions are nested, as in Sec. 8.9. Our aim in this chapter is to construct better methods that also work when there are more complicated relations between regions, e.g., (5.21).

But first I illustrate the general notation with a simple mathematical example motivated by a one-loop graph for DIS in a model theory, Fig. 10.2. There are two leading regions: R_0, where the top rung is hard and the bottom rung collinear, and R_1, where the whole loop is hard. They obey $R_0 < R_1$. The simple example is obtained by replacing the full Feynman graph by the following one-dimensional integral

$$I(Q, P, m) = \int_0^\infty dk \, \Gamma(k, Q, P, m) = \int_0^\infty dk \, \frac{Q}{Q+k+m} \frac{1}{k+P+m}. \quad (10.7)$$

The factor of Q in the numerator makes the integral dimensionless, and gives an overall leading power of Q^0.

If our general subtraction method works, then the leading-power asymptote for the graph is

$$C_{R_0} \Gamma + C_{R_1} \Gamma = T_{R_0} \Gamma + T_{R_1}(1 - T_{R_0})\Gamma. \quad (10.8)$$

We define the approximators T_R to be applied to the integrand, Γ, rather than to the integral as a whole. Each T_R sets to zero the (lower) external momentum and the internal mass of the hard scattering. Thus

$$C_{R_0}\Gamma = T_{R_0}\Gamma = \frac{Q}{Q}\frac{1}{k+P+m} = \frac{1}{k+P+m}, \tag{10.9a}$$

$$(1-T_{R_0})\Gamma = \left(\frac{Q}{Q+k+m} - 1\right)\frac{1}{k+P+m}, \tag{10.9b}$$

$$T_{R_1}\Gamma = \frac{Q}{Q+k}\frac{1}{k}, \tag{10.9c}$$

$$C_{R_1}\Gamma = T_{R_1}(1-T_{R_0})\Gamma = \left(\frac{Q}{Q+k} - 1\right)\frac{1}{k}, \tag{10.9d}$$

so that the remainder is

$$r(\Gamma) = \Gamma - C_{R_0}\Gamma - C_{R_1}\Gamma = (1-T_{R_1})(1-T_{R_0})\Gamma$$

$$= \left(\frac{Q}{Q+k+m} - 1\right)\frac{1}{k+P+m} - \left(\frac{Q}{Q+k} - 1\right)\frac{1}{k}. \tag{10.10}$$

Applying $1 - T_{R_0}$ gives a suppression by k/Q or m/Q, whichever is larger, in the factors in parentheses on the last line of (10.10). This has a minimum of m/Q, which is the desired overall error, but the error degrades as k increases towards Q.

Applying $1 - T_{R_1}$ gives a suppression by m/Q or m/k. (We assume P is of order m.) The intrinsic variable of the large region R_1 is k, and T_{R_1} is designed to give an accurate approximation when $k \sim Q$. But as k approaches R_0 the accuracy of an approximation of Γ by $T_{R_1}\Gamma$ degrades to m/k. But multiplying this error by the previously determined factor of k/Q compensates this, to leave an overall relative error of m/Q.

Notice that the error in $(1 - T_{R_1})\Gamma$ gets even worse if $k \ll m$, because T_{R_1} makes a massless approximation, replacing $1/(k+P+m)$ by $1/k$. By itself, this would give an actual divergence in the integral at $k = 0$. But the $1 - T_{R_0}$ factor applied in this same massless approximation gives a k/Q factor to kill the divergence.

10.3 Sudakov form factor

The fundamental object in our method is the approximator $T_R\Gamma$ for a region R of a graph Γ. We let λ_R be the radial variable, we let $\overline{k}_R$ be the angular variables surrounding R, and we let z_R be the intrinsic variables for R (Secs. 5.5 and 5.7). The approximator must give a good approximation to Γ in the core of the region R, i.e., where λ_R is small and the $\overline{k}_R$ variables are not close to larger regions.

In simple examples, as in Sec. 10.2, the accuracy of T_R only degrades when the intrinsic variable(s) of R approach the PSS of a smaller region. However, when we treat soft gluons, the accuracy of T_R also degrades when the angular variables $\overline{k}_R$ approach *larger* PSSs than R. This issue is responsible for complications in many QCD processes, when they are compared with simple Euclidean problems, like the OPE.

A simple case to illustrate these issues is the Sudakov form factor, i.e., the electromagnetic form factor of an elementary particle at high Q. We defined the Sudakov form

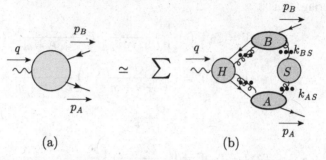

Fig. 10.3. (a) Sudakov form factor. (b) Its leading regions with loop momenta connecting the subgraphs. The dots indicate arbitrarily many gluons exchanged between the neighboring subgraphs. Note that the soft subgraph S *may be empty or may have more than one connected component*. The complete amplitude is approximated by a sum over regions and graphs when the contribution of each region is interpreted as $C_R\Gamma$.

factor and its kinematics in Sec. 5.1.1. Our aims now are to make suitable definitions of the approximators T_R, and to derive factorization. The form factor and its leading regions are shown in Fig. 10.3.

10.3.1 Factorization

We will obtain a factorization property in which the form factor F is the product of a hard factor H, collinear factors A and B for each external quark, and a soft factor S:

$$F = HABS + \text{power-suppressed},\qquad(10.11)$$

each with dependence on only some parameters of F. Later we will redefine the factors so that a square root of S is absorbed into each collinear factor. (We will accompany this by some further redefinitions of A and B.) Then S will not appear in the final factorization formula.

10.3.2 Overall motivation for factorization approach

At this point, I review the rationale for using the factorization approach in QCD. This will indicate the kinds of theorem we need to formulate.

Typically, multiple regions contribute to an amplitude or cross section when there are large momenta. In perturbative calculations, this gives rise to large logarithms which prevent a straightforward use of perturbation theory in QCD. The two logarithms per loop present in many cases like the Sudakov form factor are particularly bothersome.

Moreover, in almost all interesting cases in QCD, some momenta in leading regions have low virtualities, where the effective coupling is large, so that low-order perturbative calculations are inapplicable.

In a factorized formula like (10.11), the different factors are each concerned with a particular kind of 4-momentum. Besides dependence on external kinematic variables, each

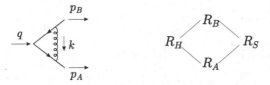

Fig. 10.4. One-loop graph for Sudakov form factor, and the hierarchy of its leading regions. In each case, the name of the region, e.g., "R_S", refers to the category of the gluon's momentum. A line connecting two regions denotes that they are ordered, with the bigger region on the left. Thus the diagram component $R_2 - R_1$ means that $R_2 > R_1$ in the sense defined in Sec. 10.1.2.

factor has dependence on one or more auxiliary parameters (like a renormalization scale). The auxiliary parameters can be roughly characterized as setting the boundaries between kinematic regions. The logarithms can be tamed by deriving evolution equations for the dependence on the auxiliary parameters. The kernels of the evolution equations are free of logarithms in the parameter whose dependence is governed by the evolution equation, and thus the kernels are susceptible to perturbative calculations (and hence *prediction* from first principles).

After the application of evolution equations, we need the individual factors, each at appropriate reference values of the auxiliary parameters. Some factors depend on low momentum scales, and are therefore genuinely non-perturbative in QCD. Others depend only on a single large scale, and therefore are perturbatively calculable in QCD. The non-perturbative quantities in QCD are typified by parton densities. They will be proved to be universal, i.e., the same parton densities appear in many different reactions. As explained in Sec. 9.13, universality underlies much of the predictive power of QCD: The non-perturbative quantities can be measured from a limited set of data, and then predictions are made for a wide variety of other experiments, with the aid of perturbative calculations for hard-scattering coefficients and evolution kernels.

10.3.3 Sudakov: regions for one- and two-loop graphs

As explained in Secs. 5.4.1 and 10.1.3, the regions for a graph can be organized as a hierarchy. To illustrate this, Figs. 10.4 and 10.5 show some important one- and two-loop graphs for the Sudakov form factor together with a representation of the hierarchies of their leading regions. A useful exercise is to check the hierarchies.

10.4 Region approximator T_R for Sudakov form factor

The definition in this section of the region approximator T_R uses the methods of Collins, Rogers, and Staśto (2008).

10.4.1 Decomposition of graph for one region

Consider a particular graph Γ for the Sudakov form factor. A leading region R corresponds to a graphical decomposition of the form of Fig. 10.3(b), with subgraphs which we label H,

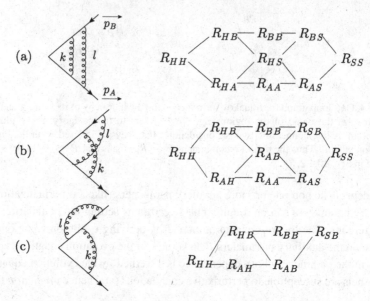

Fig. 10.5. Some two-loop graphs for the Sudakov form factor, and their leading-region hierarchies. The two-lettered code for a region, e.g., in "R_{HS}", refers to the categories of gluon k and gluon l.

A, B, and S.[5] We choose loop momenta coupling the subgraphs as follows. Momenta on external lines of the soft subgraph circulate into one collinear subgraph, round through the hard subgraph and back by the other collinear subgraph. Remaining loops involve momenta from each collinear subgraph entering the hard subgraph and circulating back to the same collinear subgraph. Thus we write the integral for the graph as

$$I = \int dk_{AS}\, dk_{BS}\, dk_{HA}\, dk_{HB} \; H(q, k_{HA} + k_{HAS}, k_{HB} + k_{HBS}, m)$$

$$\times A(p_A, k_{HA}, k_{AS})\, B(p_B, k_{HB}, k_{BS})\, S(k_{BS}, k_{AS}). \tag{10.12}$$

Here k_{AS} denotes the array of momenta flowing from the A subgraph into the S subgraph, and similarly for k_{BS}. These momenta flow through the hard subgraph, with k_{HAS} and k_{HBS} denoting how the circulating soft momenta are apportioned among lines entering H. The remaining momenta circulating between the H and A subgraphs are k_{HA}, and similarly for k_{HB}. Thus to denote the full set of momenta entering the hard subgraph from each collinear subgraph we use $k_{HA} + k_{HAS}$ and $k_{HB} + k_{HBS}$.

The soft factor is defined to include a momentum-conservation factor for each of its connected components. All loops contained entirely within the separate subgraphs do not need to be indicated explicitly. Although the integrals in (10.12) are commonly of high

---

[5] Note that the use of these symbols is different than in Figs. 10.4 and 10.5, where the symbols refer to particular categories of gluon momentum instead of subgraphs with momenta in a category.

dimension, it is possible that some or all are absent, for example when the soft subgraph is empty, or when only a single line connects a collinear graph to the hard subgraph.

Although to construct the definition of T_R we will examine properties of the graph when the values of momenta correspond to the region under consideration, we do not intend the loop momenta in (10.12) to be restricted to the region. In that sense, (10.12) is an exact expression for the whole Feynman graph. The purpose of this decomposition is simply to provide a convenient notation for use in a general definition of T_R.

When the momenta are near the PSS of R, some propagator denominators are particularly small. In general, we can make a suitable approximant by expanding in powers of small variables compared with large variables. Since we are concerned here only with the leading power of Q, the first term in the series suffices, i.e., we simply neglect the small variables compared with the large variables.

One complication now arises. As follows from the discussion in Sec. 5.10.2, there are two clashing characterizations of a collinear momentum. One is that it has energy of order Q and low virtuality. The other is that it has high center-of-mass rapidity. The distinction is particularly important when we deal with graphs with a massless gluon, as in QCD, and it is the second, more general, characterization that is more appropriate.

The use of this second definition strongly influences our construction of the region approximator T_R, since it affects the characterization of large and small variables. There can be leading contributions when some gluons are simultaneously soft and collinear, in the sense that all their momentum components are much less than Q and that their rapidities are large.

10.4.2 Definition of T_R

We consider the region R of a graph Γ associated with the decomposition (10.12). We also label momenta of particular lines by their category (soft, collinear-to-A, etc.) at the PSS for the region. The power-counting for the momentum components was given in Sec. 5.7.4.

The basic method to construct the region approximator T_R is to expand to leading power in the radial variable λ_R for the region. This will tend to introduce divergences. Some of the divergences are endpoint divergences, associated with regions R' that are smaller than R; these we will find to be canceled by the subtractions in the definition (10.4) of the region's contribution $C_R\Gamma$. Other divergences arise when we extend loop-momentum integration beyond the immediate neighborhood of R. These are essentially UV divergences removed by conventional renormalization that we include in the definition of T_R.

For simple Euclidean asymptotic problems like the OPE, there are no further divergences. But characteristic of asymptotic problems in Minkowski space with soft gluons are further divergences, which we term rapidity divergences; see the discussion around (10.35) below. We will modify the definition of T_R to cut off rapidity divergences. The evolution equations with respect to the cutoffs are essential to using the factorization theorem, and we will see important applications in Ch. 13. The only place where a modification is needed is in the approximation of soft momenta entering the collinear subgraphs. Later, in Sec. 10.11, we will reorganize the factorization formula into a form where the cutoffs on rapidity divergences can be removed.

The approximator's definition is made in three stages. The first is to extract the leading power of λ_R in the numerators and denominators of the subgraphs, with modifications to cut off rapidity divergences, and to improve properties of the hard scattering. It is implemented by defining linear projectors on loop momenta: P_{AS}, P_{BS} for soft loop momenta in the A and B subgraphs, and P_{HA}, P_{HB} for collinear and soft loop momenta in the H subgraph. Then certain adjustments of the momenta in H are implemented by non-linear functions R_{HA} and R_{HB}, so that the following replacement is made:

$$I \mapsto \int dk_{AS}\, dk_{BS}\, dk_{HA}\, dk_{HB}\; H\big(q, \hat{k}_{HA}, \hat{k}_{HB}, 0\big)$$

$$\times A\big(p_A, k_{HA}, \hat{k}_{AS}\big)\, B\big(p_B, k_{HB}, \hat{k}_{BS}\big)\, S\big(k_{BS}, k_{AS}\big), \qquad (10.13)$$

where

$$\hat{k}_{AS} = P_{AS}(k_{AS}), \qquad\qquad\qquad\qquad\qquad (10.14a)$$

$$\hat{k}_{BS} = P_{BS}(k_{BS}), \qquad\qquad\qquad\qquad\qquad (10.14b)$$

$$\hat{k}_{HA} = R_{HA}P_{HA}\big(k_{HA} + P_{AS}(k_{HAS})\big) = R_{HA}P_{HA}(k_{HA}), \qquad (10.14c)$$

$$\hat{k}_{HB} = R_{HB}P_{HB}\big(k_{HB} + P_{BS}(k_{HBS})\big) = R_{HB}P_{HB}(k_{HB}). \qquad (10.14d)$$

It will be a considerable convenience that soft momenta are approximated by exactly zero in the hard subgraph H, which is enforced by defining projectors so that $P_{HA}P_{AS} = P_{HB}P_{BS} = 0$.

The second stage of the definition of T_R is to apply corresponding approximations to the numerator factors in the lines connecting the subgraphs. The final stage is renormalization of UV divergences.

The approximator makes use of some auxiliary vectors to define particular directions in the (t, z) plane:

$$w_1 = (1, 0, \mathbf{0}_T), \qquad w_2 = (0, 1, \mathbf{0}_T), \qquad\qquad (10.15a)$$

$$n_1 = \big(1, -e^{-2y_1}, \mathbf{0}_T\big), \qquad n_2 = \big(-e^{2y_2}, 1, \mathbf{0}_T\big). \qquad (10.15b)$$

Thus w_1 and w_2 are light-like vectors corresponding to the external momenta p_A and p_B, while n_1 and n_2 are similar vectors that are slightly space-like. The rapidity parameters y_1 and y_2 are among the auxiliary parameters referred to earlier, for which evolution equations will be derived; initially they are chosen to be comparable to the rapidities y_{p_A} and y_{p_B} of the external on-shell lines. The vectors in (10.15) specify directions, and all their uses will be unchanged if any of the vectors is scaled by a positive non-zero number.

I now present the detailed definitions that make up T_R, leaving some details of the justification to Sec. 10.6.

1. *Soft to collinear-A:* Consider a momentum k_{AS} flowing from A into S. The denominator for a line in A has the form $(k_A + k_{AS}^2) - m^2 = k_A^2 - m^2 + 2k_A \cdot k_{AS} + k_{AS}^2$, where k_A

is a momentum classified as collinear-to-A. From Sec. 5.7.4, the leading power of λ_R is λ_R^2, for the terms k_A^2 and $2k_A^+k_{AS}^-$. So the basic leading-power approximation for subgraph A is to neglect all but the minus component of k_{AS}, i.e., to make the replacement $k_{AS} \mapsto (0, k_{AS}^-, \mathbf{0}_T)$. To cut off rapidity divergences we then modify the minus component slightly, and define T_R to use the following projector:

$$k_{AS} \mapsto \hat{k}_{AS} = P_{AS}(k_{AS}) = (0, 1, \mathbf{0}_T)\left(k_{AS}^- - e^{-2y_1}k_{AS}^+\right). \tag{10.16}$$

In covariant form this is

$$\hat{k}_{AS} = P_{AS}(k_{AS}) = \frac{w_2^\mu \, k_{AS} \cdot n_1}{w_2 \cdot n_1}, \tag{10.17}$$

where n_1 and w_2 are defined in (10.15), with y_1 in the definition of n_1 being a large positive rapidity appropriate to the p_A particle. But the precise value of y_1 is not critical; the effect of changes in y_1 will cancel in the complete factorization formula.

The use of k_{AS} in (10.12) treats k_{AS} as the array of loop momenta flowing from A into S. So the above definition is to be applied separately to each of the momenta in the array.

The justification of the exact form of the above projector will be given in Sec. 10.6, including the choice that n_1 is space-like.

2. *Soft to collinear-B:* A similar replacement is applied to soft momenta in the B subgraph, with the roles of plus and minus components exchanged:

$$k_{BS} \mapsto \hat{k}_{BS} = P_{BS}(k_{BS}) = \frac{w_1^\mu \, k_{BS} \cdot n_2}{w_1 \cdot n_2}. \tag{10.18}$$

Naturally y_2 in the definition of n_2 should be a large negative rapidity appropriate to p_B.

3. *Collinear-A and collinear-B to H:* In the hard subgraph H, the basic approximation is to replace momenta $k_{HA} + k_{HAS}$ from the A subgraph by their plus components and momenta $k_{HB} + k_{HBS}$ from the B subgraph by their minus components:

$$P_{HA}(k_{HA}) = \frac{w_1 \, (k_{HA} + \hat{k}_{HAS}) \cdot w_2}{w_1 \cdot w_2} = (k_{HA}^+, 0, \mathbf{0}_T), \tag{10.19a}$$

$$P_{HB}(k_{HB}) = \frac{w_2 \, (k_{HB} + \hat{k}_{HBS}) \cdot w_1}{w_2 \cdot w_1} = (0, k_{HB}^-, \mathbf{0}_T). \tag{10.19b}$$

Hence soft momenta are replaced by zero in the hard subgraph.

4. *Masses in H:* We also normally replace masses by zero in H. Under some circumstances, it is appropriate to retain masses. In that case it is normally appropriate to put on-shell the external massive quark lines of the hard subgraph by modifying P_{HA} and P_{HB}.

5. *Alternative for H:* In applications, like QCD, where the gluon is massless, there can be important contributions from gluons that are soft in the sense of having very low energy, but collinear in the sense of having rapidity comparable to that of p_A or p_B. Such gluons we call "*soft-collinear*". From the point of view of regions and approximations, we will treat them as collinear. They can be external lines of the hard scattering.

To treat them adequately, we modify the definition of the approximator for a hard subgraph: masses are left unapproximated, and the external quark lines of the hard scattering are put on-shell, but now massive. The projectors for Dirac matrix connections between collinear and hard subgraphs are modified to project onto massive wave functions.

After use of Ward identities to extract the extra collinear gluons from the hard subgraph, the modified H subgraph can be replaced by the standard one.

6. *Numerators connecting subgraphs:* We project on the leading-power part of the numerators for Dirac lines and for gluons connecting the H, A, B and S subgraphs as follows:

   (a) For the attachment of a gluon from S to A, insert the following matrix to implement a Grammer-Yennie approximation (modified from (5.51)):

$$\frac{\hat{k}_{AS}^{\mu} n_1^{\nu}}{k_{AS} \cdot n_1 + i0}. \tag{10.20}$$

Note that $k_{AS} \cdot n_1 = \hat{k}_{AS} \cdot n_1$. The $i0$ prescription is correct when k_{AS} is defined to flow *out* of the collinear subgraph. The μ index is contracted with the A subgraph and ν with S. We will see that because the *approximated* A subgraph is contracted with the *approximated* momentum $\hat{k}_{AS}$, exact Ward identities can be applied to convert the S-to-A couplings to a Wilson line in direction n_1.

Thus the following replacement is made on the product of the A and S subgraphs:

$$A(p_A, k_{AS,1}, \ldots, k_{AS,N})^{\mu_1 \cdots \mu_N} S(k_{AS,1}, \ldots, k_{AS,N})_{\mu_1 \ldots \mu_N}$$

$$\mapsto A(p_A, \hat{k}_{AS,1}, \ldots, \hat{k}_{AS,N})^{\mu_1 \cdots \mu_N}$$

$$\times \prod_{j=1}^{N} \frac{\hat{k}_{AS,j,\mu_j} n_{1,\nu_j}}{k_{AS,j} \cdot n_1 + i0} S(k_{AS,1}, \ldots, k_{AS,N})^{\nu_1 \cdots \nu_N}, \tag{10.21}$$

where the individual momenta of the array k_{AS} are denoted by $k_{AS,j}$. It can be verified that the approximation is accurate to leading power when the k_{AS} momenta are in the soft region: i.e., all components are much less than Q, their rapidities are much lower than those of the collinear-to-A lines, and they are not in the Glauber region.

   (b) Similarly, for the attachment of a gluon from S to B, insert

$$\frac{\hat{k}_{BS}^{\mu} n_2^{\nu}}{k_{BS} \cdot n_2 + i0}, \tag{10.22}$$

where the momentum is flowing out of B.

   (c) For a gluon of momentum $k_{HA} + k_{HAS}$ out of H into the A subgraph, make the insertion

$$\frac{P_{HA}(k_{HA})^{\mu} w_2^{\nu}}{k_{HA} \cdot w_2 + i0}. \tag{10.23}$$

(d) For a gluon of momentum $k_{HB} + k_{HBS}$ out of H into the B subgraph, make the insertion

$$\frac{P_{HA}(k_{HB})^\mu w_1^\nu}{k_{HB} \cdot w_1 + i0}.$$ (10.24)

(e) For a Dirac line *entering* H from B, and for a Dirac line *leaving* H to A, insert the projector $\mathcal{P}_B = \frac{1}{2}\gamma^+\gamma^-$. This and the next item are cases of the Dirac spinor projector derived for the parton model in Sec. 6.1.2.

(f) But for a quark line in the reverse direction, use $\mathcal{P}_A = \frac{1}{2}\gamma^-\gamma^+$.

(g) If a version of the approximator is used in which approximated quark momenta are massive (and on-shell), then the projectors need to be modified, but in such a way that their massless limits exist. See problem 10.8 for possible definitions.

7. *Slightly scaled H:* The approximated hard scattering will generally not obey momentum conservation:

$$\sum_j (k_{HA, j}^+, 0, \mathbf{0}_{\mathrm{T}}) + \sum_j (0, k_{HB, j}^-, \mathbf{0}_{\mathrm{T}}) \neq q. \quad \text{(Pre-rescaling)} \quad (10.25)$$

Here j labels the lines carrying the relevant momenta. To correct momentum conservation, we apply overall scaling factors separately to the plus and minus components:

$$k_{HA, j}^+ \mapsto \tilde{k}_{HA, j}^+ = k_{HA, j}^+ \frac{q^+}{\sum_{j'} k_{HA, j'}^+}, \quad (10.26a)$$

$$k_{HB, j}^- \mapsto \tilde{k}_{HB, j}^- = k_{HB, j}^- \frac{q^-}{\sum_{j'} k_{HB, j'}^-}, \quad (10.26b)$$

a replacement to be made in H alone. Since we defined q to have $q_{\mathrm{T}} = 0$, no correction of approximated transverse momenta is needed. After the rescaling, we have exact momentum conservation:

$$\sum_j (\tilde{k}_{HA, j}^+, 0, \mathbf{0}_{\mathrm{T}}) + \sum_j (0, \tilde{k}_{HB, j}^-, \mathbf{0}_{\mathrm{T}}) = q. \quad \text{(Post-rescaling)} \quad (10.27)$$

The correction factors in (10.26) differ from unity by order m^2/Q^2. This is because the sums of the unapproximated collinear momenta are the external momenta: $\sum_j k_{HA, j} = p_A$, $\sum_j k_{HB, j} = p_B$, while p_A^-/p_A^+ and p_B^+/p_B^- are of order m^2/Q^2.

8. *Renormalization of extra UV divergences:* As in our treatment of DIS in a non-gauge theory, the approximator short-circuits certain loop-momentum components, thereby inducing UV divergences beyond those renormalized in the Lagrangian. These are removed by UV counterterms defined, for example, in the $\overline{\mathrm{MS}}$ scheme with the use of dimensional regularization. After we obtain factorization, renormalization will behave much like that for the local operators used in the OPE (e.g., Collins, 1984), but now applied to the operators defining the soft and collinear factors. We will generally leave this renormalization implicit until we do actual calculations.

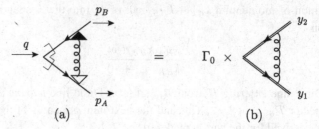

Fig. 10.6. Representation of soft term (10.29) for vertex graph, (a) before and (b) after use of Ward identities.

10.5 One-loop Sudakov form factor

We now illustrate the general definitions given in Sec. 10.4 by applying them to the one-loop graph, Fig. 10.4. The external fermions are on-shell, and the gluon has a non-zero mass m_g. But some issues will be illustrated by taking m_g to zero and/or taking the fermions off-shell.

The graph is

$$\Gamma_1 = \frac{ig^2}{(2\pi)^n} \int d^n k \, \frac{-g_{\kappa\lambda}}{(k^2 - m_g^2 + i0)} \frac{\bar{u}_A \gamma^\kappa (\not{p}_A - \not{k} + m)\gamma^\mu(-\not{p}_B - \not{k} + m)\gamma^\lambda v_B}{[(p_A - k)^2 - m^2 + i0][(p_B + k)^2 - m^2 + i0]},$$
(10.28)

where u_A and v_B are the Dirac wave functions for the outgoing quark and antiquark. Its leading regions are R_S, R_A, R_B and R_H, where the subscripts indicate the type of gluon momentum. For a compact notation, the region approximators and the region contributions are written $T_S \overset{\text{def}}{=} T_{R_S}$, $C_S \overset{\text{def}}{=} C_{R_S}$, etc.

10.5.1 Soft-gluon term C_S

The soft region R_S is a minimal region, so its term is obtained by applying the region's approximator, as defined in the list starting on p. 326:

$$C_S \Gamma_1 = T_S \Gamma_1 = \frac{ig^2}{(2\pi)^n} \int d^n k \, \frac{-g_{\kappa\lambda}}{(k^2 - m_g^2 + i0)} \frac{n_1^\kappa}{-n_1 \cdot k + i0} \frac{n_2^\lambda}{n_2 \cdot k + i0}$$

$$\times \frac{\bar{u}_A(-\not{k}_1)(\not{p}_A - \not{k}_1 + m)\mathcal{P}_B \gamma^\mu \mathcal{P}_B(-\not{p}_B - \not{k}_2 + m)\not{k}_2 v_B}{[(p_A - k_1)^2 - m^2 + i0][(p_B + k_2)^2 - m^2 + i0]}$$

$$= \frac{ig^2}{(2\pi)^n} \int d^n k \, \frac{n_1 \cdot n_2 \, \bar{u}_A \mathcal{P}_B \gamma^\mu \mathcal{P}_B v_B}{(k^2 - m_g^2 + i0)(-n_1 \cdot k + i0)(n_2 \cdot k + i0)}, \qquad (10.29)$$

which we write diagrammatically in Fig. 10.6. The hard scattering is just the factor γ^μ; it is surrounded by factors of $\mathcal{P}_B = \frac{1}{2}\gamma^+\gamma^-$, to project onto the appropriate on-shell massless Dirac wave functions. This is indicated by the hooks in Fig. 10.6(a), just as for the parton model in Fig. 6.4.

From (10.17) and (10.18), the projected gluon momenta in the collinear subgraphs are

$$k_1 = (0, k^- - e^{-2y_1}k^+, \mathbf{0}_T) \quad \text{and} \quad k_2 = (k^+ - e^{2y_2}k^-, 0, \mathbf{0}_T). \tag{10.30}$$

At the ends of the gluon line are applied the Grammer-Yennie approximants (10.20) and (10.22). The result is notated by the arrows at the ends of the gluon in Fig. 10.6(a).

To get the last line of (10.29), we applied the identities $\not{k}_2 = (\not{p}_B + \not{k}_2 + m) - (\not{p}_B + m)$ and $\not{k}_1 = (\not{p}_A - m) - (\not{p}_A - \not{k}_1 - m)$. For each of these, one term gives zero on a Dirac wave function and the other cancels the neighboring quark propagator. The result is represented in Fig. 10.6(b). On the left is a lowest-order vertex

$$\Gamma_0 = \bar{u}_A \mathcal{P}_B \gamma^\mu \mathcal{P}_B v_B. \tag{10.31}$$

On the right, the two double lines represent the $gn_1/(-n_1 \cdot k + i0)$ and $-gn_2/(n_2 \cdot k + i0)$ factors in (10.29). With two changes, these factors are just as the first-order application of the Feynman rules, Figs. 7.10 and 7.11, for Wilson lines, as in the gauge-invariant definition of a parton density, (7.40). One change is that we have two Wilson-line segments in different directions. The other is that the Wilson line in direction n_2 has a reversed sign of the coupling; physically this is because it approximates an outgoing antiquark, with the opposite charge to a quark.

We therefore identify $C_S \Gamma_1$ as Γ_0 times the one-loop value of the vacuum matrix element of two Wilson lines of opposite charge, joined at the origin:

$$\text{soft factor}_{\text{ver. 1}} = \langle 0| \, W(\infty, 0, n_2)^\dagger \, W(\infty, 0, n_1) \, |0\rangle, \tag{10.32}$$

where W is defined by

$$W(\infty, 0; n) = P\left\{ e^{-ig_0 \int_0^\infty d\lambda \, n \cdot A_{(0)\alpha}(\lambda n) \, t_\alpha} \right\}. \tag{10.33}$$

Notice that this definition uses the bare coupling and field, as needed to get the correct gauge-transformation properties. A factor of a representation matrix t_α of the gauge group appears in the exponent to give a formula that is also appropriate for a non-abelian theory. In the simpler case of an abelian gauge theory, one omits the t_α factor, and one can replace the coupling and field by their renormalized counterparts, since $g_0 A_{(0)} = g\mu^\epsilon A$ in an abelian theory. The opposite charge of the Wilson line for direction n_2 is implemented by a hermitian conjugation in (10.32).

After we formulate a factorization theorem, we will see that the formula for the one-loop soft factor, $C_S \Gamma_1$, is sufficient to determine almost completely the Wilson-line definition. However, we will modify some details of the definition. Hence we include a version subscript on the left-hand side of (10.32). The matrix element in (10.33) is a primary ingredient in the later redefinitions.

The approximations used to give $C_S \Gamma_1$ are valid in the soft region, provided we deform the integration contour out of the Glauber region. As we will show in Sec. 10.6.4, the choice of space-like vectors (10.15b) for n_1 and n_2, and of the $i0$ prescriptions in (10.29) is needed to be compatible with the contour deformation.

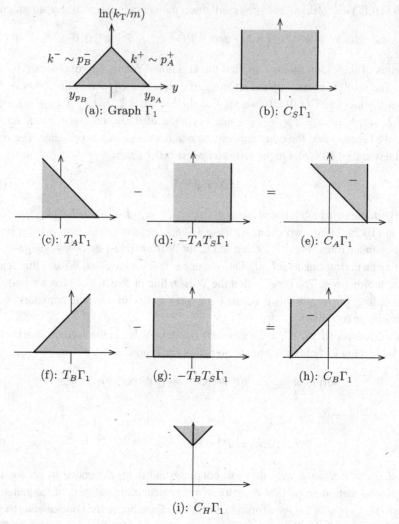

Fig. 10.7. Main regions in y and k_T for one-loop Sudakov form factor. The shaded areas indicate where there are leading-power contributions, and the thick lines show where there is a cutoff. A lack of a thick edge to a shaded area indicates that the area goes to infinity. These diagrams are for the original graph and for various terms in the decomposition of the graph by regions, with subtractions. The $-$ signs on a shaded region indicate a negative contribution. The top of the triangle in graph (a) is at $\ln(k_T/m) = \ln(Q/m)$.

10.5.2 *(Double) leading-logarithm approximation*

To understand the nature of the approximation, we make plots in the space of $\ln k_T$ and y, where y is the gluon rapidity $y = \frac{1}{2}\ln|k^+/k^-|$, and examine where the main contributions arise, both for the original graph and for terms contributing to each $C_R\Gamma_1$. These are shown in Fig. 10.7.

The variables are logarithmic in ordinary momentum components. With respect to these variables, we will find that the original integral Γ_1 has a uniform integrand in the interior of the triangle in Fig. 10.7(a). This uniform value is in fact that of the soft approximation $C_S\Gamma_1$. Outside of the triangle, the integrand falls off, so that a first approximation to the original graph is the uniform integrand times the area of the triangle, which is a coefficient times $\ln^2 Q$. This gives the double leading-logarithm approximation (LLA) to Γ_1. The edges and corners of the triangle give non-leading logarithms, and remaining contributions are in fact power-suppressed.

We will see that the soft term, $C_S\Gamma_1$, also has important contributions from outside the triangle. But we will find that these other contributions cancel corresponding parts of the terms $C_R\Gamma_1$ for other regions R; see Fig. 10.7(b–i). The total reproduces Γ_1 up to a power-suppressed remainder.

In the core of the soft region the original graph Γ_1 is correctly approximated by the soft term $C_S\Gamma_1$, and the approximation remains correct when n_1 and n_2 are replaced by light-like vectors, to give

$$\frac{ig^2}{(2\pi)^4} \int_{\text{core of soft region}} \mathrm{d}^4 k \, \frac{\bar{u}_A \mathcal{P}_B \gamma^\mu \mathcal{P}_B v_B}{(k^2 - m_g^2 + i0)(-k^- + i0)(k^+ + i0)}, \tag{10.34}$$

where we now work in four-dimensional space-time. We apply contour integration to the k^- integral,[6] which gives a non-zero result only for $k^+ > 0$. By closing the contour on the gluon pole and changing variables from k^+ to $y = \frac{1}{2}\ln|k^+/k^-|$ and from $\boldsymbol{k}_T$ to $\ln k_T$, we obtain:

$$\frac{-g^2}{4\pi^2} \int_{\text{core of soft region}} \mathrm{d}\ln k_T \, \mathrm{d}y \, \bar{u}_A \mathcal{P}_B \gamma^\mu \mathcal{P}_B v_B \, \frac{k_T^2}{k_T^2 + m_g^2}$$

$$\simeq \frac{-g^2}{4\pi^2} \int_{\text{core of soft region}} \mathrm{d}\ln k_T \, \mathrm{d}y \, \bar{u}_A \mathcal{P}_B \gamma^\mu \mathcal{P}_B v_B. \tag{10.35}$$

The right-hand form is obtained by restricting attention, for reasons that will soon be apparent, to large enough k_T that we can neglect the gluon mass.

Original graph

The result has a uniform integrand, and so we estimate the size of the original unapproximated graph by the area of the relevant part of the plane of $\ln k_T$ and y. We will find that the integrand falls off relative to (10.35) near the edges of the triangle in Fig. 10.7(a), so the area is that of the triangle. We examine the limits provided by each propagator denominator in turn.

In the gluon propagator, the gluon mass effectively cuts off the k_T integral at m_g, and this gives the lower boundary of the triangle, at $\ln(k_T/m_g) \simeq 0$. This is a fuzzy cutoff, not a sharp cutoff. Given that the dimensions of the triangle are of order $\ln(Q/m)$, the width of the fuzzy edge relative to the triangle is small, of order $1/\ln(Q/m)$.

---

[6] Strictly speaking, this application of contour integration includes values of k^- all the way to infinity, i.e., outside the soft region. To see that this is not a problem, observe that the contribution we use in later equations is from the gluon pole. The errors, i.e., the non-pole terms, are from a non-soft region which does not concern us here.

The A-quark denominator (after setting k on the gluon mass-shell from the contour integration and after setting $p_A^2 = m^2$) is

$$(p_A - k)^2 - m^2 = -2p_A^+ k^- - 2p_A^- k^+ + m_g^2. \tag{10.36}$$

We write $2p_A^+ k^-$ in terms of rapidities as $m\sqrt{k_T^2 + m_g^2}\,e^{y_{p_A} - y}$, where y_{p_A} is the rapidity of the A quark, also taken as the rapidity of the n_1 vector. The simplest soft approximation replaces the denominator by $-2p_A^+ k^-$. The second term in the denominator becomes equally important when the rapidity of the gluon is comparable to that of p_A, thereby providing a cutoff requiring $y \lesssim y_{p_A}$. Next, in the unapproximated graph, the k^- poles are all in the lower half plane if $k^+ > p_A^+$; this limits k^+ to be less than p_A^+. The m_g^2 term in (10.36) provides no stronger constraint.

Similar limits are associated with the B quark.

If the gluon mass is comparable to the quark mass, as we will assume for the moment, then the limits $k_T \gtrsim m$, $k^+ \lesssim p_A^+$, and $k^- \lesssim p_B^-$ dominate, giving the triangle in Fig. 10.7(a). The two diagonal lines give $y_{p_B} + \ln(k_T/m) \lesssim y \lesssim y_{p_A} - \ln(k_T/m)$, which intersect at $k_T \sim Q$.

But when the gluon mass is made small or zero (as in QCD perturbation theory), the range of k_T extends down, and other limits become important.

Finally, the graph has a renormalized UV divergence for $k_T \gg Q$. We assign this to the line going vertically up from the top vertex of the triangle.

The area of the triangle is $\frac{1}{2}(y_{p_A} - y_{p_B})\ln(Q^2/m^2) = \frac{1}{2}\ln^2(Q^2/m^2)$, which gives the leading-logarithm approximation

$$\text{LLA of } \Gamma_1 = \frac{-g^2 \ln^2(Q^2/m^2)}{16\pi^2}\,\bar{u}_A \mathcal{P}_B \gamma^\mu \mathcal{P}_B v_B. \tag{10.37}$$

This has two logarithms for a one-loop graph, unlike the case for ordinary renormalization-group (RG) logarithms, which are one per loop. At high energy the approximated vertex $\bar{u}_A \mathcal{P}_B \gamma^\mu \mathcal{P}_B v_B$ equals the unapproximated vertex $\bar{u}_A \gamma^\mu v_B$, up to a power-suppressed correction.

The effects of the cutoffs are important only in a finite range of y and $\ln k_T$ near the edges of the triangle. Thus they do not affect the double logarithm. At large Q^2, the sides of the triangle contribute single logarithms, while the vertices contribute constants. The vertical line above the triangle gives an RG single logarithm. Further contributions are suppressed by a power of Q.

All-orders sum of LLA

This line of argumentation can be extended to higher loops, to give the leading logarithms (Sudakov, 1956; Jackiw, 1968) for every order of perturbation theory. These form an exponential series. If the assumption is made that it is sufficient to retain the leading logarithm in each order, then one obtains the LLA for the form factor:

$$F \simeq e^{-g^2 \ln^2(Q^2/m^2)/(16\pi^2)}\,\bar{u}_A \mathcal{P}_B \gamma^\mu \mathcal{P}_B v_B. \tag{10.38}$$

We will derive this from our general factorization approach in Sec. 10.11.5. The result given above is for the case of a massive gluon with on-shell external quarks, and was first found

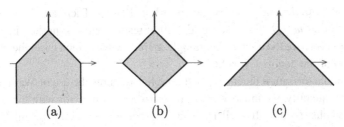

Fig. 10.8. Modifications to Fig. 10.7(a) when: (a) the gluon mass is zero; (b) the gluon mass is zero but the external quarks are off-shell; (c) the quark and the gluon masses are both zero, and the external quarks are on-shell.

by Jackiw (1968). As we will see below, the case of a massless gluon with off-shell external quarks has double the coefficient of the double logarithm, and this was what Sudakov (1956) actually calculated.

At large Q the LLA form factor drops faster than any power of Q. This obviously indicates that power-law corrections might dominate, for sufficiently large Q. However, without further information, there is no guarantee that non-leading logarithms have to fall into the same pattern of summing to a strongly decreasing function of Q. For example, as a hypothetical example, if the non-leading logarithms consisted of a single term g^2, this would be non-vanishing at large Q, and would dominate the LLA. In some analogous problems in QCD (Ch. 13) such a phenomenon does occur, a standard example being the Drell-Yan cross section at small transverse momentum; the LLA does not even get correct the qualitative behavior of the cross section. The factorization approach provides a much more systematic and powerful approach to dealing with these issues.

10.5.3 Massless gluon; off-shell external quarks

The above estimates assumed that the gluon and quark masses are comparable, and that the external quarks are on-shell. But in QCD the gluon is massless. Although a massive gluon might be considered more representative of the real physics of a theory with quark and gluon confinement, perturbative calculations definitely need a massless gluon. Moreover applications to QED require a massless photon instead. We will also need to consider vertex graphs embedded in bigger graphs, so it is also useful to understand the effect of taking the external quarks off-shell.

Figure 10.8(a) shows the effect of setting $m_g = 0$, which is to remove the lower cutoff on k_T. Thus a leading contribution occurs all the way to $k_T = 0$, or minus infinity on a logarithmic scale. As for the rapidity range at low k, the dominant restriction is caused by the rapidities of the external quarks, which give the lower vertical lines. The integral has a divergence, which is a conventional IR divergence, as in QED, with a coefficient that grows with energy like $y_{p_A} - y_{p_B}$.

The IR divergence arises from the $1/(k^- k^+)$ factor in the soft approximation. If we now set the external quarks off-shell, there is an extra term in the quark denominators. This cuts off the k_T integral at the lower end. If the external quark virtuality is of the order of the quark mass, i.e., m^2, the result is shown in Fig. 10.8(b). There are effective cutoffs at k^- and k^+ of order m^2/Q. The leading-logarithm result comes from the diamond-shaped region, which has twice the area of the triangle in the massive gluon case, thereby doubling the coefficient of the double logarithm.

The general factorization theory we will establish requires the use of Ward identities. In a real physical quantity, we must combine the off-shell form factor with the contributions from other graphs, so the off-shell form factor does not represent the final result for a physical quantity.

Finally there is the case of the on-shell form factor with all particles massless. In that case, there is no longer a cutoff caused by the rapidities of the external lines, so we have the region shown in Fig. 10.8(c), where we effectively have a doubly logarithmic divergence composed of both IR and collinear divergences.

10.5.4 Region for $C_S \Gamma_1$

In contrast to the actual denominators of the quark propagators, the approximated eikonal denominators $(n_2 \cdot k + i0)(-n_1 \cdot k + i0)$ in the soft term C_S provide cutoffs only at the rapidities of the Wilson line. As illustrated in Fig. 10.7(b), the limits on gluon rapidity, $y_2 \lesssim y \lesssim y_1$, are the same at all k_T. We choose the rapidities of n_1 and n_2 to be approximately the same as the rapidities of the external quark lines p_A and p_B. The soft term forms a good approximation at small y and k_T. It is most accurate at the center of the bottom line in Fig. 10.7(a) and (b), and in fact is equally good for even smaller k_T. The approximation degrades as one approaches the upper lines of the triangle; one can characterize these lines as where the error in the soft term is around 100%.

The soft term obviously contributes in a region where the original graph does not. This is above the triangle, and therefore where at least one of the following holds: the energy of the gluon is large, its rapidity is large, and/or its transverse momentum is large. These all concern other regions than the soft region. Compensation for the extra area for the soft term will be obtained from subtraction terms in the terms for regions bigger than the soft region.

We can apply the same area argument as we used for the LLA for the original graph. There is evidently an infinity (multiplied by $y_1 - y_2$) for the infinite range of k_T. This can be regulated dimensionally and renormalized, although we will not exhibit the calculation yet.

Our general proof will require us to understand the errors in the soft approximation more systematically. To do this we return to ordinary non-logarithmic momentum space. The PSSs forming the skeletons of the leading regions are shown in Fig. 10.9(a). The relative error in approximating the integrand is

$$\frac{|C_S I_1 - I_1|}{\|C_S I_1\|} = O\left(\frac{|k^+|}{p_A^+}, \frac{|k^-|}{p_B^-}, e^{-2(y_{p_A} - y)}, e^{-2(y - y_{p_B})}\right). \tag{10.39}'$$

Here I_1 denotes the integrand. One might expect the denominator to be just the absolute value $|C_S I_1|$. But we use the double bars, $\|C_S I_1\|$, to indicate that in a more general situation

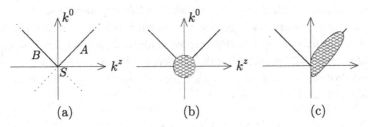

Fig. 10.9. (a) Leading PSSs for one-loop Sudakov form factor. (b) Neighborhood of S for evaluating errors in soft approximation. (c) Neighborhood of A for evaluating errors in collinear-to-A approximation. The squashing on the left indicates that we restrict attention to positive rapidity.

a modification is needed. The problem is that there may be what we can term accidental cancellations; for example, a numerator might have a variable sign, with necessarily a zero at some place in the integration. We wish to use the general order of magnitude of the integrand, for which we use the power-counting estimate of $C_S I_1$, obtained by the methods of Sec. 5.8, with avoidance of any accidental cancellations. The result is denoted $\| C_S I_1 \|$. We also use the approximated integrand $C_S I_1$ in $\| C_S I_1 \|$, rather than the original integrand I_1.

The right-hand side of (10.39) simply comes from listing the sources of error in the soft approximation, i.e., from examining the terms in the quark denominators and numerators that were neglected in making the soft approximation. The first two terms simply measure distance from the center of the soft region, viewed in the center-of-mass frame; these sources of error are roughly constant on surfaces such as those in Fig. 10.9(b) surrounding S. In a purely Euclidean asymptotic problem, this would be the whole story.

But in a Minkowski-space problem, such as ours, the errors worsen as the rapidity of the gluon gets large, and approaches the A and B lines. The errors are given quantitatively by the last two terms in (10.39). These are of order m^2/Q^2 when the gluon rapidity y is small. But when the gluon rapidity is comparable to that of one of the external quarks, the errors become of order unity.

10.5.5 Why integrate $C_S \Gamma_1$, etc., over all k?

Given that $C_S \Gamma_1$ has important contributions from a much broader range of loop momentum than has Γ_1, it is natural to want to restrict the integration to, for example, the triangular range in Fig. 10.7(a). Nevertheless we define $C_S \Gamma_1$ (and all other C_R) to have an integral over *all* loop momenta. The combination of $C_S \Gamma_1$ with terms for other leading regions will not only cancel the large excess regions, but will correct the inaccuracies in the soft approximation at the edges of the triangle. Then the sum over $C_R \Gamma_1$ will give a complete and useful representation of the leading-power part of Γ_1.

The reasons for not using cutoffs (beyond those given by the finite rapidities of the Wilson lines) are as follows. To get a systematic treatment, we need to have operator definitions for the factors in the factorization theorem. An example definition is (10.32), whose main one-loop graph gives $C_S\Gamma_1$. A cutoff on the loop momentum k would require an unpleasantly complicated operator. It is not known how to do this and combine it with the Ward identities that we use later. Proving Ward identities needs shifts in loop momenta and uses gauge-invariance properties of the operators; these are difficult to make consistent with a cutoff. Instead, without cutoffs we are led directly to simple Wilson-line operators whose gauge-invariance properties are obvious.

10.5.6 Collinear-A term C_A

We now construct $C_A\Gamma$, corresponding to the gluon being collinear to p_A. First we just apply the approximator for region R_A. Using the definitions in the list starting on p. 326 we get

$$
T_A\Gamma_1 = \frac{ig^2}{(2\pi)^n} \int d^n k \, \frac{-g_{\kappa\lambda}}{(k^2 - m_g^2 + i0)} \frac{w_2^\lambda}{w_2 \cdot k + i0}
$$

$$
\times \, \frac{\bar{u}_A \gamma^\kappa (\not{p}_A - \not{k} + m) P_B \gamma^\mu (-p_B^- \gamma^+ - \gamma^- k^+)\gamma^- k^+ P_B v_B}{[(p_A - k)^2 - m^2 + i0][2p_B^- k^+ + i0]}
$$

$$
= \frac{ig^2}{(2\pi)^n} \int d^n k \, \frac{-g_{\kappa\lambda}}{(k^2 - m_g^2 + i0)} \frac{\bar{u}_A \gamma^\kappa (\not{p}_A - \not{k} + m) P_B \gamma^\mu P_B (-w_2^\lambda) v_B}{[(p_A - k)^2 - m^2 + i0](w_2 \cdot k + i0)}. \tag{10.40}
$$

The collinear approximant changes the quark denominator $(p_B + k)^2 - m^2$ to $2p_B^- k^+$, because in the hard subgraph it replaces p_B and k by massless vectors in the minus and plus directions, and sets masses to zero. Examining the neglected terms $2p_B^+ k^-$ and k^2, with the knowledge that $2k^+ k^-$ and k_T^2 are comparable shows that the relative errors in this approximation are of order $e^{-2(y-y_{p_B})}$ and $e^{-(y-y_{p_B})} k_T/m$. Thus the approximant is accurate when the gluon rapidity is much larger than the rapidity of the B line. There is also a degradation for large $k_T \gg m$, but that concerns the hard-gluon region, to be treated later.

The region of (y, k_T) space for $T_A\Gamma_1$ is shown in Fig. 10.7(c). Since the eikonal denominator $w_2 \cdot k$ is k^+, without an additional k^- term, the integral has a rapidity divergence, where the rapidity of the gluon goes to negative infinity.

Our aim is to construct a term for the collinear-to-A region such that $C_A\Gamma_1 + C_S\Gamma_1$ is accurate over the whole of the soft and collinear-to-A regions. Observe both of the soft term and the collinear approximation contribute in each other's regions. So we compensate the double counting by subtracting

$$
T_A T_S \Gamma_1 = \frac{ig^2}{(2\pi)^n} \int d^n k \, \frac{-g_{\kappa\lambda}}{(k^2 - m_g^2 + i0)} \frac{n_1^\kappa (-w_2^\lambda) \bar{u}_A P_B \gamma^\mu P_B v_B}{(-n_1 \cdot k + i0)(w_2 \cdot k + i0)}. \tag{10.41}
$$

The $n_2/(n_2 \cdot k + i0)$ factor of $T_S\Gamma_1$ is in the hard subgraph with respect to the collinear-to-A approximator T_A. Hence applying T_A, defined by (10.19a) and (10.23), changes n_2 to the light-like vector w_2.

We therefore define the term for the A region by

$$C_A\Gamma_1 = T_A(1 - T_S)\Gamma_1$$

$$= \frac{ig^2}{(2\pi)^n} \int d^n k \frac{-g_{\kappa\lambda}}{(k^2 - m_g^2 + i0)} \left[\frac{\bar{u}_A\gamma^\kappa(\not{p}_A - \not{k} + m)}{(p_A - k)^2 - m^2 + i0} - \frac{\bar{u}_A n_1^\kappa}{-n_1 \cdot k + i0} \right]$$

$$\times \frac{(-w_2^\lambda)}{(w_2 \cdot k + i0)} \mathcal{P}_B \gamma^\mu \mathcal{P}_B v_B. \tag{10.42}$$

This results in the cancellation of the rapidity divergence, justifying our use of a light-like vector in the collinear approximant. The cancellation is because the soft approximant on the A side is accurate when the gluon has large negative rapidity relative to p_A. Thus we get a cancellation in the square-bracket term in (10.42) with the result going to zero as the gluon's rapidity goes to minus infinity.

The placement of the Dirac projectors $\mathcal{P}_B$ is also critical to making the formalism work correctly.

The result is that the $C_A\Gamma_1$ term is power-suppressed in the soft region; Fig. 10.7(e). The combination of the terms constructed so far, $C_S\Gamma_1 + C_A\Gamma_1$, gives a good approximation to Γ_1 over the whole of the soft and the collinear-A regions, with a restriction to the positive rapidity side.

We can see this by observing that the remainder is $\Gamma_1 - C_S\Gamma_1 - C_A\Gamma_1 = (1 - T_A)(1 - T_S)\Gamma_1$. The $1 - T_S$ factor gives a suppression basically by a power of $|k|/Q$ but with a degradation to $e^{-(y_{p_A}-y)}$ as we go around the soft PSS and approach the PSS A, given that y_1 is close to y_{p_A}. At this point we restrict to positive gluon rapidity, leaving negative rapidity to our treatment of $C_B\Gamma_1$. The $1 - T_A$ factor gives a suppression e^{-y}, when $k_T \lesssim m$. At the soft end of the A region, this compensates the worsening of $1 - T_S$ factor. It also gives a power-suppression over the rest of the A region, a power of k_T/Q. Thus we get surfaces of constant error for $C_S\Gamma_1 + C_A\Gamma_1$ as symbolized in Fig. 10.9(c).

The collinear-to-A term itself is suppressed in the soft region, because of the $1 - T_S$ factor, as illustrated in Fig. 10.7(e). Thus for central rapidity only the C_S term is needed to get a good approximation to Γ_1, which it was constructed to do.

Furthermore, the soft subtraction has ensured that the C_A term is also suppressed in the whole of the opposite collinear region. This is an example of a general result critical to our general treatment of overlapping regions: C_A is suppressed both in regions smaller than A, i.e., S, and in regions that overlap with it, in this case the B region.

A generally applicable argument is that in applying T_A we made the first term in the expansion of the B propagator in powers of k^- and k_T. In $T_A(1 - T_S)\Gamma_1$, the $1 - T_S$ factor gives a suppression for small k^+ and k_T from its application to the A side. Going to the B region involves extrapolating the common B-side factor to large k^-. The suppression at small k^+ and k_T continues to apply.

Effectively, once the approximator for the A region, T_A, is applied, the power-counting in the B region corresponds to that for the intersection of the two overlapping regions, i.e., $A \cap B = S$.

In contrast to these cancellations, there is a contribution in the upper region in Fig. 10.7(e), where the gluon rapidity is positive, but its k^+ is much larger than p_A^+. Such a contribution is not present in the original graph, but is an artifact of the soft subtraction, in the term $-T_A T_S \Gamma_1$, as is the divergence when $k_T \to \infty$. Strange though this contribution might appear, it will allow us to derive convenient evolution equations by differentiating with respect to the rapidity cutoffs associated with the vertical lines in Fig. 10.7(b), (e), and (h). When we add C_S, C_A, and C_B, there is a cancellation of these extra contributions for the case that $k_T \lesssim Q$. This leaves only the region $k_T \gtrsim Q$, which is the province of the hard region H, which we have yet to treat, and whose double-counting subtractions will compensate for the incorrect value of $C_S + C_A + C_B$ in the hard region. There is also an actual divergence as $k_T \to \infty$, which we remove by UV renormalization, which will correspond to conventional UV renormalization defining the operators used to construct the soft and collinear factors in the factorization property.

Note that when the transverse momentum is large, $k_T \gg Q$, there is also an important region of *negative* gluon rapidity. This is surprising given that $C_A \Gamma_1$ is intended to deal with gluons that are collinear to p_A, i.e., of positive rapidity. But the problematic region is of hard momenta, and so its full treatment will also bring in the term $C_H \Gamma_1$, whose subtraction terms will correct the apparently problematic regions.

10.5.7 Collinear-B term C_B

The collinear-to-B term is constructed exactly similarly to the collinear-to-A term:

$$
C_B \Gamma_1 = T_B (1 - T_S) \Gamma_1
$$

$$
= \frac{ig^2}{(2\pi)^n} \int d^n k \; \frac{-g_{\kappa\lambda}}{(k^2 - m_g^2 + i0)} \frac{\bar{u}_A \mathcal{P}_B \gamma^\mu \mathcal{P}_B w_1^\kappa}{-w_1 \cdot k + i0}
$$

$$
\times \left[\frac{(-\not{p}_B - \not{k} + m)\gamma^\lambda v_B}{(p_B + k)^2 - m^2 + i0} - \frac{-n_2^\lambda v_B}{n_2 \cdot k + i0} \right]. \tag{10.43}
$$

The contributing regions for this term and its components, shown in Fig. 10.7(f)–(h), are, naturally, a mirror image of those for the A region.

Just as before, the sum of the soft and collinear-to-B terms, i.e., $C_S \Gamma_1 + C_B \Gamma_1$, gives a good approximation in the combination of the S and B regions. We next observe that each of $C_A \Gamma_1$ and $C_B \Gamma_1$ is suppressed in both the central soft region and the opposite collinear region. Thus we can add all three terms to get $C_S \Gamma_1 + C_A \Gamma_1 + C_B \Gamma_1$ and the result provides a good approximation to Γ_1 over all three regions, including both positive and negative rapidity.

10.5.8 Hard term C_H

The only degradation in $C_S \Gamma_1 + C_A \Gamma_1 + C_B \Gamma_1$ as an approximation to Γ_1 occurs as we move away from the combined $S \cup A \cup B$ regions, i.e., as we go into the hard region H of large transverse momenta and of virtualities of order Q^2. We define the approximator T_H

for this region to make a massless approximation. As before, we avoid double counting in the C_H term specific to this region by applying the approximator to Γ_1 only after subtracting the contributions from smaller regions, i.e.,

$$C_H\Gamma_1 = T_H(\Gamma_1 - C_S\Gamma_1 - C_A\Gamma_1 - C_B\Gamma_1)$$

$$= T_H(1 - T_A - T_B)(1 - T_S)\Gamma_1. \tag{10.44}$$

We have seen that $C_S\Gamma_1 + C_A\Gamma_1 + C_B\Gamma_1$ gives a good approximation to Γ_1 near the combined S, A, and B regions, so that $\Gamma_1 - C_S\Gamma_1 - C_A\Gamma_1 - C_B\Gamma_1$ is power-suppressed in the distance to any of these regions. Thus the remaining contribution is when the momenta are hard, i.e., for k_T of order Q or larger, i.e., in the H region. So we define the approximator T_H for this region to set masses to zero, and to make p_A and p_B massless. It also replaces the n_1 and n_2 vectors (in the definition of T_S) by light-like versions: $n_1 \mapsto w_1 = (1, 0, \mathbf{0}_T)$, and $n_2 \mapsto w_2 = (0, 1, \mathbf{0}_T)$. The soft term $T_S\Gamma_1$ and the soft subtractions in $C_A\Gamma_1$ and $C_B\Gamma_1$ now have the same light-like vectors, so they combine to a single added term, and we get

$$C_H\Gamma_1 = \frac{ig^2}{(2\pi)^n} \int d^n k \, \frac{-1}{(k^2 + i0)} \bar{u}_A \mathcal{P}_B \left\{ \frac{\gamma^\kappa (p_A^+\gamma^- - \slashed{k})\gamma^\mu (-p_B^-\gamma^+ - \slashed{k})\gamma_\kappa}{[-2p_A^+k^- + k^2 + i0] \, [2p_B^-k^+ + k^2 + i0]} \right.$$

$$- \frac{\gamma^+(p_A^+\gamma^- - \slashed{k})}{-2p_A^+k^- + k^2 + i0}\gamma^\mu \frac{-1}{k^+ + i0} - \frac{1}{-k^- + i0}\gamma^\mu \frac{(-p_B^-\gamma^+ - \slashed{k})\gamma^-}{2p_B^-k^+ + k^2 + i0}$$

$$\left. + \frac{1}{-k^- + i0}\gamma^\mu \frac{-1}{k^+ + i0} \right\} \mathcal{P}_B v_B. \tag{10.45}$$

10.5.9 UV divergences

The original graph Γ_1 has a UV divergence. This is canceled in a complete calculation of the one-loop vertex when the correct definition is used for the current at the photon vertex. The current is a Noether current for a conserved charge, with unit coefficient when the current is expressed in terms of bare fields: $j^\mu = \bar{\psi}_0 \gamma^\mu \psi_0$. In terms of renormalized fields, it has a factor Z_2: $j^\mu = Z_2 \bar{\psi} \gamma^\mu \psi$, and this factor of Z_2 cancels the divergences in the loop calculations. This is a well-known standard result in renormalization theory.[7] This results in a non-zero anomalous dimension associated with the one-loop graph. In the full form factor calculation, we must also allow for the LSZ residue factors for the external on-shell quarks. These are also associated with Z_2, but inversely, so that the complete form factor is RG invariant.

However, the hard, collinear and soft factors in (10.11) all have their own UV renormalization and need renormalization that is different from that in the current itself. This is illustrated by the one-loop quantities computed above, $C_S\Gamma_1$, $C_A\Gamma_1$, $C_B\Gamma_1$, and $C_H\Gamma_1$. Their UV divergences are associated with new vertices: where a Wilson line attaches to an ordinary field (in $C_A\Gamma_1$ and $C_B\Gamma_1$) and where two Wilson lines attach to each other (in

---

[7] However, there are some complications beyond the ones seen in most textbooks. See Collins, Manohar, and Wise (2006) for a correct treatment.

$C_S\Gamma_1$ and the subtractions in $C_A\Gamma_1$ and $C_B\Gamma_1$). As we will see, all these divergences are logarithmic. Our ultimate definitions of the region contributions include renormalization counterterms to remove the UV divergences.

Finally, $C_H\Gamma_1$ is formed from the original graph together with subtractions for the smaller regions, all taken in the massless limit. Therefore, in the sum over all regions, i.e., $C_S\Gamma_1 + C_A\Gamma_1 + C_B\Gamma_1 + C_H\Gamma_1$, the extra UV divergences cancel to leave just the same UV divergence as in Γ_1. This is necessary if this sum is to give a correct large-Q asymptote for Γ_1.

We will treat the extra UV divergences and their renormalization in more detail later.

But for now we just examine one simple case, the UV divergence for $C_S\Gamma_1$, and indicate some interesting properties, notably that it depends on the directions of the Wilson lines, and more specifically on the hyperbolic angle between them.

In the formula (10.29) for $C_S\Gamma_1$, the integrals over the longitudinal momenta are readily performed, e.g., by contour integration over k^- followed by an elementary integral over k^+. Without the UV counterterm

$$\frac{C_S\Gamma_1}{\Gamma_0} \overset{\text{no c.t.}}{=} \frac{-g^2(2\pi\mu)^{2\epsilon}}{8\pi^2}(y_1 - y_2)\coth(y_1 - y_2) \int d^{2-2\epsilon}k_T \frac{1}{k_T^2 + m_g^2}$$

$$= \frac{-g^2}{8\pi^2}(y_1 - y_2)\coth(y_1 - y_2)\Gamma(\epsilon)\left(\frac{4\pi\mu^2}{m_g^2}\right)^\epsilon, \tag{10.46}$$

where Γ_0 is given by (10.31). The UV counterterm in the $\overline{\text{MS}}$ scheme is

$$\frac{g^2 S_\epsilon}{8\pi^2\epsilon}(y_1 - y_2)\coth(y_1 - y_2), \tag{10.47}$$

so that the renormalized $C_S\Gamma_1$ is

$$\frac{C_S\Gamma_1}{\Gamma_0} \overset{\text{renorm.}}{=} \frac{-g^2}{8\pi^2}(y_1 - y_2)\coth(y_1 - y_2)\ln\frac{\mu^2}{m_g^2}. \tag{10.48}$$

Observe the dependence on the difference in rapidities between the lines. (Lorentz invariance requires that the dependence is on the rapidity difference, not on the rapidities separately, since we can always transform to a frame in which one rapidity, y_2 say, is zero, in which case the other line's rapidity is changed to $y_1 - y_2$.)

Since $e^{-(y_1-y_2)} \sim m^2/Q^2$ at large Q, a correct leading-power approximation is to replace $\coth(y_1 - y_2)$ by unity. This leaves the remaining factor of $y_1 - y_2$. Therefore, there is a further divergence if we take the Wilson lines light-like, an explicit example of a rapidity divergence.

We call $y_1 - y_2$ the hyperbolic angle between the two vectors. The name is appropriate because if we continue y_1 and y_2 to imaginary values, with $y_1 - y_2 = i\theta$, then n_1 and n_2 are vectors in Euclidean space, and θ is the ordinary angle between them. (Actually θ is the angle between n_1 and $-n_2$.)

We have seen that $C_S\Gamma_1$ is a one-loop term in the vacuum expectation value (10.32) of a Wilson line composed of two straight line segments in directions n_1 and n_2, joined at a

Fig. 10.10. Notation for derivative of $C_S\Gamma_1$ with respect to y_1. The crossed vertex is defined as a rapidity derivative of the Wilson line, in (10.49).

cusp. Our calculation has shown that there is a UV divergence associated with the cusp and that both the divergence and the associated anomalous dimension depend on the hyperbolic angle between the two lines.

10.5.10 Evolution with respect to Wilson-line rapidity

To illustrate evolution of the soft factor with respect to the direction of a Wilson line, consider the derivative of the one-loop soft term $C_S\Gamma_1$ with respect to y_1. This is obtained by differentiating the n_1-dependent factor:

$$\frac{\partial}{\partial y_1}\left(\frac{n_1}{-n_1 \cdot k + i0}\right) = \frac{\partial}{\partial y_1}\left(\frac{(1, -e^{-2y_1}, \mathbf{0}_T)}{-k^- + e^{-2y_1}k^+ + i0}\right)$$

$$= \frac{-n_1^2 \tilde{k}}{(-n_1 \cdot k + i0)^2}, \tag{10.49}$$

where $\tilde{k} \stackrel{\text{def}}{=} (k^+, -k^-, \mathbf{0}_T)$. Let us represent this object by a vertex with a cross, as in Fig. 10.10. Then the derivative of $C_S\Gamma_1$ is

$$\frac{\partial\, C_S\Gamma_1}{\partial y_1} = \frac{ig^2}{(2\pi)^n}\int d^n k \frac{1}{(k^2 - m_g^2 + i0)}\frac{-n_1^2\tilde{k}\cdot n_2\,\Gamma_0}{(-n_1\cdot k + i0)^2\,(n_2\cdot k + i0)} + \text{UV c.t.} \tag{10.50}$$

The key to further simplifications in the full evolution equation is that the derivative with respect to y_1 restricts the integral over k to rapidities near y_1, to leading power, so that we can take the limit $y_2 \to -\infty$ without a rapidity divergence. To see this, we observe that in the integrand of (10.50), when the rapidity y of the gluon is much less than y_1, the factor $1/(-n_1 \cdot k)^2$ becomes $1/(k^-)^2 \propto e^{2y}$, which gives a suppression. So (10.50) concerns gluons of rapidity close to y_1.

Therefore in (10.50) we replace n_2 by a light-like vector w_2 in the minus direction. The numerator and denominator factors $n_2 \cdot \tilde{k}$ and $n_2 \cdot k$ both become $w_2^- k^+$, and therefore cancel, so that

$$\frac{\partial\, C_S\Gamma_1}{\partial y_1} = \frac{ig^2}{(2\pi)^n}\int d^n k \frac{1}{(k^2 - m_g^2 + i0)}\frac{2\Gamma_0}{(-e^{y_1}k^- + e^{-y_1}k^+ + i0)^2}$$

$$+ \text{UV c.t.} + O\left(e^{-2(y_1 - y_2)}\right). \tag{10.51}$$

The unsuppressed first term is independent of y_1. The k^- and k^+ integrals are easy to evaluate, giving

$$\frac{\partial\, C_S\Gamma_1}{\partial y_1} = \frac{-g^2}{(2\pi)^{n-1}} \int \frac{d^{n-2}k_T}{k_T^2 + m_g^2} \Gamma_0 + \text{UV c.t.} + O\!\left(e^{-2(y_1-y_2)}\right), \tag{10.52}$$

consistent with (10.46) and (10.48).

We will see that the evolution equation for the soft factor S in the factorization property (10.11) has the form

$$\frac{\partial \ln S}{\partial y_1} = \frac{1}{2} K(m_g, m, \mu, g(\mu)) + O\!\left(e^{-2(y_1-y_2)}\right), \tag{10.53}$$

with the kernel K being independent of y_1 and y_2. The right-hand side of (10.52) is in fact the first term in the perturbation expansion of $\frac{1}{2} K$. In accordance with the convention in Collins (1989); Collins and Soper (1981); Collins, Soper, and Sterman (1985b), a factor $\frac{1}{2}$ is defined to accompany K. The lowest-order value of K, from (10.52), is

$$K = \frac{-g^2}{4\pi^2} \ln \frac{\mu^2}{m_g^2} + O(g^4). \tag{10.54}$$

It follows from (10.53) that S depends exponentially on $y_1 - y_2$:

$$S(y_1 - y_2) = S_0 e^{\frac{1}{2}(y_1-y_2)K} \left[1 + O\!\left(e^{-2(y_1-y_2)}\right)\right], \tag{10.55}$$

with S_0 independent of $y_1 - y_2$.

The quantity K also plays a key role in the evolution of the other factors in (10.11), and analogous results hold for factorization theorems for other processes, like Drell-Yan with measured transverse momentum for the lepton pair. In using the factorization theorems, it will be necessary to use different values of the renormalization scale μ in different factors, e.g., $\mu \sim Q$ in the hard factor H, but $\mu \sim$ mass in the soft and collinear factors. Thus the RG equation for K is also important. This has the form

$$\frac{dK}{d \ln \mu} = -\gamma_K(g). \tag{10.56}$$

From (10.52), we read off the one-loop term in the anomalous dimension:

$$\gamma_K = \frac{g^2}{2\pi^2} + O(g^4), \tag{10.57}$$

which plays a central role in applications.

This anomalous dimension has two roles, of the kernel of the RGE for K, and as controlling the rapidity dependence of the anomalous dimension γ_S of the soft factor S:

$$\gamma_K = -\frac{dK}{d \ln \mu} = -2 \frac{d}{d \ln \mu} \frac{\partial \ln S}{\partial y_1} = -2 \frac{\partial}{\partial y_1} \frac{d \ln S}{d \ln \mu} = -2 \frac{\partial \gamma_S}{\partial y_1}, \tag{10.58}$$

where we have dropped power-suppressed terms of order $e^{-2(y_1-y_2)}$.

10.6 Rationale for definition of T_R

The definition of the region approximator T_R in Sec. 10.4.2 is obtained from the first term in an expansion in powers of small variables. However, the actual soft-to-collinear approximators were modified, to use space-like auxiliary vectors in (10.17) and (10.18), and to have specific $i0$ prescriptions in (10.20)–(10.24). We now justify these modifications. The modifications are unique, given some mild assumptions which are used to ensure the proofs are relatively simple.

Some of the justifications are more readily understood by referring to the one-loop example in Sec. 10.5.

The Grammer and Yennie (1973) paper gives a general approach to obtaining a leading approximation for soft gluons (and for related situations). But their approximator (in their K term) differs significantly in form from what we wrote in Sec. 10.4.2. This indicates that a variety of alternative approximators are conceivable, and we should justify a particular choice of approximator.

The Grammer-Yennie method was constructed to deal with IR divergences in QED; it concerns regions where photon momenta go to zero. In that situation IR photons do not interact with each other, even via loops of lines for electrons and any other matter fields. The Ward identities are particularly simple in an abelian theory. Not only do the IR divergences factorize from the rest of the cross section, but it was shown that the complete IR factor is the exponential of its one-loop value. The correctly computed divergence includes contributions from IR photons with rapidities comparable to that of an external charged line.

In the asymptotic problems treated in this book, what we mean by soft momenta is much broader; we include momenta whose absolute size may be large, but still much less than Q. Thus interactions of soft lines are important: the S factor in Fig. 10.3(b) is an arbitrary multigluon graph. However, we do not require the soft factor to correctly treat low-energy gluons of high rapidity; these belong in the collinear factors with other high-rapidity phenomena. The soft factor becomes a matrix element of Wilson lines, as do the collinear-to-hard gluon couplings. Furthermore the non-abelian Ward identities used in QCD are more complicated than the Ward identities in QED.

Consider our soft-to-A approximant, (10.21). In comparison, the original Grammer-Yennie approach would use no approximation of k_{AS} on the A subgraph, and would have a more complicated non-linear function in place of our denominator $k_{AS,j} \cdot n_1$. We will also need to justify the particular $i0$ prescriptions used in the denominators in (10.20)–(10.24).

10.6.1 Structure of soft and collinear approximants

The structure of all our generalized Grammer-Yennie approximants is

$$A(k)^{\mu} S(k)_{\mu} \mapsto A(\hat{k})^{\mu} \frac{u_{\mu} v^{\nu}}{u \cdot v} S(k)_{\nu} = A(\hat{k})^{\mu} \frac{\hat{k}_{\mu} v^{\nu}}{k \cdot v} S(k)_{\nu}, \tag{10.59}$$

where

$$\hat{k}^\mu = \frac{u^\mu \, k \cdot v}{u \cdot v}. \tag{10.60}$$

Here u and v are fixed vectors chosen so as to extract the leading behavior of $A \cdot S$ in the design region of the approximant. (That means, for example, that a soft-to-collinear-A approximant should give an approximation that is accurate to leading power when the momenta in S are soft and the momenta in A are collinear-to-A.) The names of the vectors in (10.59) are changed from our original formula, to indicate that we address general structural issues, allowing possible modifications of the formalism.

In general, multiple applications of (10.59) are used, one for each gluon joining A and S, as in (10.21). All the considerations in this section apply equally if the pair AS is replaced by BS, HA, or HB, merely needing a choice of appropriate auxiliary vectors u and v.

10.6.2 *Requirements on soft and collinear approximants*

To show that this form is required, and to determine further restrictions on the auxiliary vectors w_1, n_1, etc., we apply the requirements on region approximators T_R.

1. T_R should give an approximation correct to leading power at its design region R.
2. It should be compatible as necessary with contour deformations applied to the original graph.

   We have already dealt with the consequences of this requirement.
3. The conversion of the sum over graphs and regions to a Wilson-line form should be exact. Compare the derivation of the gauge-invariant parton model in Sec. 7.7.

   That is, in applying the Ward identities to Grammer-Yennie approximants, there should be no remainder terms. Typically such remainder terms are power-suppressed and hence innocuous in the design region of T_R, but can be unsuppressed elsewhere. These terms are not in principle undesirable, but they make it hard to construct complete proofs of factorization.
4. The approximant should be exact when applied to the Wilson lines derived from it.

   Ward identities applied to the approximated Wilson line give back exactly the Wilson line, as required by item 3. So the remainders between the graph and approximant must sum to zero. It avoids a probably hard subsidiary proof if the remainders are not zero term-by-term.
5. Summing the gluon attachments should actually give a Wilson line with a straight path, rather than some more general object, at least if this is possible consistently.[8]

   One can imagine more general ways of constructing gauge-invariant operators, e.g., by having Wilson lines with non-rectilinear paths, or by having an integral or sum over Wilson lines with different paths and given endpoints. All such cases are even more complicated than what we are already dealing with, so we should avoid them if possible.

---

[8] In the applications treated in Ch. 13, the definition of gauge-invariant transverse-momentum-dependent parton densities will require a minor modification to this assumption, with an extra segment of a Wilson line at infinity. The effects of the modification will cancel in the ultimate results.

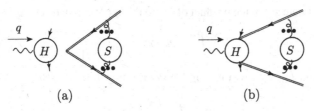

Fig. 10.11. Hard factor times (a) normal local soft factor, (b) conceivable non-local soft factor.

6. After applying T_R, the hard subgraph should not depend on soft momenta.

Close to the design region of T_R, the hard subgraph obviously has a power-suppressed dependence on soft momenta. But if the dependence is not removed exactly in the definition of T_R, there will be significant dependence of H on soft momenta, and this will introduce a complicated non-locality in the operator defining the soft factor. From the sum over Fig. 10.3(b), we will derive a factorization with soft factors defined by the vacuum matrix element of two Wilson lines joined at a point, so that the hard factor times soft factor is as in Fig. 10.11(a). But if the hard subgraph H had dependence on momenta circulating from the soft factor, the hard subgraph would give non-locality between the otherwise-joined ends of the Wilson lines, as in Fig. 10.11(b). We could avoid this by a subsidiary expansion of H *after* the use of Ward identities, but at the expense of hard-to-control remainders in diagrammatic treatments: the subtraction formalism would not correspond exactly to factorization. There would also be issues with gauge invariance of H. It is simpler if we avoid the extra step, as we will be able to.

7. An approximated momentum $\hat{k}$ is a linear function of the unapproximated momentum $\hat{k} = P(k)$. Applying the approximator a second time reproduces $\hat{k}$, i.e., $P(P(k)) = P(k)$.

One can find other requirements, but these are the ones that impinge most directly on the issues we wish to discuss. Evidently some of the requirements are not absolute, but are to prevent us from going outside known general ideas on gauge-invariant parton densities, etc. unless we are absolutely forced to.

10.6.3 General form of Grammer-Yennie-type approximation

The different cases of a Grammer-Yennie-type approximant are very similar. So to obtain its general form from the above requirements, it is sufficient to treat the case of a gluon of momentum k connecting the S to the A subgraph. The relevant approximant is the approximated A subgraph multiplied by a special factor and the unapproximated S subgraph, as in (10.59). We regard this as the approximated A subgraph (which is 1PI in the gluon) times the matrix element with a gluonic operator that defines S. To connect this to the Wilson-line formulation, the result is to be expressed by a Fourier transformation in terms of an integral over the coordinate-space gluon field.

The Wilson-line requirement implies that the coordinate-space integral is along a straight line, of some direction v, which we will identify with the same vector in (10.59) and (10.60).

That is, in coordinate space the product of S and the approximated A subgraph has the form

$$\int d\lambda \; F_A(\lambda) \, X(\lambda v), \tag{10.61}$$

for some function F_A, with $X(x)$ being the Fourier transform of $S(k)$,

$$X(x) = \int \frac{d^4k}{(2\pi)^4} \, e^{ik \cdot x} \, S(k). \tag{10.62}$$

In momentum space the product of S and approximated A is

$$\int \frac{d^4k}{(2\pi)^4} f_A(k \cdot v) S(k), \tag{10.63}$$

where

$$f_A(k \cdot v) = \int d\lambda \, e^{ik \cdot v\lambda} F_A(\lambda). \tag{10.64}$$

Hence the approximated A is a function of $v \cdot k$. Since $\hat{k}$ is a linear function of k, it is a fixed vector times $v \cdot k$. Reapplication of the approximator reproduces $\hat{k}$, so $\hat{k}$ must be of the form (10.60).

The exactness of the Ward identities in a non-abelian theory requires the vectors u to be the same at all gluons connecting S to A.

At each gluon between S and A, the approximant therefore has the form

$$A(k)^\mu S(k)_\mu \mapsto A(\hat{k})^\mu M(\hat{k})_\mu{}^\nu S(k)_\nu, \tag{10.65}$$

where M is some matrix to be determined. The approximant is exact if A^μ is obtained from a Wilson line in direction v; in that case A is some function of $k \cdot v$ times the vector v. The function is unchanged by the approximant, since $\hat{k} \cdot v = k \cdot v$. So the requirement of exactness of approximating a Wilson line gives

$$v^\mu M(\hat{k})_\mu{}^\nu = v^\nu, \tag{10.66}$$

from which we find that $M_\mu{}^\nu$ is of the form $a_\mu v^\nu / (a \cdot v)$ for some vector a. For the Ward identities to work exactly, we need $a^\mu \propto \hat{k}^\mu$. The structure in (10.59) follows.

10.6.4 Auxiliary vectors in soft approximation

In setting up the soft-to-collinear approximators, (10.21) etc., the natural expansion in small variables would make the vectors n_1 and n_2 light-like, in the plus and minus directions. But to cut off rapidity divergences, we made them non-light-like with rapidities y_1 and y_2.

We now derive the $i0$ prescription in (10.20) and (10.22), and determine that n_1 and n_2 are space-like. Examination of one-loop examples is sufficient for this. As we saw in Sec. 5.5.10, the soft approximation fails in the Glauber region, i.e., when $|k^+k^-| \ll k_{\rm T}^2$. We avoid the Glauber region by deforming the k^+ and k^- integrals away from the poles on the quark propagators. The approximators are applied on the deformed contours, so the denominators in (10.20) and (10.22) must use $i0$ prescriptions compatible with the deformed contours.

In (10.28), the simplest deformation is symmetric. Where the real parts of k^+ and k^- are in the Glauber region, we deform k^+ into the upper half plane away from the $p_B + k$ pole, and we deform k^- into the lower half plane away from the $p_A - k$ pole:

$$k^+ \mapsto k^+ + i\Delta, \qquad k^- \mapsto k^- - i\Delta, \tag{10.67}$$

where Δ is positive and of order k_T. The signs reflect that both the quark and antiquark are in the final state relative to the hard interaction, and the reversed sign between k^+ and k^- is because k flows into the A subgraph but out of the B subgraph. In (10.29), the Grammer-Yennie denominators are

$$\frac{1}{(-n_1 \cdot k + i0)(n_2 \cdot k + i0)} = \frac{1}{(-k^- + e^{-2y_1}k^+ + i0)(k^+ - e^{2y_2}k^- + i0)}. \tag{10.68}$$

Not obstructing the contour deformation determines the $k^+ + i0$ and $-k^- + i0$ parts to be as written, since e^{2y_2} and e^{-2y_1} are much less than one.

Fourier transformation of the Feynman rules for the Wilson lines shows that in coordinate space they are future-pointing, corresponding to the fact that the external quark and antiquark are in the final state.

We will also use factorization for other processes, and it is important that, if possible, we have universality of the collinear and soft factors between processes. Now, as explained by Collins and Metz (2004), other processes require an asymmetric contour deformation. As we will see in Sec. 12.14.3, in DIS we would use a contour deformation in k^+ only:

$$k^+ \mapsto k^+ + i\,O(Q), \qquad k^- \mapsto k^-. \tag{10.69}$$

The large k^+ deformation is away from final-state singularities, but k^- is generally trapped at small values by a combination of initial- and final-state singularities associated with the hadron target in DIS. This asymmetric deformation takes k from a Glauber configuration to a collinear-to-A configuration, and hence out of the soft region. But the soft approximant is to be integrated over all momenta, and it is used in a subtraction in collinear terms, so auxiliary denominators must not obstruct the contour deformation.

To get maximum universality of the soft and collinear factors, we should avoid changing the Wilson lines when we change processes, if possible. This requires (Collins and Metz, 2004) that our soft approximant for the Sudakov form factor also be compatible with the asymmetric deformation (10.69). This is achieved if the relative signs between the $i0$s and the k^+ terms in (10.68) all be the same, and therefore as written. A similar argument applies to the k^- terms. This determines all the signs in (10.68), from which we deduce that n_1 and n_2 are space-like, in agreement with our definitions.

An important advantage is that, since gluon fields commute at space-like separation, the use of space-like Wilson lines ensures automatic compatibility between the path ordering defining the Wilson lines and the time ordering used to define Green functions.

A disadvantage arises when one extends the use of the approximations to cases with emission of real gluons. Then singularities at $k \cdot n = 0$ with n space-like occur in the region of physical gluon emission. But with a time-like vector, the singularity is restricted to $k = 0$, because of the positive energy condition on a physical state. (In the rest frame of n, $k \cdot n = k^0$, which is positive for a physical state.)

If one gave up the argument about universality, one could use time-like auxiliary vectors. In the Sudakov form factor (and generally in reactions in $e^+ e^- \to$ hadrons) one could use time-like future-pointing vectors. In DIS one would still need a future-pointing vector on the struck quark side, but a *past*-pointing vector on the target side. The issues of universality in this context need further investigation.

10.6.5 Auxiliary vectors in the collinear approximants

As for the collinear-to-hard approximants, subtractions for soft regions cancel the possible rapidity divergences; we will see this as general result. Therefore it is sufficient to use light-like vectors in the collinear approximants, as given in (10.19), (10.23), and (10.24).

The $i0$ prescriptions in (10.23) and (10.24) are determined in the same way as in the soft approximants. The signs are in fact the same, and correspond to future-pointing Wilson lines. Although the Glauber region appears to have nothing to do with a collinear region, the approximators are applied to the graph as a whole with a deformed momentum contour. The momentum denominators in the collinear approximant must therefore be compatible with the contour deformation out of the Glauber region.

10.6.6 Alternative definition of the collinear-to-hard approximants

In our definitions in Sec. 10.4.2, we chose all the approximated momenta to be light-like. Thus in (10.59), the vector u is light-like. Although this is generally the most convenient choice, other choices are conceivable. However, constraints arise from other requirements. In the case of the hard-scattering factor, gauge invariance is most conveniently assured, if its external lines are on-shell. This implies that these lines are light-like given that they are massless. Practical perturbative calculations are enormously much simpler when masses are zero and external lines on-shell.

We also used a light-like Wilson-line vector in the hard scattering, i.e., w_2 in (10.19a) and w_1 in (10.19b).

A constraint now arises from the requirement that the hard factor does not depend on the soft momenta, after application of an approximator. This ensures that the hard factor completely factors from the soft factor. In the notation of (10.59), let u_{AS} and v_{AS} be the vectors for the soft-to-A approximant, and let u_{HA} and v_{HA} be the vectors for the A-to-H approximant. In this general case, the approximated momentum in H is

$$\hat{k}^\mu_{HA} = u^\mu_{HA} \frac{(k_{HA} + \hat{k}_{HAS}) \cdot v_{HA}}{u_{HA} \cdot v_{HA}} \tag{10.70}$$

Since $\hat{k}_{HAS}$ is proportional to u_{AS}, we only get independence of $\hat{k}_{HA}$ from k_{AS} if

$$u_{AS} \cdot v_{HA} = 0, \tag{10.71}$$

i.e., if the approximated soft momentum is orthogonal to the Wilson-line vector for the A-to-H connections.

This is obviously satisfied for our actual choice, in (10.17) and (10.19a), that u_{AS} and v_{HA} both equal w_2, a light-like vector in the minus direction.

What other possibilities are there? We restrict to vectors in the $(+, -)$ plane, otherwise we break azimuthal rotation symmetry in our approximators, without having a transverse vector in the process's kinematics to give a preferred transverse direction.

If v_{HA} stays light-like, this requires u_{AS} to be light-like in the same direction, which is our original choice.

Given our results on $i0$ prescriptions, the other choice is a space-like vector v_{HA}. An orthogonal vector is time-like. A simple and natural case is to put v_{HA} in the z direction in the center-of-mass frame. The corresponding Wilson line restricts gluon rapidity in the A factor to be approximately positive, which is very natural; it gives a natural cutoff of the rapidity divergence in $T_A \Gamma_1$ before subtraction. Then we would need $u_{AS} \propto q^\mu$, a not unnatural choice.

As far as I can see, this is an legitimate alternative possibility.

However, as we will see, it is generally preferable to avoid non-light-like Wilson lines whenever possible: It makes calculations easier and avoids inhomogeneous terms in evolution equations.

10.7 General derivation of region decomposition

In this section, we prove the main result needed to apply the subtraction formalism. This is that, for a general Feynman graph for any of the many processes that we consider, the remainder, (10.5), is actually power-suppressed. That is, it is a power of Q smaller than the leading power for the process (which is, for example, Q^0 for DIS structure functions). This then demonstrates (10.1), which is the key formula for our later derivations of factorization of various kinds.

The derivation uses certain properties of the region approximators T_R, so effectively we are finding and using a set of requirements on good approximators.

A general treatment involves regions in a loop-momentum space of arbitrarily high dimension, and thus necessarily has a high degree of abstraction. As we will see, a recursive, or inductive, strategy enormously simplifies the proof by reducing it to considering relations between two generic regions. These can be visualized in a space of two dimensions, and simple examples, like those in Secs. 10.2 and especially 10.5, give the main ideas for the generic situation. It would be useful to read those sections concurrently with the present section to gain better understanding, visualization, and motivation.

Even so, it will become apparent that the rigor of the derivations is insufficient. Mathematically inclined readers are strongly urged to do better; the literature on deriving factorization leaves much to be desired.

10.7.1 Results so far

So far, we have explicitly defined the main ingredients of the method. The region contributions $C_R\Gamma$ were defined in (10.4) in terms of region approximators $T_R\Gamma$. Then the asymptotic behavior of Γ is intended to be correctly given by the sum over regions: $\Sigma_R C_R\Gamma$. Explicit definitions of the region approximators were given in Sec. 10.4 for the Sudakov form factor; these definitions apply with at most minor changes to the many other processes we will treat.

10.7.2 Overall view

It is important to keep in mind the main motivations for the subtraction formalism. First, the region approximant $T_R\Gamma$ is intended to give a good approximation to Γ near the PSS R; that is,

$$\Gamma - T_R\Gamma = \mathcal{O}\left(\left(\frac{\lambda_R + m}{Q}\right)^p\right)\|\Gamma\|, \tag{10.72}$$

with some qualifications that I will explain in Sec. 10.7.3. Here, λ_R is the radial variable for region R. Naturally the approximators we use are such that the soft, collinear and hard subgraphs of a region correspond to contributions to factors in a phenomenologically useful factorization property. The error specified in (10.72) improves as λ_R decreases, but only until λ_R becomes of order m. There are additional sources of error in neglecting m with respect to Q when appropriate. So all these issues are covered by adding m to λ_R in (10.72).

The approximant contributes in regions larger than R, but with an inaccurate value. To handle the consequent double counting, we defined the region contribution $C_R\Gamma$ by (10.4), where T_R is applied after subtraction of the contributions from smaller regions. This is also intended to solve the problem that the accuracy of the approximator T_R degrades close to PSSs smaller than R: the region contribution $C_R\Gamma$ is intended to be leading power at region R but suppressed in smaller regions. Including the contributions of smaller regions, $C_R\Gamma + \sum_{R' < R} C_{R'}\Gamma$, is intended to give a correct leading-power approximation near the whole of R, including smaller regions.

As we saw in Ch. 8, this setup works quite straightforwardly to give factorization, if the relevant regions are just nested inside each other, i.e., if they have a total ordering. But, in general, the regions can have more general relations involving overlaps and non-intersection, as in (5.21). This is responsible for the main complications in the proof. They are a non-trivial generalization of those involved in dealing with overlapping divergences in renormalization theory.

The most fundamental problem solved by the subtraction formalism is that the accuracy of a region approximant $T_R\Gamma$ degrades in certain places, associated with other regions. An example is at the approach to a smaller region $R_1 < R$. As we have seen in examples, the worsening of the accuracy of $T_R\Gamma$ is compensated in the subtraction formalism. In forming

$C_R\Gamma$, T_R is applied to Γ only after subtractions are used for all the smaller regions. Then it is the sum $C_R\Gamma + \sum_{R'<R} C_{R'}\Gamma$ that gives an accurate approximation to Γ over the whole of R, including smaller regions.

Another problem is the large multiplicity of regions, as in Fig. 10.5, a problem that obviously gets much worse for even higher-order graphs. Our proofs will be inductive, i.e., recursive, and a generic step of a proof will only involve a single region and its nearest neighbors in the region hierarchy. Then the most complicated relation between regions that we need to discuss explicitly is Fig. 5.32. Most of the time, the relation we treat will be essentially of the form of Fig. 5.28. So with an appropriate viewpoint, the most general situation can be reduced to many copies of what happens in one-loop graphs, or at most two-loop graphs.

Now, our aim is to derive power-law estimates of the accuracy of a factorization statement, i.e., to obtain results that are accurate to some given power of a small ratio (e.g., m/Q). But we often have logarithmic integrals interpolating between different regions, and these worsen basic power-law estimates by some number of logarithms. So it is convenient to define the following notation:

$$f(x) = \Lambda_p(x)g(x) \quad \text{as } x \to 0, \tag{10.73}$$

which means that

$$f(x) = \mathcal{O}\!\left(x^p |\ln x|^\alpha\right) g(x) \quad \text{as } x \to 0, \tag{10.74}$$

for some value of the power α of the logarithm. That is, there are constants C, α and x_0, such that

$$|f(x)| < C\,|x|^p\,|\ln x|^\alpha\,|g(x)| \quad \text{for all } |x| < x_0. \tag{10.75}$$

Normally, p is fixed for the problem we are analyzing (e.g., graphs for the Sudakov form factor to leading power), but α depends on the graph, being up to two times the number of loops.

10.7.3 Accuracy of approximator T_R

The basic form of the accuracy of a region approximator T_R was given in (10.72). We now modify it to obtain a strictly correct error estimate which will form the basis of the rest of our work.

Basic error estimate

The accuracy of the approximator for a leading region can be read off from the accuracy of its individual components, as defined in Sec. 10.4. Since we are working to leading-power accuracy, the exponent p of the power law is $p = 1$. Often such errors involve some transverse momentum relative to Q: k_T/Q, and these commonly vanish after an integral over angle. Then the actual error is one power better: $p = 2$. We can also imagine improved region approximators with an expansion to more orders in small momentum components, with a correspondingly larger value of p. The precise value of p will not matter.

There are also non-leading regions, such as $R_{A'}$ defined in Sec. 5.4 for the one-loop Sudakov form factor. Since the graph is already non-leading in such a region, we can define the associated approximator to be zero, e.g., $T_{R_{A'}} \Gamma = 0$. But the use of the integrand Γ on the r.h.s. of error estimates such as (10.72) is then not appropriate; rather we need a value characteristic of the graph integrated over all regions. Thus we replace Γ on the r.h.s. of (10.72) by

$$\left\| \int_{\text{all}} \Gamma \right\|. \tag{10.76}$$

Here the double-bar notation has the same meaning as in (10.39). That is, it is a power-counting estimate of the size of the integral arranged to avoid dynamical cancellations. (Thus for DIS, we would write $\| W^{\mu\nu} \| = O(1)$, even though some specific components vanish.)

Correspondingly, we should use an integral for the l.h.s., but now over a range near the PSS R:

$$\int_{\text{local}} (\Gamma - T_R \Gamma). \tag{10.77}$$

Then λ_R on the r.h.s. of an error estimate should be interpreted as the maximum value of the radial variable in the range of integration. The integration is over some range of all variables, not just λ_R but also the angular and intrinsic coordinates for R. Naturally, the integral should be on a deformed contour if we need to avoid a Glauber region. Since there is the possibility of logarithmic enhancements in such integrals, we must replace the power-law estimate on the r.h.s. by

$$\Lambda_p \left(\frac{\lambda_R + m}{Q} \right). \tag{10.78}$$

Situations needing adjustment

We now quantify that for a given value of λ_R, the error estimates need modification for two situations, as can be obtained from the definitions in Sec. 10.4. First, they generally degrade when the intrinsic coordinates approach the positions of any particular smaller PSS $R_1 < R$, since then the conditions for neglecting a small momentum component with respect to a large component become weaker.

The second issue concerns lines with soft-collinear momenta, as in the example in Sec. 10.5.4. These lines have both a small energy and a high rapidity. The small energy allows them to be considered as soft, and the high rapidity allows them to be considered as collinear. Let R be a region in which the soft-collinear lines are part of the soft subgraph. Let R_2 be the larger region obtained from it by changing the category of the soft-collinear lines to the appropriate collinear category. We notate this relation by

$$R_2 \overset{\text{sc}}{>} R. \tag{10.79}$$

In terms of the underlying PSSs, this relation is defined to mean that certain collinear lines at the PSS R_2 are changed to zero momentum to obtain the PSS R.

Soft-collinear lines are at an end of their collinear range in fractional momentum. But their high rapidity implies that the approximator T_{R_2} continues to be valid, removing the degradation that would otherwise occur near the smaller region R.

In the approximator T_R, the soft-collinear lines are treated as soft, but then their high rapidity implies that the approximators where they attach to the corresponding collinear subgraph degrade in accuracy. The errors become of order $e^{-\Delta y}$, where Δy is the rapidity difference between the soft and collinear lines, with the soft line always being taken as having rapidity between the two collinear groups of the whole process.

Generalizing our proof from the example in Sec. 10.5.4, we will find that these effects combine to give correctness of the subtraction method to extract the asymptotics of the graphs.

Generic degradation near smaller PSSs

The accuracy of the approximator T_R defined in Sec. 10.4 degrades when the intrinsic coordinates appropriate for PSS R approach the positions of any particular smaller PSS $R_1 < R$. For example, in a hard subgraph, we neglect a collinear transverse momentum with respect to a large momentum component of order Q. But near R_1 we may need to replace Q by the smaller value λ_{R_1}. So in our error estimate we insert a degradation factor

$$W_{R_1,R} = 1 + \Lambda_p\left(\frac{Q}{\lambda_{R_1} + m}\right), \tag{10.80}$$

with one term for each smaller region. Here, I added 1 to the basic degradation factor, so that the factor $W_{R_1,R}$ can be applied universally: close to R_1, the $\Lambda_p(\ldots)$ term dominates, but away from R_1, it decreases, leaving $W_{R_1,R}$ to relax to unity.

Soft-collinear problem

Surrounding PSS R, consider integrating around a surface of fixed λ_R, as in Fig. 5.28. Close to each larger PSS R_2 that obeys the soft-collinear relation $R_2 \overset{\text{sc}}{>} R$, we get degradation of the approximation by a factor $V_{R_2,R}$. This factor replaces $\Lambda_p(\lambda_R/Q)$ by $\Lambda_p(e^{-\Delta y})$, where Δy is the rapidity difference between the soft-collinear lines in the soft subgraph of R and lines in the collinear subgraph of R to which they attach.

Consider next these same lines in the same momentum region in the other approximator $T_{R_2}\Gamma$. Relative to R_2, the configuration is close to the smaller region R, where there is a default degradation factor W_{R,R_2}. But the approximator applies accurately to the soft-collinear lines, so we multiply the degraded error estimate by the inverse of the large $V_{R_2,R}$ factor.

10.7.4 Overall error estimate

Putting all these components together, we have shown that the error in T_R is characterized by

$$\int_{\text{local}} (\Gamma - T_R \Gamma) = \Lambda_p \left(\frac{\lambda_R + m}{Q} \right)$$

$$\times \left[1 + \sum_{R_1 < R} W_{R_1, R} \frac{1}{1 + V_{R, R_1}} \right] \left[1 + \sum_{R_2 \overset{\text{sc}}{>} R} V_{R_2, R} \right] \left\| \int_{\text{all}} \Gamma \right\|. \quad (10.81)$$

The $1/(1 + V_{R, R_1})$ factors only appear for subregions obeying $R_1 \overset{\text{sc}}{<} R$.

10.7.5 Theorems to be proved

I now state some theorems to be proved inductively. They generalize properties we have seen in examples. The first three theorems are properties labeled by a region.

Theorem 1_R Define $\int_{\text{local}} \bar{C}_R \Gamma \overset{\text{def}}{=} \int_{\text{local}} (\Gamma - \sum_{R' < R} C_{R'} \Gamma)$ which has subtractions for smaller regions than R. It is suppressed in all regions R_1 smaller than R, but with degradation for soft-collinear situations that concern regions R or bigger:

$$\int_{\text{local at } R_1} \bar{C}_R \Gamma = \Lambda_p \left(\frac{\lambda_{R_1} + m}{Q} \right) \left[1 + \sum_{R_2 \overset{\text{sc}}{\geq} R} V_{R_2, R_1} \right] \left\| \int_{\text{all}} \Gamma \right\|. \quad (10.82)$$

Theorem 2_R The same property applies to $C_R \Gamma = T_R \bar{C}_R \Gamma$.

Theorem 3_R When we also subtract $C_R \Gamma$, there is a suppression at R, and the soft-collinear degradation only applies on regions strictly bigger than R:

$$\int_{\text{local at } R} \left(\Gamma - C_R \Gamma - \sum_{R' < R} C_{R'} \Gamma \right) = \int_{\text{local at } R} (1 - T_R) \bar{C}_R \Gamma$$

$$= \Lambda_p \left(\frac{\lambda_R + m}{Q} \right) \left[1 + \sum_{R_2 \overset{\text{sc}}{>} R} V_{R_2, R} \right] \left\| \int_{\text{all}} \Gamma \right\|. \quad (10.83)$$

The suppression is uniform over the whole of R including smaller regions.

Theorem 4 The sum of $C_R \Gamma$ over all regions approximates Γ to power-law accuracy:

$$\int_{\text{all}} \left(\Gamma - \sum_R C_R \Gamma \right) = \Lambda_p \left(\frac{m}{Q} \right) \left\| \int_{\text{all}} \Gamma \right\|. \quad (10.84)$$

10.7.6 Proofs of theorems $1_{R_{\min}}$ to $3_{R_{\min}}$

We will first prove these theorems for a minimal region, and then prove them for larger regions given that they hold for all smaller regions.

Minimal regions

For a minimal region $R_{\min}$, theorems $1_{R_{\min}}$ and $2_{R_{\min}}$ are trivial because there are no smaller regions. Theorem $3_{R_{\min}}$ follows directly from the approximation property (10.81); because of the lack of smaller regions $\bar{C}_{R_{\min}}\Gamma = \Gamma$.

Theorem 1_R

For a general region R, we make the inductive hypothesis that theorems 1–3 have already been proved for regions smaller than R. Then to prove the suppression (10.82), we partition the terms in $\bar{C}_R\Gamma$ into three sets according to the relation of the relevant regions to R_1, and then consider each set separately.

First, we note the following structural properties of $\bar{C}_R\Gamma$ that follow directly from its definition.

- $\bar{C}_R\Gamma$ is a sum of terms, each of which involves a product of $-T_{R'}$ operations applied to Γ. Each product involves a sequence of strictly ordered regions, since subtractions in the definition of any particular region contribution $C_{R'}\Gamma$ only involves yet smaller regions.
- A factor $T_{R'}$ only appears in combinations that combine to form a $C_{R'}\Gamma$ factor.

The partitioning of $\bar{C}_R\Gamma$ is as follows.

- The first set consists of terms in which all the $T_{R'}$ factors are for regions that are ordered relative to R_1. The sum gives an object of the form:

$$\sum \prod_{R''}(-T_{R''})(1 - T_{R_1})\bar{C}_{R_1}\Gamma. \tag{10.85}$$

The sum is over the ways in which can appear $T_{R''}$ factors for regions R'' bigger than R_1 (and necessarily smaller than R). The two terms in the middle parentheses account for all the terms in which $-T_{R_1}$ does not or does appear.
- The second set has at least one $-T_{R'}$ overlapping with R_1, but none that fail to intersect R_1. We group these terms by the minimal such R':

$$\sum \prod_{R''}(-T_{R''})C_{R'}\Gamma, \tag{10.86}$$

where R' overlaps R_1, i.e., the intersection $R' \cap R_1$ is non-empty and strictly smaller than both R' and R_1.
- The third set is where there is at least one $-T_{R'}$ factor for a region that does not intersect at all with R_1. We group these terms by the minimal such R':

$$\sum \prod_{R''}(-T_{R''})C_{R'}\Gamma, \tag{10.87}$$

where the R'' regions are larger than R'.

For the first set, the factor $(1 - T_{R_1})\bar{C}_{R_1}\Gamma$ is suppressed by theorem 3_{R_1}, which is true by the inductive hypothesis. But this has the soft-collinear degradation at any R_2 obeying $R_2 \overset{\text{SC}}{>} R_1$. For those R_2 that are also smaller than R, i.e., that obey $R_2 < R$, there are subtractions in (10.85). By an inductive application of theorem 3 to region R_2, we find a suppression by the $1/V_{R_2,R_1}$ factor. There remain the cases $R_2 = R$, and $R_2 \overset{\text{SC}}{>} R$, which are allowed in (10.82).

For the second set, (10.86), our treatment uses the ideas given in Sec. 10.5.6. There we found for the one-loop Sudakov form factor that the collinear term $C_A\Gamma$ was suppressed in the opposite collinear region R_B. In this term, the factor T_A acts by first projecting the loop-momentum configuration down to the intersection R_S of the two regions. Then it extrapolates in the normal coordinates for A, preserving the value of the intrinsic coordinates. A momentum close to R_B gives an intrinsic coordinate close to the endpoint R_S of the R_A PSS. We then get a suppression because of the suppression of $C_A\Gamma$ at regions smaller than R_A. This idea applies generally, by changing R_A to R', R_B to R_1, and R_S to $R' \cap R_1$. The approximator $T_{R'}$ coerces a momentum configuration near R_1 to be effectively near $T_{R' \cap R_1}$.

For the third set, R' and R_1 do not intersect at all. Again the $T_{R'}$ operation coerces the momentum configuration to be changed from R_1-like to R'-like. The lack of intersection of R' and R_1 implies that the coerced configuration is a generic one for R' and that the radial variable is of order Q. More propagators are off-shell without a change in the integration measure, so we get a suppression.

This completes the proof of theorem 1_R.

Theorem 2_R

The application of the approximator T_R does not change the suppressions and degradations in (10.82). So theorem 2_R follows.

Theorem 3_R

The l.h.s. of (10.83) differs from that of (10.82) by a factor $1 - T_R$. From the basic approximation property, (10.81), this gives a factor $\Lambda_p((\lambda_R + m)/Q)$ on the r.h.s. The suppression factors for $\bar{C}_R\Gamma$ at smaller regions on the r.h.s. of (10.82) cancel the corresponding degradation terms in (10.81), while the $1/(1 + V_{R,R_1})$ factors cancel the effect of the V_{R_2,R_1} factors in (10.82) for the case that $R_2 = R$.

This gives (10.83).

Theorem 4

Theorem 4, (10.84) is the actual theorem we need to use in proving factorization, since it states that to power-law accuracy, Γ is given by the sum of $C_R\Gamma$ over regions. It is just theorem 3 applied to the largest possible region R_H, where all momenta are hard. For this region all coordinates are intrinsic, so we must set the radial coordinate to zero: $\lambda_H = 0$. There are no larger regions, so we need no $V_{R_2,R}$ terms. Thus theorem 4 is just an application of (10.83) for $R = R_H$.

10.8 Sudakov form factor factorization: first version

The general leading region for the Sudakov form factor was depicted in Fig. 10.3(b). For each region R of each graph Γ, we defined a corresponding contribution $C_R\Gamma$, and the sum over Γ and R gives a correct leading-power approximation to the form factor:

$$F = \sum_{\Gamma,R} C_R\Gamma + \text{power-suppressed.} \tag{10.88}$$

The sum can be specified by independent sums over the region subgraphs H, A, B, and S in Fig. 10.3 (subject to the constraint that there is a match of the numbers of gluon lines connecting the different subgraphs). We must convert this sum into the factorized form of hard, collinear and soft factors, as in (10.11), with definite definitions for the factors as matrix elements of certain operators containing Wilson lines.

The basis of our method is that the region approximators T_R allow Ward identities to be applied to the connections of gluons from S to the collinear subgraphs A and B, and to the gluons from A and B to the hard subgraph H. In each case there is a factor of the gluon momentum contracted with one of the subgraphs, which we will call the destination subgraph (A, B or H respectively). It is this contraction that allows Ward identities to be used, generalizing the results of Sec. 7.7.

Elementary Ward identities in an abelian gauge theory are for ordinary Green functions or matrix elements. Relative to these cases, we have two primary complications. The first is that our Green functions have subtractions for smaller regions. The second is that the graphs for A, B, and H are restricted by certain irreducibility requirements: Each collinear subgraph A and B is one-particle-irreducible (1PI) in the soft lines, while the hard subgraph H is 1PI separately in the A lines and the B lines.

10.8.1 Statement of definitions of factors

The Ward identities entail definitions for the soft and collinear factors that we state in this section.

The soft factor is

$$S(y_1 - y_2) = \frac{\langle 0| \, W(\infty, 0, n_2)^\dagger \, W(\infty, 0, n_1) \, |0\rangle}{\text{W.L. self-energies for } n_2 \text{ and } n_1} Z_S. \tag{10.89}$$

Here the Wilson-line operators are defined in (10.33), with directions n_1 and n_2, while Z_S is a UV renormalization factor defined by, say, the $\overline{\text{MS}}$ scheme. The denominator will be defined in (10.101); it removes graphs that contribute to the numerator but that are not produced from the Ward-identity argument. Applying Lorentz invariance shows that the dependence of S and Z_S on the Wilson-line rapidities y_1 and y_2 is only on the difference $y_1 - y_2$. However, it is sometimes convenient to write separate y_1 and y_2 arguments: $S(y_1, y_2)$ instead of $S(y_1 - y_2)$.

As for the collinear factors, I first define an unsubtracted collinear factor for the A side:

$$A^{\text{unsub}}(y_{p_A} - y_{u_2}) = \frac{\langle p_A | \bar{\psi}_0(0) \, W(\infty, 0, u_2)^\dagger \, \mathcal{P}_B \, |0\rangle}{(\text{W.L. self-energies for } u_2) \, \bar{u}_A \mathcal{P}_B} Z_A^{\text{unsub}}$$

$$= \frac{\langle p_A | \bar{\psi}(0) \, W(\infty, 0, u_2)^\dagger \, \mathcal{P}_B \, |0\rangle}{(\text{W.L. self-energies for } u_2) \, \bar{u}_A \mathcal{P}_B} Z_A^{\text{unsub}} Z_2^{1/2}. \tag{10.90}$$

In the first line, the numerator has a matrix element of a *bare* quark field and a Wilson line in a space-like direction $u_2 = (-e^{2y_{u_2}}, 1, \mathbf{0_T})$. The vector u_2 is just like n_2 except for a different rapidity y_{u_2}, and we will later use a limit with $y_{u_2} \to -\infty$. There is also a UV renormalization factor. The second line is simply the first line written in terms of the renormalized quark field, as appropriate for calculations. As in the soft factor, there is a denominator to cancel Wilson-line self-energy graphs.

The numerator is actually a Dirac spinor, and contains the factor $\mathcal{P}_B = \gamma^+ \gamma^-/2$ which is used to connect the collinear and hard factors. As I now show, the numerator is just a factor times $\bar{u}_A \mathcal{P}_B$. Therefore we include in the denominator in (10.90) a factor to divide out the spinor dependence, so that the quantity A^{unsub} is a numerical-valued scalar quantity. To derive the spinor structure, we observe that the only vector variables on which the collinear factor depends are in the $(+, -)$ plane. After the use of parity invariance, the most general Dirac structure for A^{unsub} is

$$\bar{u}_A(aI + b^+ \gamma^-)\mathcal{P}_B. \tag{10.91}$$

Because of the $\mathcal{P}_B$ factor, all other combinations of Dirac matrices can either be reduced to this by anticommutation relations or give zero. By use of $\bar{u}_A(\not{p}_A - m) = 0$, it is easily checked that the most general form is actually proportional to $\bar{u}_A \mathcal{P}_B$.

An unsubtracted B factor is defined exactly similarly:

$$B^{\text{unsub}}(y_{u_1} - y_{p_B}) = \frac{\langle p_B | \, W(\infty, 0, u_1) \, \mathcal{P}_B \, \psi_0(0) \, |0\rangle}{(\text{W.L. self-energies for } u_1) \, \mathcal{P}_B v_B} Z_B^{\text{unsub}}$$

$$= \frac{\langle p_B | \, W(\infty, 0, u_1) \, \mathcal{P}_B \, \psi(0) \, |0\rangle}{(\text{W.L. self-energies for } u_1) \, \mathcal{P}_B v_B} Z_B^{\text{unsub}} Z_2^{1/2}, \tag{10.92}$$

with a Wilson line in the direction $u_1 = (1, -e^{-2y_{u_1}}, \mathbf{0_T})$.

Not only do soft and collinear factors like S, A^{unsub}, and B^{unsub} depend on the rapidities of their non-light-like Wilson line(s), but so do their renormalization factors Z_S, Z_A^{unsub}, and Z_B^{unsub}. For S and Z_S this is simply a dependence on $y_1 - y_2$, as in (10.47).

The renormalization factors Z_A^{unsub}, and Z_B^{unsub} are mass independent and so variables to parameterize their dependence on the Wilson-line rapidities must use the massless limit of p_A and p_B. Appropriate variables for Z_A^{unsub}, and Z_B^{unsub} are, respectively,

$$\zeta_{A,u_2} \overset{\text{def}}{=} 2(p_A^+)^2 e^{-2y_{u_2}} = m^2 e^{2(y_{p_A} - y_{u_2})}, \tag{10.93a}$$

$$\zeta_{B,u_1} \overset{\text{def}}{=} 2(p_B^-)^2 e^{2y_{u_1}} = m^2 e^{2(y_{u_1} - y_{p_B})}. \tag{10.93b}$$

Next we define subtracted collinear factors. Their names, A and B, are decorated with a superscript "basic" to indicate that the definitions are in a sense preliminary, since in later sections we will construct an improved factorization with modified definitions of the factors. Each subtracted collinear factor is defined by dividing the unsubtracted collinear factor by a version of the soft factor, and then taking the light-like limits u_1 and u_2 in a certain way. Thus the subtracted A factors are

$$A^{\text{basic}} = \frac{\langle p_A | \bar{\psi}_0(0) \, W(\infty, 0, w_2)^\dagger \, \mathcal{P}_B \, |0\rangle \;\; (\text{W.L. self-energies for } n_1)}{\langle 0| \, W(\infty, 0, w_2)^\dagger \, W(\infty, 0, n_1) \, |0\rangle \; \bar{u}_A \mathcal{P}_B} \, Z_A^{\text{basic}}$$

$$= \frac{\langle p_A | \bar{\psi}(0) \, W(\infty, 0, w_2)^\dagger \, \mathcal{P}_B \, |0\rangle \;\; (\text{W.L. self-energies for } n_1)}{\langle 0| \, W(\infty, 0, w_2)^\dagger \, W(\infty, 0, n_1) \, |0\rangle \; \bar{u}_A \mathcal{P}_B} \, Z_A^{\text{basic}} \, Z_2^{1/2},$$

$$(10.94a)$$

$$B^{\text{basic}} = \frac{\langle p_B | \, W(\infty, 0, w_1) \, \mathcal{P}_B \, \psi_0(0) \, |0\rangle \;\; (\text{W.L. self-energies for } n_2)}{\langle 0| \, W(\infty, 0, n_2)^\dagger \, W(\infty, 0, w_1) \, |0\rangle \; \mathcal{P}_B v_B} \, Z_B^{\text{basic}}$$

$$= \frac{\langle p_B | \, W(\infty, 0, w_1) \, \mathcal{P}_B \, \psi(0) \, |0\rangle \;\; (\text{W.L. self-energies for } n_2)}{\langle 0| \, W(\infty, 0, n_2)^\dagger \, W(\infty, 0, w_1) \, |0\rangle \; \mathcal{P}_B v_B} \, Z_B^{\text{basic}} \, Z_2^{1/2}. \quad (10.94b)$$

The above definitions agree with our one-loop calculations in (10.42) and (10.43). The renormalization factors Z_A^{basic} and Z_B^{basic} depend on $\zeta_{A,n_1}/\mu^2$ and $\zeta_{B,n_2}/\mu^2$ respectively, as well as on g and ϵ. Here the ζ variables are defined by (10.93).

We will see that the denominators (10.94) are obtained as a result of the subtractions in $C_R\Gamma$ for smaller regions; they have the effect of compensating double counting between the collinear and soft factors. Closely related to this is that we will find that rapidity divergences associated with the Wilson lines in light-like directions cancel between the numerators and denominators. In effect,

$$A^{\text{basic}} = \text{``lim''}_{y_{u_2} \to -\infty} \frac{A^{\text{unsub}}(y_{p_A} - y_{u_2})}{S(y_1 - y_{u_2})}, \quad (10.95)$$

and similarly for B^{basic}. However, there is a non-uniformity in taking the infinite rapidity limits and removing the UV regulator, which impacts calculations. As indicated by the quotation marks, the limit in (10.95) is taken in a special way to be defined in Sec. 10.8.2.

Finally, the hard factor is essentially whatever is left over, in the limit that masses are neglected:

$$H = \frac{F}{A^{\text{basic}} B^{\text{basic}} S} \Bigg|_{m_g = m = 0, \, p_A, n_1, p_B, n_2 \text{ light-like}}. \quad (10.96)$$

Originally we choose n_1 and n_2 to be vectors with approximately the rapidities of p_A and p_B. So taking the massless limit for p_A and p_B implies that we replace n_1 and n_2 by their light-like limits, i.e., w_1 and w_2. Our definition of the collinear factors implies that H includes factors of spinors $\bar{u}_A \mathcal{P}_B$ and $\mathcal{P}_B v_B$ with a Dirac matrix between them.

10.8.2 Limit of infinite rapidity Wilson lines

The limit $y_{u_2} \to -\infty$ on the Wilson-line rapidity in (10.95) needs a little care in its definition concerning the hard region of large transverse momenta: there is non-uniformity in combining the limits of infinite rapidities with the removal of a UV regulator. We use the following procedure to define A^{basic} and B^{basic}.

- For A^{unsub} and S, apply a UV regulator, e.g., dimensional regularization with $n < 4$.
- Take the limit $y_{u_2} \to -\infty$ on the r.h.s. of (10.95).
- Apply UV counterterms.
- Remove the UV regulator, e.g., take $n \to 4$.

This corresponds to our procedure for calculating $C_A \Gamma_1$ and $C_B \Gamma_1$ in (10.42) and (10.43).

If we reversed the limits, we would need to compensate by an extra hard factor, e.g.,

$$A^{\text{basic}} = \lim_{y_{u_2} \to -\infty} \left[\lim_{n \to 4} \frac{A^{\text{unsub}}(y_{p_A} - y_{u_2})}{S(y_1 - y_{u_2})} \tilde{Z}_A(\zeta_{A,n_1}/\mu^2, y_1 - y_{u_2}, g(\mu), \epsilon) \right]. \quad (10.97)$$

The factor $\tilde{Z}_A$ is to be adjusted so that we get the same results as in (10.94a). Now the non-uniformity of the limits $n \to 4$ and of infinite Wilson-line rapidities only concerns the limit of infinitely large transverse momentum; for $n < 4$, the limits can be exchanged. Thus the factor $\tilde{Z}_A$ is a pure UV factor, and can be regarded as a kind of generalized UV renormalization factor, chosen to make a renormalization prescription that agrees with the combination of $\overline{\text{MS}}$ renormalization and the opposite order of the limits.

Within the context of low-order perturbation theory, especially at one loop, the first description works; an example is in the calculation of the one-loop collinear term at (10.42).

An exactly similar procedure applies to the B factor.

10.8.3 Elements of diagrammatic Ward identities

Ward identities can be derived without perturbation theory, as properties of Green functions. From these we could try to derive identities for the factors, H, A, B, and S in Fig. 10.3, which are *modified* Green functions, with appropriate irreducibility properties and subtractions.[9] For our present work, it is considerably easier just to give a perturbative proof, valid to all orders, where we will take full account of the necessary subtractions and irreducibility properties. The general approach was seen in Sec. 7.7, where we derived a gauge-invariant parton model in a full non-abelian theory, i.e., QCD with a limited set of graphs.

Here we handle the full set of graphs, but restrict to an abelian theory in a covariant gauge. In deriving factorization, it will be important to understand which subgraphs are allowed and which are not, for A, for B, and particularly for H, in Fig. 10.3(b), given a specification of their external lines. This will modify the derivation of the Ward identities from the standard derivation, e.g., Sterman (1993, p. 334–340).

Consider one gluon from subgraph S to subgraph A, and its attachment to a quark line, as in the left-hand side of Fig. 10.12(a). The triangle at the vertex denotes the application of the soft approximation. *For the moment we ignore the subtraction terms.*

---

[9] Here Fig. 10.3(b) is treated as specifying the term $C_R \Gamma$, with the subgraphs H, A, B, S being those the specify the region R.

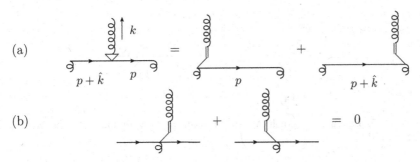

Fig. 10.12. Graphical elements of Ward identity: (a) application to line, (b) sum at vertex (in abelian gauge theory).

Let k be the gluon momentum, and let $\hat{k}$ be its approximant defined in (10.16). We apply the following identity:

$$\frac{n_1^\mu}{k \cdot n_1 + i0} \frac{i}{\not{p} - m + i0} (-ig\hat{k}) \frac{i}{\not{p} + \hat{k} - m + i0}$$

$$= \frac{i(-ign_1^\mu)}{k \cdot n_1 + i0} \left[\frac{i}{\not{p} - m + i0} - \frac{i}{\not{p} + \hat{k} - m + i0} \right]. \tag{10.98}$$

Thus one or other quark propagator is canceled, as pictured on the right-hand side of Fig. 10.12(a). The gluon is now attached to a special vertex that is at one or other end of the quark line. At this special vertex,[10] the double line denotes a factor of a Wilson-line propagator with an accompanying vertex, and the diagonal single line codes an overall sign. The sign essentially concerns the charge of the quark field.

We now sum over all places where the gluon can attach to the quark line. Now, when an S gluon attaches to an A quark, an equally allowed graph is where the S gluon attaches to the opposite side of a neighboring gluon vertex, as in Fig. 10.13. Note that the other gluon, of momentum l, may either be part of the A subgraph or the S subgraph; the argument works equally in both cases. This gives pairs of canceling terms, at each other gluon vertex on the quark, as illustrated in Fig. 10.12(b). If the quark line goes around in a loop inside the collinear graph, we get zero. But if the quark line goes out of the collinear graph, we are left with only the special vertices at the outside end(s) of the quark line. At the on-shell p_A end we in fact get zero, exactly as in the standard textbook case.[11] There remains one term, at the end of the quark line where it enters the hard scattering. The result is just as in the lowest-order case, (10.29), and is equivalent to a gluon attaching to a Wilson line.

In certain model calculations, we might use a scalar quark. In that case, we must take account of the vertex with two gluons. The necessary vertex identity is Fig. 10.14, which replaces Fig. 10.12(b) for spin-$\frac{1}{2}$ quarks. It is readily verified from the form of the two-quark–two-gluon vertex.

---

[10] In the context of diagrammatic proofs of Ward identities (e.g., Sterman, 1993, p. 351) the vertex represents the BRST transformation of the field at the end of the quark propagator, but in our work it is multiplied by an eikonal denominator.

[11] However, the details are not always made explicit in the textbooks!

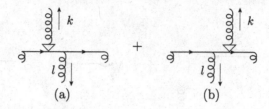

Fig. 10.13. Example of graphical structure which leads to the canceling terms in Fig. 10.12(b).

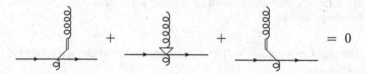

Fig. 10.14. Vertex sum as in Fig. 10.12(b), but for scalar quark.

10.8.4 Extraction of soft lines from collinear subgraphs

Now consider all the gluons entering collinear subgraph A from the soft subgraph S, continuing to omit the subtractions. We apply the Ward-identity argument of Sec. 10.8.3 to each gluon in turn, summing over allowed graphs for the A subgraph, given a particular set of external lines for the subgraph. Then we apply the same argument to the gluons from S to the other collinear subgraph B, and represent the result in Fig. 10.15(a) and (b). Each external gluon of the S subgraph now attaches to a Wilson-line factor of the form

$$\frac{i(-ign_1^{\mu})}{k_j \cdot n_1 + i0} \quad \text{on } A \text{ side}, \qquad \frac{i(ign_2^{\mu})}{k_j \cdot n_2 + i0} \quad \text{on } B \text{ side}, \qquad (10.99)$$

where k_j is the gluon momentum, defined to flow *into* the S subgraph.

We convert the result to exactly the Wilson-line form by using the following identity for the product of elementary Wilson-line propagators:

$$\prod_{j=1}^{N} \frac{i}{k_j \cdot n + i0} = \sum_{\text{permutations}} \frac{i}{k_1 \cdot n + i0} \times \frac{i}{k_1 \cdot n + k_2 \cdot n + i0}$$

$$\times \cdots \times \frac{i}{k_1 \cdot n + k_2 \cdot n + \ldots k_N \cdot n + i0}. \qquad (10.100)$$

This identity is readily proved by induction on N, and is applied separately to the parts of the diagram with $n = n_1$ and $n = n_2$. The right-hand side is exactly the product of lines resulting from the Feynman rules for a Wilson line (Sec. 7.6). Wilson-line vertex factors are exactly the $-ign_1$ and ign_2 factors in (10.99).

Next we observe that, with the region approximator T_R defined in Sec. 10.4.2, the approximated hard subgraph H is independent of the soft momenta. Thus we can contract the free ends of the Wilson lines together to give Fig. 10.15(c). The right-hand factor

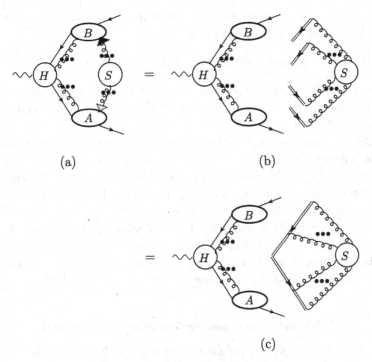

Fig. 10.15. Application of Ward identities to extract S gluons from the collinear subgraph with the soft approximation in (a). After use of Ward identities we get graph (b), and after use of (10.100), we get graph (c).

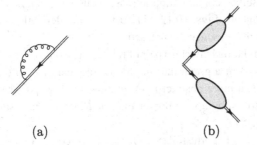

Fig. 10.16. (a) Example of Wilson-line self-energy graph. (b) Denominator of (10.89).

(summed over graphs for S) is just what we already stated as the definition (10.89) of the soft factor; there is one complication in the proof that I now explain.

Each connected component of an S subgraph joins the A and B sides. So no graph arises in Fig. 10.15(b) where a component of S just connects n_1 lines to themselves, or n_2 lines to themselves. However, such graphs do arise from the matrix element of the Wilson line, the numerator of (10.89), giving for example Fig. 10.16(a). If we were to sum over all such graphs, they would form extra factors in Fig. 10.15(b), which we call Wilson-line self-energy factors. Converting these factors to the Wilson-line form gives the general form of Fig. 10.16(b), which has the operator form

$$\text{W.L. self-energy factor} = \langle 0|W(\infty, 0, n_2)^\dagger|0\rangle \, \langle 0|W(\infty, 0, n_1)|0\rangle \,. \qquad (10.101)$$

Since these graphs are not produced by our Ward-identity argument, they must be removed from the definition of the soft factor. Thus (10.101) is the denominator in the definition (10.89) of the soft factor.

A careful examination of calculations of the self-energy factor shows that it has a divergence as the length of the Wilson line goes to infinity. No such divergence arises from graphs that connect the n_1 to the n_2 lines. So for a correct definition of the soft factor, we first replace the occurrences of "∞" in (10.89) and (10.101) by some large finite length L. Then the soft factor (10.89) is defined with a limit $L \to \infty$.

Finally, there are UV divergences in many of the relevant graphs. Just as in the textbook treatment of conventional Ward identities (e.g., Collins, 1984, Ch. 9) we define these to be canceled by UV counterterms. Just as in that case, the counterterms preserve the derivation of the Ward identities, provided that an appropriate renormalization scheme is used, like $\overline{\text{MS}}$.

10.8.5 Subtractions and the derivation of the soft factor

We have extracted soft gluons from their attachments to the collinear factors. But our derivation so far has applied to $T_R\Gamma$, i.e., to the approximator for region R of graph Γ, followed by a sum over graphs. We now examine the effect of the subtractions that convert $T_R\Gamma$ to the region term $C_R\Gamma$, defined in (10.4). These prevent double counting with the terms for smaller regions $R' < R$. Note that for a general region and graph, the subtraction terms $-C_{R'}\Gamma$ themselves contain subtractions, recursively applied. We now show how the fundamental elements, Figs. 10.12 and 10.14, in the derivation of the Ward identities continue to apply in the presence of subtractions.

We represent the relation between a pair of relevant regions in Fig. 10.17. There, diagram (a) depicts the division of a graph into the hard, collinear, and soft subgraphs associated with a region R; it is a more abstract representation of Fig. 10.3(b). In a smaller region $R' < R$, either the soft subgraph is bigger than in R, or the hard subgraph is smaller, or both, as in Fig. 10.17(b).

A generic term in $C_R\Gamma$ corresponds to a set of nested regions R_j that obey $R_1 < R_2 < \cdots < R_n < R$, and the corresponding contribution to $C_R\Gamma$ is

$$(-1)^n T_R \prod_{j=1}^{n} T_{R_j} \Gamma. \qquad (10.102)$$

The T_{R_j} operations are applied from inside out, smallest region to largest. Then $C_R\Gamma$ is the sum over possibilities for (10.102), including the case $n = 0$. This follows from the definition (10.4) of $C_R\Gamma$, exactly as in the theory of renormalization (Collins, 1984). The differences with renormalization are only in the specification of the regions and in the definitions of the region approximators.

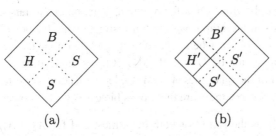

Fig. 10.17. (a) Partition of graph for Sudakov form factor by subgraphs for a region R. (b) Partition for a smaller region $R' < R$. The dotted lines indicate the boundaries of the subgraphs for the first region.

Now each region R_j corresponds to a pinch-singular surface (PSS) in the massless limit. Its approximator T_{R_j} is obtained from the leading power of the integrand expanded in powers of the radial variable λ_{R_j} for the region, with masses treated as an appropriate power of λ_{R_j}. This expansion is then slightly modified by the following replacements for soft loop momenta in the collinear subgraphs:

$$k_{AS} \cdot w_1 \mapsto k_{AS} \cdot n_1, \qquad k_{BS} \cdot w_2 \mapsto k_{BS} \cdot n_2, \tag{10.103}$$

as in (10.17) and (10.18). We now show that the Ward identities we use for extracting the soft factor continue to apply in the presence of the subtractions.

Let a gluon of momentum k from the S subgraph of R attach to an A quark. The line identity, (10.98) and Fig. 10.12(a), has the structure

$$\frac{1}{A_1} (A_1 - A_2) \frac{1}{A_2} = \frac{1}{A_2} - \frac{1}{A_1}, \tag{10.104}$$

up to an overall factor of a phase and a coupling. Here $1/A_1$ and $1/A_2$ are the quark propagators, and $A_1 - A_2$ is the vertex factor, $\hat{k}^- \gamma^+$.

Now, to get from $T_R\Gamma$ to $C_R\Gamma$ we sum (10.102) over all possibilities for nested sets of smaller regions. Each term in (10.102) has region approximator(s) applied to the graph, which contains the l.h.s. of (10.104) as a factor. Each region approximator replaces each factor in the graph by (the first term) in its expansion in powers of λ_{R_j}, supplemented by the replacements like (10.103). All of these operations can be applied equally well when the l.h.s. of (10.104) is replaced by one or other of the terms on the r.h.s. Furthermore, the same collection of operations can be applied to each of the terms in the vertex identity Fig. 10.12(b) or Fig. 10.14.

This indicates that the Ward identities that apply to $\sum_{R,\Gamma} T_R\Gamma$ are also valid in the presence of subtractions, so that the Ward-identity result should also apply to $\sum_{R,\Gamma} C_R\Gamma$. However, there is a potential problem that to use the vertex identity, we are combining terms obtained from different graphs, and these could have different regions. To see the difficulty, observe that the canceling terms at a vertex arise from different graphs, e.g., from Fig. 10.13. To make the vertex identity work in the presence of subtractions, we must use a correspondence between the regions for the different graphs. We need to determine the situations where the correspondence fails to exist, and to deal with the consequences.

Another related complication is that the region approximator T_{R_j} takes the leading power in λ_{R_j} of the factors in the graph; we must investigate what happens if an approximator gives a different power of λ_{R_j} when applied to A_1 and A_2 on the r.h.s. of (10.104).

Consider the application of T_{R_j} to (10.104). It takes the leading power in λ_{R_j} of each factor on the l.h.s. For the quantities A_1 and A_2, let the leading-most terms be $\hat{A}_1$ and $\hat{A}_2$. In the most general context, there are three possible cases for the power laws:

- The power of λ_{R_j} is the same for both quantities, and for $A_2 - A_1$. The line identity applies equally to the leading-power expansion

$$\frac{1}{\hat{A}_1} (\hat{A}_1 - \hat{A}_2) \frac{1}{\hat{A}_2} = \frac{1}{\hat{A}_2} - \frac{1}{\hat{A}_1}. \tag{10.105}$$

The left-hand side gives the effect of T_{R_j} on the left-hand side of (10.104), and the two terms on the right are the effect of applying T_{R_j} to the terms on the right-hand side of (10.104). Effectively, T_{R_j} is a linear operation that commutes with the manipulations giving the Ward identity. If T_{R_j} had been defined to make different operations on the vertex factor and the propagators, this result need not be true. The quantity on the left-hand side and the two terms on the right-hand side have the same power-counting and therefore do not change the necessary set of subregions.

  The above situation is always the case for a soft line connected to a collinear line, with the one trivial exception that one line, e.g., A_2, is an external line. Then we omit the $1/A_2$ factor, and replace A_2 by zero.

- Another possibility is that the power of λ_{R_j} for one line, A_2 say, is larger than for the other line A_1. Thus $A_2/A_1 \to 0$ in the limit of $\lambda_{R_j} \to 0$. Then the leading power of the vertex factor $A_1 - A_2$ is just $\hat{A}_1$, and we must replace (10.105) by

$$\frac{1}{\hat{A}_1} \hat{A}_1 \frac{1}{\hat{A}_2} = \frac{1}{\hat{A}_2}. \tag{10.106}$$

At the PSS R_j, the $\hat{A}_2$ line can be viewed as on-shell, and we get exactly one term on the right-hand side, just as when such a line is exactly on-shell. The term $1/\hat{A}_1$ is smaller by a power of λ_j than $1/\hat{A}_2$, and so is correctly neglected.

- A final possibility is that A_2 and A_1 are comparable, but $A_1 - A_2$ is much smaller. In that case, no subtraction associated with R_j is actually needed for the original graph. But for the individual terms on the right-hand side we do need subtractions. Even though R_j is not actually a leading region for the original graph, we add it to the catalog of leading regions.

The above treatment applies literally for scalar quarks, for then the quantities A_1 and A_2 are scalars, and the definition of their power is unambiguous. For fermions, each is a matrix, whose inverse is taken in the propagators. A slightly more complicated version of the argument leads to the same outcome.

Finally we apply the vertex identity. This relates graphs with the same set of denominators, and hence with the same subtractions. So the vertex identities continue to apply after all the subtractions for subregions have been applied.

When applying the vertex identity, we will have canceling terms obtained from applying the line identity to neighboring lines. In the above derivations we have only examined the vertex and lines in question. It is important that everything else in the graphs remains the same. For example, in defining the soft (and collinear) factors, we inserted Wilson-line denominators with non-light-like directions to cut off rapidity divergences. The success of the vertex identities depends on these non-light-like lines being the same everywhere they are encountered, e.g., always the same n_1 for a soft gluon connecting to a collinear-to-A quark.

The final result is that Ward identities apply in the presence of subtractions just as they did in the elementary case we examined where we ignored subtractions. However, we must take care to apply subtractions to the resulting factors.

So far we have extracted the soft factor. Since there are no smaller momentum classes than soft, this factor needs no subtraction. Thus we have completed the proof that the soft part of the form factor factorizes, and that the soft factor can be defined by (10.89). That is, after summing over graphs and regions, we get Fig. 10.15(c).

But subtractions are needed in the remaining parts of the graphs, and our next task is to convert them into hard and collinear factors (which will have subtractions).

10.8.6 Extraction of collinear factors from hard scattering, without effect of subtractions

We now extract the collinear gluon attachments from the hard scattering and convert them to attachments to Wilson-line operators, as in (10.94a) and (10.94b). As before, we start by examining the part of $C_R\Gamma$ without subtractions, and extract the collinear gluons one-by-one. The argument will be somewhat modified from that for soft gluons attaching to a collinear subgraph, because the allowed subgraphs for H have important restrictions by being 1PI in each set of collinear lines.

Of the two graphical elements for the Ward identity, a line identity like Fig. 10.12(a) continues to apply, with only the caveat that one of the lines p and $p + \hat{k}$, inside the H subgraph, may be set on-shell by the approximator applied to a quark line at the collinear edge of H. But for the vertex identity, Fig. 10.12(b), we can miss one of the graphs it implicates.

An example is shown in Fig. 10.18, where we sum over the possible attachments of a B gluon of momentum k to a one-loop hard subgraph. In graph (a), there is an on-shell quark to the right of the vertex with the gluon, so that one term in the line identity gives zero, as usual for an on-shell quark.[12] There is then the usual chain of cancellations, with graphs (b) and (c). But we do not have the graph where gluon k attaches one place to the right of where it is in (c), i.e., we are missing graph (d). This is because in graph (d), gluon k attaches to another B line at its lower end, so that vertex is not part of the hard subgraph;

---

[12] Note that in the general case, with a non-trivial collinear-to-B subgraph, the quark in question is on-shell not because it is an external quark, but because it is the outermost quark line of the hard scattering. Our definition of the approximator for a region replaces the (possibly off-shell) external quarks of the hard scattering by exactly on-shell quarks.

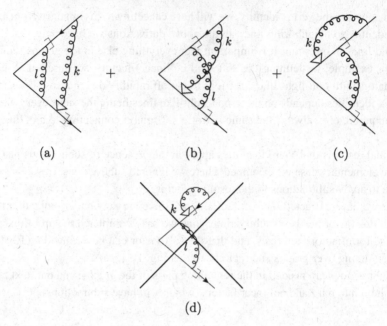

Fig. 10.18. Example of sum over attachments of gluon from collinear subgraph to hard subgraph. The gluon l is in subgraph H, and the gluon k is in subgraph B. The hooks on the quark lines indicate lines that are approximated as on-shell in the hard subgraph H. The big arrow at the bottom of line k has the same meaning as in Fig. 10.6(a), except that it uses the vector w_1 instead of n_1. Graphs (a)–(c) are summed, while graph (d) is excluded by the condition that the hard subgraph is 1PI in collinear-to-B lines.

we see here an example of the general result that a hard subgraph is 1PI in lines that are collinear to a particular direction.

The result is shown in Fig. 10.19, and it shows that the sum over attachments of gluon k to a hard subgraph has extracted the gluon from the hard scattering and attached it instead to a Wilson line. The Wilson line has exactly the form that results from the definition, (10.94b), of the collinear-to-B factor. The remaining factor is a one-loop graph for the hard subgraph without any extra gluons.

In the general case of a B gluon connecting to any H subgraph, what possibilities are there? They are when one but not the other of the two graphs in Fig. 10.13 is not allowed, given that gluon k is in the B subgraph, and that, at least on one side, the quark line is in the H subgraph. It is easily checked that there are two cases, each where one of the two subgraphs would have a collinear quark line.

One corresponds to Fig. 10.19(a), where the quark on one side of the vertex for k is in the B subgraph. This gives the expected Wilson-line vertex.

The other case is where the other gluon l and the quark line on one side are in the A subgraph, as in Fig. 10.20. We get an extra term in the sum over attachments of the k gluon, Fig. 10.20(c). This graph is in fact zero. The reason, which applies generally, is that at the attachment of gluon l the approximator picks out exactly the minus component of the

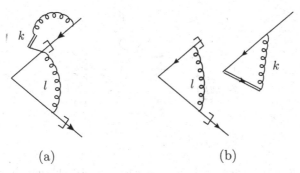

Fig. 10.19. Result of sum in Fig. 10.18: (a) in the notation of Fig. 10.12, (b) as an attachment to a Wilson line.

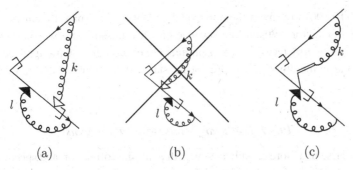

Fig. 10.20. Simplest example of the other case that the vertex cancellation in the Ward identity has a missing term. Approximators are applied for the case that k is collinear to B, and l is collinear to A.

vertex; see (10.23), where the H subgraph is contracted with $P_{HA}(k_{HA})$, which is exactly in the w_1 direction. But the vertex is now exactly at the edge of the hard subgraph where there is a quark that is exactly in the plus direction. It has a projection onto on-shell massless wave functions for the quark, by the matrix $\mathcal{P}_B$. Therefore multiplying by the vertex factor γ^- gives zero; this is essentially from the Dirac equation for a massless quark in the plus direction:

$$0 = \bar{u}_{A\,\text{massless}} \gamma^- p_A^+. \tag{10.107}$$

Although we have formulated this argument for one graph, and for Dirac quarks, the argument is actually general. It concerns an approximation where both the quark and the gluon l have been made exactly massless and collinear in the plus direction in one part of the hard subgraph. The minus component of the vertex goes to zero under an infinite boost from a rest frame.

We now repeat the above arguments for all gluons entering the H subgraph from the collinear subgraphs, first from the collinear-to-B subgraph and then from the collinear-to-A subgraph. After a sum over all graphs, we get two collinear factors times a hard factor. As with the soft factor, each collinear factor has a product of one-gluon Wilson-line factors,

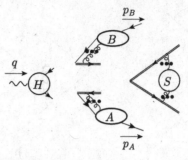

Fig. 10.21. Factorized structure for Sudakov form factor. The double lines are Wilson lines with the following rapidities: $-\infty$ for A, $+\infty$ for B, y_1 and y_2 for S. Subtractions in H, A, and B are not indicated explicitly.

and we use (10.100) to convert them to exactly a Wilson-line matrix element. Again, just as with the soft factor, Wilson-line self-energies are missing. So we must divide by a Wilson-line self-energy factor. The Wilson lines are exactly those with light-like directions that are in the numerators of the previously stated definitions of the collinear factors, (10.94).

10.8.7 Collinear factors, with subtractions

Subtractions arise in a more complicated way than for the soft factor, and specific examples in multiloop graphs can become quite elaborate.

The most general method of dealing with subtractions is to appeal to the argument given in Sec. 10.8.5, which applies quite generally. This is that subtractions apply whenever a graph would have singularities in the massless limit, and that they are obtained from the analytic structure of the denominators, together with power-counting. We showed that the Ward identities apply in the presence of subtractions.

Therefore all we have to do to convert the unsubtracted result is to apply subtractions. Without the subtractions, the arguments so far give the factorized structure shown in Fig. 10.21. We have separate hard, collinear and soft factors multiplied together. The correct formula is obtained simply by applying subtractions to the factors.

For the soft factor, as already explained, no subtractions are needed, because there are no momentum regions smaller than a soft configuration. (Beyond this we also need the Wilson-loop denominator in (10.89), to remove the Wilson-line self-energies, which do not arise from the Ward-identity argument.)

For each collinear factor we have soft subtractions and for the hard factor we have soft and collinear subtractions.

The easiest way to obtain an operator form for a subtracted collinear factor is to apply the factorization argument to the unsubtracted collinear factor, e.g., to the limit $y_{u_2} \to -\infty$ of (10.90), which has a non-light-like Wilson line, of rapidity y_{u_2}. The leading regions have the form shown in Fig. 10.22(a). These each have a collinear-to-A subgraph and a soft graph that connects the Wilson line to the collinear subgraph, by arbitrarily many

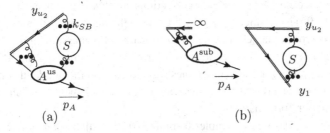

Fig. 10.22. (a) Regions for unsubtracted collinear matrix element (10.90). Here A^{us} is an abbreviation for A^{unsub}. (b) After applying Ward identities to the gluons attaching the soft subgraph to the collinear subgraph, we get this factorized form. Here A^{sub} denotes the subtracted collinear factor. Next to each Wilson line is a label indicating its rapidity.

gluons. We define the soft region with respect to u_2 rather than the overall center-of-mass. In accordance with the order of limits specified in Sec. 10.8.2, we take the limit $y_{u_2} \to -\infty$ with fixed space-time dimension $4 - 2\epsilon < 4$, so that loop corrections to the hard subgraph are power-suppressed, and we need no hard subgraph, just the connections from the collinear subgraph to the Wilson line.

Since the collinear-B part is already in a Wilson-line form, it is enough to combine the soft factor and the collinear-B factor in a new soft factor, denoted S in Fig. 10.22. The usual Ward-identity argument is applied to gluons entering the collinear-A subgraph from S. The same argument that we applied to the whole form factor now applies here, and results in a soft factor times a subtracted collinear factor:

$$A^{unsub}(y_{p_A} - y_{u_2}) = A^{sub} \times S(y_1 - y_{u_2}), \tag{10.108}$$

up to terms that are power-suppressed in the limit $y_{u_2} \to -\infty$. The soft factor is the same as in factorization for the form factor itself, *except that* the direction of the Wilson line on the B side is u_2 instead of n_2. This is depicted in Fig. 10.22(b).

Dividing by the soft factor on both sides of the above equation gives the subtracted A factor as the unsubtracted matrix element (10.90) divided by the relevant soft factor. Taking the limit $y_{u_2} \to \infty$, i.e., $u_2 \to w_2$, gives our definition of the subtracted soft factor A^{basic} in (10.94a). The subtractions are the same as in the collinear factor used for the form factor, so it has the same definition. Thus A^{basic} is to be identified with the graphical factor A both in Fig. 10.21, and in the factorization formula (10.11).

An exactly similar argument applies to the collinear-to-B factor, of course.

10.8.8 Hard factor

At this point, we have actually proved a form of factorization, (10.11), given in diagrams in Fig. 10.21, and we have given explicit definitions of the soft and collinear factors.

Now we obtain an explicit formula (10.96) for the hard factor H. The graphs for H are the same as for the form factor itself, i.e., for the reaction $\gamma^* \to q\bar{q}$, but they have subtractions for soft and collinear regions. The graphs are to be 1PI in the external quark and antiquark, since external propagator corrections are always part of a collinear subgraph. The formula (10.96) is obtained simply by observing that the power-suppressed corrections in (10.11) go to zero as masses are taken to zero. Taking the massless limit means not only setting the quark and gluon masses to zero in graphs, but also taking the light-like limits for the vectors n_1 and n_2 in the Wilson lines associated with the quark and antiquark. The one-loop expansion of (10.96) reproduces the result (10.45) which we already obtained from the subtraction formalism.

Later, we will find a slightly simpler formula (10.120), after we examine the evolution equations of the soft factor S with respect to the rapidities of its Wilson lines.

In addition to the kinematic variable Q, the hard factor depends on the renormalization scale μ. As usual, the μ dependence is governed by an RGE. So we can use the RGE to set μ of order Q, and then the hard factor would be perturbatively calculable (in a QCD problem). For the evolution, anomalous dimension are generally perturbatively calculable.

10.9 Factorization in terms of unsubtracted factors

To compensate double counting between soft and collinear regions, we implemented subtractions in the collinear factors. We then saw that after summing over graphs and regions, the subtractions were implemented by dividing out a certain factor.

We can write the factorized form factor in terms of the unsubtracted matrix elements:

$$
F \sim \lim_{\substack{y_{u_1} \to +\infty \\ y_{u_2} \to -\infty}} H \, \frac{A^{\text{unsub}}(y_{p_A} - y_{u_2}) \, B^{\text{unsub}}(y_{u_1} - y_{p_B}) \, S(y_1 - y_2)}{S(y_{u_1} - y_2) \, S(y_1 - y_{u_2})}.
\tag{10.109}
$$

Here, we have indicated the dependence of the factors on the directions of the Wilson lines. Of course, the dependence on the Wilson-line rapidities must disappear after taking the product $HABS$, at least to leading power in Q, since the Wilson lines do not appear in the original form factor. If the rapidity limits in (10.109) are taken *after* the UV regulator is removed, then the definition of the hard factor must be modified, as follows from Sec. 10.8.2.

In the definitions of A^{unsub}, B^{unsub}, and S, Wilson-line self-energies are canceled by dividing each quantity by the appropriate version of (10.101). When we combine all the factors in (10.109) the self-energies exactly cancel, since there are equal numbers of each direction of Wilson line in the numerator and denominator of (10.109).

After deriving evolution equations, it will be convenient to reorganize this formula to give it more convenient properties; see Sec. 10.11.

10.10 Evolution

We need evolution equations for the dependence of the soft and collinear factors on the rapidities of their Wilson lines. Evolution equations provide much of the predictive power of factorization.

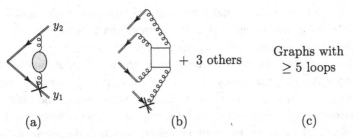

Fig. 10.23. Graphs for the connected part of the derivative of the soft factor, up to four loops. The blob on the gluon line in (a) denotes all corrections to the gluon propagator. The crossed vertex is the same as in Fig. 10.10, and is defined from (10.49). The left-hand ends of the Wilson lines are intended to be joined together, to implement the l.h.s. of (10.100).

Without the evolution equations, we would have no better predictive power than from direct perturbative calculations of the form factor, and accuracy would be particularly compromised by the two logarithms per loop. With the evolution equations including the RGEs, we can obtain all the factors in terms of quantities that are free of large logarithms.

The evolution equations given below were first obtained by Collins (1980), but by different and less general methods, and with different, but closely related, gauge-dependent definitions of the factors.

10.10.1 Evolution of basic soft factor

We start with the dependence on y_1 or y_2 of $S(y_1 - y_2)$. Deriving its evolution equation is a fairly simple generalization of our one-loop calculation in Sec. 10.5.10.

Since we are in an abelian theory, we use the identity (10.100) to write the value of a Wilson line as the product of elementary one-vertex Wilson lines. Then S is the exponential of its irreducible connected part:

$$S(y_1 - y_2) = \exp(S_{\text{conn}}). \tag{10.110}$$

Differentiating with respect to y_1 gives

$$\frac{\partial S(y_1 - y_2)}{\partial y_1} = S \frac{\partial S_{\text{conn}}}{\partial y_1}. \tag{10.111}$$

As illustrated in Fig. 10.23, graphs for $\partial S_{\text{conn}}/\partial y_1$ have one vertex for a differentiated Wilson line, just as in the lowest-order case, Fig. 10.10, together with at least one Wilson-line vertex on the other side, and any number of extra Wilson-line vertices, but no Wilson-line self-energies. Notice that the corrections at two- and three-loop order only arise from corrections to the gluon propagator.

We now perform a region analysis for $\partial S_{\text{conn}}/\partial y_1$. Because of the restriction to connected graphs and because of the differentiated vertex, this analysis is very simple. As usual, graphs for $\partial S_{\text{conn}}/\partial y_1$ can have H, A, B, and S subgraphs.[13] These subgraphs must be connected to each other, and this must occur through one or more quark loops, since the connections to the Wilson line are to single line segments, after we used (10.100).[14] Therefore, if a region for $\partial S_{\text{conn}}/\partial y_1$ has more than one of the subgraphs H, A, B, and S, we get zero, after applying a Ward identity to the sum over graphs. Exactly as in the one-loop case, the differentiation with respect to y_1 at the crossed vertex forces the gluon line at the differentiated vertex to have rapidity close to y_1; thus it is either collinear-to-A (i.e., to n_1) or hard. It follows that the only two leading regions are where the whole of $\partial S_{\text{conn}}/\partial y_1$ is collinear-to-A or where it is all hard.

Thus the situation we saw for the one-loop case in Sec. 10.5.10 immediately generalizes to all orders:

- The limit $y_2 \to -\infty$ can be taken, so that we can write the evolution equation in terms of a rapidity-independent kernel

$$K\left(m_g, m, \mu, g(\mu)\right) \stackrel{\text{def}}{=} 2 \lim_{y_2 \to -\infty} \frac{\partial S_{\text{conn}}}{\partial y_1}, \tag{10.112}$$

plus power-suppressed corrections. Thus in Fig. 10.23, the upper Wilson line can be taken light-like in the minus direction without encountering any divergence.

The above definition of K is asymmetric between the two Wilson lines of S, and we will later make a symmetric definition in (10.122), which leads to the same numerical results for calculations in a covariant gauge.
- The kernel K has an additive anomalous dimension γ_K, as in (10.56).

Hence the previously stated results (10.53) and (10.56) apply generally.

It follows that at large $y_1 - y_2$, the $y_1 - y_2$ and μ dependence of the soft factor has the form

$$S = S_0(m_g, m, \mu_0, g(\mu_0))$$
$$\times \exp\left\{ -\frac{y_1 - y_2}{2} \left[\int_{\mu_0}^{\mu} \frac{d\mu'}{\mu'} \gamma_K(g(\mu')) - K(m_g, m, \mu_0, g(\mu_0)) \right] \right\}, \tag{10.113}$$

where μ_0 is a fixed reference value of the renormalization scale, and S_0 is independent of $y_1 - y_2$. Because of power-suppressed corrections, S_0 does *not* equal the value of S when $y_1 = y_2$ and $\mu = \mu_0$.

Naturally, we could equally well have performed the differentiation with respect to y_2 instead of y_1. In that case there would be a change of sign, and the Feynman rules would have the crossed vertex in Fig. 10.23 on the opposite Wilson line. We will redefine K more symmetrically later, in Sec. 10.11.3; the redefinition also remedies a lack of gauge independence of K when one uses a non-covariant gauge.

---

[13] It should be possible to simplify this by a classification of lines by rapidity: collinear-to-B, and n_1-rest-frame.
[14] An example would be Fig. 10.23(b), when the quark loop and the lines to the lower Wilson lines are collinear-to-A, but one or both of the upper gluons are soft.

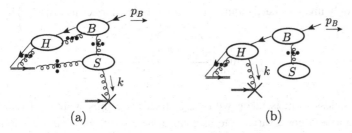

Fig. 10.24. Leading regions for $\partial B(y_1 - y_{p_B})/\partial y_1$, (10.114). *In (b), the soft subgraph has at least one gluon attachment to the main Wilson line,* but we do not show this, to avoid complicating the graph.

10.10.2 Evolution of collinear factor

We now obtain an equation for the derivative with respect to y_1 of the *unsubtracted* collinear factor $B^{\text{unsub}}(y_1, y_{p_B})$. The effect of differentiating the Wilson line is

$$\frac{\partial B^{\text{unsub}}(y_1 - y_{p_B})}{\partial y_1} = \quad\text{[figure]}\quad \tag{10.114}$$

As in Fig. 10.23, the left-hand end of the differentiated Wilson-line element is attached to the main quark-Wilson-line vertex, and we used (10.100) to allow us to treat each vertex of the Wilson line independently.

We now apply the same arguments as we used for factorization. But we simplify the argument by using a frame where n_1 has zero rapidity, so that the momentum categories are soft, hard, and collinear-to-B. A soft momentum has rapidity comparable to y_1, and there is now no separate collinear-to-A category. As usual, the momentum k at the differentiated vertex is restricted to have a rapidity close to y_1, so that it is either soft or hard. There correspond two types of leading region, shown in Fig. 10.24(a) and (b) respectively.

For the case that k is soft, graph (a), we examine the component of the soft subgraph to which is attaches, and apply Ward identities for all the gluons that couple it to the collinear subgraph. This gives a factor of exactly the kernel $\frac{1}{2}K$ for the evolution of the soft factor, and it multiplies the original collinear factor.

When k attaches to the hard subgraph, we use Ward identities to extract the collinear gluon attachments. The result is a factor times the original collinear factor. To this must be applied subtractions for the soft-gluon part. Since there are now no collinear or soft contributions to the hard factor, we can apply the massless limit to it. This gives the following evolution equation (Collins, 1980):

$$\frac{\partial B^{\text{unsub}}(y_{u_1} - y_{p_B})}{\partial y_{u_1}} = \frac{1}{2}\left[K(m, m_g, \mu) + G(\zeta_{B,u_1}, \mu)\right] B^{\text{unsub}}$$

$$+ \text{non-leading power of } \zeta_{B,u_1}, \tag{10.115}$$

where ζ_{B,u_1} is defined in (10.93b).

Since G only involves hard momenta, it can be defined in terms of B^{unsub} by a massless limit as

$$G = 2 \lim_{\substack{m \to 0 \\ m_g \to 0}} \left[\frac{\partial \ln B^{\text{unsub}}(\zeta_{p_B,u_1})}{\partial y_{u_1}} - K \right]. \tag{10.116}$$

Here the massless limit is taken with ζ_{p_B,u_1} fixed, and thus with p_B^- fixed. (Note that y_{p_B} would not be a good variable to use, since the rapidity of a massless momentum is infinite.)

If we dimensionally regulate, G decreases like a power of $\zeta_{B,u_1}/\mu^2$. But the power-suppression goes away when $n \to 4$. This gives another view of how, in defining A^{basic} and B^{basic}, we took the $y_{u_1} \to \infty$ and $y_{u_2} \to -\infty$ limits. The limits are of $A^{\text{unsub}}(y_{p_A} - y_{u_2})/S(y_1 - y_{u_2})$ and $B^{\text{unsub}}(y_{u_1} - y_{p_B})/S(y_{u_1} - y_2)$. In accordance with Sec. 10.8.2, these limits are taken with $n < 4$. With $n < 4$ the evolution equations only involve the K terms in the infinite rapidity limit. Since the u_2 (or u_1) Wilson line appears in both numerator and denominator, the evolution equation shows that the K terms cancel, so that the infinite rapidity limits exist. This is consistent with and confirms what we earlier derived by another method.

The companion equation for A has a reversed sign:

$$\frac{\partial A^{\text{unsub}}(y_{p_A} - y_{u_2})}{\partial y_{u_2}} = -\frac{1}{2} \left[K(m, m_g, \mu) + G(\zeta_{A,u_2}, \mu) \right] A^{\text{unsub}}$$

$$+ \text{ non-leading power of } Q \text{ and } \zeta_{A,u_2}, \tag{10.117}$$

where ζ_{A,u_2} was defined in (10.93a).

These equations bring under control the dependence of the collinear factors on the Wilson-line rapidities. We then use the RG to tame the logarithms of μ: to set μ to be a fixed scale in K and in the collinear factors, but to be of order Q in G and H. We will discuss this in more detail after we perform a final reorganization of the factorization formula.

10.11 Sudakov: redefinition of factors

The above formalism has some defects, particularly in its generalization to measurable cross sections in QCD:

1. The soft factor has no independent experimental consequences. It always appears multiplied by two collinear factors.

    In QCD applications of factorization, the soft factor is non-perturbative. Although the values of non-perturbative quantities are in principle predicted by QCD, our ability to actually calculate them is currently close to zero. So generally we have to measure them from experiment, and rely on universality to make predictions for the same reactions at different energies and for different reactions. But there is no experimental probe of the soft factor by itself.

2. Feynman rules for the soft factor involve non-light-like Wilson lines. Perturbative calculations of such quantities are more difficult than when at least one Wilson line is light-like. (But, of course, with light-like Wilson lines, there must be subtractions to cancel rapidity divergences.)

3. Associated with the non-light-like Wilson lines in S are power-suppressed corrections to the evolution equation (10.53).

4. The definitions of the factors involve removal of Wilson-line self-energies (10.101). However, these cancel in the complete factorization formula, which suggests a non-optimality in the formulation.

5. The removal of Wilson-line self-energies makes the factors gauge-dependent.

6. Related to this is that although the evolution kernel K defined in (10.112) is gauge independent when restricted to covariant gauges, it changes when the gauge is transformed to an axial or Coulomb gauge. See problem 10.9.

These defects are to be regarded not as errors in the formalism, but as practical problems that make the formalism more complicated to use.

We will now perform a redefinition of the soft, collinear, and hard factors to remove these defects as much as possible. A useful starting point is (10.109), where factorization is given in terms of unsubtracted collinear factors and three occurrences of the basic soft factor S with different rapidity arguments. We can use (10.113), which shows that S has exponential rapidity dependence, to reorganize the factors of S.

Then we will absorb the S factor(s) into redefined collinear factors, to give a new factorization formula with no soft factor:

$$F = HAB + \text{power-suppressed}. \tag{10.118}$$

This overcomes the lack of experimental probes of the soft factor.

The definitions of the new collinear factors are at first sight surprisingly complicated. I will first state the definitions (which supersede those proposed by Collins and Hautmann, 2000). Then I will show how they correspond to the previous factorization formula in the form (10.109). After that I will give the rationale for the new definitions; they are unique given certain reasonable requirements.

10.11.1 Collinear factors

The redefined collinear factors A and B involve an arbitrary rapidity parameter y_n. We assign y_n the physical significance of separating left- and right-moving quanta; the A factor contains the effects of right-movers and B the effects of left-movers. The new collinear factors depend on the difference in rapidity between their particle (p_A or p_B) and y_n.

We will find that the dependence of each collinear factor on y_n is governed by an exactly homogeneous evolution equation involving the kernel K. Thus we can express each collinear factor in terms of its value when its particle has the same rapidity as y_n. This gives an optimal form of factorization.

The redefined collinear factors are

$$A(m, m_g, g, \mu, y_{p_A} - y_n)$$

$$\stackrel{\text{def}}{=} \lim_{\epsilon \to 0} \lim_{\substack{y_1 \to +\infty \\ y_2 \to -\infty}} Z_A \, A^{\text{unsub, bare}}(y_{p_A} - y_2) \sqrt{\frac{S^{\text{bare}}(y_1 - y_n)}{S^{\text{bare}}(y_1 - y_2) \, S^{\text{bare}}(y_n - y_2)}}$$

$$= A^{\text{unsub}}(y_{p_A}, -\infty) \sqrt{\frac{S(+\infty, y_n)}{S(+\infty, -\infty) \, S(y_n, -\infty)}}, \tag{10.119a}$$

$$B(m, m_g, g, \mu, y_n - y_{p_B}) = B^{\text{unsub}}(+\infty, y_{p_B}) \sqrt{\frac{S(y_n, -\infty)}{S(+\infty, -\infty) \, S(+\infty, y_n)}}. \tag{10.119b}$$

As in Sec. 10.8.2, we first take the limits of infinite rapidity, and then we remove the UV regulator $\epsilon \to 0$, with the aid of renormalization factors Z_A and Z_B. This order of limits entails adjusting the renormalization coefficients relative to our previous definitions. Thus it is convenient to write the new definitions in terms of bare soft and collinear factors, i.e., quantities defined without the renormalization factors Z_S, Z_A^{unsub}, and Z_B^{unsub} used in (10.89), (10.90), and (10.92). It is convenient to use a notation with infinite rapidities for the Wilson lines, as in the third and fourth lines of (10.119). It implies the limits given on the second line.

Each of the factors on the r.h.s. of (10.119) was originally defined to have Wilson-line self-energies divided out. It can be shown that the self-energy factors cancel in the combinations used in (10.119). (The total power of self-energy factors for each direction of Wilson line is zero. The only complication is that the Wilson lines for direction n are for opposite charges, but charge-conjugation invariance can be used to show that this is irrelevant.)

I now show that the product of A and B defined in (10.119) equals the product of the soft and collinear factors in our first form of factorization, when it is expressed in terms of unsubtracted collinear factors in (10.109).

First we examine the limits $y_{u_1} \to \infty$ and $y_{u_2} \to -\infty$ in (10.109), by using the evolution equations (10.53), (10.115) and (10.117). The K terms cancel for the y_{u_1} and y_{u_2} dependence in (10.109). This leaves just the G terms from (10.115) and (10.117). These concern a hard momentum region, and are effectively absorbed in UV renormalization factors. From (10.53), we see that the y_1 and y_2 dependence also cancels in (10.109). Thus the unsubtracted collinear factors are the same in (10.109) and in the product of (10.119a) and (10.119b).

After that, we apply the solution (10.113) for S, to show that the combination of S factors in (10.109) agrees with the combination of S factors in the product of (10.119a) and (10.119b).

Hence the two forms of factorization agree.

Notice that Wilson-line self-energies cancel for each of the different types of Wilson line in (10.119), so we do not need to insert any Wilson-loop factor to cancel Wilson-line self-energies, unlike our previous definitions. In fact, the definitions above are unique given the following requirements:

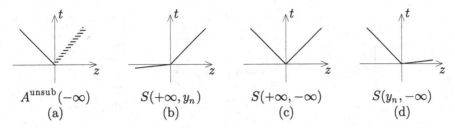

Fig. 10.25. Directions of Wilson lines in the factors in (10.119a): the solid lines are the Wilson lines (which should extend to infinity), which are either light-like or in the direction n, which is here drawn with a slightly positive rapidity y_n. The shaded part of (a) is intended to suggest the final-state quark itself, which moves in a time-like direction.

1. A collinear factor is a product of an unsubtracted collinear factor and powers of S-type objects.
2. Non-light-like Wilson lines only appear in S factors with one light-like and one non-light-like line.
3. Rapidity divergences cancel.[15]
4. Only one light-like direction y_n is used.
5. The definitions obey charge-conjugation symmetry; thus the definition of B is obtained from the definition of A, simply by changing $A^{\text{unsub, bare}}$ to $B^{\text{unsub, bare}}$ and by exchanging the roles of y_1 and y_2.
6. The factorization formula is HAB, without any soft factor.

The actual directions of the Wilson lines are shown in Fig. 10.25. In all the S objects, the two Wilson lines are at space-like separations. All the Wilson lines are either space-like or are obtained from a limit of space-like lines. Thus we do not have to be concerned with the ordering of the gauge-field operators on the Wilson lines. At least in covariant gauge, the fields commute at space-like separation. Thus the path ordering on the lines creates no conflict with the time ordering needed to define Green functions that use time-ordered fields. There is also maximum compatibility with Euclidean lattice gauge theory, which is important for attempts to compute non-perturbative collinear factors in QCD.

One perhaps unexpected feature is that the Wilson lines of rapidity y_n in the numerator and denominator of each collinear factor have opposite directions. For example, in (10.119a), y_n in the numerator factor $S(y_1 - y_n)$ corresponds to a Wilson line related to the antiquark. Therefore it has the charge of the antiquark and goes in the direction of a vector $n_B = (-e^{y_n}, e^{-y_n}, \mathbf{0}_T)$ whose minus component is positive. But y_n in the denominator factor $S(y_n - y_2)$ corresponds to a Wilson line with the charge of the quark and in the direction of a vector $n_A = (e^{y_n}, -e^{-y_n}, \mathbf{0}_T)$ whose plus component is positive. Thus the cancellation of Wilson-line self-energies for the y_n lines in (10.119a) is not as transparent as it would be if the lines were in exactly the same direction. This should be investigated.

In Sec. 10.8.2 was mentioned a non-uniformity of the limits of infinite rapidity and of $n \to 4$. For the newly defined collinear factors, we can see this from Fig. 10.26, which

---

[15] Except for regions of $k_T \to \infty$, which can be canceled by UV renormalization.

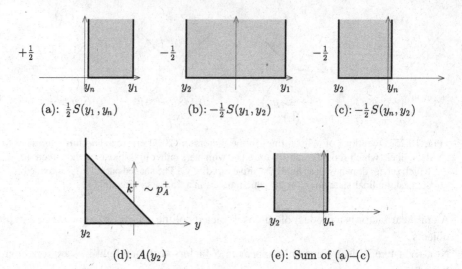

Fig. 10.26. Like Fig. 10.7, but showing the main regions for the one-loop contributions to (10.119a), with y_n chosen slightly positive. The diagrams are written before the limits $y_1 \to \infty$ and $y_2 \to -\infty$ are taken. The scale is reduced from Fig. 10.7.

shows the regions in gluon k_T and rapidity that contribute at one-loop order to the factors in (10.119a). In the region of low transverse momentum, the S terms combine to give a negative contribution running between y_2 and y_n that cancels the corresponding contribution from the one-loop term in the A term. This cancels the rapidity divergence as $y_2 \to -\infty$. But as the transverse momentum increases, the upper limit on gluon rapidity decreases in the A term, but not in the sum of the S terms. This weakens the cancellation, leaving an uncanceled contribution from a triangular region above the diagonal line in Fig. 10.26(d). With a UV regulator applied (e.g., $n < 4$) the integral is convergent at large k_T, so the limit $y_2 \to -\infty$ exists.

When the UV regulator is removed, the contribution of the triangle is a doubly logarithmic infinity, to be canceled by a UV counterterm. As in Sec. 10.8.2 the limits are applied in the order $y_2 \to -\infty$ and then $n \to 4$. Because of the doubly logarithmic divergence, the UV divergence has the two poles of $\epsilon = 2 - n/2$ per loop instead of the conventional single pole, and it is energy dependent. See (10.139).

10.11.2 Factorization and re-examination of hard factor

The new collinear factors (10.119a) and (10.119b) are obtained from the original collinear and soft factors by reorganizing the S factors. Changes are only by power-suppressed corrections. Thus the hard factor H is unchanged. But we can convert the old formula for H, (10.96), to use the new version of factorization:

$$H(Q, \mu, g(\mu)) = \lim_{\text{massless}} \frac{F}{AB} = \lim_{\text{massless}} \frac{F \, S(+\infty, -\infty)}{A^{\text{unsub}}(-\infty) B^{\text{unsub}}(+\infty)}. \qquad (10.120)$$

As before, the notation of infinite rapidity for the Wilson lines includes the definition that the infinite rapidity limit is applied before the removing the UV regulator by $\epsilon \to 0$.

10.11.3 Evolution kernel K

The final versions of the collinear factors A and B in (10.119) depend on the rapidity parameter y_n, only via the factors $S(y_1 - y_n)$ and $S(y_n - y_2)$, in the limit $y_1 \to \infty$, $y_2 \to -\infty$. So to get an equation for the dependence on y_n, we need the kernel K defined earlier.

This earlier definition was appropriate for differentiating $S(y_1 - y_2)$ with respect to y_1, and thus the diagrammatic definition was *not* symmetric between the positive and negative rapidity directions. However, since S depends on the difference of the two rapidities an equal result is obtained by differentiating with respect to the other rapidity argument, except for a sign. For use with the new collinear factors, we now make a more symmetric definition of K, and we put it into an operator form. We first define the vector $n = (e^{y_n}, -e^{-y_n}, \mathbf{0}_T)$, and define a differentiated vector

$$\delta n \stackrel{\text{def}}{=} \frac{dn}{dy_n} = (e^{y_n}, e^{-y_n}, \mathbf{0}_T). \tag{10.121}$$

Then we redefine

$$K\left(m_g, m, \mu, g(\mu)\right) \stackrel{\text{def}}{=} \frac{\partial}{\partial y_n} \ln \frac{S(y_n, -\infty)}{S(+\infty, y_n)}$$

$$= \frac{\langle 0 | T \ W(\infty, 0, w_2)^{\dagger} \ W(\infty, 0, n) \ (-i g_0) \int_0^{\infty} d\lambda \, \Delta_1(n\lambda) | 0 \rangle}{\langle 0 | T W(\infty, 0, w_2)^{\dagger} W(\infty, 0, n) | 0 \rangle}$$

$$+ \frac{\langle 0 | T \ W(\infty, 0, -n)^{\dagger} \ W(\infty, 0, w_1) \ (i g_0) \int_0^{\infty} d\lambda \, \Delta_2(-n\lambda) | 0 \rangle}{\langle 0 | T W(\infty, 0, -n)^{\dagger} W(\infty, 0, w_1) | 0 \rangle}$$

with renormalization, $\tag{10.122}$

where w_1 and w_2 are the light-like vectors defined in (10.15a), and

$$\Delta_1(x) = \delta n^{\mu} A_{\mu}^{(0)}(x) + \lambda \delta n^{\nu} n^{\mu} \frac{\partial A_{\mu}^{(0)}(x)}{\partial x^{\nu}} \qquad \text{with } x = \lambda n, \tag{10.123a}$$

$$\Delta_2(x) = \delta n^{\mu} A_{\mu}^{(0)}(x) - \lambda \delta n^{\nu} n^{\mu} \frac{\partial A_{\mu}^{(0)}(x)}{\partial x^{\nu}} \qquad \text{with } x = -\lambda n. \tag{10.123b}$$

The Feynman rules for the special vertices are given in Fig. 10.27.

See problem 10.9 for the gauge independence of K with the new definition.

10.11.4 Factorization, evolution equations: Final form

In this section, we collect all the results in their final form: the factorization formula, and the evolution equations for the dependence on the Wilson-line rapidity and on the renormalization scale. The evolution equations are the key to practical applications. We will refer back to the definitions of all the factors.

$$-ig_0 \frac{i\left(\delta n^\mu\, n \cdot k - n^\mu\, \delta n \cdot k\right)}{(k \cdot n + i0)^2} \qquad ig_0 \frac{i\left(\delta n^\mu\, n \cdot k - n^\mu\, \delta n \cdot k\right)}{(k \cdot n + i0)^2}$$

Fig. 10.27. Feynman rules for special vertices for K. See Fig. 10.23 for examples using the vertex labeled 1. The first rule agrees with that in (10.49).

The factorization equation is

$$F = H(Q, \mu, g(\mu))\, A\left(y_{p_A} - y_n, m_g, m, \mu, g(\mu)\right)$$
$$\times B\left(y_n - y_{p_B}, m_g, m, \mu, g(\mu)\right) + \text{power-suppressed},\tag{10.124}$$

where A and B are defined in (10.119) and H in (10.120).

Initially, the rapidity y_n might be taken to be zero in the overall center-of-mass frame, so that the collinear factors A and B can be characterized as giving the contribution of quanta of, respectively, positive and negative rapidities. Then both the rapidity difference arguments $y_{p_A} - y_n$ and $y_n - y_{p_B}$ are $\ln(Q/m)$. Evolution equations, that we now summarize, enable us to adjust the values of y_n differently for each collinear factor, and thereby express them in terms of values with fixed rapidity-difference arguments. Similarly, we will use RG equations to make suitable (and different) choices for the scale μ in each factor.

From the results in Sec. 10.11.3, it follows that the evolution equations with respect to y_n for the collinear factors are

$$\frac{\partial A}{\partial y_n} = -\frac{1}{2} K\left(m_g, m, \mu, g(\mu)\right) A, \tag{10.125a}$$

$$\frac{\partial B}{\partial y_n} = \frac{1}{2} K\left(m_g, m, \mu, g(\mu)\right) B, \tag{10.125b}$$

where K is defined by (10.122). It follows that the product AB that appears in the factorization formula is independent of y_n.

The RG equations have the form

$$\frac{dK}{d\ln \mu} = -\gamma_K(g(\mu)), \tag{10.126a}$$

$$\frac{dA}{d\ln \mu} = \gamma_A(\zeta_A/\mu^2, g(\mu))\, A, \tag{10.126b}$$

$$\frac{dB}{d\ln \mu} = \gamma_B(\zeta_B/\mu^2, g(\mu))\, B. \tag{10.126c}$$

The anomalous dimensions can be obtained from the renormalization counterterms for K, A and B. Now, the renormalization factors for the two collinear factors are energy dependent, for reasons explained earlier with the aid of Fig. 10.26. This causes energy dependence in the anomalous dimensions. Since the anomalous dimensions are determined by UV phenomena, they involve only the large components of quark momenta, i.e., p_A^+ and

p_B^-. So we write the energy dependence in terms of

$$\zeta_A \overset{\text{def}}{=} 2(p_A^+)^2 e^{-2y_n} = m^2 e^{2(y_{PA}-y_n)}, \qquad (10.127\text{a})$$

$$\zeta_B \overset{\text{def}}{=} 2(p_B^-)^2 e^{2y_n} = m^2 e^{2(y_n-y_{PB})}, \qquad (10.127\text{b})$$

which are versions of (10.93a) and (10.93b), but now defined relative to the single rapidity y_n. Note that these differ by power-suppressed corrections from the corresponding defini- tions in Collins and Soper (1981) and Soper (1979), which are $\zeta_{A,CS} = |4p_A \cdot n^2/n^2|$, and $\zeta_{B,CS} = |4p_B \cdot n^2/n^2|$. Note also that $\zeta_A \zeta_B = (2p_A^+ p_B^-)^2 = Q^4 \left(\frac{1}{2} + \frac{1}{2}\sqrt{1 - 4m^2/Q^2} \right)^4 \simeq Q^4$.

Since the collinear factors differ only by an exchange of plus and minus coordinates and by a charge-conjugation transformation, the anomalous dimensions γ_A and γ_B of A and B are the same.

The final ingredient we need is an equation for the energy dependence of γ_A. This is obtained by applying $d/d \ln \mu$ to (10.125a) and then exchanging the order of differentiation:

$$\frac{d}{d \ln \mu} \frac{\partial A}{\partial y_n} = \frac{1}{2} \gamma_K A - \frac{1}{2} K \gamma_A A, \qquad (10.128\text{a})$$

$$\frac{\partial}{\partial y_n} \frac{dA}{d \ln \mu} = \frac{\partial \gamma_A}{\partial y_n} A - \frac{1}{2} K \gamma_A A. \qquad (10.128\text{b})$$

Hence

$$\frac{\partial \gamma_A \left(\zeta_A/\mu^2, g(\mu) \right)}{\partial y_n} = -\frac{\partial \gamma_A}{\partial \ln \zeta_A^{1/2}} = \frac{1}{2} \gamma_K \left(g(\mu) \right), \qquad (10.129)$$

thereby completely determining the energy dependence of γ_A (and γ_B):

$$\gamma_A \left(\zeta/\mu^2, g(\mu) \right) = \gamma_B \left(\zeta/\mu^2, g(\mu) \right) = \gamma_A(1, g(\mu)) - \frac{1}{4} \gamma_K(g(\mu)) \ln \frac{\zeta}{\mu^2}. \qquad (10.130)$$

The above equations, together with the definitions of A, B, H, and K, are a complete formulation of factorization.

10.11.5 Solution

We now use the evolution equations to set the arguments of H, A and B to avoid large logarithms.

- In H, we set μ proportional to Q: $\mu = C_2 Q$.
- In A, B, and K we set μ to a fixed value μ_0, of order the particle masses.
- In A, we set $y_n = y_{PA}$.
- In B, we set $y_n = y_{PB}$.
- In γ_A and γ_B, we set the ζ/μ^2 argument to $1/C_2^2$, as with H.

For the coefficient of proportionality C_2 between μ and Q, the notation C_2 is that of Collins and Soper (1981).

It can be readily deduced from the evolution equations that

$$
\begin{aligned}
F &= H(1/C_2, g(C_2 Q))\ A\big(y_{p_A} - y_n, m_g, m, C_2 Q, g(C_2 Q)\big) \\
&\quad \times B\big(y_n - y_{p_B}, m_g, m, C_2 Q, g(C_2 Q)\big) \\
&= H(1/C_2, g(C_2 Q))\ A\big(0, m_g, m, \mu_0, g(\mu_0)\big)\ B\big(0, m_g, m, \mu_0, g(\mu_0)\big) \\
&\quad \times \exp\left\{ -\int_{\mu_0}^{C_2 Q} \frac{\mathrm{d}\mu}{\mu} \left[\ln \frac{C_2 Q}{\mu}\, \gamma_K(g(\mu)) - 2\gamma_A\big(1/C_2^2, g(\mu)\big) \right] \right\} \\
&\quad \times \exp\left[\frac{1}{2}(y_{p_A} - y_{p_B})\, K\big(m_g, m, \mu_0, g(\mu_0)\big) \right],
\end{aligned}
\tag{10.131}
$$

where power-suppressed corrections are ignored.

10.11.6 Properties and use of solution

Results of the same structure appear in many important problems in QCD (Chs. 13 and 14). So we now examine the solution (10.131) with a view to QCD applications.[16] In QCD, the effective coupling is large at small momenta, and is small at large momenta. Thus perturbative calculations are not valid for collinear factors for light particles in QCD.

By setting the renormalization scale proportional to Q in the hard scattering H, we removed large logarithms in the perturbative expansion of H. This enables effective perturbative predictions to be made for H.[17] But then the collinear factors have Q dependence; see the *first* line of (10.131).

We remedied this by using the evolution equations to give *different values of μ and y_n in the different factors*, in the lower three lines of (10.131). There, each of the collinear factors has a Q-independent value of the renormalization mass and of the rapidity difference argument. In a weak-coupling situation, this enables a perturbative calculation to be made without logarithms. In QCD, it allows us to use universality to make predictions: the same collinear factors appear at all values of Q and in all processes with the same kind of factorization. Thus determination of a collinear factor can be made from experimental data in one process at one energy, and the value used for the otherwise unknown quantity both in the same process at other energies, and in different processes.

The exponential in (10.131) shows that our solution radically differs from a straightforward use of perturbation theory, in a way that is much stronger than in cases containing only ordinary RG logarithms. The anomalous dimensions γ_K and γ_A are to be used in the weak-coupling regime, so that low-order perturbation calculations are effective. The generally biggest term in the exponent is the γ_K term; it has a logarithm relative to γ_A.

There remains the term involving K in the exponent. It gives a substantially energy-dependent factor:

$$
\exp\left[\frac{1}{2}(y_{p_A} - y_{p_B})\, K\big(m_g, m, \mu_0, g(\mu_0)\big) \right] = \left(\frac{Q^2}{m^2} \right)^{\frac{1}{2} K(m_g, m, \mu_0, g(\mu_0))}.
\tag{10.132}
$$

---

[16] But (10.131) is also useful in a QED-like theory with a coupling that is weak at all relevant scales.
[17] The coefficient C_2 can be adjusted to further optimize perturbative coefficients.

In QCD this would give a power-law dependence on Q with a non-perturbative exponent. The exponent $K(\mu_0)$ can be determined from the derivative of the amplitude with respect to energy, at one value of energy. Then the same exponent is used at all energies and in other processes. Determination of generalizations of K to other process appear as a critical element of good phenomenology (e.g., Landry *et al.*, 2003) for the Drell-Yan and other processes. It gives a substantial and characteristic energy dependence to the shape of Drell-Yan cross sections differential in transverse momentum.

If the IR coupling were weak, as in QED, the exponent K would be perturbatively calculable.

10.11.7 Asymptotic large Q behavior

The biggest term in the exponent in (10.131) is the one with γ_K. It implies that at large enough Q, the factorized formula for the form factor goes to zero faster than any power of Q; this happens both for our form factor in an abelian theory (at least if we stay in a weak-coupling regime), and for analogous quantities in an asymptotically free theory.

However, the derivation ignored power-suppressed corrections, which therefore have the potential to be asymptotically larger than the final factorized answer: the leading-power contributions have undergone a strong cancellation. Thus beyond some energy, the precise numerical result of the factorization formula is phenomenologically irrelevant.

To assess the significance of such a factorization in QCD, we observe that in e^+e^- annihilation to hadrons, the Sudakov form-factor graphs give the component of the cross section that has a pure quark-antiquark final state. But in the *total* cross section we found a cancellation of all IR-sensitive regions, with the total cross section going to a constant at large Q; see Ch. 4. This cancels the strong decrease of the Sudakov form factor in the quark-antiquark component. At high energy the cross section for $e^+e^- \to$ hadrons is dominantly highly inelastic.

In Chs. 13 and 14, we will investigate reactions where the amount of cancellation of IR-sensitive effects depends on the value of a measurable transverse-momentum variable. In these situations, a generalization of the factorization derived in this chapter will be very useful.

10.11.8 Relation of factorization to LLA

From (10.131), we see systematically how all logarithms arise. We derive the leading-logarithm approximation (LLA) as follows: (a) expand γ_K to lowest order in coupling; (b) ignore the running of the coupling; (c) neglect the other terms in the exponent; (d) set the outside H, A and B factors to their lowest-order values (i.e., unity). This reproduces (10.38), when μ_0 is of the order of particle masses.

There are important gains from the factorization formalism relative to the LLA, particularly in generalizations in QCD. In the first place the factorization formalism shows how corrections arise, and how they may be made systematically. The corrections are in the

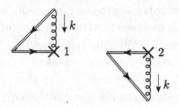

Fig. 10.28. One-loop graphs for K. The vertical heights of the graphs are adjusted to symbolize the rapidities of the light-like lines. The rules for the vertices with a cross are given in Fig. 10.27.

exponent, and also in non-logarithmic corrections to the H, A and B factors preceding the exponential.

In contrast, the logic of the LLA alone gives no information on non-leading logarithms. For example, the LLA itself does not prevent there from being an *additive* correction, e.g.,

$$g^2 \times \text{constant}, \tag{10.133}$$

which does not vanish as $Q \to \infty$. This would completely change the qualitative behavior. Such a phenomenon actually occurs for the Drell-Yan and related cross sections at zero transverse momentum. There the LLA gives a cross section that vanishes at zero transverse momentum, but the true result from a correct factorization theorem is non-zero (Collins and Soper, 1982a).

An important result is that the factorization method indicates how non-perturbative effects should affect the Q dependence in analogous QCD problems, by the factor (10.132). Of course our formal derivation stayed within perturbation theory. But the structures we use have a much more general appearance.

10.12 Calculations for Sudakov problem

In this section we show how the Feynman rules for H, A, B and K work out at one-loop order.

10.12.1 Evolution kernel K

First we calculate the evolution kernel K. From the rules given in Fig. 10.27, we have the one-loop graphs shown in Fig. 10.28. They give

$$K = \frac{ig^2\mu^{2\epsilon}}{(2\pi)^{4-2\epsilon}} \int d^{4-2\epsilon}k \, \frac{1}{(k^2 - m_g^2 + i0)(-n \cdot k + i0)^2}$$

$$\times \left[\frac{n^+ \delta n \cdot k - \delta n^+ n \cdot k}{k^+ + i0} + \frac{n^- \delta n \cdot k - \delta n^- n \cdot k}{-k^- + i0} \right] + \text{UV c.t.} + O(g^4)$$

$$= \frac{ig^2\mu^{2\epsilon}}{(2\pi)^{4-2\epsilon}} \int d^{4-2\epsilon}k \, \frac{-2n^2}{(k^2 - m_g^2 + i0)(-n \cdot k + i0)^2} + \text{UV c.t.} + O(g^4)$$

$$= -\frac{g^2 \Gamma(\epsilon)}{4\pi^2} \left(\frac{4\pi\mu^2}{m_g^2}\right)^{\epsilon} + \frac{g^2 S_\epsilon}{4\pi^2 \epsilon} + O(g^4)$$

$$= -\frac{g^2}{4\pi^2} \ln\frac{\mu^2}{m_g^2} + O(g^4). \tag{10.134}$$

(In obtaining this, note the reversal of the direction of k compared with Fig. 10.27, and remember the reversed sign of the ordinary vertex on a Wilson line that corresponds to an antiquark.) The calculation of the integral can be done by contour integration on k^- followed by an elementary integral for k^+. Then the k_T integral gives a beta function. The result agrees with our previous calculation at (10.54), but now we used our updated Feynman rules. Note that the evolution equation has no power corrections, in contrast with (10.53).

As an exercise the reader can show that the sole two-loop graph gives the $O(g^4)$ term in γ_K:

$$\gamma_K = \frac{g^2}{2\pi^2} - \frac{10}{9}\left(\frac{g^2}{4\pi^2}\right)^2 + O(g^6). \tag{10.135}$$

10.12.2 Collinear factor A

We now calculate the collinear factor A at one-loop order. This will illustrate the peculiar energy dependence of the counterterm. The graphs, obtained from the definition (10.119a), are shown in Fig. 10.29. To this is to be added a term associated with the external propagator correction.

The graphs in Fig. 10.29(a) give

$$A_{1a} = \frac{ig^2\mu^{2\epsilon}}{(2\pi)^{4-2\epsilon}} \frac{1}{\bar{u}_A \mathcal{P}_B} \int d^{4-2\epsilon}k \frac{1}{(k^2 - m_g^2 + i0)} \bar{u}_A$$

$$\times \left\{ \frac{\gamma^+(\not{p}_A - \not{k} + m)}{[(p_A - k)^2 - m^2 + i0](k^+ + i0)} + \frac{\frac{1}{2}e^{-y_n}}{(-k^- + i0)(k^+ e^{-y_n} - k^- e^{y_n} + i0)} \right.$$

$$\left. - \frac{\frac{1}{2}e^{y_n}}{(-k^- e^{y_n} + k^+ e^{-y_n} + i0)(k^+ + i0)} - \frac{\frac{1}{2}}{(-k^- + i0)(k^+ + i0)} \right\} \mathcal{P}_B, \tag{10.136}$$

to which is to be added a UV counterterm. The $e^{\pm y_n}$ factors in the exponents arise from the vertices for the Wilson lines of rapidity y_n. As usual, we use the residue theorem to perform the k^- integral. This gives

$$A_{1a} = \frac{-g^2(2\pi\mu)^{2\epsilon}}{8\pi^3} \int d^{2-2\epsilon}k_T$$

$$\times \left\{ \int_0^1 \frac{dx}{x} \left[\frac{1-x}{k_T^2 + m_g^2(1-x) + m^2x^2} + \frac{1}{-k_T^2 - m_g^2 + 2(xp_A^+ e^{-y_n})^2 + i0} \right] \right.$$

$$\left. + \int_1^\infty \frac{dx}{x} \frac{1}{-k_T^2 - m_g^2 + 2(xp_A^+ e^{-y_n})^2 + i0} \right\} + \text{UV c.t.}, \tag{10.137}$$

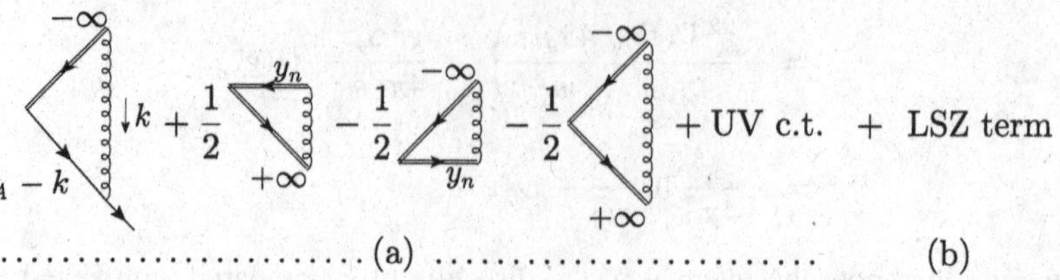

Fig. 10.29. Graphs for A at one-loop, including subtractions and the counterterm for canceling the UV divergence. Next to each double line representing a Wilson line is a label for its rapidity, $-\infty$, $+\infty$ or y_n. The factors of $\frac{1}{2}$ multiplying the Wilson-line terms arise from the one-loop expansion of the factors in the square root in (10.119a). The upper Wilson lines have the charge of an *anti*quark. The LSZ term is a self-energy graph for the on-shell quark.

where the potential divergence at $x = 0$ has canceled. Much of the x integral, including all the Wilson-line terms, can be performed by very elementary methods to give

$$A_{1a} = \frac{-g^2(2\pi\mu)^{2\epsilon}}{8\pi^3} \int \frac{d^{2-2\epsilon} k_T}{k_T^2 + m_g^2} \left\{ -\int_0^1 dx \frac{k_T^2 + m^2 x}{k_T^2 + m_g^2(1-x) + m^2 x^2} \right.$$

$$\left. + \frac{1}{2} \ln \frac{2(p_A^+ e^{-y_n})^2}{k_T^2 + m_g^2} - i\frac{\pi}{2} \right\} + \text{UV c.t.} \quad (10.138)$$

The remaining x integral is well behaved.

A simple computation of the UV counterterm in the $\overline{\text{MS}}$ scheme uses the techniques of Sec. 3.4. The UV divergence is governed by the leading large k_T behavior of the integrand, which is therefore independent of the masses:

UV c.t. $= -\overline{\text{MS}}$ pole part of integral in (10.138)

$$= -\overline{\text{MS}} \text{ pole part of } \frac{-g^2(2\pi\mu)^{2\epsilon}}{8\pi^3} \int_{k_T > \mu} \frac{d^{2-2\epsilon} k_T}{k_T^2} \left[-1 + \frac{1}{2} \ln \frac{2(p_A^+ e^{-y_n})^2}{k_T^2} - i\frac{\pi}{2} \right]$$

$$= \frac{g^2 S_\epsilon}{8\pi^2} \left[\frac{-1}{2\epsilon^2} + \frac{1}{\epsilon} \left(-1 + \frac{1}{2} \ln \frac{2(p_A^+ e^{-y_n})^2}{\mu^2} - i\frac{\pi}{2} \right) \right]. \quad (10.139)$$

As in Sec. 3.4, the use of the lower limit μ on the k_T integral gives exactly the $\overline{\text{MS}}$ pole part with its accompanying factor of S_ϵ with no further finite part. This relies on exactly our specific definition of S_ϵ in (3.18).

The k_T integral in (10.138) is readily performed. To get the complete one-loop contribution to the collinear factor, the LSZ reduction formula tells us to add half the one-loop residue of the quark propagator:

$$\frac{1}{2}\Sigma_1 = \frac{g^2}{8\pi^2} \left\{ \frac{1}{4} + \int_0^1 dx \frac{1}{2}(1-x) \ln \frac{m_g^2(1-x) + m^2 x^2}{\mu^2} + \int_0^1 dx \frac{m^2 x(1-x^2)}{m_g^2(1-x) + m^2 x^2} \right\}. \quad (10.140)$$

Then the full one-loop contribution to A at $n = 4$ is

$$A_1 = \frac{-g^2}{8\pi^2}\left\{ -\int_0^1 \frac{dx}{x}\ln\left(1 - x + x^2 m^2/m_g^2\right)\right.$$

$$+ \int_0^1 dx\, \frac{1}{2}(1+x)\ln\frac{m_g^2(1-x)+m^2 x^2}{\mu^2} - \int_0^1 dx\, \frac{m^2 x(1-x^2)}{m_g^2(1-x)+m^2 x^2}$$

$$\left. -\frac{1}{4}+\frac{1}{4}\ln^2\frac{2(p_A^+ e^{-y_n})^2}{m_g^2} - \frac{1}{4}\ln^2\frac{2(p_A^+ e^{-y_n})^2}{\mu^2} + i\frac{\pi}{2}\ln\frac{m_g^2}{\mu^2}\right\}. \qquad (10.141)$$

We can now check the evolution and RG equations. First, we see from (10.94a), and its generalization to the new definition of A, that the counterterm in (10.139) gives the one-loop contribution to $Z_A Z_2^{1/2}$. With the aid of (3.23) for Z_2, we find that

$$Z_A = 1 + \frac{g^2 S_\epsilon}{8\pi^2}\left[\frac{-1}{2\epsilon^2} + \frac{1}{\epsilon}\left(-\frac{3}{4} + \frac{1}{2}\ln\frac{2(p_A^+ e^{-y_n})^2}{\mu^2} - i\frac{\pi}{2}\right)\right] + O(g^4). \qquad (10.142)$$

From this we get the anomalous dimension:

$$\gamma_A = \frac{d\ln A}{d\ln\mu} = \frac{d\ln Z_A}{d\ln\mu}$$

$$= \frac{\partial\ln Z_A}{\partial\ln\mu} + \frac{dg^2/16\pi^2}{d\ln\mu}\frac{\partial\ln Z_A}{\partial g^2/16\pi^2}$$

$$= \frac{g^2}{8\pi^2}\left[\frac{3}{2} - \ln\frac{2(p_A^+ e^{-y_n})^2}{\mu^2} + i\pi\right] + O(g^4) \qquad \text{(at } \epsilon = 0\text{)}, \qquad (10.143)$$

where the first line uses $A = Z_A A_0$ and the RG invariance of A_0, defined in terms of bare fields, while the third line uses (3.44) for $d(g^2/16\pi^2)/d\ln\mu$. The explicit μ dependence of the single-pole counterterm was needed to get finiteness of γ_A. It is readily checked that the dependence on y_n is as predicted from (10.129) with the calculated value of γ_K from (10.135).

10.12.3 Hard factor

From the definition, (10.120), we find that the one-loop hard-scattering coefficient arises from the graphs in Fig. 10.30. This gives

$$H_1 = \frac{-ig^2\mu^{2\epsilon}}{(2\pi)^{4-2\epsilon}}\int d^{4-2\epsilon}k\, \frac{\bar{u}_A \mathcal{P}_B\, I_H(k)\, \mathcal{P}_B v_B}{k^2 + i0}, \qquad (10.144)$$

where

$$I_H(k) = \frac{\gamma^\kappa(p_A^+\gamma^- - \not{k})\gamma^\mu(-p_B^-\gamma^+ - \not{k})\gamma_\kappa}{(-2p_A^+ k^- + k^2 + i0)(2p_B^- k^+ + k^2 + i0)} + \frac{-\gamma^\mu}{(-k^- + i0)(k^+ + i0)}$$

$$- \frac{\gamma^+(p_A^+\gamma^- - \not{k})\gamma^\mu(-1)}{(-2p_A^+ k^- + k^2 + i0)(k^+ + i0)} - \frac{\gamma^\mu(-p_B^-\gamma^+ - \not{k})\gamma^-}{(-k^- + i0)(2p_B^- k^+ + k^2 + i0)}. \qquad (10.145)$$

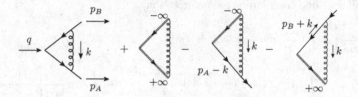

Fig. 10.30. Graphs for one-loop hard coefficient.

The factors of (-1) in two of the numerators are for the negative charges of the upper Wilson lines. To (10.144) is to be added a UV counterterm, as usual.

The integrals over k^- and k^+ can be performed analytically, to give

$$H_1 = \frac{-g^2(4\pi\mu^2)^\epsilon \bar{u}_A \mathcal{P}_B \gamma^\mu \mathcal{P}_B v_B}{8\pi^2 \Gamma(1-\epsilon)} \int_0^\infty \frac{dk_{\mathrm{T}}^2}{(k_{\mathrm{T}}^2)^{1+\epsilon}}$$

$$\times \left\{ \ln \frac{k_{\mathrm{T}}^2}{Q_E^2} + \frac{1 + (3+2\epsilon)k_{\mathrm{T}}^2/Q_E^2}{\sqrt{1+4k_{\mathrm{T}}^2/Q_E^2}} \ln \frac{\sqrt{1+4k_{\mathrm{T}}^2/Q_E^2}+1}{\sqrt{1+4k_{\mathrm{T}}^2/Q_E^2}-1} \right\}, \qquad (10.146)$$

where $Q_E^2 = -Q^2 - i0$: the integral is defined by continuing from a positive value of Q_E^2 to $-Q^2$ approaching from the appropriate side of the real axis. Observe that the Wilson-line terms combine to remove the divergence at $k_{\mathrm{T}} = 0$.

From the behavior of the integrand at large k_{T}, it can be computed that the necessary $\overline{\mathrm{MS}}$ counterterm is

$$H_{1,\text{c.t.}} = \frac{g^2 S_\epsilon \bar{u}_A \mathcal{P}_B \gamma^\mu \mathcal{P}_B v_B}{8\pi^2} \left[\frac{1}{\epsilon^2} + \frac{1}{\epsilon}\left(-\ln \frac{Q_E^2}{\mu^2} + \frac{3}{2} \right) \right]$$

$$= \frac{g^2 S_\epsilon \bar{u}_A \mathcal{P}_B \gamma^\mu \mathcal{P}_B v_B}{8\pi^2} \left[\frac{1}{\epsilon^2} + \frac{1}{\epsilon}\left(-\ln \frac{Q^2}{\mu^2} + i\pi + \frac{3}{2} \right) \right]. \qquad (10.147)$$

This is exactly equal and opposite to the sum of the one-loop contributions to Z_A and Z_B, so that for the one-loop contribution to HAB the total counterterm is zero. This corresponds to the non-renormalization theorem for matrix elements of a conserved current. Notice that the counterterm has a logarithm, just as for the collinear factors. Thus the one-loop anomalous dimension of H is also momentum dependent:

$$\gamma_H(Q^2/\mu^2, g) \overset{\text{def}}{=} \frac{d\ln H}{d\ln\mu}$$

$$= \frac{g^2}{8\pi^2}\left(2\ln \frac{Q^2}{\mu^2} - 2i\pi - 3 \right) + O(g^4)$$

$$= -\gamma_A(\zeta_A/\mu^2, g(\mu)) - \gamma_B(\zeta_B/\mu^2, g(\mu)), \qquad (10.148)$$

with the last line being a general result following from the RG invariance of the whole form factor, and hence of its factorized form HAB. Observe that the dependence of γ_H on the ratio Q^2/μ^2 can be derived from the ζ dependence of γ_A and γ_B. Thus from (10.129) we have

$$\frac{\partial \gamma_H\left(Q^2/\mu^2, g\right)}{\partial \ln(Q^2/\mu^2)} = \frac{1}{2}\gamma_K(g), \tag{10.149}$$

so that

$$\gamma_H\left(Q^2/\mu^2, g\right) = \gamma_H\left(1, g\right) + \frac{1}{2}\gamma_K(g)\ln\frac{Q^2}{\mu^2}. \tag{10.150}$$

10.13 Deduction of some non-leading logarithms

Our formalism gives a lot of information on the structure of non-leading logarithms even in the absence of explicit Feynman-graph calculations beyond lowest order. To see some of the results, we examine the perturbation series for the *logarithm of the form factor*. We keep the logarithmic dependence on Q, expressing the coefficients as polynomials in $t = \ln(-Q^2/\mu^2)$, with power corrections dropped:

$$\ln F = \frac{g^2}{4\pi^2}\left(C_{12}t^2 + C_{11}t + C_{10}\right)$$

$$+ \left(\frac{g^2}{4\pi^2}\right)^2\left(C_{24}t^4 + C_{23}t^3 + C_{22}t^2 + C_{21}t + C_{20}\right) + O(1/Q^2), \tag{10.151}$$

where the coefficients may depend on m, M and μ, but not on Q.

The leading logarithm results imply that $C_{24} = 0$. But we can deduce considerable more from the factorization formula (10.124) and the evolution equations (10.125). We do this by deducing an equation for the Q dependence of $\ln F$:

$$\frac{\partial \ln F}{\partial \ln Q} = \frac{\partial \ln J_A}{\partial \ln Q} + \frac{\partial \ln J_B}{\partial \ln Q} + \frac{\partial \ln H}{\partial \ln Q} + \text{power correction}$$

$$= K(m_g, m, g, \mu) + G(Q/\mu; g) + \text{power correction}, \tag{10.152}$$

where G is a purely UV quantity that obeys $dG/d\ln\mu = -\gamma_K$. Now, from (10.151) we have

$$\frac{\partial \ln F}{\partial \ln Q} = \frac{g^2}{4\pi^2}(4C_{12}t + 2C_{11}) + \left(\frac{g^2}{4\pi^2}\right)^2\left(6C_{23}t^2 + 4C_{22}t + 2C_{21}\right) + \dots \tag{10.153}$$

In order that G in (10.152) be independent of the masses m and M, C_{12}, C_{23} and C_{22} must be independent of m and M (and hence of μ). Furthermore, once one puts in the one-loop values, the requirement that G satisfies its RG equation implies that

$$C_{23} = -\frac{1}{36}. \tag{10.154}$$

Hence the new information for the form factor F at two loops is two logarithms down from the leading logarithm, i.e. it is in C_{22} and the less leading coefficients, C_{21} and C_{20}. The double logarithm coefficient C_{22} is related to the two-loop term in γ_K, given in (10.135);

this was the result of a relatively easy calculation. Hence

$$C_{22} = \frac{5}{36}.$$ (10.155)

The remaining information, for which a full two-loop calculation of the form factor is needed, is in the terms with one and no logarithms of Q. These are three and four logarithms down from the leading $\ln^4 Q$ term.

10.14 Comparisons with other work

In this section, I give a brief comparison between the present treatment of the Sudakov and other work on the same and related problems. I restrict attention to work that aims at something like a complete factorization theorem, rather than just obtaining a LLA.

The first treatment in a similar fashion was in Collins (1980). There I used Coulomb gauge in a frame with a time-like rest vector n, where the numerator of the gluon propagator is

$$-g^{\mu\nu} + \frac{(n^\mu k^\nu + k^\mu n^\nu)n \cdot k}{n \cdot k^2 - k^2 n^2} - \frac{k^\mu k^\nu n^2}{n \cdot k^2 - k^2 n^2}.$$ (10.156)

The collinear factors are defined by formulae like (10.90) and (10.92) except that the Wilson lines are removed, so that the matrix elements are $\langle p_A | \bar{\psi}_0(0) | 0 \rangle$ and $\langle p_B | \psi_0(0) | 0 \rangle$. Thus the rapidity of the vector n plays the same role as y_n in our final definitions (10.119). Factorization and evolution equations of a similar kind were derived, differing from those in Sec. 10.11 essentially by a change of scheme. But the old evolution equations had power-suppressed corrections, rather than being exactly homogeneous. There was also a separate soft factor, which we have now eliminated.

A treatment in covariant gauge with Wilson lines was given in Collins (1989). The collinear factors were now defined as what are here called the "unsubtracted" collinear factors (10.90) and (10.92), but with the Wilson lines now having a rapidity y_n corresponding to that in our final definitions (10.119). In this formalism, it is the soft factor that has the subtractions, which is harder to justify from a systematic approach. The evolution equations continue to have power-suppressed corrections, and the factorization formula has a separate soft factor. They also have not only the K we use, but also a G term, as in (10.117).

An earlier approach is found in Mueller (1979), but the methods are less general, particularly as regards their extension to inclusive processes in QCD.

When the methods of Collins (1980) were extended (Collins and Soper, 1981) to inclusive processes in QCD, it was found convenient to replace Coulomb gauge by a non-light-like axial gauge, where the numerator of the gluon propagator is

$$-g^{\mu\nu} + \frac{k^\mu n^\nu + n^\mu k^\nu}{k \cdot n} - \frac{k^\mu k^\mu n^2}{(k \cdot n)^2}.$$ (10.157)

This gives definitions (Collins and Soper, 1982b; Soper, 1979) of parton densities and fragmentation functions exactly like those in a *non-gauge* theory, i.e., without Wilson lines.

Essentially these are equivalent to gauge-invariant definitions with Wilson lines in direction n. Applied to the Sudakov form factor, these definitions amount to using our unsubtracted definitions (10.90) and (10.92) as the actual collinear factors in the factorization formula, but with the Wilson lines having rapidity y_n. The factorization formula still has a subtracted soft factor. Again the evolution equation for a collinear factors has power-suppressed corrections and a G term. The use of a non-light-like vector rather than a light-like vector in the collinear factors complicates calculations. The singularity in (10.157) at $k \cdot n = 0$ is defined as a principal value, which causes problems with the Glauber region; the definitions are not exactly equivalent to the definition with a Wilson line going to infinity in a definite direction. The difficulties have become particularly apparent when inclusive processes with transversely polarized beams are treated (Secs. 13.16 and 13.17). Furthermore, it was not realized that there is a need for the equivalent of what in Feynman gauge is the removal of Wilson-line self-energies. A version of this formalism was applied to semi-inclusive DIS in Meng, Olness, and Soper (1996), with a gauge-invariant version being given in Ji, Ma, and Yuan (2005).

Exercises

10.1 Show explicitly how the formulae in Sec. 8.9, like (8.70) and (8.74), give particular cases of the general formulae for the subtraction method in Sec. 10.1.

10.2 (**) *This problem refers both to material in this chapter and related material in Ch. 13.* Work out more details of the comparison with other work summarized in Sec. 10.14, and with any other papers you can find. Compare the various definitions of the collinear and soft factors. To what extent do they agree up to an allowed scheme change? Are there important differences or errors?

10.3 Assume that a solution of the form of (10.131) applies to some quantity in QCD, with the standard results for the numerical value the effective coupling as a function of μ. Deduce the form of the asymptotic large-Q behavior of the form factor. It would be appropriate to use the same one-loop value of γ_K we derived above, except for an insertion of a factor C_F. This would arise exactly as in the calculations of $e^+e^- \to$ hadrons in Sec. 4.1.

10.4 Estimate the fractional error in the LLA for the Sudakov form factor. When does the LLA give a usefully accurate approximation to the true form factor, in the following different types of theory?
  (a) In the QED-like situation where the coupling is weak over the whole range of scales involved, and the coupling is smallest in the infra-red.
  (b) In the QCD-like asymptotically free situation when the coupling is small only in the UV.
  (c) In an asymptotically free situation, like QCD, except that the masses are large, so that the largest relevant effective coupling is $g(M)$, where M is a scale characterizing the masses of the theory.

10.5 In momentum space the renormalization of the collinear factor A is by a P^+-dependent multiplicative factor. What does this correspond to in coordinate space?

10.6 In our standard definition of the soft approximations we used space-like auxiliary vectors n_1 and n_2, for maximum universality with QCD factorization theories.
   (a) Show that, for the Sudakov form factor, time-like vectors work.
   (b) Take these vectors to be (proportional) to the external particle momenta (i.e., $n_1 = p_A$ and $n_2 = p_B$). Examine the IR divergence when the gluon mass goes to zero. Show that the divergence is completely contained in the soft factor.
   (c) In contrast, examine the case that the auxiliary vectors are space-like or are not proportional to the external momenta. Use the version of the definition of H where masses are preserved, but collinear and soft subtractions are made. Show that there is a power-suppressed divergence as $m_g \to 0$. (It should be proportional to something like $(m^2/Q^2) \ln m_g^2$. The divergence is associated with the gluon mass, but the power-suppression with the quark mass.)

10.7 Verify the two-loop term in (10.135) by explicit calculation.

10.8 When masses are retained in a hard scattering, the external lines are approximated by massive on-shell lines. Show that appropriate choices of the projectors for Dirac fields are as follows.

- For a Dirac particle of momentum $\hat{k}$ *leaving* H to A: $\dfrac{\gamma^+(\hat{k}+m)}{2\hat{k}^+}$. Here the collinear function and the actual wave function $\bar{u}_A$ are on the left.
- For a Dirac antiparticle of momentum $\hat{k}$ *entering* H from A: $\dfrac{(\hat{k}-m)\gamma^+}{2\hat{k}^+}$. Here the collinear function and the actual wave function v_A are on the right.
- For a Dirac particle of momentum $\hat{k}$ *leaving* H to B: $\dfrac{\gamma^-(\hat{k}+m)}{2\hat{k}^-}$. Here the collinear function and the actual wave function $\bar{u}_B$ are on the left.
- For a Dirac antiparticle of momentum $\hat{k}$ *entering* H from B: $\dfrac{(\hat{k}-m)\gamma^-}{2\hat{k}^-}$. Here the collinear function and the actual wave function v_B are on the right.

A general projector has to project onto an on-shell wave function from a general spinor, and should be non-singular in the limit $m \to 0$.

10.9 A change of gauge condition in an abelian theory can be implemented by changing the numerator of the gluon propagator by

$$-g^{\mu\nu} \mapsto -g_{\mu\nu} + f_\mu k_\nu + k_\mu f_\nu, \tag{10.158}$$

for some vector function f of momentum. In a covariant gauge, f^μ is proportional to k^μ times a function of the scalar k^2. There are also more general gauges; such non-covariant gauges are exemplified by the Coulomb and axial gauges.

It can be proved that physical matrix elements of gauge-invariant operators are unchanged under such a change of gauge condition, i.e., that they are gauge independent.[18] Our definitions of collinear and soft factors etc (S, A^{unsub}, B^{unsub}, A^{basic}, B^{basic}, A, B, and K) involve operators that are not exactly gauge invariant, since the operators in them have open Wilson lines.

In this problem, investigate to what extent these quantities are gauge independent at the one-loop level.

As an example, you should find that K with its first definition (10.112) is gauge dependent, but with the second definition (10.122) it is gauge independent. But with a restriction to covariant gauges, even the first definition is gauge independent, and the two definitions agree.

10.10 (**) Consider those quantities that in the previous problem you found to be gauge independent at one-loop order. Try to prove gauge independence to all orders of perturbation theory.

---

[18] Note carefully that gauge invariance and gauge independence are distinct concepts.

11

DIS and related processes in QCD

In this chapter we complete our treatment of inclusive structure functions for DIS in QCD. Our analysis so far started with the parton model, and we generalized it to a factorization property, for which we found a complete proof in a non-gauge theory in Sec. 8.9. We then formulated factorization in QCD (without a proof), using gauge-invariant definitions of parton densities from Sec. 7.6. This enabled us to make low-order calculations of the perturbative hard-scattering coefficients in Ch. 9

The methods of Ch. 10 allow us to complete the work for QCD. Compared with a non-gauge theory, there is no change in the form of factorization, i.e., (8.81) and (8.83). The DGLAP evolution equations, associated with the renormalization of parton densities, are also unchanged in structure.

One change in QCD is that the operators defining the parton densities acquire Wilson lines; we also need to justify the form of the gluon density. For the proof, the enhancements relative to Sec. 8.9 are caused by the extra gluons joining the hard and collinear subgraphs in leading regions. We need generalization beyond the related work in Ch. 10 because the gauge group of QCD is non-abelian. The subtractions in the hard scattering are more complicated than those with the ladder structures appropriate to a non-gauge theory. Finally, in generalizing DIS to an off-shell Green function instead of an on-shell matrix element, we need extra parton-density-like quantities involving gauge-variant operators.

11.1 General principles

The steps to obtain factorization are:

1. List the regions as specified by PSSs in the massless limit of the theory (Ch. 5). These are labeled by subgraph decompositions like Fig. 11.1(b).
2. Find those regions that are leading, as in Sec. 5.8.
3. To leading power, write the amplitude as a sum over contributions for each region of each graph: $\sum_{R,\Gamma} C_R \Gamma$ (Sec. 10.1). Subtractions in $C_R \Gamma$ compensate double counting between regions.
4. Diagrams like Fig. 11.1(b) now acquire extra meanings:
   - The subgraph decomposition can symbolize a particular $C_R \Gamma$.
   - The diagram can imply a sum over R and Γ, and hence a sum over the Feynman graphs for each subgraph. Thus, it almost denotes the factorization property.

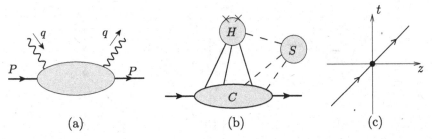

Fig. 11.1. (a) Uncut amplitude $T^{\mu\nu}$ for DIS. (b) General reduced graph for $T^{\mu\nu}$. (c) Space-time structure of its massless PSSs when $x \neq 1$.

5. The factors in $C_R\Gamma$ are defined from a power-series expansion in parameters/variables that the region R labels as small. (But renormalization, etc. is applied as needed to prevent divergences from momenta in larger regions.)
6. Finally we apply Ward identities. This, for example, extracts extra collinear gluons attaching to the hard subgraph and converts them to a Wilson-line form, as in Sec. 10.8. Methods from that section ensure that subtractions and renormalization are compatible with the Ward identities.

Note that Ward identities are not compatible with a naive region analysis, i.e., one where momentum space is partitioned into categories of hard, soft, etc., with boundaries between the regions, and where each region subgraph is defined to have its momenta restricted to the subgraph's category. But a proof of a Ward identity involves shifts of loop-momentum variables. Particularly when momenta are close to boundaries of regions, shifts of loop momenta can take them across boundaries; thus the shifted momenta can be of different categories. This was a primary motivation to define the region contributions $C_R\Gamma$ with unrestricted integrals over loop momenta.

11.2 Regions and PSSs, with uncut hadronic amplitude

As we saw in Sec. 5.3.3, the analysis of regions for DIS is simpler for the *uncut* amplitude, Fig. 11.1(a),

$$T^{\mu\nu}(q, P) = \frac{1}{4\pi} \int d^4z \, e^{iq\cdot z} \, \langle P, S| \, T \, j^{\mu}(z/2) \, j^{\nu}(-z/2) \, |P, S\rangle, \qquad (11.1)$$

from which the ordinary structure tensor is obtained as a discontinuity across the physical-region cut: $W^{\mu\nu}(q, P) = T^{\mu\nu}(\nu + i0) - T^{\mu\nu}(\nu - i0)$.

As usual, the relevant regions are determined by PSSs corresponding to physical scattering of massless particles, with a general reduced graph typified in Fig. 11.1(b). It has collinear and hard subgraphs with a possible connecting soft subgraph. The space-time structure is shown in Fig. 11.1(c): there is a short-distance scattering at the vertex for the virtual photon, while the collinear subgraph and the target hadron correspond to the diagonal (light-like) line.

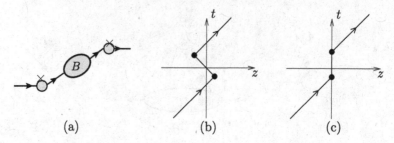

Fig. 11.2. (a) Reduced graph for $T^{\mu\nu}$ at $x \simeq 1$. Not indicated are extra collinear lines and a possible soft subgraph. (b) Space-time structure of its massless PSSs when $x = 1$. (c) Same with massive intermediate state.

11.2.1 Local averaging

This picture fails when x is close to unity, i.e., where $(P + q)^2 \simeq Q^2(1 - x)/x$ gets small. In that case we can have a reduced graph like Fig. 11.2(a), where there is an intermediate state whose mass is small compared with Q. For simplicity, a possible soft subgraph has been omitted. The corresponding PSS has a massless system going in the minus direction, Fig. 11.2(b). Possible intermediate states include a single proton (giving elastic scattering) and low-mass resonances. If we work in perturbation theory with an elementary quark target, instead of a hadron target, we have emission of soft and of final-state-collinear quanta, as in the NLO calculations in Ch. 9.

A full analysis of this region needs more sophisticated methods than we use here. Instead, we obtain the standard factorization formalism by the averaging method used in Secs. 4.1.1 and 4.4 for the total hadronic cross section for e^+e^- annihilation. In DIS we use an average in x:

$$T^{\mu\nu}[f] = \int dx \, T^{\mu\nu}(q, P) \, f(x), \tag{11.2}$$

with a smooth function $f(x)$. In the uncut amplitude, the troublesome final-state singularities all lie on one side of the real x axis, e.g.,

$$\frac{i}{Q^2(1 - x)/x - m^2 + i0} = \frac{ix/Q^2}{1 - x - m^2/Q^2 + i0}. \tag{11.3}$$

Thus we can deform[1] the integration contour away from the singularities. Then the relevant propagators are off-shell by order Q^2, and the leading regions return to the form of Fig. 11.1(b), for all x, and our standard derivations will now apply. Then the difference between $T^{\mu\nu}[f]$ and its complex conjugate gives a valid prediction for the locally averaged structure functions.

The averaging method also solves another conceptual problem. This is that in a theory with confined quarks, the evolution of the final state might be more like that of an elastic spring than of a fragile string, to use the terminology of Sec. 4.3.1. In that case a final state

---

[1] Strictly, a test function f need not be an analytic function, which makes questionable a contour deformation. But a basis set of analytic functions, e.g., Gaussians, suffices for our argument.

of a high-energy struck quark and a target remnant would evolve not to a pair of connected jets, but to a spectrum of bound states or of narrow resonances.

Standard factorization methods do not describe the bound-state structure. Thus, the true predictions of factorization are only for locally averaged structure functions. This has been verified by Einhorn (1976) in a model with elastic spring confinement: QCD in two space-time dimensions in the limit of a large number of colors. Only if the structure functions are already smooth does factorization apply point-by-point.

We already saw the need for local averaging in our NLO calculations in Ch. 9. There we found a cancellation between real and virtual emission of gluons that are soft or are final-state collinear. The cancellation was embodied in the plus distribution in the coefficient functions, e.g., (9.20). At large x, the necessary average must be done by the local averaging of the hadronic structure functions. But at smaller x, it suffices to use the integral over parton momentum in the factorization formula (8.81), provided that the parton densities are sufficiently smooth.

11.2.2 Parton-hadron duality

At large x and moderate Q^2, there are many noticeable resonances in DIS structure functions. That partonic methods can nevertheless be applied, but only to locally averaged structure functions, is an instance of the concept called parton-hadron duality. It was first found before the advent of QCD and factorization theorems in an analysis of data by Bloom and Gilman (1971). Duality carries the implication that the partonic structure and the resonance structure are parts of the same overall mechanism, rather than two distinct mechanisms to be added to each other.

For a recent review, see Melnitchouk, Ent, and Keppel (2005). One of their comparisons with recent data is shown in Fig. 11.3. As Q^2 is increased, the resonances move to the right in x, a necessary kinematic property. This is not compatible with the generally smooth scaling violations given by DGLAP evolution. Naturally the spacing of the resonances in x decreases as Q^2 increases. But there is little or no decrease in the height of the resonances, as a fraction of the structure function.

Much of the phenomenological application of duality is at low Q^2, where the region of noticeable resonances extends a long way down in x. But even at large Q^2, resonances remain, close to $x = 1$. According to duality, the smooth curves for F_2 from factorization should cross the resonance oscillations approximately midway between their peaks and troughs. However, with the MRST fit shown in Fig. 11.3, this appears *not* to be the case, at least for the larger values of Q^2. The reasons are unclear; the CTEQ and MRST curves disagree.

11.2.3 Leading and super-leading terms

We now restrict our attention to those regions that contribute at the leading power, Q^0, or larger, determined by the methods of Sec. 5.8. The basic rule is that increasing the number of lines connecting the hard and collinear subgraphs gives a suppression, as does

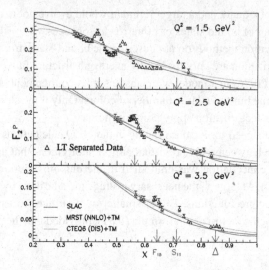

Fig. 11.3. Proton structure function F_2^p measured at Jefferson Lab Hall C. Two of the smooth curves are the results of QCD fits to other data by MRST (Martin *et al.*, 1998) and CTEQ (Lai *et al.*, 2000), with target mass corrections included by the method of Barbieri *et al.* (1976). The SLAC curve is a fit to DIS data (Whitlow *et al.*, 1992). The arrows indicate the positions of prominent resonances. Reprinted from Melnitchouk, Ent, and Keppel (2005), with permission from Elsevier.

the presence of a soft subgraph. But, just as with the Sudakov form factor in Ch. 10, there is an exception for collinear gluons of polarization in the plus direction; dealing with these is the main difficulty in our proof. The proof will be organized differently than in Ch. 10, in order to overcome the complications of working in a non-abelian gauge theory.

Of the lines entering the hard scattering H from the collinear subgraph C, let N be gluons, for which we write the polarization sum as

$$H \cdot C = H_{\mu_1 \dots \mu_N} \prod_{j=1}^{N} g^{\mu_j \nu_j} C_{\nu_1 \dots \nu_N}. \tag{11.4}$$

Let k_j be the momentum of gluon j flowing into H. The largest term in its polarization sum has $\mu_j = -$, $\nu_j = +$, and we manipulate it into a form suitable for the use of Ward identities. Accordingly, we make a Grammer-Yennie decomposition

$$g^{\mu_j \nu_j} = K^{\mu_j \nu_j} + G^{\mu_j \nu_j}, \tag{11.5}$$

where

$$K^{\mu_j \nu_j} = \frac{k_j^{\mu_j} w_2^{\nu_j}}{k_j \cdot w_2 - i0}, \quad \text{and} \quad G^{\mu_j \nu_j} = g^{\mu_j \nu_j} - \frac{k_j^{\mu_j} w_2^{\nu_j}}{k_j \cdot w_2 - i0}, \tag{11.6}$$

and the vector w_2 projects onto plus components of momentum: $w_2 = (0, 1, \mathbf{0}_T)$. Then from (11.4), we get a sum of terms which we label by saying that each of the gluons is a K gluon or a G gluon according to which term in (11.5) is used.

The denominators $k_j \cdot w_2$ introduce singularities at $k_j^+ = 0$, that have no corresponding actual singularities in H. In the final result, we will find a cancellation of these artificial singularities. We choose to equip the singularities with an $i0$ prescription appropriate for a final-state pole; it must be the same in all terms for our Ward identities to work.

(Notice a contrast with the situation for the Sudakov form factor, for which hard-scattering subgraphs often had singularities for soft and for opposite-side collinear configurations. These were canceled by subtractions for smaller regions. To ensure contour-deformation arguments for the Glauber region worked, we found that the $i0$ prescription for the denominators $k_j \cdot w_2$ had to correspond to that of the subtracted singularities in the hard-scattering subgraph.)

The normal suppression for extra collinear lines entering the hard scattering applies to the G gluons but not to the K gluons (Sec. 5.8). For a collinear gluon with radial coordinate λ, the K term has a power Q/λ relative to the G term.

Complications now arise when all the lines connecting the hard and collinear subgraphs are K gluons, because they give super-leading contributions from individual graphs, with a power Q^2/λ^2 relative to the final result. This also permits there to be a soft subgraph at leading power. There is in fact a cancellation (Labastida and Sterman, 1985) of super-leading terms in the sum over graphs. Although in a model with an abelian gluon field the cancellation of K gluons is exact, in QCD there are left-over leading-power terms (Collins and Rogers, 2008), and these are needed for factorization.

After the Grammer-Yennie decomposition, we can define two classes of contribution. The first has a pair of ordinary leading-power partons accompanied by any number of K gluons. In these situations, the lines joining the collinear and hard subgraphs are:

1. two G gluons plus any number of K gluons;
2. or: a quark and an antiquark line plus any number of K gluons;
3. or: a ghost and an antighost line plus any number of K gluons. One of the simplest graphs with such a region is shown in Fig. 11.4.

In all the above cases, there is no soft subgraph, and we have a leading-power (Q^0) contribution. Adding extra G gluons, quarks, ghosts, or a soft subgraph gives a power-suppression.

The case with collinear ghost lines does not correspond to any term in the factorization theorem. Instead we will find it combines with part of the next class of contributions to give a result that vanishes in physical quantities.

A second class of terms covers the remaining possibilities for leading and super-leading powers. In these, all the collinear lines entering the hard scattering are gluons:

1. If all of the gluons are K gluons and there is no soft subgraph, we have a super-leading contribution of order Q^2.
2. If all but one of the gluons is a K gluon, we have a super-leading contribution of order Q^1.
3. A soft subgraph contributes a suppression, but may leave the contribution leading.

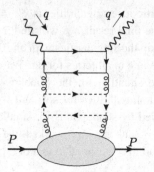

Fig. 11.4. Graph for DIS with Faddeev-Popov ghost loop.

All other cases give a power-suppression. From Sec. 5.8, the only case that a soft subgraph allows a leading-power contribution is where all the collinear attachments to H are K gluons, the external lines of the soft subgraph are gluons, and exactly one soft gluon enters H.

11.3 Factorization for DIS

To obtain factorization, we apply the steps listed in Sec. 11.1. Now that we have determined the leading and super-leading regions, it remains to apply Ward identities to sum over attachments of K gluons to the hard subgraph H. This will determine the operators defining parton densities.

As stated in Sec. 11.1, we now use Fig. 11.1(b) to refer to a generic term $C_R \Gamma$ in the sum over regions and graphs, rather than just to specify a region. We also will generally impose a sum over graphs and regions. It is important to be conscious of the shifts in meaning of such a diagram.

It is convenient to combine the soft and collinear subgraphs into a single subgraph, and then to decompose all the external gluons of the hard subgraph H into K and G gluons. *The Ward identities are applied to gluonic external lines of the H bubble, which is treated as a sum over the possible graphs and equipped with subtractions for smaller regions.*

Now the hard factor in $C(R)$ is defined to be expanded in powers of small momentum components, with retention of terms that contribute to the leading power or higher. Thus for a leading term, e.g., two G gluons plus any number of K gluons, the hard factor is simply taken with its external lines massless, on-shell at zero transverse momentum. As in our examples in previous chapters, this short-circuits integrals over k_j^- and $k_{j,\mathrm{T}}$, so that the coordinate-space fields at the edges of the collinear-soft subgraph (and hence in the parton densities) are separated in the minus component of position, as in (7.40).

In situations where higher terms in the expansion of H in powers of k^- and k_T are used, we get extra factors of these momentum components. In the operator definitions of the collinear factors, these give derivatives with respect to the various x_T and x^+ coordinates in the operators, with the derivatives taken at $x_\mathrm{T} = x^- = 0$.

11.3.1 Abelian gluon

We start with the case of a model theory with an *abelian* gluon field, since its Ward identities are simple, as in Sec. 10.8.3.

We apply a Ward identity in turn to the attachment of each K gluon to the hard subgraph H, defined as the sum over graphs with a given number of external lines, with appropriate irreducibility properties, and with subtractions for smaller regions.

We get terms for attaching the K gluon to each of the external charged lines of the hard scattering. Just as with the Sudakov form factor, the Ward identities are unaffected by the presence of subtractions.

Case of all-gluon connection

When all the external lines of H are gluons, there are no external charged lines, so that summed over graphs, the attachment of a K gluon is zero:

$$\sum \quad \begin{matrix} q \\ H \end{matrix} \quad = 0 \tag{11.7}$$

Here the solid triangle is the vertex for a K gluon, similarly to Fig. 10.6.

So we are left only with G gluons. For the leading power of Q, we keep the minimal number of gluons exchanged between the collinear and hard subgraphs. Since the case of one gluon gives exactly zero by charge-conjugation invariance, the minimum is two G gluons. Summing over attachments of K gluons to the hard subgraph gives zero. So the sum over all gluonic terms gives

$$\sum \quad \begin{matrix} q & q \\ & H \\ P & C & P \end{matrix} \quad = \quad \begin{matrix} q & q \\ & H \\ P & C & P \end{matrix} \quad + \text{power-suppressed} \tag{11.8}$$

In the main term on the r.h.s., the crosses denote what we now prove to be the vertices for the gluon density, as defined in (7.44).

Each cross starts out as a vertex $G^{\mu_j \nu_j}$ for a G gluon; (11.6). The $\mu_j = +$ case is zero, while the $\mu_j = -$ term gives a power-suppressed contribution, by the boost argument in Sec. 5.8.8. This leaves the transverse components. The two $G^{T \nu_j}$ factors are each $1/k^+$ times the vertex for the gluon field strength tensor (Fig. 7.12). One factor of $1/k^+$ gives the explicit $1/\xi P^+$ in the definition of the gluon density (equation (7.44) and Fig. 7.9). The only difference with those formulae and Feynman rules is that there is no Wilson line in the gluon density in an abelian theory. The remaining $1/k^+$ factor goes with the integral over k^+ that joins the hard and collinear subgraphs, to give the $d\xi/\xi$ factor in (8.81). According to the standard construction of a hard scattering, the external lines of the H subgraph are set on-shell with zero transverse momentum; kinematically this is just as in the parton model. As in (7.44), we wrote the gluon density factor in the form $\rho_{g,j'j}(\xi, S) f_g(\xi)$, where $\rho_{g,j'j}(\xi, S)$ is normalized to be a density matrix, i.e., it has trace unity.

As usual, renormalization is applied to the parton density. After we take the discontinuity of the uncut amplitude, we get the gluon term in the factorization theorem (8.81)

$$\int_{x-}^{1+} \frac{d\xi}{\xi} C_g^{\mu\nu;j'j}(q, \xi P; \alpha_s, \mu) \, \rho_{g,j'j}(\xi, S; \mu) \, f_g(\xi; \mu). \tag{11.9}$$

Here we have inserted two (summed) transverse spin indices. For the common case of an unpolarized target, the gluon spin density matrix $\rho_{g,j'j}$ is half the unit matrix. The normalization of the coefficient function C_g is exactly that of DIS on a transversely polarized on-shell gluonic target, with subtractions applied to cancel collinear divergences. As usual, the integral over gluon k^- and k_T is inside the standard definition of the gluon density.

Quark-antiquark plus K gluons

Since Faddeev-Popov ghosts are non-interacting in an abelian gauge theory in the gauges we use, the only other case that gives a leading contribution is where a quark and an antiquark line connect the collinear and hard subgraphs, together with any number of K gluons.

The Ward-identity argument works exactly as for collinear-to-A gluons attaching to the hard scattering in the Sudakov form factor (Sec. 10.8). There the sum over K gluons gave a Wilson line at the collinear-to-A quark entering the hard scattering.

For DIS the essential difference is that the hard scattering has both a quark and an antiquark external line, so that we get a Wilson line for each. The quark field $\psi(0)$ in the parton density therefore becomes $W(\infty, 0)\psi(0)$, while the antiquark field has a Wilson line of the opposite charge: $\bar{\psi}(w^-)W(\infty, w^-)^\dagger$. The Wilson lines have zero transverse separation, so they can be combined to give a Wilson line between the two fields: $[W(\infty, w^-)]^\dagger W(\infty, 0) = W(w^-, 0)$. The Wilson lines are in the light-like direction w_2, and the operators defining the quark density are exactly the ones in (7.40).

We must also apply the same leading-power approximations on the quark polarization as in the parton-model. Compared with the gauge-invariant parton model (Sec. 7.7), the new features are that we have arbitrarily higher-order corrections to the hard factor H, with subtractions as usual, and that the parton densities must be renormalized.

We write the overall result as

$$\tag{11.10}$$

which gives the quark term in the factorization theorem.

Cancellation of rapidity divergences

The key technical details of the full proof have involved a minor generalization of the methods we applied to the Sudakov form factor.

One notable difference is that the Wilson lines are light-like, which gives rapidity divergences graph-by-graph. But the divergences cancel in the final result. The easiest way of seeing this in general is to work in coordinate space and use the identity (7.39). Now the rapidity divergences are associated with on-shell Wilson-line denominators, and hence with a situation in which the Wilson line is infinitely long, i.e., when we integrate the vertices all the way to infinity. But (7.39) shows that the segment out to infinity cancels. As with our argument about the pinch singularities of $T^{\mu\nu}$, to use this argument in momentum space requires that we take a local average of the parton density over longitudinal momentum fraction. An example can be seen in our one-loop calculations in Sec. 9.4.3.

Because of the rapidity divergences at intermediate stages, it may be appropriate to use a non-light-like denominator $k \cdot n$ until the all the K gluons are extracted and converted to Wilson lines. After that one replaces n by a light-like vector w_2.

Overall view

We now have completed the proof of factorization in the model theory. All the standard consequences follow, including the ability to implement perturbative calculations as explained in Ch. 9.

11.3.2 Non-abelian gluon

In a non-abelian gauge theory like QCD, we have Slavnov-Taylor identities instead of simple Ward identities. They and their proof by direct diagrammatic methods are much more complicated than in the abelian case. In much of the original work on proving factorization the issues related to extracting K gluons from the hard scattering were glossed over.

Labastida and Sterman (1985) did give a diagrammatic proof of one critical result that one gets zero when all or all but one of the external lines of the hard scattering are K gluons. In Sec. 11.9, I will summarize an argument that generalizes to a non-abelian theory the Ward-identity methods for K gluons that were obtained for an abelian theory in Sec. 11.3.1.

But the proof only applies in the strictly collinear limit. Since individual contributing graphs are super-leading, this leaves open the possibility that there is a non-zero remainder of leading power. The remainder is power-suppressed with respect to the contributions of individual graphs, but not with respect to the final result. In fact, Collins and Rogers (2008) recently found by the simplest possible explicit calculation that the remainder is actually non-zero; the pure K-gluon terms contribute to the gluon density, unlike the case in an abelian gauge theory.

So more powerful methods are needed.

The ultimate result is standard factorization of the form (8.81), where each term is a coefficient convoluted with the matrix element of a gauge-invariant operator, and all the relevant operators are the ones listed in (7.40) and (7.43) (generalized to include polarization effects). Essentially identical issues arose in the short-distance OPE for moments of DIS structure functions.

One possible approach to a proof is to generalize the diagrammatic arguments of Sec. 10.8.3, as in Labastida and Sterman (1985) and Sec. 11.9.

Instead we now use an argument using BRST invariance that has been used in the renormalization of gauge-invariant local operators; see Collins (1984, Sec. 12.6).

Without the use of gauge invariance, the structure of the leading regions, Fig. 11.1(b), leads to a factorization in which there is an infinite collection of operators; each different number of gluons gives a different pdf-like object, and for each extra gluon there is an extra longitudinal-momentum argument to be convoluted with the associated hard-scattering coefficient. If this were the whole story, the formalism would have little predictive power. But, in reality, terms differing by extra gluons have the same coefficient function, and the pdf-like objects all sum to a gauge-invariant pdf, with a single longitudinal-momentum variable ξ.

BRST restrictions on operators

A natural initial idea for the proof is that because QCD is color-gauge invariant, so are all the operators defining allowed parton densities. However, the actual QCD Lagrangian is not gauge invariant, but only BRST invariant (Sec. 3.1.3).

It is useful to generalize DIS to treat an off-shell Green function corresponding to the amplitude $T^{\mu\nu}$:

$$T^{\mu\nu}_{\text{off-shell}}(q) = \frac{1}{4\pi} \int d^4z \, e^{iq \cdot z} \, \langle 0| \, T \text{ fields } j^\mu(z/2) \, j^\nu(-z/2) \, |0\rangle. \tag{11.11}$$

Here, "fields" denotes a product of two (or more) fields that are Fourier-transformed to be in a similar kinematic region to the target bra and ket, $\langle P, S|$ and $|P, S\rangle$ in (11.1). The derivation of leading regions works equally well for $T^{\mu\nu}_{\text{off-shell}}(q)$ as it does for the normal on-shell tensor. Therefore, to leading power we obtain a sum (and convolution) over coefficients and pdf-like matrix elements:

$$T^{\mu\nu}_{\text{off-shell}}(q) = \sum_i C_i \otimes \langle 0| \, T \text{ fields } \mathcal{O}_i \, |0\rangle + \text{p.s.c.} \tag{11.12}$$

We have a sum over possible operators $\mathcal{O}_i$, and a convolution with the longitudinal-momenta arguments of the operators. At this point in the argument there is the possibility that we have arbitrarily complicated multilocal operators, as pointed out above.

Now from BRST symmetry of the Lagrangian, there arises a conserved Noether current, and exactly as for an ordinary internal symmetry it follows that Green functions are BRST invariant, i.e.,

$$\delta_{\text{BRST}} \, \langle 0| \, T \text{ any fields } |0\rangle = 0. \tag{11.13}$$

(See, e.g., Collins, 1984; Nakanishi and Ojima, 1990.) The BRST variations of individual fields are given in (3.6).

We apply (11.13) to (11.11). The electromagnetic currents are gauge invariant and hence BRST invariant. Therefore

$$\langle 0| \, T \, (\delta_{\text{BRST}} \text{ fields}) \, j^\mu(z/2) \, j^\nu(-z/2) \, |0\rangle = 0. \tag{11.14}$$

Since the BRST variation adds a ghost field η (or removes an antighost field $\bar{\eta}$), the interesting cases of this equation have one more antighost than ghost fields in "fields".

Exactly the same formula must apply to the factorized form, up to possible power-suppressed terms:

$$\sum_i C_i(Q) \otimes \langle 0| \, T \, (\delta_{\text{BRST}} \text{ fields}) \, \mathcal{O}_i \, |0\rangle = \text{p.s.c.} \tag{11.15}$$

We remove the power-suppressed corrections by defining the coefficient functions to be obtained from an expansion in powers of Q and $\ln Q$, and by restricting to the leading power of Q.[2]

Using (11.13), we get

$$\sum_i C_i \otimes \langle 0| \, T \text{ fields } \delta_{\text{BRST}} \mathcal{O}_i \, |0\rangle = 0. \tag{11.16}$$

This is true no matter which set of fields is used, so the operators themselves are BRST invariant:

$$\delta_{\text{BRST}} \sum_i C_i \otimes \mathcal{O}_i = 0. \tag{11.17}$$

Factorization follows, generalized from (8.81) to apply to the off-shell amplitude $T^{\mu\nu}_{\text{off-shell}}$, and with the operators restricted to be BRST-invariant operators.

Up to here the derivation is identical to the one for the OPE, or for the renormalization of gauge-invariant operators.

Gauge-invariant operators are BRST invariant, so the important question is what other BRST-invariant operators exist. In the OPE, the operators $\mathcal{O}_i$ are local, i.e., they are polynomials in elementary fields and their derivatives all at the same space-time point. In that case, we have a theorem (Joglekar and Lee, 1976; Joglekar, 1977a, b; Nakanishi and Ojima, 1990) that all the other possible operators are one of the following classes:

A. operators that are BRST variations: $A = \delta_{\text{BRST}} A_{\text{source}}$;
B. operators that vanish by the equations of motion.

Operators in class B have vanishing matrix elements in on-shell states, but they contribute (Collins, 1984, p. 14) in time-ordered Green functions, because of the peculiarities of combining time-ordering of operators with derivatives of fields. The BRST invariance of operators in class A follows from the nilpotence of BRST transformations (up to terms vanishing by the equations of motion).

Operators in both of these classes vanish in on-shell matrix elements with physical states. This is trivial for operators vanishing by the equations of motion, and follows

---

[2] See Sect. 11.7 for variations on this expansion when quark masses may be non-negligible.

simply (Collins, 1984, p. 318) from BRST invariance of physical states for operators in class A.

A minor generalization of these results is that we also have vanishing contributions of operators of classes A and B in Green functions with gauge-invariant operators, as well as in matrix elements with physical scattering states. Equation-of-motion operators give delta functions in coordinate space in Green functions with other operators, and we can eliminate these by requiring the positions of the other operators to be away from the operators $\mathcal{O}_i$. The Green functions of BRST-variation operators with gauge-invariant (and indeed BRST-invariant) operators vanish by a simple application of (11.13).

Unfortunately, the published proofs that BRST-invariant operators are either gauge invariant or are in one of classes A or B apply as written to local operators. It is natural that the result also applies to the non-local operators we use in factorization. But, as far as I know, no proof has been given. For the purposes of the discussion, I will assume the result is true, and leave the proof (or refutation) to future research.

An example of a BRST-variation operator is

$$\delta_{\mathrm{BRST}} \left[\bar{\eta}^{\alpha}(0, x^-, \mathbf{0}_{\mathrm{T}}) \, A_{\mu}^{\beta}(0) \right] / \delta\lambda$$

$$= \left[\partial \cdot A^{\alpha}(0, x^-, \mathbf{0}_{\mathrm{T}}) \, A_{\mu}^{\beta}(0) \right] + \left[\bar{\eta}^{\alpha}(0, x^-, \mathbf{0}_{\mathrm{T}}) \, D_{\mu}^{\beta} \eta(0) \right]. \tag{11.18}$$

The free Lorentz index μ could be a $-$ index (corresponding to the A^+ component), or it could be a transverse index contracted with a transverse momentum somewhere.

I am not aware of an explicit calculation of the presence of such operators in calculations of factorization with off-shell Green functions. But there are calculations in the analogous case of the renormalization of local operators (Dixon and Taylor, 1974; Kluberg-Stern and Zuber, 1975), which showed the occurrence of non-gauge-invariant operators as counter-terms to local operators.

Gauge-invariant operators

We now have the result that all the operators needed to apply factorization in physical matrix elements are gauge invariant. We call their matrix elements parton densities.

Obvious possibilities are the operators used to define gauge-invariant parton densities in (7.40) and (7.43). In each case we have a pair of basic partonic fields ($\bar{\psi}$ and ψ or two field strength tensors) separated in the minus direction and connected by a Wilson line starting at one partonic operator and ending at the other. The representation of the gauge group in the Wilson line is the one appropriate to the partonic field. Each of the fields and the Wilson lines transforms covariantly under gauge transformations, e.g., (7.35), without derivatives, and it is then easy to deduce gauge invariance for the operators in the parton densities.

It is important to rule out other possibilities. Generalizations of this issue arise in dealing with power-law corrections where more complicated operators get used, and they also arise in treating transverse-momentum-dependent (TMD) parton densities, etc., where the Wilson lines may be non-light-like. Gauge invariance alone does not determine the path along which the gluon field is integrated in a Wilson line $W(C)$: the transformation law (7.35) involves only the endpoints of the path C, and is independent of which path is chosen between the endpoints. As we have seen with the Sudakov form factor, the path should be one such that a factorization theorem can be derived.

For our case the requirements on the operators defining the parton densities are:

1. The operator is formed out of the elementary fields of the theory, and the elementary fields correspond to the lines entering the hard scattering.
2. Since the hard scattering is expanded in powers of k^- and k_T, for each parton line entering the hard-scattering subgraph, the parton density has the corresponding momentum components integrated over. In coordinate space the operators therefore have zero relative position in x^+ and x_T. Thus the operators are localized on a line in the x^- direction.
3. By power-counting all but at most two of the elementary fields are A^+.

A simple way of dealing with this problem is to convert to light-cone gauge $A^+ = 0$. That eliminates all the extra gluons entering the hard scattering. The operators are now the same as in the elementary parton model without gauge links. Since the standard gauge links in the minus direction are unity in $A^+ = 0$ gauge, one can insert the standard gauge links and recover the standard gauge-invariant links.

But given the known problems with $A^+ = 0$ gauge, it would be nice to have a proof that does not rely on the gauge.

Since the Wilson line is restricted to a line in the x^- direction, the results of Sec. 7.5.2 show that the results now depend only on the endpoints of the path, at 0 and at $(0, x^-, 0_T)$. So we can choose the path just as we did when we first defined gauge-invariant parton densities, in Secs. 7.5.4 and 7.5.5.

We will see in Ch. 13 that the case of TMD densities shows a notable contrast, because for TMD densities the path in the Wilson line has segments at different transverse positions.

11.4 Renormalization of parton densities, DGLAP evolution

We will also need the DGLAP equations for the evolution of the parton densities:

$$\frac{d}{d \ln \mu} f_{j/H}(\xi; \mu) = \sum_{j'} \int \frac{dz}{z} 2 P_{jj'}(z, g) f_{j'/H}(\xi/z; \mu), \qquad (11.19)$$

As explained in Sec. 8.4, these equations are the RG equations for the parton densities, and the kernels can be computed from the renormalization coefficients; see (8.31)–(8.33).

Compared with that section, the main difference in the derivations for QCD is the same as for factorization in QCD compared with factorization in non-gauge theories. This is that there can be arbitrarily many K gluons connecting the collinear subgraph to the hard subgraph. For the case of renormalization, a hard subgraph is a subgraph whose loop integration gives a UV divergence. The possible operators used in renormalization are organized into the same classes: the standard gauge-invariant operators, BRST variations, and operators that vanish by the equations of motion. For the same reasons as with the local operators used in the OPE (Collins, 1984, p. 318), the renormalization matrix has a triangular form:

$$\begin{pmatrix} \mathcal{O} \\ \mathcal{A} \\ \mathcal{B} \end{pmatrix} = \begin{pmatrix} Z_{\mathcal{O}\mathcal{O}} & Z_{\mathcal{O}A} & Z_{\mathcal{O}B} \\ 0 & Z_{AA} & Z_{AB} \\ 0 & 0 & Z_{BB} \end{pmatrix} \begin{pmatrix} \mathcal{O}_{(0)} \\ \mathcal{A}_{(0)} \\ \mathcal{B}_{(0)} \end{pmatrix}. \tag{11.20}$$

Here $\mathcal{O}$ denotes the collection of gauge-invariant operators for the parton densities, while A and B denote the operators of classes A and B; see p. 409. The symbols $\mathcal{O}_{(0)}$ etc. with a subscript (0) denote the bare operators, and the unadorned symbols denote renormalized operators.

In physical matrix elements, only the operators $\mathcal{O}$ are non-zero, so that the normal DGLAP kernels can be computed from the $Z_{\mathcal{O}\mathcal{O}}$ factors alone. In physical calculations, we can therefore replace (11.20) by $\mathcal{O} = Z_{\mathcal{O}\mathcal{O}}\mathcal{O}_{(0)}$.

At one-loop order, all the necessary calculations can be performed (Sec. 9.4) with on-shell matrix elements in quark and gluon states; hence there is no need to treat the extra operators A and B. But a correct treatment of renormalization beyond one-loop order needs to take account of the presence of other operators in renormalization of the operators in off-shell Green functions. Cf. Hamberg and van Neerven (1992) and Collins and Scalise (1994).

11.5 DIS with weak interactions

So far in this chapter, we have worked with DIS with photon exchange, so that there are electromagnetic currents in the definition of $W^{\mu\nu}$. All the same methods and ideas work identically for other processes, with Z and W boson exchange. See Sec. 7.1 for an account of the structure functions and the application of the parton model, which corresponds to the LO QCD approximation. Naturally, the parton model is supplemented by an application of DGLAP evolution, and by the use of higher-order corrections to the hard scattering.

11.6 Polarized DIS, especially transverse polarization

So far, this chapter's treatment has (mostly implicitly) allowed for general polarization states for the target and the partons, so that factorization was derived in the form (8.81), with a helicity density matrix for the parton initiating the hard scattering. In a non-gauge theory, we projected this onto factorization for individual structure functions in (8.83). There F_1 and F_2 use unpolarized parton densities $f_j(\xi)$, g_1 uses the helicity densities $\Delta f_j(\xi)$, and g_2 is zero at the leading-power level (so that the transversity densities $\delta_T f_j(\xi)$ do not appear).

The derivation of these results is entirely unchanged in QCD. First, there is the classification of parton densities into unpolarized densities, helicity densities, and transversity densities (with a generalization for spin-1 gluons and for targets of spin other than $\frac{1}{2}$). The derivation of the classification in Secs. 6.4 and 6.5 used parity invariance and angular-momentum conservation about the z axis. This derivation is affected neither by inserting Wilson lines in the minus direction in the operator definitions of the parton densities nor by renormalization. As for the hard scattering, the derivation of the form of polarization dependence is unchanged from that in a non-gauge theory in Sec. 8.10. Notably there is no change in the proof in Sec. 8.10.5 that at leading power there is no contribution from transverse spin.

11.7 Quark masses

We obtained factorization by an expansion to the leading power of an appropriate large scale Q (with logarithms of Q being allowed for). This implies setting masses to zero in the hard scattering, thereby entailing the assumption that masses are all much less than Q. But in reality this is not always the case. Relevant experiments in DIS and other processes currently range from Q below 2 GeV to many hundreds of GeV, which more than spans the masses of the charm, bottom, and top quarks.

Evidently we must generalize our formulation of factorization to correctly treat heavy quarks. I will not give a complete treatment, but just summarize the results. The essential insights are in the decoupling theorem of Appelquist and Carazzone (1975) and in the work of Witten (1976) on the contributions of heavy quarks to DIS in the framework of the OPE. We saw the underlying ideas in Secs. 3.9–3.11.

The basic observation is that if the mass of a particular field in a QFT is much bigger than the momentum scale of a process, then we can drop that field from consideration with errors suppressed by a power of the heavy quark relative to the process's scale. Because of the need for renormalization, the decoupling theorem modifies this by showing that the values of renormalized parameters may need to be adjusted after dropping the heavy quarks.

Complications arise because in a reaction with a hard scale, like DIS, there are two momentum scales: Q and Λ. For example, a mass m_q may be small relative to Q, but not relative to Λ. In that case, we might want to neglect m_q with respect to Q, but also we might want to perform the opposite operation, decoupling of the quark, with respect to low-energy phenomena. Moreover, the simplest applications have errors of the order of ratios like m_q/Q, whereas we would like factorization to be valid up to power corrections in the smallest ratio Λ/Q uniformly as we vary the relative sizes of Q and the heavy quark masses.

We distinguish four cases with corresponding approximation methods:

- $m_q \gg Q$. Then we simply decouple the heavy quark.
- $m_q \sim Q$. Then we must keep the heavy quark's mass unapproximated in the hard scattering. We can apply the decoupling theorem to the parton densities, in such a way that the sum over quark flavors in the factorization theorem is restricted to the lighter quarks.
- $Q \gg m_q \gg \Lambda$. Then we can neglect m_q in the hard scattering, and we treat the quark like a light quark. But we apply a modified decoupling theorem to compute the evolved heavy quark distribution in terms of the light-parton distributions.
- $m_q \lesssim \Lambda$. The quark is a light parton, so that the methods we have derived so far are valid.

To get the best accuracy uniformly in the relative sizes of Q and the heavy quark masses, a combination of the basic approximation methods is needed, with possibly different methods being applied to different quarks.

The solution is not unique, and a variety of methods can be found in the literature, as reviewed in Thorne and Tung (2008), although not all are equally adequate. However, the variety is much less if one insists that the methods apply to all cases rather than just the limiting cases $m_q \ll Q$ and $m_q \gg Q$, and if one insists that there must be a definite gauge-invariant operator definition of every parton density. (An operator definition ensures that one actually knows the meaning of the concept of a parton density.)

I adopt the scheme of Collins, Wilczek, and Zee (1978) (CWZ), as extended to parton densities by Collins and Tung (1986); see Sec. 3.10. This involves a sequence of renormalization subschemes, parameterized by the number of "active quark flavors" n_{act}. Counterterms for graphs containing only the lightest n_{act} flavors are renormalized by the $\overline{MS}$ method, and the heavier quarks by zero-momentum subtractions, which continue to preserve gauge invariance automatically. Manifest decoupling occurs when the masses of inactive quarks are much larger than the scale of the process; graphs containing the inactive quarks are then power-suppressed, and can be simply dropped. Matching calculations between the subschemes have been performed (Chetyrkin, Kniehl, and Steinhauser, 1997, 1998; Aivazis *et al.*, 1994).

In a particular subscheme, the evolution equations, both the RGE for the QCD parameters and the DGLAP equations for the parton densities, are exactly those in the $\overline{MS}$ scheme with n_{act} quarks. One then talks of a 3-flavor scheme, a 4-flavor scheme, etc.

A straightforward generalization of the factorization property is set up by choosing the active flavors to be those for which $m_q \lesssim Q$. The sum over flavors in (8.81) etc. is then only over active flavors. Masses of heavy quarks, and especially of inactive quarks, are not neglected in the hard scattering, unless $m_q \ll Q$. This method was first proposed by Aivazis *et al.* (1994) (ACOT). It has the following consequences:

- In the hard scattering, inactive quarks can appear as internal lines.
- Parton densities for inactive quarks are suppressed by a power of Λ / m_q and are generally dropped. But this is not required.
- When the mass of some heavy quark m_q is much larger than Q, there is a power-suppression of graphs containing this quark. Such graphs may be dropped, with an error suppressed by a power of Q / m_q.
- But when the mass is comparable with Q, there is no power-suppression.
- When the mass of a quark is much less than Q, its mass can be neglected in the hard scattering; such a quark is always an active quark.

When the mass of some heavy quark (notably charm or bottom) is comparable to Q, that quark may be legitimately treated either as active or as inactive, by a change of subscheme. Equivalent accuracy is obtained provided that the mass of that quark is retained in the hard scattering, at least when the quark is internal and the hard scattering is initiated by a lighter parton (e.g., a gluon).

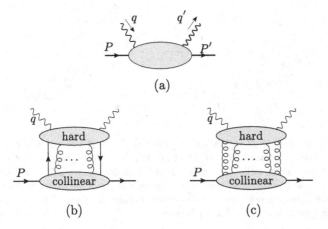

Fig. 11.5. (a) Uncut amplitude for DVCS or DIS. (b) and (c) Leading regions.

However, it is generally best if the mass of the parton initiating the hard scattering is replaced by zero. Also some kinematic modifications to the hard scattering improve its match to the physics. Tung, Kretzer, and Schmidt (2002) have provided a suitable implementation, which they call the ACOT(χ) scheme.

11.8 DVCS and DDVCS

One quite simple extension of factorization for DIS is to a quantity like the DIS structure tensor $W^{\mu\nu}$, but where the two target states have different momenta P and P' and where the current operators are time ordered:

$$T^{\mu\nu}(q, P, P') = \frac{1}{4\pi} \int d^4z \; e^{iz \cdot (q+q')/2} \langle P' | \, T \, J^\mu(z/2) \, J^\nu(-z/2) \, | P \rangle . \qquad (11.21)$$

Thus we have an uncut off-diagonal amplitude, Fig. 11.5(a), with incoming momentum q on the photon at $J^\nu(-z/2)$, and outgoing momentum $q' = q + P - P'$ on the photon at $J^\mu(z/2)$. The process is $\gamma^*(q) + P \to \gamma^*(q') + P'$.

Realizable physical processes using this amplitude are deeply virtual Compton scattering (DVCS), and double deeply virtual Compton scattering (DDVCS):

- DVCS: $e + P \to e + \gamma + P'$;
- DDVCS: $e + P \to e + \mu^+ \mu^- + P'$.

In both cases the incoming virtual photon in Fig. 11.5(a) is space-like and is exchanged with the same kind of lepton as in ordinary DIS. In DVCS the outgoing photon is real, while in DDVCS the outgoing photon is virtual and time-like, generating a lepton pair.

We obtain factorization by a minor generalization of the method used for the *uncut* amplitude (11.1) for DIS. For DDVCS, the regions have exactly the same form, and so do the leading regions.

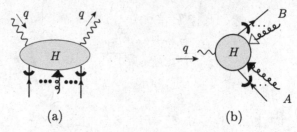

Fig. 11.6. Hard subgraph H with K gluons attached. The thick curved lines indicate where an external quark is set on-shell in H. The triangle indicates the application of a Grammer-Yennie K approximant, defined as in Sec. 10.4.2. Graph (a) is for DIS, with one collinear subgraph. Graph (b) is for e^+e^- annihilation in the simplest case of two collinear groups. Here the solid arrow denotes the approximant for collinear-A gluons and the open arrow denotes the approximant for collinear-B gluons.

But for DVCS, it is possible to have regions with a group collinear to the outgoing real photon, Fig. 5.16(a). After allowing for the usual Grammer-Yennie cancellation of K gluons, all these extra regions are power-suppressed, and the leading regions are the same as for DDVCS.

In both cases, the factorization theorem has the same form as for DIS except that the parton density is replaced by a generalized parton density (GPD) (6.90), whose definition differs from that for an ordinary parton density simply by being off-diagonal in the target state. Equation (6.90) was written for the case of super-renormalizable non-gauge theory, and for a quark. The structural modifications to treat renormalization, to insert a Wilson line, and to define a gluon density are the same as in ordinary pdfs. The DGLAP kernels have to be generalized, and include dependence on the longitudinal momentum transfer. See Diehl (2003) for a review.

11.9 Ward identities to convert K gluons to Wilson line

This section gives a graphical proof of the conversion of K gluons from attachments to a particular kind of subgraph to couplings to a Wilson line. It generalizes to non-abelian gauge theories and to other processes the work done in Ch. 10 for the Sudakov form factor in an abelian gauge theory.

11.9.1 Statement of general situation

In a gauge theory, we consider a subgraph for a particular momentum category to which Grammer-Yennie K gluons attach from a subgraph for another momentum category. One example is a hard scattering, where the K gluons come from collinear subgraph(s), Fig. 11.6. Another example is a collinear subgraph with soft K gluons attached, Fig. 11.7.

Each K-gluon attachment, notated by a triangle, represents an approximant of the form

$$H(k)_\mu \, g^{\mu\nu} \, A(k)_\nu \mapsto H(\hat{k})_\mu \, \frac{\hat{k}^\mu n_2^\nu}{k \cdot n_2} \, A(k)_\nu, \tag{11.22a}$$

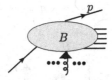

Fig. 11.7. Collinear subgraph B with K gluons attached. The subgraph is written for the amplitude for a quark to produce a jet of hadrons in which a hadron of momentum p is detected. The final state on the right consists of on-shell physical particles. The quark has come out of the hard scattering, as in $e^+ + e^-$ annihilation to hadrons; see Ch. 12. Unlike the case of Fig. 11.6, the external left-hand quark line is off-shell and includes a full propagator.

where the approximated momentum is

$$\hat{k}^\mu = n_1^\mu \frac{k \cdot n_2}{n_1 \cdot n_2}. \tag{11.22b}$$

The vector, n_1, normally light-like, is the direction of the approximated momentum (e.g., $(1, 0, \mathbf{0}_T)$ in H in DIS). The vector n_2 is either a conjugate light-like vector (e.g., $(0, 1, \mathbf{0}_T)$ in DIS) or is close to such a vector, to regulate rapidity divergences.

11.9.2 Proof for H in DIS

In this section, we treat the hard-scattering subgraph H for DIS, Fig. 11.6(a). Subtractions have been applied, as in Sec. 10.8, to remove contributions from smaller regions. We assume that the subtractions do not interfere with the Ward identities, as we saw in Sec. 10.8.

In that section the Ward identities were for an abelian theory, and used the basic diagrammatic elements listed in Sec. 10.8.3. Our task now is to generalize that argument to a non-abelian theory.

Consider first one K gluon. It has a factor of $\hat{k}$ contracted into some Green function. If this were a complete Green function, then, as explained in Sterman (1993, p. 351), we obtain a ghost attachment to each gauge-variant external line of the Green function:

$$\text{(diagram)} \quad = \quad \sum \text{(diagram)} \quad + \quad \sum \text{(diagram)} \tag{11.23}$$

On the l.h.s., a full propagator is amputated for the K gluon. On the r.h.s. the sums are over the gauge-variant external fields of the Green function, and the special vertices with a thick diagonal line are the BRST variations of the external fields. In the first term, the (amputated) gluon directly attaches to the BRST variation of the external field, exactly as in an abelian theory. The thin diagonal line denotes the remaining factor $-in_2/(k \cdot n_2)$, from (11.22a), together with a normalization factor $-i$. In the second term, the diagonal line replaces the incoming ghost line at a ghost-gluon vertex. The ghost line continues to

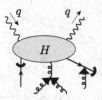

Fig. 11.8. Graphs such as these are *not* included in H because they are reducible in collinear lines.

the BRST variation of an external field. This Slavnov-Taylor identity applies in the form stated to ordinary Green functions, which are expectation values of time-ordered products of fields.

But in our case the Green functions have particular irreducibility properties. The graphs for H are irreducible in the collinear lines. Not only are the external collinear lines amputated, but any graphs in which collinear lines combine to a single line are omitted, as in Fig. 11.8.

To understand the consequences, we use the diagrammatic method of 't Hooft and Veltman (1972), which generalizes the formulation used in an abelian theory, in Sec. 10.8.3. This allows us to take account of missing items relative to the standard Slavnov-Taylor identity.

In Fig. 11.9 are shown some of the main graphical identities used in the proof in 't Hooft and Veltman (1972); this generalizes Fig. 10.12. We start by applying the line identities Fig. 11.9(a) and (b) to a K gluon. In an abelian theory, we completed the derivation of the Ward identities by applying vertex identities like Fig. 11.9(c).

But in a non-abelian theory, the ghost line is interacting, giving the second and fourth terms on the r.h.s. of Fig. 11.9(b). These require a recursive reapplication of the line identities, together with a further complication with ghost loops. After that we get a chain of cancellations at interaction vertices, from Fig. 11.9(c) and (d). With a regular Green function and a single K gluon, we get (11.23).

Now consider Fig. 11.6(a), with N K gluons of incoming approximated momenta $\hat{k}_1, \ldots, \hat{k}_N$ attaching to H. We apply the diagrammatic proof of the Slavnov-Taylor identity to the first gluon. The standard chain of cancellations occurs except at vertices where the external lines from the collinear subgraph attach. Collinear irreducibility of H prevents certain terms from occurring. The missing terms are certain BRST-variation terms in a vertex identity like Fig. 11.9(c). Now each BRST-variation term is obtained from a line identity applied to a graph where the ghost field attaches at a vertex on the immediately neighboring line. The term is missing if and only if the graph is one prohibited by the irreducibility requirements.

The first case is where a term for the BRST variation of each quark line is missing. This gives an external line coupling for the gluon, one term for each quark (or antiquark) that is exactly of the Wilson-line form; this is no different than for an abelian theory. Application of the argument multiple times will give terms of the form of Fig. 11.10 with multiple external gluon attachments.

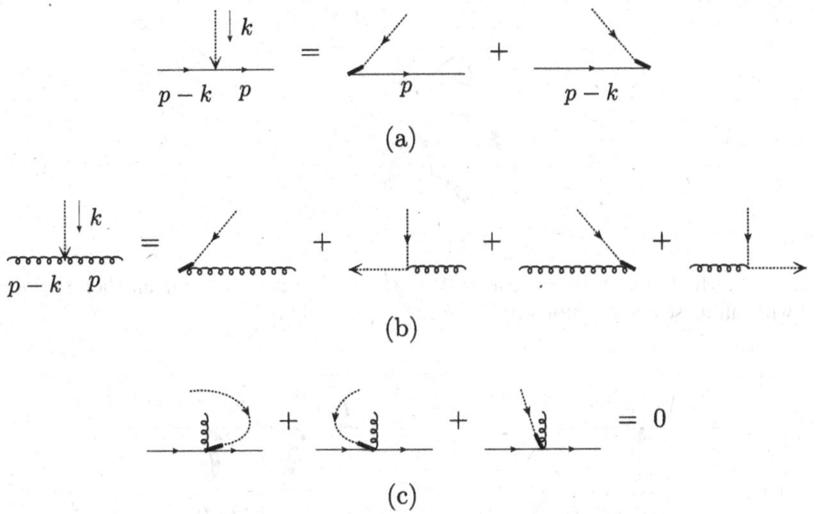

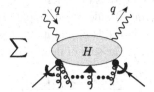

(a)

(b)

(c)

Vertex identities for gluon; identity for ghost.
(d)

Fig. 11.9. Graphical elements of Ward identity in non-abelian gauge theory. (a) Line identity for quark. On the l.h.s., the arrow on the ghost line of momentum k represents a factor of k contracted into a *gluon* vertex for the quark. On the r.h.s., the thick diagonal lines represent a vertex for the BRST transformation on the field at the end of the quark line (multiplied by a factor $-i$). (b) Line identity for gluon. (c) Vertex identity for quark-gluon vertex. There is a term for the color transformation of each field at the vertex. (d) The remaining identities can be found from 't Hooft and Veltman (1972).

Fig. 11.10. Result after applying Ward identities to some K gluons. The little black blobs denote a vertex whose form we do not need to specify.

But in a non-abelian theory, there are additional terms. First, we observe that in the recursive application of the line identity of Fig. 11.9(b) there can be a term where one of the gluons is another K gluon with momentum p. Then the corresponding term on the right of the identity is zero. This is because the l.h.s. Fig. 11.9(b), with approximated momenta, is proportional to

$$[(\hat{p} - \hat{k})^{\mu}(\hat{p} - \hat{k})^{\nu} - g^{\mu\nu}(\hat{p} - \hat{k})^2] - [\hat{p}^{\mu}\hat{p}^{\nu} - g^{\mu\nu}\hat{p}^2]. \tag{11.24}$$

So contraction with $\hat{p}_{\nu}$ makes its term zero, and no special term arises here.

Next, for a normal Slavnov-Taylor identity we have cancellations by vertex identities like Fig. 11.9(c). For a collinear-irreducible hard subgraph, there are some missing terms.

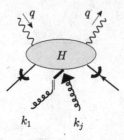

Fig. 11.11. Missing term in elementary Ward identity for connection to another K gluon. We will call these commutator terms between two K gluons.

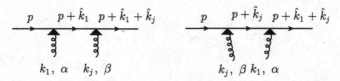

Fig. 11.12. Two ways of attaching neighboring K gluons to a quark line. α and β are color indices.

Beyond the external-line terms already noted, we have missing terms involving other K gluons. The generic case is typified by Fig. 11.9(c), whenever the ghost line directly corresponds to the k_1 gluon without any interactions and the explicit gluon is a K gluon. The missing term is the third term in Fig. 11.9(c); it would arise from attaching the two K gluons together before they enter the H subgraph, which is one of the disallowed situations. A possible notation for the general case is Fig. 11.11.

To see the meaning of the term, consider the example in Fig. 11.12. The first gluon, to which we are applying Ward identities, is k_1 and it attaches to a quark line. Next to it is another K gluon, of momentum k_j. The sum of the two (approximated) graphs is

$$\frac{i}{\not{p}+\hat{k}_1+\hat{k}_j}\left[\frac{-igt_\beta\hat{k}_j}{k_j\cdot n_2}\frac{i}{\not{p}+\hat{k}_1}\frac{-igt_\alpha\hat{k}_1}{k_1\cdot n_2}+\frac{-igt_\alpha\hat{k}_1}{k_1\cdot n_2}\frac{i}{\not{p}+\hat{k}_j}\frac{-igt_\beta\hat{k}_j}{k_j\cdot n_2}\right]\frac{i}{\not{p}}. \tag{11.25}$$

We apply the basic line identity Fig. 11.9 to gluon k_1, by writing $\hat{k}_1 = \not{p}+\hat{k}_1 - \not{p}$ in the first term and $\hat{k}_1 = (\not{p}+\hat{k}_1+\hat{k}_j) - (\not{p}+\hat{k}_j)$ in the second. We retain only the terms where the quark propagator between the gluons is canceled; this corresponds to a case of Fig. 11.11, and gives

$$\frac{i}{\not{p}+\hat{k}_1+\hat{k}_j}\frac{ig^2[t_\alpha,t_\beta]\hat{k}_j}{k_1\cdot n_2\, k_j\cdot n_2}\frac{i}{\not{p}}=\frac{i}{\not{p}+\hat{k}_1+\hat{k}_j}(-igt_\gamma\hat{k}_j)\frac{i}{\not{p}}\frac{-igf_{\alpha\beta\gamma}}{k_1\cdot n_2\, k_j\cdot n_2}. \tag{11.26}$$

This term is of the form of a vertex for one gluon times some factor associated with the gluon pair. This factor is the composite vertex in Fig. 11.11, and is the same in all these situations.

We would like to apply a Ward identity to this new object. To do this we observe that the approximated momenta are parallel, and that $k_j\cdot n_2 = \hat{k}_j\cdot n_2$. Thus we can scale the numerator and denominator at the vertex so that the gluon-quark interaction is contracted

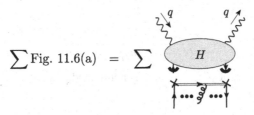

$$\sum \text{Fig. 11.6(a)} \ = \ \sum$$

Fig. 11.13. After a sum over graphs in Fig. 11.6(a), the K gluons attach to a Wilson line appropriate for a gauge-invariant parton density.

with $\hat{k}_1 + \hat{k}_j$ instead of $\hat{k}_j$, to obtain

$$\frac{i}{\not{p} + \hat{\not{k}}_1 + \hat{\not{k}}_j} \frac{-ig t_\gamma (\hat{\not{k}}_1 + \hat{\not{k}}_j)}{(k_1 + k_j) \cdot n_2} \frac{i}{\not{p}} \frac{-ig f_{\alpha\beta\gamma}}{k_1 \cdot n_2}. \tag{11.27}$$

We now have exactly a factor for an effective K gluon of momentum $k_1 + k_j$ times some other factor, and the new H factor has one K gluon less than before. So we keep reapplying the basic Ward-identity argument until all the K gluons have been extracted from H. At each intermediate stage we have a result of the form of Fig. 11.10. At the final stage we just have some external line factors, multiplying a version of H with no extra gluons at all.

However, we just know that the external-line factors are composed of many Wilson-line denominators, of a factor g for each gluon, and a product of SU(3) structure constants. But it could just be a complicated mess, not the desired Wilson line. What we do know is that the external-line factors depend only on the color of the external lines; they do not depend on the other details of H. So the same argument applies when we replace H by a Wilson line in direction n_2, and it gives the same external-line factors. Therefore the external-line factors are exactly a Wilson line, i.e., we get Fig. 11.13, our final result.

11.9.3 Unapproximated gluon momenta

Notice that because the approximated momenta of the K gluons, $\hat{k}_j$, are all parallel, the transition from (11.26) to (11.27) is exact, as are the Slavnov-Taylor-Ward identities. So the conversion to a Wilson-line form is exact, with no left-over terms.

If instead we had chosen to leave the k_j momenta unapproximated in H, as in the original Grammer-Yennie paper, then we would have extra remainder terms. These would be power-suppressed in the collinear limit for all the K gluons. But when we integrate over all momenta, as in defining a parton density, then the remainder terms would be important outside the collinear region and would have to be taken into account. This would evidently be rather complicated if we tried to do it in general.

However, the use of unapproximated momenta can help at intermediate stages when dealing with super-leading contributions. In that case we would replace the vertices in (11.26) and (11.27) by

$$(-ig t_\gamma \not{k}_j) \frac{-ig f_{\alpha\beta\gamma}}{k_1 \cdot n_2 \, k_j \cdot n_2}, \tag{11.28}$$

and

$$\frac{-igt_\gamma(\not{k}_1 + \not{k}_j)}{(k_1 + k_j) \cdot n_2} \frac{-igf_{\alpha\beta\gamma}}{k_1 \cdot n_2}. \tag{11.29}$$

The difference is the part of (11.28) which is not treated recursively as a K gluon. It has a rather symmetrical form:

$$(-igt_\gamma\gamma_\mu)\frac{-igf_{\alpha\beta\gamma}}{(k_1 + k_j) \cdot n_2}\left[\frac{k_j^\mu}{k_j \cdot n_2} - \frac{k_1^\mu}{k_1 \cdot n_2}\right]. \tag{11.30}$$

It is easily checked that, in the collinear limit, this is suppressed relative to the main effective K term (11.29) by a power of small components of k_1 and k_j relative to large components. Expression (11.30) is written as a factor of a quark-gluon vertex times a factor. It can be checked that this factor is just derived from the commutator term in Fig. 11.11, so it applies in all situations, not just to the coupling to a quark, but to a gluon or a ghost.

11.9.4 Including the case of all-gluon connections to H

Our proof of Fig. 11.13 applied directly to the case that the hard scattering is initiated by a quark and antiquark together with any number of K gluons.

The same principles apply when there are two G gluons plus any number of K gluons. But we also have to deal with the cases with super-leading contributions graph-by-graph. These are where (a) all the external lines of H are K gluons, and (b) one of the external lines is a G gluons and the rest K gluons.

When all the external lines are all K gluons, we start by applying our argument as it stands. Since there are no non-K gluons, we get exactly zero. Thus the strongest super-leading term vanishes.

But there can be a leading remainder. In general, such a remainder is to be obtained by differentiating the hard scattering with respect to the small components of the collinear momenta, e.g., k_{jT}. This would be complicated to deal with.

We rescue the situation by starting by applying the Ward-identity argument with unapproximated momenta. We start accumulating remainder terms (11.30) from the commutators. These we treat as generalizations to the G-gluon definition. Once we have two of them, we have a suppression from the Q^2 super-leading power back to an ordinary leading power. At that point, the necessary suppression is exhibited in external factors, as in (11.30). From that point on, we can restore the approximations for the k_js in the hard scattering, and drop any further accumulation of remainder terms.

We apply the same argument when there is one G gluon.

The end product of the argument is a hard scattering with either one or two external gluons containing factors appropriate for generalized G gluons. With one G gluon the hard scattering after application of the Ward-identity argument has a single external gluon; it therefore vanishes by color invariance. So we are left with two.

Application of the same argument to a product of two gluon field strength tensors joined by a Wilson line gives the same external factors. So the result should agree with the Feynman rules for an ordinary gluon density. A more explicit argument would be useful.

Fig. 11.14. Example of graph for a hard scattering that is one-particle reducible. But it is irreducible separately in the collinear group A of the lower two external lines and in the collinear group B of the upper two external lines. All external lines are amputated.

11.9.5 Generalizations

The proof in Sec. 11.9.2 generalizes in a relatively elementary way. We will deal with the modifications as needed. This notably concerns the soft-to-collinear case, which has some interesting differences, and which we will deal with in Sec. 12.8.3.

Relatively elementary generalizations concern different kinds of hard subgraph, and we summarize them here.

Trivial generalizations concern changing the nature or kinematics of the currents in Fig. 11.6. As long as they are gauge invariant (or even just BRST invariant), no change in the derivation is needed. Similarly, when we go to a collinear amplitude, we will generally have some on-shell hadrons in the initial or final state. As long as no irreducibility requirements are imposed, these give no contribution to the Ward identities.

The most important generalization is to hard scatterings where there are multiple collinear groups. The Sudakov form factor treated in Ch. 10 gives one example. The Drell-Yan process is another. The first case we will treat in QCD will be hadronic cross sections in e^+e^- annihilation, in Ch. 12.

We now discuss the simplest case of its hard-scattering amplitude as shown in Fig. 11.6(b). There is one (gauge-invariant) electromagnetic current, and two collinear groups, A and B. In the particular case shown, the partonic lines (always on-shell) are a quark and an antiquark, and any number of collinear K gluons.

The irreducibility requirements on H are now that it is irreducible in each group A and B separately; otherwise there would be an internal line of H forced to be collinear. But there is no restriction on combining lines from different collinear groups. An example is in Fig. 11.14.

For Fig. 11.6(b), we will use K gluons defined with the same collinear approximants that we defined in Sec. 10.4.2 for the Sudakov form factor. These have light-like auxiliary vectors w_1 and w_2. The external quark and antiquark lines are defined so that they are on-shell with appropriate Dirac wave functions.

We start by applying Ward identities to a first K gluon from group A. As in DIS, the restriction to graphs irreducible in the A lines implies we obtain terms associated with the other collinear-A lines. Repeated application of the Ward-identity argument converts these gluons into couplings to a Wilson line. This gives the middle graph in Fig. 11.15.

Some of the would-be cancellations involve moving the K gluons so as to uncover a graph irreducible in the B group. This is the same as for the Sudakov form factor in Sec. 10.8.6 with the one change that we can also implicate graphs involving pairs of K

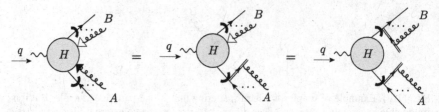

Fig. 11.15. Result of applying Ward identities to the K gluons of the A group, and then to the K gluons of the B group. The A gluons couple to a Wilson line in direction n_B, i.e., with rapidity $-\infty$. The B gluons couple to a Wilson line in direction n_A, i.e., with rapidity $+\infty$.

gluons from the B side as well as a K gluon and a quark. A typical non-canceling term of this kind is Fig. 10.20(c). But all of these terms are actually zero, because they involve a light-like collinear-B momentum from a K gluon contracted into a non-singular subgraph with momenta all in the B direction.

In the case that we use massive external quarks, the cancellation is good to order m^2/Q^2, because the quark momentum differs by this much from being light-like. This is good enough to leading power.

So there are no extra terms, and the K gluons convert to a Wilson line.

Exactly the same argument applies for the B group, finally giving Fig. 11.15. This is the same result as with the Sudakov form factor, generalized to a non-abelian theory.

Most importantly, the same argument applies with minor changes when there are multiple collinear groups, and it also applies when some of the groups consist of all gluons (up to one G gluon). Each group α has its direction defined by a light-like vector w_α, together with a conjugate light-like vector $\tilde{w}_\alpha$. We can define

$$\tilde{w}_\alpha^\mu = \frac{q^\mu}{w_\alpha \cdot q} - \frac{w_\alpha^\mu q^2}{2(w_\alpha \cdot q)^2}. \tag{11.31}$$

This is a future-pointing light-like vector with its spatial component reversed compared to w_α. It is also arranged so that $w_\alpha \cdot \tilde{w}_\alpha = 1$, which means that contracting a vector v onto $\tilde{w}_\alpha$ and w_α can be regarded as giving light-front plus and minus coordinates appropriate for the collinear group (i.e., $v^+ = v \cdot \tilde{w}_\alpha$ and $v^- = v \cdot w_\alpha$).

The approximant for a K gluon of group α to couple to the hard subgraph is a case of (11.22):

$$H(k_{\alpha i})^\mu C_\alpha(k)_\mu = H(\hat{k}_{\alpha i})^\mu \frac{\hat{k}_{\alpha i,\,\mu} \tilde{w}_\alpha^\nu}{k_{\alpha i} \cdot \tilde{w}_\alpha} C_\alpha(k)_\nu, \tag{11.32a}$$

where

$$\hat{k}_{\alpha i}^\mu = \frac{w_\alpha^\mu \, k_{\alpha i} \cdot \tilde{w}_\alpha}{w_\alpha \cdot \tilde{w}_\alpha}. \tag{11.32b}$$

The end result is that we have a hard scattering with one external on-shell line for each collinear group, and that we have a non-local vertex for each collinear group to attach to.

The non-local vertex is a natural generalization of what we have already seen in particular cases: a partonic field times a Wilson line that now goes out to infinity.

Exercises

11.1 (No stars to (***)) Examine the proofs for further weaknesses, and correct them as best you can.

11.2 ((*) to (***)) One part of our proof of factorization for DIS in QCD relied on BRST properties of operators used in defining parton densities. The published proofs of these BRST properties (Joglekar and Lee, 1976; Joglekar, 1977a, b; Nakanishi and Ojima, 1990) are restricted to *local* operators, i.e., those that are products of field operators at a single space-time point. Work through these proofs (and check them!). Do they apply more generally, to the non-local operators defining parton densities? Why? If not, construct (and publish) a correct proof.

11.3 (***) Characterize the non-gauge-invariant operators that appear in DIS on an off-shell target, with a particular emphasis on finding those that include Faddeev-Popov ghost fields.

It might be worth starting with some sample Feynman-graph calculations to get some inspiration. You should try to verify that the operators are BRST invariant.

From Fig. 11.4, you can see that the lowest-order graphs for ghost-induced hard scattering in ordinary DIS have two loops, which will probably make calculational examples hard. So you might also want to investigate a generalization of DIS in which the electromagnetic current operators are replaced by a gauge-invariant gluon operator, e.g., $G^2_{\mu\nu}$. You could also examine renormalization of the gluon density, since the counterterms generally use the operator matrix elements as those used for parton densities in factorization.

11.4 (***) Particularly if you have a sufficiently good theory of the necessary operators, it would be useful to examine renormalization, especially, one order beyond the lowest order where Faddeev-Popov ghosts appear. Examine the implications for calculations of the DGLAP kernel etc., and compare with Hamberg and van Neerven (1992) and Collins and Scalise (1994).

12

Fragmentation functions: e^+e^- annihilation to hadrons, and SIDIS

We now extend our treatment of factorization to cover the distribution of final-state hadrons in hard processes.

The simplest process is the cross section for inclusive one-hadron production in e^+e^- annihilation, Fig. 12.1,

$$e^+e^- \to \gamma^*(q) \to h(p) + X, \tag{12.1}$$

where the hadron is typically a pion, and, as usual, we work to lowest order in electroweak interactions. Kinematically the process is the same as inclusive DIS, except that certain particles are crossed between the initial and final states. We will see that the *structure* of the factorization theorem, (12.21), is the same as for DIS. The differences concern the time ordering of the process: the hard scattering is for a virtual photon to make a partonic state from a highly virtual time-like photon of momentum q. The place of a parton density is taken by a new object, called a fragmentation function; it represents the distribution of the final-state detected hadron resulting from an outgoing parton that originated in the hard scattering.

Our proof of factorization will introduce two important conceptual changes relative to DIS. The first is that we need a more detailed examination of soft gluon effects before we prove they cancel. This is because the detected hadron in the final state prevents the transition to an uncut amplitude that we used for DIS in Sec. 11.2. We will first use Ward identities to factorize the soft part, as we did for the Sudakov form factor in Ch. 10. Only after that can we employ the sum-over-cuts argument to obtain cancellation.

The second change relative to DIS arises from an issue we discussed in Ch. 4 that concerns final-state interactions. Our proof of factorization relies on the structure of leading regions as seen in Feynman graphs, which also correspond to certain regions in space-time. For our arguments to apply to full QCD, we need to assume that these regions are also correct after non-perturbative effects are included.

We will need to make the assumption that the "breakable string" picture of hadronization rather than the "unbreakable elastic spring" applies to QCD; see Sec. 4.3.1. (Of course our arguments would also apply if quarks and gluons were unconfined.) This assumption is abundantly supported by experimental data, but it is not (yet) derived from QCD. Associated with this will be a notable jump in the logic of the derivation.

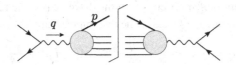

Fig. 12.1. Cross section for $e^+e^- \to h(p) + X$.

If, instead, the unbreakable spring picture had been correct, then the partonic state from the hard scattering would first form a resonance. In this case there would be no necessary correspondence between the directions of the decay products and of the partons, contrary to the measured situation with QCD jet physics.

For the DIS structure functions we evaded these issues by treating the *uncut* amplitude $T^{\mu\nu}$. But this option is no longer available to us now that we have a detected hadron in the final state.

After our treatment for the single-hadron-inclusive cross section in e^+e^- annihilation, we will have the methods necessary to treat other reactions. For example, factorization and its proof can readily be generalized to inclusive cross sections with more than one detected hadron. Another case is semi-inclusive deeply inelastic scattering (SIDIS). This is a cross section for DIS in which one (or more) particles is detected in the final state. The label SIDIS is generally applied to single particle production,

$$e + P \to e + \pi(p_B) + X, \tag{12.2}$$

although the same ideas apply to more general cases.

12.1 Structure-function analysis of one-particle inclusive cross section

12.1.1 Hadronic tensor

As with DIS, the squared amplitude for our process, Fig. 12.1, factorizes into a leptonic and a hadronic part. We use a structure-function analysis (Drell, Levy, and Yan, 1970) for the hadronic tensor, which is defined by

$$W^{\mu\nu}(q, p) \stackrel{\text{def}}{=} 4\pi^3 \sum_X \delta^{(4)}(p_X + p - q) \langle 0|j^\mu(0)|p, X, \text{ out}\rangle \langle p, X, \text{ out}|j^\nu(0)|0\rangle$$

$$= \frac{1}{4\pi} \sum_X \int d^4z \; e^{iq \cdot z} \langle 0|j^\mu(z/2)|p, X, \text{ out}\rangle \langle p, X, \text{ out}|j^\nu(-z/2)|0\rangle . \tag{12.3}$$

The primary difference compared with DIS is that the selected hadron is in the final state, so we cannot eliminate the $\sum_X$ as we did for DIS in (2.18), and we need to explicitly indicate that the states are out-states. The photon momentum q is now time-like. Our normalization conventions, both for the states and for $W^{\mu\nu}$, differ from those of Drell, Levy, and Yan (1970).

We define $Q = \sqrt{q^2}$, and we define the equivalent of the Bjorken variable by

$$x = 2p \cdot q/Q^2. \tag{12.4}$$

We interpret x as the center-of-mass energy of the detected particle relative to its maximum value $Q/2$ (when masses are neglected). The decomposition into structure functions is

$$W^{\mu\nu} = \left(-g^{\mu\nu} + \frac{q^{\mu}q^{\nu}}{q^2}\right) F_1(x, Q^2) + \frac{\left(p^{\mu} - \frac{q^{\mu}p\cdot q}{q^2}\right)\left(p^{\nu} - \frac{q^{\nu}p\cdot q}{q^2}\right)}{p\cdot q} F_2(x, Q^2), \quad (12.5)$$

which assumes current conservation, parity invariance, and that the detected hadron is either spinless or has its polarization states summed over. An F_3 structure function [cf. (7.3)] is needed if Z boson exchange is included, since then parity is violated.

The inclusive cross section is

$$E\frac{d\sigma}{d^3p} = \frac{2\alpha^2}{Q^6} L_{\mu\nu} W^{\mu\nu}, \tag{12.6}$$

where the leptonic tensor is

$$L^{\mu\nu} = l_1^{\mu}l_2^{\nu} + l_2^{\mu}l_1^{\nu} - g^{\mu\nu}l_1 \cdot l_2, \tag{12.7}$$

where l_1 and l_2 are the momenta of the incoming electron and positron, and where the electron mass is neglected. Hence

$$E\frac{d\sigma}{d^3p} = \frac{2\alpha^2}{Q^4}\sqrt{1 - \frac{4m^2}{Q^2x^2}}\left[F_1(x, Q) + \frac{x}{4}\left(1 - \frac{4m^2}{Q^2x^2}\right)\sin^2\theta F_2(x, Q)\right]$$

$$\simeq \frac{2\alpha^2}{Q^4}\left[F_1(x, Q) + \frac{x}{4}\sin^2\theta F_2(x, Q)\right]. \tag{12.8}$$

Here, m is the mass of the detected hadron, and θ is its angle relative to the electron in the center-of-mass frame. In the last line, m was neglected.

A standard presentation is

$$\frac{d\sigma}{dx\, d\cos\theta} = \frac{3}{8}(1 + \cos^2\theta)\frac{d\sigma_T}{dx} + \frac{3}{4}\sin^2\theta\frac{d\sigma_L}{dx}, \tag{12.9}$$

where

$$\frac{d\sigma_T}{dx} = \frac{4\pi\alpha^2}{3Q^2}x F_1(x, Q), \tag{12.10a}$$

$$\frac{d\sigma_L}{dx} = \frac{\pi\alpha^2}{3Q^2}\left[2x F_1(x, Q) + x^2 F_2(x, Q)\right]. \tag{12.10b}$$

12.1.2 Averaging with test function

To derive factorization, we will use certain cancellations generalizing those we saw in Sec. 4.1.1, for the total cross section for $e^+e^- \rightarrow$ hadrons, and in Sec. 11.2.1, for DIS. The cancellations involve terms that differ by whether particular lines are real or virtual, and thus by change of final state. The proof is clearest if a loop integral involving the momentum of a particular final-state particle can be routed out through the current vertex rather than back through other final-state particles. This can be done by averaging the hadronic tensor with a test function $f(q)$:

$$W^{\mu\nu}([f], p) \overset{\text{def}}{=} \int d^4q \, f(q) \, W^{\mu\nu}(q, p)$$

$$= 4\pi^3 \sum_X f(p_X + p) \langle 0 | j^\mu(0) | p, X, \text{out} \rangle \langle p, X, \text{out} | j^\nu(0) | 0 \rangle. \quad (12.11)$$

Here, the square brackets in $[f]$ act as a reminder that the argument is a whole function, and not just its value at one point. The integral over q has removed the momentum-conservation delta function from (12.3), and therefore allows the desired routing of integrals of final-state momenta.

The actual hadronic tensor is obtained by functional differentiation:

$$W^{\mu\nu}(q, p) = \frac{\delta W^{\mu\nu}([f], p)}{\delta f(q)}. \quad (12.12)$$

But the derivation of factorization only applies when the test function has a suitably slow dependence on Q, so that factorization generally only applies to a locally averaged quantity. If the actual hadronic tensor has smooth dependence on kinematic variables, the local average is unnecessary.

12.2 Statement of factorization etc. for $e^+e^- \to h(p) + X$

I now state the main results for factorization and fragmentation function evolution, all to be derived in later sections.

12.2.1 Factorization for cross section

The factorized cross section has the form

$$E \frac{d\sigma(e^+e^- \to h(p) + X)}{d^3 p} = \sum_j \int_{x-}^{1+} \frac{dz}{z^2} \, E_k \frac{d\hat\sigma_j}{d^3 k} \, d_{h/j}(z; \mu). \quad (12.13)$$

As usual, this formula is valid up to corrections suppressed by a power of $1/Q$. We use $d\hat\sigma_j$ to denote the perturbatively calculable hard-scattering factor, normalized like the differential cross section for inclusive production of an on-shell massless parton of type j and 3-momentum k. Like the hard scattering in DIS, it must contain subtractions to prevent double counting between momentum regions. The 3-momenta, p and k, of the hadron and parton are made parallel in the overall center-of-mass frame, with a ratio z: $p = zk$.

The quantity $d_{h/j}(z; \mu)$ is the fragmentation function, whose exact definition we will give later; we approximate its meaning as the number density to find hadron h in the jet initiated by parton j, with the hadron having a fraction z of the parton's momentum. The reality of the jets is evidenced by pictures like Fig. 5.10, which shows an event in DIS.

As with parton densities, this intuitive meaning only applies literally in a super-renormalizable non-gauge theory. After applying correct definitions and derivations in

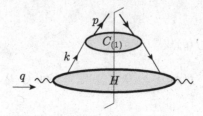

Fig. 12.2. Structure of factorization for one-hadron-inclusive cross section in e^+e^- annihilation.

QCD, the factorization formula will remain correct, but the number/probability interpretation will not be really correct. As with parton densities, there is a DGLAP evolution equation for the fragmentation function. It lets us set the scale μ of the fragmentation functions to be of order the experimentally dependent quantity Q, so that the hard scattering is calculable in fixed-order perturbation theory.

The factor $1/z^2$ in (12.13) arises because of the change of variable between the hadron momentum on the l.h.s. and the parton momentum on the r.h.s.: with neglect of masses $d^3p/E_p = z^2 d^3k/E_k$. If we were to use n space-time dimensions, the factor $1/z^2$ would be replaced by $1/z^{n-2}$.

The structure of factorization is illustrated in Fig. 12.2. The hard scattering H makes two or more final-state partons. One of these of momentum k goes into the fragmentation subgraph, which is labeled $C_{(1)}$ to be consistent with a notation used later. The details of the proof of factorization will show that Fig. 12.2 is misleadingly simple. There will be non-trivial cancellations to be proved before we obtain the factorized structure. The figure also omits reference to the Wilson lines in the definitions of the fragmentation functions.

Where fragmentation functions get used

The most basic situation for using a fragmentation function is the one-hadron-inclusive cross section in e^+e^- annihilation, just described.

Straightforward generalizations of the factorization theorem apply to semi-inclusive DIS (e.g., $e + p \to e + \pi + X$), to multiple hadron production in e^+e^- annihilation, and to production of hadrons of high transverse momentum in hadron-hadron collisions. All of these factorization theorems use fragmentation functions for the detected final-state hadrons; the same fragmentation functions for all these processes. Global fits have been performed in de Florian, Sassot, and Stratmann (2007); Albino, Kniehl, and Kramer (2008).

Terminology

The Particle Data Group (Amsler *et al.*, 2008, p. 202) uses the term "fragmentation function" to refer to both a partonic fragmentation function, as in (12.13), and to the following normalized cross section:

$$F^h(x, Q^2) = \frac{1}{\sigma_{\text{tot}}} \frac{d\sigma}{dx}(e^+e^- \to hX). \tag{12.14}$$

I find it preferable to distinguish these concepts, for the same reasons that the concepts of "structure function" and "parton density" should be distinguished in DIS. So I only use "fragmentation function" to refer to a partonic fragmentation function.

12.2.2 Renormalization and DGLAP evolution of fragmentation functions

As in the case of parton densities, the basic definitions of fragmentation functions have UV divergences. The factorization formula uses renormalized fragmentation functions obtained from bare fragmentation functions by formulae of the form

$$d_{h/j}(z; \mu) = \lim_{\epsilon \to 0} \sum_{j'} \int_{z-}^{1+} \frac{d\rho}{\rho} \, d_{(0)h/j'}(z/\rho) \, L_{j'j}(\rho; g(\mu), \epsilon). \tag{12.15}$$

(Here, as usual $\epsilon = 2 - n/2$, where n is the space-time dimension.)

There follow DGLAP evolution equations, of the same form (8.30) as for parton densities:

$$\frac{d}{d\ln\mu} d_{h/j}(z; \mu) = 2\sum_{j'} \int_{z-}^{1+} \frac{d\rho}{\rho} \, d_{h/j'}(z/\rho; \mu) \, P_{j'j}(\rho; g(\mu)). \tag{12.16}$$

See Sec. 12.10 below for a calculation that shows the LO kernels have the same value as for parton densities. The finite evolution kernels are obtained from the renormalization factors:

$$\frac{d}{d\ln\mu} L_{j'j}(\rho; g(\mu), \epsilon) = 2\sum_{j''} \int \frac{d\rho'}{\rho'} \, L_{j'j''}(\rho/\rho'; g, \epsilon) \, P_{j''j}(\rho'; g, \epsilon). \tag{12.17}$$

Just as with parton densities, the convolutions in the above equations turn into multiplications for moments. Let us define

$$\tilde{d}_{h/j}(n; \mu) = \int_0^1 dz \, z^{n-1} d_{h/j}(z; \mu), \quad \tilde{L}_{j'j}(n; \mu) = \int_0^1 d\rho \, \rho^{n-1} L_{j'j}(\rho; \mu), \tag{12.18}$$

etc. Then the renormalization and DGLAP equations are

$$\tilde{d}_{h/j}(n; \mu) = \lim_{\epsilon \to 0} \sum_{j'} \tilde{d}_{(0)h/j'}(n) \, \tilde{L}_{j'j}(n; g(\mu), \epsilon), \tag{12.19}$$

$$\frac{d}{d\ln\mu} \tilde{d}_{h/j}(n; \mu) = 2\sum_{j'} \tilde{d}_{h/j'}(n; \mu) \, \tilde{P}_{j'j}(n; g(\mu)). \tag{12.20}$$

12.2.3 Factorization for hadronic tensor

We convert (12.13) to a factorization for $W^{\mu\nu}$:

$$W^{\mu\nu}(p, q) = \sum_j \int_{x-}^{1+} \frac{dz}{z^2} \, d_{h/j}(z; \mu) \, C_j^{\mu\nu}(\hat{k}, q; g(\mu), \mu). \tag{12.21}$$

The hard-scattering tensor $C_j^{\mu\nu}$ is just like $W^{\mu\nu}$, except that it is at the partonic level, and is defined with double-counting subtractions to remove non-short-distance contributions.

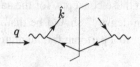

Fig. 12.3. Lowest-order partonic graph for $e^+e^- \to \text{quark}_j(\hat{k}) + X$.

It uses an approximated parton momentum

$$\hat{k} = (p^+/z, 0, \mathbf{0}_{\mathrm{T}}), \tag{12.22}$$

where we use light-front coordinates such that $p = (p^+, m^2/2p^+, \mathbf{0}_{\mathrm{T}})$.

12.2.4 Projection onto structure functions

We define partonic structure functions $\hat{F}_{ij}$ by applying (12.5) to $C_j^{\mu\nu}$:

$$C_j^{\mu\nu} = \left(-g^{\mu\nu} + \frac{q^\mu q^\nu}{q^2}\right) \hat{F}_{1j}(x/z, Q^2) + \frac{\left(\hat{k}^\mu - \frac{q^\mu \hat{k}\cdot q}{q^2}\right)\left(\hat{k}^\nu - \frac{q^\nu \hat{k}\cdot q}{q^2}\right)}{\hat{k}\cdot q} \hat{F}_{2j}(x/z, Q^2). \tag{12.23}$$

Hence we get factorization formulae for the structure functions:

$$F_1(x, Q^2) = \sum_j \int_{x-}^{1+} \frac{\mathrm{d}z}{z^2} d_{h/j}(z; \mu) \, \hat{F}_{1j}(x/z, Q^2), \tag{12.24a}$$

$$F_2(x, Q^2) = \sum_j \int_{x-}^{1+} \frac{\mathrm{d}z}{z^3} d_{h/j}(z; \mu) \, \hat{F}_{2j}(x/z, Q^2), \tag{12.24b}$$

and similarly for the transverse and longitudinal cross sections:

$$\frac{\mathrm{d}\sigma_T}{\mathrm{d}x} = \sum_j \int_{x-}^{1+} \frac{\mathrm{d}z}{z} d_{h/j}(z; \mu) \frac{\mathrm{d}\hat{\sigma}_{T,j}(x/z)}{\mathrm{d}(x/z)}, \tag{12.25a}$$

$$\frac{\mathrm{d}\sigma_L}{\mathrm{d}x} = \sum_j \int_{x-}^{1+} \frac{\mathrm{d}z}{z} d_{h/j}(z; \mu) \frac{\mathrm{d}\hat{\sigma}_{L,j}(x/z)}{\mathrm{d}(x/z)}. \tag{12.25b}$$

12.3 LO calculation

Even without a proof of factorization we can see how to get the lowest order of the hard scattering. As in DIS, we just need the LO calculation of one-parton-inclusive scattering, from the graph of Fig. 12.3:

$$\begin{aligned}
C_j^{\mu\nu} &= \frac{e_j^2 N_c}{4\pi} \int \frac{\mathrm{d}^3 k_2}{(2\pi)^3 2|k_2|} (2\pi)^4 \delta^{(4)}(q - \hat{k} - k_2) \operatorname{Tr} \gamma^\mu \hat{\not{k}} \gamma^\nu \not{k}_2 \\
&= e_j^2 N_c \delta(x/z - 1)\left[-g^{\mu\nu} - \frac{2}{Q^2}(2\hat{k}^\mu \hat{k}^\mu - q^\mu \hat{k}^\nu - \hat{k}^\mu q^\nu)\right]. \tag{12.26}
\end{aligned}$$

We have included a factor N_c, since we always sum over parton color in the final state. (The fragmentation function will have a color average.) In the first line, we integrate over the momentum k_2 of what we can term the unobserved parton, the one not associated with the observed hadron. In the calculation of the hard scattering the external partons are set on-shell, whereas the actual parton momenta are off-shell. In particular, $\hat{k}$ denotes the approximated on-shell momentum given in (12.22).

We deduce the partonic structure functions

$$\hat{F}_{1j} = e_j^2 N_c \delta(x/z - 1) + O(\alpha_s), \quad \hat{F}_{2j} = -2e_j^2 N_c \delta(x/z - 1) + O(\alpha_s), \quad (12.27)$$

which apply to both quarks and antiquarks.

From the factorization formula, we deduce that at the hadronic level

$$F_1 = \frac{1}{x} \sum_{\text{quarks } j} e_j^2 N_c \left(d_{h/j}(x) + d_{h/\bar{j}}(x) \right) + O(\alpha_s), \quad (12.28a)$$

$$F_2 = \frac{-2}{x^2} \sum_{\text{quarks } j} e_j^2 N_c \left(d_{h/j}(x) + d_{h/\bar{j}}(x) \right) + O(\alpha_s). \quad (12.28b)$$

We see that at lowest order an analog of the Callan-Gross relation applies: $F_2 = -2x^{-1}F_1$. Therefore the angular distribution of the hadron is given by the same $1 + \cos^2 \theta$ factor as for the elementary $e^+e^- \to q\bar{q}$ process:

$$E\frac{d\sigma}{d^3 p} = \frac{\alpha^2}{Q^4 x}(1 + \cos^2 \theta) \sum_{\text{quarks } j} e_j^2 N_c \left(d_{h/j}(x) + d_{h/\bar{j}}(x) \right) + O(\alpha_s). \quad (12.29)$$

From the inclusive cross section formula (12.8) and the total cross section formulae in Sec. 4.1, we find that the normalized distribution in x directly reflects the values of the fragmentation functions, weighted by quark charge squared:

$$\frac{d\sigma(e^+e^- \to hX)/dx}{\sigma(e^+e^- \to \text{hadrons})} = \frac{\sum_{\text{quarks } j} e_j^2 \left(d_{h/j}(x) + d_{h/\bar{j}}(x) \right)}{\sum_{\text{quarks } j} e_j^2} + O(\alpha_s). \quad (12.30)$$

12.4 Introduction to fragmentation functions

The intuitive idea of a fragmentation function to represent the number density for hadrons in the jet induced by a parton is quite natural. Light-front quantization, which we studied in Sec. 6.6, gives a natural first attempt at a formal definition of fragmentation functions that directly implements the desired distribution. We now present these definitions, which are quite simple. They provide an orientation for the more complicated results in QCD.

Since the number interpretation depends on the use of the canonical commutation relations for bare fields, the definitions in this section are for bare fragmentation functions. Bare quantities are denoted by a subscript "(0)". For a statement of the renormalization properties, see Sec. 12.2.2.

12.4.1 Kinematics

Our implementation uses two different coordinate frames, called the hadron and parton frames, with components denoted by subscripts h and p respectively. Thus for a vector V, we write

$$V_h = \left(V_h^+, V_h^-, \boldsymbol{P}_{h\mathrm{T}}\right), \quad V_p = \left(V_p^+, V_p^-, \boldsymbol{P}_{p\mathrm{T}}\right), \tag{12.31}$$

in the hadron and parton frames. We will arrange that the plus components are the same in both frames, so that for this component we can drop the frame's subscripts: $V_h^+ = V_p^+ = V^+$.

The hadron frame is the one already used where the detected hadron has zero transverse momentum: $p_h = \left(p^+, m^2/(2p^+), \boldsymbol{0}_\mathrm{T}\right)$. The actual parton momentum, as in Fig. 12.2, has non-zero transverse momentum:

$$k_h = \left(k^+, k_h^-, \boldsymbol{k}_{h\mathrm{T}}\right) = \left(p^+/z, k_h^-, \boldsymbol{k}_{h\mathrm{T}}\right). \tag{12.32}$$

Of course, the approximated parton has zero transverse momentum: $\hat{k}_h = (k^+, 0, \boldsymbol{0}_\mathrm{T}) = (p^+/z, 0, \boldsymbol{0}_\mathrm{T})$.

For defining the number density of hadrons in the jet induced by a parton of momentum k, we need a frame in which it is the parton that has zero transverse momentum, and the hadron has non-zero transverse momentum. For this we use the parton frame, defined to be obtained from the hadron frame by the following Lorentz transformation:

$$V_p^+ = V_h^+,$$

$$V_p^- = \frac{k_{h\mathrm{T}}^2}{2(k^+)^2} V_h^+ + V_h^- - \frac{\boldsymbol{k}_{h\mathrm{T}}}{k^+} \cdot \boldsymbol{V}_{h\mathrm{T}}, \tag{12.33}$$

$$\boldsymbol{V}_{p\mathrm{T}} = -\frac{\boldsymbol{k}_{h\mathrm{T}}}{k^+} V_h^+ + \boldsymbol{V}_{h\mathrm{T}},$$

a Lorentz transformation that changes as k varies. Hence the parton-frame components are

$$k_p = \left(k^+, k_p^-, \boldsymbol{0}_\mathrm{T}\right) = \left(k^+, k_h^- - k_{h\mathrm{T}}^2/(2k^+), \boldsymbol{0}_\mathrm{T}\right), \tag{12.34a}$$

$$p_p = \left(zk^+, \frac{m^2 + k_{h\mathrm{T}}^2 z^2}{2zk^+}, -z\boldsymbol{k}_{h\mathrm{T}}\right). \tag{12.34b}$$

Note carefully the factor of z and the reversed sign between the parton transverse momentum in the one frame and the hadron transverse momentum in the other frame.

Although the parton frame is a natural one for defining fragmentation functions as number densities, it is inconvenient for derivations of factorization. The problem is that, in a physical process, there is an integral over parton momentum, and so the parton-frame axes are not fixed. Neither parton momenta nor the resulting parton-frame axes can be determined from experimentally measured quantities. Therefore we will express the definitions of fragmentation functions in hadron-frame coordinates. In the derivation of factorization, we will use a hadron frame defined in terms of measured quantities.

12.4.2 General definition of fragmentation function

We define the fragmentation function as the number density for finding a hadron of flavor h in a parton of flavor j, given the hadron's fractional plus momentum, and its transverse momentum:

$$
d_{(0)h/j}(z, \boldsymbol{p}_{p\mathrm{T}}) \langle j; k_1 | j; k_2 \rangle \overset{\text{def}}{=} \frac{\mathrm{Tr_{color}}}{N_{c,j}} \langle j; k_1 | j; k_2 \rangle \frac{\mathrm{d}N_{j/h}}{\mathrm{d}z\, \mathrm{d}^{n-2}\boldsymbol{p}_{p\mathrm{T}}}
$$

$$
= \frac{\mathrm{Tr_{color}}}{N_{c,j}} \frac{1}{2z(2\pi)^{n-1}} \sum_X \langle j, k_1 | p, X, \text{out} \rangle \langle p, X, \text{out} | j, k_2 \rangle .
$$

$$
(12.35)
$$

Here $|j, k_1\rangle$ denotes a partonic state created by a light-front creation operator, i.e., $a^\dagger_{j, k^+_{1,p}, \boldsymbol{k}_{1,p\mathrm{T}}} |0\rangle$. A number density is obtained from matrix elements of normalized states for partons and hadrons. Since we wish to use momentum eigenstates, we bring in an off-diagonal matrix element in (12.35). To cover the generalization to QCD, we define the fragmentation function to include an average over parton color. We let $N_{c,j}$ be the number of colors for field ϕ_j; in QCD it is 3 for a quark and 8 for a gluon. Then the color average is implemented as $1/N_{c,j}$ times a trace over color indices for the parton.

Next we use formulae like (6.64a) to express the annihilation and creation operators in terms of fields, and use methods similar to those used for the pdf in Sec. 6.7.3.

12.4.3 Scalar quark

For a scalar quark we get

$$
d_{(0)h/j}(z, \boldsymbol{p}_{p\mathrm{T}}) = \frac{\mathrm{Tr_{color}}}{N_{c,j}} \sum_X \frac{k^+}{z} \int \frac{\mathrm{d}x^-_p\, \mathrm{d}^{n-2}\boldsymbol{x}_{p\mathrm{T}}}{(2\pi)^{n-1}} e^{ik^+x^-_p}
$$

$$
\times \langle 0 | \phi^{(0)}_j(x/2) | p, X, \text{out} \rangle \langle p, X, \text{out} | \phi^{(0)}_j(-x/2)^\dagger | 0 \rangle , \qquad (12.36)
$$

where $x_p = (0, x^-_p, \boldsymbol{x}_{p\mathrm{T}})$. In the hadron frame, we get

$$
d_{(0)h/j}(z, -z\boldsymbol{k}_{h\mathrm{T}}) = \frac{\mathrm{Tr_{color}}}{N_{c,j}} \sum_X \frac{k^+}{z} \int \frac{\mathrm{d}x^-_h\, \mathrm{d}^{n-2}\boldsymbol{x}_{h\mathrm{T}}}{(2\pi)^{n-1}} e^{ik^+x^-_h - i\boldsymbol{k}_{h\mathrm{T}}\cdot\boldsymbol{x}_{h\mathrm{T}}}
$$

$$
\times \langle 0 | \phi^{(0)}_j(x/2) | p, X, \text{out} \rangle \langle p, X, \text{out} | \phi^{(0)}_j(-x/2)^\dagger | 0 \rangle
$$

$$
= \frac{\mathrm{Tr_{color}}}{N_{c,j}} \frac{k^+}{z} \int \frac{\mathrm{d}k^-_h}{(2\pi)^n}
$$

$$
(12.37)
$$

where the vector x has hadron-frame components $x_h = (0, x_h^-, \mathbf{x}_{hT})$. The integrated fragmentation function is

$$d_{(0)h/j}(z) \overset{\text{def}}{=} \int d^{n-2}\mathbf{p}_{pT}\, d_{h/j}(z, \mathbf{p}_{pT})$$

$$= \frac{\text{Tr}_{\text{color}}}{N_{c,j}} \sum_X k^+ z^{n-3} \int \frac{dx_h^-}{2\pi} e^{ik^+ x_h^-}$$

$$\times \langle 0|\phi_j^{(0)}(x/2)|p, X, \text{out}\rangle \langle p, X, \text{out}|\phi_j^{(0)}(-x/2)^\dagger|0\rangle$$

$$= \frac{\text{Tr}_{\text{color}}}{N_{c,j}} k^+ z^{n-3} \int \frac{dk_h^-\, d^{n-2}\mathbf{k}_{hT}}{(2\pi)^n} \quad \text{} \tag{12.38}$$

where now $x_h = (0, x_h^-, \mathbf{0}_T)$. The factor z^{n-3} is perhaps unexpected, given the corresponding formula (6.124) for a parton density. It arises from the normalization of the hadron state, and then from the change of variable from the parton to the hadron frame.

12.4.4 Unpolarized Dirac quark

For a Dirac field, the derivations are readily modified to give

$$d_{(0)h/j}(z, \mathbf{p}_{pT}) = \frac{\text{Tr}_{\text{color}} \text{Tr}_{\text{Dirac}}}{N_{c,j}} \frac{1}{4} \sum_X \frac{1}{z} \int \frac{dx_h^-\, d^{n-2}\mathbf{x}_{hT}}{(2\pi)^{n-1}} e^{ik^+ x_h^- - i\mathbf{k}_T \cdot \mathbf{x}_{hT}}$$

$$\times \langle 0|\gamma^+ \psi_j^{(0)}(x/2)|p, X, \text{out}\rangle \langle p, X, \text{out}|\bar{\psi}_j^{(0)}(-x/2)|0\rangle$$

$$= \frac{\text{Tr}_{\text{color}} \text{Tr}_{\text{Dirac}}}{N_{c,j}} \frac{1}{4} \frac{1}{z} \int \frac{dk_h^-}{(2\pi)^n} \gamma^+ \quad \text{} \tag{12.39}$$

Here we assume we work with an unpolarized situation, where there is a spin average on the quark; this gives a factor $\text{Tr}_{\text{Dirac}}/2$. There is another factor of $\frac{1}{2}$ that arises in the same way as for the pdf for a Dirac parton. The corresponding integrated fragmentation function is

$$d_{(0)h/j}(z) = \frac{\text{Tr}_{\text{color}} \text{Tr}_{\text{Dirac}}}{N_{c,j}} \frac{1}{4} \sum_X z^{n-3} \int \frac{dx_h^-}{2\pi} e^{ik^+ x_h^-}$$

$$\times \gamma^+ \langle 0|\psi_j^{(0)}(x/2)|p, X, \text{out}\rangle \langle p, X, \text{out}|\bar{\psi}_j^{(0)}(-x/2)|0\rangle$$

$$= \frac{\text{Tr}_{\text{color}} \text{Tr}_{\text{Dirac}}}{N_{c,j}} \frac{1}{4} z^{n-3} \int \frac{dk_h^-\, d^{n-2}\mathbf{k}_{hT}}{(2\pi)^n} \gamma^+ \quad \text{} \tag{12.40}$$

12.4.5 Unpolarized Dirac antiquark

The same approach works for fragmentation functions in an antiquark. The result is that (12.40) can be used to define the fragmentation function of an antiquark, just with the positions of the ψ and $\bar{\psi}$ field exchanged, and the natural change in the flow of indices.

12.4.6 Renormalization

In a renormalizable theory, the integral over k^- and $\mathbf{k}_T$ for an integrated fragmentation function has a UV divergence quite similar to that of a parton density. Renormalization works in the same way as for parton densities (Secs. 8.3 and 11.4). This leads to the statement of renormalization already given in (12.15).

12.4.7 Polarized Dirac quark

The hard scattering can generate a polarized quark, whose state is parameterized (Sec. 6.4.1) by a helicity λ and a transverse Bloch vector $\mathbf{s}_T$. To deal with the most general case, we make the following replacement in the definition of the fragmentation function (integrated or unintegrated):

$$\frac{1}{4}\mathop{\mathrm{Tr}}_{\mathrm{Dirac}} \gamma^+ \ldots \mapsto \frac{1}{4}\mathop{\mathrm{Tr}}_{\mathrm{Dirac}} \gamma^+(1 + \gamma_5\lambda - \gamma_5\boldsymbol{\gamma}_T \cdot \mathbf{s}_T)\ldots \qquad \text{(quark)}, \qquad (12.41a)$$

$$\frac{1}{4}\mathop{\mathrm{Tr}}_{\mathrm{Dirac}} \gamma^+ \ldots \mapsto \frac{1}{4}\mathop{\mathrm{Tr}}_{\mathrm{Dirac}} \gamma^+(1 - \gamma_5\lambda - \gamma_5\boldsymbol{\gamma}_T \cdot \mathbf{s}_T)\ldots \qquad \text{(antiquark)}, \qquad (12.41b)$$

These projections are applied in the *hadron frame*. They arise from the wave functions used in light-front quantization, which correspond to those for massless Dirac particles. They are obtained from (A.27), which is surrounded by factors of γ^+ from the formula (6.64b) that gives the light-front annihilation operator. This reverses the sign of the helicity term relative to (A.27). Note also the reversal of the sign of the helicity terms between the quark and antiquark cases. The definitions of λ and $\mathbf{s}_T$ are normalized to have a maximum value of unity (for a pure state).

The effects of polarization depend on the situation:

- For an integrated fragmentation function, the situation is like that for the pdfs. From a combination of parity invariance and conservation of angular momentum about the z axis, we find the following.
  - If the measured hadron is spinless, like a pion, or if its polarization is not detected, then there is no dependence on λ and $\mathbf{s}_T$. Only the unpolarized fragmentation function is non-zero; our original definition suffices. This is the most common case.
  - If the measured hadron has spin $\frac{1}{2}$, then there are polarized fragmentation functions comparable to the Δf and $\delta_T f$ parton densities, with the analogous interpretations.
  - It is possible to generalize the definition of the fragmentation function to have two (or more) nearby measured hadrons instead of one. In dihadron fragmentation the

azimuthal distribution of the hadrons can be correlated with the transverse spin of the quark, and an appropriate fragmentation function is defined (Collins, Heppelmann, and Ladinsky, 1994), for which measurements can be found in Airapetian *et al.* (2008); Vossen *et al.* (2009); Wollny (2009).

– For a spin-1 gluon, further possibilities arise which have not been explored in the literature.

• For a k_T-dependent unintegrated fragmentation function, angular momentum can be taken up by the azimuthal dependence of the measured hadron. The possibilities are described by several fragmentation functions:

– The unpolarized fragmentation function defined above, which gives a uniform distribution in the azimuthal angle of k_T or $p_{p\,T}$, i.e., this fragmentation function depends only on the size of the transverse momentum.

– The Collins function. This is obtained from the $\gamma^+\gamma_5\boldsymbol{\gamma}_T \cdot \boldsymbol{s}_T$ part of the trace (Collins, 1993). It gives a characteristic $\sin\phi$ dependence, where ϕ is the azimuthal angle of the hadron relative to the quark spin. See Boer (2008) for a review of recent theoretical and experimental work.

– Other possibilities involving a detected hardron polarization. These have undergone little investigation.

See Sec. 13.4.1 for more details.

12.4.8 Sum rules and symmetry properties

Momentum sum rule

In a fragmentation function, we have a state created by a light-front creation operator: $a^\dagger_{j,k^+,k_T} |0\rangle$, and we project it onto a particular final state $|p, X, \text{out}\rangle$. The total plus momentum in the final state is k^+. We can measure the plus momentum in the final state by integrating the fragmentation function over all p, with a weight p^+, and summing over all hadron types. Dividing by k^+ gives the momentum sum-rule:

$$\sum_h \int_0^1 \mathrm{d}z \, z d_{(0)h/j}(z) = \sum_h \int_0^1 \mathrm{d}z \, z d_{h/j}(z) = 1. \tag{12.42}$$

The derivation applies to the bare quantities. As with the sum rules for parton densities, it implies that there is no UV divergence in the sum over h of the second moment, and therefore that if we use a suitable renormalization scheme, like $\overline{\text{MS}}$, the sum rule applies also to the renormalized fragmentation functions, as indicated above.

Flavor relations

There is one fragmentation function for each combination of hadron type and parton type. In QCD, even with just pions, and light quarks and antiquarks and gluons, this gives 21 fragmentation functions. But many of these can be related by applying isospin transformations and charge conjugation transformations. This leaves just four independent

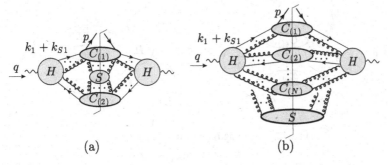

Fig. 12.4. Leading regions in gauge theory for one-hadron-inclusive cross section. The subgraphs H are the hard subgraphs, and there can be any number (greater than 2) of collinear subgraphs $C_{(1)}$, $C_{(2)}$, ..., of which two examples are shown. The soft subgraph may have any number of connected components, including zero. In (b), each gluon from the soft subgraph may connect to any collinear subgraph.

fragmentation functions

$$d_{\pi^+/u}(z) = d_{\pi^-/d}(z) = d_{\pi^-/\bar{u}}(z) = d_{\pi^+/\bar{d}}(z), \tag{12.43a}$$

$$d_{\pi^-/u}(z) = d_{\pi^+/d}(z) = d_{\pi^+/\bar{u}}(z) = d_{\pi^-/\bar{d}}(z), \tag{12.43b}$$

$$d_{\pi^+/s}(z) = d_{\pi^0/s}(z) = d_{\pi^-/s}(z) = d_{\pi^+/\bar{s}}(z) = d_{\pi^0/\bar{s}}(z) = d_{\pi^-/\bar{s}}(z), \tag{12.43c}$$

$$d_{\pi^+/g}(z) = d_{\pi^0/g}(z) = d_{\pi^-/g}(z). \tag{12.43d}$$

From these we obtain π^0 fragmentation functions of u and d quarks:

$$d_{\pi^0/u}(z) = d_{\pi^0/\bar{u}}(z) = d_{\pi^0/d}(z) = d_{\pi^0/\bar{d}}(z) = \frac{1}{2}[d_{\pi^+/u}(z) + d_{\pi^-/u}(z)]. \tag{12.43e}$$

by a Clebsch-Gordan decomposition of the final state. Only isospin $\frac{1}{2}$ and $\frac{3}{2}$ are possible for the X part of the state when the initiating parton has $I = \frac{1}{2}$ and the detected hadron has $I = 1$.

12.5 Leading regions and issues in a gauge theory

In a gauge theory, like QCD, the power-counting rules of Ch. 5 show that the general leading region (and its associated PSS) has the form specified by Fig. 12.4. Each electromagnetic vertex is part of a hard-scattering subgraph H, and out of each H exit two or more groups of lines of high energy and low virtuality. These groups each go into a collinear subgraph which crosses the final-state cut. To the collinear subgraphs may be connected a soft subgraph S (which consists of any number of connected components and can be absent).

We define the first collinear subgraph $C_{(1)}$ to be the one attached to the detected hadron. Consequently the direction for the corresponding collinear singularity of the region's PSS is fixed by p. The distinctness of the collinear configurations implies that the different collinear groups are treated as being at wide angle to each other. However, the angles of the other collinear groups are to be integrated over. When two or more of the directions get close, the

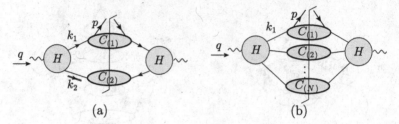

Fig. 12.5. Leading regions in *non*-gauge theory.

originally separate collinear configurations combine into a single collinear configuration, i.e., the corresponding jets merge into a single jet. In defining the contribution $C_R\Gamma$ for a region R with N collinear groups, there are subtractions for smaller regions, in particular for regions with fewer collinear groups. This results in a suppression of $C_R\Gamma$ where the angles of two or more collinear groups approach each other. Note that before subtractions there is a logarithmic enhancement after angular integration between a separated jet configuration and a merged jet.

The basic situation is more easily visualized by the leading regions in a model theory *without* gauge fields, in Fig. 12.5. Then there is no soft subgraph, and each collinear subgraph is initiated by a single parton, which can correspond to any of the fields in the theory. For a leading power, only one collinear line on each side of the final-state cut initiates each collinear subgraph, whose final state is essentially a jet.

To return to QCD, we have any number of extra gluons joining each collinear subgraph to the hard subgraphs, and we have a possible soft subgraph with gluonic couplings to any of the collinear subgraphs. Basically all these extra gluons have the polarization that we characterize as a Grammer-Yennie K gluon, as defined in Sec. 11.2.3.

Initially, each diagram like Fig. 12.4 codes a particular region of momentum space for some generic graph. According to the subtractive methods of Ch. 10, we reinterpret the diagram as an actual Feynman graph with integrals over *all* internal momenta, but with approximations applied that are appropriate for the region, and with subtractions to prevent double counting. Thus to leading power, the complete hadronic tensor $W^{\mu\nu}$ of (12.3) is a sum over all cases of Fig. 12.4.

12.5.1 Complication 1: super-leading regions

But one annoying extra possibility is generated by collinear groups that are purely gluon initiated. The amplitude between a photon and two gluons is prohibited by charge-conjugation invariance, so the smallest case is three collinear groups (Fig. 12.6). Just as we found with DIS in Secs. 11.2.3 and 11.3, there are graph-by-graph super-leading contributions, when all the external gluons of a collinear subgraph are K gluons. The Ward-identity arguments of Sec. 11.9 show that there is a cancellation after a sum over all graphs for the hard scattering. The result is that the gluons attaching to the hard subgraphs combine to give the operator defining the collinear factor, with its Wilson line.

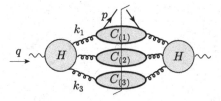

Fig. 12.6. A possible leading region in gauge theory, where all the jets are gluon initiated. Possible extra collinear gluons and a soft subgraph are not shown, for simplicity.

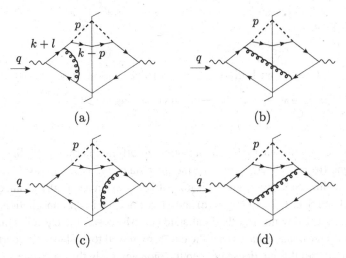

Fig. 12.7. To illustrate soft cancellation. The quarks and the pion (modeled by an elementary scalar field) form two collinear groups. The gluon is soft, and emitted off an internal line.

12.5.2 Complication 2: soft gluons and sum-over-cuts

Another complication is that to get the expected factorization theorem we need to show the soft factor cancels. The basic tool for getting the necessary cancellations is a sum-over-cuts. In the case of DIS, this was implemented in the conversion from a cut hadronic tensor $W^{\mu\nu}$ to the corresponding uncut quantity $T^{\mu\nu}$, in Sec. 11.2. But this argument is insufficient for our current process, because the final-state cut is restricted by being anchored at the detected hadron. This prevents the basic sum-over-cuts argument from combining some of cut graphs relevant to the cancellation of soft gluon effects, and we will need a more powerful argument to be given later.

As a simple example, consider the graphs in Figs. 12.7 and 12.8. We have chosen a model in which the pion is replaced by an elementary scalar field with a Yukawa coupling to the quarks. The pion analog may or may not be color singlet. We choose kinematics in which there is a quark and an antiquark jet, with the quark collinear to the pion, and then we add a soft gluon.

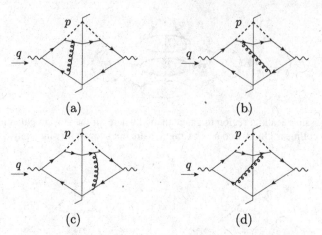

(a)   (b)

(c)   (d)

+ Graphs with gluon attached to the "pion" line (------).

Fig. 12.8. The same as Fig. 12.7, but with soft gluon emission off the external quark in the upper jet.

First consider graphs where the gluon is emitted off the internal quark line (Fig. 12.7). Graphs (a) and (b) are related by moving the final-state cut, as are graphs (c) and (d). But cancellation within each pair of graphs is not sufficient to get a soft gluon cancellation; see problem 12.4 and Sec. 12.5.3. Graphs (c) and (d) are needed in the cancellation, but are not related to (a) and (b) by moving the final-state cut. Moreover, the cancellation will need, among others, a term that behaves as if the cut were moved to the leftmost quark-antiquark pair in graph (a); but this cut does not keep the pion analog in the final state.

To demonstrate cancellation of the soft subgraphs, we first factorize the soft subgraph, just as for the Sudakov form factor in Ch. 10. In our example this requires us to sum all ways of attaching the soft gluon to the lines in the collinear graph, and thus to include the diagrams with external-line emission of the soft gluon (Fig. 12.8). After applying the soft approximation and Ward identities, we get a factorized result:

$$\int \frac{\mathrm{d}^4 q_S}{(2\pi)^4} \quad \underset{q-q_S}{} \quad \times \quad \sum_{\text{cuts}} {}^{q_S} \qquad\qquad (12.44)$$

Here the first factor is just as in a non-gauge theory, and is a product of two collinear factors. The soft gluon is in second factor, where it is attached to eikonalized quark lines (i.e., Wilson lines). The sum-over-cuts argument applies to the soft factor:

$$\sum_{\text{cuts}} {}^{q_S} \qquad = \qquad + \qquad\qquad (12.45)$$

and as with the corresponding argument for the total cross section in Sec. 4.4, we will see in Sec. 12.7 that this implies that the integration is not trapped in the soft region.

12.5.3 *KLN theorem is not sufficient*

The above summary indicates that the cancellation of soft gluons is obtained somewhat indirectly: first the soft part is factorized, and then the sum-over-cuts argument is applied. Now the soft-gluon issue is a generalization of the IR-divergence problem in QED. It is therefore tempting to suppose that a more direct proof of cancellation can be done by appealing to the theorem of Kinoshita (1962) and Lee and Nauenberg (1964) (KLN theorem). This theorem is well known in applications to QED, and it also applies to the e^+e^- total cross section discussed in Sec. 4.1. The KLN theorem (together with minor generalizations) applies when the canceling terms are related by a sum-over-cuts of individual graphs.

With a massless gluon, the KLN theorem does indeed show that the actual IR divergences cancel. These arise from where the gluon momentum goes to zero, and therefore only from graphs like Fig. 12.8, where the IR gluons are emitted from external lines. In this case the sum-over-cuts argument succeeds. But for merely soft gluons we use a much broader range of gluon momenta: any that are much less than Q and central in rapidity. Then internal line emission, Fig. 12.7, is also important.

To see this explicitly, we apply power-counting from Ch. 5 to the loop momenta defined in Fig. 12.7(a). We characterize the relative transverse momentum of the upper collinear lines by λ, so that the pion-quark invariant mass is of order λ^2. We let the size of the soft momentum l be of order λ_S in all components. The size of the denominator of an internal line carrying soft and collinear momenta is then

$$(k + l)^2 - m_q^2 = O(\lambda^2) + O(Q\lambda_S). \tag{12.46}$$

For lines next to an external line attachment, the λ^2 term is missing:

$$(k - p + l)^2 - m_q^2 = O(Q\lambda_S), \tag{12.47}$$

because the momentum $k - p$ is exactly on-shell.

IR gluon: $\lambda_S \ll \lambda^2/Q$

When λ_S is sufficiently small, much less than λ^2/Q, graphs with emission from external lines give logarithmic power-counting, from the external line factors (12.47) and the gluon line. This gives an IR divergence in the graphs of Fig. 12.8, if the gluon is massless. Internal line emission, Fig. 12.7, is suppressed because a denominator of order $Q\lambda_S$ is replaced by a much larger denominator of order λ^2; see (12.46).

Hence in this region only external line emission is important, and the KLN theorem applies.

Harder gluon: $\lambda_S \gg \lambda^2/Q$

When the contrary situation holds, i.e., $\lambda_S \gg \lambda^2/Q$, internal-line emission, Fig. 12.7, dominates. In this case those collinear denominators that carry soft momentum are of order $\lambda_S Q$; collinear denominators without a soft momentum have the much smaller value λ^2.

Relative to a graph without the soft gluon, internal line emission has two collinear denominators of order $\lambda_S Q$, and we get logarithmic power-counting, and hence a contribution at leading power. But for external line emission, a second collinear denominator λ^2 is replaced by an extra larger $\lambda_S Q$ denominator, which is much larger and therefore leads to a suppression.

$$\textit{Borderline: } \lambda_S \sim \lambda^2/Q$$

In the intermediate range of gluon momentum both internal and external emission are equally important.

Combination is simple

The Ward-identity argument for soft gluons combines all the above contributions and applies independently of the relative size of λ_S and λ. We get a coherent sum over gluon emission from the whole jet, and the result is as if the gluon were emitted from a single quark, as in the right-hand factor in (12.44). In this factor we have uniform power-counting independently of the relative size of λ_S and λ.

12.6 Which gauge to use in a proof?

Our characterization of leading regions was appropriate to Feynman gauge. But, as we saw in Sec. 5.5, the situation is gauge dependent.

The most notable effect is in the axial gauge, $n \cdot A = 0$, for which numerator of the gluon propagator is (5.40). We choose the gauge-fixing vector n proportional to q, i.e., to be at rest in the overall center-of-mass frame (Collins and Sterman, 1981). Then the enhancement of gluons connecting a collinear subgraph to a hard subgraph is removed. Regions with extra gluons connecting collinear to hard subgraphs, as in Fig. 12.4, are now power-suppressed. Regions with gluon-generated jets, Fig. 12.6, are merely leading instead of super-leading. To see this, we observe that in the rest frame of a collinear momentum, the gauge-fixing vector n is approximately a light-like vector w_2, and that the same light-like vector can be used in the Grammer-Yennie K term, (11.6), when the gluon attaches to the hard scattering. The K term then gives zero, because the $w_2^{\nu_j}$ factor in (11.6) contracts with a gluon propagator which is being treated as if it were in $w_2 \cdot A = 0$ gauge.

More formally, consider a collinear subgraph for a region R, and let w_1 be the light-like vector for the subgraph's momenta on the PSS for the region. We let w_2 be the conjugate light-like vector used in the Grammer-Yennie decomposition (11.6) for the attachment of a collinear gluon to the hard subgraph. Define light-front coordinates such that $w_1^\mu = \delta_+^\mu$ and $w_2^\mu = \delta_-^\mu$. Then for a collinear momentum k, its contraction with the gauge-fixing vector is $k \cdot n \simeq k^+ n^- = k \cdot w_2 n \cdot w_1$. By scaling n we can set $n^- = 1$, and therefore to leading power, $k \cdot n \simeq k \cdot w_2$, i.e., we can replace n by w_2 on collinear gluons. The w_2 factor in the K-gluon definition gives zero when contracted into the approximated propagator.

In the $n \cdot A = 0$ gauge, a Wilson line in direction n is simply unity. So if this Wilson line were used in the definition of a fragmentation function or parton density, the Wilson

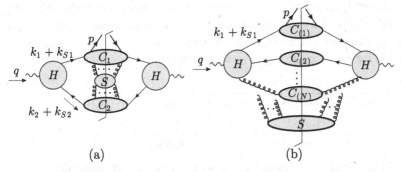

Fig. 12.9. Leading regions in gauge theory for one-hadron-inclusive cross section, now in $n \cdot A = 0$ gauge for case of (a) two and (b) more collinear groups.

line could be ignored in the $n \cdot A = 0$ gauge. Therefore such a definition has the same form as in a non-gauge theory.

But we still have a soft subgraph, unlike the case in a non-gauge theory. The resulting leading regions are illustrated in Fig. 12.9. A further advantage of an axial gauge is that Ward identities in a non-abelian theory are simple: they are essentially the same as in an abelian theory, without the terms involving Faddeev-Popov ghosts.

On all of these points, the axial gauge is superior to the Feynman gauge. However, one issue reverses the situation. This is that the necessary contour deformation out of the Glauber region is often obstructed in axial gauge.

Recall that the Glauber region is where the longitudinal momentum of a gluon is much less than its transverse momentum. The Ward-identity method to extract soft gluons from a collinear subgraph uses an approximation of the Grammer-Yennie type. But the approximation fails in the Glauber region. In Feynman gauge, the normal causal structure of Feynman denominators allows a deformation out of the Glauber region; the argument in Sec. 5.5.10 applies to the process we are considering, since all the partons are outgoing from a single interaction point.

But in the axial gauge there are unphysical singularities $1/k \cdot n$ in the gluon propagator, and these have the potential to obstruct the contour deformation.

Consider, for example, the vertex graph Fig. 5.21 for the Sudakov form factor, which also appears as a subgraph in hadron production in $e^+ e^-$ annihilation. In Feynman gauge, we get out of the Glauber regions the gluon momentum k, by using any deformation of the form

$$k^+ \mapsto k^+ + iC, \quad k^- \mapsto k^- - iD, \qquad (12.48)$$

where both C and D are non-negative, and at least one is positive. In an axial gauge, we have $1/k \cdot n$ singularities, which are principal value, and therefore prevent any deformation on $k \cdot n = k^+ n^- + k^- n^+$. The non-deformation can be satisfied by setting $C/D = n^+/n^-$, which requires that n be time-like. If we use the center-of-mass frame, and choose $n \propto q$, then we need $C = D$, so that on the deformed contour the imaginary parts of k obey $\Im k^0 = 0$ and $\Im k^3 > 0$. The gauge condition in use here is $A^0 = 0$.

More generally, as in Fig. 12.9, we can have extra collinear subgraphs, and multiple soft gluon loops. Consider a soft momentum k routed out through one collinear subgraph and back through another. Let the light-like directions for the collinear subgraphs be w_1 and w_2. Then in the collinear subgraphs we have propagators like

$$\frac{1}{(k_1 - k)^2 - m^2 + i0} = \frac{1}{-2k \cdot w_1 \times \text{positive} + \cdots + i0}, \tag{12.49a}$$

$$\frac{1}{(k_2 + k)^2 - m^2 + i0} = \frac{1}{2k \cdot w_2 \times \text{positive} + \cdots + i0}. \tag{12.49b}$$

We avoid these singularities by deforming with imaginary parts such that $\Im k \cdot w_1 < 0$ and $\Im k \cdot w_2 > 0$. In addition, we require $\Im k^0 = 0$ to avoid the axial gauge singularities, and then $\Im k^3 > 0$. The Coulomb gauge, (10.156), is also compatible with such a deformation: its unphysical singularities are not operative in the Glauber region, where $k^\pm \ll k_T$.

While this all appears to work for inclusive processes in e^+e^- annihilation, it does not help in other processes we consider. For example, for SIDIS (Sec. 12.14) the avoidance of poles in collinear propagators will restrict the deformations incompatibly with the axial gauge. In an equivalent of (12.48), we will need D to be zero: k^+, but not k^-, is not to be deformed. In the Drell-Yan and other reactions in hadron-hadron collisions (Ch. 14) we will need more elaborate arguments. Different parts of the proof will need different deformations, some on plus-components and some on minus-components of loop momenta; such different deformations are prevented by an axial gauge.

The offending singularities for the axial gauge are in a gauge-dependent part of the propagator, so their effects should cancel in the final result for a gauge-invariant amplitude. What is not clear is how to make this demonstration while at the same time preserving the rest of a factorization proof, and to the best of my knowledge has not been done.

So we conclude that for fully general and reliable proofs one should use Feynman gauge. Even if we can avoid the problems of axial (or Coulomb) gauge for a subset of processes, the methods of proof do not extend to other important cases.

In summary, here is a list of possible approaches, including some not mentioned above:

- Time-like axial gauge. This gave the first good approximation to a proof (Collins and Sterman, 1981) of factorization in e^+e^- annihilation. The leading regions and the Ward identities are relatively simple, but the Glauber region cannot be handled for more general processes.
- Coulomb gauge. The extra singularities in the gluon propagator are $1/k^2$ instead of $1/n \cdot k$, which improves the Glauber problem, but not sufficiently for a general process. In addition, there are complications in setting up the Coulomb gauge in a non-abelian theory.
- A non-covariant generalization of the Feynman-type gauges proposed by Sterman (1978, Sec. V). The extra gluon connections between collinear and hard subgraphs are suppressed, but the unphysical singularities are now off the real axis, so at a first approximation this allows a contour deformation out of the Glauber region. But the singularities

are such as not to allow all the needed contour deformations, and definitely not in semi-inclusive DIS.

- Light-cone gauge, where n is light-like. This suppresses extra connections from a collinear to a hard subgraph, but only for one collinear direction. So this gauge is of use in ordinary fully inclusive DIS, where the only direction of collinearity is that of the target hadron. But the light-cone gauge does not help when there are multiple directions of collinearity. This gauge also has problems with rapidity divergences, which fail to cancel in transverse-momentum-dependent parton densities and fragmentation functions (Ch. 13).
- Covariant gauge.
- A more recent and very popular approach is that of soft-collinear effective theory (Fleming, 2009). See Sec. 15.9 below for my comments.

12.7 Unitarity sum over jets/sum over cuts

Before we go to the proof of factorization for the process (12.1), it is useful to examine in its simplest setting a method that is repeatedly used in proofs of factorization. The method is that of summing over cuts of a region subgraph (for a collinear or soft region) when there are no detected final-state particles. The result is that momenta in the region subgraph are not trapped in regions of low virtuality after the sum-over-cuts, whereas momentum integrations are trapped for individual cut diagrams.

In this section, we examine the simplest case, which is for the collinear subgraphs $C_{(2)}, \ldots, C_{(N)}$ in the leading regions, Fig. 12.5, for a non-gauge theory. Once we have proved the momentum integrals are not trapped in these subgraphs, we can absorb these subgraphs in the hard subgraph, so that the effective leading regions are those in Fig. 12.2. It has a single collinear subgraph, $C_{(1)}$, the one containing the detected hadron. Precisely because the cut of $C_{(1)}$ is anchored to include the detected hadron, the sum-over-cuts argument fails for this one subgraph.

We combine as a group the sum of region graphs that are related by moving the final-state cut to cross different parts of the collinear subgraphs in Fig. 12.5. The cuts can be applied independently to each collinear subgraph $C_{(\alpha)}$. We use a version of the Cutkosky (1960) rules to relate the sum-over-cuts of $C_{(\alpha)}$ to corresponding uncut amplitudes:

$$\quad (12.50)$$

In the overall center-of-mass frame, we write the momenta of the subgraph's lines as $l_{\alpha i} = (E_{\alpha i}, l_{\alpha i})$, with the index i labeling the lines.

First we convert the Feynman graphs to time-ordered graphs (exactly as for x^+ perturbation theory in Sec. 7.2). The integrand in each term in time-ordered perturbation theory corresponds to a sequence of intermediate states with energy denominators:

$$I(b) = \prod_{a=1}^{b-1} \left[\frac{i}{E - E_a + i\epsilon} (ig) \right] 2\pi \delta(E - E_b) \prod_{c=b+1}^{n} \left[(-ig) \frac{-i}{E - E_c - i\epsilon} \right]. \quad (12.51)$$

Here there is a set of n intermediate states, with state number b being cut. The part to the right of the final-state cut is a complex-conjugated amplitude. The energy of state a is E_a when all its lines are on-shell, and the external energy entering from the hard scattering is $E = k_\alpha^0$. Multiplying the above formula is a common factor that depends only on the 3-momenta of the lines, and is independent of the position of the cut.

It is easily proved that the sum-over-cuts gives

$$\sum_{b=1}^{n} I(b) = (ig)^{n-1} \prod_{a=1}^{n} \frac{i}{E - E_a + i\epsilon} + (-ig)^{n-1} \prod_{c=1}^{n} \frac{-i}{E - E_c - i\epsilon}. \tag{12.52}$$

The proof is made by using the identity

$$2\pi\,\delta(E - E_b) = \frac{i}{E - E_b + i\epsilon} + \frac{-i}{E - E_b - i\epsilon}, \tag{12.53}$$

and showing that in the sum over b, there is a cancellation of all the resulting terms except for the two on the r.h.s. of (12.52). Converting back to Feynman perturbation theory gives (12.50).

We now use the same idea as with the e^+e^- total cross section in Sec. 4.4. This uses the property that all the poles in E are on one side of the real axis in each uncut collinear subgraph on the r.h.s. of (12.52). We arrange to deform the integration over E, so that the contour no longer goes close to the poles. The propagators are now far off-shell, and we can treat the collinear graphs $C_{(2)}$ to $C_{(N)}$ as part of the hard scattering. *Effectively* the leading regions are of the form of Fig. 12.2, where the only collinear subgraph is the one containing the detected hadron.

To implement the contour deformation, we use the averaged hadronic tensor defined in Sec. 12.1.2. This enables us to route the total momentum k_α of each collinear subgraph through the hard scattering and out at the virtual photon vertex. The averaging function is then $f(\sum_{\alpha=1}^{N} k_\alpha)$, and we do not have to route each k_α back through another collinear subgraph. The averaging function is slowly varying as a function of Q, and the hard scattering involves dominantly highly virtual momenta, so neither obstructs the contour deformation.

In reality, the hard subgraph can have collinear and soft singularities, but these are suppressed by subtractions. When we deform the E integration for a collinear subgraph, singularities in the hard scattering must be crossed. But the resulting contributions are power-suppressed because of the subtractions, and we therefore ignore them.

12.8 Factorization for $e^+e^- \to h(p) + X$ in gauge theory

The proof of factorization in Feynman gauge uses the methods of Ch. 10 supplemented by the non-abelian Ward-identity results for K gluons in QCD given in Sec. 11.9. Given these techniques and the associated graphical notation, the proof can be given quite quickly. Any issues about the accuracy of the proof really concern the earlier work.

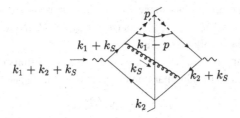

Fig. 12.10. Momentum routing for Fig. 12.7. The gluon is treated as soft, and the quarks as collinear in one of two directions.

A new feature relative to DIS is that the cancellation of the soft subgraphs occurs at a late stage. Therefore much of the proof will apply equally to situations (Ch. 13) where the soft cancellation does not happen because of more stringent conditions on the final state.

The proof starts from the leading regions symbolized diagrammatically in Fig. 12.4.

12.8.1 Definition of approximators

We now apply the principles for making the approximator for a region that we formulated in Secs. 10.4 and 10.6.

Momentum routing

Soft loop momenta flow between the soft subgraph into collinear subgraphs. We define them to flow outwards to the electromagnetic vertex. As illustrated in Fig. 12.10, to avoid routing soft momenta back through the final state of a collinear graph, we route them out of the electromagnetic vertex. This takes advantage of our definition (12.11) of averaging the hadronic tensor with a test function. We label the collinear subgraphs by α: $1 \leq \alpha \leq N$; we let the momenta of the lines from H to C_α be $k_{\alpha i}$.

Light-like auxiliary vectors w_α and $\tilde{w}_\alpha$ for collinear subgraph $C_{(\alpha)}$

Auxiliary vectors w_α and $\tilde{w}_\alpha$ are defined as follows:

- For each collinear subgraph $C_{(\alpha)}$, we define a characteristic momentum p_α. For the collinear subgraph $C_{(1)}$ that contains the detected hadron, we use $p_1 = p$, the momentum of the detected hadron. For the other collinear subgraphs, p_α is the total final-state momentum of the collinear subgraph, i.e., $p_\alpha = \sum_i k_{\alpha i}$.
- For each $C_{(\alpha)}$, the corresponding light-like direction is

$$w_\alpha^\mu = \frac{Q}{\sqrt{2} p_\alpha \cdot q} \left[\frac{p_\alpha^\mu - q^\mu p_\alpha \cdot q / q^2}{\sqrt{1 - p_\alpha^2 Q^2 / (p_\alpha \cdot q)^2}} + \frac{q^\mu p_\alpha \cdot q}{q^2} \right]. \tag{12.54}$$

In the center-of-mass frame, the direction of the 3-vector part is that of p_α.
- The conjugate auxiliary vector is defined by

$$\tilde{w}_\alpha^\mu = \frac{q^\mu}{w_\alpha \cdot q} - \frac{w_\alpha^\mu q^2}{2(w_\alpha \cdot q)^2}. \tag{12.55}$$

- For each collinear group, the auxiliary vectors can be regarded as defining light-front coordinates in which $(w_\alpha^\mu)_{\text{frame }\alpha} = \delta_+^\mu$ and $(\tilde{w}_\alpha^\mu)_{\text{frame }\alpha} = \delta_-^\mu$. Note carefully that this is a different frame for each collinear group.
- For use in the soft approximation, we generalize the definitions (10.15b) of n_1 and n_2. For each α, we choose a rapidity parameter y_α, and define a space-like vector by

$$n_\alpha^\mu = w_\alpha^\mu - e^{-2y_\alpha} \tilde{w}_\alpha^\mu. \tag{12.56}$$

As in Sec. 10.4.2, the letter w denotes a light-like vector and n denotes a non-light-like vector.

Approximators

We split the gluons connecting different subgraphs into K and G terms, as in Sec. 11.2.3:

- For a K gluon attaching collinear subgraph α to a hard subgraph H, we copy (11.32).

$$H^\mu(k_{\alpha i}) \, C_{(\alpha),\mu}(k_{\alpha i}) = H^\mu(\hat{k}_{\alpha i}) \frac{\hat{k}_{\alpha i,\mu} \tilde{w}_\alpha^\nu}{k_{\alpha i} \cdot \tilde{w}_\alpha + i0} C_{(\alpha),\nu}(k_{\alpha i}), \tag{12.57a}$$

where

$$\hat{k}_{\alpha i}^\mu = \frac{w_\alpha^\mu \, k_{\alpha i} \cdot \tilde{w}_\alpha}{w_\alpha \cdot \tilde{w}_\alpha}. \tag{12.57b}$$

This projects the gluon's momentum onto direction w_α.

- For G gluons and quarks, we project the momentum by (12.57b).
- For a quark exiting the hard scattering to collinear subgraph α, we project its Dirac spinor onto a massless on-shell wave-function multiplying H by inserting a factor $\gamma^-\gamma^+/2$. This is made relative to light-front coordinates defined by w_α and $\tilde{w}_\alpha$. In covariant form, the projector is $\psi_\alpha \bar{\psi}_\alpha/(2w_\alpha \cdot \tilde{w}_\alpha)$.
- For an *antiquark* exiting the hard scattering, we use the projector $\bar{\psi}_\alpha \psi_\alpha/(2w_\alpha \cdot \tilde{w}_\alpha)$.
- At the coupling of a K gluon of S to collinear subgraph $C_{(\alpha)}$, we denote the line's momentum (out of $C_{(\alpha)}$) by $k_{S\alpha i}$, and define the soft approximant by

$$C_{(\alpha)}^\mu(k_{S\alpha i}) \, S_\mu(k_{S\alpha i}) = C_{(\alpha)}^\mu(\hat{k}_{S\alpha i}) \frac{\hat{k}_{S\alpha i,\mu} n_\alpha^\nu}{k_{S\alpha i} \cdot n_\alpha + i0} S_\nu(k_{S\alpha i}), \tag{12.58a}$$

where

$$\hat{k}_{S\alpha i}^\mu = \frac{\tilde{w}_\alpha^\mu \, k_{S\alpha i} \cdot n_\alpha}{\tilde{w}_\alpha \cdot n_\alpha}. \tag{12.58b}$$

This projects the soft momentum onto our defined conjugate direction for $C_{(\alpha)}$, and uses the non-light-like vector n_α to cut off the rapidity divergence that would otherwise occur at $k_{S\alpha i} \cdot w_\alpha = 0$.

- The approximated momenta are also used in the test function $f(q) = f(p + p_X)$ in (12.11). Thus the approximant changes $f(p + p_X)$ to $f(\sum_\alpha \hat{k}_\alpha)$, where $\hat{k}_\alpha$ is the total approximated momentum for collinear subgraph $C_{(\alpha)}$.

- Finally we redefine the hard-scattering factor by an extra factor:

$$H = \text{basic definition of } H \times \prod_{\alpha=2}^{N} \left(\frac{|\hat{k}_\alpha|}{|k_\alpha|} \right)^{n-2}. \tag{12.59}$$

Here k_α is the total final-state momentum of collinear subgraph $C_{(\alpha)}$, and n is the space-time dimension.

At this point (12.59) is a totally unobvious redefinition. Note that, in the collinear limit, $\hat{k}_\alpha \rightarrow k_\alpha$, and the extra factor goes to unity. Hence the redefinition is one that is allowed; it is a change of factorization scheme. Appropriate versions of the redefinition will also appear, applied to smaller hard subgraphs, in the double-counting subtractions defining the region terms $C_R\Gamma$. The result is that the redefinition does affect the correctness of the factorization formula, but only the precise definition of the factors. Notice also that the redefinition involves only collinear but not soft momenta.

The rationale for the redefinition will appear in Sec. 12.8.5, where to get the most desirable form of factorization we will change the variables from k_α to $\hat{k}_\alpha$. The redefinition factor will cancel a part of the Jacobian for the change of variable. We will see that we only apply the redefinition to collinear subgraphs without an observed hadron, i.e., the case $\alpha = 1$ is omitted from the product in (12.59).

For the Sudakov form factor, we also rescaled the external momenta of H; see p. 329. We do not need to do this at this point in our case.

With these definitions, the approximated hard subgraph does not depend on the soft momenta, because a soft momentum in collinear subgraph α is approximated to be in direction $\tilde{w}_\alpha$. This gives a zero contribution in the projection (12.57b) onto an approximated momentum in H, because $\tilde{w}$ is light-like.

12.8.2 Extraction of collinear gluons from hard subgraph

We first extract the collinear gluons from the hard subgraphs.

The necessary result was given in Sec. 11.9.5, and stated graphically for the case of two collinear groups in Fig. 11.15. We apply this result on both sides of the final-state cut in Fig. 12.4, to obtain Fig. 12.11.

For example, suppose collinear subgraph $C_{(1)}$ has a quark entering it from the hard subgraph. When the accompanying K gluons are extracted from H, they couple to a Wilson line at the end of the quark line. On the left side of the final-state cut, the color matrices in the Wilson line are those appropriate to make a gauge-invariant operator with the $\bar{\psi}$ field that creates the quark entering $C_{(1)}$. The Wilson line extends out to infinity in the direction $\tilde{w}_1$. It represents a source of the opposite color to the parton initiating the collinear subgraph. Thus the Wilson line is an approximation to the rest of the event, seen as recoiling against collinear system $C_{(1)}$. In the mathematics, we get the Wilson line from graphs *omitted* from H because of the irreducibility requirements.

For a collinear subgraph initiated by an antiquark or a G gluon, the Wilson line has the corresponding color representation, and similarly on the right of the cut.

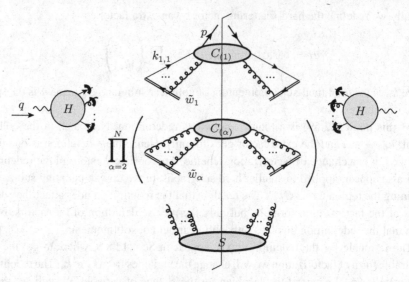

Fig. 12.11. Result of extraction of collinear K gluons from hard subgraphs in Fig. 12.4. There are Wilson lines for each subgraph in the directions shown. The soft subgraph S still couples to the collinear subgraphs. The little thick arcs on the hard subgraphs indicate on-shell partonic lines.

In accordance with our definitions, soft subtractions are applied in each collinear subgraph, and these remove rapidity divergences.

12.8.3 Factorization of soft subgraph

Similarly, we apply the Ward-identity argument to the connections of the soft subgraph to each collinear subgraph, to obtain Fig. 12.12.

Now in the extraction of collinear K gluons from a hard subgraph H, the external lines of H are on-shell. So the Wilson lines in Fig. 12.11 arose from the lack of collinear-reducible graphs in H. But for soft K gluons attaching to C_α, the irreducibility only concerns the K gluons themselves. As we saw in Sec. 11.9.5, these by themselves give no contribution; from the terms "missing" in the Ward identity due to irreducibility requirement we obtain commutator terms that are themselves of the K-gluon form. The on-shell external lines of $C_{(\alpha)}$ in the final state give no contribution to the Ward identity.

On each side of the final-state cut, each $C_{(\alpha)}$ has an external off-shell quark (or G gluon), with a Wilson line to make a gauge-invariant operator. But graphs are missing where external soft gluons directly couple to the Wilson line, since these did not come out of the argument that derived the Wilson lines. However, an approximated soft gluon gives zero when it attaches to the Wilson line, because the vertex for the Wilson line is proportional to $\tilde{w}_\alpha$. The K gluon specified in (12.58) multiplies this by $\hat{k}_{S\alpha i}$, and gives zero since $\hat{k}_{S\alpha i}$ is in the light-like direction $\tilde{w}_\alpha$.

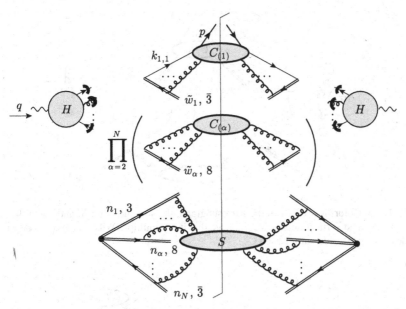

Fig. 12.12. Result of extraction of soft K gluons from collinear subgraphs in Fig. 12.11. The soft factor has a Wilson line for each external parton of the hard scattering, with the appropriate color charge, e.g., ("3", "8"). There is a non-trivial flow of color indices between the hard subgraph, the soft subgraph, and the collinear subgraphs; see Fig. 12.13 below.

So we can add to Fig. 12.11 all graphs where the soft gluons attach to the Wilson lines of $C_{(\alpha)}$. After applying the usual Ward-identity argument, we find that the soft gluons are moved to Wilson-line factors external to $C_{(\alpha)}$, as in Fig. 12.12.

12.8.4 Color flow

The Wilson lines are matrices in color space, and their color representation and color flow need attention. Each Wilson line of S has the color representation corresponding to the color charge of the outgoing parton initiating the associated jet.

The Wilson line for the gluons attaching S to $C_{(\alpha)}$ interposes itself between H and $C_{(\alpha)}$, as indicated in Fig. 12.13. We make a concrete illustration from the particular graph given in Fig. 12.14(a), which shows an extract from a particular diagram for the process we are analyzing. Diagram (b) shows one of the unapproximated graphs that are combined to give diagram (a) (after the use of approximators and Ward identities). Corresponding to diagram (a) is the formula

$$C_{(1)}(k_{1,1}, k_{1,2})_\mu \frac{i}{k_{1,2} \cdot \tilde{w}_1 + i0}(ig\tilde{w}_1^\mu t_\alpha)(-ign_1^\nu t_\beta)S(k_S)_\nu \frac{i}{k_S \cdot n_1 + i0}H(\tilde{w} \cdot (k_1 + k_2)).$$

$$(12.60)$$

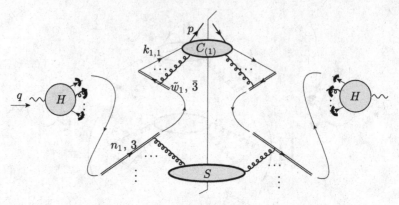

Fig. 12.13. Color flow between the hard subgraph, the soft subgraph S, and the collinear subgraphs. The vertical dots near S indicate that it has multiple Wilson lines, one on each side of the final-state cut for each $C_{(\alpha)}$.

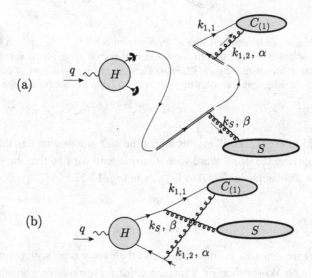

Fig. 12.14. (a) Specific example of Fig. 12.13. Each gluon is labeled with its momentum and color index. (b) A graph that, after approximants and Ward identities are applied, would contribute to (a).

The color charge (quark or antiquark) on a Wilson line is coded both in the ordering of the color matrices and in the sign of the vertices.

Now that all soft gluon lines have been extracted from the collinear subgraphs $C_{(\alpha)}$, each $C_{(\alpha)}$ becomes diagonal in color. This implies that we can rearrange the color flow as in Fig. 12.15. After rearrangement, there is an average over the color of each $C_{(\alpha)}$, i.e., a trace over colors divided by $N_{c\alpha}$, the number of colors for the parton initiating $C_{(\alpha)}$. That is, $N_{c\alpha}$ is 3 for the case of a quark or antiquark, but 8 for a gluon.

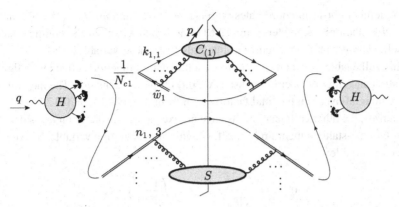

Fig. 12.15. Color flow of Fig. 12.13 after use of color-singlet property of C_1, with N_{c1} being the number of colors for the primary parton initiating C_1, which is a quark in the case shown, with $N_{c1} = 3$.

Then the color flow for the Wilson lines of S is direct from a Wilson line on the left of the cut to the Wilson line on the right; the collinear subgraphs are completely factored out. *But the color flow remains entangled with the color structure of the hard scattering.*

12.8.5 Factorization

We now have a factorized form for the hadronic tensor averaged with a function $f(q)$:

$$
W^{\mu\nu}([f], p) = \sum_{N \geq 2} \prod_{\alpha=1}^{N} \left[\sum_{j_\alpha} \int \frac{d^n k_\alpha}{(2\pi)^n} C_{(\alpha), j_\alpha}(k_\alpha) \right] \int \frac{d^n k_S}{(2\pi)^n} S_{j_1, \ldots, j_N}(k_S)
$$

$$
\times H_{j_1, \ldots, j_\alpha}(\hat{k}_1, \ldots, \hat{k}_N) \, f(\hat{k}_1 + \ldots + \hat{k}_N) \prod_{\alpha=2}^{N} \left(\frac{|\hat{k}_\alpha|}{|k_\alpha|} \right)^{n-2}. \tag{12.61}
$$

Here j_α is the flavor of the parton initiating collinear subgraph $C_{(\alpha)}$, k_α is the total momentum entering $C_{(\alpha)}$, and k_S is the total momentum entering the soft subgraph S; these momenta are also the total final-state momenta of the subgraphs. The hard subgraph H depends on the parton flavors and on the approximated momenta. Recall that the approximated $\hat{k}_\alpha$ is k_α projected by (12.57b) in a light-like direction appropriate for collinear subgraph $C_{(\alpha)}$. The dependence of S on parton flavors j_α is only through the color of the partons (3 v. $\bar{3}$ v. 8).

In the final form of factorization, (12.13), we treat the hard scattering as behaving like a cross section at the partonic level. For this purpose it is convenient to change variables in (12.61) from k_α to $\hat{k}_\alpha$, and so we need to take account of the Jacobian of the transformation.

For the first collinear group, the axes w_1 and $\tilde{w}_1$ are determined by the detected hadron, so we simply write

$$
d^n k_1 = dk_1^+ \, dk_1^- \, d^{n-2} k_{1\,T} = d\hat{k}_1^+ \, dk_1^- \, d^{n-2} k_{1\,T}. \tag{12.62}
$$

Here we define light-front coordinates with respect to w_1 and $\tilde{w}_1$: $k_1^+ = k_1 \cdot \tilde{w}_1$ and $k_1^- = k_1 \cdot w_1$. Now the only dependence in (12.61) on k_1^- and $\boldsymbol{k}_{1T}$ is in the collinear subgraph $C_{(1)}$ itself. So we can short-circuit the integrals over these variables, restricting them to $C_{(1)}$. This will enable us to convert $C_{(1)}$ to a fragmentation function defined with light-front annihilation and creation operators, just as we did for parton densities. The integral over $\hat{k}_1$ gives the integral over z in the final factorization formulae, (12.13) and (12.21).

In contrast, the other collinear graphs have no fixed axes; the direction w_α is determined by the total final-state momentum k_α itself, which is an integration variable. So we perform a change of variable:

$$d^n k_\alpha = \frac{d^{n-1}\hat{k}_\alpha}{2|\hat{k}_\alpha|} 2k_\alpha^+ \, dk_\alpha^- \left(\frac{|k_\alpha|}{|\hat{k}_\alpha|} \right)^{n-2}, \tag{12.63}$$

where the light-front variables are now those *local to* $C_{(\alpha)}$:

$$\hat{k}_\alpha^+ = k_\alpha^+ = k_\alpha \cdot \tilde{w}_\alpha = \frac{k_\alpha^0 + |k_\alpha|}{\sqrt{2}} = \sqrt{2}|\hat{k}_\alpha|, \tag{12.64a}$$

$$k_\alpha^- = k_\alpha \cdot w_\alpha = \frac{k_\alpha^0 - |k_\alpha|}{\sqrt{2}}. \tag{12.64b}$$

The first factor on the r.h.s. of (12.63) is just the Lorentz-invariant phase space for a massless parton out of the hard scattering. We will use the integral over k_α^- to convert $C_{(\alpha)}$ into a light-cone object, similarly to the fragmentation function. The hard scattering is independent of k_α^-. However, the final factor in (12.63) arises from the Jacobian for the change of variables, and it introduces extra dependence on k_α^-.

We now see the reason for introducing the extra factor in the definition (12.59) of the hard factor. It is to cancel the inverse factor in (12.63). At this point we functionally differentiate with respect to the test function f, to obtain a factorized form

$$W^{\mu\nu}(q, p) = \int d\hat{k}_1^+ \sum_{j_1} \left[\int \frac{dk_1^- \, d^{n-2}\boldsymbol{k}_{1T}}{(2\pi)^n} C_{(1), j_1}(k_1, p) \right]$$

$$\times \prod_{\alpha=2}^N \sum_{j_\alpha} \int \frac{d^{n-1}\hat{k}_\alpha}{2|\hat{k}_\alpha|(2\pi)^{n-1}} \prod_{\alpha=2}^N \left[2k_\alpha^+ \int \frac{dk_\alpha^-}{2\pi} C_{(\alpha), j_\alpha}(k_\alpha) \right]$$

$$\times \int \frac{d^n k_S}{(2\pi)^4} S_{j_1,\dots j_\alpha}(k_S)(2\pi)^n \delta^{(n)}\left(q - \sum_{\alpha=1}^N \hat{k}_\alpha \right) H_{j_1,\dots j_\alpha}(\hat{k}_1, \dots, \hat{k}_N). \tag{12.65}$$

12.8.6 Soft cancellation

The soft factor in (12.65) and Fig. 12.12 has an unrestricted sum over cuts, and an unweighted integral over its external momentum k_S. I will now show that the result is zero. One possible argument uses the methods of Sec. 12.7. But instead, we use a simpler argument relying on properties of the Wilson lines in the soft factor's operator definition.

There is a slight complication because of UV divergences. As we know from Ch. 10, there is a multiplicative renormalization to make finite virtual graphs for the soft factor. Because of the integral over all k_S, in (12.65), real-emission graphs for the soft factor also acquire logarithmic UV divergences, which need to be renormalized. Our proof that the soft factor is unity will initially apply to the bare soft factor. The result for the bare factor implies that the UV divergences also cancel between real and virtual corrections.

On each side of the final-state cut we have a product of Wilson lines all going out from the same point to infinity. Because of the integral over *all* k_S the common point is the same on both sides of the cut. Thus the integrated soft factor is

$$S_{0,\text{integrated}} \stackrel{\text{def}}{=} \int d^n k_S \, S = \left\langle 0 | M_1 T \left(\prod_\alpha W_{R(j_\alpha)}(+\infty, 0; n_\alpha) \right) \right.$$

$$\left. \times \bar{T} \left(\prod_\beta W_{R(j_\beta)}(+\infty, 0; n_\beta)^\dagger \right) M_2 | 0 \right\rangle. \qquad (12.66)$$

Here M_1 and M_2 are the color matrices coupling the soft factor to the hard factor. The Wilson line $W_{R(j_\alpha)}(+\infty, 0; n_\alpha)$ goes from the origin towards infinity in direction n_α and has the color representation $R(j_\alpha)$ corresponding to collinear subgraph $C_{(\alpha)}$ with its parton of flavor j_α. The color index at its right-hand end (at infinity) couples to the corresponding index of the conjugate Wilson line $W_{R(j_\alpha)}(+\infty, 0; n_\alpha)^\dagger$.

The Wilson lines are space-like, so the time-ordering and anti-time-ordering prescriptions give the same results, since gluon fields commute at space-like separation. Then the Wilson lines cancel: $W_{R(j_\alpha)}(+\infty, 0; n_\alpha) W_{R(j_\alpha)}(+\infty, 0; n_\alpha)^\dagger = 1$, and the bare soft factor (12.66) is unity. Although there are UV divergences in individual cut Feynman graphs, the divergences cancel in the sum over all graphs and cuts. Therefore no overall minimal-subtraction counterterms are needed, and the renormalized soft factor is also unity.

As explained in a similar case in Ch. 10, the collinear factors are equipped with soft subtractions, and the collinear factor is the product of an unsubtracted collinear factor and a soft factor related to (12.66). Again the complete soft factor is integrated over all momentum, so it gives unity; thus the collinear factors are effectively unsubtracted.

12.8.7 Collinear cancellation

For each of the collinear factors $C_{(2)}, \dots, C_{(N)}$ without a measured hadron, we have an unrestricted sum over cuts, and then an integral over k_α^-:

$$2k_\alpha^+ \int dk_\alpha^- \quad \underset{\tilde{w}_\alpha}{\overset{C_{(\alpha)}}{\cdots}} \qquad (12.67)$$

Here, $k_\alpha^\pm$ refers to components defined in local light-front coordinates, (12.64).

We now show that there is a suppression of (12.67) in the collinear region it was designed to treat. At first sight, it would be sufficient to appeal to the sum-over-cuts argument of Sec. 12.7. Indeed this argument works in a non-gauge theory, where we just have cut graphs for a full parton propagator, as in (12.50). We found that in each of the terms on the r.h.s. of (12.50) the k_α^- contour is not trapped. Thus we could deform the integral over k_α^- to a semicircle at infinity, to obtain a purely UV contribution.

This argument is broken by the Wilson lines in (12.67), for a somewhat non-trivial reason. The Wilson lines on the left and right of the final-state cut carry any value of momentum; loop momenta circulate freely through them. To make the sum-over-cuts argument work, we must also include a double line to continue the Wilson line across the final-state cut:

$$(12.68)$$

But to agree with the calculation in (12.67), the cut Wilson line must carry exactly zero 4-momentum, i.e., a cut Wilson line with momentum k' must be defined to have the value $(2\pi)^n \delta^{(n)}(k')$. In contrast, to apply the sum-over-cuts argument, the cut line needs to obey

$$= \frac{i}{k' \cdot \tilde{w}_\alpha + i0} + \frac{-i}{k' \cdot \tilde{w}_\alpha - i0} \qquad (12.69)$$

$$= 2\pi \delta(k' \cdot \tilde{w}_\alpha)$$

That is, although the delta function forces one component of k', viz., $k' \cdot \tilde{w}_\alpha$, to be zero, the other components can have any value.

We solve this problem in two stages. First we show that if the quantity defined in (12.67) is integrated over $\boldsymbol{k}_{\alpha T}$, then the cut Wilson line can be treated as being given by (12.69). The resulting quantity, (12.68), is one to which the sum-over-cuts argument can be applied, so that it gives no trap of the integration momentum in a non-UV region. The second stage of the argument is to deduce that when the collinear quantity without the transverse-momentum integral is inserted in the factorization formula, there is a power-suppression of the collinear region.

When we use an integral over $\boldsymbol{k}_{\alpha T}$ as well as k_α^- applied to the cut graph in (12.67), we can choose to route these momenta through the cut Wilson line. As we explained, the cut Wilson line at this point is effectively replaced by $(2\pi)^n \delta^{(n)}(k')$. We apply $n-1$ dimensions of the delta functions to the integrations over $\boldsymbol{k}_{\alpha T}$ and k_α^-. The rest of the Wilson line does not depend on these two variables. There remains in the cut Wilson line the factor $2\pi \delta(k' \cdot \tilde{w}_\alpha) = 2\pi \delta(k'^+)$, which is just the rule (12.69) that we needed to use the sum-over-cuts argument. What we have just shown is that the integral of (12.67) over $\boldsymbol{k}_{\alpha T}$ is exactly the cut diagram (12.68) without any external integral at all. What was an external integral over $\boldsymbol{k}_{\alpha\,\mathrm{T}}$ and k_α^- is now an internal loop integral routed from the $C_{(\alpha)}$ part of the graph back across the cut Wilson line.

The resulting quantity, which we will call $D_{(\alpha)}$, depends only on k_α^+, and the sum-over-cuts argument, as in Sec. 12.7, converts it to a difference of uncut amplitudes,

$$
D_{(\alpha)} \;=\; \left[\;\raisebox{-1em}{\parbox{5cm}{\centering $C_{(\alpha)}$ \\[1.2em] $\tilde{w}_\alpha$}}\;\right] \;+\; \left[\;\raisebox{-1em}{\parbox{5cm}{\centering $C_{(\alpha)}$ \\[1.2em] $\tilde{w}_\alpha$}}\;\right] \tag{12.70}
$$

and for each term there is no trapping of the momentum integration in the collinear region.

But this result is not sufficient for our purposes, since we defined coordinates for the collinear subgraph with respect to its final-state momentum (without the Wilson line). Thus $\boldsymbol{k}_{\alpha\,\mathrm{T}}$ is actually fixed at zero in the factor $C_{(\alpha)}$ in the factorization result (12.65). Now, at an exact collinear limit, transverse momenta are zero. The cancellation of collinear singularities in the integrated quantity (12.70) is between terms that have final states that differ by a shift in transverse momentum, which vanishes in the collinear limit. In the factorization formula (12.65) there is an integral over the center-of-mass angles of the collinear graphs $C_{(2)}, \dots,$ $C_{(N)}$, and the remaining angular dependence is smooth dependence in the hard factor. This is sufficient to get the desired collinear cancellation.

The overall result is that the only genuinely collinear factor is $C_{(1)}$, which contains the detected hadron. The rest of the graph can be treated as making a hard subgraph. A more direct proof would be desirable.

12.8.8 Definitions of fragmentation functions

From the approximant for attaching the subgraph $C_{(1)}$ to the hard subgraph, we obtain the operator definitions of the fragmentation functions. The operators are the same as in parton densities (Sec. 7.5), since the approximants are the same. The normalizations are the same as in non-gauge theories, and therefore correspond to a number density interpretation.

We define the bare fragmentation function for a quark of flavor j by

$$
d_{(0)\,h/j}(z) = \frac{\mathrm{Tr}_{\mathrm{color}}}{N_{c,j}} \frac{\mathrm{Tr}_{\mathrm{Dirac}}}{4} \sum_X z^{n-3} \int \frac{\mathrm{d}x^-}{2\pi} e^{ik^+ x^-}
$$

$$
\times\, \gamma^+ \langle 0 |\, \bar{T}\, W(\infty, x^-/2; \tilde{w}_1)\, \psi_j^{(0)}(x/2)\, | p, X, \text{out} \rangle
$$

$$
\times\, \langle p, X, \text{out} |\, T\, \bar{\psi}_j^{(0)}(-x/2)\, W(\infty, -x^-/2; \tilde{w}_1)^\dagger\, | 0 \rangle
$$

$$
= \frac{\mathrm{Tr}_{\mathrm{color}}}{N_{c,j}} \frac{\mathrm{Tr}_{\mathrm{Dirac}}}{4} z^{n-3} \int \frac{\mathrm{d}k^-\, \mathrm{d}^{n-2}\boldsymbol{k}_{\mathrm{T}}}{(2\pi)^n} \gamma^+ \raisebox{-1em}{\parbox{4cm}{\centering p \\[0.3em] $C_{(1)}$ \\[1em] $\tilde{w}_1$}} \tag{12.71}
$$

where $x^\mu = (0, x^-, \boldsymbol{0}_{\mathrm{T}})$. The Fourier transform implements the integral over $\boldsymbol{k}_{1\,\mathrm{T}}$ and k_1^-, and we now drop the subscript "1" on the external momentum of $C_{(1)}$.

This definition differs from that in a non-gauge theory, (12.40), only by having a Wilson line going out to future infinity in the light-like direction $\tilde{w}_1$ from the quark field. In principle, there are rapidity divergences associated with the light-like Wilson line and these are to be canceled by appropriate subtractions, which amount to a soft factor in the fragmentation function. But according to Sec. 12.8.6 this soft factor is unity, so that rapidity divergences cancel. This happens since we integrated over all transverse momentum.

When we go to the physical dimension $n = 4$, there are UV divergences, which we define to be renormalized away. Thus the final definition of the finite renormalized fragmentation function is given by (12.15), with the QCD definitions of bare fragmentation functions.

The antiquark fragmentation function is defined similarly.

The bare gluon fragmentation function is

$$
d_{(0)\,h/g}(z) = \frac{z^{n-3}}{N_{c,\,\text{gluon}}(n-2)k^+} \sum_X \int \frac{dx^-}{2\pi} e^{ik^+x^-}
$$

$$
\times (-g_{\lambda\lambda'}) \langle 0 |\, G^{+\lambda}_{(0),\,b}(-x/2) \left[W_A(\infty, -x^-/2; \tilde{w}_1)^\dagger \right]_{bc} |p, X,\,\text{out}\rangle
$$

$$
\times \langle p, X,\,\text{out}|\, \left[W_A(\infty, x^-/2; \tilde{w}_1) \right]_{cd}\, G^{+\lambda'}_{(0),\,d}(-x/2) |0\rangle , \qquad (12.72)
$$

with again renormalization to be applied by (12.15). The field strength tensor $G^{+\lambda}$ is used for the same reason explained in Sec. 11.3 for the gluon density: it corresponds to a collinear G gluon, (11.6), attaching to the hard scattering. Since $G^{++} = 0$, the only terms in the Lorentz trace are for transverse indices. Then the overall factor $1/(n-2)$ in (12.72) gives an average over transverse gluon polarizations. The Wilson lines W_A are of course in the adjoint representation appropriate for gluons.

Feynman rules for the above definitions can be read off those for parton densities, with minor obvious variations. Renormalization is applied, leading to DGLAP equations, as stated in Sec. 12.2.2, with the derivations being like those for parton densities, Sec. 11.4.

12.8.9 Final state in fragmentation function

The final state $|p, X,\,\text{out}\rangle$ in the fragmentation functions has non-zero color, because the field that creates it has color. This is obviously a wrong situation non-perturbatively in a confining theory. A full resolution of the issue has not appeared in the literature. But the following remarks suggest a possible approach.

The Wilson line represents a color source moving in the opposite direction to the parton initiating the fragmentation function. The color source is non-dynamical, moving along a fixed line, so let us call it a pseudo-parton. In the definition of a fragmentation function we treat the operator on the right as creating a state consisting of a parton and a pseudo-parton in an overall color singlet state. The pseudo-parton propagates to future infinity in the opposite direction to the jet that we can consider as being initiated by the regular parton. Then the final state consists of the ordinary hadrons in the jet, and at the opposite end a pseudo-meson consisting of the pseudo-parton and a regular parton. There are in addition some hadrons of intermediate rapidity. In some sense we consider the state space of QCD to include states of pseudo-partons.

12.8.10 Final result for factorization

Using these definitions of the renormalized fragmentation functions, together with the cancellation in the soft factor, we convert the factorization formula (12.65) to the form already stated in (12.13), (12.21), and (12.24). The factorization formula has the same form as in a non-gauge theory. However, the derivation was much more complicated.

12.9 Use of perturbative calculations

To apply the factorization formalism phenomenologically, we need perturbative calculations of the hard-scattering coefficients and of the evolution kernels of the fragmentation functions. These are independent of the choice of the detected particle. So, as explained in Sec. 9.3.1 for DIS, a convenient method of calculation is to choose the detected particle to be an on-shell quark or gluon, and then to perform low-order perturbative calculations of the hadronic tensor $W^{\mu\nu}$ and of the fragmentation functions. The factorization formula and the evolution equations allow us to deduce the hard-scattering coefficients and the evolution kernels.

We perform these calculations with masses set equal to zero, and with dimensional regularization applied. There are soft and collinear divergences at the physical space-time dimension, but the divergences cancel in the hard scattering and the evolution kernels.

12.10 One-loop renormalization of fragmentation function

In this section I summarize one-loop calculations of the fragmentation functions for massless partons. We will deduce the renormalization of the fragmentation functions, from which follows the DGLAP kernels. At one-loop order, these are in fact equal to those for the parton densities. The calculations will also be used in the subtractions in calculations of the hard-scattering coefficients, Sec. 12.11.

12.10.1 Quark in gluon

There is one graph, Fig. 12.16, for the fragmentation function of a gluon into a quark. From the Feynman rules (cf. Fig. 7.12) for (12.72) we get

$$
\frac{g^2}{16\pi^2} d_{q/g}^{[1]}(z) = \frac{g^2 \mu^{2\epsilon} z^{1-2\epsilon}}{N_{c,g}(2-2\epsilon)(2\pi)^{4-2\epsilon}} \int dk^- \, d^{2-2\epsilon} k_{\mathrm{T}} \, \frac{(2\pi)\delta((k-p)^2)}{(k^2)^2}
$$

$$
\times \mathrm{Tr}\, t_\alpha t_\alpha \, \mathrm{Tr}\Big[-k^+ \slashed{p}\gamma^\mu (\slashed{k}-\slashed{p})\gamma_\mu + \slashed{p}\slashed{k}(\slashed{k}-\slashed{p})\gamma^+
$$

$$
+ \slashed{p}\gamma^+ (\slashed{k}-\slashed{p})\slashed{k} - \frac{k^2}{k^+}\slashed{p}\gamma^+ (\slashed{k}-\slashed{p})\gamma^+ \Big] + \text{UV counterterm}
$$

$$
= \frac{g^2 T_F (4\pi\mu^2/z^2)^\epsilon}{8\pi^2 \Gamma(1-\epsilon)} \left[1 - \frac{2z(1-z)}{1-\epsilon} \right] \int_0^\infty dk_{\mathrm{T}}^2 (k_{\mathrm{T}}^2)^{-1-\epsilon} + \text{UV c.t.} \quad (12.73)
$$

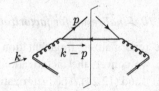

Fig. 12.16. One-loop Feynman graph for fragmentation function of quark in gluon.

Here the superscript "[1]" denotes "one-loop", and the group-theory factor is $T_F = \frac{1}{2}$ in QCD. The integral in the last line is zero. Being scale-free, it has a cancellation between equal and opposite divergences at zero and infinite k_T. The counterterm is computed from the large k_T part, giving

$$\frac{g^2}{16\pi^2} L^{[1]}_{q/g}(z) = -\frac{g^2 T_F [z^2 + (1-z)^2]}{8\pi^2} \frac{S_\epsilon}{\epsilon}, \tag{12.74}$$

in the notation of (12.15), with S_ϵ given in (A.41).

This is also the UV-renormalized value of the fragmentation function:

$$\frac{g^2}{16\pi^2} d^{[1]}_{q/g}(z) = \frac{-g^2 T_F [z^2 + (1-z)^2]}{8\pi^2} \frac{S_\epsilon}{\epsilon}. \tag{12.75}$$

This exhibits the collinear divergence in the massless fragmentation function, and will be used in a subtraction term for the hard-scattering coefficient $C^{\mu\nu}_g$.

12.10.2 Quark in quark

The one-loop graphs for the fragmentation function of a quark into a quark are shown in Fig. 12.17. All are diagonal in quark flavor.

Graph (a)

This is

$$\frac{g^2}{16\pi^2} d^{[1,a]}_{q(j')/q(j)}(z) = \frac{g^2 \mu^{2\epsilon} \delta_{j'j} z^{1-2\epsilon}}{4N_{c,q} (2\pi)^{4-2\epsilon}} \int dk^- \, d^{2-2\epsilon} k_T \, (2\pi) \delta\big((k-p)^2\big)$$

$$\times \frac{-\operatorname{Tr} \gamma^+ \slashed{k} \gamma^\mu \slashed{p} \gamma_\mu \slashed{k}}{(k^2)^2} \operatorname{Tr} t_\alpha t_\alpha + \text{UV counterterm}$$

$$= \frac{g^2 C_F \delta_{j'j} (4\pi \mu^2)^\epsilon}{8\pi^2} \frac{(1-\epsilon)(1-z)z^{-2\epsilon}}{\Gamma(1-\epsilon)}$$

$$\times \int_0^\infty dk_T^2 (k_T^2)^{-1-\epsilon} + \text{UV c.t.} \tag{12.76}$$

The minus sign in the Dirac trace is from the numerator of the gluon propagator. Again, the value of the integral is zero, while canceling its UV divergence gives the counterterm:

$$\text{UV c.t. of (a)} = -\frac{g^2 C_F \delta_{j'j} (1-z)}{8\pi^2} \frac{S_\epsilon}{\epsilon}. \tag{12.77}$$

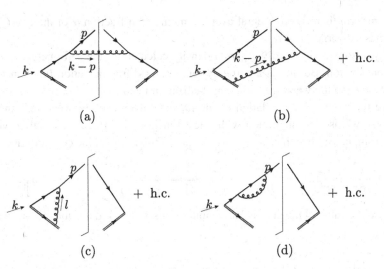

Fig. 12.17. One-loop Feynman graphs for fragmentation function of quark in quark; "h.c." means "hermitian conjugate".

Graph (b)

To make the relation to the Feynman rules for the Wilson line explicit, I write graph (b) in more detail, listing the individual propagators and vertices.

$$
\frac{g^2}{16\pi^2} d^{[1,b]}_{q(j')/q(j)}(z) = \frac{g^2 \mu^{2\epsilon} \, \delta_{j'j} \, z^{1-2\epsilon}}{4 N_{c,q} (2\pi)^{4-2\epsilon}} \int dk^- \, d^{2-2\epsilon} k_T
$$

$$
\times \; \underset{\text{Dirac color}}{\text{Tr Tr}} \; \gamma^+ \frac{i}{k^+ - p^+} (i g t_\alpha g^{\mu+}) \, \slashed{p} \, (-i g t_\alpha)^\dagger \gamma^\nu \frac{-i\slashed{k}}{k^2 - i0}
$$

$$
\times \; (-g_{\mu\nu})(2\pi)\delta\big((k-p)^2\big) + \text{UV counterterm}
$$

$$
= \frac{g^2 C_F \, \delta_{j'j} \, (4\pi\mu^2)^\epsilon z^{-2\epsilon}}{8\pi^2 \Gamma(1-\epsilon)} \frac{z}{1-z} \int_0^\infty dk_T^2 (k_T^2)^{-1-\epsilon} + \text{UV c.t.} \quad (12.78)
$$

Notice that the Wilson line has a vertex $i g t_\alpha$ rather than $-i g t_\alpha$, because it corresponds to an outgoing *anti*-triplet object. There are reversed $i0$s on the right of the final-state cut, as usual. The rapidity divergence of this graph manifests itself in the $1/(1-z)$ singularity, which gives a divergence when the graph is integrated with a test function $f(z)$. The hermitian-conjugate graph gives the same value. The UV counterterm for the two graphs is

$$
\text{UV c.t. of (b)+(b)}^\dagger = -\frac{g^2 \delta_{j'j} C_F}{4\pi^2} \frac{z}{1-z} \frac{S_\epsilon}{\epsilon}, \quad (12.79)
$$

Remainder of calculation

The virtual graph (c) involving the Wilson line has the same expression as the corresponding graph in the parton-density calculation (9.11). The value of the graph is again zero. In the

UV counterterm there is an integral over the momentum fraction α of the quark, and the integral has a divergence at $\alpha = 1$.

Finally, also exactly as in the parton-density calculation, the self-energy graph gives a zero contribution. But to get a correctly renormalized fragmentation function we add a contribution from the wave-function renormalization factor.

To see the expected cancellation of the rapidity divergence between real and virtual corrections, we use an integration with a test function, as in Sec. 9.4.4, after which the $1/(1-z)$ singularity becomes a plus distribution. The result for the counterterm is

$$\frac{g^2}{16\pi^2}L^{[1]}_{q/q}(z) = -\frac{g^2}{16\pi^2}C_F\frac{S_\epsilon}{\epsilon}\left[-\frac{4}{(1-z)_+} + 2 + 2z - 3\delta(z-1)\right]. \quad (12.80)$$

This is also the value of the graphs plus counterterms, to be used in subtractions in the hard scattering.

12.10.3 Gluon in quark, and gluon in gluon

The one-loop fragmentation functions to a gluon, $d^{[1]}_{g/q}(z)$ and $d^{[1]}_{g/g}(z)$, can be computed similarly. The calculation is left as an exercise.

12.10.4 DGLAP kernels

The one-loop renormalization counterterms are exactly the same for the fragmentation functions as those we calculated in Sec. 9.4 for the parton densities. It follows that the DGLAP evolution kernels are the same, so that the values in (9.6), (9.23), (9.24), and (9.25) also apply to fragmentation functions.

This relation does not hold at higher order. For the two-loop values see Furmanski and Petronzio (1980); Curci, Furmanski, and Petronzio (1980); Floratos, Kounnas, and Lacaze (1981); Kalinowski, Konishi, and Taylor (1981); Kalinowski *et al.* (1981). Note that there are misprints in the published version of Furmanski and Petronzio (1980).

12.11 One-loop coefficient functions

I now summarize a calculation of the one-loop coefficient functions for $e^+e^- \to hX$. The calculation is phenomenologically significant, and it will also illustrate the principles to be applied.

Thus let $W^{\mu\nu}_{\text{partonic } j}$ be defined like the hadronic tensor, but with the detected particle being an (on-shell) massless parton of flavor j, so that the process is $e^+e^- \to jX$. This object exists if we restrict ourselves to perturbation theory and use dimensional regularization to regulate the collinear and soft divergences. Factorization (12.21) gives

$$W^{\mu\nu}_{\text{partonic } j}(p,q) = \sum_{j'}\int_{x-}^{1+}\frac{dz}{z^2}\,d_{j/j'}(z;\mu)\,C^{\mu\nu}_{j'}(\hat{k},q;g(\mu),\mu). \quad (12.81)$$

Both $W^{\mu\nu}_{\text{partonic } j}$ and the partonic fragmentation functions can be computed from Feynman rules. From the expansion to one loop, we find

$$W^{[1]\,\mu\nu}_{\text{partonic } j}(p, q), = \sum_{j'} \frac{1}{x} d^{[1]}_{j/j'}(x; \mu)\, \tilde{C}^{[0]\,\mu\nu}_{j'}(p/x, q) + C^{[1]\,\mu\nu}_{j}(p, q), \qquad (12.82)$$

where the superscripts "[0]" and "[1]" denote the order of perturbation theory, and $\tilde{C}^{[0]}$ denotes the lowest-order coefficient function (12.26) *without* its $\delta(x/z - 1)$ factor. From this we see that the one-loop coefficient function is its unsubtracted counterpart $W^{[1]\,\mu\nu}_{\text{partonic } j}$ minus a one-loop partonic fragmentation function, as calculated in Sec. 12.10. The formula is easily converted to one for the structure functions F_1 and F_2.

For $W^{[1]\,\mu\nu}_{\text{partonic } j}(p, q)$, the graphs are exactly the same as for the calculation of the e^+e^- annihilation total cross section in Sec. 4.2. We simply have to remove the integral over the momentum of the detected parton, adjusting the normalization to be that of $W^{\mu\nu}$.

Thus for the case of the inclusive production of a quark, consider the real-gluon-emission graphs of Fig. 4.8(a) and (b), which previously gave (4.27). Now, we replace the 3-body phase-space integral (A.44) by

$$\frac{1}{4\pi} \int \prod_{i=2}^{3} \frac{d^{3-2\epsilon} k_i}{(2\pi)^{3-2\epsilon} 2|k_i|} (2\pi)^{4-2\epsilon} \delta^{(4-2\epsilon)}(q - p - k_2 - k_3) f(\boldsymbol{p}, \boldsymbol{k}_2, \boldsymbol{k}_3)$$

$$= \frac{(4\pi)^\epsilon Q^{-2\epsilon}}{32\pi^2 \Gamma(1 - \epsilon)} \text{Ang. avg.} \int_0^1 d\alpha \big[(1 - x)\alpha(1 - \alpha)\big]^{-\epsilon} f(\boldsymbol{p}, \boldsymbol{k}_2, \boldsymbol{k}_3). \qquad (12.83)$$

Here the quark momentum k_1 is replaced by p, and the scalar variables in (4.27) are written as $y_1 = 1 - x$, $y_2 = x\alpha$, and $y_3 = x(1 - \alpha)$. There also appears the factor $1/(4\pi)$ from the definition (12.3) of $W^{\mu\nu}$.

One way of simplifying the calculation is to use scalar projections of the hadronic tensor, $-g_{\mu\nu}W^{\mu\nu}$ and $p_\mu p_\nu W^{\mu\nu}$, from which can be deduced results for the structure functions and for $d\sigma_T / dx$ and $d\sigma_L / dx$. At the end of the calculation, the variable x will be replaced by x/z for use in the factorization formulae.

The details of the calculation are left as an exercise. The results, in the $\overline{\text{MS}}$ scheme, are given in Rijken and van Neerven (1997), where also the NNLO coefficients are calculated. The NLO coefficients were first calculated by Baier and Fey (1979); Altarelli *et al.* (1979).

The variables in (12.83) were suitable for the quark coefficient function. The antiquark coefficient function is equal. The gluonic coefficient function is obtained by applying the variables α and x to a different permutation of partonic momenta: $y_1 = x\alpha$, $y_2 = x(1 - \alpha)$, and $y_3 = 1 - x$. As in Fig. 4.8, the labels 1, 2, and 3 refer to the quark, antiquark, and gluon respectively.

12.12 Non-perturbative effects and factorization

Gupta and Quinn (1982) pointed out a problem with factorization in the case that QCD is replaced by a theory in which all the quarks are heavy. Initially a quark-antiquark pair that is produced in e^+e^- annihilation at large Q/M goes outward at almost the speed of light. If the quark and antiquark were to hadronize into a jet of color-singlet hadrons, there would need to be production of quark-antiquark pairs in the color flux tube joining the pair. But since all quarks are heavy, this is a slow weak-coupling process, governed by $\alpha_s(M)$. At the same, the gluonic non-perturbative interaction is still effective, and will tend to bring the quark and antiquark back. In the language of Sec. 4.3.1, the elastic-spring picture would likely be a better approximation than the breakable-string picture that appears to be valid in real QCD with its light quarks. This would break factorization for the inclusive hadron cross section; for example the direction of the jet and the hadrons in it would not correspond to the direction of a parton produced at short distances.

Now our proof of factorization used the structure of momentum regions that is seen in perturbation theory. So an important issue of principle is to what extent non-perturbative effects change the results. This is a far-from-completely understood subject. It would seem best to consider the process in coordinate space. Then the breakable-string picture would appear to be compatible with preserving the factorization structure seen in perturbation theory.

This led Gupta and Quinn to an interesting question. Suppose an experimental test were made of a perturbatively calculated jet cross section or of a factorized hadron-production process such as we treated in this chapter, and suppose that experiment and theory substantially disagreed. Would this count as evidence against QCD? Gupta and Quinn argued cogently that it would not, by itself, falsify QCD. The reason is that they could show a counterexample where the theoretical methods are violated non-perturbatively without any problem with the perturbative calculations.

What would actually be falsified would be the combination of QCD and the (mostly implicit) assumptions about non-perturbative physics used in deriving factorization etc.

In the time since Gupta and Quinn (1982), there have be many successful comparisons of QCD predictions with data. So we should not count all of these successes as successful predictions of QCD itself. An isolated single experiment in this area does not test QCD. Some of the results should be counted as establishing the breakable-string picture. Then the other experiments can be regarded as QCD tests.

At the present time, one must regard QCD as being very well established. Failure of a comparison between QCD predictions and experiment is highly unlikely to impinge on QCD itself. Depending on the situation, much more likely situations would involve any or all of: (a) problems with the experiment itself, (b) problems with more exotic QCD methods, and (c) physics beyond the Standard Model.

12.13 Generalizations

Although the last part of the proof in Sec. 12.8 was specific to the one-particle-inclusive cross section, the bulk of it applies to much more general situations in e^+e^- annihilation.

12.13.1 Multiparticle cross sections

Consider an inclusive cross section differential in more than one hadron. We first suppose the particles are all at wide angles with respect to each other. In that case, in the

Fig. 12.18. Three-jet configuration with registered particles (thick lines) in two of the jets. The line lengths indicate momenta.

leading-region analysis, each of the particles arises from a different collinear subgraph. An example of such a final-state configuration is shown in Fig. 12.18.

We simply apply the same method of proof as for the single-particle-inclusive cross section. For each measured particle its collinear subgraph becomes a fragmentation function, and we have the factorization property

$$\frac{d\sigma}{\prod_{\alpha=1}^{N_p}(d^3\boldsymbol{p}_\alpha/E_{p_\alpha})} = \prod_{\alpha=1}^{N_p}\left[\sum_{j_\alpha}\int\frac{dz_\alpha}{z_\alpha^2}d_{h_\alpha/j_\alpha}(z_\alpha)\right]\frac{d\hat{\sigma}_{\text{partonic, subtracted}}}{\prod_{\alpha=1}^{N_p}(d^3\boldsymbol{k}_\alpha/E_{k_\alpha})}, \qquad (12.84)$$

with hadron and parton 3-momenta related by $\boldsymbol{p}_\alpha = z_\alpha \boldsymbol{k}_\alpha$. We treat this as a partonic cross section convoluted with a number density for the partons to make the measured hadrons. As usual, the partonic cross section is subtracted.

Each of the fragmentation functions contains an integral over its parent parton's minus and transverse momentum (defined with respect to the hadron in the fragmentation function). As in the one-particle-inclusive cross section, we use approximated parton kinematics for the hard scattering. That works when the hadrons are at wide angle, since it is equivalent to a small shift in the hadronic momenta. For example a transverse momentum of order Λ corresponds to an angular shift of order Λ/Q.

(The reader may point out that the integral over partonic momenta also extends to large minus and transverse momenta, where the approximation is always bad. A reminder is needed that within the subtraction method, all that is necessary is that the approximant be accurate to order k_T/Q and k^-/Q for its design region. As the distance of momenta from the skeleton of some region R increases, so does the error in the region's approximant T_R. But, as illustrated in Sec. 10.2, the increasing errors are compensated by the terms for larger regions together with their double-counting subtraction terms.)

12.13.2 Back-to-back region

But when the detected hadrons are almost back-to-back, as in Fig. 12.19, the neglect of partonic transverse momentum in the hard scattering is no longer correct, even in the collinear region. The situation therefore needs a somewhat different kind of factorization, which we will treat in Ch. 13.

Alternatively, we can integrate over the angle between the measured particles, averaging over the back-to-back region with a suitably broad function. At this point, the neglect of

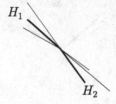

Fig. 12.19. Two-jet configuration with a registered particle (thick lines) in each jet.

partonic transverse momenta in the averaging function regains its accuracy in the collinear region, so that we can continue to use integrated fragmentation functions.

12.13.3 Multiparticle fragmentation

Another simple generalization is when two (or more) measured hadrons are approximately parallel. Then they come out of a single collinear subgraph.

This situation is dealt with by an elementary generalization (Konishi, Ukawa, and Veneziano, 1978) of the definitions of fragmentation functions. For example, consider the case of two measured hadrons of momenta p_1 and p_2. In the final state in a definition like (12.35), we replace $|p, X, \text{out}\rangle \langle p, X, \text{out}|$ by $|p_1, p_2, X, \text{out}\rangle \langle p_1, p_2, X, \text{out}|$. At the partonic end of the fragmentation function, nothing changes. So all the issues about renormalization, DGLAP evolution, and the construction of a hard-scattering coefficient function are unchanged. The fragmentation function becomes a function of more variables, representing the kinematics of p_1 and p_2 relative to the parton.

A significant use of this idea is in transverse-spin physics. With fragmentation to a single pion, there is no polarization dependence of the fragmentation function; only the unpolarized fragmentation functions are non-zero. But with two-particle fragmentation, a transversely polarized quark can give an azimuthal dependence of the form

$$A + B \cos\phi. \tag{12.85}$$

Here ϕ is the angle in the transverse plane between the transverse-spin vector of the quark and the normal to the plane of the two measured pions. The coefficient A is proportional to the ordinary unpolarized fragmentation function, while B is proportional to a kind of polarized fragmentation function that was proposed in Collins, Heppelmann, and Ladinsky (1994). It can be probed in e^+e^- annihilation, because there is a correlation of the transverse spins of the quark and antiquark in the lowest-order graph. Therefore the polarized dihadron fragmentation function appears in the factorization theorem for e^+e^- annihilation to four pions, with the pions grouped in two small-angle pairs (Artru and Collins, 1996). The function also appears in factorization for DIS with two measured hadrons in the final state when the target hadron is transversely polarized.

Data on the two-hadron fragmentation function have recently become available: Airapetian *et al.* (2008); Vossen *et al.* (2009); Wollny (2009). Fits have been made by Bacchetta *et al.* (2009).

12.13.4 Jet cross sections

For the e^+e^- annihilation total cross section to hadrons, we used a sum over all final states to get a perturbatively calculable IR-safe cross section. Similarly, in inclusive cross sections, we used a similar sum to obtain cancellations of IR-sensitive parts of the soft factor and of those collinear factors that did not couple to measured hadrons.

The cancellations involve collinear and soft interactions. In the exact collinear (or soft) limit, these interactions cause transitions between different final states with the same momentum. This suggests a general strategy to obtaining perturbatively calculability by defining IR-safe jet cross sections. These are computed from the angular pattern of energy flow, and do not depend on how the energy is split among particles.

A simple example is a calorimetric cross section (Sterman, 1996) in e^+e^- annihilation. Here we use a cross section weighted by a suitable function S of the momenta of the particles in the final state:

$$\sigma_S = \sum_n \int d\tau_n \frac{d\sigma}{d\tau_n} S_n(p_1/Q, \ldots, p_n/Q), \qquad (12.86)$$

where n is the number of hadrons in the final state, and $d\tau_n$ represents the element of n-body phase-space. The weight function S_n is defined for any n-body configuration.

The cancellations needed for IR safety of the cross section σ_S occur if (Sterman, 1996) "the weight function does not distinguish between states in which one set of collinear particles is substituted for another set with the same total momentum, or when zero-momentum particles are absorbed or emitted". Mathematically this is formulated as follows:

- The weighting functions are smooth.
- They are symmetric functions of their arguments.
- For massless momenta, they obey

$$S_n(p_1/Q, \ldots, p_i/Q, \ldots, p_{n-1}/Q, \lambda p_i/Q)$$
$$= S_{n-1}(p_1/Q, \ldots, p_i(1+\lambda)/Q, \ldots, p_{n-1}/Q), \qquad (12.87)$$

with λ being any real parameter $\lambda \geq 0$.

The weighting functions are defined to be functions of momenta scaled by Q. This matches the Libby-Sterman analysis, since the cancellations needed for IR safety occur in a fixed region of the scaled momenta. Smoothness of the weighting functions is needed because the necessary cancellations occur in a neighborhood of the massless PSS configurations.

In practice, two rather different approaches are used instead, which more directly probe the jet structure of final states. One is to define global measures of the jet structure of a final state. A classic example is thrust,[1] defined on an n-particle state by

$$T \overset{\text{def}}{=} \frac{1}{\sum_i |p_i|} \max_{\hat{n}} \sum_{i=1}^n |p_i \cdot \hat{n}|, \qquad (12.88)$$

---

[1] The definition given here is the current standard one, and is based on a slightly different definition by Farhi (1977).

with the maximum being over unit 3-vectors $\hat{n}$ in the overall center-of-mass frame. The direction that gives the maximum is called the thrust axis. Thrust has a maximum value of unity, when a final state has a perfect 2-jet configuration, i.e., some of the momenta are exactly aligned in one direction and the others are exactly aligned in the opposite direction. A spherically uniform distribution of momenta gives $T = \frac{1}{2}$.

Applying this definition of thrust as a weight function gives the average value of thrust, a measure of the average 2-jet-likeness of final states. Commonly a cross section *differential* in thrust is measured, e.g., Bethke *et al.* (2009). Showing that the differential thrust distribution is IR safe requires a generalization of the previous discussion.

Perhaps the most common approach is to define jets directly by grouping measured hadrons into clusters by some "jet algorithm". The clusters are labeled as jets, and cross sections differential in jet momenta are measured. In a leading-order approximation, a jet's momentum is close to its parent parton's momentum as in Fig. 2.3. It is quite non-trivial to determine whether a particular jet algorithm is IR safe. An important practical constraint is that the algorithm should be suitable for implementation both in experimental analyses and in theoretical calculations. See Salam (2010) for a recent review.

12.14 Semi-inclusive deeply inelastic scattering

Another classic process where fragmentation functions appear is semi-inclusive deeply inelastic scattering (SIDIS), i.e., DIS differential in one (or more) hadrons in the final state, e.g., $e(l) + P \rightarrow e(l') + \pi(p_h) + X$, Fig. 12.20.

12.14.1 Kinematics and structure functions

For the kinematics, we need to supplement the variables for DIS by a specification of the momentum of the outgoing hadron. In the Breit frame, we write

$$q = \left(-xP^+, \frac{Q^2}{2xP^+}, \mathbf{0}_{\mathrm{T}} \right), \tag{12.89a}$$

$$P = \left(P^+, \frac{M^2}{2P^+}, \mathbf{0}_{\mathrm{T}} \right), \tag{12.89b}$$

$$p_h = \left(\frac{p_{h\mathrm{T}}^2 + m_h^2}{2p_h^-}, p_h^-, \boldsymbol{p}_{h\mathrm{T}} \right). \tag{12.89c}$$

For the independent scalar variables of the hadronic part of the cross section, we use x and Q as usual, together with

$$z = \frac{P \cdot p_h}{P \cdot q} \simeq \frac{p_h^-}{q^-}, \qquad |\boldsymbol{p}_{h\mathrm{T}}|^2 \simeq z^2 Q^2 + 2zq \cdot p_h, \tag{12.90}$$

and the azimuthal angle ϕ_h of $\boldsymbol{p}_{h\mathrm{T}}$. The approximations in (12.90) are valid when masses can be neglected. The standard specification of the angle is given by the Trento convention (Bacchetta *et al.*, 2004); the angle is relative to the lepton plane: Fig. 12.21.

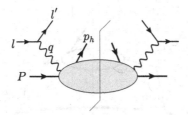

Fig. 12.20. SIDIS cross section.

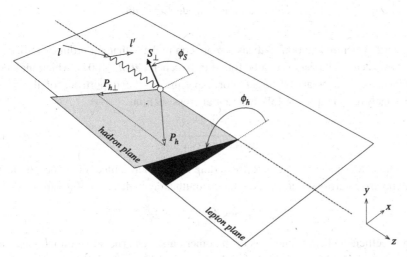

Fig. 12.21. SIDIS kinematics, from Bacchetta *et al.* (2004). This diagram is given in the target rest frame, and gives the Trento convention for defining the azimuthal angles of the measured outgoing hadron and of the target's spin vector. (Copyright (2004) by The American Physical Society.)

The significance of the z variable is given in the parton-model approximation applied to Fig. 12.22. Viewed in the Breit frame, the outgoing quark is approximately light-like, $k + q \simeq (0, q^-, \mathbf{0}_\mathrm{T})$. Thus the experimentally measured variable z approximates the fractional momentum of the detected hadron relative to its parent quark, just as x approximates the fractional momentum of the struck quark k relative to the target hadron.

Given the basic meaning of a fragmentation function as a number density for a hadron in a parton-induced jet, we immediately deduce a parton-model formula for the cross section integrated over the transverse momentum $\boldsymbol{p}_{h\,\mathrm{T}}$ of the detected hadron:

$$\frac{\mathrm{d}\sigma}{\mathrm{d}x\,\mathrm{d}y\,\mathrm{d}z} \simeq \frac{4\pi\alpha^2}{yQ^2}(1 - y + y^2/2) \sum_j e_j^2 f_j(x) d_{h/j}(z). \quad \text{(Parton model)} \qquad (12.91)$$

This is obtained by appending fragmentation functions to the parton model for inclusive DIS, from (2.22) and (2.29), with neglect of masses. As usual $y = q \cdot P / l \cdot P \simeq Q^2/(xs)$.

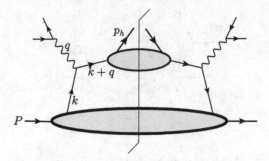

Fig. 12.22. Parton model for SIDIS.

A general structure function analysis is more complicated than for ordinary DIS, because of the extra vector. The details can be found in Bacchetta *et al.* (2007), which includes the important case of a polarized target. For experiments, the importance of the structure function analysis is that in the full differential cross section,

$$\frac{d\sigma}{dx\,dy\,dz\,dp_{h\,T}\,d\phi_h},\tag{12.92}$$

the azimuthal dependence is restricted to certain trigonometric functions. The simplest case is the unpolarized cross section, where the azimuthal dependence is of the form

$$A + B\cos\phi_h + C\cos 2\phi_h,\tag{12.93}$$

with the coefficients being functions of the other variables. The situation is more complicated for the polarized case, generalizing the same idea. The extra terms are each associated with certain polarized parton densities and fragmentation functions.

12.14.2 Leading regions

To derive factorization, we use the same sequence of steps as for e^+e^- annihilation. An important difference is in how a contour deformation is made to get out of the Glauber region; this will have particularly notable consequences when we treat situations needing transverse-momentum-dependent parton densities and fragmentation functions in Ch. 13.

The leading regions have a hard scattering on each side of the final-state cut, and the virtual photon is attached to the hard scattering. There are at least two collinear graphs, one of which includes the target. There may be a soft subgraph connecting by gluons to any of the collinear subgraphs. Each of the collinear subgraphs connects to each hard-subgraph amplitude by a primary parton line plus any number of Grammer-Yennie K-gluons. All this follows from the usual power-counting.

There are two classes of collinear subgraph: the target subgraph and what we will call hard-jet subgraphs. The "hard-jet" terminology associates them with the final state of a hard scattering. In the laboratory frame the associated regions have large transverse momentum, of order Q. In the brick-wall frame the associated regions either have large transverse

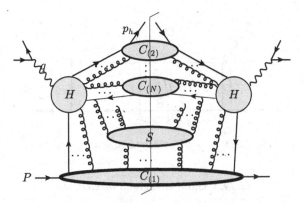

Fig. 12.23. Typical leading region for SIDIS, when the measured final-state hadron is in the "current fragmentation region". Three collinear subgraphs are shown; the minimum is two.

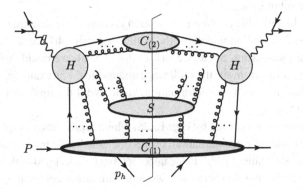

Fig. 12.24. Like Fig. 12.23, but when the measured final-state hadron is in the "target fragmentation region". Two collinear subgraphs are shown, but extra hard-jet subgraphs are also possible.

momentum or they have large minus momentum (appropriate for the parton model with a single hard jet).

Depending on its kinematics, the measured hadron either comes from a hard-jet subgraph, the target subgraph, or the soft subgraph.

When the hadron is from a hard jet, it is said to be in the "current fragmentation region", as shown in Fig. 12.23. When it is from the target subgraph, it is said to be in the "target fragmentation region", as shown in Fig. 12.24. The hadron comes from the target or soft subgraphs only when z is small, so that p_h^- is small. We can characterize a canonical situation for a target-collinear hadron by $z \sim m^2/Q^2$ and a soft hadron by $z \sim m/Q$.

For the rest of this section we will only be concerned with the current fragmentation region, $z \gg m/Q$. Target-collinear hadrons will be briefly discussed in Sec. 12.15. We will not treat soft measured hadrons at all.

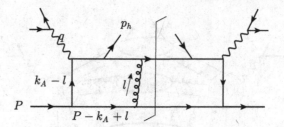

Fig. 12.25. Illustrating Glauber region for (SI)DIS.

12.14.3 Glauber region and (SI)DIS

The Ward-identity argument needed to factor the soft subgraph requires that the contour of integration over soft momenta avoid the Glauber region, i.e., that there be no pinch in the Glauber region. The general conditions for a pinch in a Glauber region were given in Sec. 5.6. They concern a pinch of the smallest components of soft momenta when they flow through collinear subgraphs.

The ability to do a suitable contour deformation is completely determined by examining low-order graphs. In e^+e^- annihilation, the collinear subgraphs are all in the final state relative to the hard scattering. So, as we saw in Sec. 10.6.4, we could deform exchanged Glauber momenta away from all the collinear singularities. For example, in an exchange between two collinear groups, the deformation would avoid the final-state singularities of both collinear groups.

In DIS, the situation is typified by Fig. 12.25, which represents a graph in a model for SIDIS. In contrast with e^+e^- annihilation, only a one-sided deformation works, to deform l^+ away from final-state singularities in the upper jet, but l^- is trapped in the target-collinear subgraph. Consider the region where the lower lines are target-collinear, with momenta of order $(Q, \lambda^2/Q, \lambda)$, the upper lines are the opposite collinear: $(\lambda^2/Q, Q, \lambda)$, while the gluon is Glauber: $l \sim (\lambda^2/Q, \lambda^2/Q, \lambda)$. The target lines trap l^- between initial- and final-state poles:

$$\frac{1}{[(k_A - l)^2 - m^2 + i0]\,[(P - k_A + l)^2 - m^2 + i0]}$$

$$\simeq \frac{1}{[-2k_A^+ l^- + \cdots + i0]\,[2(P^+ - k_A^+)l^- + \cdots + i0]}. \tag{12.94}$$

The approximation is valid when k_A is target collinear and l is soft or Glauber. The dots indicate terms that do not depend on l^-. The trapping of the l^- contour at values of order λ^2/Q is because on one line the direction of l is with the flow of plus momentum and on the other the line it is against. As to l^+, in the Glauber region the only significant dependence is in the upper two lines, where it always is in the opposite direction relative to the large outgoing collinear minus momentum. Thus l^+ is not trapped; we can deform the contour to much larger values of l^+. The deformation only stops when we get out of the Glauber region.

In order to make the region approximators, we must choose the auxiliary vectors (Sec. 10.6.4) for the K gluons so as not to obstruct the deformation. The space-like future-pointing vectors chosen in that section continue to work here.

12.14.4 Factorization for SIDIS

Once the contours are out of the Glauber region, we can copy the factorization proof for e^+e^- annihilation. We have applied approximants and subtractions for the contributions of each region of each graph, and sum over possibilities.

We first use Ward identities to extract collinear K gluons from the hard scattering, converting them to Wilson lines. Similarly we extract the soft gluons from the collinear subgraphs.

The sum-over-cuts argument applies to all the collinear subgraphs except for the target subgraph and the one to which the measured hadron attaches. Provided we average over a broad range of transverse momentum $\boldsymbol{p}_{h\,\mathrm{T}}$ for the measured hadron, we can also apply the sum-over-cuts argument to the soft subgraph. We are left with a hard subgraph convoluted with a parton density and a fragmentation function, which arise for the same reasons as in our factorization arguments separately for DIS and e^+e^- annihilation.

The resulting factorization theorem is conveniently stated in terms of a cross section differential in Lorentz-invariant phase-space, in a form generalizing (12.13):

$$E\frac{d\sigma(e + P \to e + p_h + X)}{d^3 \boldsymbol{p}_h} = \sum_{jj'} \int_{x_-}^{1+} \frac{d\xi}{\xi} \int_{z_-}^{1+} \frac{d\zeta}{\zeta^2}\, E_k \frac{d\hat{\sigma}_{jj'}}{d^3 \boldsymbol{k}}\, d_{h/j'}(\zeta; \mu) f_{j/P}(\xi; \mu).$$

(12.95)

The hard scattering is for a parton of flavor j and massless momentum $\xi\hat{P}$ to scatter inclusively to a parton of flavor j' and massless momentum $\hat{p}_h/\zeta$. Here the hatted momenta have the same 3-momenta as the unhatted momenta in the brick-wall frame, and have their energies set to make the momenta massless.

One can convert the above to a formula for the structure functions.

12.15 Target fragmentation region: fracture functions

When the detected final-state hadron is in the target-collinear region, the leading regions have the form shown in Fig. 12.24. There is a hard subgraph H, a target-collinear subgraph $C_{(1)}$, one or more hard-jet subgraphs $C_{(2)}, \ldots, C_{(N)}$ and a soft subgraph S. These are exactly like those for ordinary DIS, except that the target-collinear subgraph now contains the detected hadron. To specify the longitudinal kinematics of p_h, it is convenient not to use z, but instead a target-relative variable

$$x_h \overset{\text{def}}{=} \frac{p_h^+}{P^+} \simeq \frac{p_h \cdot q}{P \cdot q}.$$

(12.96)

Factorization can be derived by the same arguments we have already described. First, the soft subgraph can be factored out and then shown to cancel. The K gluons from the

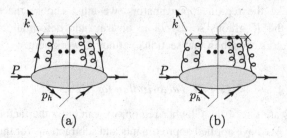

Fig. 12.26. Diagrammatic representation of gauge-invariant fracture functions. The partonic part, with its Wilson line, is the same as for parton densities in Fig. 7.9.

collinear subgraphs can be extracted from the hard subgraph, to give Wilson lines, after which the hard-jet subgraphs are of the form in which the integrations can be taken out of their collinear regions. We end up with exactly the same structure as in DIS. We have a hard scattering of identically the same form as in DIS, and it is convoluted with a target-collinear quantity whose definition is the same as a parton density, except that the final state is required to contain the detected hadron. These quantities are called "extended fracture functions" (Grazzini, Trentadue, and Veneziano, 1998). For the case of a bare quark fracture function, the definition, generalized from the quark density defined in (7.40), is (Berera and Soper, 1996)

$$M_{(0)\,jh/P}(x, x_h, \boldsymbol{p}_{h\,\mathrm{T}}) = \int \frac{\mathrm{d}w^-}{2\pi} e^{-ixP^+w^-} \sum_X \langle P|\overline{\psi}_j^{(0)}(0, w^-, \boldsymbol{0}_{\mathrm{T}})W(\infty, w^-)^\dagger|p_h, X, \mathrm{out}\rangle$$

$$\times \langle p_h, X, \mathrm{out}|\frac{\gamma^+}{2}W(\infty, 0)\psi_j^{(0)}(0)|P\rangle_{\mathrm{c}}, \tag{12.97}$$

where the Wilson line (in the minus direction) was defined in (7.41). A similar modification to the definition of the gluon density gives the gluon fracture function. These definitions are shown diagrammatically in Fig. 12.26. In the unpolarized case, there is no preferred axis in the transverse plane, so the dependence on the transverse momentum of the detected hadron is only through its size.

Since the parton kinematics are treated identically to those of parton densities, the Feynman rules at the parton end are the same as for parton densities. Hence renormalization of extended fracture functions has the same form (8.11) as for parton densities. The DGLAP equations therefore have the same form as (8.30) for parton densities:

$$\frac{\mathrm{d}}{\mathrm{d}\ln\mu}M_{jh/P}(x, x_h, \boldsymbol{p}_{h\,\mathrm{T}}; \mu) = 2\sum_{j'} \int \frac{\mathrm{d}z}{z} P_{jj'}(z; g)M_{j'h/P}(x/z, x_h, \boldsymbol{p}_{h\,\mathrm{T}}; \mu). \tag{12.98}$$

There is a kinematic constraint $x + x_h \leq 1$, given by energy positivity for the unobserved part $|X\rangle$ of the hadronic final state.

The above functions are officially called "extended fracture functions", even though the term "fracture functions" would be more natural. However the latter term was already defined (Trentadue and Veneziano, 1994) to refer to similar quantities defined with an integral over all $p_{h\,T}$. Because of the integral, these quantities also include contributions from the current fragmentation region and have more complicated evolution equations. It seems better to use only extended fracture functions.

The extended fracture functions can be notated as parton densities differential in p_h:

$$M_{jh/P}(x, x_h, p_{h\,T}) = (2\pi)^3 2 E_{p_h} \frac{d f_{jh/P}}{d^3 p_h}. \tag{12.99}$$

One way of stating the factorization theorem is to project the SIDIS cross section onto structure functions. These are like F_2 etc. for DIS, but now differential in x_h and $p_{h\,T}$:

$$\frac{d F_2(x, Q^2; x_h, p_{h\,T})}{d x_h\, d^2 p_{h\,T}}. \tag{12.100}$$

Then factorization is a simple generalization of the version (8.83) for DIS:

$$\frac{d F_1(x, Q^2; x_h, p_{h\,T})}{d x_h\, d^2 p_{h\,T}} = \sum_j \int_{x-}^{1+} \frac{d\xi}{\xi} \hat{F}_{1j}(Q/\mu, x/\xi; \alpha_s) \frac{M_{jh/P}(\xi, x_h, p_{h\,T})}{16\pi^3 x_h}, \tag{12.101a}$$

$$\frac{d F_1(x, Q^2; x_h, p_{h\,T})}{d x_h\, d^2 p_{h\,T}} = \sum_j \int_{x-}^{1+} d\xi\, \hat{F}_{2j}(Q/\mu, x/\xi; \alpha_s) \frac{M_{jh/P}(\xi, x_h, p_{h\,T})}{16\pi^3 x_h}, \tag{12.101b}$$

valid up to power-suppressed corrections. The hard-scattering coefficients are the same as in ordinary DIS.

The primary phenomenological applications are to diffractive DIS on protons. This concerns the case that x_h is close to unity (with, necessarily, $x \ll 1$), and that the detected hadron is also a proton. See Chekanov *et al.* (2010) for recent results. In this case the extended fragmentation functions are commonly referred to as "diffractive parton densities".

Quite elementary extensions of these ideas can be applied to cross sections differential in more final-state hadrons. One example is the dijet cross section in diffractive DIS (Aktas *et al.*, 2007a), which is differential in one proton in the target fragmentation region and in two hard jets. Diffractive parton densities can be obtained from a fit to ordinary diffractive DIS, without the dijet condition. Then the cross section for diffractive dijet DIS is predicted with the aid of standard perturbative calculations for the hard scattering. The success of the prediction confirms the experimental validity of the factorization approach.

Exercises

12.1 (**) Find and prove any extensions to the Ward-identity arguments of Ch. 11 that are needed to apply them to the processes treated in this chapter.

12.2 (*****) Construct a good formalism for the evolution of states in space-time from a quark state to a hadronic state. Ideally, this should be a rigorous formalism from which you can derive from first principles that partonic states evolve to jet-like configurations. Publish your results.

   Undoubtedly I have stated this (very difficult) problem quite badly, and part of the answer should be to formulate this problem more appropriately. A good solution to this problem should answer the issues raised, for example, by Gupta and Quinn (1982). See problem 5.1 for some results that may be of use.

12.3 (**) In Sec. 12.12, I discussed whether non-perturbative effects can ruin factorization in inclusive cross sections in e^+e^- annihilation. Give a more detailed and explicit account of these issues by critically using the methods and results of Einhorn (1976, 1977).

In these papers Einhorn made approximate calculations in the model of large-N_c QCD in two space-time dimensions. The final states given by this model, both in e^+e^- annihilation and in DIS, are a series of closely spaced narrow resonances. Thus the model consistently realizes an approximately unbreakable elastic-spring picture. Einhorn found contrasting results relative to the parton model for different kinds of cross section: DIS, the total cross section for e^+e^- annihilation, and the single-hadron-inclusive process $e^+e^- \to \pi + X$.

Use these results to illustrate the non-perturbative properties that either preserve or violate factorization in the different reactions.

12.4 (**) Investigate the soft gluon cancellation in Figs. 12.7 and 12.8. Assume that the quark is approximately parallel to the modeled pion, that the antiquark moves in approximately the opposite direction, and that the gluon is soft. Show that there is a cancellation between all the graphs in Figs. 12.7 and 12.8. But show that the cancellation does not work if it is restricted to subsets of graphs related by sums-over-cuts, i.e.,

- between Fig. 12.7(a) and (b),
- between Fig. 12.7(c) and (d),
- between the graphs of Fig. 12.8 alone.

For sufficiently soft gluons, internal emission will be suppressed, and there will be a cancellation in Fig. 12.8 alone. But when the gluon momentum l becomes comparable to or larger than M^2/Q, internal line emission is important. Here M denotes the invariant mass of the upper jet.

12.5 (**) Complete the proofs sketched in Sec. 12.8. Deal properly and explicitly with the issues of subtractions and of the necessary Ward identities in non-abelian gauge theories.

12.6 Obtain explicit factorization formulae for the differential cross sections for more complicated inclusive cross sections, e.g., $e^+ + e^- \to H_1 + H_2 + X$, $e^+ + e^- \to H_1 + H_2 + H_3 + X$. Assume here that the observed hadrons are at wide angles with respect to each other (and are not close to back-to-back in the two-hadron case). More general situations can be considered, of course. But that will lead you into other topics, such as those in Ch. 13.

12.7 Check and complete the one-loop calculations in Sec. 12.10.

12.8 Complete the one-loop calculations in Sec. 12.11. Verify that your results agree with Rijken and van Neerven (1997).

13

TMD factorization

An appealing interpretation of a parton density is that it is a number density of partons in a target hadron. As we saw in Sec. 6.7, a parton density in a simple theory is an expectation value of a light-front number operator, integrated over transverse momentum. A similar interpretation applies to fragmentation functions: Sec. 12.4.

As explained in Secs. 6.8 and 12.4, it is equally natural to define unintegrated, or transverse-momentum-dependent (TMD), parton densities and fragmentation functions, simply by omitting the integral over transverse momentum. In a sense, the TMD functions are more fundamental and present more information on non-perturbative phenomena than do the ordinary integrated functions. Therefore it is useful to find situations where TMD functions are needed.

In this chapter, I treat two characteristic cases. One is two-particle-inclusive e^+e^- annihilation when the detected hadrons are close to back-to-back. This process needs TMD fragmentation functions. Then I will extend this work to semi-inclusive DIS (SIDIS) with a detected hadron of low transverse momentum. In SIDIS, TMD parton densities are needed as well as fragmentation functions. A further extension to the Drell-Yan process at low transverse momentum will be covered in Sec. 14.5.

There are substantial complications in QCD. Although the discussion about light-front quantization and the associated definitions of number densities gives a general motivation, it does not work correctly in QCD (or any other gauge theory). The actual definitions are whatever is appropriate to consistently obtain a valid factorization theorem.

The generally used jargon is that factorization with integrated pdfs and fragmentation functions is called "collinear factorization", while factorization with the unintegrated functions is called "k_T factorization". For the second case, I prefer "TMD factorization". Its overall structure generalizes the results for the Sudakov form factor in Ch. 10.

13.1 Overview of two-particle-inclusive e^+e^- annihilation

The definition of an ordinary integrated pdf or fragmentation function arises from the approximants used in deriving factorization. There are two parts to an approximant. One is in the actual amplitude for the hard scattering, where we neglect transverse and

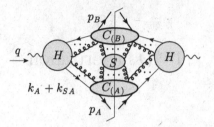

Fig. 13.1. A leading region in gauge theory for two-hadron-inclusive cross section in e^+e^- annihilation. Like Fig. 12.4(a), but with an extra detected hadron.

minus components of momentum with respect to Q. The other is in the kinematics of some groups of the final-state particles, as when some components of a jet or soft momentum are neglected relative to the large component of momentum in some other jet subgraph.

It is the second part of the approximant that determines the definition of a pdf or fragmentation function, and it can fail, even when the more fundamental first part of the approximant remains valid. Consider, for example, the collinear momentum k_A entering one jet subgraph in Fig. 12.4(a) or Fig. 12.5(a), and complete the loop by circulating it through the other jet subgraph. Neglecting k_A^- and $k_{A\,T}$ in the second jet amounts to changing the kinematics of the jet. If the jet is not observed, this gives a legitimate approximation for the inclusive cross section, as we showed more formally by routing the momentum out through the virtual photon, and by applying the approximant to the external test function in (12.11).

But the situation is quite different if, instead, we consider two-particle-inclusive annihilation $e^+e^- \rightarrow H_A H_B + X$ and choose the measured hadrons to be close to back-to-back. The leading regions are shown in Fig. 13.1, and a corresponding 2-jet final state was sketched in Fig. 12.19. Neglecting k_A^- in the second jet is still legitimate, because k_A^- is small and is neglected with respect to a large minus momentum in the unobserved part of the second jet.

But the neglect of $k_{A\,T}$ in the second jet is no longer justified. The neglect shifts the second detected hadron transversely by an amount that can be comparable (or even larger) than its transverse momentum relative to the jet. Exactly similar considerations apply to the approximant for the soft factor.

Therefore, a valid approximant must preserve the exact values of collinear and soft transverse momenta when they flow through other collinear subgraphs. A similar idea applies if we use a cross section averaged with a test function. Then we route loop integrals over soft and collinear subgraphs out through the photon vertex, and the approximant must preserve transverse momenta in the test function, unlike the definition in Sec. 12.8.1.

One direct consequence is that the relevant fragmentation functions are TMD functions, rather than the integrated functions. Another consequence is that the soft factor no longer cancels. In one-particle-inclusive annihilation, we defined the soft factor with an independent integral over all momenta for its final state, thereby enabling the proof of cancellation in Sec. 12.8.6. But this fails when the transverse-momentum integral is coupled to the other factors. Our treatment must include the uncanceled soft factor, just as for the Sudakov form factor.

One simplification does occur, and this is that the leading regions in the back-to-back case only have two collinear subgraphs, as in Fig. 13.1. To understand this consider a region like Fig. 12.4(b), with three or more collinear groups, and for which a 3-jet final state was sketched in Fig. 12.18. We have two detected hadrons which are almost back-to-back, and so the directions of their parent jets are also almost back-to-back. Now, the propagators in the hard subgraphs are power-counted as having denominators of order Q^2. But as the directions of collinear subgraphs approach each other to give a 2-jet configuration, some denominators get much smaller, to approach collinear singularities. The neighborhood of these singularities therefore dominates the cross section. In the contribution to the cross section from a region with 3 (or more) collinear subgraphs, the 2-jet region is of course subtracted out, thereby giving a power-suppression relative to the 2-jet regions. Therefore, as claimed, the leading regions are restricted to those of Fig. 13.1, when the detected hadrons are close to back-to-back.

Deviations from the exact back-to-back configuration of the hadrons are controlled by transverse momentum within the two collinear subgraphs (and in the soft subgraph). Thus they are controlled by transverse momentum generated in fragmentation functions. This suggests a general pattern: TMD functions are needed whenever the directions of detected hadrons match a parton configuration that does not allow for extra jets.

13.2 Kinematics, coordinate frames, and structure functions

Much of the derivation uses the same elements that we already used in Ch. 12 and in earlier chapters. We focus on the changes.

In this section, we specify the kinematics and then define a hadronic tensor $W^{\mu\nu}$ for our process, together with corresponding structure functions. Let p_A and p_B be the momenta of the detected hadrons in $e^+e^- \to H_A H_B + X$, and let q be the momentum of the virtual photon. It is convenient to use two different coordinate frames:

- A *photon frame*, in which the photon has zero transverse momentum. This is chosen as a center-of-mass (CM) frame, supplemented by a condition on the direction of the z axis. It is a frame most directly related to an actual experiment, and is best suited for the analysis of the hard scattering.
- A *hadron frame*, in which the hadrons are back-to-back in the $\pm z$ directions. This matches the hadron frame used in (12.37) and (12.38) for defining fragmentation functions in momentum space.

Subscripts γ and h denote components of a vector in the two frames.

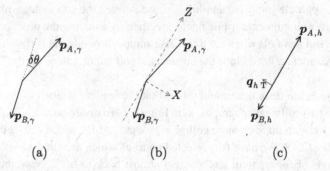

Fig. 13.2. Kinematics of two hadrons in final state: (a) center-of-mass frame; (b) center-of-mass frame with dashed lines to indicate Z and X axes used to define the structure functions in (13.9); (c) hadron frame of (13.4). *The incoming leptons in (a) and (b) can be out of the plane defined by the hadrons.*

13.2.1 Photon frame

Our standard photon frame, illustrated in Fig. 13.2(a), is a CM frame, where the momenta in ordinary Cartesian coordinates are

$$q_\gamma = (Q, \mathbf{0}), \tag{13.1a}$$

$$p_{A,\gamma} = (E_{A,\gamma}, \mathbf{p}_{A,\gamma}) \simeq |\mathbf{p}_{A,\gamma}| (1, \mathbf{n}_{A,\gamma}), \tag{13.1b}$$

$$p_{B,\gamma} = (E_{B,\gamma}, \mathbf{p}_{B,\gamma}) \simeq |\mathbf{p}_{B,\gamma}| (1, \mathbf{n}_{B,\gamma}). \tag{13.1c}$$

Here $\mathbf{n}_{A,\gamma}$ and $\mathbf{n}_{B,\gamma}$ are unit vectors for the directions of the hadrons. In the second form for $p_{A,\gamma}$ and $p_{B,\gamma}$, we neglected masses. To parameterize the deviation from the exact back-to-back configuration, we let $\delta\theta$ be the angle between $\mathbf{p}_{A,\gamma}$ and $-\mathbf{p}_{B,\gamma}$.

Although some issues can be treated with coordinate axes fixed in the laboratory, independent of the detected hadrons, we will find it convenient to use light-front coordinates with a z axis defined from the hadron directions. The spatial axes can be defined covariantly by normalized 4-vectors whose energy components in the CM frame are zero. For the z and x axes we choose

$$Z_\gamma^\mu = \frac{(0, \mathbf{n}_{A,\gamma} - \mathbf{n}_{B,\gamma})}{|\mathbf{n}_{A,\gamma} - \mathbf{n}_{B,\gamma}|}, \qquad X_\gamma^\mu = \frac{(0, \mathbf{n}_{A,\gamma} + \mathbf{n}_{B,\gamma})}{|\mathbf{n}_{A,\gamma} + \mathbf{n}_{B,\gamma}|}. \tag{13.2}$$

As shown in Fig. 13.2(b), the z axis bisects the angle between $\mathbf{p}_{A,\gamma}$ and $-\mathbf{p}_{B,\gamma}$, and the x axis is orthogonal to it in the hadron-hadron plane. Thus in the back-to-back region, Z characterizes the jet axis, and X characterizes the transverse direction of the hadron pair. The y axis is the remaining axis in a right-handed system. The time axis can be defined by $T^\mu = q^\mu/Q$.

Then we define photon-frame light-front coordinates for a vector V by

$$V_\gamma^\pm \overset{\text{def}}{=} \frac{V \cdot (T \mp Z)}{\sqrt{2}}. \tag{13.3}$$

13.2.2 Hadron frame

In the hadron frame, illustrated in Fig. 13.2(c), the detected hadrons are exactly back-to-back, but the virtual photon has a generally non-zero transverse momentum. We choose the positive z axis to be the direction of p_A, and define light-front coordinates for this frame by

$$q_h = (q_h^+, q_h^-, \mathbf{q}_{h\,T}), \tag{13.4a}$$

$$p_{A,h} = (p_{A,h}^+, m_A^2/2p_{A,h}^+, \mathbf{0}_T) \simeq (p_{A,h}^+, 0, \mathbf{0}_T), \tag{13.4b}$$

$$p_{B,h} = (m_B^2/2p_{B,h}^-, p_{B,h}^-, \mathbf{0}_T) \simeq (0, p_{B,h}^-, \mathbf{0}_T). \tag{13.4c}$$

We define scaling variables by

$$z_A = \frac{p_{A,h}^+}{q_h^+} \simeq \frac{p_A \cdot p_B}{q \cdot p_B}, \qquad z_B = \frac{p_{B,h}^-}{q_h^-} \simeq \frac{p_A \cdot p_B}{q \cdot p_A}. \tag{13.5}$$

The photon transverse momentum in the photon frame measures how much the hadrons deviate from the back-to-back configuration in the CM frame:

$$q_{h\,T}^2 = 2q_h^+ q_h^- - Q^2 \simeq \frac{2p_A \cdot q \, p_B \cdot q}{p_A \cdot p_B} - Q^2 = Q^2 \tan^2 \frac{\delta\theta}{2}. \tag{13.6}$$

Formulae for z_A, z_B, and $q_{h\,T}$ in terms of Lorentz invariants can also be obtained with retention of hadron masses, but I will not present them. Our definition of the hadron frame is non-unique, in that it can be changed by a boost in the z direction, which will not affect our derivations. If necessary, the frame can be fixed by requiring the photon to have zero rapidity, i.e., $q_h^+ = q_h^-$. In the general case

$$q_h^\pm = e^{\pm y} \frac{Q}{\sqrt{2}\cos(\delta\theta/2)}. \tag{13.7}$$

13.2.3 Lorentz transformation between photon and hadron frames

The Lorentz transformation between the photon and hadron frames is

$$\begin{aligned}
V_h = L\left(V_\gamma^+, V_\gamma^-, \mathbf{V}_{\gamma\,T}\right) = \Bigg(& e^y \left[V_\gamma^+ \frac{\kappa+1}{2} + V_\gamma^- \frac{\kappa-1}{2} + \frac{\mathbf{V}_{\gamma\,T} \cdot \mathbf{q}_{h\,T}}{Q\sqrt{2}} \right], \\
& e^{-y} \left[V_\gamma^+ \frac{\kappa-1}{2} + V_\gamma^- \frac{\kappa+1}{2} + \frac{\mathbf{V}_{\gamma\,T} \cdot \mathbf{q}_{h\,T}}{Q\sqrt{2}} \right], \\
& \mathbf{V}_{\gamma\,T} + \mathbf{q}_{h\,T} \left[\frac{V_\gamma^+}{Q\sqrt{2}} + \frac{V_\gamma^-}{Q\sqrt{2}} + \frac{\mathbf{V}_{\gamma\,T} \cdot \mathbf{q}_{h\,T}}{Q^2(\kappa+1)} \right] \Bigg),
\end{aligned} \tag{13.8}$$

where $\kappa = \sqrt{1 + q_{h\,T}^2/Q^2} \simeq 1/\cos(\delta\theta/2)$, and y is the rapidity of q in the hadron frame, i.e., $y = \ln(q_h^+/q_h^-)$. Note carefully that although the components of V on the right-hand side of this equation are in the photon frame, the transverse vector $\mathbf{q}_{h\,T}$ is for the photon in the hadron frame. Note also that with the x and z axes defined in Fig. 13.2 q_h has a negative x component: $q_h^x = -Q \tan(\delta\theta/2)$, $q_h^y = 0$.

13.2.4 Structure function analysis

We make a structure function analysis by the method that Lam and Tung (1978) used for the Drell-Yan process. It starts from a hadronic tensor $W^{\mu\nu}$, which obeys current conservation, $q_\mu W^{\mu\nu} = W^{\mu\nu} q_\nu = 0$, is symmetric under $\mu \longleftrightarrow \nu$, and obeys parity conservation. When the detected hadrons have zero spin or their polarization is not measured, we have

$$W^{\mu\nu}(q, p_A, p_B) \overset{\text{def}}{=} 4\pi^3 \sum_X \delta^{(4)}(p_X + p_A + p_B - q)$$

$$\times \langle 0| j^\mu(0)|p_A, p_B, X, \text{out}\rangle \langle p_A, p_B, X, \text{out}| j^\nu(0)|0\rangle$$

$$= (-\tilde{g}^{\mu\nu} - Z^\mu Z^\nu) W_{\text{T}} + Z^\mu Z^\nu W_{\text{L}} - (X^\mu Z^\nu + Z^\mu X^\nu) W_\Delta$$

$$+ (-\tilde{g}^{\mu\nu} - 2X^\mu X^\nu - Z^\mu Z^\nu) W_{\Delta\Delta}, \tag{13.9}$$

where the structure functions W_{T}, etc., are functions of Lorentz invariants. We define $\tilde{g}^{\mu\nu} = g^{\mu\nu} - q^\mu q^\nu/Q^2$, and the orthogonal unit vectors Z and X were defined in (13.2).

The structure function decomposition (13.9) and the associated cross section formulae can be readily generalized to include the case of Z exchange or that the hadrons are polarized. But to explain the principles, we avoid these complications.

The names of the structure functions (T, L, Δ, and $\Delta\Delta$) characterize the corresponding polarization state of a spin-1 particle of momentum q: T is for an azimuthally symmetric transverse polarization around Z, L is for longitudinal polarization, Δ is for one unit of helicity flip in the density matrix, and $\Delta\Delta$ is for two units of helicity flip. Each gives a characteristic term in the angular dependence of the cross section:

$$E_A E_B \frac{d\sigma}{d^3 p_A\, d^3 p_B} = \frac{\alpha^2}{16\pi^3 Q^4} \left[(1 + \cos^2\theta)\, W_{\text{T}} + \sin^2\theta\, W_{\text{L}} \right.$$

$$\left. + \sin 2\theta \cos\phi\, W_\Delta + \sin^2\theta \cos 2\phi\, W_{\Delta\Delta} \right]. \tag{13.10}$$

Here θ is the polar angle of the leptons with respect to the Z direction, and ϕ is the azimuthal angle around Z, with the direction X corresponding to $\phi = 0$. The angular dependence corresponds to the angular momentum associated with each structure function.

Some confusion about the azimuthal angle can be avoided by realizing that there are actually two azimuthal angles that can be measured from the two hadrons, but that the cross section (13.10) only depends on one of them. In the overall CM frame with the incoming lepton beams along the z axis, these angles can be characterized as (a) the azimuthal angle of the overall jet axis Z^μ relative to some fixed axis, and (b) the azimuthal angle of the hadron plane relative to the plane that contains Z^μ and the leptons. The dependence is on the second angle, but not the first. The reason for this is that because the leptons are unpolarized, the initial state has nothing to allow an intrinsic azimuthal axis to be defined.

There are "kinematic zeros" in W_Δ and $W_{\Delta\Delta}$ at $q_{h\,T} = 0$, since the dependence on the direction X arises only from the transverse momentum $\boldsymbol{q}_{h\,T}$. So when $q_{h\,T} \to 0$, W_Δ is proportional to $q_{h\,T}$ and $W_{\Delta\Delta}$ is proportional to $q_{h\,T}^2$.

13.3 Region analysis

We now start the derivation of a factorization property suitable for the case of relatively low transverse momentum, i.e., $q_{h\,T} \ll Q$. Later we will combine this with a more standard factorization for large transverse momentum to give a result valid for all $q_{h\,T}$. We will assume throughout that the hadron energies in the CM frame are comparable with Q. That is, we do not treat the case of very small values for the scaling variables z_A and z_B.

The strategy was already explained in Chs. 10 and 12. One feature critical to a proper derivation is the use of the integral of the hadronic tensor with a test function, as in (12.11); this allows a clean understanding of the accuracy of the region approximants. Another feature is a shift between hadron and photon frames in defining the hadron scattering; this will give consistency of parton kinematics between fragmentation functions and perturbative calculations of the hard scattering.

13.3.1 Only two jets

Since we assume that the observed hadrons H_A and H_B have energies of order Q, they are part of jet subgraphs, not of the soft subgraph, in the leading regions. As already explained in Sec. 13.1, the leading regions when $q_{h\,T} \ll Q$ have only have two jet subgraphs, as in Fig. 13.1; regions with three or more jet subgraphs are suppressed by a power of $q_{h\,T}/Q$.

13.3.2 Region approximators

In the subtraction formalism, Sec. 10.1, the contribution of a particular region R of a graph is obtained by applying an approximator T_R to the graph. But it is applied only after subtractions are made for smaller regions, to avoid double-counting problems.

For reasons already encountered in Secs. 12.7 and 12.8, we apply the approximators not to the hadronic tensor $W^{\mu\nu}$ itself, but to an integral of it with a test function. The integral, $W^{\mu\nu}([f], p_A, p_B)$, is defined just as in (12.11) for the one-particle-inclusive case. The argument of the test function is the sum of the collinear and soft momenta in the final state: $f(k_A + k_B + k_S)$. Region approximants are applied to internal virtual lines of collinear and hard subgraphs, and to the argument of the test function.

If, instead, the approximant were used for the unintegrated $W^{\mu\nu}$, it would be applied to soft momenta circulating through final states of collinear subgraphs. The errors associated with approximants that directly change the final-state momenta are hard to control.

The approximant for a soft momenta in a collinear subgraph is unchanged from that in Sec. 10.4.2 for the Sudakov form factor. The approximant is also unchanged from the one for single-particle-inclusive e^+e^- annihilation in Sec. 12.8.1, except for the choice of the directions defining the auxiliary vectors. These are now derived from the momenta of the observed hadrons, and we apply the definitions in the hadron frame using the light-front coordinates defined in (13.4).

The approximants for soft and collinear momenta in the hard subgraphs have an apparently small but very significant change compared with (10.19) for the Sudakov form factor. This concerns the frames used to specify the light-like auxiliary vectors. We now define the projectors for collinear momenta into the hard subgraph by

$$P_{HA}(k_A) = \frac{w_{HA}\,k_A \cdot w_B}{w_{HA} \cdot w_B}, \qquad P_{HB}(k_B) = \frac{w_{HB}\,k_B \cdot w_A}{w_{HB} \cdot w_A}. \tag{13.11}$$

Here w_A and w_B are light-like vectors defined in the hadron frame: $w_{A,h} = (1, 0, \mathbf{0}_T)$ and $w_{B,h} = (0, 1, \mathbf{0}_T)$. They correspond to vectors used in the soft-to-collinear approximants and in the definitions of fragmentation functions. But for reasons to be explained below, the other vectors are defined in photon frame: $w_{HA,\gamma} = (1, 0, \mathbf{0}_T)$ and $w_{HB,\gamma} = (0, 1, \mathbf{0}_T)$.

As with the Sudakov form factor, a momentum from a collinear subgraph may include a circulating soft component. This is approximated, to be in direction w_B in collinear subgraph $C_{(A)}$, and in direction w_A in collinear subgraph $C_{(B)}$. From (13.11), these circulating soft momenta are replaced by zero in the hard scattering, as for the Sudakov form factor.

The reason for the new definitions of the projectors is that we normally perform perturbative calculations of the hard scattering in the photon frame, where the virtual photon has zero transverse momentum. Thus the calculations correspond to the elastic process $e^+e^- \to q\bar{q}$ in its CM frame. Therefore we arrange that in the photon frame the approximated quark momenta are in the plus and minus directions. This complication did not arise for the Sudakov form factor, since it is an elastic process, for which the photon and hadron frames coincide.

In hadron-frame components, the approximated momenta are

$$P_{HA}(k_A)_h = k_{A,h}^+ \left(1, \; e^{-2y}\frac{\kappa - 1}{\kappa + 1}, \; e^{-y}\frac{q_{h\,T}\sqrt{2}}{Q(\kappa + 1)}\right), \tag{13.12a}$$

$$P_{HB}(k_B)_h = k_{B,h}^- \left(e^{2y}\frac{\kappa - 1}{\kappa + 1}, \; 1, \; e^{y}\frac{q_{h\,T}\sqrt{2}}{Q(\kappa + 1)}\right). \tag{13.12b}$$

Note that these leave unchanged the "large components", i.e., $k_{A,h}^+$ and $k_{B,h}^-$. The photon-frame components are

$$P_{HA}(k_A)_\gamma = \frac{2e^{-y}k_{A,h}^+}{1 + \kappa}(1, 0, \mathbf{0}_T), \qquad P_{HB}(k_B)_\gamma = \frac{2e^{y}k_{B,h}^-}{1 + \kappa}(0, 1, \mathbf{0}_T). \tag{13.13}$$

These formulae apply not just to the total collinear momenta entering the hard subgraph, but equally to the individual momenta on particular external lines of H. Let these momenta be indicated by an index j: k_{Aj}, k_{Bj}. Then, by momentum conservation at the hard scattering, the virtual photon's momentum in the photon frame is changed to

$$\hat{q}_\gamma = \left(\sum_j k_{Aj,h}^+ e^{-y}, \sum_j k_{Bj,h}^- e^y, \mathbf{0}_T \right) \frac{2}{\kappa + 1}. \tag{13.14}$$

To restore the original value of q, we will define an approximant on the test function, in (13.18) below. In effect, this approximant will change the momentum of the final state relative to q.

Dirac projectors on the external lines of the hard subgraphs need to be modified from those defined in Sec. 10.4.2. For quark lines between C_A and H we use

$$\mathcal{P}_A \overset{\text{def}}{=} \frac{\gamma \cdot w_{HA} \gamma \cdot w_B}{2 w_B \cdot w_{HA}}, \qquad \overline{\mathcal{P}}_A \overset{\text{def}}{=} \frac{\gamma \cdot w_B \gamma \cdot w_{HA}}{2 w_B \cdot w_{HA}}, \tag{13.15}$$

and for quark lines between C_B and H we use

$$\mathcal{P}_B \overset{\text{def}}{=} \frac{\gamma \cdot w_{HB} \gamma \cdot w_A}{2 w_A \cdot w_{HB}}, \qquad \overline{\mathcal{P}}_B \overset{\text{def}}{=} \frac{\gamma \cdot w_A \gamma \cdot w_{HB}}{2 w_A \cdot w_{HB}}. \tag{13.16}$$

On the side next to the hard scattering, the factors $\gamma \cdot w_{HA}$ and $\gamma \cdot w_{HB}$ project onto wave functions for the (approximated) massless on-shell quarks. On the side next to the collinear subgraphs, the factors $\gamma \cdot w_A$ and $\gamma \cdot w_B$ project onto the components of the Dirac fields that are used in the hadron-frame definitions of fragmentation functions. As usual, we use the Dirac conjugation notation: $\overline{\Gamma} \overset{\text{def}}{=} \gamma^0 \Gamma^\dagger \gamma^0$.

In the approximant in the test function we must preserve the exact transverse momentum of the collinear and soft partons, since we wish to obtain a cross section differential in q_{hT}.

Previously, in the one-particle-inclusive cross section, the approximator made the replacement

$$f(k_A + k_B + k_S) \mapsto f(k_A^+, k_B^-, \mathbf{0}_T), \quad \text{(previous)} \tag{13.17}$$

where we neglected not only the small longitudinal components k_A^- and k_B^+, but also all the transverse momenta. For TMD factorization, we must change the approximant to retain the transverse momenta. But to keep the longitudinal components consistent with those required by momentum conservation in the hard scattering, we apply a scaling to the plus and minus components of k_A and k_B. We therefore define the approximant on the test function in terms of hadron-frame momenta by

$$f(k_{A,h} + k_{B,h} + k_{S,h}) \mapsto f\left(k_{A,h}^+ \frac{2\kappa}{\kappa + 1}, k_{B,h}^- \frac{2\kappa}{\kappa + 1}, k_{A,hT} + k_{B,hT} + k_{S,hT} \right). \tag{13.18}$$

The scaling factor $2\kappa/(\kappa + 1) = 2/(1 + \cos(\delta\theta/2))$ is, of course, unity in the limit that q_{hT} is zero. It is chosen so that after the next step of functional differentiation, $\hat{q}_\gamma$ in (13.14) reproduces q, i.e., the hard scattering has the original value of q.

To find the actual hadronic tensor $W^{\mu\nu}(q, p_A, p_B)$, we functionally differentiate the integrated tensor with respect to the test function f:

$$W^{\mu\nu}(q, p_A, p_B) = \frac{\delta W^{\mu\nu}([f], p_A, p_B)}{\delta f(q)}. \tag{13.19}$$

The result is that the approximated parton momenta, in (13.18), sum to q. Relative to an unapproximated graph, the transverse momenta are unchanged, but the longitudinal momenta are shifted by amounts that are power-suppressed in the design region of the approximator. This results in power-suppressed errors in the hadronic tensor itself, provided that scale of the $q_h^\pm$ dependence of the hadronic tensor is Q rather than a smaller scale.

As usual, the internal integrations are over all momenta. Outside the design region R of approximator T_R, the accuracy of the approximation degrades. But this is handled by terms for larger regions than R, combined with the double-counting subtractions in the subtraction method.

13.3.3 Ward identities

There is no change in the Ward identities that extract K gluons from hard and collinear subgraphs and that led to Fig. 12.12. There are now only two collinear subgraphs, so the hard scattering only has two external lines, and each collinear subgraph has a detected hadron.

At this point the color flow is as shown in Fig. 12.13. We now disentangle the color flow between the various factors. As before, the collinear factors are color-singlet, so we convert to the form of Fig. 12.15, where the sums over the color indices of the hard scattering bypass the collinear factors, which are now defined with a color average.

In our two-collinear-subgraph case, the entangled hard-soft combination has color sums of the form

$$H_{ab} S_{ab;a'b'} H^*_{a'b'}, \tag{13.20}$$

with repeated indices summed. Since the hard-scattering amplitudes are color-singlet, we replace this by

$$\left(H_{ab} H^*_{ab}\right) \frac{1}{N_c} S_{cc;dd}. \tag{13.21}$$

Here there is a color trace for the hard scattering, the same as in a cross section with a sum over final-state color, while the soft factor is color averaged. So from Fig. 12.12 we obtain Fig. 13.3, where each of the collinear and soft factors has a color average. The collinear and hard factors are still linked by a Dirac trace that we will analyze later.

As usual, a sum over graphs and regions converts Fig. 13.3 to a factorization formula. The operator definitions for the factors are determined by the approximants, and there are appropriate double-counting subtractions in the factors. The factorized form is

$$W^{\mu\nu} = 4\pi^3 z_A z_B \sum_f \int d^2 k_{A\mathrm{T}} \, d^2 k_{B\mathrm{T}} \, S(q_{h\mathrm{T}} - k_{A\mathrm{T}} - k_{B\mathrm{T}})$$

$$\times \mathrm{Tr}\, \mathcal{P}_A \, C_A(z_A, k_{A\mathrm{T}}; f) \overline{\mathcal{P}}_A \, H^\nu_f \, \mathcal{P}_B \, C_B(z_B, k_{B\mathrm{T}}; \bar{f}) \overline{\mathcal{P}}_B \, \overline{H}^\mu_f(Q), \tag{13.22}$$

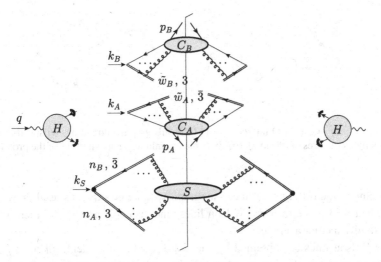

Fig. 13.3. Same as Fig. 12.12, but for two-hadron-inclusive cross section in the back-to-back region. The color flow has been reorganized: the collinear and soft factors have color averages, and there is a color trace between the hard-scattering amplitudes on the left and the right of the final-state cut.

where all transverse momenta are in the hadron frame, the Dirac projectors were defined in (13.15) and (13.16), and the factors S, C_A, and C_B will be defined below. The sum over f is over the flavors of quark and antiquark that can enter the C_A factor, i.e., u, $\bar{u}$, d, $\bar{d}$, etc.; the opposite flavor is used for C_B.

The following steps give the above formula.

1. We perform the functional differentiation, (13.19), for the approximated $W^{\mu\nu}$. This sets the transverse momentum in the soft factor equal to $q_{h\mathrm{T}} - k_{A,h\mathrm{T}} - k_{B,h\mathrm{T}}$. It also sets $k^+_{A,h} = q^+_h$ and $k^-_{B,h} = q^-_h$ in the collinear factors C_A and C_B.
2. The approximation removed dependence of the test function on $k^-_{A,h}$, $k^+_{B,h}$, and $k^{\pm}_{S,h}$. So the integrals over these variables are "short-circuited" and included in the definitions of C_A, C_B, and S.
3. The soft factor is $S = Z_S S_{(0)}$, which is a UV-renormalization factor $Z_S(y_A - y_B, g, \epsilon)$ times a bare soft factor

$$S_{(0)}(k_{S\mathrm{T}}) = \frac{1}{N_c} \int \frac{dk^+_S \, dk^-_S}{(2\pi)^{4-2\epsilon}} \qquad\qquad\qquad\qquad\qquad (13.23)$$

The Wilson lines are in non-light-like directions defined as in Sec. 12.8.1, but now using the hadron frame: $n_{A,h} = (1, -e^{-2y_A}, \mathbf{0}_\mathrm{T})$, $n_{B,h} = (-e^{2y_B}, 1, \mathbf{0}_\mathrm{T})$. The rapidities and light-front coordinates for the two collinear subgraphs are in the hadron frame, rather

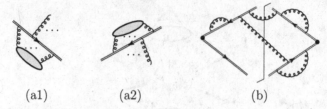

(a1)   (a2)   (b)

Fig. 13.4. (a1) and (a2) Omitted from (13.23) are graphs containing Wilson-line self-interaction structures of this kind. (b) Example of omitted graph with four of the prohibited structures.

than being in the different collinear-subgraph-specific coordinates used in Sec. 12.8.1. There is a color trace over the Wilson lines, and, as explained above, there is a factor $1/N_c$ to give a color average.

The Wilson lines are obtained by the Ward-identity argument, given in Sec. 11.9. But this does not produce graphs that contain a subgraph connecting only to one of the straight-line segments of the Wilson line. Thus graphs containing structures like those in Fig. 13.4 are omitted; this is indicated by the subscript "no SI" (for "no self-interaction"). More details are given in Sec. 13.3.4.

4. The collinear factor C_A is defined with the integral

$$\frac{k_{A,h}^+}{p_{A,h}^+} \int \frac{dk_{A,h}^-}{(2\pi)^{4-2\epsilon}}, \tag{13.24}$$

as appropriate for a fragmentation function; see (12.39).

5. The longitudinal momentum fraction in C_A is then

$$\frac{p_{A,h}^+}{k_{A,h}^+} = z_A \frac{2}{\kappa + 1}. \tag{13.25}$$

Now the factor $2/(\kappa + 1)$ goes to unity in the limit $q_{h\,T}/Q \to 0$, and the approximations used elsewhere in the derivation are only valid only to leading power in $q_{h\,T}/Q$. So we replaced the momentum fraction by z_A in (13.22).

Corrections will handled by methods appropriate to the large transverse momentum region, in Sec. 13.12.

6. As in the Sudakov form factor, soft subtractions are applied to the collinear factors.

7. The same methods give the other collinear factor C_B.

8. A prefactor of $z_A z_B$ compensates factors of $1/z_A$ and $1/z_B$ in the definitions of the fragmentation functions, as in (13.24). Those factors normalize the fragmentation functions, like number densities.

9. The Dirac and color traces are explicit in (13.22) rather than being absorbed into the fragmentation functions.

As for determining the Wilson lines and implementing the soft subtractions in the collinear factors: In Sec. 13.6, we will use Fourier transforms on transverse momenta to convert the convolution in transverse momentum in (13.22) to a product in transverse position space. After that, the arguments we used in Ch. 10 for the Sudakov form factor apply in the same way here to determine appropriate directions for the Wilson lines, and to find optimal definitions of the factors.

We will determine the allowed polarization dependence of the fragmentation functions, after which we will determine the angular distribution, with an interesting correction to the standard $1 + \cos^2\theta$ form.

13.3.4 Wilson-line self-interactions

As already remarked, Wilson-line self-interactions, Fig. 13.4, are omitted from the definition of the soft factor, (13.23). We met exactly the same issue in an abelian theory, e.g., in (10.89). But in an abelian theory, the Wilson lines could be simplified, by the use of (10.100), to replace the sum over gluon attachments to a Wilson line by a product of single Wilson-line propagators. Then the self-interactions of the Wilson lines could be factored out. A simple factorization of Wilson-line self-interactions does not work in a non-abelian theory.

An immediate consequence is that despite being defined from a matrix element of a Wilson-line operator, the soft factor depends on the gauge used to formulate the theory; a gauge-dependent set of graphs is omitted. Similar issues apply to the collinear factors.

These problems will be solved by a reorganization of the factorization formula in Sec. 13.7, just as for the Sudakov form factor in Sec. 10.11.

13.4 Collinear factors

There are two parts to our treatment of the collinear factors, leading to definitions of fragmentation functions. In this section, we treat quark polarization and the azimuthal dependence of the fragmentation functions. The second part, in Sec. 13.7, concerns the Wilson lines.

13.4.1 Quark fragmentation, including polarization

The Dirac projectors for the leading power restrict the collinear factor C_A to terms proportional to γ^-, $\gamma^-\gamma_5$, and $\gamma^-\gamma^i$ (with i a *transverse* index). Given that we choose the observed hadrons to be spinless, the $\gamma^-\gamma_5$ term is prohibited by parity invariance. For an *integrated* fragmentation function, the $\gamma^-\gamma^i$ term is prohibited by invariance under rotations about the z axis. But for an unintegrated fragmentation function, the quantity C_A can have a term $\gamma^-\gamma^i k^i_{A,h\,T}$ times a function of the size of $k_{A,h\,T}$.

We now recall our results in Sec. 12.4.7, and use them to convert (13.22) to use fragmentation functions that allow for quark polarization. Given the longitudinal and transverse polarization of the initial quark, we must project the trace of C_A as indicated in (12.41a).

The first term on the r.h.s. of (12.41a) gives the unpolarized fragmentation function, which is independent of the azimuthal angle of the quark's transverse momentum. The second term goes with a factor that is zero by parity invariance. The trace with the third term picks out the $\gamma^-\gamma^i k^i_{A\,T}$ term in C_A,

$$-\operatorname{Tr}\gamma^- \sum_{i=1,2} \gamma^i k^i_{A,h\,T}\gamma^+ \sum_{j=1,2}\gamma_5\gamma^j s^j_T$$

$$= 4i(k^x_{A,h\,T}s^y_T - k^y_{A,h\,T}s^x_T) = 4i|k_{A,h\,T}||s_T|\sin(\phi_s - \phi_k), \qquad (13.26)$$

where ϕ_s and ϕ_k are the azimuthal angles of s_T and $k_{A,h\,T}$. The result is a characteristic angular dependence, which we will relate to the angular dependence of the two-hadron-inclusive cross section. In this equation, superscripts x and y are used for transverse components. The transverse spin vector s_T of the quark will be obtained from calculations of the hard scattering in the photon frame. But it can be verified from the form of the Dirac projector $\mathcal{P}_A$ that the same numerical vector can be applied in the hadron frame.

The basic derivation just given applies in a model field theory without gauge fields. In a gauge theory, Wilson lines need to be attached to the quark fields. Since the Wilson lines are chosen to be in the (t, z) plane, they do not affect any azimuthal dependence. So we can apply the same Dirac trace in full QCD.

The factorization formula (13.22) has transverse momentum for the quark, and zero transverse momentum for the hadron. But the number-density interpretation is in terms of a transverse momentum of the hadron relative to the quark, as given by (12.34a), so fragmentation functions are treated as functions of z_j and $p_{j\mathrm{T}}$, where

$$p_{j\mathrm{T}} = -z_j k_{j,h\mathrm{T}}. \qquad (13.27)$$

Don't forget the minus sign in this relation!

The resulting fragmentation function $d_{h/f}(z, p_{\mathrm{T}})$ depends on the azimuthal angle of the transverse momentum. We decompose it into azimuth-independent fragmentation functions, which we normalize by the Trento conventions (Bacchetta *et al.*, 2004),

$$d_{h/f}(z, p_{\mathrm{T}}) = D_{1, h/f}(z, p_{\mathrm{T}}) + H^\perp_{1, h/f}(z, p_{\mathrm{T}})\frac{p^x_{h\mathrm{T}}s^y_T - p^y_{h\mathrm{T}}s^x_T}{zM_h}$$

$$= D_{1, h/f}(z, p_{\mathrm{T}}) + H^\perp_{1, h/f}(z, p_{\mathrm{T}})\frac{|s_T||p_{h\mathrm{T}}|}{zM_h}\sin(\phi_h - \phi_s), \qquad (13.28)$$

where the notations D_1 and $H^\perp_1$ are those of Mulders and Tangerman (1996), M_h is the mass of the detected hadron, and ϕ_h and ϕ_s are the azimuths of the hadron and the quark spin. Note that the ϵ tensor in Bacchetta *et al.* (2004) has the opposite sign to the one used in this book, so I have avoided using it in (13.28).

We therefore have two TMD fragmentation functions. Without the Wilson lines, the unpolarized one is defined by (12.39), but we now use the symbol D_1 instead of d. The polarized fragmentation function is

$$
H^{\perp}_{1,h/f}(z, p_{\mathrm{T}}) \frac{p^x_{\mathrm{T}} s^y_{\mathrm{T}} - p^y_{\mathrm{T}} s^x_{\mathrm{T}}}{z M_h}
$$

$$
\overset{\text{prelim}}{=} -\frac{\mathrm{Tr}_{\text{color}}\, \mathrm{Tr}_{\text{Dirac}}}{N_{c,f}}\, \frac{1}{4} \sum_X \frac{1}{z} \int \frac{\mathrm{d}x^-\, \mathrm{d}^{n-2}x_{\mathrm{T}}}{(2\pi)^{n-1}} e^{ik^+_h x^- - ik_{h\mathrm{T}} \cdot x_{\mathrm{T}}}
$$

$$
\times \gamma^+ \gamma_5 \boldsymbol{\gamma}_{\mathrm{T}} \cdot \boldsymbol{s}_{\mathrm{T}}\, \langle 0|\psi^{(0)}_j(x/2)|p, X, \text{out}\rangle\, \langle p, X, \text{out}|\bar{\psi}^{(0)}_j(-x/2)|0\rangle
$$

$$
= -\frac{\mathrm{Tr}_{\text{color}}\, \mathrm{Tr}_{\text{Dirac}}}{N_{c,j}}\, \frac{1}{4}\, \frac{1}{z} \int \frac{\mathrm{d}k^-}{(2\pi)^n}\, \gamma^+ \gamma_5 \boldsymbol{\gamma}_{\mathrm{T}} \cdot \boldsymbol{s}_{\mathrm{T}} \qquad (13.29)
$$

where the overall minus sign is from the third term on the r.h.s. of (12.41a). $H^{\perp}_{1,h/f}$ is commonly called the Collins function (Collins, 1993). Its physical importance is that it gives a correlation between the azimuthal distribution of a hadron and the transverse spin of its parent quark. It therefore provides a measure of the quark polarization. The "prelim" designation of this definition is a reminder that we have not yet included Wilson lines.

13.4.2 Antiquark fragmentation

Exchanging the quark and antiquark lines in the above definition gives the Collins function for an antiquark. From (12.41) it follows that no change of sign is needed.

13.4.3 Coordinate systems for C_A and C_B factors

The above definitions apply to the fragmentation functions corresponding to C_A in (13.22). But an exchange of the roles of the plus and minus axes, i.e., a reversal of the z axis is necessary for the C_B factor. Since the x, y, and z axes form a right-handed coordinate system, certain signs will reverse in obtaining the polarized fragmentation function.

13.4.4 Positivity and Collins function

In the absence of Wilson lines, the fragmentation function is positive, as follows from the general definition (12.35). This must apply for any polarization state of the quark in (13.28). Hence the Collins function is restricted to obey

$$
\frac{|H^{\perp}_{1,h/f}(z, p_{\mathrm{T}})||\boldsymbol{p}_{h\mathrm{T}}|}{z M_h} \leq D_{1,h/f}(z, p_{\mathrm{T}}). \qquad (13.30)
$$

When we use the full QCD definitions with subtractions to prevent double counting with the soft factor, we may find some violation of this constraint.

13.5 Initial version of factorization with TMD fragmentation

13.5.1 Factorization

We now express (13.22) in terms of the fragmentation functions:

$$
W^{\mu\nu} \overset{\text{prelim}}{=} \frac{8\pi^3 z_A z_B}{Q^2} \sum_f \int \mathrm{d}^2 k_{A,h\,\mathrm{T}} \, \mathrm{d}^2 k_{B,h\,\mathrm{T}} \, S(q_{h\,\mathrm{T}} - k_{A\mathrm{T}} - k_{B\mathrm{T}})
$$

$$
\times D_{1,\,H_A/f}(z_A, z_A k_{A,h\,\mathrm{T}}) \, D_{1,\,H_B/\bar{f}}(z_B, z_B k_{B,h\,\mathrm{T}})
$$

$$
\times \mathrm{Tr}\, k_{A,\gamma}^+ \gamma^- (1 - \gamma_5 \boldsymbol{\gamma}_\mathrm{T} \cdot \boldsymbol{a}_{A,\gamma\,\mathrm{T}}) \, H_f^\nu(Q) \, k_{B,\gamma}^- \gamma^+ (1 - \gamma_5 \boldsymbol{\gamma}_\mathrm{T} \cdot \boldsymbol{a}_{B,\gamma\,\mathrm{T}}) \, \overline{H}_f^\mu(Q).
$$

$$(13.31)$$

The "prelim" notation is used because we will modify the definitions of the factors to get our final factorization formula. The Collins function appears in the transverse vectors $\boldsymbol{a}_{A\mathrm{T}}$ and $\boldsymbol{a}_{B\mathrm{T}}$ defined by

$$
\boldsymbol{a}_{A,\gamma\,\mathrm{T}} \overset{\text{def}}{=} \left(+ k_{A,h\,\mathrm{T}}^y, \; - k_{A,h\,\mathrm{T}}^x \right) \alpha_A, \tag{13.32a}
$$

$$
\boldsymbol{a}_{B,\gamma\,\mathrm{T}} \overset{\text{def}}{=} \left(- k_{B,h\,\mathrm{T}}^y, \; + k_{B,h\,\mathrm{T}}^x \right) \alpha_B, \tag{13.32b}
$$

where the scalar coefficients are

$$
\alpha(z, z k_{h\,\mathrm{T}}; h/f) \overset{\text{def}}{=} \frac{H_{1,\,h/f}^\perp(z, z k_{h\,\mathrm{T}})}{M_h D_{1,\,h/f}(z, z k_{h\,\mathrm{T}})}, \tag{13.33}
$$

and the reversed sign between the definitions of $\boldsymbol{a}_{A,\gamma\,\mathrm{T}}$ and $\boldsymbol{a}_{B,\gamma\,\mathrm{T}}$ allows for the reversed z axis in the definitions of the fragmentation functions between the two collinear subgraphs.

The scalar coefficients α_A and α_B have the dimensions of inverse mass, and quantify the Collins function relative to unpolarized fragmentation. The vectors $\boldsymbol{a}_{j,\gamma\,\mathrm{T}}$ are the analyzing power of single-particle fragmentation for measuring the transverse spin of a quark. The transverse momenta for the quarks on the r.h.s. of (13.32) are in the hadron frame. But the numerical values of the resulting transverse vectors $\boldsymbol{a}_{j,\gamma\,\mathrm{T}}$ on the l.h.s. are treated as photon-frame vectors to be combined with the calculation of the hard scattering, performed in the photon frame.

Note: In (13.31) there is a γ_5 factor multiplying each $\boldsymbol{a}_{j\,\mathrm{T}}$ vector, thereby allowing the interpretation in terms of an analyzing power for transverse spin. But the formulae can also be expressed without the γ_5 in terms of transverse momenta, which is a convenience in calculations with loop graphs with dimensional regularization.

13.5.2 Lowest-order (LO) calculation

The hard-scattering factor in (13.31) is easily calculated at LO. We now use the photon frame, in which the quark labeled A goes in the $+z$ direction and has energy $Q/2$. The LO hard scattering is

$$\operatorname{Tr} k_A^+ \gamma^- (1 - \gamma_5 \boldsymbol{\gamma}_T \cdot \boldsymbol{a}_{AT}) H_f^\nu(Q) \, k_B^- \gamma^+ (1 - \gamma_5 \boldsymbol{\gamma}_T \cdot \boldsymbol{a}_{BT}) \, \overline{H}_f^\mu(Q)$$

$$\overset{\text{LO}}{=} e_f^2 \operatorname{Tr} k_A^+ \gamma^- (1 - \gamma_5 \boldsymbol{\gamma}_T \cdot \boldsymbol{a}_{AT}) \gamma^\nu \, k_B^- \gamma^+ (1 - \gamma_5 \boldsymbol{\gamma}_T \cdot \boldsymbol{a}_{BT}) \gamma^\mu$$

$$= 2 e_f^2 Q^2 \left[\delta_T^{\mu\nu} + (\delta_T^{\mu\nu} \boldsymbol{a}_{AT} \cdot \boldsymbol{a}_{BT} - a_{AT}^\mu a_{BT}^\nu - a_{BT}^\mu a_{AT}^\nu) \right]$$

$$= 2 e_f^2 Q^2 \left[\delta_T^{\mu\nu} + \alpha_A \alpha_B (\delta_T^{\mu\nu} \boldsymbol{k}_{AT} \cdot \boldsymbol{k}_{BT} - k_{AT}^\mu k_{BT}^\nu - k_{BT}^\mu k_{AT}^\nu) \right], \qquad (13.34)$$

where $\delta_T^{\mu\nu}$ is a transverse Kronecker delta, the same as $-\tilde{g}^{\mu\nu} - Z^\mu Z^\nu$ in the structure function definition (13.9). Note that the dependence on $\boldsymbol{a}_{AT}$ and $\boldsymbol{a}_{BT}$ is only on a product of both. This implies that the quark and antiquark are individually unpolarized, but that their spins are correlated; the spin state is thus an entangled state. In the last line of (13.34), we used the definitions (13.32).

To find the results for the structure functions, defined in (13.9), we insert (13.34) in the factorization formula, and integrate over quark transverse momentum. The unpolarized term in (13.34) contributes to W_T only, giving the well-known $1 + \cos^2\theta$ distribution associated with a spin-$\frac{1}{2}$ quark. Comparison of the spin-dependent tensor in (13.34) with the structure function decomposition shows that it gives a contribution to the $W_{\Delta\Delta}$ structure function. This gives rise to a characteristic $\cos 2\phi$ azimuthal dependence in the cross section, (13.10).

13.5.3 Lack of single-quark polarization

The lack of transverse polarization of each single quark is actually a result valid to all orders of perturbation theory. The general proof uses chirality conservation in massless perturbation theory, and is made by the argument associated with (8.84) in DIS.

13.6 Factorization and transverse coordinate space

In this section, I will restrict attention to the unpolarized term in the factorization formula. The extension to the remaining term is left as an exercise (problem 13.6).

By using a Fourier transform, we can diagonalize the convolutions over transverse momentum in the factorization formula, (13.31), and in the evolution equations to be discussed later. So we define

$$\tilde{S}(b_T) = \int d^{2-2\epsilon} k_T \, e^{i k_T \cdot b_T} S(k_T), \qquad (13.35a)$$

$$\tilde{D}_{1,\,h/f}(z, b_T) = \int d^{2-2\epsilon} k_T \, e^{i k_T \cdot b_T} D_{1,\,h/f}(z, z k_T), \qquad (13.35b)$$

etc. The normalizations differ from those in Collins and Soper (1982b). The lack of a $1/(2\pi)^{2-2\epsilon}$ normally associated with k_T integral is because this factor is already in the definition of a fragmentation function. Although the phenomenological use of factorization is in four space-time dimensions, the above formulae are written in a general space-time dimension, because we will also use them in dimensionally regulated perturbative calculations. All the transverse vectors are in the hadron frame.

Applying the limit $b_T \to 0$ naively, gives the integrated fragmentation function up to normalization factor:

$$\tilde{D}_{1,h/f}(z, 0_T) \overset{?}{=} \frac{1}{z^2} d_{h/f}(z) \text{ (at } \epsilon = 0). \tag{13.36}$$

The factor $1/z^2$ (or $z^{-2+2\epsilon}$ in a general space-time dimension $4 - 2\epsilon$) is from the scaling between parton and hadron transverse momentum, (13.27). The above result applies in a super-renormalizable non-gauge theory, and is the equivalent of (6.75) for a parton density.

Applying (13.35b) to the definition (12.39) gives

$$\tilde{D}_{1,h/f}(z, b_T) \overset{\text{prelim}}{=} \frac{\text{Tr}_{\text{color}}}{N_{c,j}} \frac{\text{Tr}_{\text{Dirac}}}{4} \sum_X \frac{1}{z} \int \frac{dx^-}{2\pi} e^{ik^+x^-}$$

$$\times \langle 0|\gamma^+\psi_j^{(0)}(x/2)|p, X, \text{out}\rangle \langle p, X, \text{out}|\bar{\psi}_j^{(0)}(-x/2)|0\rangle$$

$$= \frac{\text{Tr}_{\text{color}}}{N_{c,j}} \frac{\text{Tr}_{\text{Dirac}}}{4} \frac{1}{z} \int \frac{dk^- \, d^{n-2}k_T}{(2\pi)^n} e^{ik_T \cdot b_T} \gamma^+ \quad \tag{13.37}$$

where the vector x is $(0, x^-, b_T)$. Thus the transverse coordinate b_T in the Fourier transform is exactly the transverse separation of the quark and antiquark fields. *The "prelim" notation alerts us that we have not yet explicitly treated the Wilson-line issues in the definition.* The orientation of the diagram corresponds to hadron p_A in Fig. 13.1.

Then the factorization formula becomes

$$W^{\mu\nu} \overset{\text{prelim}}{=} \frac{8\pi^3 z_A z_B}{Q^2} \sum_f \text{Tr} \, k_{A,\gamma}^+ \gamma^- \, H_f^\nu(Q) k_{B,\gamma}^- \gamma^+ \, \overline{H}_f^\mu(Q)$$

$$\times \int \frac{d^{2-2\epsilon} b_T}{(2\pi)^{2-2\epsilon}} e^{-iq_{hT} \cdot b_T} \tilde{S}(b_T) \tilde{D}_{1, H_A/f}(z_A, b_T) \, \tilde{D}_{1, H_B/\bar{f}}(z_B, b_T)$$

$$+ \text{polarized terms.} \tag{13.38}$$

13.7 Final version of factorization for e^+e^- annihilation

After the Fourier transform into transverse coordinate space, the factorization structure in (13.38) has the same multiplicative structure as the Sudakov form factor in Ch. 10. We therefore apply the same manipulations as we used there to obtain an improved scheme for factorization for our process.

One defining property of this scheme is that a square root of the soft factor is absorbed into a redefinition of the TMD fragmentation functions, so no soft factor is needed in the factorization formula itself. This is appropriate, since the non-perturbative part of the soft factor always appears multiplied by two collinear factors, so that it cannot be independently determined from data.

A second defining property of the scheme is that in the definitions of fragmentation functions, (13.42) below, as many Wilson lines as possible are made light-like. A non-light-like Wilson line appears only in a matrix of a certain elementary soft factor where it is multiplied with a light-like Wilson line. This will have the consequence, just as with the Sudakov form factor, that the evolution equations for the TMD functions are homogeneous. It also makes calculations of integrals simpler.

We will also formulate a further kind of factorization that will determine the TMD functions for small b_T in terms of ordinary integrated fragmentation functions. Later, we will add in a correction term to factorization for large q_{hT}/Q. After that we will have a complete formalism suitable for phenomenological use, with certain functions needing to be obtained by fits to data.

The use of transverse coordinate space simplifies many formulae. An equivalent formalism with transverse momentum variables would involve many convolutions (and their inverses).

The results in Ch. 10 were derived for an abelian gauge theory, and gave definitions for the factors with Wilson lines in certain directions. *In the following treatment, I will only briefly sketch the necessary generalizations of the proofs to extend the results to a non-abelian theory.* The general subtractive method still applies, and the eikonal denominators in the Grammer-Yennie method are the same as before. It is an urgent problem to completely fill in the details of the proofs (problem 13.7).

13.7.1 Definitions of TMD functions

As with the Sudakov form factor in Ch. 10, a basic entity in implementing factorization and subtractions is the bare soft factor defined in (13.23). Its Wilson lines are in non-light-like directions n_A and n_B, whose rapidities are y_A and y_B. They are chosen in the hadron frame as in (10.32). That is, they are space-like (Collins and Metz, 2004), and initially their rapidities approximately correspond to those of the hadrons p_A and p_B. The color charges of the Wilson lines correspond to those of the quark and antiquark. The Fourier transform to transverse coordinate space gives

$$\tilde{S}_{(0)}(b_T; y_A, y_B) = \frac{1}{N_c} \langle 0| \; W(b_T/2; \infty, n_B)^\dagger_{ca} \; W(b_T/2; \infty, n_A)_{ad}$$

$$\times \; W(-b_T/2; \infty, n_B)_{bc} \; W(-b_T/2; \infty, n_A)^\dagger_{db} \; |0\rangle \Bigg|_{\text{No SI}}, \qquad (13.39)$$

where Wilson-line self-interactions are again omitted, and the Wilson line rooted at position x is

$$W(x; \infty, n)_{ab} = P\left\{e^{-ig_0 \int_0^\infty d\lambda \, n \cdot A_{(0)\alpha}(x+\lambda n)t_\alpha}\right\}_{ab}. \qquad (13.40)$$

Here the index a corresponds to the start of the line at point x, and the index b corresponds to the end at infinity.

The soft factor has dependence on all parameters of the theory, notably, coupling, masses and renormalization scale, in addition to the parameters indicated explicitly.

In just the same way as we did for the Sudakov form factor, we reorganize the definitions of the factors in the factorization formula, (13.31). After the initial derivation, the collinear factors are matrix elements defined as in the basic formula, (12.35), for a fragmentation function, but with the following modifications:

- In the fragmentation function to hadron H_A, a Wilson line is attached to each of the quark and antiquark fields. Each Wilson line goes to positive infinity in the direction w_B corresponding to the opposite hadron. In the fragmentation function to hadron H_B, the Wilson lines are in direction w_A.
- Wilson-line self-interactions, Fig. 13.4, are omitted.
- Soft subtractions and UV renormalization are applied.

A component of the results is the unsubtracted TMD fragmentation function for hadron H_A:

$$\tilde{D}_{1,\,H_A/f}^{\text{unsub}}(z_A, b_{\text{T}}, y_{p_A} - y_B)$$

$$\stackrel{\text{def}}{=} \frac{\text{Tr}_{\text{color}}}{N_{c,j}} \frac{\text{Tr}_{\text{Dirac}}}{4} \sum_X \frac{1}{z_A} \int \frac{dx^-}{2\pi} e^{ik_A^+ x^-} \langle 0|\gamma^+ W(x/2; \infty; n_B)\psi_f(x/2)|p, X, \text{ out}\rangle$$

$$\times \langle p, X, \text{ out}|\bar{\psi}_f(-x/2)W(-x/2; \infty; n_B)^\dagger|0\rangle\Big|_{\text{No SI}}$$

$$= \frac{\text{Tr}_{\text{color}}}{N_{c,j}} \frac{\text{Tr}_{\text{Dirac}}}{4} \frac{\gamma^+}{z_A} \int \frac{dk_A^- \, d^{2-2\epsilon}k_{A\,\text{T}}}{(2\pi)^{4-2\epsilon}} e^{ik_{A\text{T}}\cdot b_{\text{T}}} \quad \xrightarrow{\;k_A\;} \qquad (13.41)$$

where the vector x in the first line is $(0, x^-, b_{\text{T}})$. We do not equip this fragmentation function with a UV-renormalization factor, leaving that to the final definition in (13.42).

Exactly as for the Sudakov form factor, in (10.119a), we combine soft factors into the collinear factors, to make the final definition of the fragmentation function for hadron H_A:

$$\tilde{D}_{1,\,H_A/f}(z_A, b_{\text{T}}; \zeta_A; \mu)$$

$$\stackrel{\text{def}}{=} \lim_{\substack{y_A \to +\infty \\ y_B \to -\infty}} \tilde{D}_{1,\,H_A/f}^{\text{unsub}}(z_A, b_{\text{T}}; y_{p_A} - y_B)\sqrt{\frac{\tilde{S}_{(0)}(b_{\text{T}}; y_A, y_n)}{\tilde{S}_{(0)}(b_{\text{T}}; y_A, y_B)\,\tilde{S}_{(0)}(b_{\text{T}}; y_n, y_B)}}$$

$$\times \text{ UV-renormalization factor}$$

$$= \tilde{D}_{1,\,H_A/f}^{\text{unsub}}(z_A, b_{\text{T}}; y_{p_A} - (-\infty))\sqrt{\frac{\tilde{S}_{(0)}(b_{\text{T}}; +\infty, y_n)}{\tilde{S}_{(0)}(b_{\text{T}}; +\infty, -\infty)\,\tilde{S}_{(0)}(b_{\text{T}}; y_n, -\infty)}}\,Z_D Z_2.$$

$$(13.42)$$

As in (10.119a), y_n is an arbitrary rapidity value, used to specify non-light-like Wilson lines. It will have the function in the factorization formula of separating left- and right-moving quanta. We use the notation with infinite-rapidity Wilson lines to imply the appropriate limit operations. The fragmentation function depends on the rapidity difference $y_{p_A} - y_n$. But for the corresponding argument of the fragmentation function we use the following variable:

$$\zeta_A \overset{\text{def}}{=} 2(k_{A,h}^+)^2 e^{-2y_n} = \frac{2(p_{A,h}^+)^2 e^{-2y_n}}{z_A^2} = \frac{m_A^2}{z_A^2} e^{2(y_{p_A} - y_n)}. \tag{13.43}$$

This is a convenient variable for use in renormalization. Compare (10.127a).

As explained following (10.119a), UV renormalization and removal of the UV regulator (i.e., space-time dimension $n \to 4$) are to be applied *after* taking the limits of infinite rapidity for the Wilson lines. In the case of the integrated fragmentation function there was an integral over all external parton transverse momentum, and the associated UV divergence gave a non-trivial RG equation of the DGLAP type. But for the TMD function, this integral is absent, and UV divergences are only in virtual corrections, essentially the same as in the collinear factors for the Sudakov form factor. Since $\tilde{D}^{\text{unsub}}$ is defined with renormalized quark fields, the multiplicative UV-renormalization factor in (13.42) is written as $Z_D Z_2$, where Z_2 is the wave-function renormalization factor of the quark field. Then Z_D is the ratio of the renormalized fragmentation function to the unrenormalized fragmentation function defined with *bare* fields. The anomalous dimension will be obtained from Z_D.

A fragmentation function for the other hadron is defined like (13.42), just with the roles of the plus and minus coordinates exchanged, and with exchange of the labels A and B. Instead of ζ_A we use

$$\zeta_B \overset{\text{def}}{=} 2(k_{B,h}^-)^2 e^{2y_n} = \frac{2(p_{B,h}^-)^2 e^{2y_n}}{z_B^2} = \frac{m_B^2}{z_B^2} e^{2(y_n - y_{p_B})}. \tag{13.44}$$

Note that with the values of $k_{A,h}^+$ and $k_{B,h}^-$ specified in Sec. 13.3.2, and with neglect of power-suppressed corrections, we have

$$\zeta_A \zeta_B = \frac{Q^4}{\cos^4(\delta\theta/2)} \quad \text{(original values)}. \tag{13.45}$$

But we will be able to clean up the factorization formula by changing the values of ζ_A and ζ_B slightly.

13.7.2 Wilson-line self-interactions in final definitions

For the same reasons as for the soft factor, Wilson-line self-interactions were omitted from the definition of the unsubtracted fragmentation function, (13.41). We now show that in the final definition of the complete fragmentation function, (13.42), we can replace the unsubtracted collinear and soft factors by versions with Wilson-line self-interactions allowed.

In an abelian theory, Wilson-line self-interactions just gave an overall factor, e.g., in (10.89). Thus it was straightforward to show that the self-interactions cancel in the combinations relevant to (10.119). Thus we could retain Wilson-line self-energies on the r.h.s. of this equation, and obtain gauge independence.

For the non-abelian case, the steps are not so direct. This is because we can no longer use (10.100) to disentangle the different gluons attaching to a Wilson line. Instead we use a factorization theorem for each of the factors on the r.h.s. of (13.42), in the limit that $y_A \to \infty$ and $y_B \to -\infty$; cf. Sec. 10.8.7. Each factor then becomes the product of a hard factor, a soft factor, and two collinear factors. For each Wilson line in (13.42), we obtain a particular collinear factor after this new factorization, and it is these collinear factors that we treat as the Wilson-line self-interactions.

To see that the Wilson-line self-interactions cancel in (13.42), we simply count the number of appearances of each kind of self-interaction factor in the complete expression. For example, the Wilson-line self-interaction factor for the y_B Wilson line of D^{unsub} is canceled by the two y_B Wilson-line self-interaction factors from the two factors of $\tilde{S}$ in the denominator of the square-root factor.

The factorization of Wilson-line self-interaction contributions is correct for each correlation function up to errors suppressed by exponentials of the differences in Wilson-line rapidities, e.g., $e^{-(y_A - y_B)}$. Thus, when the infinite rapidity limits are taken in (13.42), these errors become exactly zero. The result for the abelian Sudakov form factor is just a special case.

An immediate and important advantage is that the collinear factors defined in (13.42) are gauge invariant. Hence the results of calculations are independent of the choice of gauge fixing, unlike the case for the individual factors on the r.h.s. of (13.42); see problem 10.9.

It is true that the Wilson lines do not quite join at infinity. For example, in (13.39) we have segments at different transverse positions. To get an exactly gauge-invariant operator, two links must be inserted at infinity in a transverse direction. But the factorization argument for cancellation of Wilson-line self-interactions also applies to the transverse links at infinity.

13.7.3 Factorization

To complete the factorization formula with the redefined fragmentation functions, we anticipate a result from Sec. 13.12 below that gives an additive correction term Y to the structure derived so far. Our derivation has been appropriate for small $q_{h\,T}$: we have neglected not only terms suppressed by a power of a hadronic mass divided by Q, but also terms suppressed by a power of $q_{h\,T}/Q$. But when $q_{h\,T}$ is of order Q, conventional factorization with integrated fragmentation functions is valid. So in Sec. 13.12, we will show how to formulate a large-$q_{h\,T}$ correction term Y. The resulting factorization formula is

$$
W^{\mu\nu} = \frac{8\pi^3 z_A z_B}{Q^2} \sum_f \text{Tr}\, k_{A,\gamma}^+ \gamma^- H_f^\nu(Q) k_{B,\gamma}^- \gamma^+ \overline{H}_f^\mu(Q)
$$

$$
\times \int \frac{\mathrm{d}^{2-2\epsilon} \boldsymbol{b}_{\mathrm{T}}}{(2\pi)^{2-2\epsilon}} e^{-i q_{h\mathrm{T}} \cdot \boldsymbol{b}_{\mathrm{T}}} \tilde{D}_{1,\,H_A/f}(z_A, b_{\mathrm{T}}; Q^2 e^{-2y_n}) \tilde{D}_{1,\,H_B/\bar{f}}(z_B, b_{\mathrm{T}}; Q^2 e^{2y_n})
$$

$$
+ \text{polarized terms} + \text{large-}q_{h\mathrm{T}} \text{ correction, } Y. \tag{13.46}
$$

As in earlier factorization formulae, we use "polarized terms" to indicate term(s) that involve the entangled transverse-spin state. The values of the ζ_A and ζ_B arguments of the fragmentation functions are changed from the values in (13.43) and (13.44), thereby removing the $\cos(\delta\theta/2)$ in (13.45). This will simplify the formulae used in phenomenology. Since the effect is of order $q_{h\,T}^2/Q^2$, it is comparable to errors in the approximants used in the derivation, and does not affect the correctness of the formula. A correct Y term will cancel $q_{h\,T}^2/Q^2$ errors in the TMD term, including those associated with the changes of the values of ζ_A and ζ_B to those in (13.46). Hence the overall result for $W^{\mu\nu}$ is valid for all $q_{h\,T}$, small and large, with relative errors suppressed by a power of mass divided by Q.

In the TMD part of this and previous formulae, the collinear, soft and hard factors appear in the same way as in the Sudakov form factor in Ch. 10. In particular, a soft factor is absent in the final formula. The changes relative to Ch. 10 simply accommodate that the factors correspond to scattering amplitudes times conjugate amplitudes and that the Wilson lines etc. are shifted transversely between the amplitude and the conjugate. The collinear factors have the operators and normalizations appropriate for fragmentation; the Dirac structure is unchanged from a non-gauge theory. Relative to the definitions in a non-gauge theory, only the Wilson-line factors are new.

13.8 Evolution equations for TMD fragmentation functions

When initially conceived, a TMD fragmentation function was simply the number density of hadrons in a parton-induced jet. With its complete definition, it acquires dependence on both a renormalization scale and on ζ. We can regard this dependence as the effect of recoil against emission of soft gluons into approximately a range determined by μ and ζ. Appropriate values of these parameters are energy dependent.

To regain predictive power and effective universality together with some extra predictions, we now derive evolution equations: an RG equation for the μ dependence and a Collins-Soper (CS) equation for the rapidity dependence. The overall structure is the same as for the Sudakov form factor.

13.8.1 CS evolution of TMD fragmentation function

The CS equation has the form

$$\frac{\partial \ln \tilde{D}_{1,\,H_A/f}(z_A, b_T; \zeta_A, \ldots)}{\partial \ln \sqrt{\zeta_A}} = \tilde{K}(b_T; \mu). \tag{13.47}$$

The derivative is equivalent to a derivative with respect to $-y_n$. Since the only dependence on y_n is in the S factors in (13.42), we have

$$\tilde{K}(b_T; \ldots) = \frac{\partial}{\partial y_n} \left[\frac{1}{2} \ln \tilde{S}_{(0)}(b_T; y_n, -\infty) - \frac{1}{2} \ln \tilde{S}_{(0)}(b_T; +\infty, y_n) \right] + \text{UV counterterm}$$

$$= \frac{1}{2\tilde{S}_{(0)}(b_T; y_n, -\infty)} \frac{\partial \tilde{S}_{(0)}(b_T; y_n, -\infty)}{\partial y_n}$$

$$- \frac{1}{2\tilde{S}_{(0)}(b_T; +\infty, y_n)} \frac{\partial \tilde{S}_{(0)}(b_T; +\infty, y_n)}{\partial y_n} + \text{UV c.t.} \tag{13.48}$$

This is normalized to be like K in Ch. 10. Since K is derived from the soft factor, it is the same for all quark and antiquark fragmentation functions. It would be different for the fragmentation of a gluon, which is a color octet.

Now in a differentiated soft factor, e.g., $\partial S(b_T, y_A, y_B)/\partial y_A$, there is a sum of terms in each of which one Wilson-line vertex and its neighboring line are differentiated with respect to rapidity. The resulting vertex is the same as in Fig. 10.27 for the Sudakov form factor, but with the insertion of the appropriate color matrix. Exactly as for the Sudakov form factor, the momentum at the differentiated vertex is, to leading power, close in rapidity to that of the parent Wilson line. Factorization then gives the original soft factor times a factor associated with differentiated vertex. Taking the rapidity difference of the Wilson lines to infinity then removes the power-suppressed corrections to factorization, and leaves a kernel $\tilde{K}$ that depends on b_T and on the parameters of the theory (μ, $g(\mu)$, etc.), but not on the Wilson-line rapidities.

13.8.2 RG evolution of K

As with the Sudakov form factor, the evolution kernel $\tilde{K}$ is renormalized by adding a counterterm, and this gives an additive anomalous dimension. The UV divergence only arises from virtual graphs, so it has no b_T dependence. The RG equation for $\tilde{K}$ then has the form

$$\frac{d\tilde{K}}{d\ln\mu} = -\gamma_K(g(\mu)). \tag{13.49}$$

13.8.3 RG evolution of TMD fragmentation function

Since the UV divergences of the TMD fragmentation function $\tilde{D}$ arise only from virtual graphs, the associated RG equation arises from the overall Z_D factor in (13.42). Thus the RG equation for $\tilde{D}$ has the form

$$\frac{d\ln\tilde{D}_{1, H_A/f}(z_A, b_T; \zeta_A, \dots)}{d\ln\mu} = \gamma_D(g(\mu); \zeta_A/\mu^2). \tag{13.50}$$

Unlike the DGLAP equation, there is no convolution with the longitudinal momentum fraction z_A. However, as with the Sudakov form factor (see Sec. 10.11.4) the anomalous dimension does depend on the longitudinal momentum of the quark, via the variable ζ_A defined in (13.43).

We obtain the energy dependence of γ_D, by the same proof as for the Sudakov form factor in Sec. 10.11.4, and obtain

$$\gamma_D(g(\mu); \zeta_A/\mu^2) = \gamma_D(g(\mu); 1) - \frac{1}{2}\gamma_K(g(\mu))\ln\frac{\zeta_A}{\mu^2}. \tag{13.51}$$

13.9 Flavor dependence of CS and RG evolution

In the most common applications, TMD fragmentation functions (and also TMD parton densities) are used only for Dirac quarks. But TMD functions can be defined for any kind of

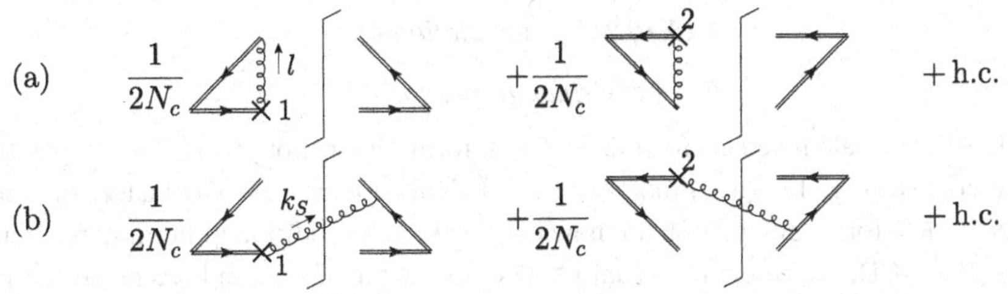

Fig. 13.5. Lowest-order graphs for K: (a) virtual gluon, (b) real gluon. Note the addition of the hermitian conjugate (h.c.) graphs, where the differentiated vertex is on the opposite side of the final-state cut. The empty Wilson lines on the right in the virtual graphs give a unit factor for zeroth-order Wilson lines. The overall factor of $1/N_c$ is from (13.39), and the factor of $\frac{1}{2}$ is from (13.48).

parton, both for the gluon in QCD and for other partonic fields in hypothesized extensions of QCD.

The kernel K and its anomalous dimension γ_K arise from the Wilson-line soft factors, and thus depend only on the color representation of the parton. Thus there are separate versions of these for the gluon, which we could denote K_8 and γ_{K8}. The lowest-order values are obtained from those for ordinary quarks by changing C_F to C_A.

But in a supersymmetric extension of QCD, we would not need any extra functions. The extra fields in such a theory are squarks and gluinos. A squark is a scalar triplet, so it would use the same values of K and γ_K as ordinary quarks. This applies both to the perturbative and non-perturbative parts, of course. A gluino is a spin-$\frac{1}{2}$ octet, so it would need the same K_8 and γ_{K8} as a gluon.

In contrast, the anomalous dimension γ_D would differ between all these types of parton.

13.10 Analysis of CS kernel K: perturbative and non-perturbative

13.10.1 Feynman rules for K

The definition of K in (13.48) involves differentiation of Wilson lines in $\tilde{S}$ with respect to rapidity. The basic vertex needed for the derivatives are obtained from differentiating one Wilson-line vertex and its neighboring line, exactly as in Fig. 10.27, but with an appropriate color matrix.

At lowest order, the resulting graphs are shown in Fig. 13.5, a simple generalization of those for the Sudakov form factor.

In QCD, we no longer have the simplification that we had in an abelian theory where each Wilson line is a graphical exponential of its first-order term. Therefore higher-order graphs for K are more complicated than the simple connected graphs shown in Fig. 10.23 for the abelian theory. It is left as a research exercise to search for any corresponding simplification in a non-abelian theory.

13.10.2 LO calculation of K

LO virtual graphs for K

In (10.134), we calculated K for the Sudakov form factor. For the *virtual* graphs in the present context, Fig. 13.5(a), almost the same calculation applies. The factors of $\frac{1}{2}$ in the definition of K for fragmentation are now canceled by the addition of hermitian conjugate graphs. For QCD, we insert the usual QCD color factor of C_F, and we make the gluon massless.

In momentum space, there is no final-state momentum for the virtual graphs, so for the contribution of these graphs in (13.23) we insert a factor $(2\pi)^{4-2\epsilon}\delta^{(4-2\epsilon)}(k_S)$. After the integration given in (13.23), we have a factor $\delta^{(2-2\epsilon)}(k_T)$ relative to the Sudakov form factor case. The Fourier transform to transverse coordinate space converts this to unity, to give

$$\tilde{K}_{1V} = \frac{-g^2 C_F (4\pi\mu^2)^\epsilon}{4\pi^3} \int d^{2-2\epsilon} l_T \frac{1}{l_T^2} + \frac{g^2 C_F S_\epsilon}{4\pi^2 \epsilon}. \tag{13.52}$$

Here, the UV counterterm is in the $\overline{\text{MS}}$ scheme, and the l_T integral is left explicit. This exhibits a negative IR divergence at $l_T = 0$, which will cancel against a positive divergence from the real emission graphs. The subscript "$1V$" denotes "1-loop virtual".

LO real graphs for K

In Feynman gauge, each of the two graphs in Fig. 13.5(b) gives an equal result, as do their hermitian conjugates; this is checked by explicit calculation. So the complete result is given by multiplying one graph (including its explicit factor of $\frac{1}{2}$) by 4. With the rule in Fig. 10.27 for the differentiated line, we get

$$\tilde{K}_{1R} = \frac{-4g^2 C_F \mu^{2\epsilon}}{(2\pi)^{4-2\epsilon}} \int d^{4-2\epsilon} k_S \, e^{ik_{ST}\cdot b_T} \frac{(2\pi)\delta(k_S^2)\theta(k_S^0)}{(k_S^- e^{y_n} - k_S^+ e^{-y_n} + i0)^2}$$

$$= \frac{g^2 C_F (4\pi^2\mu^2)^\epsilon}{4\pi^3} \int d^{2-2\epsilon} k_T \frac{e^{ik_T\cdot b_T}}{k_T^2}. \tag{13.53}$$

LO total for K

The IR divergence at zero transverse momentum cancels between the real and virtual graphs:

$$\tilde{K}_1 = \frac{g^2 C_F (4\pi^2\mu^2)^\epsilon}{4\pi^3} \int d^{2-2\epsilon} k_T \frac{e^{ik_T\cdot b_T} - 1}{k_T^2} + \frac{g^2 C_F S_\epsilon}{4\pi^2 \epsilon}. \tag{13.54}$$

To perform the k_T integral, we use (A.45), and get

$$\tilde{K}_1 = \frac{g^2 C_F}{4\pi^2} \left[(\pi\mu^2 b_T^2)^\epsilon \Gamma(-\epsilon) + \frac{S_\epsilon}{\epsilon} \right] \overset{\epsilon=0}{=} -\frac{g^2 C_F}{4\pi^2} \left[\ln(\mu^2 b_T^2) - \ln 4 + 2\gamma_E \right]. \tag{13.55}$$

The anomalous dimension, from (13.49), is therefore

$$\gamma_K = \frac{g^2 C_F}{2\pi^2} + O(g^4). \tag{13.56}$$

13.10.3 Analysis of b_T dependence for K

The b_T dependence of TMD functions and of $\tilde{K}$ determines the transverse-momentum dependence of cross section, and so it is important to understand to what extent the b_T dependence can be predicted by perturbative calculations.

At large b_T the dependence is non-perturbative, simply because the operators are separated by a large distance. There the functions must be obtained by analyzing experimental data, given the present lack of non-perturbative calculations. In contrast, at small b_T we can employ perturbative methods, as we will now see. Since the boundary between non-perturbative and perturbative regions is quite vague, we will also need a method to combine for a single function perturbative predictions and non-perturbative fits.

Now in the definition of the soft factor $\tilde{S}(b_T)$, there is an integral over the external momentum k_S, with the Fourier-transform factor $e^{ik_{ST} \cdot b_T}$ providing, roughly speaking, an upper cutoff at $k_{ST} \sim 1/b_T$. We have the same situation as in the e^+e^- annihilation cross section that this gives an IR-safe quantity. The same applies to $\tilde{K}$, which is a derivative of $\tilde{S}$, as is evidenced by the canceled IR singularity in (13.54).

So to calculate $\tilde{K}(b_T)$ at small b_T we simply apply an RG transformation to set μ of order $1/b_T$:

$$\tilde{K}(b_T; \mu, g(\mu), m(\mu)) \simeq \tilde{K}\left(b_T; \frac{C_1}{b_T}, g\left(\frac{C_1}{b_T}\right), 0\right) - \int_{C_1/b_T}^{\mu} \frac{d\mu'}{\mu'} \gamma_K(g(\mu')). \quad (13.57)$$

On the r.h.s., K has its renormalization mass proportional to $1/b_T$, which eliminates large logarithms. The constant of proportionality, C_1, can be used to optimize the accuracy of perturbative calculations. In the $\overline{\text{MS}}$ scheme, a value not far from unity is appropriate. Then, for example, truncating perturbation theory for $\tilde{K}$ to the first-order term gives an error of order the first term omitted, i.e., $O(g^4)$. We have also chosen to neglect quark masses relative to $1/b_T$, as is appropriate for light quarks.

In applications, we will set μ equal to the value used in calculating the hard scattering perturbatively, i.e., of order Q, unambiguously in a perturbative domain. As b_T is varied with μ fixed, the largest contribution to the r.h.s. of (13.57) comes from the integral over the anomalous dimension.

Evidently the accuracy of a perturbative estimate worsens as b_T increases, and for large enough b_T, presumably around 0.5 fm, perturbation theory becomes inapplicable. In this case, we can perform a transformation to a value μ_0 of the renormalization mass μ_0 that stays fixed when we change the experimental energy Q and hence change μ. Thus we write

$$\tilde{K}(b_T; \mu, g(\mu), m(\mu)) = \tilde{K}(b_T; \mu_0, g(\mu_0), m(\mu_0)) - \int_{\mu_0}^{\mu} \frac{d\mu'}{\mu'} \gamma_K(g(\mu')). \quad (13.58)$$

This demonstrates an important result: the non-perturbative information in K is contained in a single universal function of b_T. (The universality is between all processes using TMD functions for color-triplet partons.) As we will see, this function can be measured from the derivative with respect to energy of a suitable cross section. In principle, the derivative can be taken at one energy, thereby allowing a prediction of the cross section at other energies.

Consequences in transverse-momentum space

When we Fourier-transform back to transverse momentum, perturbative calculations at small b_T conveniently combine two types of perturbatively calculable information. First, the singularity of $\tilde{K}(b_T)$ at small b_T determines the shape of $K(k_T)$ at large k_T. Second, the value of $\tilde{K}$ at small b_T determines the integral of $K(k_T)$ over all k_T up to about $1/b_T$. Similar remarks apply to the TMD fragmentation function.

13.10.4 Matching perturbative and non-perturbative b_T dependence

To combine information on b_T dependence from perturbative calculations valid at small enough b_T with a non-perturbative part that must be determined by a fit to experimental data, a matching procedure was formulated by Collins and Soper (1982a). First a parameter b_{max} is chosen which has the interpretation of the maximum distance at which perturbation theory is to be trusted. One value (Landry *et al.*, 2003) that has been used is $b_{max} = 0.5\,\text{GeV}^{-1} = 0.1\,\text{fm}$.

Then a function $b_*(b_T)$ is defined with the properties that at small b_T it is the same as b_T, and that at large b_T it is no larger than b_{max}. The standard choice is

$$b_* \stackrel{\text{def}}{=} \frac{b_T}{\sqrt{1 + b_T^2/b_{max}^2}}. \tag{13.59}$$

Changes in the form of this function or in the value of b_{max} do not affect the physical cross section, but only the way in which non-perturbative phenomena are parameterized.

We now write $\tilde{K}(b_T) = \tilde{K}(b_*) + $ correction term. The idea is that $\tilde{K}(b_*)$ is always in a situation where perturbation theory is appropriate, and the correction term is only important at large b_T. Therefore we write

$$\tilde{K}(b_T; \mu, \ldots) = \tilde{K}(b_*; \mu, g(\mu), m(\mu))$$
$$+ \left[\tilde{K}(b_T; \mu, g(\mu), m(\mu)) - \tilde{K}(b_*; \mu, g(\mu), m(\mu)) \right]$$
$$= \tilde{K}(b_*; C_1/b_*, g(C_1/b_*), 0) - \int_{C_1/b_*}^{\mu} \frac{d\mu'}{\mu'} \gamma_K(g(\mu'))$$
$$+ \left[\tilde{K}(b_T; \mu_0, g(\mu_0), m(\mu_0)) - \tilde{K}(b_*; \mu_0, g(\mu_0), m(\mu_0)) \right]$$
$$= \tilde{K}(b_*; C_1/b_*, g(C_1/b_*), 0) - \int_{C_1/b_*}^{\mu} \frac{d\mu'}{\mu'} \gamma_K(g(\mu')) - g_K(b_T), \tag{13.60}$$

where the correction term is denoted $-g_K$. Phenomenologically it is a function of one variable b_T, to be fit to data. It is RG invariant, since it is a difference of $\tilde{K}$ at two values of its position argument. The correction term vanishes as $b_T \to 0$. Recent fits use a quadratic ansatz:

$$g_K(b_T) = \frac{1}{2}g_2 b_T^2.$$
(13.61)

The measured value of g_2 is correlated with the value of b_{max}, and with assumptions about other non-perturbative functions; see Sec. 14.5.3. Landry *et al.* (2003) and Konychev and Nadolsky (2006) found

$$g_2 = \begin{cases} 0.68 \, {}^{+0.01}_{-0.02} \, \text{GeV}^2 & \text{with } b_{max} = 0.5 \, \text{GeV}^{-1}, \\ 0.17 \pm 0.02 \, \text{GeV}^2 & \text{with } b_{max} = 1 \, \text{GeV}^{-1}. \end{cases}$$
(13.62)

(The second number is my average of two fits.) Then

$$g_K(b_T) \simeq \begin{cases} 0.34 \, b_T^2 & \text{with } b_{max} = 0.5 \, \text{GeV}^{-1}, \\ 0.08 \, b_T^2 & \text{with } b_{max} = 1 \, \text{GeV}^{-1}. \end{cases}$$
(13.63)

The fits are made with truncated perturbative approximations for both $\tilde{K}(b_*; C_1/b_*, g(C_1/b_*), 0)$ and $\gamma_K(g(\mu'))$. Because the coupling increases with decreasing scale, the approximations lose accuracy when applied at larger values of b_* compared with lower values. A phenomenological fit for the function $g_K(b_T)$ in (13.60) effectively includes an allowance for errors in truncated perturbation theory for scales near b_{max}.

Thus one should *not* expect the numerical values of $g_K(b_T)$ to be stable against the inclusion of yet higher-order perturbative estimates of $\tilde{K}$ and γ_K. Only the total value of $\tilde{K}(b_T)$ should be stable against improvements in perturbative calculations.

Note that the numerical results quoted from Landry *et al.* (2003) were from fits to Drell-Yan data. In describing them here for their application to e^+e^- annihilation, we are using a result to be explained later that the soft function is the same in the two reactions.

13.11 Relation of TMD to integrated fragmentation function

We now generalize the methods of Secs. 13.10.3 and 13.10.4 to analyze the dependence of the fragmentation function on b_T, and to formulate perturbative and non-perturbative parts.

13.11.1 Perturbative small-b_T dependence

First, we analyze the small-b_T region. We show that in contrast to the case of the evolution kernel K, the perturbative calculation for a TMD fragmentation function at small b_T does not determine the fragmentation function absolutely, but only expresses it, by a factorization property, in terms of a perturbative coefficient convoluted with integrated fragmentation functions. The integrated fragmentation functions themselves must still be obtained from experiment. This result does give notable predictive power since a function of two kinematic variables is expressed in terms of a non-perturbative function of one variable.

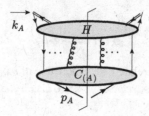

Fig. 13.6. Leading regions for TMD fragmentation function at small b_T.

See Sec. 13.11.2 for how to combine this with non-perturbative information for large b_T.

To motivate that a relation should exist between a TMD fragmentation function at small b_T and an integrated fragmentation function, we recall the discussion at the beginning of Sec. 13.6. There we showed that in super-renormalizable non-gauge model theories, a TMD fragmentation function at zero b_T equals the corresponding integrated fragmentation function (up to a standard normalization factor); this is the parton-model result. In QCD, this relation fails, because of the need to renormalize fragmentation functions and because of complications associated with the soft factors in (13.42).

I will now formulate a corrected relation. It is a factorization formula involving a coefficient function whose lowest-order value is unity, corresponding to the parton-model result.

Region analysis

We start with a region analysis for the unsubtracted unintegrated TMD fragmentation function (13.41). Leading regions involve hard, collinear and soft subgraphs. The soft subgraphs connect the collinear subgraphs, and there are collinear factors associated with the detected hadron and with the Wilson line. The hard factor is associated with the external quark–Wilson-line vertices. Its lowest-order term is just these vertices. There can be higher-order hard subgraphs with highly virtual loops, and there can be further hard subgraphs with production of final-state jets of high transverse momentum, the transverse momenta being limited basically by the large value of $1/b_T$. The structure is essentially the same as we encountered for simple inclusive cross sections in e^+e^- annihilation, around Fig. 12.4. As in that case, the sum/integral over final states in each extra jet is fully inclusive, so after a sum-over-cuts, as in Sec. 12.8.7, the subgraphs for the extra high-transverse-momentum jets are effectively far off-shell and count as part of the hard part.

But now a difference arises, that the definition of the subtracted TMD fragmentation function (13.42) includes subtractions to remove the opposite (p_B-associated) collinear region and the soft region. So the only remaining effective regions have a subgraph collinear to p_A and a possible hard subgraph, as shown in Fig. 13.6. The parton-model result applies when the only hard subgraphs are trivial, i.e., when all the parton lines in the graph on the last line of (13.41) are hadron-collinear.

Factorization for TMD at small b_T

We now apply the usual factorization argument to Fig. 13.6, summed over all cases and with double-counting subtractions. The extra gluons entering the hard part are converted to Wilson lines by Ward identities. Since the integration over transverse momenta in the collinear part is limited solely by $1/b_T$ in the hard part, the collinear factor gives an *integrated* fragmentation function. We get

$$\tilde{D}_{1,H_A/f}(z_A, b_T; \zeta_A; \mu)$$

$$= \sum_j \int_{z_A}^1 \frac{d\hat{z}}{\hat{z}^{3-2\epsilon}} \, d_{H_A/j}(\hat{z}; \mu) \, \tilde{C}_{j/f}(z_A/\hat{z}, b_T; \zeta_A, \mu, g(\mu)) + O[(mb_T)^p]. \quad (13.64)$$

The error term is suppressed by some power of transverse position. The sum over j is over all types of parton, including gluons and antiquarks. The coefficient function $\tilde{C}_{f/j}$ is calculated, as usual for a hard factor, from graphs with external on-shell parton of type j, with double-counting subtractions that cancel all collinear contributions.

Since the on-shell parton has infinite rapidity, we convert the dependence on the rapidity y_n to a dependence of C on an energy variable ζ_A, defined by (13.43). The lowest-order coefficient is

$$\tilde{C}_{j/f}(z_A/\hat{z}, b_T; \zeta_A, \mu, g(\mu)) = \delta_{jf}\delta(z_A/\hat{z} - 1) + O(g^2). \quad (13.65)$$

The integral in (13.64) has a measure $d\hat{z}/\hat{z}^{3-2\epsilon}$, rather than the $d\hat{z}/\hat{z}$ that we would get for a corresponding formula with a parton density. This arises from the different powers of z in the normalizations of the definitions of the TMD and integrated fragmentation function, (12.39) and (12.40). Although the formula is phenomenologically applied at $\epsilon = 0$, it was written for a general ϵ. This allows the factorization formula also to be applied in perturbative calculations, where intermediate stages use dimensional regularization.

Logarithms in coefficient

Evolution equations for the coefficient can be derived from the CS and RG evolution equations for the TMD fragmentation function and the DGLAP equation for the integrated fragmentation function. They show that the dependence of the coefficient on μ and on ζ_A is logarithmic in each order of perturbation theory. Hence by dimensional analysis the dependence on b_T is also logarithmic. Thus the functional form of the b_T dependence in each order of perturbation theory is a polynomial in $\ln b_T^2$. From the evolution equations it can be seen that the order of the polynomial is $2L$ where L is the number of loops: i.e., there are two logarithms per loop, giving leading logarithms characteristic of the Sudakov form factor. Fourier transformation gives $1/k_T^2$ times a polynomial in $\ln k_T^2$, and the order of this polynomial is $2L - 1$.

NLO calculations

To see how actual calculations work and for the values of the coefficients, see Sec. 13.14.

TMD fragmentation function at large k_T

Fourier transformation of (13.64) gives factorization for the TMD fragmentation function at large transverse momentum:

$$D_{1, H_A/f}(z_A, k_T; \zeta_A; \mu)$$

$$= \sum_j \int_{z_A}^1 \frac{\mathrm{d}\hat{z}}{\hat{z}^{3-2\epsilon}} d_{H_A/j}(\hat{z}; \mu)\, C_{j/f}\left(\frac{z_A}{\hat{z}}, k_T; \zeta_A, \mu, g(\mu)\right) + O\left[\left(\frac{m}{k_T}\right)^p \frac{1}{k_T^2}\right]. \quad (13.66)$$

13.11.2 Matching perturbative and non-perturbative b_T dependence for TMD fragmentation

To combine the perturbative information on fragmentation at small b_T with non-perturbative information (to be fitted to data) at large b_T, we copy the method applied in Sec. 13.10.4 to the kernel K.

Intrinsic transverse momentum dependence and energy dependence

A complication is that in addition to the kinematic variables z and $\boldsymbol{b}_T$, the TMD fragmentation function depends on two parameters $y_{p_A} - y_n$ and μ, which can be thought of as cutoffs on internal gluon momenta. The CS and RG equations control dependence on these parameters. In an application, we will normally set μ of order the large kinematic variable Q, to allow a useful perturbative calculation of the hard scattering; we might choose y_n to be zero in the overall CM frame. Thus the values of $y_{p_A} - y_n$ and μ change, depending on the kinematics of the process being considered.

So we solve the evolution equations to gives the TMD fragmentation function in terms of its value at fixed reference values of ζ_A and μ:

$$\tilde{D}_{1, H_A/f}(z_A, b_T; \zeta_A; \mu) = \tilde{D}_{1, H_A/f}\left(z_A, b_T; m_A^2/z_A^2; \mu_0\right) \exp\left\{ \ln \frac{\sqrt{\zeta_A} z_A}{m_A} \tilde{K}(b_T; \mu_0) \right.$$

$$\left. + \int_{\mu_0}^\mu \frac{\mathrm{d}\mu'}{\mu'} \left[\gamma_D(g(\mu'); 1) - \ln \frac{\sqrt{\zeta_A}}{\mu'} \gamma_K(g(\mu')) \right] \right\}. \quad (13.67)$$

Here, the reference value of ζ_A was chosen to correspond to $y_{p_A} = y_n$. Some other value could equally well be used, but this value is appropriate as the limit of where the detected hadron is moving to the right in the rest frame of the vector n. As for the reference value μ_0 of the renormalization scale, it should be in the perturbative region, so that low-order perturbative calculations of γ_D and γ_K are useful. Notice that the μ dependence gives an overall normalization change, but does not affect the shape of the b_T dependence of the fragmentation function.

The dependence on ζ_A involves the function $\tilde{K}(b_T)$, so it gives energy dependence to the shape of the transverse momentum distribution. We characterize the result as follows. The function $\tilde{D}$ at its reference value of the parameters can be thought of as the Fourier transform of an intrinsic transverse momentum distribution of a hadron in its parent parton, essentially a parton-model concept. But this is multiplied by $e^{\ln(\sqrt{\zeta_A}z_A/m_A)\tilde{K}(b_T;\mu_0)}$. This is the effect of energy-dependent recoil against the emission of soft gluons. In momentum space we can treat its effects as the result of convoluting the intrinsic distribution with $\ln(\sqrt{\zeta_A}z_A/m_A)$ factors of the Fourier transformation of $e^{\tilde{K}(b_T;\mu_0)}$. Thus we can treat $e^{\tilde{K}(b_T;\mu_0)}$ as giving the distribution of gluon emission per unit rapidity, with the emission being uniform in rapidity. Note that perturbative calculations and fits indicate that $\tilde{K}(b_T)$ is basically negative, and that it becomes very negative at large b_T, so that the Fourier transform is well behaved.

Matching perturbative and non-perturbative parts

To match the perturbative and non-perturbative parts of $\tilde{D}$, we again use the quantity b_* defined in (13.59). Generalizing (13.60) we formulate an intrinsically non-perturbative part by the following decomposition:

$$\tilde{D}_{1,H_A/f}(z_A, b_T; \zeta_A; \mu)$$

$$= \tilde{D}_{1,H_A/f}(z_A, b_*; \zeta_A; \mu) \left[\frac{\tilde{D}_{1,H_A/f}(z_A, b_T; \zeta_A; \mu)}{\tilde{D}_{1,H_A/f}(z_A, b_*; \zeta_A; \mu)} \right]$$

$$= \tilde{D}_{1,H_A/f}(z_A, b_*; \zeta_A; \mu) \exp\left[-g_{H_A/f}(z_A, b_T) - \ln\frac{\sqrt{\zeta_A}z_A}{m_A} g_K(b_T) \right]. \tag{13.68}$$

In the second line we simply separated out a factor of $\tilde{D}$ at b_*, which we will calculate perturbatively. We then evolved the fragmentation functions in the brackets to the reference values of ζ_A and μ. The effects of the anomalous dimension γ_D cancel between numerator and denominator, while from CS evolution there survived only the "non-perturbative" part of $\tilde{K}$, i.e., g_K, defined in (13.60). The remaining factor we chose to write as an exponential $e^{-g_{H_A/j}(z_A,b_T)}$, which we can label as the non-perturbative part of the intrinsic transverse momentum distribution (Fourier transformed).

The Fourier transform of (13.68) into momentum space should be well behaved under conditions when the factorization formula is used, i.e., when ζ_A is large enough, probably bigger than a few GeV^2. This implies that the function g_K should go to positive infinity as $b_T \to \infty$. Typical fits assume that this behavior is proportional to one or two powers of b_T, i.e., that an exponential or Gaussian is appropriate. The constraints on the other function, $g_{H_A/j}$, are less severe. If its power law is the same as g_K, then there will be a problem when ζ_A is too low, since then the exponent would grow indefinitely at large b_T. This is not in principle a problem, since we should only use parton densities when a factorization formula is valid, i.e., only for ζ_A above some lower limit.

See Landry *et al.* (2003) and Konychev and Nadolsky (2006) for fits in the completely analogous case of TMD quark densities in a hadron.

To use the perturbative small-b_T result from (13.64), we now evolve the b_* factor (13.68) to a situation with no large kinematic ratios in the coefficient function $\tilde{C}$, whose logarithms would prevent the effective use of perturbation theory. We therefore choose to replace μ_0 in (13.67) by

$$\mu_b = \frac{C_1}{b_*(b_T)},\tag{13.69}$$

and we replace the reference value m_A^2/z_A^2 for ζ_A by μ_b^2. Then

$$\tilde{D}_{1,\,H_A/f}(z_A, b_T; \zeta_A; \mu)$$

$$= \sum_j \int_{z_A}^1 \frac{\mathrm{d}\hat{z}}{\hat{z}^{3-2\epsilon}}\, d_{H_A/j}(\hat{z}; \mu_b)\, \tilde{C}_{j/f}\left(z_A/\hat{z}, b_*; \mu_b^2, \mu_b, g(\mu_b)\right)$$

$$\times \exp\left[-g_{H_A/f}(z_A, b_T) - \ln\frac{\sqrt{\zeta_A}z_A}{m_A} g_K(b_T)\right]$$

$$\times \exp\left\{\ln\frac{\sqrt{\zeta_A}}{\mu_b}\tilde{K}(b_*; \mu_b) + \int_{\mu_b}^\mu \frac{\mathrm{d}\mu'}{\mu'}\left[\gamma_D(g(\mu'); 1) - \ln\frac{\sqrt{\zeta_A}}{\mu'}\gamma_K(g(\mu'))\right]\right\}.\tag{13.70}$$

This is probably the best formula for calculating and fitting TMD fragmentation functions; see (13.81) for its use in a factorization formula.[1] Besides the integrated fragmentation functions, which can be measured from simpler inclusive processes, there are further non-perturbative functions $g_{H_A/j}(z_A, b_T)$ and $g_K(b_T)$ that must be obtained by fits to data. The first of these functions requires essentially the same amount and kind of data to determine as we would need to determine TMD fragmentation functions if the simple parton model were valid, without any QCD modifications. The second function $g_K(b_T)$ depends only on a single variable, and can be obtained from the energy dependence of the process. Many predictions can be made with the aid of g_K, since it is independent of z_A, and also since exactly the same function appears in many other processes with TMD functions, both for fragmentation and for parton densities.

All remaining quantities are perturbative, and can therefore be predicted to useful accuracy from first principles by low-order Feynman-graph calculations. For this to work, the lower limit on μ_b, i.e., $C_1/b_{\max}$, should be at an energy scale where the use of perturbation theory is appropriate, say about 2 GeV. However, this is typically a fairly low scale, where the errors in truncated perturbation theory are substantially larger than in the calculation of the hard scattering at a scale of tens or hundreds of GeV. It is worth noting that because of the form of (13.70) these errors can dominantly be compensated by adjustments of the non-perturbative functions. That is, actual fits for the non-perturbative functions automatically compensate the largest higher-order terms in $\tilde{K}$ and $\tilde{C}$.

---

[1] In this application, ϵ is set to zero in the factor $1/\hat{z}^{3-2\epsilon}$. The ϵ dependence is retained in (13.70) so that the formula can also be related to regulated perturbative calculations.

13.12 Correction term for large $q_{h\mathrm{T}}$

The TMD factorization formalism described above applies when $q_{h\mathrm{T}}$ is treated as a small variable. Approximations were made that have errors of order a power of $q_{h\mathrm{T}}/Q$. When $q_{h\mathrm{T}}$ is of order Q, the conventional formalism, with its integrated fragmentation functions, is valid: Ch. 12. Notably, the large $q_{h\mathrm{T}}$ then arises from hard scattering with three or more final-state partons, whereas the TMD formalism associates $q_{h\mathrm{T}}$ with parton transverse momenta in the TMD fragmentation functions.

The TMD formalism loses accuracy at large $q_{h\mathrm{T}}$, with fractional errors we characterize as $(q_{h\mathrm{T}}/Q)^{\alpha}$. The other formalism loses accuracy at small $q_{h\mathrm{T}}$ with fractional errors $(m/q_{h\mathrm{T}})^{\beta}$, where m denotes a typical hadronic scale. In these estimates, α and β are positive powers for the first neglected terms in the region approximants, either 1 or 2 in reality, and we then can reduce these powers slightly so as to obtain errors valid in the presence of logarithmic corrections. We assume throughout that we do not let $q_{h\mathrm{T}}$ increase beyond order Q.

To work with the whole range of $q_{h\mathrm{T}}$ it is necessary to find a way of combining the two formalisms without loss of accuracy.

A simple-minded approach would be to use the TMD formalism for $q_{h\mathrm{T}}$ below some scale $Q_0 \ll Q$, and to use the conventional formalism above that scale. But this would substantially degrade the accuracy of the predictions. For example, suppose the error exponents are $\alpha = \beta = 1$. Then the worst fractional error in the use of the TMD formalism would be Q_0/Q, while that for the conventional formalism would be m/Q_0, both at $q_{h\mathrm{T}} = Q_0$. Globally optimal errors would be obtained with Q_0 proportional to the geometric mean of Q and m, for a fractional error of order $\sqrt{m/Q}$, i.e., with an error exponent 0.5 instead of unity.

Using the general principles of subtraction methods, Collins and Soper (1982a) devised a method that in principle gives m/Q errors for all $q_{h\mathrm{T}}$. Their idea was to treat the TMD term as a first approximation to the cross section (or structure function). It is obtained by applying a TMD approximator, T_{TMD}, to the structure function:

$$L = T_{\mathrm{TMD}} W^{\mu\nu}. \tag{13.71}$$

(Hidden inside the action of T_{TMD} are all the details of the extraction of the hard factor, the definitions of the TMD fragmentation function, etc.) This "low-transverse momentum term" L gives all but the Y term on the r.h.s. of (13.46).

The fractional error in the approximant is power-suppressed in $q_{h\mathrm{T}}/Q$:

$$|W - T_{\mathrm{TMD}} W| = O((q_{h\mathrm{T}}/Q)^{\alpha} |W|). \tag{13.72}$$

We define the correction term Y in (13.46) by applying an approximator for ordinary collinear factorization to the remainder:

$$Y \stackrel{\mathrm{def}}{=} T_{\mathrm{coll}}(W^{\mu\nu} - L) \tag{13.73}$$

The fractional errors in this approximation are suppressed by a factor $(m/q_{h\mathrm{T}})^{\beta}$. Although this degrades as $q_{h\mathrm{T}}$ gets small, it is applied to a quantity that itself is getting small. Therefore the sum of L and Y, i.e., the whole r.h.s. of (13.46), is a uniformly good approximation, i.e., $W - L - Y$ is power-suppressed:

$$|W - L - Y| = |(1 - T_{\text{coll}})(W - L)|$$
$$= O\big((m/q_{h\,\text{T}})^\beta |W - L|\big)$$
$$= O\big((m/q_{h\,\text{T}})^\beta (q_{h\,\text{T}}/Q)^\alpha |W|\big)$$
$$= O\big((m/Q)^{\min(\alpha,\beta)} |W|\big). \tag{13.74}$$

The above error estimate applies when $q_{h\,\text{T}}$ is less than of order Q. However, although there is a kinematic limit at when $q_{h\,\text{T}}$ gets larger than Q, this kinematic limit is not respected by the low-$q_{h\,\text{T}}$ term. Its large-$q_{h\,\text{T}}$ behavior represents only a kind of extrapolation of the low-$q_{h\,\text{T}}$ behavior. Once one gets close to or beyond the kinematic limit, the error between W and L increases far beyond 100%. An appropriate solution is to redefine L with a cutoff to restrict the values of $q_{h\,\text{T}}$ to which it is applied. That is, L is replaced by

$$L_F = F(q_{h\,\text{T}}/Q)\, T_{\text{TMD}} W. \tag{13.75}$$

Here the function $F(q_{h\,\text{T}}/Q)$ is chosen so that it is unity at $q_{h\,\text{T}} = 0$ and zero for large $q_{h\,\text{T}}$. A possible choice would be a theta function $F(q_{h\,\text{T}}/Q) = \theta(Q - q_{h\,\text{T}})$. A better choice would be a smooth function. Since the kinematic limit is dependent on the momentum fractions z_A and z_B, it would be appropriate to give corresponding dependence to the function F.

The cutoff function should be inserted in (13.46), multiplying its first term, and an appropriate redefinition of Y must be made:

$$Y_F \overset{\text{def}}{=} T_{\text{coll}}(W - L_F). \tag{13.76}$$

If L and Y were computed exactly, the choice of cutoff function would be unimportant. But actual estimates of L and Y involve truncations of perturbation expansions, so the cutoff function F should be chosen to minimize errors, as well as these can be understood.

Other procedures are possible, for example as proposed by Arnold and Kauffman (1991). The overall aim is to minimize the likely errors of calculations.

13.13 Using TMD factorization

To use the factorization formalism we exploit the CS and RG evolution equations to change the values of μ and y_n in each factor separately, so that:

- perturbatively calculated quantities are applied in a region where their coefficients have no large logarithms;
- non-perturbative quantities are applied with fixed values of μ and Δy, so that the functions that need to be fitted to data have the minimum number of variables.

The resulting formula, (13.81) below, is suitable for data fitting and for using the results of perturbative calculations. However, this formula is quite complicated. So I show the factorization result in two other forms to exhibit the overall structure.

13.13.1 Three views of factorization

Main factorization formula

First is the main factorization formula, presented earlier (13.46), which follows most immediately from the derivation of factorization. It directly exhibits the low-q_{hT} part of the $W^{\mu\nu}$ in terms of TMD fragmentation functions. It is equivalent to a convolution of TMD fragmentation functions.

Factorization with fixed fragmentation functions

The fragmentation functions have dependence on auxiliary parameters as well as the momentum fractions and the transverse coordinates and momenta. We can exhibit factorization in terms of TMD densities at fixed reference values of the auxiliary parameters, by the use of (13.67), obtained from solving the CS and RG equations. This gives

$$
W^{\mu\nu} = \frac{8\pi^3 z_A z_B}{Q^2} \sum_f H_f^{\mu\nu}\left(Q; g(\mu_Q), \mu_Q\right) \int \frac{d^2 \boldsymbol{b}_T}{(2\pi)^2} e^{-i\boldsymbol{q}_{hT}\cdot\boldsymbol{b}_T} e^{-S(b_T; Q; \mu_Q, \mu_0)}
$$

$$
\times \tilde{D}_{1,\,H_A/f}\left(z_A, b_T; \frac{m_A^2}{z_A^2}; \mu_0\right) \tilde{D}_{1,\,H_B/\bar{f}}\left(z_B, b_T; \frac{m_B^2}{z_B^2}; \mu_0\right)
$$

$$
+ \text{polarized terms} + \text{large-}q_{hT} \text{ correction, } Y. \tag{13.77}
$$

We now have fixed fragmentation functions combined with an allowance for recoil against energy-dependent gluon emission, in the factor

$$
e^{-S(b_T; Q; \mu_Q, \mu_0)} \overset{\text{def}}{=} \exp\left\{\ln \frac{Q^2 z_A z_B}{m_A m_B} \tilde{K}(b_T; \mu_0)\right\}
$$

$$
\times \exp\left\{\int_{\mu_0}^{\mu_Q} \frac{d\mu'}{\mu'}\left[2\gamma_D(g(\mu'); 1) - \ln \frac{Q^2}{(\mu')^2} \gamma_K(g(\mu'))\right]\right\}, \tag{13.78}
$$

where the first factor gives an energy-dependent shape to the TMD distribution, but the second factor only affects the normalization. Observe that all dependence on y_n has disappeared.

In (13.77) the renormalization scale in the hard factor $H_f^{\mu\nu}$ is chosen to be proportional to Q,

$$
\mu_Q = C_2 Q. \tag{13.79}
$$

This is used to minimize logarithms in perturbative calculations of $H_f^{\mu\nu}$, which is obtained as a (Dirac and color) trace over on-shell hard-scattering amplitudes:

$$
H_f^{\mu\nu}(Q; g(\mu), \mu) = \text{Tr}\, k_A^+ \gamma^- H_f^\nu(Q) k_B^- \gamma^+ \overline{H}_f^\mu(Q). \tag{13.80}
$$

Factorization with maximum perturbative content

Finally, we apply the small-b_T perturbative expansion of the TMD fragmentation functions in the form of (13.70), where it is combined with functions to parameterize the non-perturbative large b_T dependence. This gives

$$
W^{\mu\nu} = \frac{8\pi^3 z_A z_B}{Q^2} \sum_{f, j_A, j_B} H_f^{\mu\nu}(Q; g(\mu_Q), \mu_Q) \int \frac{d^2 \boldsymbol{b}_T}{(2\pi)^2} e^{-i\boldsymbol{q}_{hT} \cdot \boldsymbol{b}_T}
$$

$$
\times \int_{z_A}^1 \frac{d\hat{z}_A}{\hat{z}_A^3} d_{H_A/j_A}(\hat{z}_A; \mu_b) \, \tilde{C}_{j_A/f}\left(\frac{z_A}{\hat{z}_A}, b_*; \mu_b^2, \mu_b, g(\mu_b)\right)
$$

$$
\times \int_{z_B}^1 \frac{d\hat{z}_B}{\hat{z}_B^3} d_{H_B/j_B}(\hat{z}_B; \mu_b) \, \tilde{C}_{j_B/\bar{f}}\left(\frac{z_B}{\hat{z}_B}, b_*; \mu_b^2, \mu_b, g(\mu_b)\right)
$$

$$
\times \exp\left[-g_{H_A/f}(z_A, b_T) - g_{H_B/\bar{f}}(z_B, b_T)\right]
$$

$$
\times \exp\left[-\ln \frac{Q^2 z_A z_B}{m_A m_B} g_K(b_T) + \ln \frac{Q^2}{\mu_b^2} \tilde{K}(b_*; \mu_b)\right]
$$

$$
\times \exp\left\{\int_{\mu_b}^{\mu_Q} \frac{d\mu'}{\mu'} \left[2\gamma_D(g(\mu'); 1) - \ln \frac{Q^2}{(\mu')^2} \gamma_K(g(\mu'))\right]\right\}
$$

$$
+ \text{polarized terms} + \text{large-}q_{hT} \text{ correction}, Y. \qquad (13.81)
$$

The second and third lines are the part contributed by the integrated fragmentation functions. At lowest order

$$
\text{Lines 2 and 3 of (13.81)} \stackrel{\text{LO}}{=} \frac{d_{H_A/f}(z_A; \mu_b) \, d_{H_B/\bar{f}}(z_B; \mu_b)}{z_A^2 z_B^2} + O(\alpha_s(\mu_b)). \qquad (13.82)
$$

The fourth line of (13.81) gives the non-perturbative contribution to the non-evolving part of the transverse distributions. Finally, the last two lines give the effect of gluon radiation, perturbative and non-perturbative.

The overall non-perturbative factor has the form

$$
\exp\left[-g_{H_A/f}(z_A, b_T) - g_{H_A/\bar{f}}(z_B, b_T) - \ln \frac{Q^2 z_A z_B}{m_A m_B} g_K(b_T)\right]. \qquad (13.83)
$$

This represents the TMD part that must (currently) be obtained by fitting to data. It concerns the region of large b_T. To avoid interfering with the results of valid perturbative calculations, the functions in the exponent should decrease like a power of b_T at *small* b_T. The choice giving the logarithm in (13.83) (and the corresponding logarithm in (13.81)) was explained below (13.67).

13.13.2 Arbitrariness in renormalization scales and b_{max}

In the perturbative parts of (13.81), there are choices of the scales μ_b and μ_Q, with an arbitrariness parameterized by the coefficients C_1 and C_2. As is usual with a choice of renormalization/factorization scale, if all the factors were calculated exactly, then the result for $W^{\mu\nu}$ would be independent of C_1 and C_2. This follows simply from the CS and RG evolution equations treated exactly. But the perturbative calculations of the various quantities needed, H, $\tilde{C}$, $\tilde{K}$, γ_D and γ_K, are always truncated finite-order calculations. So there is residual dependence on C_1 and C_2 due to truncation errors. This dependence is small if the truncation errors are small. The coefficients should be chosen to minimize truncation errors, which can only be done approximately in the absence of exact calculations. My own approach to estimating appropriate values of renormalization scales is summarized in Sec. 3.4.

The remaining arbitrary parameter is $b_{\max}$, which roughly characterizes the boundary between the non-perturbative and perturbative domains for b_T dependence. The dependence on $b_{\max}$ arises from the dependence of the definition of $b_*(b_T)$ on $b_{\max}$.

In (13.81), there is explicit dependence on b_* only in the $\tilde{C}$ factors and in $\tilde{K}$. There is also dependence via the dependence of many quantities on μ_b; see (13.69). But we have already seen that the μ_b dependence cancels up to perturbative truncation errors. The explicit dependence on b_* is in places where it can be exactly compensated by a change in the functional form of the non-perturbative functions $g_{H_A/f}$, $g_{H_B/f}$, and g_K.

Therefore a change of $b_{\max}$ in no way affects the fundamental validity of the formalism, but only the extent to which perturbation theory is used to predict the b_T dependence. However, fits of the non-perturbative functions are normally made by postulating particular functional forms, e.g., (14.38) below, with a small number of parameters. In principle, if such a parameterization is accurate at one value of $b_{\max}$, it will become invalid when $b_{\max}$ is changed. How much of a practical issue this is, needs an examination of actual fits. See Sec. 14.5.3 for further comments.

13.13.3 Fitting data, etc.

It would be interesting to see how (13.81) compares to actual experimental data. However, the most developed phenomenology is for the Drell-Yan process, which we will treat later. See Sec. 14.5 for the factorization formalism, which has the same general structure as for the two-hadron-inclusive cross section in e^+e^- annihilation. A review of the phenomenology for the Drell-Yan process is given in Sec. 14.5.3.

13.13.4 Leading-logarithm approximation

One method of analyzing a process with a large scale is to determine in each order of perturbation theory the term with the highest power of a logarithm. These leading logarithms can often be derived analytically. The leading-logarithm approximation (LLA) is the sum of these terms, and it is often treated as a useful approximation to the exact result, because it sums the biggest terms in the perturbation expansion. In a strict LLA, the coupling is treated as fixed. In the b_T-space integrand, (13.81), the leading logarithms for large Qb_T are the two per loop associated with the leading order γ_K. Relative to a pure LO result, we have a factor

$$W_{\mathrm{LLA}}(b; Q) = \exp\left(-\frac{g^2(\mu)}{\pi^2} C_F \int_{1/b}^{Q} \frac{d\mu'}{\mu'} \ln\frac{Q}{\mu'}\right) d_{H_A/f}(z_A; \mu)\, d_{H_B/\bar{f}}(z_B; \mu).$$

$$= \exp\left(-\frac{g^2(\mu)}{2\pi^2} C_F \ln^2(Qb)\right) d_{H_A/f}(z_A; \mu)\, d_{H_B/\bar{f}}(z_B; \mu). \qquad (13.84)$$

This exhibits all the main qualitative features just described. The value of μ and hence the value of the coupling are not determined within the LLA. The choice of an appropriate value needs some intuition. One natural choice is that $\mu = 1/b_{\mathrm{peak}}$, where b_{peak} is the maximum of $bW(b, Q)$, so μ is the solution of $\ln(Q/\mu) = 2\pi^2/(g^2(\mu)C_F)$.

The LLA can also be obtained in transverse momentum space, for example by Fourier transformation of each term in the LLA in b_{T} space. This gives

$$\frac{d\sigma}{d^2 q_{h\mathrm{T}}} \propto \frac{g^2 C_F \ln\frac{Q}{q_{h\mathrm{T}}} \exp\left(-\frac{g^2(\mu)}{2\pi^2} C_F \ln^2(Q/q_{h\mathrm{T}})\right)}{q_{h\mathrm{T}}^2} d_{H_A/f}(z_A; \mu)\, d_{H_B/\bar{f}}(z_B; \mu)$$

$$= \frac{g^2 C_F \ln\frac{Q}{q_{h\mathrm{T}}}}{q_{h\mathrm{T}}^2} \left(\frac{q_{h\mathrm{T}}}{Q}\right)^{\frac{g^2(\mu)}{2\pi^2} C_F \ln(Q/q_{h\mathrm{T}})} d_{H_A/f}(z_A; \mu)\, d_{H_B/\bar{f}}(z_B; \mu). \qquad (13.85)$$

This last formula serves as an excellent warning about the inadequacies of the leading-logarithm method, despite its widespread use and tacit acceptance. Without the logarithms, the cross section diverges like $1/q_{h\mathrm{T}}^2$ as $q_{h\mathrm{T}} \to 0$. But with the resummed logarithms, the cross section decreases to zero faster than any power of $q_{h\mathrm{T}}$, as exhibited on the second line. This contradicts the correct result, which is that the cross section is finite and non-zero at $q_{h\mathrm{T}} = 0$. Even the LLA in b_{T} space implies this result.

The LLA can indeed provide some semi-quantitative information when the logarithms are not too large, by focusing attention on the largest terms in the perturbation expansion. One of the dangers of taking the LLA too literally is indicated by the Fourier transformation. Even if the LLA in b_{T} space were appropriate, the LLA in $q_{h\mathrm{T}}$ space need not be.

In general, there is no justification for using the LLA beyond some limited domain where $\frac{g^2(\mu)}{2\pi^2} C_F \ln^2(Q/q_{h\mathrm{T}}) \lesssim 1$. In contrast the derivation of the TMD factorization theorem is intended to be valid all the way down to $q_{h\mathrm{T}} = 0$.

13.13.5 Resummation methodology

The LLA presents some quantity like $W(b)$ or a cross section as a sum over all orders of perturbation theory, with each order being calculated as some analytically tractable approximation to full perturbation theory. This is called a "resummation" of perturbation theory.

In the literature can be found many generalizations of such resummations, for example to allow for a running coupling. Indeed TMD factorization formulae like (13.81) are often claimed to be resummation formulae: the starting point in this viewpoint is the *normal, collinear* factorization formula valid at large transverse momentum, i.e., at $q_{h\mathrm{T}} \sim Q$. In the hard-scattering coefficient H, higher-order terms contain logarithms of Q/q_{hT}, as in (13.85). Of course, when q_{hT} is too small, the logarithms prevent the reliable use of fixed-order perturbation theory, and resummation tries to overcome this problem.

If the logarithms are large but not too large, the use of resummation is reasonable. However, the justification for using *collinear* factorization as a starting point breaks down if one takes q_{hT} too small. Now the first part of the derivation was a region analysis of amplitudes, and this remains valid for arbitrarily small q_{hT}, provided that Q stays large. However, there is a failure of the approximations that led to the hard scattering and to the definitions of integrated parton densities and fragmentation functions. Parts of the approximations neglect partonic transverse momentum and virtuality, not just relative to Q, but also relative to q_{hT}. The partonic transverse momenta at issue are those intrinsically associated with the hadronic mass scale, so the actual (fractional) errors in collinear factorization are a power of M/q_{hT}.

Generally collinear factorization also applies to the *integral* of the cross section over q_{hT}, since the relevant errors in the approximations merely redistribute the cross section as a function q_{hT}.

In deriving TMD factorization, we have carefully preserved transverse momentum kinematics, and so the errors become a power of M/Q instead of M/q_{hT}. TMD factorization then applies all the way down to zero q_{hT}.

A less abstract way to see the problems with applying collinear factorization (resummed or not) at small q_{hT} is from the existence of an unphysical $1/q_{hT}^2$ singularity at $q_{hT} = 0$ in each order of the perturbative expansion of the hard-scattering factor in collinear factorization. Each of the summed terms in LLA is representative of the singularity. But the singularity (with its associated logarithms) arises from emission of collinear and soft emission gluons from parent partons that are *exactly* massless and on-shell. In physical reality such partons do not exist.

There is a further problem with an LLA such as (13.84) or (13.85), that the terms alternate in sign and exponentiate to a result much smaller than the first term when $b_T \gg 1/Q$. or $q_{hT} \ll Q$. Without further knowledge, one could not exclude that some non-leading logarithm might be outside the exponential form, e.g., a single term α_s^{10}/q_{hT}^2 *added* to an exponential series such as in (13.85). This term is of such high order that in practice it would probably not be calculable. Without an accompanying exponential of even higher-order terms, this high order would completely dominate the LLA sum.

Essentially full TMD factorization does ensure that higher-order terms can be organized so that there is an exponential factor. But the exponentials are the rather different ones in (13.81), and give rather different behavior than the LLA at zero q_{hT}.

13.14 NLO calculation of TMD fragmentation function at small b_T and at large k_T

To calculate the coefficient functions for the small-b_T fragmentation functions, we make the usual observation that the coefficient functions are independent of the type of the detected hadron. Thus we can (a) replace the hadron by a parton in IR-regulated massless QCD,

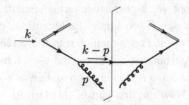

Fig. 13.7. One-loop fragmentation of quark to a gluon of *physical* polarization. The diagram applies equally to TMD and integrated fragmentation functions.

(b) compute both the TMD and the integrated fragmentation functions in strict fixed-order perturbation, and then (c) deduce the coefficient functions at that order from the perturbative expansion of (13.64).

For each of the functions, let the expansion in powers of the *renormalized* coupling be notated like

$$\tilde{C}_{j/f} = \sum_{n=0}^{\infty} \left(\frac{g^2}{16\pi^2} \right)^n \tilde{C}_{j/f}^{[n]}. \tag{13.86}$$

We write factorization (13.64) in a convolution notation as $\tilde{D} = d \otimes C$.

Then the first-order terms give

$$d^{[0]} \otimes \tilde{C}^{[1]} = \tilde{D}^{[1]} - d^{[1]} \otimes \tilde{C}^{[0]}. \tag{13.87}$$

The lowest-order coefficient $\tilde{C}^{[0]}$ is given in (13.65), and the lowest-order integrated fragmentation function is

$$d_{j/j'}^{[0]}(z) = \delta_{jj'}\delta(z-1), \tag{13.88}$$

so that

$$\tilde{C}_{j/f}^{[1]}(z, \boldsymbol{b}_{\mathrm{T}}) = \tilde{D}_{j/f}^{[1]}(z, \boldsymbol{b}_{\mathrm{T}}) - \frac{d_{j/f}^{[1]}(z)}{z^{2-2\epsilon}}, \tag{13.89}$$

where the denominator in the last term arises from the $\hat{z}^{3-2\epsilon}$ denominator of the measure in the convolution, (13.64), which in turn arises from the different powers of z in the definitions of TMD and integrated fragmentation functions, e.g., (12.39) and (12.40).

In the above formulae, j and f represent any parton type. We will compute the one-loop corrections for the cases that f is any flavor of quark, since these are the relevant ones in TMD factorization of the two-particle-inclusive cross section.

13.14.1 Gluon from quark at $O(g^2)$

The sole graph we need to calculate the fragmentation of a quark to a gluon at $O(g^2)$ is shown in Fig. 13.7. The hadron-frame momentum of the gluon is $p_h = (p_h^+, 0, \boldsymbol{0}_{\mathrm{T}}) = (zk_h^+, 0, \boldsymbol{0}_{\mathrm{T}})$, and we restrict to a sum over physical polarizations, chosen to be in the transverse plane. Then we have no graphs in which the gluon connects to a Wilson line.

For the dimensionally regulated TMD fragmentation function, we have

$$\frac{g^2}{16\pi^2} \tilde{D}_{g/q}^{[1]}(z, \boldsymbol{b}_T) = \frac{g^2 \mu^{2\epsilon} C_F}{(2\pi)^{4-2\epsilon} z} \int dk^- \, d^{2-2\epsilon} \boldsymbol{k}_T \, e^{i k_T \cdot \boldsymbol{b}_T} 2\pi \delta\big((k - p^2)\big)$$

$$\times \frac{\frac{1}{4} \text{Tr} \sum_j \gamma^+ \not{k} \gamma^j (\not{k} - \not{p}) \gamma^j \not{k}}{(k^2)^2}$$

$$= \frac{g^2 (4\pi^2 \mu^2)^\epsilon C_F}{8\pi^3} \int \frac{d^{2-2\epsilon} \boldsymbol{k}_T \, e^{i k_T \cdot \boldsymbol{b}_T}}{k_T^2} \left[\frac{1 + (1-z)^2 - \epsilon z^2}{z^3} \right]. \quad (13.90)$$

In the first line, the sum over j is over all transverse indices. There is a (collinear) divergence at $k_T = 0$, which is regulated if $\epsilon < 0$.

For the integrated function, the same formula applies, except that (a) the factor $1/z$ in the definition (12.39) is changed to $z^{1-2\epsilon}$, as in (12.40), (b) $\boldsymbol{b}_T$ is set to zero, and (c) an $\overline{\text{MS}}$ renormalization counterterm is used to cancel the resulting UV divergence:

$$\frac{g^2}{16\pi^2} d_{g/q}^{[1]}(z) = \frac{g^2 (4\pi^2 \mu^2)^\epsilon C_F}{8\pi^3} \int \frac{d^{2-2\epsilon} \boldsymbol{k}_T}{k_T^2} \left[\frac{1 + (1-z)^2 - \epsilon z^2}{z^{1+2\epsilon}} \right]$$

$$- \frac{g^2 C_F S_\epsilon}{8\pi^2 \epsilon} \left[\frac{1 + (1-z)^2}{z} \right]. \quad (13.91)$$

Using (13.89), we find the one-loop coefficient function

$$\frac{g^2}{16\pi^2} \tilde{C}_{g/q}^{[1]}(z, \boldsymbol{b}_T)$$

$$= \frac{g^2 (4\pi^2 \mu^2)^\epsilon C_F}{8\pi^3} \int \frac{d^{2-2\epsilon} \boldsymbol{k}_T (e^{i k_T \cdot \boldsymbol{b}_T} - 1)}{k_T^2} \left\{ \frac{1 + (1-z)^2 - \epsilon z^2}{z^3} \right\}$$

$$+ \frac{g^2 C_F S_\epsilon}{8\pi^2 \epsilon} \left[\frac{1 + (1-z)^2}{z^{3-2\epsilon}} \right]$$

$$= \frac{g^2 C_F}{8\pi^2} \left(\pi b_T^2 \mu^2 \right)^\epsilon \Gamma(-\epsilon) \left\{ \frac{1 + (1-z)^2 - \epsilon z^2}{z^3} \right\} + \frac{g^2 C_F S_\epsilon}{8\pi^2 \epsilon} \left[\frac{1 + (1-z)^2}{z^{3-2\epsilon}} \right]$$

$$\stackrel{\epsilon=0}{=} \frac{g^2 C_F}{8\pi^2 z^3} \left\{ 2\big[1 + (1-z)^2\big] \left[\ln \frac{2z}{\mu b_T} - \gamma_E \right] + z^2 \right\}. \quad (13.92)$$

In the first line, the collinear divergence at $k_T = 0$ is exactly canceled, and in the second line the UV divergence $k_T = \infty$ is renormalized. The integral was performed using (A.45).

Notice that if we applied $\int d^{2-2\epsilon} \boldsymbol{k}_T / k_T^2 = 0$ in the first line, then we would be left with the IR-divergent integral $\int d^{2-2\epsilon} \boldsymbol{k}_T \, e^{i k_T \cdot \boldsymbol{b}_T} / k_T^2$. The $\overline{\text{MS}}$ counterterm would appear to be canceling the IR divergence (strictly a collinear divergence). Although this method of IR cancellation corresponds to much actual calculational practice, it does not reflect the correct conceptual treatment.

After Fourier transformation of (13.92) back to momentum space, the behavior of the TMD at large transverse momentum is determined by the singularity in the Fourier conjugate variable, i.e., by the logarithm of b_T. Much more simply, one just inverts the

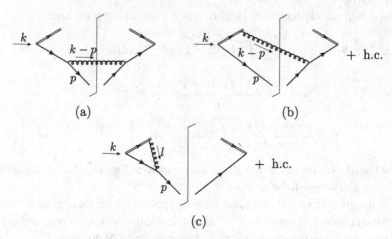

Fig. 13.8. One-loop fragmentation of quark from quark. For the TMD functions, there are also the Wilson-line terms shown in Fig. 13.9.

Fourier transform in the first line of (13.92), to obtain

$$\frac{g^2}{16\pi^2} C_{g/q}^{[1]}(z, \mathbf{k}_{\mathrm{T}}) = \frac{g^2 C_F}{8\pi^3} \frac{1 + (1-z)^2}{k_{\mathrm{T}}^2 z^3} \qquad \text{(at large } k_{\mathrm{T}}\text{)}. \qquad (13.93)$$

13.14.2 Quark from quark at $O(g^2)$

For the quark-to-quark fragmentation function, we use the graphs shown in Fig. 13.8. These need some explanation. The Wilson line is in the light-like direction $w_B = (0, 1, \mathbf{0}_{\mathrm{T}})$. It is now the outgoing quark that is detected and that has zero transverse momentum. Since the gluon has non-zero transverse momentum, its physical polarizations are no longer exactly in the transverse plane, and its coupling to the Wilson line is non-zero. It is convenient to calculate using a sum over all gluon polarizations, physical and unphysical, with the polarization sum $-g_{\kappa\lambda}$. The unphysical part cancels between graphs, by a standard textbook argument. Although quark self-energy graphs contribute to the actual one-loop fragmentation functions, they cancel in the difference used to compute the coefficient function in (13.89).

For the TMD quark fragmentation function, the graphs shown are for the $\tilde{D}^{\mathrm{unsub}}$ factor in the definition (13.42). We must add the one-loop contribution of the soft factor part of the definition, i.e., the graphs in Fig. 13.9, and these last graphs are to be multiplied by the lowest-order fragmentation function of a quark to a quark, i.e., $\delta(z-1)$. They cancel a rapidity divergence that will manifest itself in a singularity at $z = 1$ in Fig. 13.8. We also add renormalization counterterms to cancel UV divergences in all the virtual graphs, i.e., for Figs. 13.8(c) and 13.9(b).

For the integrated fragmentation function, we need only the graphs of Fig. 13.8, which now all have an unrestricted integral over transverse momentum. With this unrestricted integral, the rapidity divergences cancel between real and virtual gluon emission. We also

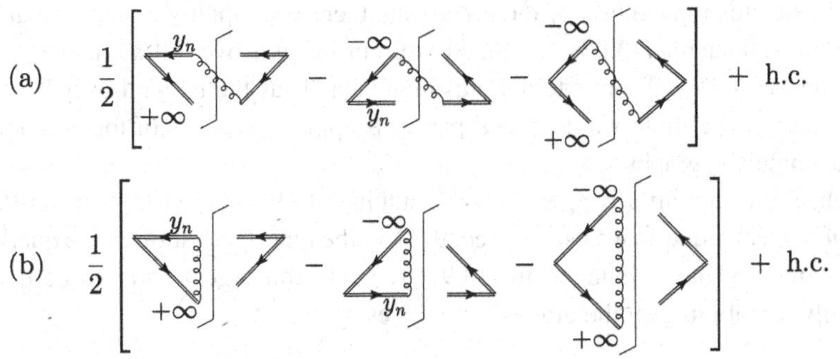

Fig. 13.9. One-loop graphs for soft-factor contributions to quark-to-quark fragmentation. The labels next to the Wilson lines indicate their rapidities. The graphs are to be multiplied by the zeroth-order fragmentation function $\delta(z-1)$.

need counterterms to cancel the UV divergences from the integral to infinite transverse momentum.

Since the graphs of Fig. 13.8 are doing double duty, for the two kinds of fragmentation function, I first summarize the overall calculational structure to obtain $\tilde{C}_{q/q}^{[1]}$:

Fig. 13.8 for TMD f.f. + (Fig. $13.9 \times \delta(z-1)$) + $\overline{\text{MS}}$ c.t.

$$- z^{-2+2\epsilon} \big[\text{Fig. 13.8 for integrated f.f.} + \overline{\text{MS}} \text{ c.t.} \big]. \quad (13.94)$$

The contribution of Fig. 13.8(a) to the dimensionally regulated TMD fragmentation function is straightforwardly

$$\frac{g^2 (4\pi^2 \mu^2)^\epsilon C_F}{8\pi^3} \frac{(1-z)(1-\epsilon)}{z^2} \int d^{2-2\epsilon} \boldsymbol{k}_{\text{T}} \frac{e^{i \boldsymbol{k}_{\text{T}} \cdot \boldsymbol{b}_{\text{T}}}}{k_{\text{T}}^2}. \quad (13.95)$$

Graph (b) (including its hermitian conjugate) is

$$\frac{g^2 (4\pi^2 \mu^2)^\epsilon C_F}{8\pi^3} \frac{2}{z(1-z)} \int d^{2-2\epsilon} \boldsymbol{k}_{\text{T}} \frac{e^{i \boldsymbol{k}_{\text{T}} \cdot \boldsymbol{b}_{\text{T}}}}{k_{\text{T}}^2}. \quad (13.96)$$

This has a singularity at $z = 1$. In the integral in the factorization formula, the singularity is at $z_A/\hat{z} = 1$, an endpoint of the integration over $\hat{z}$. The singularity is from a rapidity divergence associated with the light-like Wilson line. The rapidity divergence is canceled by the contribution from Fig. 13.9(a). This contribution is calculated almost identically to the corresponding term for the Sudakov form factor, and corresponds to the last three terms in the braces in (10.136). The differences are that (a) the gluon propagator is cut, (b) we add a hermitian conjugate term, (c) there is a group theory factor, and (d) we set masses to zero, obtaining:

$$-\frac{g^2 (4\pi^2 \mu^2)^\epsilon C_F}{8\pi^3} 2\delta(z-1) \int d^{2-2\epsilon} \boldsymbol{k}_{\text{T}} e^{i \boldsymbol{k}_{\text{T}} \cdot \boldsymbol{b}_{\text{T}}} \int_0^\infty \frac{dl^+}{l^+} \Re \frac{1}{k_{\text{T}}^2 - 2(l^+)^2 e^{-2y_n} - i0}. \quad (13.97)$$

Here l^+ is the plus momentum of the gluon, and there is a rapidity divergence at $l^+ = 0$. Because the soft factor (13.23) is defined with an integral over all k_S^+ and k_S^-, there is no dependence of (13.97) on external plus momenta. Thus the integral over l^+ ranges to infinity rather than a finite value. A real part $\Re$ is applied, because of the addition of the hermitian conjugate graphs.

To cancel the rapidity divergences, we combine (13.96) and (13.97) using the same distributional technique as we used in Sec. 9.4.4 in the renormalization of the quark parton density. After that the l^+ integral in (13.97) is made convergent and can be performed analytically. Combining all the graphs so far gives

$$\frac{g^2(4\pi^2\mu^2)^\epsilon C_F}{8\pi^3} \int d^{2-2\epsilon}k_T \frac{e^{ik_T \cdot b_T}}{k_T^2}$$

$$\times \left[\left(\frac{2}{1-z}\right)_+ + \frac{2}{z} + \frac{(1-z)(1-\epsilon)}{z^2} + \delta(z-1)\ln\frac{2(k^+)^2 e^{-2y_n}}{k_T^2} \right]. \qquad (13.98)$$

The IR/collinear divergence at $k_T = 0$ will cancel against the contribution of the integrated fragmentation function. But we will not display this explicitly. Instead we will proceed calculationally. All the remaining graphs, i.e., not only the virtual graphs for TMD fragmentation function, i.e., Figs. 13.8(c) and 13.9(b), but also all the graphs for the integrated fragmentation function, give zero, because they have scale-free transverse-momentum integrals.

So it remains to add the UV counterterms, whose total contribution is

$$\frac{g^2 C_F}{8\pi^2} \left\{ \delta(z-1) \left[-\frac{S_\epsilon}{\epsilon^2} + \frac{S_\epsilon}{\epsilon}\left(\ln\frac{2(k^+)^2 e^{-2y_n}}{\mu^2} - 2 \right) \right] \right.$$

$$\left. + z^{-2+2\epsilon} \frac{S_\epsilon}{\epsilon} \left[\left(\frac{2}{1-z}\right)_+ - 1 - z + 2\delta(z-1) \right] \right\}. \qquad (13.99)$$

The first line has the counterterms for the TMD fragmentation function's virtual graphs; their calculation is the same as for the Sudakov form factor (10.139), except for a group-theory factor C_F and except for multiplication by 2 and removal of the imaginary part. The second line has the counterterms for the integrated fragmentation function; these are the same as the DGLAP kernel, but without the contribution associated with the quark self-energy graph.

Then we perform the k_T integrals analytically and add everything together at $\epsilon = 0$, to obtain

$$\frac{g^2}{16\pi^2} \tilde{C}^{[1]}_{j'/j}(z, b_T) \stackrel{\epsilon=0}{=} \frac{g^2 C_F \delta_{j'j}}{8\pi^2} \left(2\left[\left(\frac{2}{1-z}\right)_+ + \frac{1}{z^2} + \frac{1}{z} \right]\left[\ln\frac{2z}{\mu b_T} - \gamma_E \right] \right.$$

$$+ \frac{1}{z^2} - \frac{1}{z} + \delta(z-1)\left\{ -\frac{1}{2}\left[\ln(\mu^2 b_T^2) - 2(\ln 2 - \gamma_E) \right]^2 \right.$$

$$\left. \left. - \left[\ln(\mu^2 b_T^2) - 2(\ln 2 - \gamma_E) \right]\ln\frac{2(k^+)^2 e^{-2y_n}}{\mu^2} \right\} \right). \qquad (13.100)$$

For generality, we have allowed arbitrary quark flavors j and j', with, of course, a Kronecker delta between them. Notice that there are *two* logarithms of b_T in this one-loop calculation, associated with the presence of a Sudakov form factor.

Correspondingly on Fourier transformation back to momentum space, there is a logarithm of k_T in the large-k_T behavior:

$$\frac{g^2}{16\pi^2} C^{[1]}_{j'/j}(z, k_T) \overset{\epsilon=0}{=} \frac{g^2 C_F \delta_{j'j}}{8\pi^3} \frac{1}{k_T^2}$$

$$\times \left[\left(\frac{2}{1-z}\right)_+ + \frac{1}{z} + \frac{1}{z^2} + \delta(z-1) \ln \frac{2(k^+)^2 e^{-2y_n}}{k_T^2} \right], \quad (13.101)$$

obtained most easily from (13.98).

13.14.3 Failure of positivity

As initially defined, a TMD fragmentation function had the meaning of a number density of a hadron in a parton. This would imply that the coefficient function $C(z, k_T)$ is also positive. However, the $\ln k_T$ in (13.101) ensures that the coefficient becomes negative (at $z = 1$) when k_T is larger than $\sqrt{2} k^+ e^{-y_n}$, which we normally choose to be approximately the overall CM energy Q. There is a subsidiary positivity problem that the distribution $1/(1-z)_+$ is not positive, because it is defined with a subtraction. When (13.101) is convoluted with an integrated fragmentation function, to get the TMD fragmentation function, there is a combination of positive and negative terms. But at sufficiently large k_T the negative delta-function term dominates, and positivity is violated.

Note that the quark-to-gluon coefficient (13.93) has no such problem, because it has neither a logarithm nor a plus distribution.

The resolution of the problem starts by the observation that we were forced to modify the definition of the TMD fragmentation function from its naive one. We made subtractions, notably to remove the contribution of rapidity divergences. Since we used subtractions rather than a cutoff, we can get a negative value, just as in our implementation in Sec. 3.4 of renormalization by subtraction of an asymptote.

The real positivity requirement is on cross sections. The TMD functions occur by themselves only in a factorization theorem for $q_{hT} \ll Q$, where small values of parton transverse momenta dominate. In that region, the logarithm at issue is indeed positive.

For large q_{hT} the TMD factorization formula represents only part of an estimate of the cross section. To compensate the error, we devised a correction term Y in Sec. 13.12. It corrects the cross section to the one obtained from standard collinear factorization, and is available to compensate the negativity in one individual term.

Obtaining a positive physical cross section from a combination of terms of opposite sign can be dangerous numerically, since the negative term can be larger than the final answer. It would not be a real issue if we could calculate exactly all the coefficients involved, to all orders of perturbation theory. But there can practical difficulties with low-order estimates.

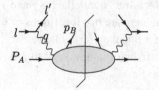

Fig. 13.10. SIDIS cross section.

This suggests that modifications of the basic formalism would be useful; see Sec. 13.12 and Arnold and Kauffman (1991). Any modification should agree with the TMD form of factorization at small q_{hT} and should agree with normal collinear factorization at large q_{hT}. But creativity in combining and/or matching the two kinds of contribution without double counting is appropriate.

13.14.4 Other cases

By charge-conjugation invariance, the above coefficients are unchanged if the quark (of any flavor) is changed to an antiquark.

At one loop, there is zero coefficient to get a quark from an antiquark, or vice versa. This process needs a minimum of two loops.

The coefficients for quark from gluon and gluon from gluon are left as an exercise. These cases are *currently* of lesser experimental importance, since the main currently studied reactions sensitive to TMD fragmentation are those where the hard scattering involves quarks (and antiquarks). These reactions are e^+e^- annihilation, SIDIS and Drell-Yan.

13.15 SIDIS and TMD parton densities

So far, we treated TMD factorization for reactions in e^+e^- annihilation, where TMD fragmentation functions were used. We now extend these ideas to a process that needs parton densities, specifically semi-inclusive DIS (SIDIS). Another process that uses TMD parton densities is the Drell-Yan process to be treated in Ch. 14 along with the complications in obtaining factorization in hadron-hadron collisions.

The results for SIDIS are a straightforward generalization of those for e^+e^- annihilation, so it is mainly necessary to explain the changes.

13.15.1 Kinematics

Semi-inclusive deeply inelastic scattering (SIDIS) is DIS with inclusive measurement of one hadron as well as a lepton in the final state: $e(l) + H_A(P_A) \rightarrow e(l') + H_B(p_B) + X$, Fig. 13.10. We choose the outgoing lepton to be in the DIS region, so that the reaction has large Q, and we also choose the hadron H_B to be in a region where it can be a fragmentation product of one of the jets produced by the hard scattering.

We have already examined this reaction in Sec. 12.14, but without a treatment appropriate for small transverse momentum. Here we write the momenta of the incoming hadron and the detected outgoing hadron as P_A and p_B instead of P and p_h. This notation is consistent with the rest of this chapter, and avoids confusion with use of a subscript h to denote components in the hadron frame. As in Sec. 13.2, we use two coordinate frames: the photon frame and the hadron frame.

The photon frame was used in (12.89), now notated

$$q_\gamma = \left(-x P_{A,\gamma}^+, \ \frac{Q^2}{2x P_{A,\gamma}^+}, \ \mathbf{0}_{\mathrm{T}} \right), \tag{13.102a}$$

$$P_{A,\gamma} = \left(P_{A,\gamma}^+, \ \frac{M_A^2}{2 P_{A,\gamma}^+}, \ \mathbf{0}_{\mathrm{T}} \right), \tag{13.102b}$$

$$p_{B,\gamma} = \left(\frac{p_{B,\gamma\,\mathrm{T}}^2 + M_B^2}{2 p_{B,\gamma}^-}, \ p_{B,\gamma}^-, \ \boldsymbol{p}_{B,\gamma\,\mathrm{T}} \right). \tag{13.102c}$$

Lorentz scalars for the process are x, Q, $z \stackrel{\text{def}}{=} P_A \cdot p_B / P_A \cdot q$, $|\boldsymbol{p}_{B,\gamma\,\mathrm{T}}|$, and the azimuthal angle $\phi_{B,\gamma}$ of $\boldsymbol{p}_{B,\gamma\,\mathrm{T}}$. Thus $p_{B,\gamma}^- \simeq Q^2 z / (2x P_{A,\gamma}^+)$.

In the hadron frame, both the hadrons have zero transverse momentum:

$$q_h = \left(q_h^+, \ q_h^-, \ \boldsymbol{q}_{h\,\mathrm{T}} \right), \tag{13.103a}$$

$$P_{A,h} = \left(P_{A,h}^+, \ M_A^2 / 2 P_{A,h}^+, \ \mathbf{0}_{\mathrm{T}} \right), \tag{13.103b}$$

$$p_{B,h} = \left(m_B^2 / 2 p_{B,h}^-, \ p_{B,h}^-, \ \mathbf{0}_{\mathrm{T}} \right). \tag{13.103c}$$

Since

$$q_{h\,\mathrm{T}}^2 = 2 q_h^+ q_h^- + Q^2 \simeq \frac{2 p_B \cdot q \ P_A \cdot q}{P_A \cdot p_B} + Q^2, \tag{13.104}$$

we use $\boldsymbol{q}_{h\,\mathrm{T}} = -\boldsymbol{p}_{B,\gamma\,\mathrm{T}}/z$ in the zero-mass limit, and we define the Lorentz transformation between the frames to be

$$\left(V_h^+, \ V_h^-, \ \boldsymbol{V}_{h\,\mathrm{T}} \right) = L \left(V_\gamma^+, \ V_\gamma^-, \ \boldsymbol{V}_{\gamma\,\mathrm{T}} \right)$$

$$= \left(V_\gamma^+ + \frac{2x^2 (P_{A,\gamma}^+)^2 q_{h\,\mathrm{T}}^2 V_\gamma^-}{Q^4} + \frac{2x P_{A,\gamma}^+ \boldsymbol{q}_{h\,\mathrm{T}} \cdot \boldsymbol{V}_{\gamma\,\mathrm{T}}}{Q^2}, \right.$$

$$\left. V_\gamma^-, \ \boldsymbol{V}_{\gamma\,\mathrm{T}} + \boldsymbol{q}_{h\,\mathrm{T}} \frac{2x P_{A,\gamma}^+ V_\gamma^-}{Q^2} \right), \tag{13.105}$$

in an approximation valid *when hadron masses are neglected*. (The formula with hadron masses is more complicated.) Note that the large components of P_A and p_B are unchanged between the frames: $p_{B,h}^- = p_{B,\gamma}^-$ and $P_{A,h}^+ = P_{A,\gamma}^+$ (the last up to a mass-suppressed correction).

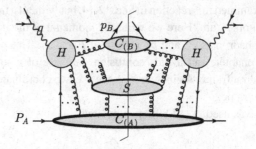

Fig. 13.11. Leading region for SIDIS, for low transverse momentum.

13.15.2 Overall structure of proof

An analysis giving TMD factorization in a CSS-style formalism was first given by Meng, Olness, and Soper (1996), but only when the energy of the outgoing hadron was integrated over. For the unintegrated cross section, a treatment was given by Nadolsky, Stump, and Yuan (2000), and by Ji, Ma, and Yuan (2005). These treatments ignored important quark polarization effects that are absent with integrated parton densities and fragmentation functions. The formalism with polarization effects was provided by Ji, Ma, and Yuan (2004). A list of the necessary structure functions is presented in Kotzinian (1995); Diehl and Sapeta (2005); Bacchetta *et al.* (2007).

Kinematically, SIDIS differs from two-particle-inclusive e^+e^- annihilation simply by crossing one hadron from the final to the initial state, and vice versa for one lepton, thereby making the photon space-like instead of time-like. The graphical specification of the leading regions therefore looks very similar, i.e., Fig. 13.11 for the $q_{hT} \ll Q$ case instead of Fig. 13.1.

The same pattern of factorization proof works as for two-particle-inclusive e^+e^- annihilation.

An important change concerns the Glauber region. Previously we simply copied the treatment for the Sudakov form factor, in Sec. 5.5.10. For a Glauber gluon connected to the upper collinear subgraph $C_{(B)}$ in Fig. 13.1, we deformed plus momentum away from final-state poles in $C_{(B)}$, and for a Glauber gluon connected to $C_{(A)}$, we deformed minus momentum away from final-state poles in $C_{(A)}$.

But for SIDIS, there are both initial- and final-state poles in $C_{(A)}$, as in Fig. 12.25. Luckily, as we saw after that figure, it is sufficient to deform away from the final-state poles in $C_{(B)}$, i.e., to make a one-sided deformation.

After the use of region approximators and Ward identities, we get soft and collinear factors whose operators involve Wilson-line factors. Since the deformation to get out of the Glauber region is in the same direction as for e^+e^- annihilation, we can use the same (future-pointing) directions of the Wilson lines, which must not obstruct the contour deformations. Hence the fragmentation function associated with collinear subgraph $C_{(B)}$ is identical to the one in e^+e^- annihilation.

13.15.3 Unpolarized TMD quark density

For the target-collinear subgraph we use a parton density instead of a fragmentation function. Its definition (in transverse coordinate space) is the natural modification of (13.42), again applied in the hadron frame:

$$\tilde{f}_{f/H_A}(x, \boldsymbol{b}_{\mathrm{T}}; \zeta_A; \mu)$$

$$\stackrel{\text{def}}{=} \lim_{\substack{y_A \to +\infty \\ y_B \to -\infty}} \tilde{f}_{f/H_A}^{\text{unsub}}(x, \boldsymbol{b}_{\mathrm{T}}; y_{P_A} - y_B) \sqrt{\frac{\tilde{S}_{(0)}(\boldsymbol{b}_{\mathrm{T}}, y_A, y_n)}{\tilde{S}_{(0)}(\boldsymbol{b}_{\mathrm{T}}, y_A, y_B)\, \tilde{S}_{(0)}(\boldsymbol{b}_{\mathrm{T}}, y_n, y_B)}}$$

$$\times \text{ UV renormalization factor}$$

$$= \tilde{f}_{f/H_A}^{\text{unsub}}(x, \boldsymbol{b}_{\mathrm{T}}; y_{P_A} - (-\infty)) \sqrt{\frac{\tilde{S}_{(0)}(\boldsymbol{b}_{\mathrm{T}}; +\infty, y_n)}{\tilde{S}_{(0)}(\boldsymbol{b}_{\mathrm{T}}; +\infty, -\infty)\, \tilde{S}_{(0)}(\boldsymbol{b}_{\mathrm{T}}; y_n, -\infty)}}$$

$$\times Z_f Z_2. \tag{13.106}$$

The soft factors are exactly the same as for the fragmentation function, but the renormalization factor Z_f may differ. The definition of ζ_A is now

$$\zeta_A \stackrel{\text{def}}{=} 2(k_{A,h}^+)^2 e^{-2y_n} = 2x^2 (P_{A,h}^+)^2 e^{-2y_n} = M_A^2 x^2 e^{2(y_{P_A} - y_n)}, \tag{13.107}$$

and the unsubtracted pdf is

$$\tilde{f}_{f/H_A}^{\text{unsub}}(x, \boldsymbol{b}_{\mathrm{T}}; y_{P_A} - y_B) = \underset{\text{color}}{\text{Tr}} \int \frac{dw^-}{2\pi} e^{-ixP_A^+ w^-}$$

$$\times \langle P_A | \bar{\psi}_f(w/2) W(w/2; \infty; n_B)^{\dagger} \frac{\gamma^+}{2} W(-w/2; \infty; n_B) \psi_f(-w/2) | P_A \rangle_c$$

$$= \underset{\text{color Dirac}}{\text{Tr}\ \text{Tr}} \frac{\gamma^+}{2} \int \frac{dk_A^-\, d^{n-2}\boldsymbol{k}_{A\,\mathrm{T}}}{(2\pi)^n} e^{-i\boldsymbol{k}_{A\mathrm{T}} \cdot \boldsymbol{b}_{\mathrm{T}}} \tag{13.108}$$

where the vector w in the second line is $(0, w^-, \boldsymbol{b}_{\mathrm{T}})$. The soft factor is defined by (13.39), and the Wilson lines by (13.40). As in Sec. 13.7.2 for fragmentation functions, strictly gauge-invariant operators in the definitions of $S_{(0)}$ and $\tilde{f}^{\text{unsub}}$ would need transverse links to join the Wilson lines at infinity. But, as shown there, their effects cancel in the parton density defined by (13.106).

13.15.4 TMD parton densities: evolution, b_{T} dependence, relation to integrated density

The soft factors in (13.106) are the same as for fragmentation functions. Therefore the CSS evolution equation for the dependence of the TMD quark density on ζ_A is exactly the same as for TMD quark fragmentation, (13.47), with the same kernel K. Therefore, its anomalous dimension $\gamma_K(g)$ and its non-perturbative part, $g_K(b_T)$ in (13.60), are unchanged.

However, in the RG equation, like (13.50), the anomalous dimension, γ_f, of a TMD quark density may be different, since the quark momentum is reversed.

Because of changed normalizations, the small b_T expansion of the quark density is slightly changed from (13.64):

$$\tilde{f}_{f/H_A}(x, b_T; \zeta_A; \mu) = \sum_j \int_{x-}^{1+} \frac{d\hat{x}}{\hat{x}}\, \tilde{C}_{f/j}(x/\hat{x}, b_T; \zeta_A, \mu, g(\mu))\, f_{j/H_A}(\hat{x}; \mu) + O\!\left[(m b_T)^p\right].$$

(13.109)

The coefficient function $\tilde{C}$ need not be the same as for fragmentation.

The analysis of the large-b_T behavior of the TMD density follows as in Sec. 13.11.2. Then the appropriate version of (13.70) giving the separation of perturbative and non-perturbative parts is

$$\tilde{f}_{f/H_A}(x, b_T; \zeta_A; \mu)$$
$$= \sum_j \int_{x-}^{1+} \frac{d\hat{x}}{\hat{x}}\, \tilde{C}_{f/j}\!\left(x/\hat{x}, b_*; \mu_b^2, \mu_b, g(\mu_b)\right)\, f_{j/H_A}(\hat{x}; \mu_b)$$
$$\times \exp\!\left[-g_{f/H_A}(x, b_T) - \ln \frac{\sqrt{\zeta_A}}{M_A x}\, g_K(b_T)\right]$$
$$\times \exp\!\left\{\ln \frac{\sqrt{\zeta_A}}{\mu_b}\, \tilde{K}(b_*; \mu_b) + \int_{\mu_b}^{\mu} \frac{d\mu'}{\mu'}\!\left[\gamma_f(g(\mu'); 1) - \ln \frac{\sqrt{\zeta_A}}{\mu'}\, \gamma_K(g(\mu'))\right]\right\},$$

(13.110)

where $b_*(b_T)$ and μ_b are defined by (13.59) and (13.69). Of the non-perturbative functions, $g_{j/H_A}(x, b_T)$ is specific to parton densities, and cannot be predicted from any measurements of fragmentation functions. But $g_K(b_T)$ is the same as for fragmentation functions.

13.15.5 Hadronic tensor and kinematics of hard scattering

To determine the factorization formula, we follow the same methods as for e^+e^- annihilation. First, we define the hadronic tensor

$$W^{\mu\nu}(q, P_A, p_B) \overset{\text{def}}{=} \sum_X \delta^{(4)}(P_A + q - p_B - p_X)$$
$$\times \langle P_A | j^\mu(0) | p_B, X, \text{out} \rangle\, \langle p_B, X, \text{out} | j^\nu(0) | P_A \rangle.$$ (13.111)

For each leading region R, an approximator T_R is defined, as in Sec. 13.3.2, generally using hadron-frame coordinates. The only modification is in the approximant for collinear partons at the hard scattering H, to match the different transformation to the photon frame. Consider unapproximated parton momenta: k_A from the target subgraph to H, and k_B from H to the other collinear subgraph. They have components $k_{A,h} = (k_{A,h}^+, k_{A,h}^-, \mathbf{k}_{A,h\mathrm{T}})$ and $k_{B,h} = (k_{B,h}^+, k_{B,h}^-, \mathbf{k}_{B,h\mathrm{T}})$. Then we define the approximated momenta by

$$P_{HA}(k_A)_h = (k_{A,h}^+, 0, \mathbf{0}_\mathrm{T}), \tag{13.112a}$$

$$P_{HB}(k_B)_h = k_{B,h}^- \left(\frac{q_{h\mathrm{T}}^2}{2(q_h^-)^2}, 1, \frac{\mathbf{q}_{h\mathrm{T}}}{q_h^-} \right). \tag{13.112b}$$

From the transformation (13.105), the photon-frame components are

$$P_{HA}(k_A)_\gamma = (k_{A,h}^+, 0, \mathbf{0}_\mathrm{T}), \qquad P_{HB}(k_B)_\gamma = (0, k_{B,h}^-, \mathbf{0}_\mathrm{T}). \tag{13.113}$$

It can be verified that this approximator is unique given the following requirements:

- The total transverse momentum at the hard scattering is unchanged by the approximant. Thus let α and β label the lines between the collinear and hard subgraphs, and let $\hat{k}_{A,\alpha}$ and $\hat{k}_{B,\beta}$ be the approximated momenta. Then from (13.112)

$$\sum_\beta \hat{k}_{B,\beta,h\mathrm{T}} - \sum_\alpha \hat{k}_{A,\alpha,h\mathrm{T}} = \sum_\beta \frac{k_{B,\beta,h}^- \mathbf{q}_{h\mathrm{T}}}{q_h^-} = \mathbf{q}_{h\mathrm{T}}, \tag{13.114}$$

  where the last equality follows by momentum conservation in the approximated hard scattering.
- The approximated momenta have no transverse components in the photon frame.
- The approximated momenta are massless and on-shell.
- Fractional longitudinal momenta for the partons are the same for the approximated and unapproximated momenta. Thus $\hat{k}_{A,\alpha,h}^+ = k_{A,\alpha,h}^+$ and $\hat{k}_{B,\beta,h}^- = k_{B,\beta,h}^-$.

The first requirement defines what we mean by TMD factorization, while the second and third requirements are how we normally perform a parton-model approximation. The last requirement could be relaxed, but there is no need to; it has the convenience that the longitudinal momentum arguments of the parton densities and fragmentation functions are the standard ones. This follows because the approximated large components of parton momentum obey $\hat{k}_{A,\alpha,\gamma}^+ = k_{A,\alpha,h}^+$ and $\hat{k}_{B,\beta,\gamma}^- = k_{B,\beta,h}^-$. Then momentum conservation, $q = \sum_\alpha \hat{k}_{A,\alpha} + \sum_\beta \hat{k}_{B,\beta}$, in the approximated hard scattering gives

$$\hat{k}_{A,\gamma} = \left(-x P_{A,\gamma}^+, 0, \mathbf{0}_\mathrm{T} \right), \qquad \hat{k}_{B,\gamma} = \left(0, \frac{Q^2}{2x P_{A,\gamma}^+}, \mathbf{0}_\mathrm{T} \right). \tag{13.115}$$

Hence $\sum_\alpha k_{A,\alpha,h}^+ / P_{A,h}^+ = x$, and $p_{B,h}^- / \sum_\beta k_{B,\beta,h}^- = z$ (up to m^2/Q^2 corrections).

13.15.6 Factorization

The resulting factorization formula is

$$W^{\mu\nu} = \frac{2z}{Q^2} \sum_f \text{Tr} \frac{k_{A,\gamma}^+ \gamma^-}{2} H_f^\nu(Q; g(\mu), \mu) k_{B,\gamma}^- \gamma^+ H_f^\mu(Q; g(\mu), \mu)^\dagger$$

$$\times \int \frac{\text{d}^2 \boldsymbol{b}_T}{(2\pi)^2} e^{-i\boldsymbol{q}_T \cdot \boldsymbol{b}_T} \tilde{f}_{f/H_A}(x, \boldsymbol{b}_T; \zeta_A) \tilde{D}_{1, H_B/\bar{f}}(z, \boldsymbol{b}_T; Q^4/\zeta_A)$$

$$+ \text{ polarized terms} + \text{large } q_{h\,T} \text{ correction}, Y. \tag{13.116}$$

Here, the ζ_B argument of the fragmentation function is set to Q^4/ζ_A, corresponding to a similar choice in (13.46). The overall factor $2z/Q^2$ is obtained from the details of the integrals over loop momenta, given the definition of $W^{\mu\nu}$. The hard-scattering factor is the part of the first line of (13.116) after the summation sign. It is normalized to correspond to DIS on an on-shell massless quark in the photon frame. The vertex factor H_f is equipped with soft and collinear subtractions as usual.

13.16 Polarization issues

The explicit TMD factorization term in (13.116) has an unpolarized quark entering the hard scattering, and no sensitivity to the polarization of the quark leaving the hard scattering. The TMD quark density is intended to be defined with an unpolarized initial-state hadron.

There is an interesting set of extensions when one allows for polarization effects. The details get quite complicated, with many structure functions, parton densities and fragmentation functions. A comprehensive list is found in Diehl and Sapeta (2005), but without taking account of the full CSS-style formalism.

The main ideas are quite simple, however. There is a number density of each flavor of parton in a parent hadrons, and the parton has a helicity density matrix. Similarly, the fragmentation function can be sensitive to the polarization state of the outgoing quark (Sec. 13.4.1). In all cases the polarization state of a quark or of a spin-$\frac{1}{2}$ hadron can be described by a three-dimensional Bloch vector (e.g., a helicity λ and a transverse spin S_T), and the spin dependence is linear in the Bloch vector.

The complications arise in enumerating the list of TMD parton densities. In the case of integrated parton densities, rotation and parity invariance restrict the parton densities to an unpolarized density, and helicity and transversity distributions (Sec. 6.5); but with a transverse momentum, the number of possibilities increases substantially.

In the following we let λ and S_T be the helicity and transverse spin of the target, normalized to maximum values of unity, and we let x and k_T be the longitudinal momentum fraction and the transverse momentum of the quark. As summarized by Bacchetta *et al.* (2007) and Mulders and Tangerman (1996), we have the following eight densities for a quark in a spin-$\frac{1}{2}$ hadron.

- In an unpolarized hadron:
  - There is a number density of quarks, $f_1(x, k_T)$ in the Mulders-Tangerman notation.
  - The quark can have a transverse polarization proportional to $\epsilon_{ij} k_T^j / M$, where ϵ_{ij} is the two-dimensional antisymmetric tensor. The coefficient $h_1^\perp(x, k_T)$ is called the Boer-Mulders function.
- In a longitudinally polarized hadron with normalized helicity λ:
  - The quark may have a longitudinal polarization proportional to that of the hadron. The coefficient is $g_{1L}(x, k_T)$.
  - The quark may have a transverse polarization proportional to $\lambda k_T / M$. The coefficient is $h_{1L}^\perp(x, k_T)$.
- In a transversely polarized hadron with normalized spin S_T:
  - There may be a contribution to the number density proportional to $\epsilon_{ij} k_T^i S_T^j / M$. The coefficient, $f_{1T}^\perp(x, k_T)$, is called the Sivers function (Sivers, 1990).
  - The quark may have a contribution to its transverse polarization proportional to that of the hadron. The coefficient is $h_1(x, k_T)$.
  - The quark may have a contribution to its transverse polarization proportional to $S_T^j (k_T^j k_T^i - \delta^{ji} k_T^2 / 2)/M^2$. The coefficient, $h_{1T}^\perp(x, k_T)$, is called the pretzelosity distribution.
  - The quark may have a longitudinal polarization proportional to $k_T \cdot S_T / M$. The coefficient is $g_{1T}(x, k_T)$.

The various combinations of pdf and fragmentation contribute to different combinations of structure functions, and contribute to the SIDIS cross section with characteristic angular dependencies listed in Diehl and Sapeta (2005). The longitudinal spin densities are obtained by replacing the trace with γ^+ in (13.108) by a trace with $\gamma^+ \gamma_5$, and the transverse spin densities by replacing γ^+ by $\gamma^+ \gamma^i \gamma_5$.

As for quark fragmentation to an unpolarized hadron, there are (see Sec. 13.4.1) the ordinary number density and the Collins function, which is a final-state analog of the Boer-Mulders function. These allow the cross section to depend on all eight of the TMD densities listed above (Diehl and Sapeta, 2005, Eq. (40)).

See Boer (2009) for the use of the polarized TMD fragmentation functions in $e^+ e^-$ annihilation.

13.17 Implications of time-reversal invariance

Some interesting insights into the nature of QCD factorization and its consequences have resulted from the observation that the Sivers and Boer-Mulders functions have the property called "time-reversal odd", T-odd, for short. As we will see, this means that when we apply a PT transformation we find a reversal of sign. If Wilson lines were ignored in the definitions of these functions, each would be its own negative, and therefore zero. In a gauge theory we do have Wilson lines, and the PT transformation changes them to be past-pointing instead of future-pointing.

As we will see in Ch. 14, parton densities defined with past-pointing Wilson lines are needed for the Drell-Yan process. Thus there is a change of sign between SIDIS and DY (Collins, 2002) for the T-odd functions, i.e., for the Sivers and Boer-Mulders functions.

13.17.1 Sivers function

I now derive (Collins, 1993) the T-odd property of the Sivers function. Rather than a time-reversal transformation, it is convenient to apply a PT transformation, since it leaves momenta of physical states unchanged. It does, however, exchange in-states and out-states, which does not matter for the vacuum and for one-particle states.

Let $\mathcal{PT}$ denote the anti-unitary operator implementing PT transformation on state space. From standard QFT textbooks, we know that the transformation of a quark field is

$$(\mathcal{PT})^\dagger \psi(w)\mathcal{PT} = PT\psi(-w),\qquad(13.117)$$

where PT is a unitary Dirac matrix such that

$$(PT)^{-1}(\gamma^\mu)^* PT = \gamma^\mu.\qquad(13.118)$$

There is a possible phase in the transformation (13.117), but it will not affect our proofs. Also from the textbooks, we know that $\mathcal{PT}$ reverses the spin-vector of a single particle state for a spin-$\frac{1}{2}$ particle:

$$\mathcal{PT}\,|p, S\rangle = \text{phase factor}\,|p, -S\rangle.\qquad(13.119)$$

A bilinear in the quark fields transforms as

$$(\mathcal{PT})^\dagger \bar{\psi}(y)\Gamma\psi(z)\mathcal{PT} = \bar{\psi}(-y)(PT)^\dagger \Gamma^* PT\psi(-z),\qquad(13.120)$$

where the $*$ arises because $\mathcal{PT}$ is an antilinear operator. In the case $\Gamma = \gamma^+$, as in a quark number density, we get a positive sign: $(PT)^\dagger(\gamma^+)^* PT = \gamma^+$. For the cases used for spin densities, i.e., $\Gamma = \gamma^+\gamma_5$ and $\Gamma = \gamma^+\gamma_T^i\gamma_5$, we get a minus sign, which implements the reversal of spin by $\mathcal{PT}$.

Consider now the application of $\mathcal{PT}$ to the operator in a basic parton number density, where we initially work without a Wilson line:

$$\langle P, S| \bar{\psi}_j(w/2) \frac{\gamma^+}{2} \psi_j(-w/2)\,|P, S\rangle$$

$$= \langle P, S| \mathcal{PT}(\mathcal{PT})^{-1} \bar{\psi}_j(w/2) \frac{\gamma^+}{2} \psi_j(-w/2)\, \mathcal{PT}(\mathcal{PT})^{-1}\,|P, S\rangle$$

$$= \langle P, -S| \bar{\psi}_j(-w/2) \frac{\gamma^+}{2} \psi_j(w/2)\,|P, -S\rangle^*$$

$$= \langle P, -S| \bar{\psi}_j(w/2) \frac{\gamma^+}{2} \psi_j(-w/2)\,|P, -S\rangle.\qquad(13.121)$$

The complex conjugate in line 3 arises because of the antilinearity of the $\mathcal{PT}$ operator:

$$\langle f|(\mathcal{PT})^\dagger|g\rangle = \langle g|\mathcal{PT}|f\rangle = \langle f'|g\rangle^*, \tag{13.122}$$

where $|f'\rangle = \mathcal{PT}|f\rangle$.

Suppose the number density of quarks of some flavor were defined from the matrix element in (13.121), which has no Wilson line. We write the number density in a polarized target as

$$f(x, k_{\mathrm{T}}) + \frac{\epsilon_{ij}k_{\mathrm{T}}^i S_{\mathrm{T}}^j}{M} f_{1T}^\perp(x, k_{\mathrm{T}}), \tag{13.123}$$

where $f_{1T}^\perp(x, k_{\mathrm{T}})$ is the Sivers function. From (13.121) it follows the number density is unchanged when the spin vector of the target is reversed, and therefore that the Sivers function vanishes.

This argument is correct in a non-gauge theory. But in QCD (and any other gauge theory), there is a Wilson line going out to infinity in some light-like direction (or approximately light-like direction) from one quark field, and coming back to the other quark field. For the parton densities used for SIDIS, the lines go to *future* infinity. Let us insert this Wilson line in the left-hand side of (13.121). Then the PT transformation to get the right-hand side of (13.121) reverses the positions of the fields, so that on the right-hand side, the Wilson line goes to *past* infinity. We must conclude not that the Sivers function is zero, but that the Sivers function for SIDIS has the opposite sign to a Sivers function with past-pointing Wilson lines.

We will see that, in the Drell-Yan process, proving factorization requires that the TMD parton densities have past-pointing Wilson lines. Thus the Sivers function reverses sign between the two processes:

$$f_{1T,\mathrm{SIDIS}}^\perp(x, k_{\mathrm{T}}) = -f_{1T,\mathrm{DY}}^\perp(x, k_{\mathrm{T}}), \tag{13.124}$$

while the ordinary unpolarized parton density, $f(x, k_{\mathrm{T}})$, is numerically the same for SIDIS and DY.

The reversal of sign of the Sivers function is a notable violation of the initially intuitive idea that parton densities are universal between processes. In a sense, we already have such violations because of the renormalization-scale dependence of parton densities, and because of the process-dependent directions of Wilson lines in TMD densities.

All of these situations concern controlled and calculable violations of universality: the parton densities (and fragmentation functions) in different reactions and at different energies can be related to each other.

13.17.2 Boer-Mulders function

We generalize (13.121) to measurements of quark polarization by replacing $\gamma^+/2$ by the matrix appropriate to a helicity or transversity. In this case, the right-hand side acquires a minus sign. It follows that the Boer-Mulders function is T-odd, since this function is the transverse spin density of a quark in an unpolarized hadron. The function therefore also reverses sign between SIDIS and Drell-Yan

13.17.3 Other cases

All the other parton densities listed in Sec. 13.16 are T-even. Either they involve no polarization at all, or they involve both a quark polarization and a hadron polarization.

13.17.4 Integrated parton densities

In the definition of integrated densities, the Wilson line goes straight from one quark field to the other, without a detour to infinity. So the Wilson line is unchanged after a PT transformation. So the non-zero integrated parton densities must all be the T-even ones, even in a gauge theory. But this restriction is already implied by rotation and parity invariance, which gave us the simple restriction to a simple number density, a helicity density and a transversity density.

13.17.5 Soft factors and K

The above arguments all apply to the basic operator for a quark density, i.e., to the first factor in definition (13.106). This is multiplied by a particular combination of soft factors. Now the directions of the Wilson lines in the definition (13.39) of the soft factor must match those in the unsubtracted parton density, in order that all the necessary subtractions and the cancellations of rapidity divergences work. So after a PT transformation, the future-pointing Wilson lines in each soft factor S must be replaced by past-pointing Wilson lines.

The value of each S factor is unchanged under this transformation. This is proved by applying the same argument as (13.121) but to the matrix element in (13.39).

Hence the CS and RG evolution equations, including the values of their kernels, are unchanged when the Wilson lines are changed from future to past pointing.

13.17.6 Fragmentation

We have found two types of TMD parton density that are related by a PT transformation and that differ by whether the Wilson lines go to future or past infinity. Naturally, one can ask whether a similar situation arises for fragmentation functions. The answer is in fact negative, as we will now see.

In both the cases treated so far, e^+e^- annihilation and SIDIS, we used future-pointing Wilson lines in the definitions of the fragmentation functions. A PT transformation would indeed convert the Wilson lines to past pointing. But it would also transform out-states to in-states:

$$\sum_X \mathrm{Tr}\,\gamma^+ \, \langle 0|\psi(w/2)|p, X,\ \mathrm{out}\rangle \, \langle p, X,\ \mathrm{out}|\bar{\psi}(-w/2)|0\rangle$$

$$= \sum_X \mathrm{Tr}\,\gamma^+ \, \langle 0|\psi(w/2)|p, X,\ \mathrm{in}\rangle \, \langle p, X,\ \mathrm{in}|\bar{\psi}(-w/2)|0\rangle \,. \qquad (13.125)$$

Since in-states with two or more particles are not the same as the out-states with the same labels, but are related by the S matrix, the right-hand side of this equation cannot be equated to a matrix element used to define some fragmentation function.

So PT transformations give no useful information here. Although certain fragmentation functions like the Collins function involve only one spin and are naively T-odd, they can be non-vanishing even in a non-gauge model, unlike the case for a T-odd parton density. To better understand this difference, we insert a complete set of final states between the operators defining a parton density:

$$\sum_X \langle P, S|\bar{\psi}_j(w/2)|X,\ \mathrm{out}\rangle \, \frac{\gamma^+}{2} \, \langle X,\ \mathrm{out}|\,\psi_j(-w/2)|P, S\rangle \,. \qquad (13.126)$$

Although a PT transformation changes the intermediate states to in-states, we can use completeness in the sum/integral over *all* basis states to convert them back to out-states:

$$\sum_X |X,\ \mathrm{in}\rangle \, \langle X,\ \mathrm{in}| = \sum_X |X,\ \mathrm{out}\rangle \, \langle X,\ \mathrm{out}| \,. \qquad (13.127)$$

This argument does not apply to the inclusive sum in a fragmentation function where one particle is detected and therefore not summed over.

Exercises

13.1 Very carefully check all the signs in the derivation and use of the Collins function, notably in (13.31) (13.32), and (13.34).

13.2 (**) Complete problem 10.2 of Ch. 10.

13.3 (***) Find other work on the evolution of TMD parton densities, and try to extend problem 10.2 to it. Such work includes that resulting in the CCFM equation (Ciafaloni, 1988; Catani, Fiorani, and Marchesini, 1990a, b; Marchesini, 1995). Note that the CCFM equation has an apparently radically different structure to the evolution equation described in the present chapter. It nevertheless refers to TMD parton densities, so there should be a relation.

13.4 (**) Find and prove any extensions to the Ward-identity arguments in Ch. 11 that are needed to apply them to the processes treated in this chapter.

13.5 Show that the two-dimensional Fourier transform of an azimuthally symmetric function, defined by (13.35a), can be expressed as a one-dimensional integral:

$$\tilde{S}(b) = 2\pi \int_0^\infty \mathrm{d}k\, k\, J_0(kb)\, S(k), \tag{13.128a}$$

$$S(k) = \frac{1}{2\pi} \int_0^\infty \mathrm{d}b\, b\, J_0(kb)\, \tilde{S}(b), \tag{13.128b}$$

where J_0 is the Bessel function of order zero. This result is used in numerical work.

13.6 (**)
  (a) Generalize the treatment of CSS evolution to include the part of the factorization formula with the Collins function in two-particle-inclusive e^+e^- annihilation.
  (b) Repeat for semi-inclusive DIS, and for the DY process, where the relevant functions also include the Sivers and the Boer-Mulders functions.
  See Idilbi *et al.* (2004) for a solution. You may wish to extend their work.

13.7 (****) Complete the proofs of all the results in this chapter, notably those concerning the application of the subtraction formalism to processes in a non-abelian gauge theory with TMD functions, and the expression of these functions in terms of operator matrix elements with Wilson lines.

13.8 (***) Suppose that, contrary to the argument of Collins and Metz (2004), time-like rather than space-like Wilson lines were used in the definitions of the TMD functions. Determine whether this gives actual problems, and *under what circumstances*. Consider a variety of processes for which TMD functions are appropriate, including two-particle-inclusive e^+e^- annihilation, SIDIS, and DY.
  Notes:
  • Time-like Wilson lines appear to have the advantage of better resembling actual recoil-less partons, at least in e^+e^- annihilation, where the partons have time-like momenta.
  • But in SIDIS and DY each struck parton is space-like, at least as regards its momentum.
  • With time-like Wilson lines, you need to examine very carefully the Collins-Metz arguments about universality.

13.9 (***) If possible, find a simple elegant form for the Feynman rules for computing K beyond lowest order.

13.10 (**) Extend the methods to take account of heavy quarks. Publish the result if you are the first to solve this problem.

13.11 (***) The final definition of the TMD fragmentation function (13.42) involves a product of an unsubtracted fragmentation function and several Wilson-line factors. If possible, express Feynman graphs for this quantity as graphs for the unsubtracted fragmentation function with a systematic subtraction procedure applied. Again, publish the result if you are the first to solve this problem.

13.12 (**) Obtain the coefficients for the small-b_T coefficients for the TMD fragmentation functions of gluons; that is, extend the calculations in Sec. 13.14 from quark to gluon fragmentation.

13.13 (***) The formalism presented in this chapter uses TMD fragmentation functions and/or pdfs for the "low-q_{hT}" terms, and ordinary integrated fragmentation functions and/or pdfs for the large-q_{hT} correction. Try to obtain a more unified formalism in which everything is done with TMD functions.

14

Inclusive processes in hadron-hadron collisions

In this chapter, I treat inclusive hard processes in hadron-hadron collisions. These give some of the most important practical applications of factorization. But the actual derivation has substantial extra difficulties, compared with other processes we have examined.

Technically the extra difficulties concern the Glauber region. In e^+e^- annihilation or SIDIS, we deformed loop momenta out of the Glauber region in individual (cut) graphs. But this is no longer possible in hadron-hadron collisions. This situation results from interactions between the spectator parts of the beam hadrons, as I will illustrate by an example in Sec. 14.3. To get factorization, we will need a sum over cuts of the graphs, which in turn entails a sum over different unobserved final states in an inclusive cross section. The technical details will be explained in Sec. 14.4 for the case of the Drell-Yan process.

After that we will obtain factorization, including the version using TMD parton densities. I will summarize the situation for more general reactions with detected hadrons of high transverse momentum. There is a surprising lack of detailed published proofs. Although the statements of factorization are essentially trivial generalizations of those for Drell-Yan, there are underlying complications in the physics which makes the justification of the generalizations quite non-trivial.

This work also leads us to the frontiers of the factorization approach, beyond which more general methods are needed, e.g., in diffractive hadron-hadron scattering.

14.1 Overview

The actual statements of factorization are quite simple, but they hide physical and conceptual complications, many of which we have already seen. Examples of the processes we can consider are:

- the Drell-Yan process, i.e., the inclusive production of high-mass lepton pairs, $H_A + H_B \to \mu^+ \mu^- + X$;
- inclusive production of one or more hadrons of high transverse momentum, $H_A + H_B \to H_C + X$, $H_A + H_B \to H_C + H_D + X$, etc.;
- production of jets of high transverse momentum;
- generalizations of Drell-Yan to the production of electroweak bosons, both within the Standard Model and in conjectured extensions;
- production of hadrons containing heavy quarks (charm, bottom, top).

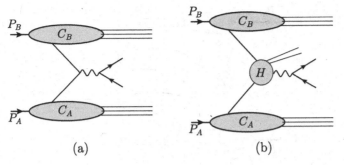

Fig. 14.1. Structure of factorization for the Drell-Yan process: (a) lowest-order hard scattering; (b) more general hard scattering.

The commonality is that in the leading regions there is a hard scattering. This can be thought of as the core of the process: a reaction involving short distances that determines the signature property of the reaction, e.g., a high-mass virtual photon or a high-transverse-momentum jet.

Leading regions, as in Fig. 14.3 for the Drell-Yan process, involve collinear subgraphs for the observed initial- and final-state hadrons and a soft subgraph, as well as the hard subgraph. To get factorization we need to use Ward identities to extract extra collinear gluons from the hard subgraph, and we need to show that either the sum of soft subgraphs cancels or it can be absorbed into the collinear factors. After that, we get the situation represented in Fig. 14.1 for Drell-Yan. The hard scattering can be treated as a on-shell partonic scattering of a kind appropriate to the chosen reaction. It is initiated by one parton out of each initial-state hadron. The collinear factors associated with the initial-state hadrons behave as number distributions for the partons initiating the hard scattering. For the Drell-Yan cross section integrated over transverse momentum, the factorization property is then

$$\frac{d\sigma}{dQ^2\,dy\,d\Omega} = \sum_{ij} \int_0^1 d\xi_a \int_0^1 d\xi_b\, f_{i/H_A}(\xi_a) f_{j/H_B}(\xi_b) \frac{d\hat\sigma(\xi_a, \xi_b, i, j)}{dQ^2\,dy\,d\Omega}, \qquad (14.1)$$

where y is the rapidity of the lepton pair, and Ω is the polar angle of one of the leptons. Very often the cross section is presented after integration over lepton-pair angle. There is an integral over the parton fractional momenta ξ_a and ξ_b, and a sum over parton flavor.

For the cross section differential also in the transverse momentum, we have

$$\frac{d\sigma}{d^4q\,d\Omega} = \sum_{ij} \int_0^1 d\xi_a \int_0^1 d\xi_b\, f_{i/H_A}(\xi_a) f_{j/H_B}(\xi_b) \frac{d\hat\sigma(\xi_a, \xi_b, i, j)}{d^4q\,d\Omega}, \qquad (14.2)$$

where now the partonic cross section is fully differential in q. This factorization is appropriate when $q_T \sim Q$. However, as we have seen in Ch. 13 for other kinematically similar processes, the approximations needed at the hard scattering need to be changed when the transverse momentum of the lepton pair q_T is much less than Q. In that case, we need a more general factorization with TMD parton densities: see (14.31).

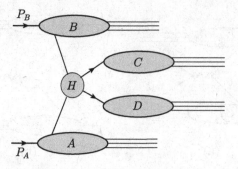

Fig. 14.2. Structure of factorization for production of hadrons at high transverse momentum.

We can interpret all these factorization properties in the parton-model sense as a partonic cross section times single-particle densities for the partons, with a sum and integral over the partonic configurations. But this interpretation must not be treated literally. For example, the hard-scattering function $d\hat{\sigma}$ is defined with subtractions. Similarly the parton densities are not genuine number densities, because of the details of the definition of parton densities in renormalizable gauge theories.

The simplicity of the interpretation of the factorization should also not obscure that substantial conceptual and technical complications are needed to derive factorization. These leave their symptoms in the non-trivial evolution equations, especially for TMD parton densities.

For other reactions, similarly simple factorization formulae can be written. For example, consider the inclusive production of a hadron of high transverse momentum, $H_A + H_B \to H_C + X$. The factorization structure is now that of Fig. 14.2. Here the final state of the hard scattering is itself completely partonic, and we need a fragmentation function to give the density of hadrons in one of the outgoing partons:

$$E_{p_C} \frac{d\sigma}{d^3 \boldsymbol{p}_C} = \sum_{ijc} \int_0^1 d\xi_a \int_0^1 d\xi_b \int_0^1 dz \, f_{i/H_A}(\xi_a) f_{j/H_B}(\xi_b) d_{H_C/c}(z)$$

$$\times \frac{1}{z^2} |\boldsymbol{k}_c| \frac{d\hat{\sigma}(\xi_a, \xi_b; k_c, i, j, c)}{d^3 \boldsymbol{k}_c}, \tag{14.3}$$

where we now have an inclusive partonic hard scattering to make a parton of type c of on-shell momentum k_c. Here the hadron and parton 3-momenta are related by $z = (E_{p_c} + |\boldsymbol{p}_c|)/(2|\boldsymbol{k}_c|)$, given the standard light-front definition of a fragmentation function, with this relation being applied in the overall center-of-mass (CM) frame.

14.2 Drell-Yan process: kinematics etc.

The Drell-Yan process is hadro-production of high-mass lepton pairs, e.g., $H_A + H_B \to \mu^+\mu^- + X$. The classic case is production of $\mu^+\mu^-$ or e^+e^- through a virtual photon, but the same ideas apply to production of any kind of lepton pair through an electroweak gauge boson (γ, W or Z), as well as to many standard mechanisms for making Higgs bosons and to many generalizations in proposed extensions of the Standard Model.

Kinematically, it differs from two-hadron-inclusive production in e^+e^- annihilation or from SIDIS by a crossing transformation: both leptons are now in the final state and the two detected hadrons are the initial state. The kinematic variables and the structure function analysis are minor generalizations of those for the previous two processes, as is the general analysis of the leading regions and the power-counting. As with those processes, we will use two coordinate frames: a hadron frame and a photon frame.

Hadron frame

We let P_A and P_B be the momenta of the incoming hadrons, and we let q be the momentum of the lepton pair. In a hadron frame, we write

$$P_{A,h} = \left(P_{A,h}^+, \frac{M_A^2}{2P_{A,h}^+}, \mathbf{0}_T \right), \tag{14.4a}$$

$$P_{B,h} = \left(\frac{M_B^2}{2P_{B,h}^-}, P_{B,h}^-, \mathbf{0}_T \right), \tag{14.4b}$$

$$q_h = \left(q_h^+, q_h^-, \mathbf{q}_{hT} \right). \tag{14.4c}$$

The rapidity of the lepton pair is $y = \frac{1}{2}\ln(q_h^+/q_h^-)$, which we normally apply in the overall CM frame, where $P_{A,h}^+ \doteq P_{B,h}^-$. The invariant mass of the lepton pair is $Q = \sqrt{q^2}$.

Photon frame

We define the photon frame to be obtained from the hadron frame by a boost along the z until the lepton pair has zero rapidity, and then a transverse boost to put the lepton pair at rest. This gives exactly the Lorentz transformation used for e^+e^- annihilation, i.e., (13.8). The momenta of the lepton pair and the hadrons are given in (13.1). With masses neglected, the z axis of the photon frame is again midway in angle between $\mathbf{P}_A$ and $-\mathbf{P}_B$, as in Fig. 13.2(b). This frame was defined by Collins and Soper (1977).

Hadronic tensor

The hadronic tensor for the Drell-Yan process is defined as

$$W^{\mu\nu} = s \int d^4z\, e^{-iq \cdot z} \langle P_A, P_B, \text{in}|\, j^\mu(z)\, j^\nu(0)\, |P_A, P_B, \text{in}\rangle. \tag{14.5}$$

The structure functions were formulated by Lam and Tung (1978) for the case that the hadrons are unpolarized and j^μ is the electromagnetic current. See Mirkes (1992) for the case of W bosons with unpolarized beams, and Ralston and Soper (1979) and Donohue and Gottlieb (1981) for the case of the electromagnetic current with polarized beams.

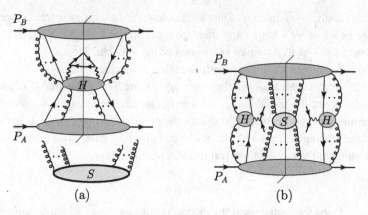

Fig. 14.3. Leading regions for the Drell-Yan process: (a) when $q_{h\,\mathrm{T}}$ is integrated over or is large, (b) when $q_{h\,\mathrm{T}} \ll Q$. The soft subgraph connects to the collinear subgraphs.

Scaling limit

The scaling limit we consider is where $s = (P_A + P_B)^2 \to \infty$, with q^+/P_A^+ and q^-/P_B^- of a fixed order of magnitude. As to the transverse momentum, there are three cases:

1. The classic case, where $\boldsymbol{q}_{h\,\mathrm{T}}$ is integrated over. The natural factorization formula uses integrated parton densities.
2. A variation, where $q_{h\,\mathrm{T}}$ is large, of order Q.
3. The cross section differential in $\boldsymbol{q}_{h\,\mathrm{T}}$, particularly for $q_{h\,\mathrm{T}} \ll Q$, where the cross section is largest. Factorization then uses TMD parton densities.

The first two cases can be unified by considering the cross section integrated over $\boldsymbol{q}_{h\,\mathrm{T}}$ with a weighting function $f(\boldsymbol{q}_{h\,\mathrm{T}}/Q)$:

$$\frac{\mathrm{d}\sigma[f]}{\mathrm{d}q^+\,\mathrm{d}q^-\,\mathrm{d}\Omega} = \int \mathrm{d}^2\boldsymbol{q}_{h\,\mathrm{T}}\ f(\boldsymbol{q}_{h\,\mathrm{T}}/Q)\frac{\mathrm{d}\sigma}{\mathrm{d}^4q\,\mathrm{d}\Omega}. \tag{14.6}$$

The lepton angle Ω is taken with respect to our chosen photon frame. From (14.6), the $q_{h\,\mathrm{T}}$-integrated cross section is obtained by setting $f = 1$ for all $\boldsymbol{q}_{h\,\mathrm{T}}$. The differential cross section is obtained by functional differentiation.

Leading regions

The leading regions for the process are shown in Fig. 14.3(a). There is a hard-scattering subgraph out of which comes the virtual photon coupled to the lepton pair. There are collinear subgraphs associated with the two beams and a possible soft subgraph. The hard subgraph may include extra high-k_{T} partons going into the final state. These extra partons manifest themselves as high-k_{T} jets, which are treated as unobserved in the inclusive Drell-Yan cross section. In principle, these high-k_{T} partons ought each to be attached to their individual collinear subgraphs. To avoid notational complications, they are not indicated in the diagram. This is appropriate since in an inclusive cross section we expect to use the argument of Sec. 12.7 to eliminate these extra collinear factors after a sum over the relevant

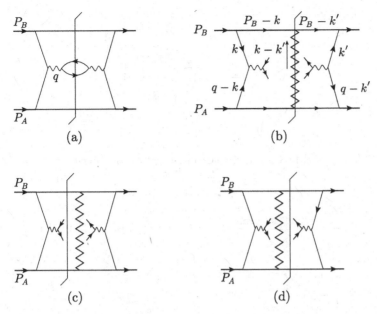

Fig. 14.4. (a) Pure parton-model contribution to Drell-Yan with single spectator. (b) With addition of cut multiperipheral ladder (Fig. 14.5) to fill in rapidity gap. (c) and (d) Two other cuts of graph (b) with a diffractive final state.

cuts. There are also the usual extra K gluons between the collinear subgraphs and the hard subgraph.

In the case that $q_{h\,T} \ll Q$, the leading regions are shown in Fig. 14.3(b). As with other processes with low transverse momentum (Ch. 13) there are no extra high-k_T jets, i.e., jets with transverse momentum of order Q. These graphs are a subset of those for the cross section integrated over $q_{h\,T}$.

14.3 Glauber region example

The issues with the Glauber region are conveniently illustrated by the Feynman-graph model for the Drell-Yan process shown in Fig. 14.4.

The model is simplified to treat the hadron as being composed of exactly two constituents. Then graph (a) is the lowest-order basic parton-model approximation, without any soft subgraph, and with a lowest-order hard scattering. If this literally represented the actual physics, then we could apply the usual parton-model approximator to the hard scattering. This would directly give the TMD factorization formula (14.31) below, but simplified to have the LO hard factor and without any rapidity or scale dependence to the parton densities. Integrating over $q_{h\,T}$ would then give the integrated DY cross section as a hard factor times two integrated parton densities.

However, by itself this graph gives a hadronic final state consisting of the fast-moving remnants of the two beams, with a large rapidity gap between them. The rapidity gap's

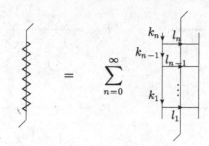

Fig. 14.5. Ladder graph, with cut and sum over rungs.

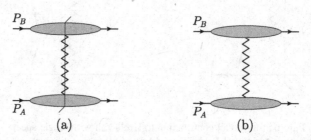

Fig. 14.6. Ladder model (a) for total hadron-hadron cross section, and (b) for elastic hadron-hadron amplitude.

size would be approximately $\ln(s/M^2)$, roughly the difference in rapidity between the two beams. Such a rapidity gap is only present in a small fraction of actual DY events (Abe *et al.*, 1997). Moreover in QCD, the two parts of the final state would have fractional charge, because they are each obtained by subtracting a quark or antiquark from a beam hadron.

A simple but influential model giving a more realistic kinematic structure to the final state is the multiperipheral model (Gribov, 2009, Ch. 9) illustrated in Fig. 14.4(b). Exchanged between the spectator partons is a ladder graph, as defined in Fig. 14.5, with a sum over the number of rungs. It is not necessary to specify exactly the nature of the lines in the ladder. To match reality, the lines should probably represent some effective degrees of freedom appropriate for non-perturbative QCD, and we will not need to specify the details at all precisely. A similar ladder can be exchanged between intact hadrons, without a hard scattering, to give a model for the total and elastic hadron-hadron cross sections, Fig. 14.6. The exchanged ladder sum was used (Gribov, 2009, Ch. 9) as an elementary model for what is called Reggeon exchange, with the exchange dominating the high-energy behavior being called the "pomeron". It is useful to use the term "pomeron" to denote the sum over ladder graphs.

For the model, it is assumed that all the lines have a mass of order a typical hadronic mass M, and that the virtualities of the displayed hadronic/partonic lines are of order M^2, as is appropriate for modeling the non-perturbative regime of the strong interaction.

14.3.1 Energy-dependence of exchanged ladder

Details of unreferenced results in this section can be found, for example, in Gribov (2009).

The final-state particles are ordered in rapidity: $y_1 > y_2 > \cdots > y_n$, between the rapidities of the two beams. The orders of magnitude of these momenta are then $l_j \sim M(e^{y_j}, e^{-y_j}, 1)$. In the case that the rapidities are strongly ordered, $e^{y_j} \gg e^{y_{j+1}}$, the plus momentum of each l_j mostly comes from below, and the minus momentum from above, so that the momenta on the sides of the ladder obey

$$k_j \sim M(e^{y_{j+1}}, e^{-y_j}, 1). \tag{14.7}$$

If either of k_j^+ or k_j^- were larger, the excess would have to flow along the sides of the ladder (above and below respectively), as is shown by momentum conservation at the vertices with the rungs. This would give these other lines much higher virtuality, which we have ruled out by the definition of the model. It follows that (when the rapidities are strongly ordered), the vertical lines have Glauber-like momenta: $k_j^2 \simeq -k_{jT}^2$.

The integrals over the final-state momenta are transverse-momentum integrals times integrals over rapidity:

$$\int \frac{\mathrm{d}^3 l_j}{2E_{l_j}(2\pi)^3} \cdots = \int \mathrm{d}y_j \int \frac{\mathrm{d}^2 l_{jT}}{2(2\pi)^3} \cdots \tag{14.8}$$

When the rapidities are strongly ordered, the integrand depends only on transverse momenta, to leading power. This enables us to estimate the energy dependence of the ladder graphs.

When the number of rungs is zero, only the sides of the ladder exist, and at high CM energy the exchanged system gives, relative to graph (a), a power $s^{J_1+J_2-2}$, where J_1 and J_2 are the spins of the exchanged fields. Although there is no suppression for gluon exchange, an exchange of quarks (as would be appropriate for getting color-singlet final-state particles) would give a power $1/s$, i.e., a power-suppression relative to the parton-model graph.

Now, the rapidity integral for an n-rung ladder gives energy dependence approximated by

$$\int_0^{\Delta y} \mathrm{d}y_1 \int_0^{y_1} \mathrm{d}y_2 \ldots \int_0^{y_{n-1}} \mathrm{d}y_n = \frac{(\Delta y)^n}{n!}, \tag{14.9}$$

where Δy is the rapidity difference between the two beams, i.e., $\Delta y \simeq \ln(s/M^2)$. If, as is appropriate, we assign a general order of magnitude λ to the transverse momentum integral per rung, then the ladder sum gives

$$\sum_{n=0}^{\infty} \frac{(\lambda \Delta y)^n}{n!} = e^{\lambda \Delta y} \simeq \left(\frac{s}{M^2}\right)^\lambda. \tag{14.10}$$

This increases the power of s relative to the no-rung case, to give a total power $s^{\alpha-1} = s^{\lambda+J_1+J_2-2}$. (Note that each term is positive, so λ is positive.) Since we are modeling a non-perturbative part of QCD, λ is not small. In the model, we have calculated a contribution to the cross section, necessarily positive. But we will find a cancellation with graphs (c) and (d) where the ladder is uncut, so that the Drell-Yan cross section is just the parton-model value, from graph (a). To give the cancellation, the contribution of graphs (c) and (d) is necessarily negative.

The ladder model can also be applied to ordinary soft cross sections, Fig. 14.6, in which case α corresponds to the "intercept" of the exchanged pomeron. The pomeron intercept is measured to be approximately unity, to give an approximately constant total cross section.[1]

14.3.2 Cancellation after sum over cuts

To show the cancellation of Fig. 14.4(b)–(d), we start by performing the integrals over the plus and minus components of k and k' for graph (b). In the region we are considering, the lines k and k' are collinear to P_B, while the lines $q - k$ and $q - k'$ are collinear to P_A. Thus k^+ and k'^+ are of order M^2/P_B^- and therefore in the lower half of the graph they are negligible compared with the large components of plus momenta, which are of order P_A^+. Similarly $q^- - k^-$ and $q^- - k'^-$ are of order M^2/P_A^+, and can be neglected in the top half of the graph. In the top half, we therefore make the replacements $k^-, k'^- \mapsto q^-$.

We will work with the case that the end rungs of the ladder have strongly ordered rapidity relative to the hadrons: $e^{y_1} \ll e^{y_{P_A}}$, and $e^{y_{P_B}} \ll e^{y_n}$. Then the dependence of the sides of the ladder on $k^\pm$ and $k'^\pm$ can be neglected. Of course, there is a significant region where the end rungs are collinear to the hadrons. But in that case we should consider the rungs as part of the collinear subgraphs, with a more general collinear subgraph, as in Fig. 14.7(a) below. To better capture the correct concept of pomeron exchange we should redefine the exchanged entity to have such collinear contributions removed, perhaps by some subtractive technique. We will not investigate this issue here, although it is interesting and needs investigation. For the purposes of a motivational example, the strongly ordered case is sufficient.

For simplicity of presenting the results, we will take all the lines to have equal mass. This is not essential.

After the approximations, the only dependence on k^+ is in the two lines k and $P_B - k$. We perform the integral by closing the k^+ contour on the pole of the "final-state" line $P_B - k$:

$$\int \frac{\mathrm{d}k^+}{2\pi} \frac{i}{(2q^-k^+ - E_T^2 + i0)} \frac{i}{[2(-k^+ + P_B^+)(P_B^- - q^-) - E_T^2 + i0]}$$

$$= \frac{1}{2(P_B^- - q^-)} \frac{i}{(2q^-k_{\text{on-shell}}^+ - E_T^2 + i0)}$$

$$= \int \frac{\mathrm{d}k^+}{2\pi} \frac{i}{(2q^-k^+ - E_T^2 + i0)} 2\pi\delta\big(2(-k^+ + P_B^+)(P_B^- - q^-) - E_T^2\big), \qquad (14.11)$$

---

[1] If the basic ladder gives an exponent significantly larger than $2 - J_1 - J_2$, then it would give cross sections substantially above the Froissart bound. We should then imagine that the ladder represents "bare pomeron" exchange and that the calculation of the true cross section involves multiple bare pomerons.

where we have made the approximation $k^- \mapsto q^-$, and have defined $E_T^2 = k_T^2 + m^2$. In the second line,

$$k_{\text{on-shell}}^+ = P_B^+ - \frac{E_T^2}{2(P_B^- - q^-)}. \tag{14.12}$$

The effect is to set the line $P_B - k$ on-shell.

Graphically let us denote on-shell lines by a cross. Then after similarly performing the integrations over k'^+, k^-, and k'^-, we find that

$$\text{Fig. 14.4(b)} = \qquad\qquad \tag{14.13}$$

to leading-power accuracy in the region we are considering.

Exactly similar calculations can be done on the other graphs, Fig. 14.4(c) and (d). In those graphs, the final-state cut goes through two of the spectator lines, and these are set on-shell from the beginning. The total of the three graphs is therefore

$$\tag{14.14}$$

The left- and rightmost factors are the same in all these graphs:

$$\tag{14.15}$$

and they equal the corresponding factor in the pure parton-model graph Fig. 14.4(a).

The pomeron factor is therefore a sum over all the kinematically allowed cuts of the ladder graphs, with on-shell external lines:

$$\tag{14.16}$$

This is zero by a standard theorem, which we used in Sec. 12.7. Note that, because of our choice of kinematics for the final-state partons, the only non-zero cut that goes through the pomeron is where all the rungs are cut, as in the first of the graphs.

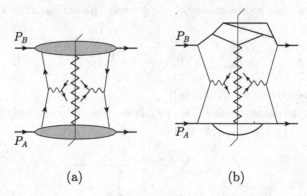

(a)   (b)

Fig. 14.7. (a) General class of parton-model graphs supplemented by ladder-graph exchange. (b) Example where generalization of the argument for cancellation of Glauber region is particularly non-trivial compared with Fig. 14.4(b)–(d).

The pure parton-model graph Fig. 14.4(a) gives what is commonly termed a diffractive final state: two isolated particles (or groups of particle) separated by a large rapidity gap. The effect of the cancellation after the sum over cuts is that the graphs with an uncut pomeron reduce the diffractive part of the cross section and replace it by a contribution from the cut pomeron graph in which the rapidity gap is filled. In reality, only about one or two percent of Drell-Yan events are diffractive (Abe *et al.*, 1997). Hence exchanges of the kind modeled in Fig. 14.4 are a substantial effect in QCD. (The data quoted are actually for production of W bosons, a minor generalization of the standard Drell-Yan process.)

14.3.3 More general view

The above example indicates that in the Drell-Yan process (and actually more generally in hard processes in hadron-hadron collisions) the Glauber region is handled by a sum over final-state cuts, restricted to those compatible with the specification of the cross section. The cancellation applies only to the inclusive cross section.

In the model, we made restrictions that the spectator part of each hadron consisted of a single line, and that the rungs of the ladders were strongly ordered in rapidity. In fact, the argument generalizes (DeTar, Ellis, and Landshoff, 1975; Cardy and Winbow, 1974). The key point is that to get a Glauber pinch, the results of Sec. 5.11 show that one must have an exchange attached to both the spectator parts of the hadrons, as roughly indicated in Fig. 14.7(a). Once the exchanged system is in the relevant region, the cancellation only depends on general properties, not on the detailed structure of the exchanged pomeron-like object. The argument applies as it stands if the zigzag line is replaced by a gluon, for example. Given the general structure of the argument, we expect that it applies non-perturbatively, to the actual final-state interactions of QCD.

One cut of an example of a more complicated graph to which the general argument applies is shown in Fig. 14.7(b).

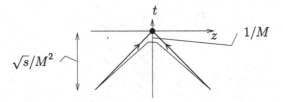

Fig. 14.8. Space-time location of side lines of ladder graph. The slightly time-like thick lines represent the trajectories of the incoming hadrons, and the space-like thin line represents where the ladder's side line is. This diagram is not to scale: the collinear ends should be much further away.

In our example, our choice of kinematic region and of approximation was such that the contour could be deformed to infinity, and the contour at infinity gave zero. In the more general case, to be treated in Secs. 14.4.2 and 14.4.3, we may get a non-zero result on the deformed contour: the contour might be obstructed before it gets to infinity (e.g., by a pole in an exchanged gluon line), or the integrand might not fall rapidly enough in k^+ and k^-. But such a contribution corresponds to some region other than the Glauber region, e.g., a normal soft region where k^+ and k^- are comparable to k_T. That is sufficient to allow us to derive factorization, by our standard methods. What matters is that there is a cancellation of the contributions from the singularities obstructing the contour deformation.

14.3.4 Space-time structure

We now show that the cancellation has a useful but non-trivial interpretation in coordinate space. At first sight, the fact that we obtain a cancellation by setting certain lines on-shell suggests that these lines have a long lifetime and that the cancellation therefore concerns interactions that happen long after the hard scattering occurs, and thus too late to affect the inclusive cross section. If this were the case, then we could imagine making a general proof by working with time-ordered perturbation theory in the overall CM frame. Then we could use a unitarity argument like that used in Sec. 12.7 where we showed a cancellation from a sum over the final states of a jet.

I show that a more powerful argument is needed by determining space-time properties of a (cut or uncut) ladder graph, as used in Fig. 14.4(b)–(d). The result is illustrated in Fig. 14.8. Now all the vertical lines of the ladder have virtuality of order M^2 in the region we consider. Therefore, in the rest frame of each of these lines, the lifetime of the corresponding state is of order $1/M$. But this is boosted, so in the CM frame, the lifetime for a line of rapidity y is $e^{|y|}/M$. For a collinear line, this gives a time scale $\sqrt{s}/M^2$. But for a central line, without a boost, the scale remains at $1/M$. These time scales and the corresponding distances give the separation between the ends of the corresponding lines. Naturally, the positions of the vertices are integrated over, so the estimates give typical values, not exact values.

Next we show that the vertices along the sides are at space-like separation. We do this by examining the correspondence with light-front perturbation theory, but using two

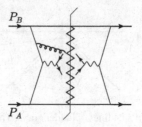

Fig. 14.9. Pomeron/ladder exchange with gluon connection to an active quark.

versions, both x^-- and x^+-ordered perturbation theory. For x^--ordered perturbation theory, there is minus momentum flowing down the left side of the ladder from the top of the graphs, especially graph (b). These values are positive, in order to give the positive minus components of momentum for the rungs. In x^--ordered perturbation theory, this implies that the vertices on the left of the final-state cut are ordered from the top to the bottom in order of increasing x^-.

But the same argument applied to x^+-ordered perturbation theory implies the reverse ordering, from bottom to top, for x^+. Thus the difference in position of the ends of one of the side lines has the opposite sign for the plus and minus components. Hence the ends have a space-like separation, as illustrated by the lower thin line in Fig. 14.8.

A similar argument actually also applies to the partons that initiate the hard scattering. But although these lines are space-like, they also both have high rapidity, and are therefore close to light-like. Fitting all this information together, in Fig. 14.8, shows that the central rungs of the ladder are initiated *before* the hard scattering.

Thus we cannot argue that the ladder is literally a final-state effect, so the simplest argument using time-ordered perturbation in the CM frame is not powerful enough to show the result we need. A correct argument will in fact use both relativistic causality and the topological structure of the graphs with Glauber exchanges.

It is worth noticing that the relevant physical coordinate-space separation of the central part of the ladder from the hard scattering is a normal hadronic scale, i.e., of order 1 fm. Moreover, there is a transverse separation by the same order of magnitude.

That the side lines are space-like gives by itself a reason that there is no causal influence of the ladder on the hard scattering. One can perhaps rationalize this by asserting that the central rungs, particularly, correspond to pre-existing virtual fluctuations in the vacuum, which become instantiated because an appropriate collision happens nearby.

One could try to evade the lack of causal influence, by connecting a gluon (or other line) between the central part of the ladder and one of the active parton lines, as in Fig. 14.9. If all the lines, both in the ladder and in the upper collinear subgraph, have an unchanged virtuality from the previous situation, then the components of the momentum of the extra line have the sizes $l = (l^+, l^-, l_{\mathrm{T}}) \sim (M^2/Q, Me^{-y}, M)$, where y gives the rapidity of the part of the ladder that the gluon attaches to. This is actually a Glauber momentum. But at the active-parton end, the extra gluon attaches at a place which does not give a Glauber pinch. Therefore we can deform the integration out of the region we were originally discussing.

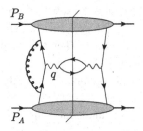

Fig. 14.10. Initial-state interaction of active partons.

14.4 Factorization for Drell-Yan

For the general treatment of the Glauber region, I will follow the proof given by Collins, Soper, and Sterman (1988), but with some important improvements and with correction of errors. That reference supersedes earlier work (Bodwin, 1985; Collins, Soper, and Sterman, 1985a).

14.4.1 Overview

Most of the proof of factorization for the Drell-Yan process follows the same pattern as for other processes we have treated, so we need not repeat those details. The differences concern the Glauber region and the consequences for the directions of Wilson lines. The steps to obtain factorization are as follows.

1. Perform the region decomposition, as in Fig. 14.3, and apply approximants and sub-tractions as usual.
2. For each graphical decomposition for a region, we include the sum over all allowed final-state cuts.
3. To get out of a Glauber-region contribution for the soft subgraph, we move the contours in a direction we characterize as "away from initial-state singularities". This is the appropriate direction for avoiding the Glauber region when gluons are attached to initial-state lines, e.g., Fig. 14.10.
4. But for a general graph the contour deformation entails crossing certain final-state singularities in the collinear subgraphs. These give non-factorizing extra terms which we prove to cancel after the sum-over-cuts. The proof is made by demonstrating that after the sum-over-cuts there are no singularities obstructing the contour deformation. An example of the cancellation was seen in Sec. 14.3.
5. We apply a Grammer-Yennie approximant where needed. To be compatible with the contour deformation out of the Glauber region, the eikonal denominators correspond to past-pointing Wilson lines. This contrasts with our treatment of e^+e^- annihilation and (SI)DIS, where future-pointing Wilson lines worked.
6. The usual Ward-identity arguments give factorization into a hard factor, two parton densities and a soft factor.
7. For the $q_{hT} \ll Q$ case, we then have TMD factorization.

8. For the $q_{h\,T}$-integrated cross section, the soft factor has initial-state Wilson lines. We apply the time-reversal argument of Sec. 13.17 to show that the soft factor is equal to the one with final-state Wilson lines. Then the usual unitarity cancellation applies, after which we get normal factorization.

9. We also use the the time-reversal argument to relate the parton densities for Drell-Yan to those for (SI)DIS.

10. For most parton densities, the numerical values are the same for the two versions. But as explained in Sec. 13.17, certain TMD densities, the Sivers function and the Boer-Mulders function, are T-odd and reverse sign between Drell-Yan and SIDIS.

The treatment of the Glauber region impinges on important issues concerning the physics of soft hadronic interactions. Some of the physics issues manifest themselves in the predicted reversal of the sign of the T-odd distributions. Others manifest themselves in an outright failure of the standard factorization structure in certain natural generalizations of the Drell-Yan process when conditions are imposed in the target fragmentation region. This failure is found both theoretically (Henyey and Savit, 1974; Landshoff and Polkinghorne, 1971) and experimentally (Abe *et al.*, 1997; Aktas *et al.*, 2007b), even though in DIS with a comparable final-state condition factorization does hold (Collins, 1998b).

14.4.2 Separation of collinear subgraph and the rest

We start by consider leading regions, which each correspond to a graph of the form of Fig. 14.3(a) or (b). They involve a convolution of a collinear factor for each incoming hadron, a soft factor and a hard factor. At the hard scattering, let us apply the usual approximants, and let us apply subtractions for smaller regions. Then we sum over collinear attachments to the hard subgraph, by the usual Ward-identity argument. The necessary eikonal lines are past-pointing, corresponding to initial-state poles.

We do not yet apply the full approximant where the soft lines attach to the collinear subgraphs, since we wish to display the nature of the contour deformation out of the Glauber region. Only after the deformation will the standard soft approximation apply.

It is convenient to write the result as a product $C_A R$, where C_A is the collinear factor attached to P_A, and R is everything else. (Thus R includes the hard subgraph, the soft subgraph and the opposite collinear subgraph.) Since we have already extracted the extra collinear-to-A gluons from the hard subgraph, there is only a single collinear-to-A line connecting C_A to R on each side of the final-state cut. All the extra gluons displayed in Fig. 14.11 are therefore part of the soft subgraph. Next we perform the sum over final-state cuts, organized as follows. We start with the vertices at which the soft gluons enter C_A, and we let V denote a choice of which of these vertices are to the left of the cut and which are to the right. The sum-over-cuts is partitioned by the value of V. Given V, we sum over the set $\mathcal{A}(V)$ of compatible cuts of C_A, and over the set $\mathcal{R}(V)$ of compatible cuts of R. These sets can be summed over independently, given that as regards collinear lines we have on

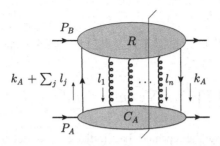

Fig. 14.11. Separation of region decomposition into a collinear-A part C_A and everything else, R. The displayed gluon lines are part of the soft subgraph, and are included in the definition of subgraph R.

each side of the cut one definite collinear line from C_A to R. This gives

$$G_L = \int \frac{dk_A^+ \, d^2 \mathbf{k}_{AT}}{(2\pi)^3} \int \prod_{j=1}^n \frac{d^4 l_j}{(2\pi)^4} \sum_V \sum_{F_A \in \mathcal{A}(V)} C_A^{F_A} \left(k_A, \{l_j\}\right)^{\mu_1 \dots \mu_n}$$

$$\times \sum_{F_R \in \mathcal{R}(V)} R^{F_R} \left(k_A^+, \mathbf{k}_{AT}, \{l_j\}\right)_{\mu_1 \dots \mu_n}. \tag{14.17}$$

Here $C_A^{F_A}$ denotes C_A with cut F_A, and similarly for R. *The collinear subgraph C_A is defined to include its external collinear lines $k_A + \sum_{j=1}^n l_j$ and k_A, but to exclude the soft gluons l_j.* The soft lines are all in R.

The momenta are organized as follows: l_j are the momenta entering C_A from the soft subgraph, while k_A^+ and $\mathbf{k}_{AT}$ are the collinear loop momentum components from C_A entering the hard scattering. Since the minus component k_A^- is approximated by zero in the hard scattering, its integral is considered to be included in the definition of C_A. As for routing the soft momentum loops, we choose them as in Fig. 14.11: the collinear line on the left of the cut has momentum $k_A + \sum_{j=1}^n l_j$ outgoing from C_A, and the collinear line on the right has momentum k_A incoming to C_A.

We next apply to the soft lines two parts of the soft approximants that remain valid in the Glauber part of the soft region. The first is that we keep only the $\mu_j = +$ components of the gluon polarizations, since these correspond to the large components for the collinear subgraph:

$$C_A^{F_A} \left(k_A, \{l_j\}\right)^{\mu_1 \dots \mu_n} R^{F_R} \left(k_A^+, \mathbf{k}_{AT}, \{l_j\}\right)_{\mu_1 \dots \mu_n}$$

$$\mapsto C_A^{F_A} \left(k_A, \{l_j\}\right)^{+\dots} R^{F_R} \left(k_A^+, \mathbf{k}_{AT}, \{l_j\}\right)_{+\dots}. \tag{14.18}$$

The second part of the approximant is to drop the plus component of each soft momentum in C_A, because in the soft region each l_j^+ is much smaller than the order-Q components of collinear momenta. Thus in C_A we replace each l_j by

$$\tilde{l}_j = (0, l_j^-, \mathbf{l}_{jT}). \tag{14.19}$$

This all gives

$$G_{L,1} = \int \frac{\mathrm{d}k_A^+ \, \mathrm{d}^2 k_{A\mathrm{T}}}{(2\pi)^3} \int \prod_{j=1}^n \frac{\mathrm{d}l_j^- \, \mathrm{d}^2 l_{j\mathrm{T}}}{(2\pi)^3} \sum_V \sum_{F_A \in \mathcal{A}(V)} C_A^{F_A}(k_A, \{\tilde{l}_j\})$$

$$\times \int \prod_{j=1}^n \frac{\mathrm{d}l_j^+}{2\pi} \sum_{F_R \in \mathcal{R}(V)} R^{F_R}(k_A^+, k_{A\mathrm{T}}, \{l_j\}). \tag{14.20}$$

It is convenient not to write the repeated fixed indices, so we define the indexless symbols for the factors by $C_A = C_A^{+\cdots}$ and $R = R_{+\cdots} = R^{-\cdots}$.

At this point, the integral over soft momenta still includes the Glauber region. As we know from earlier chapters, this implies that the remaining part of the soft approximant cannot yet be applied. This is the approximation of neglecting the transverse components of l_j in the collinear subgraph, i.e., to replace $\tilde{l}_j$ by

$$\hat{l}_j = (0, l_j^-, \mathbf{0}_\mathrm{T}). \tag{14.21}$$

After we have justified a contour deformation on l_j^- out of the Glauber region, we can apply this last approximation. Then we will be able to apply the Grammer-Yennie method to factor the soft lines from C_A. (A similar argument will apply to the soft lines connecting to the collinear-to-B subgraph, which at the moment is inside the R factor.)

However, the use of (14.21) is not yet valid because in the Glauber region l_j^- is particularly small compared with $l_{j\mathrm{T}}$.

14.4.3 Contour deformation

In (14.20), the integrals over l_j^+ are confined to R, while the integrals over l_j^- and k_A^- are confined to C_A. This suggests writing C_A and R in terms of light-front perturbation theory, *but in two opposite versions*, x^--ordered perturbation theory for R, x^+-ordered perturbation theory for C_A.

x^+ ordering for C_A

To obtain the x^+-ordered form of C_A, we perform all the internal k^- integrals, as in Sec. 7.2. We write $C_A^{F_A}$ as a sum over x^+ orderings T of its vertices. With each ordering, there is a set of intermediate states, each with its energy denominator (actually a k^- denominator), and an on-shell final state with a delta function to make its momentum physical.

We classify the intermediate states as to whether they are earlier or later than the vertex that annihilates the active parton, and as to whether they are on the left or right of the final-state cut. Then the sum over x^+ orderings is given as

$$C_A^{F_A} = \sum_T I_T'(\{\hat{l}_j\})^* F_T(\{\hat{l}_j\}) I_T(\{\hat{l}_j\}) \times \text{vertices}. \tag{14.22}$$

Here I_T contains those factors for intermediate states that are *earlier* than the active-parton vertex and that are on the left of the cut; these we treat as being initial-state interactions. Similarly I_T', with a complex conjugation, is for the initial-state interactions that are on the

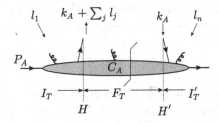

Fig. 14.12. Classification of initial- and final-state interactions. x^+ is assumed to increase from the left to the final-state cut and then decrease again. The soft gluon lines are short to symbolize that their propagators are excluded from the subgraph C_A.

right of the cut. Everything else is later than the active-parton vertices, and we put it in the factor F_T, which we label as "final state". This classification is illustrated in Fig. 14.12.

In this diagram we arrange the vertices from left to right so that there is x^+ ordering on the left of the cut and x^+ anti-ordering on the right. Correspondingly we define an ordering on the vertices and the intermediate states used in x^+-ordered perturbation theory: thus, if j is a vertex and ξ is an intermediate state, then $j < \xi$ means that j is to the left of ξ in Fig. 14.12. We let H and H' be the vertices at the end of the collinear parton lines. Then we define an intermediate state ξ to be in the final state if $H < \xi < H'$, and to be in the initial-state if $\xi < H$ or $H' < \xi$.

We have already extracted extra gluons from the hard scattering, converting them to attachments to Wilson lines attached to the primary parton lines of the collinear graphs (e.g., at H and H' in C_A). Since we use past-pointing Wilson lines, they are all in the initial-state factors I_T and $I_T'^{\,*}$; hence the Wilson lines will not affect the sum-over-cuts argument that we will apply to F_T.

Explicit expressions for I_T, F_T and $I_T'^{\,*}$, given the position of the final-state cut F_A, are

$$I_T\left(\{\hat{l}_j\}\right) = \prod_{\substack{\text{states } \xi: \\ \xi < H}} \frac{1}{P_A^- + \sum_{\substack{\text{vertices } j: \\ j < \xi}} l_j^- - \sum_{\substack{\text{lines } L: \\ L \in \xi}} X_L + i0}, \tag{14.23a}$$

$$I_T'\left(\{\hat{l}_j\}\right)^* = \prod_{\substack{\text{states } \xi: \\ H' < \xi}} \frac{1}{P_A^- - \sum_{\substack{\text{vertices } j: \\ j > \xi}} l_j^- - \sum_{\substack{\text{lines } L: \\ L \in \xi}} X_L - i0}, \tag{14.23b}$$

$$F_T\left(\{\hat{l}_j\}\right) = \int_{-\infty}^{\infty} \frac{dk_A^-}{2\pi} \prod_{\substack{\text{states } \xi: \\ F_A < \xi < H'}} \frac{1}{P_A^- - k_A^- - \sum_{\substack{\text{vertices } j: \\ j > \xi}} l_j^- - \sum_{\substack{\text{lines } L: \\ L \in \xi}} X_L - i0}$$

$$\times 2\pi\, \delta\!\left(P_A^- - k_A^- - \sum_{\substack{\text{vertices } j: \\ j > F_A}} l_j^- - \sum_{\substack{\text{lines } L: \\ L \in \xi}} X_L \right)$$

$$\times \prod_{\substack{\text{states } \xi: \\ H < \xi < F_A}} \frac{1}{P_A^- - k_A^- - \sum_{\substack{\text{vertices } j: \\ j > \xi}} l_j^- - \sum_{\substack{\text{lines } L: \\ L \in \xi}} X_L + i0}. \tag{14.23c}$$

Here ξ denotes an intermediate state, which contains a certain number of lines labeled by L. For each line L, the quantity X_L is its on-shell value of minus momentum:

$$X_L = \frac{k_{LT}^2 + m_L^2}{2k_L^+}. \tag{14.24}$$

By the rules of x^+-ordered perturbation theory, the initial-state factors I_T and I_T' contain no dependence on k_A^-, so the integral over k_A^- is confined to the final-state factor F_T.

Normally, there would be a factor i for each intermediate state on the left of the final-state cut, and a factor with the opposite sign $-i$ for each state on the right. However, at the end of each state there is a vertex, and interaction vertices also have opposite signs between their occurrences on opposite sides of the cut. So there is no difference in sign between a state-vertex pair on the left and the right of the cut. Therefore we omit the i and $-i$ that go with the states and with the vertices; hence the factor for the vertices in (14.22) is independent of the placement of the cut.

The initial-state denominators in I_T and $I_T'^*$ all give poles in the lower half plane for those l_j^-s that enter at initial-state vertices. Thus to avoid these poles we deform l_j^- into the upper half plane. But the final-state poles and delta function obstruct this deformation. Our aim will be to show that the obstructions cancel after a sum-over-cuts, so that we can deform the integrations over all the l_j^- momenta into the upper half plane. Thus we can avoid all Glauber configurations, where some of the l_j^- values are small.

Independence of R on soft-vertex position

We will be able to obtain this result if we can sum over all cuts F_A of $C_A^{F_A}$ independently of the remaining parts of (14.20). But the allowed cuts for the remainder factor, in $\sum_{F_R \in \mathcal{R}(V)} R^{F_R}$, depend on V, which labels the placement of the soft-gluon vertices in C_A relative to the cut. We solve this problem by showing that the remainder factor,

$$\int \prod_{j=1}^{n} \frac{dl_j^+}{2\pi} \sum_{F_R \in \mathcal{R}(V)} R^{F_R}\left(k_A^+, k_{AT}, \{l_j\}\right), \tag{14.25}$$

is in fact independent of V.

A proof of this result (Collins, Soper, and Sterman, 1988) can be made by examining x^--ordered perturbation theory for R. (This is in contrast to the x^+-ordered perturbation theory that we used for C_A.) Instead I will show here an argument that uses commutators for the gluon field, and that is therefore more suggestive of the underlying physics issues.

For a given placement of the ends of the soft gluons relative to the cut, i.e., for a given value of V, the quantity in (14.25) is obtained from a Fourier transform of a matrix element of fields of the form

$$\langle P_B | \bar{T} \left\{ \bar{\psi}(0) \prod_{j>F} A(x_j) \right\} T \left\{ \prod_{j<F} A(x_j)\psi(y) \right\} | P_B \rangle. \tag{14.26}$$

The soft-gluon fields diagrammatically on the right of the cut are in the anti-time-ordered part; these are in the left-hand part of the matrix element. Conversely, the fields on the left of the cut are in the time-ordered part that is in the right-hand part of the matrix element.

The quark fields are those for the parton density for hadron H_B, and they are on fixed sides of the cut. Implicit in (14.26) are sums over all ways of inserting interactions; this includes the sum-over-cuts compatible with a given placement V of the explicit gluon fields in (14.26).

Now we made the approximation to neglect l_j^+ in C_A, so that the integrals over l_j^+ are confined to R, as in (14.25). It follows that the fields in (14.26) have zero separation in the x_j^- coordinates. With a generally non-zero separation in the transverse direction, the A fields are at space-like separation, and therefore they all commute with each other, given that we use Feynman gauge. Hence the ordering of the fields does not affect the value of (14.26). This gives the desired result that the R factor in (14.25) is independent of V, as was to be proved.

Sum-over-cuts of collinear subgraph

Given this result, we need to analyze the sum of the collinear-to-A factor C_A over all cuts, i.e., to analyze

$$\sum_V \sum_{F_A \in \mathcal{A}(V)} C_A^{F_A}(k_A, \{\hat{l}_j\}) = \sum_{\text{all } F_A} C_A^{F_A}(k_A, \{\hat{l}_j\}), \tag{14.27}$$

with the approximated momenta. It is in fact sufficient to take a fixed ordering of the vertices and states in Fig. 14.12, and to sum over allowed placements of the cut F_A relative to the vertices. The active-parton vertices H and H' are always to the left and right (respectively) of the cut, and we have already seen that with the formulae (14.23) for the initial- and final-state factors, the vertex factors in (14.22) are independent of the placement of the cut F_A. Thus we just need to sum F_T in (14.23c) over the placement of the cut.

Let there be N states ξ_f in F_T, which we label by an index $f = 1, \ldots, N$, and let the on-shell minus momentum for state f be

$$D_f = \sum_{\substack{\text{lines } L: \\ L \in \xi_f}} X_L. \tag{14.28}$$

Then we need to calculate

$$\int_{-\infty}^{\infty} \frac{dk_A^-}{2\pi} \sum_{c=1}^{N} \left\{ \prod_{f=c+1}^{N} \frac{1}{P_A^- - k_A^- - \sum_{j>f} l_j^- - D_f - i0} \right.$$

$$\left. \times 2\pi \, \delta\!\left(P_A^- - k_A^- - \sum_{j>c} l_j^- - D_c\right) \prod_{f=1}^{c-1} \frac{1}{P_A^- - k_A^- - \sum_{j>f} l_j^- - D_f + i0} \right\} \tag{14.29}$$

times a vertex factor. For the case $N = 1$, i.e., when there are no final-state interactions at all, we simply get unity. For larger N, we use the unitarity identity that the integrand equals

$$i \prod_{f=1}^{N} \frac{1}{P_A^- - k_A^- - \sum_{j>f} l_j^- - D_f - i0} - i \prod_{f=1}^{N} \frac{1}{P_A^- - k_A^- - \sum_{j>f} l_j^- - D_f + i0}. \tag{14.30}$$

The integral over k_A^- for each term separately is zero, since each term has its singularities on one side of the real axis.

Therefore, after the sum-over-cuts, the F_T factor becomes unity. All that remains for the total of C_A in (14.27) are those terms in which all the interactions, including the soft vertices, are in the initial state, whether on the left or the right of the final-state cut. This just leaves initial-state poles, so that to get out of the Glauber region, we can deform l_j^- into the upper half plane.

14.4.4 Soft approximation and Grammer-Yennie method

We now switch back to Feynman perturbation theory. In the previous section, we showed that we can avoid the Glauber region for soft momenta by deforming their integrals away from initial-state poles. There has therefore been a cancellation of the final-state poles that would otherwise obstruct the deformation, given that we make an inclusive sum over at least the spectator part of the hadronic final state.

On the deformed contour, the usual soft approximation applies, where we neglect l_{jT} as well as l_j^+ in the collinear-to-A subgraph. Then we apply the appropriate version of the same argument to soft connections to the opposite collinear subgraph. After that we apply the usual Grammer-Yennie method and Ward identities to obtain a factorized form for the cross section or, equivalently, for the hadronic tensor.

14.4.5 Factorization for low-q_{hT} cross section

In the case of the low-q_{hT} cross section, we now have exactly the same structure as we found in Ch. 13 for e^+e^- annihilation and SIDIS, with a product of collinear, soft, and hard factors. So all the same steps that lead to a factorization formula can be used. We will examine the consequences in Sec. 14.5.

It is interesting that the treatment of the Glauber region was originally formulated (Collins, Soper, and Sterman, 1988) only in the context of situations using integrated parton densities, e.g., the cross section integrated over $\boldsymbol{q}_{hT}$. In fact, the argument works equally well for the TMD case. What enables it to work is that we now have a complete Feynman-gauge formalism for TMD factorization and TMD parton densities. The Feynman gauge is important in giving the analytic properties that we relied on to deform contours out of the Glauber region.

14.4.6 Factorization for integrated cross section

We can also treat the Drell-Yan cross section integrated over all transverse momentum, or at large q_{hT}. The usual argument gives us a standard factorization formula with integrated parton densities.

However, there is one step that needs enhancement for the Drell-Yan process. This is in proving that there is a cancellation of the soft factor. This is an integrated soft factor, like the one we encountered in Sec. 12.8.6 for an inclusive cross section in e^+e^- annihilation.

There we used a unitarity-type argument applied to a soft factor defined with *future*-pointing Wilson lines. But for the Drell-Yan process, we get past-pointing Wilson lines. We already solved this problem in Sec. 13.17.5, where we used time-reversal invariance to show that the two kinds of soft factor are equal. After that, the cancellation of the integrated soft factor follows from Sec. 12.8.6.

Finally we obtain factorization in the standard form already stated in (14.1) and (14.2).

14.4.7 Possible use of "physical" gauges

In much early work on factorization theorems in QCD (e.g., Ellis *et al.*, 1979; Libby and Sterman, 1978a; Lepage and Brodsky, 1980; Collins and Sterman, 1981) so-called physical gauges were often used. Such gauges include the various kinds of axial gauge, with the gauge condition $n \cdot A = 0$, and the Coulomb gauge.

These gauges only have physical polarizations for the gluon, unlike the Feynman gauge with its extra unphysical states. This leads to a number of advantages including the absence of regions that give graph-by-graph super-leading contributions, Sec. 11.2.3.

Unfortunately all such gauges, at least all the known ones, have unphysical singularities in the gluon propagator and the singularities break manifest Lorentz invariance. The proof of factorization relied critically on having only physical singularities for the deformation out of the Glauber region, and especially for the treatment of final-state interactions.

In the presence of unphysical singularities, it is possible, in individual graphs, to have signals that propagate faster than light. These can correlate the two hadrons and the two active partons before the hard collision, and if uncanceled they lead to a breakdown of factorization. Of course, all such effects must cancel in a physical cross section. But the presence of unphysical singularities complicates the proof.

An example of how non-factorization could arise from initial-state interactions was given in Bodwin, Brodsky, and Lepage (1981), where a non-abelian phase factor was calculated from exchange of Glauber gluons between the incoming active partons in the Drell-Yan process. Our proof shows that we can deform the momenta out of the Glauber region, and then apply Ward identities to obtain factorization. Therefore, while the phases exist and give a contribution to the cross section, their effects do not break factorization; rather they get moved into the parton densities.

14.4.8 Issues remaining

The proof given above captures many of the physics issues involved. But an attentive and critical reader can surely raise some questions about the proof's completeness, and there are interesting problems in trying to do better.

For example, the proof relied strictly on the momentum categories for the standard leading regions as seen in perturbation theory. In particular, for a soft gluon connecting to a collinear subgraph, there is a large rapidity difference between the lines at the ends of the gluon. But the multiperipheral model used in Sec. 14.3 suggests that there is a different but related possibility that is relevant for non-perturbative hadronic interactions. This is that there is an exchange where the rapidity is graduated along the exchange without any large jumps. The overall momentum transfer along the exchange is not only soft but in fact Glauber. But without any rapidity gaps, it is difficult to apply the argument we gave as it stands. Very quickly one gets into a situation of trying to develop a good and deductive QCD version of Regge theory.

Another issue is that our characterization of leading regions was incomplete in Feynman gauge. There can be regions with extra disconnected hard-scattering subgraphs, appearing, so to speak, in parallel with the standard hard scattering, and transversely separated from it in coordinate space. These are induced by gluons, and they are not power-suppressed because the gluons can have polarization in the direction of their momentum, which gives an enhancement we have seen in many contexts. Such extra hard scatterings cancel after a sum over graphs for the hard scattering (Labastida and Sterman, 1985).

It is quite easy to state factorization intuitively in terms of parton probability densities in each hadron, convoluted with a parton hard scattering. The current versions of genuine proofs are evidently formidable. Much of the difficulty is genuine. In going to hadron-hadron scattering and in examining more detailed cross sections, one is approaching the frontier of where ordinary factorization fails, and some more general approach is needed. For one kind of indication as to where this frontier is, see Bomhof, Mulders, and Pijlman (2004) and Collins and Qiu (2007).

14.5 TMD pdfs and Drell-Yan process

We now have all the ingredients to obtain TMD factorization for the Drell-Yan process $q_{h\,\mathrm{T}} \ll Q$. The leading regions were shown in Fig. 14.3(b). We now know that after summing over final-state cuts of graphs, we can deform contours out of the Glauber region, away from initial-state poles. Therefore the methods of Ch. 13 apply to obtain factorization, provided that the parton densities and the soft function are defined with *past*-pointing Wilson lines. The coordinate frames and hadronic tensor were defined in Sec. 14.2.

14.5.1 Factorization

The result is a TMD factorization formula for the hadronic tensor $W^{\mu\nu}$, defined in (14.5). Factorization is like (13.46) for e^+e^- annihilation, or (13.116) for SIDIS, but using parton densities:

$$W^{\mu\nu} = \frac{8\pi^2 s}{Q^2} \sum_f C_f^{\mu\nu}(\hat{k}_A, \hat{k}_B) \int d^2 \boldsymbol{b}_\mathrm{T} \, e^{i q_{h\mathrm{T}} \cdot \boldsymbol{b}_\mathrm{T}} \, \tilde{f}_{f/H_A}(x_A, \boldsymbol{b}_\mathrm{T}; \zeta_A) \, \tilde{f}_{\bar{f}/H_B}(x_B, \boldsymbol{b}_\mathrm{T}; \zeta_B)$$

$$+ \text{polarized terms} + \text{large } q_{h\mathrm{T}} \text{ correction, } Y. \tag{14.31}$$

Some details of this formula will be explained in more detail in Sec. 14.5.2. It uses the Drell-Yan versions of parton densities, defined with past-pointing Wilson lines. The PT-transformation argument of Sec. 13.17 shows that these parton densities are numerically equal to those in SIDIS, except that T-odd densities in the polarization part are reversed in sign. In the parton densities, the fractional momentum arguments are

$$x_A = \frac{Qe^y}{\sqrt{s}}, \quad x_B = \frac{Qe^{-y}}{\sqrt{s}}, \tag{14.32}$$

where y is the CM rapidity of the lepton pair. The ζ arguments are as in (13.107), i.e.,

$$\zeta_A = 2x_A^2(P_{A,h}^+)^2 e^{-2y_n} = M_A^2 x_A^2 e^{2(y_{P_A} - y_n)}, \tag{14.33a}$$

$$\zeta_B = 2x_B^2(P_{B,h}^-)^2 e^{2y_n} = M_B^2 x_B^2 e^{2(y_n - y_{P_B})}, \tag{14.33b}$$

where y_n is the rapidity parameter of the parton densities (13.106).

For phenomenological use, one can take account of the CS and RG evolution equations by applying to (14.31) the same steps as for two-particle-inclusive e^+e^- annihilation. This gives a formula like (13.81).

14.5.2 Kinematics and approximations in TMD factorization

In the derivation of the TMD term in (14.31), the approximations used concern the hard scattering and the momentum-conservation delta function.

For the hard scattering, we use the tensor $C_f^{\mu\nu}$ for the on-shell partonic reaction $f\bar{f} \to \gamma^*$. Its normalization is that of a partonic scattering amplitude squared. Thus in lowest order,

$$C_{f,\mathrm{LO}}^{\mu\nu} = \frac{e_f^2}{N_c}(\hat{k}_A^\mu \hat{k}_B^\nu + \hat{k}_B^\mu \hat{k}_A^\nu - g^{\mu\nu}\hat{k}_A \cdot \hat{k}_B), \tag{14.34}$$

where the factor $1/N_c = 1/3$ results from the average over color. In higher order, there are the usual soft and collinear subtractions. The approximated external momenta of the hard scattering are chosen so that in the photon frame they have zero transverse momentum:

$$\hat{k}_{A,\gamma} = (q_\gamma^+, 0, \mathbf{0}_\mathrm{T}) = (Q/\sqrt{2}, 0, \mathbf{0}_\mathrm{T}), \quad \hat{k}_{B,\gamma} = (0, q_\gamma^-, \mathbf{0}_\mathrm{T}) = (0, Q/\sqrt{2}, \mathbf{0}_\mathrm{T}). \tag{14.35}$$

Then in the hadron frame

$$\hat{k}_{A,h} = \left(\frac{e^y Q(\kappa+1)}{2\sqrt{2}}, \frac{e^{-y}Q(\kappa-1)}{2\sqrt{2}}, \frac{\mathbf{q}_{h\mathrm{T}}}{2}\right), \tag{14.36a}$$

$$\hat{k}_{B,h} = \left(\frac{e^y Q(\kappa-1)}{2\sqrt{2}}, \frac{e^{-y}Q(\kappa+1)}{2\sqrt{2}}, \frac{\mathbf{q}_{h\mathrm{T}}}{2}\right), \tag{14.36b}$$

where $\kappa = \sqrt{1 + q_{h\mathrm{T}}^2/Q^2}$. The hadron-frame components can be useful in performing a structure function decomposition of the hard scattering.

These approximated parton momenta apply *only* to the hard scattering. In the momentum-conservation delta function, we make instead the replacement

$$\delta(q_h - k_{A,h} - k_{B,h}) \mapsto \delta\left(q_h - (k_{A,h}^+\kappa, k_{B,h}^-\kappa, \mathbf{k}_{A,h\mathrm{T}} + \mathbf{k}_{B,h\mathrm{T}})\right). \tag{14.37}$$

Here, we keep the exact values of transverse momenta, as is required to correctly treat the cross section at low $q_{h\,T}$. But for the plus and minus components we made an approximation that is valid to leading power in $q_{h\,T}/Q$. The factors of κ are arranged so that the fractional momenta in the parton densities are the variables defined in (14.32), so that $k_{A,h}^+/P_{A,h}^+ = x_A$ and $k_{B,h}^-/P_{B,h}^- = x_B$, where errors of order M_j^2/s are ignored.

Observe the mismatch between the values of $k_{A,h}^+$ and $k_{B,h}^-$ used in the hard scattering and those used in the parton densities. The reader might therefore be tempted to try to remedy this, for example, by changing the the terms in (14.37) that involve $k_{A,h}^+$ and $k_{B,h}^-$. However, such a change would not help if one stays within the parton-density framework. The reason is that to get conventional light-front-style parton densities, one must short-circuit the integrals over the opposite components, i.e., $k_{A,h}^-$ and $k_{B,h}^+$, so that each integral is internal to its parton density. Provided that the parton densities are not rapidly varying as a function of x_A and x_B, this leads to an intrinsic error in the approximation of order $q_{h\,T}^2/Q^2$. Changing (14.37) can only correct part of the error. A correct treatment needs to deal with the production of extra jets, recoil against which is the source of events with large $q_{h\,T}$. This is the province of the Y term in (14.31). With a correct Y term included, (14.31) is correct up to mass-suppressed corrections for all $q_{h\,T}$.

The exact form of the approximations on the longitudinal momenta in the TMD term in (14.31) was chosen to be fairly simple and to agree with previously stated results (Collins, Soper, and Sterman, 1985b).

It is of course possible that the parton densities are rapidly varying enough that the short-circuiting of $k_{A,h}^-$ and $k_{B,h}^+$ is a bad approximation even at small $q_{h\,T}$. But in that case one cannot use ordinary parton densities, even of the TMD type. One must use more general quantities (Collins, Rogers, and Staśto, 2008; Watt, Martin, and Ryskin, 2003, 2004) that are functions of the full 4-momentum of a parton.

14.5.3 Fitting data

A representative of the state of the art (Landry *et al.*, 2003) for fitting the Drell-Yan process is shown in Fig. 14.13.

As is usual, there is an interesting combination of fitting and prediction. The general principles are as follows.

- Obtain the *integrated* parton densities by global fits to reactions that do not need TMD factorization.
- From perturbative calculations estimate the perturbative parts of the Drell-Yan version of (13.81). This determines the integrand primarily at $b_T \lesssim b_{max}$.
- Compare the Drell-Yan version of (13.81) with data at moderate Q, and adjust the non-perturbative functions to give a fit. If the data are at one value of energy, the function g_K will not yet be separately determined.
- There are some predictions already at this point, since the non-perturbative factor, generalizing (13.83), is a product of a function of x_A and a function of x_B, rather than a more general function.

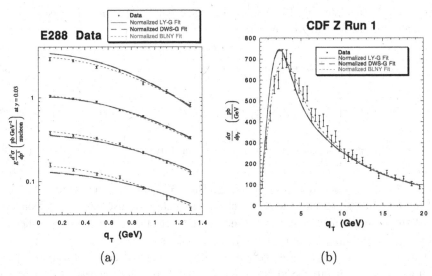

Fig. 14.13. From Landry *et al.* (2003) (with change of axis label). Copyright (2003) by The American Physical Society. Results from fitting TMD parton densities for the Drell-Yan process. (a) For $\mu^+\mu^-$ production in the E288 experiment (Ito *et al.*, 1981) at $\sqrt{s} = 27.4\,\text{GeV}$. From top to bottom, the curves and data are for $5 < Q < 6\,\text{GeV}$, $6 < Q < 7\,\text{GeV}$, $7 < Q < 8\,\text{GeV}$, $8 < Q < 9\,\text{GeV}$. (b) For production of e^+e^- pairs in the Z-boson region at CDF at $\sqrt{s} = 1800\,\text{GeV}$ (Affolder *et al.*, 2000). See the text for a description of the different curves. The fits are made with $b_{\text{max}} = 0.5\,\text{GeV}^{-1}$.

- Repeat the fit at a higher value of $\sqrt{s}$ to determine the coefficient, g_K, of $\ln Q^2$ in the non-perturbative factor.
- Since g_K is independent of x_A, x_B and the flavors of quark and hadron, this last fit can be performed for one value of x_A and x_B. The cross section is predicted for all other values.
- The cross section is then predicted for all other energies Q. (Of course, Q must be high enough for factorization to be valid.)

In practice, errors in fitted data (and in the use of low-order perturbative calculations) limit the accuracy of predictions. So when new data become available at a higher energy, not only is there a test of whether the new data agree with predictions within errors, but the new data are also used to tune up the fits. A test of the combination of QCD and factorization is by the quality of the global fit. One problem is that to make fits, the non-perturbative functions are typically replaced by some assumed (plausible) form with a few parameters. A lack of a good fit may simply be due to the use of an unsuitable parameterization.

A further complication is that from standard Drell-Yan data it is hard to obtain a complete flavor separation of the non-perturbative functions $g_{j/H}(x, b_T)$. This is probably most systematically solved by a global fit to data from all three processes (Drell-Yan, SIDIS, and e^+e^- annihilation). The flavor relations listed in Sec. 12.4.8 will considerably assist the fits for fragmentation functions.

As for Drell-Yan data, Fig. 14.13 shows fits corresponding to three choices of parameterization of the non-perturbative factor:

$$\text{DWS:} \quad \exp\left[-\left(g_1 + g_2 \ln \frac{Q}{2Q_0}\right) b_T^2\right], \tag{14.38a}$$

$$\text{GY:} \quad \exp\left\{-\left[g_1 + g_2 \ln \frac{Q}{2Q_0}\right] b_T^2 - [g_1 g_3 \ln(100 x_A x_B)] b_T\right\}, \tag{14.38b}$$

$$\text{BLNY:} \quad \exp\left\{-\left[g_1 + g_2 \ln \frac{Q}{2Q_0} + g_1 g_3 \ln(100 x_A x_B)\right] b_T^2\right\}. \tag{14.38c}$$

Here g_1, g_2, and g_3 are numerical parameters. The DWS ansatz is quadratic in b_T, corresponding to a Gaussian transverse momentum distribution; it also has no x dependence. The GY form supplements this by an x-dependent term that is *linear* in b_T rather than quadratic. Since $\ln(x_A x_B) = \ln x_A + \ln x_B$, this ansatz is of the general form of the Drell-Yan equivalent of (13.83); that is, the x dependence in the exponent is a sum of separate terms for x_A and x_B. Finally the BLNY ansatz is like GY, but with quadratic b_T dependence for all its terms.

From Fig. 14.13, we see that the last ansatz, BLNY, provides a good fit to the data. It has the parameters

$$g_1 = 0.21 {}^{+0.01}_{-0.01} \text{ GeV}^2, \quad g_2 = 0.68 {}^{+0.01}_{-0.02} \text{ GeV}^2, \quad g_3 = -0.6 {}^{+0.05}_{-0.04} \text{ GeV}^2. \tag{14.39}$$

The two plots in Fig. 14.13 are at very different energies, and the primary change is a strong broadening of the transverse-momentum distribution from low to high energy. Note the very different scales of transverse momentum for the two plots, and that the left-hand plot uses a logarithmic scale for the cross section, whereas the right-hand plot uses a linear scale. Note also that the zero at the origin of the right-hand plot is an artifact of the different variable used for the cross section, which is $d\sigma / d P_T$ rather than $d\sigma / d^2 P_T$ for the left-hand plot. In both cases the cross section $d\sigma / d^2 P_T$ differential in the two-dimensional transverse momentum is non-zero at zero transverse momentum. See Landry *et al.* (2003) for a comparison of these same fits with other data.

The numerical fitted values of the coefficients in (14.38) depend strongly on the value of the cutoff parameter b_{max}. For example, instead of the value $b_{max} = 0.5 \text{ GeV}^{-1}$ used in Fig. 14.13, a later fit in Konychev and Nadolsky (2006) used a much larger value, $b_{max} = 1.5 \text{ GeV}^{-1}$. With the same functional form, (14.38c), they found $g_2 \simeq 0.2$, a factor of 3 less than given in (14.39). This corresponds to strikingly different large-b_T behavior. However, the large-b_T asymptote is unimportant after we perform the Fourier transform to transverse-momentum space, for the cross section, as can be seen from the plots of the complete b_T-space integrand in Fig. 14.14. What matters is the integrand at 2 GeV^{-1} and smaller. There the curves for fits with different values b_{max} are in reasonable agreement. (The short dashed curve in Fig. 14.14(a) refers to an earlier fit that is not relevant here.) For higher energy, Fig. 14.14(b), the important values of b_T migrate down, so that the details of the large-b_T asymptote are even less important.

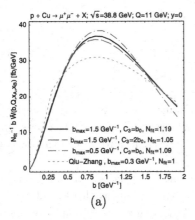
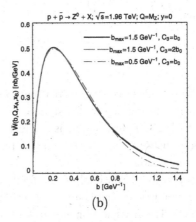

Fig. 14.14. b_T-space integrand for the equivalent of (13.81) for the Drell-Yan process. The results of fits (Konychev and Nadolsky, 2006) using different values of b_{max} are shown: (a) for $\sqrt{s} = 38.8$ GeV, (b) for Z production $\sqrt{s} = 1.96$ TeV. In these plots C_3 corresponds to what is called C_1 in this chapter, and $b_0 = 2e^{-\gamma_E} \simeq 1.123$. Reprinted from Konychev and Nadolsky (2006), with permission from Elsevier.

Given that the functional form of the non-perturbative functions was not changed when b_{max} was changed, b_{max} can be treated as a parameter to be fitted to data. A good fit implies that a good match is found between the perturbative prediction of the integrand and its continuation to large b_T. In fact Konychev and Nadolsky (2006) found that the larger values, notably 1.5 GeV^{-1}, are preferred.

It is a concern that $b_{max} = 1.5$ GeV^{-1} is rather large to trust perturbation theory, since it corresponds to a low momentum. But 1.5 GeV$^{-1} = 0.3$ fm, which is somewhat smaller than the size of a proton. Thus it is reasonable that such a distance is in the range of perturbative quark-gluon physics. An increase by another factor of 3 would not be reasonable. In contrast, this argument suggests that using $b_{max} = 0.5$ GeV$^{-1} = 0.1$ fm, as in Landry *et al.* (2003), is excessively conservative, especially given that the non-perturbative functions can absorb errors in using perturbation theory around b_{max}.

There are some general features of the b_T-space integrand, which enable us to gain a useful semi-quantitative understanding of the main properties of its Fourier transform, which determines the cross section as a function of q_{hT}. For this purpose we define the integrand as

$$W(b_T; Q) = \text{(the Drell-Yan version of) lines 2–6 of (13.81).} \qquad (14.40)$$

The plots in Fig. 14.14 show this factor multiplied by b_T, as is appropriate in its use in a one-dimensional radial integral, as in (13.128b).

Now W is positive everywhere in the fits. This is not absolutely guaranteed, since any positivity constraints on parton densities apply only to their momentum-space versions.

We next notice in the plots that, beyond some value of b_T, the integrand decreases with b_T, and that the plot narrows as Q increases. This arises from both the perturbative and non-perturbative parts of the exponents. When $b_T \gg 1/Q$, the biggest part of the perturbative exponent is

$$-2 \int_{\mu_b}^{\mu_Q} \frac{d\mu'}{\mu'} \ln \frac{Q}{\mu'} \gamma_K \big(g(\mu')\big) . \tag{14.41}$$

For fixed Q, this becomes increasingly more negative as b_T increases, the sign of this term being determined by the sign of the lowest-order calculation of γ_K. Then when Q itself is increased, the decrease is stronger. This matches the behavior of the corresponding non-perturbative term $-2g_K(b_T) \ln Q$, which at large b_T is negative to avoid a pathological Fourier transform.

We now examine the implications for the cross section. Assistance is provided by converting the two-dimensional Fourier transformation into a one-dimensional Bessel transform, (13.128b). The cross section at zero transverse momentum is the area under the curve of $b_T W(b_T)$, times an overall kinematic factor. The narrowing of the $b_T W(b_T)$ curve with increasing Q therefore shows that the cross section at $q_{hT} = 0$ decreases with increasing Q (at fixed x_A and x_B). At large Q, the large-b_T tail has decreased substantially, and therefore has a small effect on this cross section. So the precise values of the non-perturbative functions play a negligible role, and the perturbative part of (13.81) governs the cross section even at $q_{hT} = 0$. However, it is not finite-order perturbation theory for the cross section that is relevant by itself. Perturbation theory for the exponent in $W(b_T)$, especially for γ_K, is critical.

When we increase q_{hT}, the cross section decreases. In the Bessel transform this occurs because of the oscillations in the Bessel function $J_0(q_{hT} b_T)$. The width of the q_{hT} peak can be estimated as the value of q_{hT} where the first half-oscillation of $J_0(q_{hT} b_T)$ fits under the peak of the $b_T W(b_T)$ curve. Let $b_{\text{peak}}(Q)$ be the position of the maximum of $b_T W(b_T)$. Then the half-width of the q_{hT} distribution is very roughly $1/b_{\text{peak}}(Q)$, which agrees with Figs. 14.13 and 14.14. The width evidently increases substantially with Q.

We obtain this broadened distribution by recoil against gluon emission into an increasing kinematic range. Notably there is soft, non-perturbative gluon radiation uniformly in the available rapidity range. Even at fairly low energy, as in Fig. 14.13(a), the width, around 1 GeV, is much larger than the naivest expectation (around 300 MeV) based on elementary ideas of Fermi motion of bound quarks in hadrons.

Finally the behavior for large q_{hT}, of the order of Q, is governed by the sharpest features in $W(b_T)$, which come from perturbative logarithms of b_T, with $b_T \sim 1/Q$.

14.5.4 Further issues

Although the above formalism has had substantial success, a long-standing problem has been to account for the measured angular distribution of the Drell-Yan process pairs (Badier *et al.*, 1981; Falciano *et al.*, 1986; Guanziroli *et al.*, 1988; Conway *et al.*, 1989). There is a substantial $\sin^2 \theta \cos(2\phi)$ term in addition to the $1 + \cos^2 \theta$ term that is expected from unpolarized $q\bar{q}$ annihilation. It should be noted that the measurement needs particular care, because of the effects of detector acceptance (Bianconi *et al.*, 2009).

Now the standard applications of factorization to the Drell-Yan process have assumed that quarks in an unpolarized hadron are themselves unpolarized. However, in Sec. 13.16 we saw that there is a transverse polarization of a quark correlated with its transverse momentum, as described by the Boer-Mulders function. Applying this to the annihilating quark and antiquark gives a non-zero contribution to the structure function that gives the apparent anomaly in the angular distribution.

A recent measurement of the angular dependence, together with separation of a Boer-Mulders term and conventional pQCD term for large transverse momentum can be found in Zhu *et al.* (2009). Another recent fit can be found in Lu and Schmidt (2010). See Barone, Melis, and Prokudin (2009) for a recent analysis of the Boer-Mulders function in SIDIS.

14.6 Calculations with initial-state partons

Applications of factorization for the Drell-Yan process involve hadronic incoming states, with the parton densities containing non-perturbative physics. However, calculations for the hard scattering are typically made starting from perturbative calculations of the Drell-Yan cross section with partonic beams. Given that factorization is valid independently of the nature of the beams, the hard scattering can be obtained by dividing by perturbative calculations of the parton densities in partonic targets. (After the expansion in powers of coupling, this formalism gives a subtractive calculation of the hard scattering: a Feynman-graph calculation of the cross section with contributions associated with parton densities subtracted off.)

When the partons are massless the calculations of the partonic cross section have mass divergences that are canceled by mass divergences in the subtracted parton-density terms, giving an IR-safe hard-scattering coefficient. We have seen examples of such calculations in Ch. 9.

Although it is common and calculationally simplest in QCD perturbative calculations to make all the partons massless, with the divergences being dimensionally regulated, this is not essential. The principles just described apply equally to calculations with all the partons given a mass. Then the massless limit need only be taken at the end of a calculation for the hard scattering.

Naturally, if, for example, one wishes to calculate production of a quark whose mass is large, then it is inappropriate to neglect its mass at any stage: the heavy quark mass in this situation is either comparable to or actually sets the large scale Q of the hard scattering. But for the present discussion, let us treat only a situation in which the quarks are light, of masses much less than Q.

An interesting issue arises when some but not all of the partons are given a mass. This is natural in QCD, since the gluon mass is required to be zero by non-abelian gauge invariance. Then one can perturbatively calculate a Drell-Yan cross section with incoming quarks which have a non-zero mass, while keeping the gluon mass exactly zero. This will regulate collinear divergences involving quarks, but will leave IR divergences. In an NLO calculation, these are much as in QED. But in higher order there will be collinear divergences associated with gluonic self-interactions.

In this situation the calculation has a danger of giving uncanceled IR divergences (Catani, Ciafaloni, and Marchesini, 1986) in a hard scattering, perhaps not for the Drell-Yan process which is completely inclusive in the hadronic final state, but for other processes. The problem is that IR divergences occur at all beam energies, even when the beams are non-relativistic, whereas the intricate cancellations needed to get factorization for the Drell-Yan process require the relativistic limit. The relativistic limit implied that influences of one incoming hadron on the other cannot travel fast enough to correlate the two active partons in such a way as to break factorization. So factorization by itself does not imply that all IR divergences cancel, but only those that are leading power in Q. One can imagine a divergence proportional to

$$\frac{m^2}{Q^2} \times \frac{1}{\epsilon},$$ (14.42)

where ϵ is the dimensional-regularization parameter. Should one count this as a power-suppressed correction because of the m^2/Q^2 factor, or is it infinite, because one should take the physical limit $\epsilon \to 0$, with quark masses non-zero? Of course, if one also had a massive gluon, then one would replace $1/\epsilon$ by something like $\ln(Q^2/m_g^2)$: the gluon mass would provide a physical IR cutoff. In that case, (14.42) would be unambiguously power-suppressed as $Q \to \infty$.

Of course, in real QCD, confinement should give a physical non-zero IR cutoff. But this is not present in pure perturbative calculations.

It has been proposed that when calculations are made with heavy quarks, whose mass is not always negligible with respect to Q, it would be a legitimate method to preserve heavy quark masses in the hard scattering, including the case of incoming quarks. The above argument indicates that this is a bad idea.

See also Aybat and Sterman (2009) for work on the cancellation of soft gluons when the initial state is partonic.

14.7 Production of hadrons

Our proof of factorization for the Drell-Yan process depended quite essentially on the cross section being completely inclusive in the hadronic part of the final state. So one can anticipate further complications if one wants to generalize the result to production of hadrons, e.g., $H_A + H_B \to H_C + X$ and $H_A + H_B \to H_C + H_D + X$, where the detected hadrons have large transverse momentum. The most common experimentally investigated case is where jets of large transverse momentum are measured in the final state.

It is easy to state factorization properties as obvious and natural generalizations of the ones already proved, e.g., (14.3). They involve a parton density for each incoming hadron, and a fragmentation function for each detected final-state hadron, all convoluted with a partonic hard-scattering cross section. Many examples of such factorization properties are in regular and successful phenomenological use, and there is an industry of calculating important higher-order corrections to the hard scatterings.

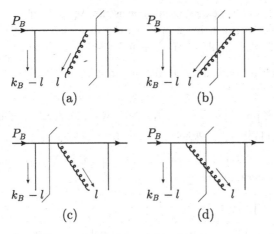

Fig. 14.15. Example of diagram for R: (a) and (b) with gluon vertex to left of final-state cut, (c) and (d) with vertex to right of cut.

But there is a relative lack of detailed proofs of factorization in these cases, to justify factorization from fundamental QCD. One of notable notable exception is Nayak, Qiu, and Sterman (2005). This work applies to factorization of a kind that uses ordinary integrated parton densities.

One can consider cross sections which are sensitive to partonic transverse momentum, for example $H_A + H_B \to H_C + H_D + X$ when the final-state hadrons H_C and H_D are close to back-to-back azimuthally. It is not too hard to formulate apparently suitable TMD factorization properties, as a natural generalization of those valid for the Drell-Yan and SIDIS processes.

However, unlike the Drell-Yan case, these TMD factorization properties appear to fail (Collins and Qiu, 2007; Rogers and Mulders, 2010). A failure of factorization in these areas in situations where it is intuitively plausible implies that there is a possibility of interesting new areas of QCD physics that are in need of investigation.

Exercises

14.1 This problem is about the Glauber region treatment of Secs. 14.4.2 and 14.4.3. Verify in a simple example, e.g., Fig. 14.15, that $\sum_{F_R \in \mathcal{R}(V)} R^{F_R}$ in (14.20) is independent of the choice V of where soft gluons are placed relative to the final-state cut. There is no need to assume any particular momentum region, but only that the plus components of the external momenta are integrated over; in Fig. 14.15, these components are k_B^+ and l^+.

For this example the result to be proved is that

$$\text{Graph (a)} + \text{Graph (b)} = \text{Graph (c)} + \text{Graph (d)}. \tag{14.43}$$

For the purposes of this exercise, you can assume the solid lines correspond to scalar fields, with the hadron being a color singlet that couples to the quark fields by a ϕ^3-type vertex. You should use Feynman gauge for the gluon.

14.2 Does the result in problem 14.1 continue to hold in a "physical gauge" like axial, light-cone or Coulomb gauge?

14.3 (***) Find and prove any extensions to the Ward-identity arguments of Ch. 11 that are needed to apply to the processes treated in this chapter.

14.4 (*) Repeat the calculations of Sec. 13.14, but for TMD parton densities instead of TMD fragmentation functions.

15

Introduction to more advanced topics

This book has covered many of the primary topics in perturbative QCD, with a focus on certain inclusive processes for which particularly systematic treatments are available. It should provide the reader with a sound conceptual framework for further study and research. However, hadronic interactions form a vast subject, and there is an enormous literature where perturbatively based methods have been applied.

This chapter gives a summary of a selection of important areas of further application of perturbative QCD.

One common theme, a prerequisite for actual perturbative calculations, is that the reactions have in some sense a controlling hard subprocess, occurring on a short distance scale, i.e., a distance scale significantly less than 1 fm, or, more-or-less equivalently, a momentum transfer significantly larger than the typical hadronic scale of a few hundred MeV.

Another recurring idea, perhaps the closest to a unifying motif, is the idea that one should try to separate (factor) phenomena on different scales of distance and momentum. This refers not just to scales of different virtuality, but also to a separation of phenomena at widely different rapidities. A characteristic here is that almost scattering processes examined in high-energy physics are ultra-relativistic. Thus time dilation and Lorentz contraction of fast-moving hadrons by themselves provide a wide range of distance scales. For example at the Tevatron collider we have proton and antiproton beams of energy almost 1 TeV. This allows the measurement of hard processes with momentum scales of several hundred GeV. Therefore distances as small as 10^{-3} fm can be probed. Now the intrinsic distance scale of phenomena in a proton is about 1 fm in its rest frame. So time dilation of the beams indicates that there are phenomena relevant to the collisions occurring on the much larger scale of 10^3 fm. Thus relevant distance scales span 6 orders of magnitude (the *square* of E/M). Such a big ratio allows for many simplifications and useful approximations, and not just those that directly impinge on the applicability of perturbative methods.

This train of thought leads to one common (but not universal) theme, that of light-front methods. Most systematically, one can represent the states of fast-moving hadronic systems in terms of their light-front wave functions. As we have seen throughout this book, one cannot take the elementary formulations of light-front quantization etc. literally; many of the basic ideas must be considerably distorted to be applied correctly in QCD. Nevertheless this area gives concepts and methodology that underlie much of the work.

The significance of light-front methods goes beyond that of perturbative applications to relatively short-distance phenomena. There is a close relation to phenomena in soft hadronic physics (Gribov, 1973, 2009). This is an area often characterized as the domain of Regge theory. Although Regge theory was extremely influential in the pre-QCD era, and although one can still see its effects on current research, there is not yet a properly established connection with QCD from first principles. This is an area that deserves more investigation now that QCD is a very mature subject.

Many of the topics listed in this chapter concern some of the most difficult parts of QCD. It is not surprising therefore that their justification from fundamental principles is not always sufficient. It is generally difficult for an outsider, even for one experienced in perturbative QCD, to acquire an full understanding of these areas from the published literature. Whether or not scepticism in any particular case is justified, I will leave to the future to decide.

15.1 Light-front wave functions and exclusive scattering at large momentum transfer

One natural application of hard-scattering methods is to elastic scattering at large momentum transfer. The classic early reference is Lepage and Brodsky (1980). Standard examples include elastic hadron-hadron scattering $H_A + H_B \rightarrow H_C + H_D$ at wide angle, and electromagnetic form factors of hadrons at large momentum transfer.

The standard methods of region analysis apply: Sec. 5.9.3. An obvious kind of region was shown in Fig. 5.34(a), where essentially the partonic content of each external hadron collapses to a small configuration at a single hard scattering. The wide-angle hadronic scattering is then controlled by a kinematically equivalent scattering of valence quarks from each hadron. If we follow the same logic as for inclusive scattering, the non-perturbative factors are light-front wave functions (with an integral over transverse momentum). They are obtained from matrix elements of light-front annihilation operators between a single hadron and the vacuum, e.g.,

$$\langle 0 | b_{k_1,\lambda_1} b_{k_2,\lambda_2} b_{k_3,\lambda_3} | P \rangle , \tag{15.1}$$

where the operators are as in Secs. 6.6 and 6.7. One expects the usual QCD complications, of course.

But because the hard-scattering subgraph has more external partons than in inclusive scattering, the cross sections fall with a higher power of the hard scale Q than corresponding inclusive cross sections. So it is hard to probe very large Q experimentally. In addition, this strong decrease allows the possibility of other regions contributing with either the same power law or a less-suppressed power. See Sec. 5.9.3 for a brief discussion of one example, the Landshoff process. See the citations to Lepage and Brodsky (1980) and Landshoff (1974) for subsequent work.

15.2 Exclusive diffraction: generalized parton densities

A related topic concerns exclusive processes in large-Q inelastic lepton-hadron scattering. We examined the leading regions for such processes in Sec. 5.3.6. Standard examples presented there were deeply virtual Compton scattering, double deeply virtual Compton scattering, and exclusive production of mesons. The hadronic parts of these reactions are $\gamma^*(q) + P \to \gamma + P$, $\gamma^*(q) + P \to \gamma^*(q') + P$, and $\gamma^* + P \to M + P$, respectively.

Experimentally, these processes are often investigated at small Bjorken x (where the cross section is largest), so they also take on the characteristics of diffractive scattering.

In the normal case that the momentum transfer from the target-hadron end is small, the appropriate factorization property uses what are called "generalized parton densities" (GPDs). These are defined exactly like parton densities, except that the hadronic matrix element is off-diagonal, (6.90). See Sec. 11.8 for a further discussion. In exclusive production of mesons, a light-front wave function of the meson is needed, the same quantity that appears in elastic scattering of the meson. See Diehl (2003) for a good review.

15.3 Small-x, BFKL, perturbative Regge physics

In DIS much work deals with the region of small x. There is considerable experimental data from the HERA collider, where the high center-of-mass energy allowed ep collisions to go to small x while maintaining Q in a perturbative region, e.g., Q of a few GeV with x as small as 10^{-5}. The standard treatment of DIS involves the limit of large Q at fixed x, so the small-x regime introduces another large ratio in addition to the ratio of the hard scale to the hadron mass, Q/M.

In the small-x region, the ideas of Regge theory become relevant. Regge theory concerns asymptotic behavior where s is large and momentum transfer is fixed. This includes the total hadronic cross section at large s.

Now DIS structure functions correspond to a cross section for scattering of a virtual photon on a hadron: $\gamma^* P \to X$. At small x, the mass of both the photon and the hadron are much less than their center-of-mass energy, which is $Q\sqrt{(1-x)/x}$. When Q is in a perturbative region, one can hope that Regge theory can be usefully approximated by perturbative methods. Investigations of a Regge limit in perturbation theory for non-abelian gauge theories led to the equation of Balitsky, Fadin, Kuraev, and Lipatov (BFKL) (Fadin, Kuraev, and Lipatov, 1975; Balitsky and Lipatov, 1978). For a review, see Lipatov (1997).

This and a number of closely allied developments have had many applications, in situations where a Regge limit is appropriate. If the DGLAP equation is regarded as governing the Q dependence of DIS structure functions and parton densities, then the BFKL equation governs the x dependence, at small x.

For partonic scattering at high energy and small angle, the BFKL equation gives a ladder structure that is very similar to the multiperipheral model we mentioned in Sec. 14.3. However, the actual Feynman graphs that give the leading behavior are gauge dependent and need not be actual ladder graphs. Primarily the derivations use the leading-logarithm method, and therefore concern the situation where the gluons are strongly ordered in rapidity. But important work concerns NLO corrections.

There is interesting work by Balitsky (e.g., Balitsky, 1999), who relates the BFKL equation to the evolution of Wilson-line matrix elements with respect to the rapidity of the Wilson-line directions; thus his work is related to our treatment of TMD functions in Ch. 13.

One characteristic of the BFKL equation is that it implies that its approximation to the pomeron has an intercept well above unity. The pomeron was originally characterized as the Regge exchange that gives the highest power of energy in elastic hadronic scattering. Its intercept $\alpha(0)$ gives a total hadron-hadron cross section proportional to $s^{\alpha(0)-1}$. But the Froissart bound requires that the cross section rise at most like $\ln^2 s$. Phenomenologically hadronic total cross sections do rise slowly. Thus something with an intercept far above unity cannot be the true pomeron. However, there does appear to be a transition in DIS between soft pomeron behavior at low Q, with approximately constant $\gamma^* p$ cross sections, and a "hard pomeron" behavior at higher Q, with a substantial rise with energy (e.g., H1 Collaboration, 2010)

Related issues concern the CCFM equation (Ciafaloni, 1988; Catani, Fiorani, and Marchesini, 1990a, b; Marchesini, 1995) for parton densities, etc. at small x.

15.4 Resummation, etc.

The basic method of using perturbation theory in QCD for a quantity with a large momentum scale Q is to use the RG to set the renormalization mass μ of order Q. This removes large logarithms of Q. But in many cases there are other parameters which can also give large logarithms. One way of viewing the problem is to observe that the quantity being calculated depends on multiple momentum scales rather than just Q.

One example is the hard-scattering coefficient in ordinary "collinear" factorization for the Drell-Yan process when $q_T \ll Q$. Among many other examples are processes at small Bjorken x (Sec. 15.3), and at large x (Sec. 15.8 below).

The most fundamental method of dealing with such situations is to formulate an appropriately improved factorization theorem, such as we did using TMD factorization for the Drell-Yan cross section in Ch. 14. After that the various perturbative coefficients are all single-scale quantities.

Another very common method is that of resummation. There one analyzes the source of the large logarithms. It is often not too hard to determine the leading logarithms to all orders of perturbation theory, even without a more complete treatment. This avoids exact Feynman-graph calculations at very high order. Then one sums the large higher-order corrections.

The vast literature on this subject can be sampled by searching for papers with titles containing "resummation" or "resummed".

Resummation is at its most useful when the logarithms are not too large, since it can provide an efficient way to improve the accuracy of perturbative calculations. One important example is in the use of resummed calculations of jet shapes (Gehrmann, Luisoni, and Stenzel, 2008) in e^+e^- annihilation to obtain accurate estimates of the strong coupling (Bethke *et al.*, 2009).

The method gets much harder to justify when the logarithms are large. For example, in the Drell-Yan process at small transverse momentum, the errors in the approximations giving collinear factorization include terms that are a power of M/q_T. When the transverse momentum is of order a hadronic mass, the derivation does not apply. TMD factorization solves this problem, with the outcome that more non-perturbative information is needed in the transverse momentum distributions of partons and of soft-gluon emission. In the intermediate region $M \ll q_T \ll Q$, the full TMD factorization property can be used to derive a resummation formula. In the version of TMD factorization given in (13.81), a resummation result can be obtained by omitting the non-perturbative factors in the fourth and fifth lines.

15.5 Methods for efficient high-order calculations

In many realistic applications of perturbative QCD, calculations of high-order graphs are needed. For the LHC, calculations of parton-parton scattering with many partons in the final state are used, preferably at one-loop order. Examples of the calculations are in Berger *et al.* (2009).

It is readily evident that such calculations are very complex, particularly when performed in the most direct way from the standard Feynman rules. A 3-gluon vertex has 6 terms, so that a graph with n such vertices has 6^n terms. Straightforward calculations by hand become very lengthy or impractical. Of course, intensive use of computers helps. It also helps if calculations are restricted to massless on-shell amplitudes as much as possible.

But it is also observed that the final results of a calculation are often much simpler than intermediate results, and certainly much simpler than one expects from the complications in individual graphs. This suggests that there are much better methods. See Bern, Dixon, and Kosower (2007) for a review of much of the work in this direction, with further references.

15.6 Monte-Carlo event generators

The analysis of the regions for Feynman graphs for processes with a hard scattering gives much more information on the detailed structure of the final state than we used in factorization theorems for inclusive cross sections. A contrasting approach is provided by Monte-Carlo event generators, e.g., PYTHIA (Sjostrand, Mrenna, and Skands, 2006, 2008) and HERWIG (Bahr *et al.*, 2008). These are computer programs which simulate actual collisions. That is, they generate complete events with a distribution that is intended to be a useful approximation to the distribution of events in actual collisions.

Modern collider experiments generate events with many final-state particles, and the detectors are sensitive to most of the final state. The acceptance and efficiency of the detectors is quite complicated, and the signatures of many interesting signals (e.g., the Higgs particle) involve properties of whole groups of final-state particles. Therefore understanding the nature of a physics signal is greatly assisted by having a realistic simulation of the complete final state. Monte-Carlo event generators are therefore an essential tool in the analysis of experimental data in high-energy physics, not to mention the planning of future experiments.

Event generators also evade another problem. This is that the number of Feynman graphs rises with the order N of the Feynman graph roughly like $N!$. The difficulty of computing each single graph also rises with the order. Although modern methods ameliorate this somewhat, there is a formidable computational problem in directly computing production of final states with many particles. The situation is worse in QCD because straightforward perturbation theory is not useful without interesting reorganizations: factorization, renormalization group, etc. A Monte-Carlo event generator provides an approximation to the production of N particles that uses computational resources linear in N, instead of its factorial, thereby giving a dramatic improvement over unassisted perturbation theory.

The price is, of course, the approximation and the difficulty of justifying it.

To understand how the methods used by the event generators arise, consider our treatment in Sec. 8.9 of factorization in a non-gauge theory. There the dominant structures were generalized ladder graphs. We can extend these ideas to analyze the structure of the jets in the final state, obtaining a structure of ladders within ladders. If we take, as is appropriate, a fixed order for each rung, we have a small number of graphs in each order, and the number of rungs is proportional to N. The structure readily maps to the linear-in-N structure in an event generator. The use of Monte-Carlo methods, i.e., probabilistic methods, is the most sensible for numerical calculations of high-dimensional integrals and maps perfectly onto how data appears in a scattering experiments.

In QCD, the ladder structure only arises after a sum implemented by Ward identities from graphs with non-local attachments of gluons. The kinematics of final states, with soft gluons filling in rapidity gaps, is also much more complicated than in a non-gauge theory.

The theory of the event generators (see Sjostrand, Mrenna, and Skands, 2006; Bahr *et al.*, 2008; Sjostrand, 2009) is based on the ideas used in ordinary factorization theorems for inclusive processes. But a full justification, which I am not sure really exists, needs to go much further. One symptom of this is in the kinematic approximations used in deriving factorization for inclusive processes. At various points we change the kinematics of partons going into the final state from their actual values, but in such a way that the inclusive cross section is not affected (at leading power). But this is not adequate for an event generator where a complete description of the final state is to be given. An event that does not obey conservation of 4-momentum is not useful in this context. Prescriptions are needed to correct this (Bengtsson and Sjöstrand, 1988), and these do not fully match how inclusive factorization theorems are derived.

Furthermore, in hadron-hadron collisions, generating complete final states goes beyond a situation in which factorization in its basic form is valid. Event generators incorporate modelling of the soft final state and this can be regarded as a model of the spectator-spectator interactions that we examined (in the context of another very simple and naive model) in Sec. 14.3.

There has naturally been much work on Monte-Carlo event generators that I cannot review here. They represent an interesting way of combining the results of perturbative calculations with other elements including modelling of non-perturbative physics to give a very useful approximation to real QCD.

15.7 Heavy quarks

At various points in this book, I have mentioned the issues that arise when heavy quark masses are not small compared with the hard-scattering scale Q. Many situations can be dealt with by minor modifications of the standard factorization method; see for example Krämer, Olness, and Soper (2000).

But there are other situations that require different techniques. One of the most important is the analysis of the decays of hadrons containing heavy quark constituents, notably *B* mesons. This is the domain of heavy-quark effective theory (HQET). An account of HQET is found in Manohar and Wise (2000).

15.8 Large *x*

Limitations on the validity of a basic factorization theorem often arise near kinematic limits. An important case is DIS at $x \to 1$. There the spectator part of a typical leading region becomes soft instead of collinear, and therefore indicates that a change in the analysis is needed. Essentially the same considerations apply to any other inclusive process in a kinematic region where the initiating partons of a conventional hard scattering must have $x \to 1$. Similar issues arise in fragmentation as $z \to 1$. Because of the restricted kinematics of the spectator system, more accurate treatment of the kinematics is needed than in the conventional factorization.

Cross sections decrease quite rapidly as $x \to 1$ because parton densities decrease roughly as $(1 - x)^3$ or a higher power. This decrease affects the accuracy of conventional factorization methods. One indication of this is in the NLO correction (9.54) for DIS, where the plus distribution implements a cancellation between real and virtual gluon emission. Where the parton densities decrease rapidly this cancellation becomes inaccurate, giving large logarithms of $1 - x$.

Recent work can be traced from Almeida, Sterman, and Vogelsang (2009).

15.9 Soft-collinear effective theory (SCET)

In recent years a new approach to perturbative QCD has been developed under the name soft-collinear effective theory (SCET) (Bauer *et al.*, 2001; Bauer and Stewart, 2001). See Fleming (2009) for a recent overview. Historically SCET arose as a generalization of heavy-quark effective theory; see Sec. 15.7.

The overall philosophy of SCET is like that of the Wilsonian renormalization group. This is to integrate out certain ranges of momentum modes for the fields of QCD and to replace them with effective fields. In SCET momentum space is divided into many bins in each of which the integrating-out is to be done. A problem that needs to be addressed in any such method is how to deal with a momentum that lies just outside a boundary of an integrated-out region.[1] There is no small parameter to expand in, unlike the case of momenta far from the boundary.

In the Wilsonian RG this problem is overcome by using an infinite set of operators. But this rather obscures the underlying simplicity of the situation, where one has a simple factorization of coefficients times a limited set of operators.

---

[1] Compare the discussion in the first few paragraphs of Sec. 13.12.

In reality, the integrating-out in SCET is performed by integrating over all momenta, with subtractions to enforce the region conditions. This is similar to what was done in this book in Ch. 10. However, I have not been able to penetrate the SCET literature to properly understand its rationale. I just refer the reader to the literature cited above.

15.10 Higher twist: power corrections

In deriving factorization we made approximations that used the leading power of an expansion in small variables like masses relative to a hard scale Q. It is natural to ask what can be done with non-leading powers.

The basic techniques do apply to non-leading powers. In fact, the earliest of the factorization theorems, the operator product expansion (OPE), does treat all powers, leading and non-leading, in a uniform formalism. The OPE (Collins, 1984, ch. 10) expresses a suitable matrix element in a limit of large *Euclidean* momentum q as a sum of q-dependent coefficients times q-independent operator matrix elements, e.g.,

$$\int d^4x \, e^{iq\cdot x} \, \langle P|j(x)j(0)|P\rangle = \sum_i C_i(q) \, \langle P|\mathcal{O}_i|P\rangle . \tag{15.2}$$

The power law for the q dependence is controlled by the dimension of the operators.[2]

When the OPE is applied to moments of DIS structure functions, the normal leading power corresponds to operators $\mathcal{O}_i$ that obey

$$\text{dimension} - \text{spin} = 2. \tag{15.3}$$

This quantity is called twist, and non-leading powers have a higher value of twist. It has therefore become a standard jargon to use "higher twist" to refer to any power-suppressed correction. Leading-power factorization for inclusive processes is then labeled "twist-2"; integer moments of integrated parton densities are exactly matrix elements of twist-2 operators.

For work on higher-twist corrections to factorization see Qiu and Sterman (1991b).

But the vast majority of applications avoid the use of higher-twist corrections, trying to stay in kinematic regions where the leading-power formalism is sufficient. There are several reasons.

One is that the relevant generalizations of parton densities use multiparton operators, and these depend on more than one fractional momentum variable. The more non-leading the power, the larger the number of variables that is needed. But it is hard to extract such a multivariable function from data. This contrasts with the twist-2 case, where, in the parton-model approximation, DIS structure functions are simple linear combinations of quark densities.

One can only do better if one is in a special situation where the non-leading power terms are particularly simple.

---

[2] Here I ignore the effects of anomalous dimensions.

A second reason for not using power corrections is a fundamental limitation on the accuracy of perturbative calculations in QCD. Consider the perturbation series for an IR-safe quantity

$$F = \sum_{n=0}^{\infty} c_n \alpha_s(Q)^n,$$ (15.4)

with purely numeric coefficients. Suppose, as is the general expectation, that this is an asymptotic series with large-order behavior

$$c_n \alpha_s^n \sim (a n \alpha_s)^n \text{ as } n \to \infty.$$ (15.5)

We estimate the error in a truncated perturbative expansion by the first term omitted, as is appropriate for an asymptotic series. Then the minimum error in a perturbative calculation is the smallest term in the series. So, (15.5) implies that the minimum error is from the term with

$$n \propto \frac{1}{a \alpha_s(Q)}.$$ (15.6)

Then the minimum error itself is roughly

$$\exp\left(-\frac{\text{constant}}{\alpha_s(Q)}\right) \sim \exp\left(-\text{constant} \ln Q^2\right) = O(Q^{-p}),$$ (15.7)

for some positive constant p. Any higher-twist correction with a more negative power of Q is smaller than the minimum error in the perturbative calculation of the leading-twist term. It is therefore phenomenologically useless.

Another severe complication arises for higher-twist corrections in hadron-hadron collisions. Factorization has independent parton densities for each beam hadron. To obtain this independence, we needed a cancellation of interactions between the two hadrons. The proof of the cancellation, Sec. 14.4, relied on causality in the ultra-relativistic limit. In a non-relativistic situation the active partons could get correlated before the hard scattering. Such effects generally contribute to higher-twist corrections. Therefore initial-state interactions require that the non-perturbative functions in corrections of sufficiently higher twist are properties of the whole two-hadron state, rather than being multiparton correlation functions in individual hadrons.

This issue does not affect terms suppressed by $1/Q$ and $1/Q^2$ relative to the leading-power terms (Qiu and Sterman, 1991a, b). So twist-3 and twist-4 terms can be investigated in a generalized factorization framework.

Appendix A

Notations, conventions, standard mathematical results

In this appendix, I have collected the definitions of notations and conventions that I use. In addition, I have collected some standard numerical results and formulae that are frequently used in practical QCD calculations. For normalization conventions and the like, I generally follow the conventions of the Particle Data Group (PDG) (Amsler *et al.*, 2008).

In some cases it may be quite difficult to discover some of these formulae in the literature, and the reader wishing to check them may find it easier to rederive them than do a literature search.

A.1 General notations

1. I use $\stackrel{\text{def}}{=}$ to denote the definition of a symbol, as in $Q^2 \stackrel{\text{def}}{=} -(l - l')^2$.
2. I use $\stackrel{\text{prelim}}{=}$ to indicate a preliminary early version of a definition, that is to be corrected later.

$$\text{Quantity} \stackrel{\text{prelim}}{=} \text{preliminary candidate definition.} \tag{A.1}$$

3. I use $\stackrel{?}{=}$ to indicate an incorrect result: $F_{\text{L}} \stackrel{?}{=} 0$. Typically, this represents a result true only in some simplified situation.
4. I use $[A, B]_+$, with a subscript $+$, to denote an anticommutator: $[A, B]_+ \stackrel{\text{def}}{=} AB + BA$.
5. A hat over a symbol, e.g., $\hat{k}$, generally indicates that some parton-type approximator has been applied. It is also used to denote a hard scattering or a kinematic variable at a partonic level.
6. Generally a tilde, as in $\tilde{f}$, indicates a Mellin or a Fourier transform. It has some other rarer uses, e.g., the wave function renormalization $\tilde{Z}$ for the Faddeev-Popov ghost field.
7. In Feynman graphs, the normal association of line types is:

⟶	quark or lepton	⟹	Wilson line
⟿	gluon	---▶---	scalar
······▶······	Faddeev-Popov ghost	∿∿∿	photon, W, Z

A.2 Units, and conversion factors

1. Generally I use units with $\hbar = c = \epsilon_0 = 1$, with energy in GeV. To convert to standard units, factors of $\hbar$, c, etc. need to be inserted according to the demands of dimensional analysis, after which the following conversion factors are useful.
2. $\hbar c = 0.197\,327\,0\,\text{GeV fm}$.
3. $(\hbar c)^2 = 0.389\,379\,3\,\text{GeV}^2\,\text{mbarn}$, where 1 barn $= 10^{-28}\,\text{m}^2$.

4. The fine-structure constant is $\alpha = e^2/(4\pi) \simeq 1/137.036$, with e being the size of the charge of the electron. In SI units, $\alpha = e^2/(4\pi\hbar c\epsilon_0)$.

A.3 Acronyms and abbreviations

Common acronyms and abbreviations are:

1PI	one-particle irreducible
2PI	two-particle irreducible
ACOT	Aivazis-Collins-Olness-Tung
BFKL	Balitsky-Fadin-Kuraev-Lipatov
BJL	Bjorken-Johnson-Low
BNL	Brookhaven National Laboratory
BRST	Becchi-Rouet-Stora-Tyutin
CCFM	Catani-Ciafaloni-Fiorani-Marchesini
CERN	European Organization for Nuclear Research
CKM	Cabibbo-Kobayashi-Maskawa
CM	center-of-mass
CS	Collins-Soper
CSS	Collins-Soper-Sterman
CWZ	Collins-Wilczek-Zee
DDVCS	double deeply virtual Compton scattering
DESY	Deutsches Elektronen-Synchrotron
DGLAP	Dokshitzer-Gribov-Lipatov-Altarelli-Parisi
DIS	deeply inelastic scattering
DVCS	deeply virtual Compton scattering
DY	Drell-Yan
ELO	extended leading order
ENLO	extended next-to-leading order
FNAL	Fermi National Accelerator Laboratory
GPD	generalized parton density
HERA	Hadron-Electron Ring Accelerator (at DESY)
HQET	heavy-quark effective theory
IR	infra-red
KLN	Kinoshita-Lee-Nauenberg
LEET	low-energy effective theory
LEP	Large Electron Positron collider (at CERN)
LHC	Large Hadron Collider (at CERN)
l.h.s.	left-hand side (of equation)
LLA	leading-logarithm approximation
LO	leading order
LSZ	Lehmann-Symanzik-Zimmermann
MNS	Maki-Nakagawa-Sakata
$\overline{\text{MS}}$	modified minimal subtraction (renormalization scheme)
NLO	next-to-leading order
NNLO	next-to-next-to-leading order
OPE	operator product expansion
pdf	parton distribution function (or parton density function)
PDG	Particle Data Group
pQCD	perturbative QCD
p.s.c.	power-suppressed correction
PSS	pinch-singular surface

QCD   quantum chromodynamics
QED   quantum electrodynamics
QFT   quantum field theory
RG   renormalization group
RGE   renormalization-group equation
RHIC   Relativistic Heavy Ion Collider (at BNL)
r.h.s. right-hand side (of equation)
SCET   soft-collinear effective theory
SIDIS   semi-inclusive deeply inelastic scattering
SLAC   Stanford Linear Accelerator Center
SM   Standard Model
TMD   transverse momentum dependent
Tr   trace
UV   ultra-violet
VEV   vacuum expectation value

A.4 Vectors, metric, etc.

1. 3-vectors are written in boldface: $\boldsymbol{x}$.
2. In ordinary coordinates, Lorentz 4-vectors are written as, e.g., $x^\mu = (t, x, y, z) = (t, \boldsymbol{x})$, with a right-handed coordinate system.
3. The metric is $g_{\mu\nu} = \operatorname{diag}(1, -1, -1, -1)$.
4. The fully antisymmetric tensor $\epsilon_{\kappa\lambda\mu\nu}$ is normalized to $\epsilon_{0123} = 1$. With raised indices it has the opposite sign: $\epsilon^{0123} = -1$.
5. Light-front coordinates (App. B) are defined by $x^\pm = (t \pm z)/\sqrt{2}$. A vector is written $x^\mu = (x^+, x^-, \boldsymbol{x}_{\mathrm{T}})$.
6. The 2-dimensional antisymmetric tensor ϵ_{ij} obeys $\epsilon_{12} = \epsilon^{12} = 1$.
7. Rapidity for a 4-momentum is defined by $y \overset{\text{def}}{=} \frac{1}{2} \ln \left| \dfrac{p^+}{p^-} \right|$.
8. I make a clear distinction between contravariant vectors, with upper indices, and covariant vectors, with lower indices. See App. B for further details.
9. An on-shell momentum $p^\mu = (E, \boldsymbol{p})$ for a particle of mass m obeys $p^2 = m^2$, and so $E = E_p \overset{\text{def}}{=} \sqrt{\boldsymbol{p}^2 + m^2}$.
10. Hence for an on-shell particle

$$p^\pm = e^{\pm y}\sqrt{(p_{\mathrm{T}}^2 + m^2)/2}, \tag{A.2}$$

when the transverse momentum is $\boldsymbol{p}_{\mathrm{T}} = (p^1, p^2)$.

A.5 Renormalization group (RG)

1. I consistently write renormalization group equations (RGEs) in terms of a derivative with respect to $\ln \mu$.
2. Then the anomalous dimension γ_G of a quantity G is defined as

$$\gamma_G = -\frac{\mathrm{d} \ln G}{\mathrm{d} \ln \mu}. \tag{A.3}$$

Note the minus sign. This corresponds to the natural use of the term "anomalous dimension" where there is a fixed point in the coupling.

3. However, certain quantities do not have the minus sign that might otherwise be expected, notably γ_m in (3.48), β in (3.44) etc., and the DGLAP kernels in (8.30).

4. Furthermore, the definitions of β and the DGLAP kernels are conventionally made in terms of derivatives with respect to $\ln \mu^2$, so our definitions in terms of $d/d \ln \mu$ acquire factors of half.

A.6 Lorentz, vector, color, etc. sub- and superscripts

Generally, the symbols for indices of various kinds are taken from different ranges of letters:

1. Lorentz: μ, etc.
2. 3-vector: i, j, etc.
3. Dirac: ρ, etc.
4. Color, in adjoint representation: α, etc.
5. Color, in fundamental representation: a, etc.
6. Flavor, in adjoint representation: A, etc.
7. Flavor, in fundamental representation: f, etc.
8. Symbols for momenta tend to be taken from the list k, l, p, q, etc.
9. Symbols for coordinates tend to be from the end of the roman alphabet: x, y, z.

Note that there are so many symbols needed that it is not always possible to be consistent. Also symbols may be overloaded: e.g., a common sub- or superscript index (notably e, i, ρ, and δ) may have a different, standardized meaning when not used as a sub- or superscript.

A.7 Polarization and spin

Note: there is no agreement in the literature on the normalization of quantities defined in this section.

1. The Pauli-Lubański (Lubański, 1942a, b) spin vector is the operator

$$W_\mu \stackrel{\text{def}}{=} \frac{1}{2} \epsilon_{\mu\alpha\beta\gamma} J^{\alpha\beta} P^\gamma, \tag{A.4}$$

where P^γ is the momentum operator, and $J^{\alpha\beta}$ are the generators of the Lorentz group, normalized to obey commutation relations

$$[J^{\mu\nu}, J^{\alpha\beta}] = i\left(-g^{\mu\alpha} J^{\nu\beta} + g^{\nu\alpha} J^{\mu\beta} + g^{\mu\beta} J^{\nu\alpha} - g^{\nu\beta} J^{\mu\alpha}\right), \tag{A.5}$$

$$[J^{\mu\nu}, P^\alpha] = i\left(-g^{\mu\alpha} P^\nu + g^{\nu\alpha} P^\mu\right), \tag{A.6}$$

$$[P^\alpha, P^\beta] = 0. \tag{A.7}$$

2. The most general state – pure or mixed – of a particle of momentum p can be written in terms of a spin density matrix $\rho_{\alpha\alpha'}$, with α and α' being labels for the possible helicities of the particle.[1] The expectation value of an operator in such a state is

$$\langle p, \rho | \text{ op } | p, \rho \rangle \stackrel{\text{def}}{=} \sum_{\alpha, \alpha'} \rho_{\alpha, \alpha'} \langle p, \alpha' | \text{ op } | p, \alpha \rangle. \tag{A.8}$$

The basis states $|p, \alpha\rangle$ have definite momentum p and helicity α. The density matrix ρ is Hermitian, it has trace unity, and all its eigenvalues are non-negative. An unpolarized state of a particle of spin s has $\rho_{\alpha,\alpha'} = \delta_{\alpha,\alpha'}/(2s + 1)$.
3. Helicity is a particle's spin angular momentum projected on its direction of motion. Thus for a spin-$\frac{1}{2}$ particle its possible values are $\pm\frac{1}{2}$.

---

[1] Another basis could be chosen for the spin states, but the helicity basis is most convenient for our purposes.

4. The helicity basis states are simultaneous eigenvectors of the momentum operators and a suitable projection of the Pauli-Lubański vector.

5. To specify a general spin state of a spin-$\frac{1}{2}$ particle, it is also possible to use a Bloch vector $\boldsymbol{b}$, which is a real-valued 3-vector obeying $|\boldsymbol{b}| \leq 1$. The correspondence to a 2×2 density matrix is

$$\rho = \frac{1}{2}(1 + \boldsymbol{b} \cdot \boldsymbol{\sigma}). \tag{A.9}$$

6. For a spin-$\frac{1}{2}$ particle moving in the $+z$ direction, we write the Bloch vector as

$$\boldsymbol{b} = (\boldsymbol{b}_{\mathrm{T}}, \lambda). \tag{A.10}$$

Here λ is twice the average helicity of a state, and $\boldsymbol{b}_{\mathrm{T}}$ is twice the average transverse spin. We call these normalized helicity and transverse spin; their maximum values are unity.

7. The spin vector S^μ of a single-particle state is twice the expectation value of the Pauli-Lubański vector:

$$S^\mu = 2 \langle \psi(p), \rho | W^\mu | \psi(p), \rho \rangle. \tag{A.11}$$

The factor of 2 is to agree with a standard normalization (Amsler *et al.*, 2008) of S^μ. Here $|\psi(p), \rho\rangle$ denotes a normalized state whose momentum is closely centered on p, and whose helicity density matrix is ρ.

8. In the rest frame of a spin-$\frac{1}{2}$ particle, the Bloch vector corresponds exactly to the Bloch vector concept in non-relativistic spin physics, and $S^\mu = M(0, \boldsymbol{b})$. Thus S^μ is a Lorentz-covariant generalization of the Bloch vector.

9. If the particle is moving in the z direction with 4-momentum $p = (p^0, 0, 0, p^z)$, the spin and the Bloch vectors are related by

$$S = \left(S^0, S^x, S^y, S^z\right) = \left(\lambda p^z \operatorname{sign}(p^z), M b_{\mathrm{T}}^x, M b_{\mathrm{T}}^y, \lambda p^0 \operatorname{sign}(p^z)\right). \tag{A.12}$$

The factors of $\operatorname{sign} p^z$ show that the $(\boldsymbol{b}_{\mathrm{T}}, \lambda)$ representation is not ideal for a non-relativistic particle. But the factor of M with the transverse components shows that the spin vector cannot correctly represent the general spin state of a massless spin-$\frac{1}{2}$ particle.

10. For a massive spin-$\frac{1}{2}$ particle of definite momentum p, the most general spin state is determined by the spin vector S^μ. For a spin-$\frac{1}{2}$ particle of mass M, the spin vector obeys
    (a) $S \cdot p = 0$.
    (b) For a general state $0 \geq S \cdot S \geq -M^2$.
    (c) For a pure state $S \cdot S = -M^2$.
    The helicity density matrix can be deduced from the spin vector S, and therefore we also write the matrix element in (A.8) as

$$\langle p, S| \operatorname{op} |p, S \rangle = \langle p, \rho(S)| \operatorname{op} |p, \rho(S) \rangle. \tag{A.13}$$

A.8 Structure functions

Definitions of structure functions for various processes are as follows:

1. F_1, F_2, g_1, and g_2 for electromagnetic DIS, in (2.20).
2. For unpolarized weak interaction DIS, in (7.3).
3. For one-particle-inclusive $e^+ e^-$ annihilation, in (12.5).
4. For two-particle-inclusive $e^+ e^-$ annihilation, in (13.9).
5. For Drell-Yan, see Lam and Tung (1978); Mirkes (1992); Ralston and Soper (1979); Donohue and Gottlieb (1981).

A.9 States, cross sections, integrals over particle momentum

1. The normalization of single particle states is

$$\delta_{pp'} \stackrel{\text{def}}{=} \langle p' | p \rangle = (2\pi)^3 2E_p \, \delta^{(3)}(p - p')$$
$$= (2\pi)^3 2p^+ \, \delta(p^+ - p'^+) \delta^{(2)}(p_T - p'_T)$$
$$= (2\pi)^3 2 \, \delta(y - y') \delta^{(2)}(p_T - p'_T). \tag{A.14}$$

In the last two lines, light-front coordinates and rapidity were used, as defined in Sec. A.4.

2. The Lorentz-invariant integral over particle momentum is

$$\sum_p \cdots \stackrel{\text{def}}{=} \int \frac{d^3 p}{(2\pi)^3 2E_p} \cdots = \int \frac{dp^+ d^2 p_T}{2p^+(2\pi)^3} \cdots = \int \frac{dy \, d^2 p_T}{2(2\pi)^3} \cdots \tag{A.15}$$

Notice that the formula with ordinary Cartesian coordinates is explicitly dependent on the particle mass, in E_p, but the formulae with light-front coordinates or rapidity are not.

3. The differential cross section for a $2 \to n$ process with incoming momenta p_1 and p_2, and outgoing momenta $q_1, \ldots, q_n$ is

$$d\sigma = (2\pi)^4 \delta^{(4)}\left(p_1 + p_2 - \sum_j q_j\right) \prod_{j=1}^n \frac{d^3 q_j}{(2\pi)^3 2E_{q_j}} \frac{|\mathcal{M}(p_1, p_2; q_1, \ldots, q_n)|^2}{4\sqrt{(p_1 \cdot p_2)^2 - m_1^2 m_2^2}}. \tag{A.16}$$

The matrix element $\mathcal{M}$ is normalized so that it corresponds to an amputated, on-shell, connected Green function (supplemented by residue factors from the LSZ reduction formula beyond tree approximation), with the overall $(2\pi)^4 \delta^{(4)}(p_1 + p_2 - \sum_j q_j)$ factor for momentum conservation removed. See Sterman (1993) for details.

4. The integral over "final-state phase space" is defined by

$$\int dfsps \cdots = \prod_{j=1}^n \int \frac{d^3 q_j}{(2\pi)^3 2E_{q_j}} (2\pi)^4 \delta^{(4)}\left(p_1 + p_2 - \sum_j q_j\right) \cdots \tag{A.17}$$

A.10 Dirac, or gamma, matrices

Here I summarize results on Dirac matrices. They can be gleaned from a standard QFT textbook. When there are competing conventions, I normally follow Sterman (1993).

1. The anticommutator is $[\gamma^\mu, \gamma^\nu]_+ = 2g^{\mu\nu} I$, where I is a unit matrix.
2. The hermiticity relation is $(\gamma^\mu)^\dagger = \gamma_\mu$.
3. $\gamma_5 \stackrel{\text{def}}{=} i\gamma^0\gamma^1\gamma^2\gamma^3 = \frac{1}{4!}i\gamma^\kappa\gamma^\lambda\gamma^\mu\gamma^\nu \epsilon_{\kappa\lambda\mu\nu}$, where $\epsilon_{\kappa\lambda\mu\nu}$ is the totally antisymmetric tensor obeying $\epsilon_{0123} = 1$.
4. In the antisymmetric combination $\sigma^{\mu\nu} \stackrel{\text{def}}{=} \frac{i}{2}[\gamma^\mu, \gamma^\nu]$ only 6 cases are independent, in 4 space-time dimensions.
5. When the normal space-time dimension is 4, the dimensionally regulated Dirac matrices (in $n = 4 - 2\epsilon$ space-time dimensions) are normalized to have $\text{Tr } I = 4$ for all n.
6. The contraction of γ^μ and a vector is written $\not{v} \stackrel{\text{def}}{=} \gamma^\mu v_\mu$.
7. The Dirac conjugate of a matrix is defined by $\overline{\Gamma} \stackrel{\text{def}}{=} \gamma^0 \Gamma^\dagger \gamma^0$. The basic matrices obey $\overline{\gamma^\mu} = \gamma^\mu$ and $\overline{\gamma_5} = -\gamma_5$.

8. Useful identities:

$$\text{Tr odd number of } \gamma^\mu s = 0, \tag{A.18}$$

$$\text{Tr} \, \gamma^\mu \gamma^\nu = 4g^{\mu\nu}, \tag{A.19}$$

$$\text{Tr} \, \gamma_5 \gamma^\kappa \gamma^\lambda \gamma^\mu \gamma^\nu = 4i\epsilon^{\kappa\lambda\mu\nu} = -4i\epsilon_{\kappa\lambda\mu\nu}, \tag{A.20}$$

$$\gamma^\mu \gamma_\mu = (4 - 2\epsilon)I, \tag{A.21}$$

$$\gamma^\mu \gamma^\nu \gamma_\mu = -(2 - 2\epsilon)\gamma^\nu. \tag{A.22}$$

9. In 4 space-time dimensions, Dirac matrices are 4×4, and a 16-dimensional basis for them is given by $1, \gamma^\mu, \sigma^{\mu\nu}, \gamma^\mu \gamma_5, \gamma_5$.

10. Thus a general 4×4 matrix M can be written as

$$\Gamma = S + \gamma_5 P + \gamma_\mu V^\mu + \gamma_\mu \gamma_5 A^\mu + \frac{1}{2}\sigma_{\mu\nu} T^{\mu\nu}, \tag{A.23}$$

where we assume we are in 4 space-time dimensions. (Otherwise generalization is needed.) If Γ obeys the normal Lorentz-transformation properties of a matrix on Dirac spinor space, then the coefficients S, P, V^μ, A^μ, and $T^{\mu\nu}$ have respectively the transformation rules of: scalar, pseudo-scalar, vector, axial-vector, and second rank antisymmetric tensor. The factor of $\frac{1}{2}$ in the tensor term is introduced because both $\sigma_{\mu\nu}$ and $T^{\mu\nu}$ are antisymmetric, so that each independent term appears twice in the sum over μ and ν.

The coefficients can be obtained from Γ as

$$S = \tfrac{1}{4}\text{Tr}\,\Gamma, \qquad P = \tfrac{1}{4}\text{Tr}\,\Gamma\gamma_5, \qquad V^\mu = \tfrac{1}{4}\text{Tr}\,\Gamma\gamma^\mu,$$
$$A^\mu = \tfrac{1}{4}\text{Tr}\,\Gamma\gamma_5\gamma^\mu, \qquad T^{\mu\nu} = \tfrac{1}{4}\text{Tr}\,\Gamma\sigma^{\mu\nu}. \tag{A.24}$$

11. In cross sections we encounter combinations $u\bar{u}$ and $v\bar{v}$ of Dirac spinors for on-shell particles. An average over independent spin states for a Dirac particle of mass M gives

$$\tfrac{1}{2}\sum_{\text{spin}} u\bar{u} = \tfrac{1}{2}(\slashed{p} + M), \qquad \tfrac{1}{2}\sum_{\text{spin}} v\bar{v} = \tfrac{1}{2}(\slashed{p} - M). \tag{A.25}$$

For a Dirac particle with non-trivial spin, we have instead

$$(\slashed{p} + M)\tfrac{1}{2}\left(1 + \gamma_5\slashed{S}/M\right), \qquad (\slashed{p} - M)\tfrac{1}{2}\left(1 + \gamma_5\slashed{S}/M\right), \tag{A.26}$$

where S is the particle's spin vector, normalized (Amsler *et al.*, 2008) to a maximum of $-S^2 \leq M^2$. In the case of a massless particle we use a helicity variable λ and a transverse spin variable b_T, normalized to have a maximum values unity, $\lambda^2 + |b_T|^2 \leq 1$ (Sec. A.7) to give

$$\tfrac{1}{2}\slashed{p}\left(1 - \lambda\gamma_5 - \sum_{j=1,2}\gamma_5\gamma^j b_T^j\right) \quad \text{for quark,} \tag{A.27a}$$

$$\tfrac{1}{2}\slashed{p}\left(1 + \lambda\gamma_5 - \sum_{j=1,2}\gamma_5\gamma^j b_T^j\right) \quad \text{for antiquark.} \tag{A.27b}$$

A.11 Group theory

1. For SU(3), the definition of the structure constants $f_{\alpha\beta\gamma}$ and the representation matrices in the fundamental (i.e., triplet) representation t_α are the standard ones, with $t_\alpha = \lambda_\alpha/2$. Here λ_α are the Gell-Mann matrices, as defined in Amsler *et al.* (2008, p. 338).

2. The commutation relations are $[t_\alpha, t_\beta] = if_{\alpha\beta\gamma}t_\gamma$.

3. $f_{\alpha\beta\gamma}$ are totally antisymmetric.
4. Combinations of representation matrices and structure constants:

$$\text{Tr}(t_\alpha t_\beta) = T_F \delta_{\alpha\beta}, \tag{A.28}$$

$$t_\alpha t_\alpha = C_F I, \tag{A.29}$$

$$f_{\alpha\gamma\delta} f_{\beta\gamma\delta} = C_A \delta_{\alpha\beta}, \tag{A.30}$$

where repeated indices are summed, I is the unit matrix, and the t_αs are in the fundamental representation. Useful values with standard conventions:

Symbol	SU(n)	SU(3)
T_F	$\frac{1}{2}$	$\frac{1}{2}$
C_F	$\frac{n^2-1}{2n}$	$\frac{4}{3}$
C_A	n	3

$$\tag{A.31}$$

5. Combinations useful in calculations:

$$t_\beta t_\alpha t_\beta = t_\alpha \, (C_F - \tfrac{1}{2}C_A), \tag{A.32}$$

$$f_{\delta\alpha\epsilon} f_{\epsilon\beta\phi} f_{\phi\gamma\delta} = -\tfrac{1}{2} C_A f_{\alpha\beta\gamma}. \tag{A.33}$$

A.12 Dimensional regularization and $\overline{\text{MS}}$: basics

See Collins (1984, Ch. 4) for a systematic mathematical treatment of dimensional regularization.

1. The space-time dimension is $n = 4 - 2\epsilon$.
2. Rotationally symmetric Euclidean integral in d dimensions:

$$\int d^d k \, f(k^2) = \frac{\pi^{d/2}}{\Gamma(d/2)} \int_0^\infty dk^2 \, (k^2)^{d/2-1} \, f(k^2). \tag{A.34}$$

This is often used for the transverse dimensions, with $d = 2 - 2\epsilon$.
3. The Lorentz-invariant integral over particle momentum is

$$\sum_p \cdots \overset{\text{def}}{=} \int \frac{d^{3-2\epsilon} p}{(2\pi)^{3-2\epsilon} 2E_p} \cdots \tag{A.35}$$

4. Decomposition of integration over a spatial $3 - 2\epsilon$-dimensional variable into integrals over radius, a polar angle, and an azimuthal angle:

$$\int d^{3-2\epsilon} k \, f(k) = \int_0^\infty dk \, k^{2-2\epsilon} \int_{-1}^1 d\cos\theta \, (\sin\theta)^{-2\epsilon} \int d\Omega_T \, f(k), \tag{A.36}$$

where $d\Omega_T$ represents an integral over a $1 - 2\epsilon$-dimensional angle in the transverse dimensions, which would be $d\phi$ in a 3-dimensional space, i.e., at $\epsilon = 0$. The normalization of the angular integral is

$$\int d\Omega_T = \frac{2\pi^{1-\epsilon}}{\Gamma(1-\epsilon)}. \tag{A.37}$$

These results can be proved by decomposing k into a z component $k\cos\theta$ and a $2 - 2\epsilon$-dimensional transverse vector, and then using (A.34) to get the normalization of the azimuthal integral. See Sec. A.14 for the Gamma function.

5. The normalization of single particle states is

$$\langle \mathbf{p}' | \mathbf{p} \rangle = (2\pi)^{3-2\epsilon} 2E_p \, \delta^{(3-2\epsilon)}(\mathbf{p} - \mathbf{p}') = (2\pi)^{3-2\epsilon} 2p^+ \delta(p^+ - p'^+) \delta^{(2-2\epsilon)}(\mathbf{p}_T - \mathbf{p}'_T). \tag{A.38}$$

6. Loop-momentum integrals are

$$\int \frac{d^{4-2\epsilon}k}{(2\pi)^{4-2\epsilon}} \cdots \tag{A.39}$$

7. Momentum-conservation delta functions are

$$(2\pi)^{4-2\epsilon} \delta^{(4-2\epsilon)}(k_1 + \ldots). \tag{A.40}$$

8. Dirac matrices are defined to obey $\text{Tr}\, I = 4$ for all n.
9. $\overline{\text{MS}}$ definition:
   (a) The lowest-order bare coupling is defined to be $g_0 = \mu^\epsilon g$, with g dimensionless for all n.
   (b) Counterterms have a factor S_ϵ for each loop, where

$$S_\epsilon = (4\pi)^\epsilon / \Gamma(1 - \epsilon). \tag{A.41}$$

See (3.16) and (3.17) for examples. This definition differs from the more conventional one, $S_\epsilon = \left(4\pi e^{-\gamma_E}\right)^\epsilon \simeq (7.056)^\epsilon$, but only by terms of order ϵ^2. It can be shown that differences of order ϵ^2 do not affect the values of ordinary renormalized Green functions at any order (problem 3.3). However, the definition given here is preferable for $\overline{\text{MS}}$ renormalization of the collinear factors defined in Chs. 10 and 13.

A.13 Dimensional regularization: standard integrals

1. "Scale-invariant" integrals, i.e., integrals of a power of the integration momentum are zero:

$$\int d^n k (k^2)^{-\alpha} = 0. \tag{A.42}$$

2. Rotationally invariant phase-space integrals for massless particles:
   (a) Two bodies:

$$\int \prod_{i=1}^{2} \frac{d^{3-2\epsilon} k_i}{(2\pi)^{3-2\epsilon} 2|k_i|} (2\pi)^{4-2\epsilon} \delta^{(4-2\epsilon)}(q - k_1 - k_2) f(k_1, k_2)$$

$$= \frac{Q^{-2\epsilon}}{2^{4-4\epsilon} \pi^{1/2-\epsilon} \Gamma(\frac{3}{2} - \epsilon)} \times \text{angular average of } f\left(\tfrac{1}{2}Q\mathbf{n}, -\tfrac{1}{2}Q\mathbf{n}\right), \tag{A.43}$$

   in the center-of-mass, with $Q = \sqrt{q^2}$.
   (b) Three bodies:

$$\int \prod_{i=1}^{3} \frac{d^{3-2\epsilon} k_i}{(2\pi)^{3-2\epsilon} 2|k_i|} (2\pi)^{4-2\epsilon} \delta^{(4-2\epsilon)}(q - k_1 - k_2 - k_3) f(k_1, k_2, k_3)$$

$$= \frac{Q^{2-4\epsilon}}{2^{8-6\epsilon} \pi^{5/2-2\epsilon} \Gamma(\frac{3}{2} - \epsilon) \Gamma(1 - \epsilon)}$$

$$\times \text{ang. avg.} \int_0^1 \prod_{i=1}^{3} dy_i \, \delta\left(1 - \sum y_i\right) (y_1 y_2 y_3)^{-\epsilon} f(k_1, k_2, k_3). \tag{A.44}$$

Here the spatial momenta k_1, k_2, k_3, add up to 0 in the center-of-mass frame, and the sizes are given by dimensionless variables y_i defined by $|k_i| = (1 - y_i)Q/2$.

3. Integral used in Fourier transformations on transverse momenta:

$$\int \frac{e^{ik_T \cdot b_T}}{(k_T^2)^\alpha} d^{2-2\epsilon} k_T = \left(\frac{b_T^2}{4\pi}\right)^{\epsilon+\alpha-1} \frac{\pi^\alpha \Gamma(1 - \epsilon - \alpha)}{\Gamma(\alpha)}. \tag{A.45}$$

A proof can be made by converting the k_T integral to a Gaussian, by the use of $(k_T^2)^{-\alpha} = \frac{1}{\Gamma(\alpha)} \int_0^\infty x^{\alpha-1} e^{-xk_T^2} dx$.

4. For the case that the integrand has one or more powers of $\ln k_T^2$, the result is found by differentiating (A.45) with respect to α.

A.14 Properties of Γ function

1. Definition:

$$\Gamma(z) \stackrel{\text{def}}{=} \int_0^\infty dt\, t^{z-1} e^{-t}. \tag{A.46}$$

2. Integer values: $\Gamma(n + 1) = n!$.
3. $\Gamma(z + 1) = z\Gamma(z)$.
4. Expansion about $z = 0$:

$$\Gamma(z) = \frac{1}{z} e^{-\gamma_E z} \left[1 + \frac{\pi^2}{12} z^2 + O(z^3)\right], \tag{A.47}$$

where $\gamma_E = 0.5772\ldots$ is the Euler constant.

5. Expansion about $z = \frac{1}{2}$:

$$\Gamma(\tfrac{1}{2} + z) = \pi^{1/2} e^{-(\gamma_E + \ln 4)z} \left[1 + \frac{\pi^2}{4} z^2 + O(z^3)\right]. \tag{A.48}$$

6. We often use

$$\int_0^1 dx\, x^{\alpha-1}(1 - x)^{\beta-1} = \frac{\Gamma(\alpha)\Gamma(\beta)}{\Gamma(\alpha + \beta)}, \tag{A.49}$$

$$\int_0^\infty dx\, \frac{x^{\alpha-1}}{(A + x)^\beta} = A^{\alpha-\beta} \frac{\Gamma(\alpha)\Gamma(\beta - \alpha)}{\Gamma(\beta)}. \tag{A.50}$$

These and other useful formulae can be found in or deduced from results in Abramowitz and Stegun (1964). Some commonly used integrals have integrands with factors of logarithms of x, $1 - x$ or $A + x$ relative to (A.49) or (A.50); these can be found by differentiation with respect to α or β.

A.15 Plus distributions, etc.

We define the general plus distribution $(\ln^n(1 - x)/(1 - x))_+$ by its integral with an arbitrary smooth test function $f(x)$:

$$\int_0^1 dx \left(\frac{\ln^n(1 - x)}{1 - x}\right)_+ f(x) \stackrel{\text{def}}{=} \int_0^1 dx\, \frac{[f(x) - f(1)]\ln^n(1 - x)}{1 - x}. \tag{A.51}$$

When $n = 0$, and there is a smooth function (e.g., a polynomial) in the numerator, we will also write

$$\int_0^1 dx \, \frac{A(x)}{(1-x)_+} f(x) \stackrel{\text{def}}{=} \int_0^1 dx \, \frac{A(x)f(x) - A(1)f(1)}{1-x}. \qquad (A.52)$$

In calculations of structure functions with dimensionally regulated divergences, we find integrals in which plus distributions appear as a limit of regulated integrals. The following derivation shows both a result that is useful in itself, and a general method. The factor $[z/(1-z)]^\epsilon$ in the integrand arises in the phase-space integral for DIS: Sec. 9.9. The integral is regulated if $\epsilon < 0$.

$$\int_0^1 dz \, \frac{z^\epsilon}{(1-z)^{1+\epsilon}} f(z) = \int_0^1 dz \, \frac{z^\epsilon f(z) - f(1)}{(1-z)^{1+\epsilon}} + f(1) \int_0^1 dz \, \frac{1}{(1-z)^{1+\epsilon}}$$

$$= \int_0^1 dz \left\{ \frac{f(z) - f(1)}{1-z} + \epsilon \frac{f(z)\ln z - [f(z) - f(1)]\ln(1-z)}{1-z} \right\}$$

$$+ O(\epsilon^2) - \frac{f(1)}{\epsilon}. \qquad (A.53)$$

The expansion in powers of ϵ in the second line is allowed because the subtracted integrand is well behaved as $\epsilon \to 0$.

This can be treated as an expansion of $z^\epsilon/(1-z)^{1+\epsilon}$ in powers of ϵ, interpreted in the standard sense of the limit of a generalized function/distribution:

$$\frac{z^\epsilon}{(1-z)^{1+\epsilon}} = -\frac{\delta(z-1)}{\epsilon} + \frac{1}{(1-z)_+} + \epsilon \left[\frac{\ln z}{1-z} - \left(\frac{\ln(1-z)}{1-z} \right)_+ \right] + O(\epsilon^2). \qquad (A.54)$$

A.16 Feynman parameters

$$\frac{1}{A^\alpha B^\beta} = \frac{\Gamma(\alpha + \beta)}{\Gamma(\alpha)\Gamma(\beta)} \int_0^1 dx \, \frac{x^{\alpha-1}(1-x)^{\beta-1}}{[Ax + B(1-x)]^{\alpha+\beta}}. \qquad (A.55)$$

A.17 Orders of magnitude, estimation, etc.

We will frequently need to estimate sizes of Feynman graphs, the sizes of errors in approximations, etc. A correct use of appropriate mathematical notation keeps the arguments precise and reliable; I use the definitions given by Knuth (1976). As Knuth points out, it is quite common to misuse the definitions, and this results in a loss of precision of the arguments.

A.17.1 "Order at most": big-O

The most commonly used notation is

$$f(Q) = O\big(g(Q)\big) \quad \text{when } Q \to \infty, \qquad (A.56)$$

which means that there is a constant C such that

$$\left| \frac{f(Q)}{g(Q)} \right| < C \text{ for all large enough } Q. \qquad (A.57)$$

It is often useful to replace the limit by some more precise specification of the range of Q (or whatever other variable is used). An example would be

$$\frac{\sin x}{\sqrt{x^2 - 1}} = O\left(\frac{1}{x}\right) \text{ for } x \geq 2. \tag{A.58}$$

Although this notation is commonly used to indicate that the left-hand side is asymptotically of the order of magnitude of the right-hand side, this is not actually a correct usage. For this case Knuth's Θ notation should be used: Sec. A.17.2. The big-O notation is most appropriate when stating error estimates, for example, since the standard definition allows the left-hand side to have zeros, as in (A.58), or to go to zero relative to the right-hand side, as in

$$\frac{1}{x^2} = O\left(\frac{1}{x}\right) \text{ as } x \to \infty. \tag{A.59}$$

A.17.2 "Exact order": Θ

Power-counting and error estimates are often made using what we often call order-of-magnitude estimates. We replace an exact quantity by a crude approximation that is valid up to a factor. For this we use the symbol "Θ":

$$f(Q) = \Theta(g(Q)) \text{ when } Q \to \infty, \tag{A.60}$$

which means that there are two positive *non-zero* constants C_1 and C_2 such that

$$C_1 < \left|\frac{f(Q)}{g(Q)}\right| < C_2 \text{ for all large enough } Q. \tag{A.61}$$

(The use of this definition requires that $g(Q)$ is non-zero for large Q.)

An example of the use of this notation would be if we added 2 to the $\sin x$ in (A.58). The numerator of the fraction now oscillates between 1 and 3, instead of between -1 and 1, so that we have

$$\frac{2 + \sin x}{\sqrt{x^2 - 1}} = \Theta\left(\frac{1}{x}\right) \text{ for } x \geq 2. \tag{A.62}$$

This is a typical use in estimation of integrals: the right-hand side can be integrated analytically, the left-hand side at best with difficulty.

We will frequently apply this notation to denominators of Feynman propagators, in which case it is important that (A.60) also applies to the reciprocal functions. That is, (A.60) implies that

$$\frac{1}{f(Q)} = \Theta\left(\frac{1}{g(Q)}\right) \text{ when } Q \to \infty. \tag{A.63}$$

A.17.3 Little-o

Sometimes we simply wish to state that something becomes arbitrarily much smaller than something else in a limit, without wishing to say by how much. In that case we use the little-o notation

$$f(Q) = o(g(Q)) \text{ when } Q \to \infty, \tag{A.64}$$

which means simply that

$$\frac{f(Q)}{g(Q)} \to 0 \text{ when } Q \to \infty. \tag{A.65}$$

Unlike the previous cases, it makes no sense to specify a range of Q; only the limit matters. However, if there is another parameter involved, it makes sense to specify that (A.65) applies uniformly in the other parameter. See below for an example.

A.17.4 Asymptotic equality: $\sim$

This notation is frequently used when the Θ notation should be used. The standard definition is the much stronger statement that

$$f(Q) \sim g(Q) \text{ when } Q \to \infty \tag{A.66}$$

means

$$\lim_{Q \to \infty} \frac{f(Q)}{g(Q)} = 1. \tag{A.67}$$

Both this and the Θ notation have essential uses, so that it is important not to confuse them.

A.17.5 Uniformity

Frequently we will obtain order-of-magnitude estimates of some function that has parameters. (Often the function is the difference between some exact quantity and an approximation.) It is important to know whether the estimates can be made independent of the parameters.

For example, define

$$f_1(Q; a) = \frac{1}{a^2 + Q^2}. \tag{A.68}$$

Then as $Q \to \infty$,

$$f_1(Q; a) = O(1/Q^2). \tag{A.69}$$

We can set the quantity C in the definition of $O(\ldots)$, (A.57), to be unity (or larger), independently of the parameter a. Moreover, the application of (A.57) works with the same minimum value of Q for all a. In that case we say that (A.69) holds uniformly in a.

But if instead we used

$$f_2(Q; a) = \frac{1}{1 + a^2 Q^2}, \tag{A.70}$$

then we could still say that

$$f_2(Q; a) = O(1/Q^2). \tag{A.71}$$

But this would not be uniform in a. When a is made small, the quantity C in (A.57) has to be made large. A symptom of this non-uniformity is that when $a = 0$, $f_2 = O(1)$ instead of $O(1/Q^2)$.

Appendix B
Light-front coordinates, rapidity, etc.

The use of light-front variables, rapidity and pseudo-rapidity is very common in treating high-energy scattering, particularly in hadron-hadron and lepton-hadron collisions. The essential features of these collisions that make these variables of utility are the presence of ultra-relativistic particles and a preferred axis.

B.1 Definition

Light-front coordinates are defined by a change of variables from the usual (t, x, y, z) [or $(0, 1, 2, 3)$] coordinates. Given a vector V^μ, its light-front components are defined by

$$V^+ = \frac{V^0 + V^3}{\sqrt{2}}, \quad V^- = \frac{V^0 - V^3}{\sqrt{2}}, \quad V_\mathrm{T} = (V^1, V^2), \tag{B.1}$$

and I will write the components in the order $V^\mu = (V^+, V^-, V_\mathrm{T})$. Some authors prefer to omit the $1/\sqrt{2}$ factor in (B.1), but among the reasons not to is that the change of variable from ordinary coordinates has unit Jacobian. Thus the element of volume is simply

$$\mathrm{d}^4 k = \mathrm{d}k^+ \, \mathrm{d}k^- \, \mathrm{d}^2 k_\mathrm{T}. \tag{B.2}$$

What are the motivations for defining such coordinates, which evidently depend on a particular choice of the z axis? One is that these coordinates transform very simply under boosts along the z axis. Another is that when a vector is highly boosted along the z axis, light-front coordinates nicely show what are the large and small components of momentum. Typically one uses light-front coordinates in a situation like high-energy hadron scattering. In that situation, there is a natural choice of an axis, the collision axis, and one frequently needs to transform between different frames related by boosts along the axis. Commonly used frames include the rest frame of one of the incoming particles, the overall center-of-mass frame, and the center-of-mass frame of a partonic subprocess.

It can easily be verified that Lorentz-invariant scalar products have the form

$$V \cdot W = V^+ W^- + V^- W^+ - V_\mathrm{T} \cdot W_\mathrm{T},$$

$$V \cdot V = 2V^+ V^- - V_\mathrm{T}^2. \tag{B.3}$$

It follows that the metric tensor has as its non-zero components $g_{+-} = g_{-+} = 1$, $g_{ij} = -\delta_{ij}$, where the indices i and j refer to the two transverse coordinates.

It is important to make a distinction between contravariant vectors, whose indices are superscripts, and covariant vectors, whose indices are subscripts. Indices are contracted by the Einstein summation convention only between upper and lower indices, as in $g_{\mu\nu} V^\mu W^\nu$. Contravariant and covariant vectors are transformed into each other by the metric tensor, e.g., $V_\mu = g_{\mu\nu} V^\nu$. It is readily checked that the components of the metric tensor do not change their values when both indices are changed from contravariant to covariant, $g^{+-} = g^{-+} = 1$, $g^{ij} = -\delta_{ij}$, but that the mixed tensor g^μ_ν is just a Kronecker delta.

We will choose to treat ordinary coordinate vectors and momentum vectors as naturally contravariant. Derivatives with respect to these are then naturally covariant:

$$\partial_\mu f(x) \stackrel{\text{def}}{=} \frac{\partial f}{\partial x^\mu}, \tag{B.4}$$

so that a Taylor expansion can be written without any metric tensor: $f(a + x) = f(a) + a^\mu \partial_\mu f(a) + O(a^2)$. Notice also, for example, that $\partial_+ f = \partial f / \partial x^+$. Thus the corresponding contravariant derivative (with upstairs indices) has the slightly counterintuitive of being with respect to the opposite coordinate, $\partial^+ f = \partial f / \partial x^-$, and similarly for $\partial^- f$.

B.2 Boosts

Let us make a boost in the z direction to make a new vector V'^μ. In the ordinary (t, x, y, z) components we have the well-known formulae

$$V'^0 = \frac{V^0 + vV^z}{\sqrt{1 - v^2}}, \quad V'^z = \frac{vV^0 + V^z}{\sqrt{1 - v^2}}, \quad V'^x = V^x, \quad V'^y = V^y. \tag{B.5}$$

It is easy to derive the following for the light-front components:

$$V'^+ = V^+ e^\psi, \quad V'^- = V^- e^{-\psi}, \quad V'_\text{T} = V_\text{T}, \tag{B.6}$$

where the hyperbolic angle ψ is $\frac{1}{2} \ln \frac{1+v}{1-v}$, so that $v = \tanh \psi$.

Notice that if we apply two boosts of parameters ψ_1 and ψ_2 the result is a boost $\psi_1 + \psi_2$. This is clearly simpler than the corresponding result expressed in terms of velocities.

B.3 Rapidity

B.3.1 Boost of particle momentum

Consider a particle of mass m that is obtained by a boost ψ in the z direction from the particle's rest frame. Its momentum is

$$p^\mu = \left(p^+, \frac{m^2}{2p^+}, 0_\text{T} \right) = \left(\frac{m}{\sqrt{2}} e^\psi, \frac{m}{\sqrt{2}} e^{-\psi}, 0_\text{T} \right). \tag{B.7}$$

Notice that if the boost is very large (positive or negative), only one of the two non-zero light-front components of p^μ is large; the other component becomes small. With the usual coordinates, two of the components, p^0 and p^z, become large.

Suppose next that we have two such particles, p_1 and p_2, with the boost for particle 1 being much larger than that for particle 2. Then in the scalar product of the two momenta only one component of each momentum dominates the result, so that for example $(p_1 + p_2)^2 \simeq 2p_1^+ p_2^-$. This implies that, when analyzing the sizes of scalar products of highly boosted particles, it is simpler to use light-front components than to use conventional components.

B.3.2 Definition of rapidity

Since the ratio p^+/p^- gives a measure $e^{2\psi}$ of the boost from the rest frame, we are led to the following definition of a quantity called "rapidity":

$$y = \frac{1}{2} \ln \frac{p^+}{p^-} = \frac{1}{2} \ln \frac{E + p^z}{E - p^z}, \tag{B.8}$$

which can be applied to a particle of non-zero transverse momentum. The 4-momentum of a particle of rapidity y and transverse momentum $\boldsymbol{p}_T$ is

$$
p^\mu = \left(e^y \sqrt{\frac{m^2 + p_T^2}{2}}, \; e^{-y} \sqrt{\frac{m^2 + p_T^2}{2}}, \; \boldsymbol{p}_T \right),
\tag{B.9}
$$

with $\sqrt{m^2 + p_T^2}$ being called the transverse mass m_T of the particle. It can be checked that the scalar product of two momenta is

$$
p_1 \cdot p_2 = m_{1T} m_{2T} \cosh(y_1 - y_2) - \boldsymbol{p}_{T1} \cdot \boldsymbol{p}_{T2}.
\tag{B.10}
$$

In the case where the transverse momenta are negligible, this reduces to $m^2 \cosh(y_1 - y_2)$, which is like the formula for the product of two Euclidean vectors $\boldsymbol{p}_{T1} \cdot \boldsymbol{p}_{T2} = p_1 p_2 \cos\theta$, with the trigonometric cosine being replaced by the hyperbolic cosine.

B.3.3 Transformation under boosts

Under a boost in the z direction, rapidity transforms additively:

$$
y \mapsto y' = y + \psi.
\tag{B.11}
$$

This implies that in situations where we have a frequent need to work with boosts along the z axis it is economical to label the momentum of a particle by its rapidity and transverse momentum, rather than to use 3-momentum.

B.3.4 Phase-space integration

The standard Lorentz-invariant phase-space integration measure for an on-shell particle of mass m is readily converted to light-front coordinates, or to rapidity and transverse momentum:

$$
\frac{d^3\boldsymbol{p}}{2E_p(2\pi)^3} = \frac{dp^+ \, d^2\boldsymbol{p}_T}{2p^+(2\pi)^3} = \frac{dy \, d^2\boldsymbol{p}_T}{2(2\pi)^3},
\tag{B.12}
$$

where $E_p = \sqrt{\boldsymbol{p}^2 + m^2}$. Observe that the light-front version does not depend explicitly on the mass, and that there is a restriction to only positive p^+.

B.3.5 Non-relativistic limit

For a *non*-relativistic particle, rapidity is the same as velocity along the z axis, for then

$$
y = \frac{1}{2} \ln \frac{E + p^z}{E - p^z} = \frac{1}{2} \ln \frac{1 + v^z}{1 - v^z} \simeq v^z. \qquad (v_z \ll 1)
\tag{B.13}
$$

Non-relativistic velocities transform additively under boosts, and the non-linear change of variable from velocity to rapidity allows this additive rule (B.11) to apply to relativistic particles (but only in one direction of boost).

One way of seeing this is as follows. The relativistic law for addition of velocities in one dimension is

$$
\beta_{13} = \frac{\beta_{12} + \beta_{23}}{1 + \beta_{12}\beta_{23}},
\tag{B.14}
$$

where β_{12} is the velocity of some object 1 measured in the rest frame of object 2, etc. This formula is reminiscent of the following property of hyperbolic tangents:

$$\tanh(A + B) = \frac{\tanh A + \tanh B}{1 + \tanh A \tanh B}.$$ (B.15)

So to obtain a linear addition law, we should write $\beta_{12} = \tanh A_{12}$, etc. Then the rule (B.14) for the addition of velocities becomes simply $A_{13} = A_{12} + A_{23}$. The A variables are exactly relative rapidities, since

$$v^z = \frac{p^z}{E} = \frac{p^+ - p^-}{p^+ + p^-} = \tanh y.$$ (B.16)

B.3.6 Relative velocity

Rapidity is the natural relativistic velocity variable. Suppose we have a proton and a pion with the same rapidity at $p_T = 0$. Then they have no relative velocity; to see this, one just boosts to the rest frame of one of the particles. But these same particles have very different energies: $E_p = \frac{m_p}{m_\pi} E_\pi$.

B.4 Pseudo-rapidity

As I will now explain, the rapidity of a particle can easily be measured in a situation where its mass is negligible, for then it is simply related to the polar angle of the particle.

First let us define the pseudo-rapidity of a particle by

$$\eta = -\ln \tan \frac{\theta}{2},$$ (B.17)

where θ is the angle of the 3-momentum of the particle relative to the $+z$ axis. It is easy to derive an expression for rapidity in terms of pseudo-rapidity and transverse momentum:

$$y = \ln \frac{\sqrt{m^2 + p_T^2 \cosh^2 \eta} + p_T \sinh \eta}{\sqrt{m^2 + p_T^2}}.$$ (B.18)

In the limit that $m \ll p_T$, $y \to \eta$. This accounts both for the name "pseudo-rapidity" and for the ubiquitous use of pseudo-rapidity in high-transverse-momentum physics. Angles, and hence pseudo-rapidity, are easy to measure. But it is really the rapidity that is of physical significance: for example, the distribution of particles in a minimum bias event is approximately uniform in rapidity over the kinematic range available.

The distinction between rapidity and pseudo-rapidity is very clear when one examines the kinematic limits on the two variables. In a collision of a given energy, there is a limit to the energy of the particles that can be produced. This can easily be translated to limits on the rapidities of the produced particles of a given mass. But there is no limit on the pseudo-rapidity, since a particle can be physically produced at zero angle (or at 180°), where its pseudo-rapidity is infinite. The particles for which the distinction is very significant are those for which the transverse momentum is substantially less than the mass. Note: (B.18) implies that $|y| < |\eta|$ always.

B.5 Rapidity distributions in high-energy collisions

In the most common events in high-energy hadronic collisions (the so-called "minimum bias events"), the distribution of final-state hadrons is approximately uniform in rapidity (Alner *et al.*,

1986; Abe *et al.*, 1990; ATLAS Collaboration, 2010; Khachatryan *et al.*, 2010). That is, the distribution of final-state hadrons is approximately invariant under boosts in the z direction.

In contrast, the distribution in angle $dN / d\Omega$ is strongly peaked at forward and backward angles. This follows from the Jacobian between $\cos\theta$ and pseudo-rapidity:

$$\frac{d\eta}{d\cos\theta} = \cosh^2\eta = \frac{1}{\sin^2\theta}. \tag{B.19}$$

It follows that rapidity and transverse momentum are appropriate variables for analyzing data, and that detector elements should be approximately uniformly spaced in rapidity. (What is physically possible is to make a detector uniform in the pseudo-rapidity discussed in Sec. B.4.) This is in contrast to the situation for e^+e^- collisions where most of the interest is in events generated via annihilation into an electro-weak boson. Such events are much closer to uniform in solid angle than uniform in rapidity.

Appendix C
Summary of primary results

The important definitions and results are, quite naturally, spread throughout the book. However, it is frequently convenient for reference purposes to have all these equations collected together. This will be the purpose of this appendix.

Lagrangian and Feynman rules of QCD

Without regard to renormalization see (2.1) for the gauge-invariant Lagrangian, and see Sec. 3.1.2 and Fig. 3.1 for the gauge-fixed Lagrangian and the Feynman rules. See (3.6) for the BRST transformations.

With counterterms, etc., see (3.13), (3.14), and (3.15) for the Lagrangian, and Fig. 3.2 for the counterterm vertices.

The full Standard Model Lagrangian is given in (2.30).

Definition of $\overline{\text{MS}}$

See Sec. 3.2.6 for the definition of the $\overline{\text{MS}}$ renormalization scheme.

Renormalization counterterms, RG coefficients

The results for one-loop renormalization counterterms in Z_2, m_0, Z_3, $\check{Z}$, and g_0 are in (3.23), (3.24), (3.25), (3.26), and (3.31). The higher orders are left as an exercise to be derived from the RG coefficients (problem 3.2).

The renormalization group coefficients (β, etc.) are given in Sec. 3.7.

Information on relating schemes with different numbers of active quark flavors is given in Sec. 3.10.

Light-front perturbation theory, etc.

The rules for light-front perturbation theory are given in Sec. 7.2.3.

Light-front wave functions are defined in Sec. 7.3.

Parton densities

The operator definition of an unintegrated quark density in the parton-model framework is given in (6.31), while the antiquark density is given in (6.33). The corresponding polarized densities are given in (6.35) and (6.36). The unintegrated (TMD) quark density is defined in (6.79). Isospin and charge-conjugation relations are listed in Sec. 6.9.7.

Feynman rules for the above densities, still in a pre-QCD framework, are given in (6.110) and in Fig. 6.7. For an unintegrated (TMD) density, see Fig. 6.8.

Gauge-invariant unrenormalized integrated parton densities are defined in (7.40) (quark), (7.43) (unpolarized gluon), (7.44) (polarized gluon). Feynman rules for these are given in Figs. 7.9, 7.10, 7.11, and 7.12.

Our convention for the renormalization factors for parton densities is given in (8.11). Our convention for the DGLAP kernel is in (8.30).

One-loop results for the DGLAP kernels are given in (9.6) (quark in gluon), (9.23) (quark in quark), (9.24) (gluon in gluon), and (9.25) (gluon in quark).

For TMD parton densities in QCD there are a number of extra polarization-dependent functions which are defined in Sec. 13.16. These include the Boer-Mulders function, the Sivers function, and the pretzelosity distribution.

Fragmentation functions

The basic non-QCD definitions of fragmentation functions are given in Sec. 12.4.

In QCD, unrenormalized fragmentation functions are defined in (12.71) (quark) and (12.72) (gluon). The situation on the DGLAP kernels for the renormalized QCD fragmentation functions is summarized in Sec. 12.10.4.

Flavor relations are given in (12.43).

Definitions and results for deeply inelastic lepton scattering

The kinematics of DIS are defined in Sec. 2.3.2, and the structure functions in the electromagnetic case are defined in (2.20) and (2.21). Structure functions are related to the DIS cross section in (2.22).

The parton-model result for DIS is given in (2.28) and (2.29) for the unpolarized case, and in (6.25) when target polarization is included.

In the case of charged-current neutrino and antineutrino scattering on an unpolarized target, the structure functions are defined in (7.3) and the parton-model formula is given in (7.5) and (7.6).

e^+e^- annihilation total cross section

The results of perturbative calculations of the ratio R for the total cross section for $e^+e^- \to$ hadrons are given in (4.34).

Power-counting and region analysis

The power that corresponds to a general region for a hard process is given in (5.75), (5.76), (5.77), (5.78).

Factorization formulae for DIS

Factorization for DIS structure functions is stated in (8.83).

The one-loop coefficient function is in (9.43) (gluon) and (9.54) (quark). With a quark mass, the gluonic coefficients are given in (9.55) and (9.58). References to the currently known higher-order terms are given in Sec. 9.12.

Factorization for Sudakov form factor

The final form of the collinear factors for the Sudakov form factor is defined in (10.119), with the evolution kernel defined in (10.122). Factorization and evolution formulae are given in Sec. 10.11.4, with a solution in (10.131) for the factorized form factor.

The two-loop result for γ_K is given in (10.135). The one-loop hard factor is given in (10.146).

One-particle-inclusive e^+e^- annihilation

For the one-particle-inclusive e^+e^- annihilation process, the hadronic tensor and structure functions are defined in Sec. 12.1, and formulae for the cross section are given.

The factorization formula for the cross section is given in (12.13). Factorization formulae for the hadronic tensor and for the structure functions are given in Secs. 12.2.3 and 12.2.4.

The LO coefficient functions are in Sec. 12.3.

The NLO coefficient functions are discussed in Sec. 12.11, with references to the results.

Semi-inclusive DIS (SIDIS)

Factorization for the SIDIS cross section is stated in the form with integrated parton densities and fragmentation functions in (12.95).

Two-particle annihilation in e^+e^- annihilation

The kinematics, the hadronic tensor, and cross section formulae for two-particle annihilation in e^+e^- annihilation are given in Sec. 13.2.

TMD factorization for two-particle annihilation in e^+e^- annihilation

For TMD factorization, the bare soft factor is defined in (13.39). The unpolarized TMD fragmentation function is defined in (13.42), with the unsubtracted fragmentation function defined in (13.41) as an operator matrix element.

TMD factorization for two-particle annihilation in e^+e^- annihilation is stated in (13.46). The CSS and RG evolution equations are given in Sec. 13.8.

The one-loop results for the CSS evolution kernel K and its anomalous dimension γ_K are given in (13.55) and (13.56). The separation of the non-perturbative part of K is performed in Sec. 13.10.4, especially at (13.60).

The result of solving all the evolution equations is presented in three forms in Sec. 13.13.

The results of calculations of the NLO term for the small-b_T behavior of TMD fragmentation functions are given in (13.92) (gluon from quark) and (13.100) (quark from quark).

The corresponding results for large k_T are in (13.93) (gluon from quark) and (13.101) (quark from quark).

TMD factorization for SIDIS

The TMD quark densities are defined in (13.106), the unsubtracted density being defined in (13.108) as an operator matrix element. These definitions are the ones with future-pointing Wilson lines to be used for SIDIS.

TMD factorization for SIDIS is stated in (13.116).

TMD factorization for Drell-Yan

TMD factorization for Drell-Yan is stated in (14.31). The results of fits to the TMD non-perturbative functions are reviewed in Sec. 14.5.3.

References

To find unpublished preprints and reports, try a search at the SPIRES database, http://www.slac.stanford.edu/spires/hep/. Many are available in scanned form. See http://arxiv.org/ for e-prints with only an arXiv number. These www addresses were accurate in 2010.

Abazov V. M., *et al.* (2008). Measurement of the inclusive jet cross section in $p\bar{p}$ collisions at $\sqrt{s} = 1.96\,\mathrm{TeV}$. *Phys. Rev. Lett.* **101**, 062001. arXiv:0802.2400.

Abe F., *et al.* (1990). Pseudorapidity distributions of charged particles produced in $p\bar{p}$ interactions at $\sqrt{s} = 630\,\mathrm{GeV}$ and $1800\,\mathrm{GeV}$. *Phys. Rev.* **D41**, 2330.

Abe F., *et al.* (1996). Inclusive jet cross section in $\bar{p}p$ collisions at $\sqrt{s} = 1.8\,\mathrm{TeV}$. *Phys. Rev. Lett.* **77**, 438–443. arXiv:hep-ex/9601008.

Abe F., *et al.* (1997). Observation of diffractive W boson production at the Tevatron. *Phys. Rev. Lett.* **78**, 2698–2703. arXiv:hep-ex/9703010.

Abramowitz M., Stegun I. A. (1964). *Handbook of Mathematical Functions with Formulas, Graphs, and Mathematical Tables*. New York: Dover.

Abulencia A., *et al.* (2007). Measurement of the inclusive jet cross section using the k_T algorithm in $p\bar{p}$ collisions at $\sqrt{s} = 1.96\,\mathrm{TeV}$ with the CDF II detector. *Phys. Rev.* **D75**, 092006. arXiv:hep-ex/0701051.

Adloff C., *et al.* (2003). Measurement and QCD analysis of neutral and charged current cross sections at HERA. *Eur. Phys. J.* **C30**, 1–32. arXiv:hep-ex/0304003.

Affolder A. A., *et al.* (2000). The transverse momentum and total cross section of e^+e^- pairs in the Z boson region from $p\bar{p}$ collisions at $\sqrt{s} = 1.8\,\mathrm{TeV}$. *Phys. Rev. Lett.* **84**, 845–850. arXiv:hep-ex/0001021.

Airapetian A., *et al.* (2005). First measurement of the tensor structure function b_1 of the deuteron. *Phys. Rev. Lett.* **95**, 242001. arXiv:hep-ex/0506018.

Airapetian A., *et al.* (2007). Precise determination of the spin structure function g_1 of the proton, deuteron and neutron. *Phys. Rev.* **D75**, 012007. arXiv:hep-ex/0609039.

Airapetian A., *et al.* (2008). Evidence for a transverse single-spin asymmetry in leptoproduction of $\pi^+\pi^-$ pairs. *JHEP* **06**, 017. arXiv:0803.2367.

Aivazis M. A. G., *et al.* (1994). Leptoproduction of heavy quarks. 2. A unified QCD formulation of charged and neutral current processes from fixed target to collider energies. *Phys. Rev.* **D50**, 3102–3118. arXiv:hep-ph/9312319.

Aktas A., *et al.* (2007a). Dijet cross sections and parton densities in diffractive DIS at HERA. *JHEP* **10**, 042. arXiv:0708.3217.

Aktas A., *et al.* (2007b). Tests of QCD factorisation in the diffractive production of dijets in deep-inelastic scattering and photoproduction at HERA. *Eur. Phys. J.* **C51**, 549–568. arXiv:hep-ex/0703022.

Albino S., Kniehl B. A., Kramer G. (2008). AKK update: improvements from new theoretical input and experimental data. *Nucl. Phys.* **B803**, 42–104. arXiv:0803.2768.

Alekhin S., Melnikov K., Petriello F. (2006). Fixed target Drell-Yan data and NNLO QCD fits of parton distribution functions. *Phys. Rev.* **D74**, 054033. arXiv:hep-ph/0606237.

Allanach B. C., *et al.* (2006). Les Houches "Physics at TeV colliders 2005" Beyond the Standard Model working group: summary report. arXiv:hep-ph/0602198.

Almeida L. G., Sterman G., Vogelsang W. (2009). Threshold resummation for dihadron production in hadronic collisions. *Phys. Rev.* **D80**, 074016. arXiv:0907.1234.

Alner G. J., *et al.* (1986). Scaling of pseudorapidity distributions at c.m. energies up to 0.9 TeV. *Z. Phys.* **C33**, 1–6.

Altarelli G., Parisi G. (1977). Asymptotic freedom in parton language. *Nucl. Phys.* **B126**, 298–318.

Altarelli G., *et al.* (1979). Processes involving fragmentation functions beyond the leading order in QCD. *Nucl. Phys.* **B160**, 301–329.

Amsler C., *et al.* (2008). Review of particle physics. *Phys. Lett.* **B667**, 1–1339.

Anastasiou C., *et al.* (2003). Dilepton rapidity distribution in the Drell-Yan process at next-to-next-to-leading order in QCD. *Phys. Rev. Lett.* **91**, 182002. arXiv:hep-ph/0306192.

Anastasiou C., *et al.* (2004). High-precision QCD at hadron colliders: electroweak gauge boson rapidity distributions at next-to-next-to leading order. *Phys. Rev.* **D69**, 094008. arXiv:hep-ph/0312266.

Andersson B. (1998). *The Lund Model*. Cambridge: Cambridge University Press.

Appelquist T., Carazzone J. (1975). Infrared singularities and massive fields. *Phys. Rev.* **D11**, 2856–2861.

Arnold P. B., Kauffman R. P. (1991). W and Z production at next-to-leading order: from large q_T to small. *Nucl. Phys.* **B349**, 381–413.

Artru X., Collins J. C. (1996). Measuring transverse spin correlations by 4-particle correlations in $e^+e^- \to 2$ jets. *Z. Phys.* **C69**, 277–286. arXiv:hep-ph/9504220.

Artru X., Mekhfi M. (1990). Transversely polarized parton densities, their evolution and their measurement. *Z. Phys.* **C45**, 669–676.

ATLAS Collaboration (2010). Charged-particle multiplicities in pp interactions at $\sqrt{s} = 900$ GeV measured with the ATLAS detector at the LHC. *Phys. Lett.* **B688**, 21–42. arXiv:1003.3124.

Aybat S. M., Sterman G. (2009). Soft-gluon cancellation, phases and factorization with initial-state partons. *Phys. Lett.* **B671**, 46–50. arXiv:0811.0246.

Bacchetta A., *et al.* (2004). Single-spin asymmetries: the Trento conventions. *Phys. Rev.* **D70**, 117504. arXiv:hep-ph/0410050.

Bacchetta A., *et al.* (2007). Semi-inclusive deep inelastic scattering at small transverse momentum. *JHEP* **02**, 093. arXiv:hep-ph/0611265.

Bacchetta A., *et al.* (2008). Matches and mismatches in the descriptions of semi-inclusive processes at low and high transverse momentum. *JHEP* **08**, 023. arXiv:0803.0227.

Bacchetta A., *et al.* (2009). Asymmetries involving dihadron fragmentation functions: from DIS to e^+e^- annihilation. *Phys. Rev.* **D79**, 034029. arXiv:0812.0611.

Badier J., *et al.* (1981). Angular distributions in the dimuon hadronic production at 150 GeV/c. *Zeit. Phys.* **C11**, 195–202.

Bahr M., *et al.* (2008). Herwig++ physics and manual. *Eur. Phys. J.* **C58**, 639–707. arXiv:0803.0883.

Baier R., Fey K. (1979). Finite corrections to quark fragmentation functions in perturbative QCD. *Z. Phys.* **C2**, 339–349.

Baikov P. A., Chetyrkin K. G., Kuhn J. H. (2008). Order α_s^4 QCD corrections to Z and τ decays. *Phys. Rev. Lett.* **101**, 012002. arXiv:0801.1821.

Bakker B. L. G., Leader E., Trueman T. L. (2004). A critique of the angular momentum sum rules and a new angular momentum sum rule. *Phys. Rev.* **D70**, 114001. arXiv:hep-ph/0406139.

Balitsky I. (1999). Factorization and high-energy effective action. *Phys. Rev.* **D60**, 014020. arXiv:hep-ph/9812311.

Balitsky I. I., Lipatov L. N. (1978). The pomeranchuk singularity in quantum chromodynamics. *Sov. J. Nucl. Phys.* **28**, 822–829.

Barbieri R., et al. (1976). Mass corrections to scaling in deep inelastic processes. *Nucl. Phys.* **B117**, 50–76.

Bardakci K., Halpern M. B. (1968). Theories at infinite momentum. *Phys. Rev.* **176**, 1686–1699.

Bardeen W. A., et al. (1978). Deep-inelastic scattering beyond the leading order in asymptotically free gauge theories. *Phys. Rev.* **D18**, 3998–4017.

Barone V., Drago A., Ratcliffe P. G. (2002). Transverse polarisation of quarks in hadrons. *Phys. Rept.* **359**, 1–168. arXiv:hep-ph/0104283.

Barone V., Melis S., Prokudin A. (2009). The Boer-Mulders effect in unpolarized SIDIS: an analysis of the COMPASS and HERMES data on the $\cos 2\phi$ asymmetry. *Phys. Rev.* **D81**, 114026. arXiv:0912.5194.

Bassetto A., Dalbosco M., Soldati R. (1987). Renormalization of the Yang-Mills theories in the light cone gauge. *Phys. Rev.* **D36**, 3138–3147.

Bassetto A., et al. (1985). Yang-Mills theories in the light-cone gauge. *Phys. Rev.* **D31**, 2012–2019.

Bauer C. W., Stewart I. W. (2001). Invariant operators in collinear effective theory. *Phys. Lett.* **B516**, 134–142. arXiv:hep-ph/0107001.

Bauer C. W., et al. (2001). An effective field theory for collinear and soft gluons: heavy to light decays. *Phys. Rev.* **D63**, 114020. arXiv:hep-ph/0011336.

Becchi C., Rouet A., Stora R. (1975). Renormalization of the abelian Higgs-Kibble model. *Commun. Math. Phys.* **42**, 127–162.

Becchi C., Rouet A., Stora R. (1976). Renormalization of gauge theories. *Annals Phys.* **98**, 287–321.

Belitsky A. V., Müller D., Kirchner A. (2002). Theory of deeply virtual Compton scattering on the nucleon. *Nucl. Phys.* **B629**, 323–392. arXiv:hep-ph/0112108.

Bengtsson M., Sjöstrand T. (1988). Parton showers in leptoproduction events. *Z. Phys.* **C37**, 465–476.

Berera A., Soper D. E. (1996). Behavior of diffractive parton distribution functions. *Phys. Rev.* **D53**, 6162–6179. arXiv:hep-ph/9509239.

Berger C. F., et al. (2009). Next-to-leading order QCD predictions for W+3-jet distributions at hadron colliders. *Phys. Rev.* **D80**, 074036. arXiv:0907.1984.

Berger E. R., Diehl M., Pire B. (2002). Timelike Compton scattering: exclusive photoproduction of lepton pairs. *Eur. Phys. J.* **C23**, 675–689. arXiv:hep-ph/0110062.

Bern Z., Dixon L. J., Kosower D. A. (2007). On-shell methods in perturbative QCD. *Annals Phys.* **322**, 1587–1634. arXiv:0704.2798.

Bernreuther W. (1983a). Heavy quark effects on the parameters of quantum chromodynamics defined by minimal subtraction. *Z. Phys.* **C20**, 331–333.

Bernreuther W. (1983b). Decoupling of heavy quarks in quantum chromodynamics. *Ann. Phys.* **151**, 127–162.

Bernreuther W., Wetzel W. (1982). Decoupling of heavy quarks in the minimal subtraction scheme. *Nucl. Phys.* **B197**, 228–236. Erratum: **B513**, 758 (1998).

Bethke S. (2009). The 2009 world average of α_s. *Eur. Phys. J.* **C64**, 689–703. arXiv:0908.1135.

Bethke S., et al. (2009). Determination of the strong coupling α_s from hadronic event shapes and NNLO QCD predictions using JADE data. *Eur. Phys. J.* **C64**, 351–360. arXiv:0810.1389.

Bianconi A., et al. (2009). Effects of azimuth-symmetric acceptance cutoffs on the measured asymmetry in unpolarized Drell-Yan fixed target experiments. arXiv:0911.5493.

Binosi D., *et al.* (2009). JaxoDraw: a graphical user interface for drawing Feynman diagrams. Version 2.0 release notes. *Comput. Phys. Commun.* **180**, 1709–1715. Available from: http://jaxodraw.sourceforge.net/, arXiv:0811.4113.

Binoth T., *et al.* (2008). NLO QCD corrections to tri-boson production. *JHEP* **06**, 082. arXiv:0804.0350.

Bjorken J. D. (1966). Applications of the chiral U(6) $\otimes$ U(6) algebra of current densities. *Phys. Rev.* **148**, 1467–1478.

Bjorken J. D., Paschos E. A. (1969). Inelastic electron-proton and γ-proton scattering and the structure of the nucleon. *Phys. Rev.* **185**, 1975–1982.

Bloom E. D., Gilman F. J. (1971). Scaling and the behavior of nucleon resonances in inelastic electron-nucleon scattering. *Phys. Rev.* **D4**, 2901–2916.

Blümlein J., Robaschik D. (2000). On the structure of the virtual Compton amplitude in the generalized Bjorken region: integral relations. *Nucl. Phys.* **B581**, 449–473. arXiv:hep-ph/0002071.

Bodwin G. T. (1985). Factorization of the Drell-Yan cross-section in perturbation theory. *Phys. Rev.* **D31**, 2616–2642. Erratum: **D34**, 3932 (1986).

Bodwin G. T., Brodsky S. J., Lepage G. P. (1981). Initial state interactions and the Drell-Yan process. *Phys. Rev. Lett.* **47**, 1799–1803.

Boer D. (2008). Transversity asymmetries. Talk given at Transversity 2008: 2nd International Workshop on Transverse Polarization Phenomena in Hard Processes, Ferrara, Italy, 28–31 May 2008. arXiv:0808.2886.

Boer D. (2009). Angular dependences in inclusive two-hadron production at BELLE. *Nucl. Phys.* **B806**, 23–67. arXiv:0804.2408.

Bogoliubov N. N., Shirkov D. V. (1959). *Introduction to the Theory of Quantized Fields*. New York: Wiley-Interscience.

Bomhof C. J., Mulders P. J., Pijlman F. (2004). Gauge link structure in quark-quark correlators in hard processes. *Phys. Lett.* **B596**, 277–286. arXiv:hep-ph/0406099.

Born M., Heisenberg W., Jordan P. (1926). On quantum mechanics II. *Zeit. f. Phys.* **35**, 557–615.

Born M., Jordan P. (1925). On quantum mechanics. *Zeit. f. Phys.* **34**, 858–888.

Bouchiat C., Fayet P., Meyer P. (1971). Galilean invariance in the infinite momentum frame and the parton model. *Nucl. Phys.* **B34**, 157–176.

Brock R., *et al.* (1995). Handbook of perturbative QCD: version 1.0. *Rev. Mod. Phys.* **67**, 157–248.

Brodsky S. J., Farrar G. R. (1973). Scaling laws at large transverse momentum. *Phys. Rev. Lett.* **31**, 1153–1156.

Brodsky S. J., Hwang D.-S., Schmidt I. (2002). Final-state interactions and single-spin asymmetries in semi-inclusive deep inelastic scattering. *Phys. Lett.* **B530**, 99–107. arXiv:hep-ph/0201296.

Brodsky S. J., Lepage G. P. (1989). Exclusive processes in quantum chromodynamics. *Adv. Ser. Direct. High Energy Phys.* **5**, 93–240.

Brodsky S. J., Pauli H.-C., Pinsky S. S. (1998). Quantum chromodynamics and other field theories on the light cone. *Phys. Rept.* **301**, 299–486. arXiv:hep-ph/9705477.

Brodsky S. J., *et al.* (1994). Diffractive leptoproduction of vector mesons in QCD. *Phys. Rev.* **D50**, 3134–3144. arXiv:hep-ph/9402283.

Brodsky S. J., *et al.* (2001). Light-cone representation of the spin and orbital angular momentum of relativistic composite systems. *Nucl. Phys.* **B593**, 311–335. arXiv:hep-th/0003082.

Brown L. S. (1992). *Quantum Field Theory*. Cambridge: Cambridge University Press.

Buras A. J., *et al.* (1977). Asymptotic freedom beyond the leading order. *Nucl. Phys.* **B131**, 308–326.

Callan C. G., Gross D. J. (1969). High-energy electroproduction and the constitution of the electric current. *Phys. Rev. Lett.* **22**, 156–159.

Callan C. G., Gross D. J. (1973). Bjorken scaling in quantum field theory. *Phys. Rev.* **D8**, 4383–4394.

Cardy J. L., Winbow G. A. (1974). The absence of final state interaction corrections to the Drell-Yan formula for massive lepton pair production. *Phys. Lett.* **B52**, 95.

Catani S., Ciafaloni M., Marchesini G. (1986). Noncancelling infrared divergences in QCD coherent state. *Nucl. Phys.* **B264**, 588–620.

Catani S., Fiorani F., Marchesini G. (1990a). QCD coherence in initial state radiation. *Phys. Lett.* **B234**, 339–345.

Catani S., Fiorani F., Marchesini G. (1990b). Small-x behavior of initial state radiation in perturbative QCD. *Nucl. Phys.* **B336**, 18–85.

Chang S.-J., Ma S.-K. (1969). Feynman rules and quantum electrodynamics at infinite momentum. *Phys. Rev.* **180**, 1506–1513.

Chekanov S., *et al.* (2005). An NLO QCD analysis of inclusive cross-section and jet-production data from the ZEUS experiment. *Eur. Phys. J.* **C42**, 1–16. arXiv:hep-ph/0503274.

Chekanov S., *et al.* (2010). A QCD analysis of ZEUS diffractive data. *Nucl. Phys.* **B831**, 1–25. arXiv:0911.4119.

Chetyrkin K. G., Harlander R. V., Kuhn J. H. (2000). Quartic mass corrections to R_{had} at $O(\alpha_s^3)$. *Nucl. Phys.* **B586**, 56–72. Erratum: **B634**, 413–414 (2002). arXiv:hep-ph/0005139.

Chetyrkin K. G., Kniehl B. A., Steinhauser M. (1997). Strong coupling constant with flavour thresholds at four loops in the $\overline{MS}$ scheme. *Phys. Rev. Lett.* **79**, 2184–2187. arXiv:hep-ph/9706430.

Chetyrkin K. G., Kniehl B. A., Steinhauser M. (1998). Decoupling relations to $\mathcal{O}(\alpha_s^3)$ and their connection to low-energy theorems. *Nucl. Phys.* **B510**, 61–87. arXiv:hep-ph/9708255.

Ciafaloni M. (1988). Coherence effects in initial jets at small Q^2/s. *Nucl. Phys.* **B296**, 49–74.

Coleman S., Gross D. J. (1973). Price of asymptotic freedom. *Phys. Rev. Lett.* **31**, 851–854.

Coleman S., Norton R. E. (1965). Singularities in the physical region. *Nuovo Cim.* **38**, 438–442.

Coleman S., Weinberg E. (1973). Radiative corrections as the origin of spontaneous symmetry breaking. *Phys. Rev.* **D7**, 1888–1910.

Collins J. C. (1974). Structure of counterterms in dimensional regularization. *Nucl. Phys.* **B80**, 341–348.

Collins J. C. (1980). Algorithm to compute corrections to the Sudakov form-factor. *Phys. Rev.* **D22**, 1478–1489.

Collins J. C. (1984). *Renormalization*. Cambridge: Cambridge University Press.

Collins J. C. (1989). Sudakov form factors. *Adv. Ser. Direct. High Energy Phys.* **5**, 573–614. arXiv:hep-ph/0312336.

Collins J. C. (1993). Fragmentation of transversely polarized quarks probed in transverse momentum distributions. *Nucl. Phys.* **B396**, 161–182. arXiv:hep-ph/9208213.

Collins J. C. (1998a). Hard-scattering factorization with heavy quarks: a general treatment. *Phys. Rev.* **D58**, 094002. arXiv:hep-ph/9806259.

Collins J. C. (1998b). Proof of factorization for diffractive hard scattering. *Phys. Rev.* **D57**, 3051–3056. Erratum: **D61**, 019902 (2000). arXiv:hep-ph/9709499.

Collins J. C. (2002). Leading-twist single-transverse-spin asymmetries: Drell-Yan and deep-inelastic scattering. *Phys. Lett.* **B536**, 43–48. arXiv:hep-ph/0204004.

Collins J. C., Frankfurt L., Strikman M. (1997). Factorization for hard exclusive electroproduction of mesons in QCD. *Phys. Rev.* **D56**, 2982–3006. arXiv:hep-ph/9611433.

Collins J. C., Hautmann F. (2000). Infrared divergences and non-lightlike eikonal lines in Sudakov processes. *Phys. Lett.* **B472**, 129–134. arXiv:hep-ph/9908467.

Collins J. C., Heppelmann S. F., Ladinsky G. A. (1994). Measuring transversity densities in singly polarized hadron-hadron and lepton-hadron collisions. *Nucl. Phys.* **B420**, 565–582. arXiv:hep-ph/9305309.

Collins J. C., Jung H. (2005). Need for fully unintegrated parton densities. arXiv:hep-ph/0508280.

Collins J. C., Manohar A. V., Wise M. B. (2006). Renormalization of the vector current in QED. *Phys. Rev.* **D73**, 105019. arXiv:hep-th/0512187.

Collins J. C., Metz A. (2004). Universality of soft and collinear factors in hard-scattering factorization. *Phys. Rev. Lett.* **93**, 252001. arXiv:hep-ph/0408249.

Collins J. C., Qiu J.-W. (2007). k_T factorization is violated in production of high-transverse-momentum particles in hadron-hadron collisions. *Phys. Rev.* **D75**, 114014. arXiv:0705.2141.

Collins J. C., Rogers T. C. (2008). The gluon distribution function and factorization in Feynman gauge. *Phys. Rev.* **D78**, 054012. arXiv:0805.1752.

Collins J. C., Rogers T. C., Staśto A. M. (2008). Fully unintegrated parton correlation functions and factorization in lowest order hard scattering. *Phys. Rev.* **D77**, 085009. arXiv:0708.2833.

Collins J. C., Scalise R. J. (1994). The renormalization of composite operators in Yang-Mills theories using general covariant gauge. *Phys. Rev.* **D50**, 4117–4136. arXiv:hep-ph/9403231.

Collins J. C., Soper D. E. (1977). Angular distribution of dileptons in high-energy hadron collisions. *Phys. Rev.* **D16**, 2219–2225.

Collins J. C., Soper D. E. (1981). Back-to-back jets in QCD. *Nucl. Phys.* **B193**, 381–443. Erratum: **B213**, 545 (1983).

Collins J. C., Soper D. E. (1982a). Back-to-back jets: Fourier transform from b to k_T. *Nucl. Phys.* **B197**, 446–476.

Collins J. C., Soper D. E. (1982b). Parton distribution and decay functions. *Nucl. Phys.* **B194**, 445–492.

Collins J. C., Soper D. E., Sterman G. (1985a). Factorization for short distance hadron-hadron scattering. *Nucl. Phys.* **B261**, 104–142.

Collins J. C., Soper D. E., Sterman G. (1985b). Transverse momentum distribution in Drell-Yan pair and W and Z boson production. *Nucl. Phys.* **B250**, 199–224.

Collins J. C., Soper D. E., Sterman G. (1988). Soft gluons and factorization. *Nucl. Phys.* **B308**, 833–856.

Collins J. C., Sterman G. (1981). Soft partons in QCD. *Nucl. Phys.* **B185**, 172–188.

Collins J. C., Tung W.-K. (1986). Calculating heavy quark distributions. *Nucl. Phys.* **B278**, 934–950.

Collins J. C., Wilczek F., Zee A. (1978). Low-energy manifestations of heavy particles: Application to the neutral current. *Phys. Rev.* **D18**, 242–247.

Collins J. C., Zu X. (2005). Initial state parton showers beyond leading order. *JHEP* **03**, 059. arXiv:hep-ph/0411332.

Connes A., Kreimer D. (2000). Renormalization in quantum field theory and the Riemann-Hilbert problem I: the Hopf algebra structure of graphs and the main theorem. *Comm. Math. Phys.* **210**, 249–273. arXiv:hep-th/9912092.

Connes A., Kreimer D. (2002). Insertion and elimination: the doubly infinite Lie algebra of Feynman graphs. *Annales Henri Poincaré* **3**, 411–433. arXiv:hep-th/0201157.

Conway J. S., *et al.* (1989). Experimental study of muon pairs produced by 252-GeV pions on tungsten. *Phys. Rev.* **D39**, 92–122.

Curci G., Furmanski W., Petronzio R. (1980). Evolution of parton densities beyond leading order: the nonsinglet case. *Nucl. Phys.* **B175**, 27–92.

Cutkosky R. E. (1960). Singularities and discontinuities of Feynman amplitudes. *J. Math. Phys.* **1**, 429–433.

Czakon M. (2005). The four-loop QCD β-function and anomalous dimensions. *Nucl. Phys.* **B710**, 485–498. arXiv:hep-ph/0411261.

Dashen R. F., Gross D. J. (1981). The relationship between lattice and continuum definitions of the gauge theory coupling. *Phys. Rev.* **D23**, 2340.

de Florian D., Sassot R., Stratmann M. (2007). Global analysis of fragmentation functions for protons and charged hadrons. *Phys. Rev.* **D76**, 074033. arXiv:0707.1506.

DeGrand T., Detar C. E. (2006). *Lattice Methods for Quantum Chromodynamics*. Singapore: World Scientific.

Del Debbio L., *et al.* (2007). Neural network determination of parton distributions: the nonsinglet case. *JHEP* **03**, 039. arXiv:hep-ph/0701127.

DeTar C. E., Ellis S. D., Landshoff P. V. (1975). Final state interactions in large transverse momentum lepton and hadron production. *Nucl. Phys.* **B87**, 176.

Diehl M. (2003). Generalized parton distributions. *Phys. Rept.* **388**, 41–277. arXiv:hep-ph/0307382.

Diehl M., Sapeta S. (2005). On the analysis of lepton scattering on longitudinally or transversely polarized protons. *Eur. Phys. J.* **C41**, 515–533. arXiv:hep-ph/0503023.

Dine M. (2000). TASI lectures on the strong CP problem. arXiv:hep-ph/0011376.

Dirac P. A. M. (1926). The fundamental equations of quantum mechanics. *Proc. Roy. Soc. A* **109**, 642–653.

Dirac P. A. M. (1949). Forms of relativistic dynamics. *Rev. Mod. Phys.* **21**, 392–399.

Dissertori G., Knowles I. G., Schmelling M. (2003). *Quantum Chromodynamics: High Energy Experiments and Theory*. Oxford: Oxford University Press.

Dittmaier S., Kabelschacht A., Kasprzik T. (2008). Polarized QED splittings of massive fermions and dipole subtraction for non-collinear-safe observables. *Nucl. Phys.* **B800**, 146–189. arXiv:0802.1405.

Dixon J. A., Taylor J. C. (1974). Renormalization of Wilson operators in gauge theories. *Nucl. Phys.* **B78**, 552–560.

Dokshitzer Y. L. (1977). Calculation of the structure functions for deep inelastic scattering and e^+e^- annihilation by perturbation theory in quantum chromodynamics. *Sov. Phys. JETP* **46**, 641–653.

Donohue J. T., Gottlieb S. A. (1981). Dilepton production from collisions of polarized spin-1/2 hadrons. I. General kinematic analysis. *Phys. Rev.* **D23**, 2577–2580.

Doplicher S., Haag R., Roberts J. E. (1974). Local observables and particle statistics. 2. *Commun. Math. Phys.* **35**, 49–85.

Drell S. D., Levy D. J., Yan T.-M. (1970). A theory of deep inelastic lepton-nucleon scattering and lepton-pair annihilation processes. III. Deep inelastic electron-positron annihilation. *Phys. Rev.* **D1**, 1617–1639.

Drell S. D., Yan T.-M. (1970). Massive lepton pair production in hadron-hadron collisions at high-energies. *Phys. Rev. Lett.* **25**, 316–320.

Drühl K., Haag R., Roberts J. E. (1970). On parastatistics. *Commun. Math. Phys.* **18**, 204–226.

Eden R. J., *et al.* (1966). *The Analytic S-matrix*. Cambridge: Cambridge University Press.

Efremov A. V. (1978). Polarization in high P_T and cumulative hadron production. *Sov. J. Nucl. Phys.* **28**, 83.

Einhorn M. B. (1976). Confinement, form factors, and deep-inelastic scattering in two-dimensional quantum chromodynamics. *Phys. Rev.* **D14**, 3451–3471.

Einhorn M. B. (1977). Failure of the parton model in inclusive electron-positron annihilation. *Phys. Rev.* **D15**, 3037–3043.

Elitzur S. (1975). Impossibility of spontaneously breaking local symmetries. *Phys. Rev.* **D12**, 3978–3982.

Ellis R. K., Stirling W. J., Webber B. R. (1996). *QCD and Collider Physics*. Cambridge: Cambridge University Press.

Ellis R. K., *et al.* (1979). Perturbation theory and the parton model in QCD. *Nucl. Phys.* **B152**, 285–329.

Eskola K. J., Paukkunen H., Salgado C. A. (2009). EPS09 – a new generation of NLO and LO nuclear parton distribution functions. *JHEP* **04**, 065. arXiv:0902.4154.

Faddeev L. D., Jackiw R. (1988). Hamiltonian reduction of unconstrained and constrained systems. *Phys. Rev. Lett.* **60**, 1692–1694.

Faddeev L. D., Popov V. N. (1967). Feynman diagrams for the Yang-Mills field. *Phys. Lett.* **B25**, 29–30.

Fadin V. S., Kuraev E. A., Lipatov L. N. (1975). On the pomeranchuk singularity in asymptotically free theories. *Phys. Lett.* **B60**, 50–52.

Falciano S., *et al.* (1986). Angular distributions of muon pairs produced by 194 GeV/c negative pions. *Z. Phys.* **C31**, 513–526.

Farhi E. (1977). Quantum chromodynamics test for jets. *Phys. Rev. Lett.* **39**, 1587–1588.

Fetter A. L., Walecka J. D. (1980). *Quantum Theory of Many-Particle Systems*. New York: McGraw-Hill.

Feynman R. P. (1972). *Photon-Hadron Interactions*. Reading, MA: Benjamin.

Fleming S. (2009). Soft collinear effective theory: an overview. *PoS* **EFT09**, 002 arXiv:0907.3897.

Floratos E. G., Kounnas C., Lacaze R. (1981). Higher order QCD effects in inclusive annihilation and deep inelastic scattering. *Nucl. Phys.* **B192**, 417–462.

Floratos E. G., Lacaze R., Kounnas C. (1981). Space and timelike cut vertices in QCD beyond the leading order. 2. The singlet sector. *Phys. Lett.* **B98**, 285–290.

Floratos E. G., Ross D. A., Sachrajda C. T. (1979). Higher order effects in asymptotically free gauge theories. 2. Flavor singlet Wilson operators and coefficient functions. *Nucl. Phys.* **B152**, 493–520.

Frederix R., Gehrmann T., Greiner N. (2008). Automation of the dipole subtraction method in MadGraph/MadEvent. *JHEP* **09**, 122. arXiv:0808.2128.

Fritzsch H., Gell-Mann M. (1972). Current algebra: quarks and what else? In *Proceedings of XVI International Conference on High-Energy Physics, Chicago 1972* (J.D. Jackson, A. Roberts, eds.), 135–165. arXiv:hep-ph/0208010.

Fritzsch H., Gell-Mann M., Leutwyler H. (1973). Advantages of the color octet gluon picture. *Phys. Lett.* **B47**, 365–368.

Furmanski W., Petronzio R. (1980). Singlet parton densities beyond leading order. *Phys. Lett.* **B97**, 437–442.

Gastmans R., Wu T. T. (1990). *The Ubiquitous Photon: Helicity Method for QED and QCD*. Oxford: Oxford University Press.

Gehrmann T., Luisoni G., Stenzel H. (2008). Matching NLLA + NNLO for event shape distributions. *Phys. Lett.* **B664**, 265–273. arXiv:0803.0695.

Gell-Mann M. (1962). Symmetries of baryons and mesons. *Phys. Rev.* **125**, 1067–1084.

Gell-Mann M. (1964). A schematic model of baryons and mesons. *Phys. Lett.* **8**, 214–215.

Gonzalez-Arroyo A., Lopez C. (1980). Second order contributions to the structure functions in deep inelastic scattering. 3. The singlet case. *Nucl. Phys.* **B166**, 429–459.

Grammer, G. J., Yennie D. R. (1973). Improved treatment for the infrared divergence problem in quantum electrodynamics. *Phys. Rev.* **D8**, 4332–4344.

Grazzini M., Trentadue L., Veneziano G. (1998). Fracture functions from cut vertices. *Nucl. Phys.* **B519**, 394–404. arXiv:hep-ph/9709452.

Gribov V. N. (1973). Space-time description of hadron interactions at high energies. arXiv:hep-ph/0006158.

Gribov V. N. (2009). *Strong Interactions of Hadrons at High Energies*. Cambridge: Cambridge University Press.

Gribov V. N., Lipatov L. N. (1972). Deep inelastic ep scattering in perturbation theory. *Sov. J. Nucl. Phys.* **15**, 438–450.

Gross D. J., Wilczek F. (1973a). Ultraviolet behavior of non-Abelian gauge theories. *Phys. Rev. Lett.* **30**, 1343–1346.

Gross D. J., Wilczek F. (1973b). Asymptotically free gauge theories. 1. *Phys. Rev.* **D8**, 3633–3652.

Guanziroli M., *et al.* (1988). Angular distributions of muon pairs produced by negative pions on deuterium and tungsten. *Z. Phys.* **C37**, 545–556.

Guidal M., Vanderhaeghen M. (2003). Double deeply virtual Compton scattering off the nucleon. *Phys. Rev. Lett.* **90**, 012001. arXiv:hep-ph/0208275.

Gupta S., Quinn H. R. (1982). Heavy quarks and perturbative QCD calculations. *Phys. Rev.* **D25**, 838.

H1 Collaboration. (2010). Diffractive electroproduction of ρ and ϕ mesons at HERA. *JHEP* **05**, 032. arXiv:0910.5831.

H1 website. (2010). Available from: http://www-h1.desy.de.

Halzen F., Martin A. D. (1984). *Quarks and Leptons: An Introductory Course in Modern Particle Physics*. New York: Wiley.

Hamberg R., van Neerven W. L. (1992). The correct renormalization of the gluon operator in a covariant gauge. *Nucl. Phys.* **B379**, 143–171.

Hasegawa K., Moch S., Uwer P. (2008). Automating dipole subtraction. *Nucl. Phys. Proc. Suppl.* **183**, 268–273. arXiv:0807.3701.

Hasenfratz A., Hasenfratz P. (1980). The connection between the Λ parameters of lattice and continuum QCD. *Phys. Lett.* **B93**, 165.

Heinzl T. (2001). Light-cone quantization: foundations and applications. *Lect. Notes Phys.* **572**, 55–142. arXiv:hep-th/0008096.

Heinzl T. (2003). Light-cone zero modes revisited. arXiv:hep-th/0310165.

Heinzl T. (2007). A novel approach to light-front perturbation theory. *Phys. Rev.* **D75**, 025013. arXiv:hep-ph/0610293.

Heinzl T., Ilderton A. (2007). Noncommutativity from spectral flow. *J. Phys.* **A40**, 9097–9125. arXiv:0704.3547.

Heinzl T., Werner E. (1994). Light front quantization as an initial boundary value problem. *Z. Phys.* **C62**, 521–532. arXiv:hep-th/9311108.

Henyey F., Savit R. (1974). Final state interactions in the parton model and massive lepton pair production. *Phys. Lett.* **B52**, 71.

Hobbs T., Melnitchouk W. (2008). Finite-Q^2 corrections to parity-violating DIS. *Phys. Rev.* **D77**, 114023. arXiv:0801.4791.

Hofstadter R. (1956). Electron scattering and nuclear structure. *Rev. Mod. Phys.* **28**, 214–254.

Hofstadter R., Bumiller F., Yearian M. R. (1958). Electromagnetic structure of the proton and neutron. *Rev. Mod. Phys.* **30**, 482–497.

Hoodbhoy P., Jaffe R. L., Manohar A. (1989). Novel effects in deep inelastic scattering from spin 1 hadrons. *Nucl. Phys.* **B312**, 571–588.

Idilbi A., *et al.* (2004). Collins-Soper equation for the energy evolution of transverse-momentum and spin dependent parton distributions. *Phys. Rev.* **D70**, 074021. arXiv:hep-ph/0406302.

Ito A. S., *et al.* (1981). Measurement of the continuum of dimuons produced in high-energy proton-nucleus collisions. *Phys. Rev.* **D23**, 604–633.

Itzykson C., Zuber J.-B. (1980). *Quantum Field Theory*. New York: McGraw-Hill.

Jackiw R. (1968). Dynamics at high momentum and the vertex function of spinor electrodynamics. *Ann. Phys.* **48**, 292–321.

Jaffe R. L. (1983). Parton distribution functions for twist four. *Nucl. Phys.* **B229**, 205–230.

Jaffe R. L., Ji X.-D. (1991). Chiral odd parton distributions and polarized Drell-Yan. *Phys. Rev. Lett.* **67**, 552–555.

Ji X.-D. (1993). The nucleon structure functions from deep inelastic scattering with electroweak currents. *Nucl. Phys.* **B402**, 217–250.

Ji X.-D., Ma J.-P., Yuan F. (2004). QCD factorization for spin-dependent cross sections in DIS and Drell-Yan processes at low transverse momentum. *Phys. Lett.* **B597**, 299–308. arXiv:hep-ph/0405085.

Ji X.-D., Ma J.-P., Yuan F. (2005). QCD factorization for semi-inclusive deep-inelastic scattering at low transverse momentum. *Phys. Rev.* **D71**, 034005. arXiv:hep-ph/0404183.

Joglekar S. D. (1977a). Local operator products in gauge theories. 1. *Ann. Phys.* **108**, 233–287.

Joglekar S. D. (1977b). Local operator products in gauge theories. 2. *Ann. Phys.* **109**, 210–241.

Joglekar S. D., Lee B. W. (1976). General theory of renormalization of gauge invariant operators. *Ann. Phys.* **97**, 160–215.

Johnson K., Low F. E. (1966). Current algebras in a simple model. *Prog. Theor. Phys. Suppl.* **37**, 74–93.

Kalinowski J., Konishi K., Taylor T. R. (1981). Jet calculus beyond leading logarithms. *Nucl. Phys.* **B181**, 221–252.

Kalinowski J., et al. (1981). Resolving QCD jets beyond leading order: quark decay probabilities. *Nucl. Phys.* **B181**, 253–276.

Khachatryan V., et al. (2010). Transverse momentum and pseudorapidity distributions of charged hadrons in pp collisions at $\sqrt{s} = 0.9$ and 2.36 TeV. *JHEP* **02**, 041. arXiv:1002.0621.

Khriplovich I. B. (1970). Greens functions in theories with a non-abelian gauge group. *Sov. J. Nucl. Phys.* **10**, 235. *Yad. Fiz.* **10**, 409 (1969).

Kinoshita T. (1962). Mass singularities of Feynman amplitudes. *J. Math. Phys.* **3**, 650–677.

Kluberg-Stern H., Zuber J. B. (1975). Ward identities and some clues to the renormalization of gauge-invariant operators. *Phys. Rev.* **D12**, 467–481.

Knuth D. E. (1976). Big omicron and big omega and big theta. *ACM SIGACT News* **8**, 18–24.

Kogut J. B., Soper D. E. (1970). Quantum electrodynamics in the infinite momentum frame. *Phys. Rev.* **D1**, 2901–2913.

Konishi K., Ukawa A., Veneziano G. (1978). A simple algorithm for QCD jets. *Phys. Lett.* **B78**, 243–248.

Konychev A. V., Nadolsky P. M. (2006). Universality of the Collins-Soper-Sterman nonperturbative function in gauge boson production. *Phys. Lett.* **B633**, 710–714. arXiv:hep-ph/0506225.

Kotzinian A. (1995). New quark distributions and semi-inclusive electroproduction on the polarized nucleons. *Nucl. Phys.* **B441**, 234–256. arXiv:hep-ph/9412283.

Krämer M., Olness F. I., Soper D. E. (2000). Treatment of heavy quarks in deeply inelastic scattering. *Phys. Rev.* **D62**, 096007. arXiv:hep-ph/0003035.

Kretzer S., et al. (2004). CTEQ6 parton distributions with heavy quark mass effects. *Phys. Rev.* **D69**, 114005. arXiv:hep-ph/0307022.

Labastida J. M. F., Sterman G. (1985). Inclusive hadron-hadron scattering in the Feynman gauge. *Nucl. Phys.* **B254**, 425–440.

Lai H. L., et al. (2000). Global QCD analysis of parton structure of the nucleon: CTEQ5 parton distributions. *Eur. Phys. J.* **C12**, 375–392. arXiv:hep-ph/9903282.

Lam C. S., Tung W.-K. (1978). Systematic approach to inclusive lepton pair production in hadronic collisions. *Phys. Rev.* **D18**, 2447–2461.

Landry F., et al. (2003). Tevatron Run-1 Z boson data and Collins-Soper-Sterman resummation formalism. *Phys. Rev.* **D67**, 073016. arXiv:hep-ph/0212159.

Landshoff P. V. (1974). Model for elastic scattering at wide angle. *Phys. Rev.* **D10**, 1024–1030.

Landshoff P. V., Polkinghorne J. C. (1971). Two high energy processes involving detected final state particles. *Nucl. Phys.* **B33**, 221–238. Erratum: **B36**, 642 (1972).

Larin S. A., Vermaseren J. A. M. (1993). The three-loop QCD β function and anomalous dimensions. *Phys. Lett.* **B303**, 334–336. arXiv:hep-ph/9302208.

Leader E., Predazzi E. (1982). *An Introduction to Gauge Theories and the 'New Physics'*. Cambridge: Cambridge University Press.

Lee T. D., Nauenberg M. (1964). Degenerate systems and mass singularities. *Phys. Rev.* **133**, B1549–B1562.

Leibbrandt G. (1987). Introduction to noncovariant gauges. *Rev. Mod. Phys.* **59**, 1067–1119.

Lepage G. P., Brodsky S. J. (1980). Exclusive processes in perturbative quantum chromodynamics. *Phys. Rev.* **D22**, 2157–2198.

Libby S. B., Sterman G. (1978a). Jet and lepton-pair production in high-energy lepton-hadron and hadron-hadron scattering. *Phys. Rev.* **D18**, 3252–3268.

Libby S. B., Sterman G. (1978b). Mass divergences in two-particle inelastic scattering. *Phys. Rev.* **D18**, 4737–4745.

Liberati S., Maccione L. (2009). Lorentz violation: motivation and new constraints. *Ann. Rev. Nucl. Part. Sci.* **59**, 245–267. arXiv:0906.0681.

Ligterink N. E., Bakker B. L. G. (1995). Equivalence of light front and covariant field theory. *Phys. Rev.* **D52**, 5954–5979. arXiv:hep-ph/9412315.

Lipatov L. N. (1997). Small-x physics in perturbative QCD. *Phys. Rept.* **286**, 131–198. arXiv:hep-ph/9610276.

Lu Z., Schmidt I. (2010). Updating Boer-Mulders functions from unpolarized pd and pp Drell-Yan data. *Phys. Rev.* **D81**, 034023. arXiv:0912.2031.

Lubański J. K. (1942a). Sur la théorie des particules élémentaires de spin quelconque. I. *Physica* **9**, 310–324.

Lubański J. K. (1942b). Sur la théorie des particules élémentaires de spin quelconque. II. *Physica* **9**, 325–338.

Manohar A. V. (1998). Large N QCD. arXiv:hep-ph/9802419.

Manohar A. V., Wise M. B. (2000). *Heavy Quark Physics*. Cambridge: Cambridge University Press.

Marchesini G. (1995). QCD coherence in the structure function and associated distributions at small x. *Nucl. Phys.* **B445**, 49–80. arXiv:hep-ph/9412327.

Martin A. D., *et al.* (1998). Parton distributions: a new global analysis. *Eur. Phys. J.* **C4**, 463–496. arXiv:hep-ph/9803445.

Martin A. D., *et al.* (2007). Update of parton distributions at NNLO. *Phys. Lett.* **B652**, 292–299. arXiv:0706.0459.

Melnitchouk W., Ent R., Keppel C. (2005). Quark-hadron duality in electron scattering. *Phys. Rept.* **406**, 127–301. arXiv:hep-ph/0501217.

Meng R., Olness F. I., Soper D. E. (1996). Semi-inclusive deeply inelastic scattering at small q_T. *Phys. Rev.* **D54**, 1919–1935. arXiv:hep-ph/9511311.

Mirkes E. (1992). Angular decay distribution of leptons from W bosons at NLO in hadronic collisions. *Nucl. Phys.* **B387**, 3–85.

Moch S., Vermaseren J. A. M. (2000). Deep-inelastic structure functions at two loops. *Nucl. Phys.* **B573**, 853–907. arXiv:hep-ph/9912355.

Moch S., Vermaseren J. A. M., Vogt A. (2004). The three-loop splitting functions in QCD: the non-singlet case. *Nucl. Phys.* **B688**, 101–134. arXiv:hep-ph/0403192.

Mueller A. H. (1979). On the asymptotic behavior of the Sudakov form factor. *Phys. Rev.* **D20**, 2037.

Mulders P. J., Tangerman R. D. (1996). The complete tree-level result up to order $1/Q$ for polarized deep-inelastic leptoproduction. *Nucl. Phys.* **B461**, 197–237. arXiv:hep-ph/9510301.

Müller D., *et al.* (1994). Wave functions, evolution equations and evolution kernels from light-ray operators of QCD. *Fortschr. Phys.* **42**, 101–141. arXiv:hep-ph/9812448.

Nachtmann O. (1973). Positivity constraints for anomalous dimensions. *Nucl. Phys.* **B63**, 237–247.

Nadolsky P., Stump D. R., Yuan C. P. (2000). Semi-inclusive hadron production at HERA: The effect of QCD gluon resummation. *Phys. Rev.* **D61**, 014003. arXiv:hep-ph/9906280.

Nakanishi N., Ojima I. (1990). *Covariant Operator Formalism of Gauge Theories and Quantum Gravity.* Singapore: World Scientific.

Nakanishi N., Yabuki H. (1977). Null-plane quantization and Haag's theorem. *Lett. Math. Phys.* **1**, 371–374.

Nakanishi N., Yamawaki K. (1977). A consistent formulation of the null-plane quantum field theory. *Nucl. Phys.* **B122**, 15–28.

Narison S. (2002). *QCD as a Theory of Hadrons.* Cambridge: Cambridge University Press.

Nayak G. C., Qiu J.-W., Sterman G. (2005). Fragmentation, non-relativistic QCD, and NNLO factorization analysis in heavy quarkonium production. *Phys. Rev.* **D72**, 114012. arXiv:hep-ph/0509021.

Pais A. (1986). *Inward Bound.* Oxford: Oxford University Press.

Perkins D. H. (2000). *Introduction to High Energy Physics.* 4th edn. Cambridge: Cambridge University Press.

Peskin M. E., Schroeder D. V. (1995). *An Introduction to Quantum Field Theory.* Reading, MA: Addison-Wesley.

Poggio E. C., Quinn H. R., Weinberg S. (1976). Smearing the quark model. *Phys. Rev.* **D13**, 1958–1968.

Polchinski J. (1984). Renormalization and effective lagrangians. *Nucl. Phys.* **B231**, 269–295.

Politzer H. D. (1973). Reliable perturbative results for strong interactions? *Phys. Rev. Lett.* **30**, 1346–1349.

Qiu J.-W., Sterman G. (1991a). Power corrections in hadronic scattering (I): Leading $1/Q^2$ corrections to the Drell-Yan cross-section. *Nucl. Phys.* **B353**, 105–136.

Qiu J.-W., Sterman G. (1991b). Power corrections in hadronic scattering (II): Factorization. *Nucl. Phys.* **B353**, 137–164.

Quigg C. (1997). *Gauge Theories of the Strong, Weak, and Electromagnetic Interactions.* Boulder, Colorado: Westview Press.

Ralston J. P., Soper D. E. (1979). Production of dimuons from high-energy polarized proton-proton collisions. *Nucl. Phys.* **B152**, 109–124.

Rijken P. J., van Neerven W. L. (1997). Higher order QCD corrections to the transverse and longitudinal fragmentation functions in electron-positron annihilation. *Nucl. Phys.* **B487**, 233–282. arXiv:hep-ph/9609377.

Rogers T. C., Mulders P. J. (2010). No generalized transverse momentum dependent factorization in hadroproduction of high transverse momentum hadrons. *Phys. Rev.* **D81**, 094006 arXiv:1001.2977.

Salam A. (1968). In *Proceedings of the 8th Nobel Symposium.* Stockholm: Almqvist and Wiksell.

Salam G. P. (2010). Towards jetography. *Eur. Phys. J.* **C67**, 637–686. arXiv:0906.1833.

Schienbein I., *et al.* (2009). Parton distribution function nuclear corrections for charged lepton and neutrino deep inelastic scattering processes. *Phys. Rev.* **D80**, 094004. arXiv:0907.2357.

Seymour M. H., Tevlin C. (2008). TeVJet: a general framework for the calculation of jet observables in NLO QCD. arXiv:0803.2231.

Sivers D. W. (1990). Single spin production asymmetries from the hard scattering of point-like constituents. *Phys. Rev.* **D41**, 83–90.

Sjöstrand T. (2009). Monte Carlo tools. arXiv:0911.5286.

Sjöstrand T., Mrenna S., Skands P. Z. (2006). PYTHIA 6.4 physics and manual. *JHEP* **05**, 026. arXiv:hep-ph/0603175.

Sjöstrand T., Mrenna S., Skands P. Z. (2008). A brief introduction to PYTHIA 8.1. *Comput. Phys. Commun.* **178**, 852–867. arXiv:0710.3820.

Slavnov A. A. (1972). Ward identities in gauge theories. *Theor. Math. Phys.* **10**, 99–107.

Soper D. E. (1977). The parton model and the Bethe-Salpeter wave function. *Phys. Rev.* **D15**, 1141–1149.

Soper D. E. (1979). Partons and their transverse momenta in QCD. *Phys. Rev. Lett.* **43**, 1847–1851.

Srednicki M. (2007). *Quantum Field Theory*. Cambridge: Cambridge University Press.

Srivastava P. P., Brodsky S. J. (2001). Light-front quantized QCD in light-cone gauge. *Phys. Rev.* **D64**, 045006. arXiv:hep-ph/0011372.

Steinhardt P. J. (1980). Problems of quantization in the infinite momentum frame. *Ann. Phys.* **128**, 425–447.

Sterman G. (1978). Mass divergences in annihilation processes. I. Origin and nature of divergences in cut vacuum polarization diagrams. *Phys. Rev.* **D17**, 2773–2788.

Sterman G. (1993). *An Introduction to Quantum Field Theory*. Cambridge: Cambridge University Press.

Sterman G. (1996). Partons, factorization and resummation. In *QCD and beyond*. Singapore: World Scientific, 327–408. arXiv:hep-ph/9606312.

Sudakov V. V. (1956). Vertex parts at very high-energies in quantum electrodynamics. *Sov. Phys. JETP* **3**, 65–71.

't Hooft G. (1973). Dimensional regularization and the renormalization group. *Nucl. Phys.* **B61**, 455–468.

't Hooft G. (1974). A two-dimensional model for mesons. *Nucl. Phys.* **B75**, 461–470.

't Hooft G. (1999). When was asymptotic freedom discovered? or The rehabilitation of quantum field theory. *Nucl. Phys. Proc. Suppl.* **74**, 413–425. arXiv:hep-th/9808154.

't Hooft G., Veltman M. J. G. (1972). Combinatorics of gauge fields. *Nucl. Phys.* **B50**, 318–353.

Tarasov O. V., Vladimirov A. A., Zharkov A. Y. (1980). The Gell-Mann-Low function of QCD in the three-loop approximation. *Phys. Lett.* **B93**, 429–432.

Taylor J. C. (1971). Ward identities and charge renormalization of the Yang-Mills field. *Nucl. Phys.* **B33**, 436–444.

Thorne R. S., Tung W. K. (2008). PQCD formulations with heavy quark masses and global analysis. arXiv:0809.0714.

Tkachov F. V. (1994). Theory of asymptotic operation. A summary of basic principles. *Sov. J. Part. Nucl.* **25**, 649. arXiv:hep-ph/9701272.

Treiman S. B., Jackiw R., Gross D. J. (1972). *Lectures on Current Algebra and Its Applications*. Princeton, NJ: Princeton University Press.

Trentadue L., Veneziano G. (1994). Fracture functions: an improved description of inclusive hard processes in QCD. *Phys. Lett.* **B323**, 201–211.

Tung W.-K., Kretzer S., Schmidt C. (2002). Open heavy flavor production in QCD: Conceptual framework and implementation issues. *J. Phys.* **G28**, 983–996. arXiv:hep-ph/0110247.

Tung W.-K., et al. (2007). Heavy quark mass effects in deep inelastic scattering and global QCD analysis. *JHEP* **02**, 053. arXiv:hep-ph/0611254.

Tyutin I. V. (1975). Gauge invariance in field theory and statistical physics in operator formalism. Originally appeared in 1975 as preprint LEBEDEV-75-39. arXiv:0812.0580.

van Ritbergen T., Vermaseren J. A. M., Larin S. A. (1997). The four-loop β function in quantum chromodynamics. *Phys. Lett.* **B400**, 379–384. arXiv:hep-ph/9701390.

Vanyashin V. S., Terentyev M. V. (1965). Vacuum polarization of a charged vector field. *Sov. Phys. JETP* **21**, 375–380. *Zh.E.T.F.* **48**, 565–573 (1965).

Vermaseren J. A. M., Vogt A., Moch S. (2005). The third-order QCD corrections to deep-inelastic scattering by photon exchange. *Nucl. Phys.* **B724**, 3–182. arXiv:hep-ph/0504242.

Vogt A., Moch S., Vermaseren J. A. M. (2004). The three-loop splitting functions in QCD: the singlet case. *Nucl. Phys.* **B691**, 129–181. arXiv:hep-ph/0404111.

Vossen A., *et al.* (2011). Observation of the interference fragmentation function for charged pion pairs in e^+e^- annihilation near $\sqrt{s} = 10.58$ GeV, *Phys. Rev. Lett.* **107**, 072004. arXiv:1104.2425.

Wandzura S., Wilczek F. (1977). Sum rules for spin dependent electroproduction: test of relativistic constituent quarks. *Phys. Lett.* **B72**, 195–198.

Watt G., Martin A. D., Ryskin M. G. (2003). Unintegrated parton distributions and inclusive jet production at HERA. *Eur. Phys. J.* **C31**, 73–89. arXiv:hep-ph/0306169.

Watt G., Martin A. D., Ryskin M. G. (2004). Unintegrated parton distributions and electroweak boson production at hadron colliders. *Phys. Rev.* **D70**, 014012. arXiv:hep-ph/0309096.

Weinberg S. (1966). Dynamics at infinite momentum. *Phys. Rev.* **150**, 1313–1318.

Weinberg S. (1967). A model of leptons. *Phys. Rev. Lett.* **19**, 1264–1266.

Weinberg S. (1973a). Current algebra and gauge theories. 1. *Phys. Rev.* **D8**, 605–625.

Weinberg S. (1973b). Current algebra and gauge theories. 2. Nonabelian gluons. *Phys. Rev.* **D8**, 4482–4498.

Weinberg S. (1989). The cosmological constant problem. *Rev. Mod. Phys.* **61**, 1–23.

Weinberg S. (1995). *The Quantum Theory of Fields, Vol. I, Foundations*. Cambridge: Cambridge University Press.

Weinberg S. (1996). *The Quantum Theory of Fields, Vol. II, Modern Applications*. Cambridge: Cambridge University Press.

Whitlow L. W., *et al.* (1992). Precise measurements of the proton and deuteron structure functions from a global analysis of the SLAC deep inelastic electron scattering cross-sections. *Phys. Lett.* **B282**, 475–482.

Wilson K. G. (1973). Quantum field theory models in less than 4 dimensions. *Phys. Rev.* **D7**, 2911–2926.

Witten E. (1976). Heavy quark contributions to deep inelastic scattering. *Nucl. Phys.* **B104**, 445–476.

Wollny H. for the COMPASS collaboration. (2009). Transversity signal in two hadron pair production in COMPASS. arXiv:0907.0961.

Yamawaki K. (1998). Zero-mode problem on the light front. arXiv:hep-th/9802037.

Yan T.-M. (1973). Quantum field theories in the infinite momentum frame. 4. Scattering matrix of vector and Dirac fields and perturbation theory. *Phys. Rev.* **D7**, 1780–1800.

Yang C.-N., Mills R. L. (1954). Conservation of isotopic spin and isotopic gauge invariance. *Phys. Rev.* **96**, 191–195.

Zhu L. Y., *et al.* (2009). Measurement of angular distributions of Drell-Yan dimuons in $p + p$ interactions at 800 GeV/c. *Phys. Rev. Lett.* **102**, 182001. arXiv:0811.4589.

Zijlstra E. B., van Neerven W. L. (1992). Order α_s^2 QCD corrections to the deep inelastic proton structure functions F_2 and F_L. *Nucl. Phys.* **B383**, 525–574.

Zweig G. (1964a). An SU(3) model for strong interaction symmetry and its breaking. CERN-TH-401.

Zweig G. (1964b). An SU(3) model for strong interaction symmetry and its breaking. 2. CERN-TH-412.

Zweig G. (1980). Origins of the quark model. Invited talk given at 4th Int. Conf. on Baryon Resonances, Toronto, Canada, Jul. 14–16, 1980.

Index

Printed in the United States
by Baker & Taylor Publisher Services